The Lost Elements

The Lost Elements

The Periodic Table's Shadow Side

Marco Fontani, Mariagrazia Costa, and Mary Virginia Orna

OXFORD
UNIVERSITY PRESS

OXFORD
UNIVERSITY PRESS

Oxford University Press is a department of the University of Oxford.
It furthers the University's objective of excellence in research, scholarship,
and education by publishing worldwide.

Oxford New York
Auckland Cape Town Dar es Salaam Hong Kong Karachi
Kuala Lumpur Madrid Melbourne Mexico City Nairobi
New Delhi Shanghai Taipei Toronto

With offices in
Argentina Austria Brazil Chile Czech Republic France Greece
Guatemala Hungary Italy Japan Poland Portugal Singapore
South Korea Switzerland Thailand Turkey Ukraine Vietnam

Oxford is a registered trademark of Oxford University Press
in the UK and certain other countries.

Published in the United States of America by
Oxford University Press
198 Madison Avenue, New York, NY 10016

Library of Congress Cataloging-in-Publication Data

Fontani, Marco, 1969–
The lost elements : the periodic table's shadow side / Marco Fontani, Mariagrazia Costa,
and Mary Virginia Orna.
pages cm
Includes bibliographical references and indexes.
ISBN 978–0–19–938334–4 (alk. paper)
1. Chemical elements. 2. Chemical elements—History. 3. Periodic law—History.
4. Chemistry—Nomenclature—History. 5. Chemistry—Humor. I. Costa, Mariagrazia.
II. Orna, Mary Virginia. III. Title. IV. Title: Periodic table's shadow side.
QD467.F66 2014
546'.8—dc23
2014009191

3 5 7 9 8 6 4
Printed in the United States of America
on acid-free paper

It is appropriate to recall the dedication
that, many years ago, I wrote on the manuscript of my bachelor's thesis;
for this reason, I renew it with affection:
"To my parents: you were the lions; I have only roared."

—Marco Fontani

To my beloved nephews,
may they all be attracted to Science.

—Mariagrazia Costa

To Maria Lucia (Grazia) Pulaccini,
my finest teacher and most beloved mentor.

—Mary Virginia Orna

ENTE CASSA DI RISPARMIO DI FIRENZE

CONTENTS

Preface xv
Acknowledgments xvii
Note to the Reader xxi
Introduction xxiii
Why Collect into One Volume the Discoveries of Elements that Have Been Shown to be Erroneous or Have Been Forgotten? xxiii
How "an Element" Became a "Chemical Element" xxiv
Is There Any Order to the Discoveries of the Elements? xxviii
The Development of the Periodic Table xxx

Part I: Before 1789: Early Errors and Early Elements
Prologue to Part I 1
I.1. The Beginning of a Long Series of Scientific Blunders 3
I.1.1. Terra Nobilis 3
I.1.2. Siderum and Hydrosiderum 3
I.1.3. Sydneium or Australium 4
I.1.4. The Element That Breathes 6
I.1.5. The Birth of Homeopathy 7
I.2. The Elements Hidden by Alternative Names 12
I.2.1. Metallum Problematicum or Tellurium 12
I.2.2. Ochroite or Cerium 13
I.2.3. Ceresium or Palladium 13
I.2.4. Erythronium, Panchromium or Vanadium 14

Part II: 1789–1869: From Lavoisier to Mendeleev: The First Errors at the Dawn of the Concept of the Chemical Element
Prologue to Part II 19
II.1. Analytical Methodology from Lavoisier to Mendeleev 21
II.1.1. Blowpipe Analysis 22
II.1.2. Qualitative and Quantitative Analysis 23
II.1.3. Electrolysis 23
II.1.4. Emission Spectroscopy 24
II.2. The Elements of the Kingdom of Naples 27
II.2.1. Ruprecht and Tondi: Two Metallurgists Without Metals 27
II.2.2. Playing Bingo with Five Elements 28
II.2.3. The Extraction Procedure of the New Metals 29
II.2.4. Right or Wrong, Was Tondi the Victim of a Sworn Enemy? 29
II.2.5. The Elements that Replaced Those of Tondi 30

II.2.6. Possible Present-Day Interpretations 31
II.2.7. Revolution Offers a Second Career Possibility 33
II.3. Austrium: One Element, Two Elements, Three Elements, and Finally, Zero Elements 36
II.3.1. The First Fleeting Attempt to Name an Element Austrium 36
II.3.2. Austrium: A Posthumous Element 36
II.3.3. The "Austrian Element" of a Czech Chemist 38
II.3.4. A Third "Split" for Bohuslav Brauner 41
II.4. The Return of the Olympians: Silene, Aridium, Saturnum, Pelopium, Dianium, Neptunium, and Plutonium 43
II.4.1. Silene 43
II.4.2. Aridium 44
II.4.3. Saturnum 45
II.4.4. Pelopium 46
II.4.5. Dianium 47
II.4.6. Neptunium 48
II.4.7. Plutonium 49
II.5. The Unfortunate Affair of a Student of Kant: A Career Soldier, but a Chemist by Passion 53
II.5.1. Niccolanum 53
II.5.2. The Road from Oblivion 54
II.6. André-Marie Ampère Bursts onto the Chemistry Scene 56
II.6.1. "Phtore" 56
II.7. Cadmium: "Bone of Contention" Among Chemical Elements 59
II.7.1. A Related Discovery Increases the Confusion: Vestium 60
II.8. A Fireproof Family of Chemists 62
II.8.1. Chemistry as the Common Denominator 62
II.8.2. The Most Improbable of the Chemical Elements 63
II.9. A Bridge of False Hopes Between Divinity and False Elements 65
II.9.1. Crodonium 65
II.9.2. Wodanium 66
II.9.3. False Elements Exchanged for Another False Element 67
II.9.4. Ptene 69
II.9.5. Donarium 70
II.10. Gahnium, Polinium, and Pluranium 73
II.10.1. Gahnium 73
II.10.2. Polinium and Pluranium 73
II.11. Aberdonia and the "Sweet" Map of Oblivion 77
II.11.1. Donium 77
II.11.2. Treenium 78
II.11.3. The Discovery of an Already Known Element? 79
II.11.4. The Sweet Epilogue Leaves a Bitter Taste in the Mouth 80
II.12. The Brief Parentheses of Four Misleading Elements 82
II.12.1. The Fleeting Existence of Thalium 82
II.12.2. The Meteoric Appearance and Disappearance of Comesium 83
II.12.3. The Mysterious Nature of Ouralium 84
II.12.4. The Brief History of Idunium 85

II.13. Two Imaginary Elements: Sulphurium and Sulfenium 87
II.13.1. Sulphurium 87
II.13.2. The Ancient Modernity of Sulfenium 88
II.14. The Astronomer "Left in the Dark" 91
II.14.1. "Light" as a Means of Chemical Investigation 91
II.14.2. A New Family of Elements from an Old Family of Astronomers 92
II.14.3. Neptunium Is Tempting to a Lot of People 93
II.14.4. Conclusion 93
II.15. Bythium and δ: Two Elements That Arose (and Vanished) via Electrolysis 95
II.16. The Ghosts of Unnamed Elements 97
II.16.1. 1799: The Element of Fernandez 97
II.16.2. 1852: The Element of Friedrich August Genth 97
II.16.3. 1852: The Element of Carl Anton Hjalmar Sjögren 98
II.16.4. 1861: The Element of the Brothers August and Friedrich Wilhelm Dupré 98
II.16.5. 1862: The Element of Charles Frederick Chandler 99
II.16.6. 1864: The Elements of William Nylander and Carl Bischoff 101
II.16.7. 1869: The Element of Oscar Loew 102
II.16.8. 1878: The Elements of William Balthasar Gerland 102
II.16.9. 1883: The Element of Theodor Eduard Wilm 103
II.16.10. 1897: The Elements of Gethen G. Boucher and F. Ruddock 104
II.16.11. 1904: The Radium Foil of George Frederick Kunz 104
II.16.12. 1908: The Element of Clare de Brereton Evans 105
II.16.13. 1913: The Element of H. C. Holtz 107

Part III: 1869–1913: From the Periodic Table to Moseley's Law: Rips and Tears in Mendeleev's Net
Prologue to Part III 109
III.1. The Forerunners of Celtium and Hafnium: Ostranium, Norium, Jargonium, Nigrium, Euxenium, Asium, and Oceanium 111
III.2. The Discoveries of the Rare Earths Approach Their End: Philippium, Element X, Decipium, Mosandrium, Rogerium, and Columbium 119
III.2.1. Philippium and Element X 119
III.2.2. Mosandrium 121
III.2.3. Decipium and the Complexity of Didymium 122
III.2.4. Rogerium and Columbium 124
III.2.5. Conclusion 124
III.3. Lavœsium and Davyum: The Rise and Fall of Two Metals with Illustrious Names 128
III.3.1. The Discovery of Lavœsium 128
III.3.2. A Residue of Work on Platinum: Davyum 129
III.3.3. Lavœsium Falls into Oblivion 130
III.3.4. Davyum's Long Agony 131
III.3.5. Conclusion 132
III.4. The Complex Events Surrounding Two "Scandinavian" Metals: Norwegium and Wasium 135

III.4.1. The Announcement of the Discovery of Norwegium 136
III.4.2. Norwegium 136
III.4.3. A Second Claimant 137
III.4.4. The "Launching" of Wasium 138
III.4.5. The "Shipwreck" of Wasium 138
III.4.6. The Epilogue to Norwegium 139
III.5. Vesbium: An Element from the Center of the Earth 143
III.6. The Curious Case of the Triple Discovery of Actinium 147
III.6.1. The First Announcement of the Discovery of Actinium 147
III.6.2. Confessions of a Violinist 149
III.6.3. Did the Search for Neoactinium Really Delay the Discovery of Francium? 150
III.6.4. A Cold Shower at the End of a Career 152
III.7. The Improbable Elements of a Country Gentleman 155
III.8. A Bridge Between the Protochemistry of the Pharaohs and the Arab World: Masrium 158
III.9. The Demon Hidden in the Rare Earths 161
III.9.1. Provincial America Suits the Great Physicist Just Fine 161
III.9.2. The Son of a Protestant Pastor Discovers a Demon 162
III.9.3. The Tragic Conclusion 163
III.10. Dim Lights and Dark Shadows Around "Lucium" 165
III.10.1. Preview of the Discovery 165
III.10.2. The Discovery of the First "Patented" Element 165
III.10.3. The Interventions of Crookes, Fresenius, and Shapleigh 167
III.10.4. Who Was Manipulating Lucium's Strings from Behind the Scenes? 168
III.11. In the Beginning There Was Didymium... and Then Chaos Among the Rare Earths 171
III.11.1. Didymium: An Awkward Lodger in the *f*-Family 171
III.11.2. The Splitting of Didymium: Praeseodidymium and Neodidymium 173
III.11.3. A "Colorful" War: Glaucodidymium OR Glaucodymium 173
III.11.4. Claude-Henri Gorceix and Bohuslav Brauner Intervene in the Chaos 174
III.12. Sir William Ramsay: The Most "Noble" of Chemists 178
III.12.1. The First Discoveries 178
III.12.2. A Wrong Track 180
III.12.3. Anomalous Argon: The Element That Would Not Fit 181
III.12.4. A Pause in Research 182
III.12.5. Radioactivity and the Discovery of Niton 183
III.12.6. A Harvest of Laurels at the Conclusion of His Career 186
III.12.7. Postscript: Krypton II 187
III.13. Confederate and Union Stars in the Periodic Table 191
III.13.1. Introduction 191
III.13.2. Carolinium (and Berzelium) 192
III.13.3. Conclusion 195
III.14. Two Elements from the Depths of Provincial Americana 198
III.15. The Early Successes of the Young Urbain 200
III.15.1. Bauxium 201
III.15.2. From Monium to Victorium and in Pursuit of Ionium and Incognitum 202

III.15.3. The Element E or X_2 206
III.15.4. The Meta Elements 208
III.15.5. The Elements of Paul Emile (François) Lecoq de Boisbaudran and of Eugène-Anatole Demarçay 210
III.15.6. The Terbium-I, Terbium-II, and Terbium-III of Welsbach 213
III.16. The Setting of the Element of the "Rising Sun" 219
III.17. The Times Have Changed: From Canadium to Quebecium 224
III.17.1. Who Is Pierre Demers? 226

Part IV: 1914–1939: From Nuclear Classification to the First Accelerators: Chemists' Paradise Lost... (and Physicists' Paradise Regained)
Prologue to Part IV 231
IV.1. From the Eclipse of Aldebaranium and Cassiopeium to the Priority Conflict Between Celtium and Hafnium 233
IV.1.1. A Collective History: The Rare Earths 233
IV.1.2. The Lights of Paris Hide the Stars 233
IV.1.3. Celtium 235
IV.1.4. Neo-Celtium 235
IV.1.5. Celtium Doesn't Have a Leg to Stand On 238
IV.2. From the Presumed Inert Elements to Those Lost in the Dead Sea 246
IV.2.1. The Atomic Theory of James Moir and the Subelements X and Zoïkon 246
IV.2.2. The Harmonization of the Elements and the Inert Elements 247
IV.2.3. From England to Prague on the Trail of Element Number 75 249
IV.2.4. On the Banks of the Dead Sea: The First Investigations for the Identification of Element 87 251
IV.2.5. Alkalinium 252
IV.2.6. Alkalinium's Epilogue 257
IV.3. A Success "Transmuted" into Failure 260
IV.3.1. Brevium 260
IV.3.2. Lisonium and Lisottonium 262
IV.3.3. Radio-Brevium and the Missed Discovery of Nuclear Fission 264
IV.3.4. Brevium's Last Gasp 265
IV.4. From Pleochroic Haloes to the Birth of the Earth 267
IV.4.1. The Origins of the Irish Physicist 267
IV.4.2. Radioactivity Makes Dating of the Earth Possible 268
IV.4.3. Hibernium: An Elusive Element 270
IV.5. If Anyone Has a Sheep, Wolfram Will Eat It 272
IV.5.1. The Neighbors of Molybdenum and Tungsten 274
IV.6. When It Comes to New Discoveries, the More You Err, You End Up Erring More 278
IV.7. The Radioactive Element of the Hot Springs 284
IV.8. Moseleyum: The Twofold Attempt to Honor a Hero 286
IV.9. The Inorganic Evolution of Element 61: Florentium, Illinium, Cyclonium and Finally Promethium 289
IV.9.1. Florentium, the Metal of the Florentines 291
IV.9.2. The Americans Discover Illinium 295

IV.9.3. Integrity Comes with a Price Tag 299
IV.9.4. Florentium Ends Up in Court 301
IV.9.5. Cyclonium 302
IV.9.6. The Retraction of the Discovery of Florentium 303
IV.9.7. Conclusion 304
IV.9.8. Epilogue 306
IV.10. Masurium: An X-Ray Mystery 310
IV.10.1. The Discovery of Rhenium and Masurium 310
IV.10.2. No More Mention of Masurium 312
IV.10.3. Panormium and Trinacrium 312
IV.10.4. The Ignored and Underrated "Chemikerin" and Her Fission Hypothesis 315
IV.10.5. Declining Years: Sympathy for Nazism 317
IV.11. The Twilight of the Naturally Occurring Elements: Moldavium, Sequanium, and Dor 320
IV.11.1. Eka-Cæsium: From Russia to Moldavia, through Virginia 321
IV.11.2. A Digression on X-Ray Wavelength: Precision, Units, and Conversion Factors 325
IV.11.3. Eka-Rhenium: Cum Caesar in Galliam Venit, Alterius Factionis Principes Erant Haedui Alterius Sequani. . . 326
IV.11.4. Alabamine and Virginium 327
IV.11.5. Eka-Iodine Assumes the Fanciful Name of Dor 331
IV.11.6. Conclusion 333
IV.12. A Cocktail of Chemistry and Espionage: Helvetium, Anglo-Helvetium, and a Pair of Indian Elements 337
IV.12.1. Rajendralal De and His Twin Elements: Gourium and Dakin 337
IV.12.2. Walter Minder and Helvetium 339
IV.12.3. Alice Leigh-Smith and Anglo-Helvetium 340
IV.12.4. C. W. Martin and the "Elusive" Parentheses of Leptine 343
IV.12.5. Academic Conflicts with Hulubei, Paneth, and Karlik 343
IV.12.6. Conclusion 345
IV.13. Is Failure a Severe Master? 348
IV.13.1. Eline 348
IV.13.2. Verium 348

Part V : 1939–Present: Beyond Uranium, to the Stars
Prologue to Part V 351
V.1. The Obsession of Physicists with the Frontier: The Case of Ausonium and Hesperium, Littorium and Mussolinium 353
V.2. *Finis Materiae* 361
V.2.1. The Island of Nuclear Stability 365
V.2.2. Unfortunate Episodes in the Attribution of the Names of the Elements Between 101 and 109 366
V.2.3. From Atoms to the Stars 367
V.3. The Search for Primordial Superheavy Elements: Between Scientific Rigor and Atomic Fantasy 370
V.4. Names, Names, and Names Again: From A to Zunzenium 375

V.4.1. The Elements from Neptunium to Mendelevium Seen from Both Sides of the Iron Curtain 375
V.4.2. The Step Longer than Its Leg: Nobelium 381
V.4.3. Chaos Surrounds Lawrencium, Rutherfordium, Dubnium, and Seaborgium 383
V.5. Do We Have to Live With Fantasy? Hawkingium and Zunzenium 391
V.6. Naming the Last Five Arrivals in the Great "Family of the Transuranium Elements" 394

Part VI: No Place for Them in the Periodic Table: Bizarre Elements
Prologue to Part VI 401
VI.1. Inorganic Evolution: From Proto-Elements to Extinct Elements 403
VI.1.1. A Step Backward: Prime Matter, *Andronia*, and *Thelyke* 403
VI.1.2. Pantogen 405
VI.1.3. Protyle 406
VI.1.4. Other Theories of Chemical Evolution 408
VI.1.5. The Asteroid Elements 410
VI.1.6. The Painful Finale 412
VI.2. Dazzling Traces of False Suns 415
VI.2.1. The Mirage of the Source of Stellar Energy 415
VI.2.2. The Curious Appearance of Kosmium and Neokosmium 417
VI.3. From the Nonexistent Elements of Mendeleev to the Puzzle of the Existence of the Ether 419
VI.3.1. Coronium and Its Aftermath 421
VI.3.2. The Geocoronium Hypothesis 422
VI.3.3. Etherium: Elementary Gas or Subatomic Particle? 423
VI.4. Anodium and Cathodium 426
VI.5. The Exotic Damarium 428
VI.6. Subtle Is the Air: The Case of Asterium 431
VI.7. Clairvoyance as a Means of Investigating Some "Occult Elements" 435
VI.7.1. A Clairvoyant Investigates the Structure of New and Old Atoms and Their Position in the Periodic Table 435
VI.7.2. The Last Years of the Three Clairvoyants 440
VI.8. William Harkins's Element Zero: Neutronium 443
VI.8.1. A Place in the Periodic Table for the Element Without a Nuclear Charge 444
VI.8.2. From the Nuclear "Alphabet" to the Hypothesis of Neutronium 444
VI.8.3. William Draper Harkins: A Versatile and Obstinate Chemist 446

Part VII : Modern Alchemy: The Dream to Transmute the Elements Has Always Been with Us
Prologue to Part VII: Alchemy Then and Now 449
VII.1. A Piece of Research Gone Up in Smoke: Decomposition of Tungsten into Helium 451
VII.2. Transmutations of Mercury into Gold 453
VII.3. Transmutations of Silver into Gold 457
VII.4. Transmutation of Ores 461

VII.5. Other Transmutations 464
VII.6. Biological Transmutation 465
VII.7. The Transmutation of Hydrogen into Helium and Neon 468
VII.8. Radiochemistry: A Child of Both Physics and Chemistry 470
VII.8.1. Willy Marckwald Makes His Mark: The Polonium Controversy 471
VII.8.2. William Ramsay "Out of his Element" 472
VII.8.3. Tellurium X 473
VII.9. Transmutation of Lead into Mercury 476
VII.10. Some Like It "Cold" 478
VII.11. Is Cold Fusion Hot Again? 480
Epilogue 480

Appendix: Chronological Finder's Guide for the Lost Elements 483

Bibliography 493
About The Authors 499
Name Index 501
Lost Element Name Index 515
Subject Index 521

PREFACE

I have not read as truly interesting a book as this one in decades—dip into it, open it on any page, and you are immediately drawn into a tale of human ambition, folly, and ... ingenuity. Mostly chemical, too. Two pages later, there's another, even more fascinating story. Primo Levi would have loved this book. There is in it material for a dozen operas. Or is it reality shows?

Why? This question tugs at me. Why have chemists (and, in time, physicists) focused so much on the discovery of the elements? When the heart of chemistry, especially today, but even in the past, is in discovering the semi-infinite variety of molecules and compounds that they can form, why all this nervous energy and hard labor devoted to finding the building blocks, when the soaring bridge, mosquito, or antibiotic constructed from those pieces is so much more valuable, both materially and spiritually?

As I reflect on the obsessions that drove those people who sought what turned out to be spurious elements, who spent years at good chemistry (you will learn here of Lorenzo Fernandes's and Giorgio Piccardi's 56,142 fractional crystallizations of 1,200 kg of rare earth oxalates over 17 years in their search for *florentium*), I am led to think of the following potential motives:

1. The desire in us (both religious and scientific in its origins) to get to the beginning of things, to the fundamental *idea* of the element. Even if we know (or believe) that reductionism may be destructive in practice—that the way to the fundamental strips away the beauty of what people have created even though lacking knowledge of the fundamental—we really do want to know what "the natural body or bodies, one or many, of which all things consist" (Davis, 1931) are.
2. In his satire, "The Dunciad," Alexander Pope had the goddess of Dullness expose a new king to "vapours blue" and then tell him, inter alia:

> Hence the fool's Paradise, the statesman's scheme,
> The air-built castle, and the golden dream,
> The maid's romantic wish, the chemist's flame,
> And poet's vision of eternal fame.

Delusions of fame are the bane of humanity. I think of my old copy of what we called "The Rubber Book," the encyclopedia we saved money to buy volume by volume—these had simplistic, categorical attributions of discovery. As did handbooks of a 100 years ago. As do chemistry webpages today. How nice it would be to have your name in them! In the dull confines of a smelly laboratory, a scientist could aspire to embark on the chemical equivalent of the great European voyages of discovery.

And, if you found an element, you could also name it. Maybe it wouldn't seem so selfish then—maybe the name of your town or country would do nicely. A human weakness, one that shows no signs of abating in the 21st century.

3. What a challenge to the chemist's analytical prowess was the establishment of a new element! To isolate the tiny residue that is truly different, after many transformations wrought on it. Then to reduce it, in the old way, with hydrogen, to a speck of metal. Or, later, to look at its spectrum. The craftsmanship, the good hard chemical labor, in the service of a paradigmatic search for something new, pervaded the style of the inorganic chemists who searched for new elements. They could not yet see into the way the atoms were arranged in their compounds; their transformations were all they had. They were right to be proud of their skills.

In this lovingly researched book you have the dead ends, the voyages of discovery whose end is certain shipwreck. And you have here a superlative antidote to the hagiographical seduction of the stories, often just as complex in detail, of the reliable identification of new elements. Although some, such as Ramsay's wonderful identification of the noble gases, are retold here, these have been admirably recounted elsewhere. In "The Lost Elements," the failures speak to us. Completely lacking in the false condescension of "How stupid can you be?," the byways recounted in this book turn into lovely meandering paths, leading to an understanding of how chemistry really works.

—Roald Hoffmann

ACKNOWLEDGMENTS

In such a huge volume, it is impossible to remember adequately the enormous and largely hidden contributions of so many persons who, in conversations or with answers to our questions, have supplied information and suggested new lines of investigation.

Most importantly, this book has been produced with the financial support of Ente Cassa di Risparmio di Firenze (ECRF), Florence, Italy, for which the authors are sincerely grateful.

It has taken more than a dozen of years of hard work to bring this volume to fruition. It was not a continuous task, but definitely a difficult and exciting one. Over such a long period of time, many people were involved in our surveys, and if, in these official acknowledgments, someone has been omitted, we deeply apologize.

First of all, many thanks to Francesca Salvianti for her patient proofreading and for her many suggestions which greatly improved the text. Special thanks also go to these tireless friends: Alessandra Beni, Barbara Nardi, Massimiliano Nardi, Alessandro Ciandella, and Serena Bedini. Professor Juergen Heinrich Maar of the University of Florianópolis was a lively presence during the writing of this volume. Marco Fontani's colleagues at the University of Siena, all equally involved in this task, deserve a great big thank-you: Rosanna Meiattini, Marisa Pietrobono, Paola Fontani, Enrico Bellucci, Stefania Casati, Simonetta Ferri, Arnaldo Cinquantini, Franco Laschi, Donato Donati, Piero Zanello, and Emanuela Grigiotti.

The writing of the manuscript took so much time that one of the authors (Marco Fontani), soon after receiving his Ph.D., moved to the University of Florence, where he wishes to thank his new colleagues: Emily Bulukin, Chiara Carbonera, Piero Sarti-Fantoni, Ivano Bertini (1940–2012), and Paolo Manzelli. The personnel at the Polo Scientifico di Sesto Fiorentino Library, Sabina Cavicchi, Marzia Fiorini, Roberto Bongi, Laura Guarnieri, and Serena Terzani, and at the library of the College of New Rochelle, Carolyn Reid and Kathy Mannino, deserve special thanks; they were able to find, with indomitable enthusiasm, all the material we needed, overcoming the numerous difficulties created by so many and such unusual requests.

If Evan Melhado and Paul W. Bohn, Linda Stahnke and Susan Fagan of the University of Illinois, and Mary Jo Nye of Oregon State University were the first (1995) to help us, in recent times Emeriti Professors William H. Waggoner and Gilbert P. Haight greatly enriched the text with personal recollections.

The scientific, economic, and moral support of Luigi Campanella—President of the Italian Chemical Society—was indispensable in every respect in the production of the Italian Edition (2009). Special thanks to the Committee of the International Congress of History of Chemistry, Maria Elvira Callapez, Isabel Malaquias, and Ernst Homburg, and, on the Italian scene, to Ferdinando Abbri, Prince Paolo Amat di San Filippo, Giuliano Dall'Olio, Raffaella Seligardi, Marco Taddia, Andrea Karachalios, and Patrizia Papini of the Gruppo Nazionale di Storia e Fondamenti della Chimica.

Special acknowledgments go to Anna Grandolini of the Academy of Sciences of the XL and to Professor Rocco Capasso of the Italian Society for the Advancement of Science. Professors Masanori Kaji, Tokyo Institute of Technology, and H. Kenji Yoshihara, Japan Isotopes Data Institute, have courteously supplied us with abundant material concerning the discovery of *nipponium*. We also thank Peter van der Krogt, who published a specialized webpage dedicated to the chemical elements and their history; it was a mutual exchange of knowledge and a privilege for us when he published some our original papers.

Personal memories were decidedly indispensable for this volume. Gianluigi Calzetta recalled the last years of Luigi Rolla (1882–1960), whereas Enrico Franceschi, Antonella Tassara, Rocco Longo, and Aldo Iandelli (1912–2008) of the University of Genoa told us many other stories about him. Fruitful talks with Anna, Maria Stella (1924–2012) and Oretta Piccardi, daughters of Professor Georgio Piccardi (1895–1972), revealed an unusual aspect of the *florentium* story. Other personal memories of *florentium*'s chemists were collected with the collaboration of Maria Angela Maggiorelli-Canneri (1925–2011), daughter of Professor Giovanni Canneri (1897–1964) and through the courtesy of Gianluigi Fernandes, son of Dr. Lorenzo Fernandes (1902–77). Mario Galli and Fausto Fumi and Mrs. Fumi of Genoa told us about the vicissitudes related to the epilogue of *masurium* with the aim of honoring the memory of their teacher Carlo Perrier (1886–1948), co-discoverer of technetium. Natalina Angeli (1920-2007), Verdelaura Angeli (b. 1915), and Giovanni Batista di Giusto established a solid link to the memory of the great chemist Angelo Angelo (1864–1931). Marie-France Le Fel, Christopher Leigh-Smith, Carla Grazioli (of the British Council) and Mirta Alessi (of Study & Cultures, UK) supplied important information on Alice Leigh-Smith (1907–87) and *anglo-helvetium*. We thank Dr. Franziska Rogger, head archivist of the Fachbereichsbibliothek Bülplatz at Bern, who provided us with a great amount of archival material. The descendants of Georges Urbain (1872–1938)—Daniel Urbain, Michel Wagner, and Jean-Rémy Bost—have brought into clear focus the mighty image of the versatile discoverer of *neo-ytterbium, lutecium*, and *celtium*. We also thank José Claro-Gomes, Bernadette Bensaude-Vincent (Université de Paris), and Paul Ernest Léon Caro (Académie des Sciences) for their helpful information and their personal letters about Urbain. Dr. M. Ghiara of the University of Naples gave us all the existing documentation on the presumed element *vesbium*, while Dr. Edgar Swinne sent us copies of many scientific articles by his father, Richard Swinne (1885–1939), on the naturally occurring superheavy elements coming from a mysterious interstellar powder captured ages ago in the Arctic ice.

We have benefited from the courteous and information-packed letters of Oliver Sacks and George B. Kauffman (California State University, Fresno). Our friendly conversations with Barbara Scardigli-Foster, Camilla Cyriax, Marquis Benedetto Barbagli-Petrucci, Franco Samoggia, and Katharina Gemperle have enriched our knowledge and delighted us. We are likewise indebted to Francesco Michelazzo, Pierre Demers, Bruna Peri, Paola Passadore, Eleonora K. Kochetova, Boris I. Pokrovskii, Elisa Novaresi, Gayle de Maria, Marco Nucci, Francesca di Poppa, Boris Noskov, Anna Simonini, Antonella Leone, Ray Goerler, Richard E. Rice, Dario Ventra, and Gioiosa Bindi-Brogi. We are also grateful for our profitable and numerous meetings with Nicholas Long (Imperial College, London), and the kind correspondence with Didier Trubert (Institut de Physique Nucléaire) and Janis Stradins (Latvian Academy of Sciences). Emerita Professor Ruth Lewin Sime supplied unique clues and critical information enabling a complete study of Fermi's attempt to synthesize the first transuranium elements. Darleane C. Hoffman, Jean-Pierre and Madeleine Adloff supplied hard-to-find references, for which we are very grateful.

We wish to remember those persons who, during the long gestational period of this book, passed away: Professors Michele Della Corte (1915–99) and Jacob A. Marinsky (1918–2005), discoverer of promethium, both wonderful speakers and amiable gentlemen. When, in 1998, one of the authors (Marco Fontani) was in Boston, he had the rare good fortune to reminisce with Glenn T. Seaborg. The past Rector of the University of Florence Enzo Ferroni (1921–2007) was a font of useful information; Jean Busey Yntema (1898–2002) and her children Douwe B. Yntema (1923–2000) and Mary Katherine, supplied memories on the history of *illinium*, as did the youngest son of Professor B Smith Hopkins (1873–1952), B Smith Hopkins Jr. (1912–2003). Franco Berruti, director of the Insitute for Chemicals and Fuels from Alternative Sources (Canada), Nicoletta Barbarita of the Candian Embassy in Rome, and Joan Gosnell of Southern Methodist University supplied us with other details about *illinium*. Retiring professor Franco Rasetti (1901–2001), from his volunteer "exile" in Belgium, wrote numerous letters concerning *ausonium* and *hesperium*. Madame Eve Curie-Labouisse (1904–2007), daughter of Marie Curie, and her nephew Pierre Joliot-Curie, related to us the "complex alchemies" of *catium* at the Institut du Radium.

We wish to thank all those who generously supplied us with photographs, especially the family of Jaroslav Heyrovský, the Auer von Welsbach Museum, the American Theosophical Society, and the Chemical Heritage Foundation.

The authors would like to profoundly thank Professors James Marshall (University of North Texas) and Gregory S. Girolami (University of Illinois at Urbana-Champaign), as well as our anonymous reviewers, for their unstinting and meticulous reading of the manuscript and for their many suggestions, criticisms, scholarly observations, and corrections, which vastly improved the final form of this volume. Thanks are due also to Vera V. Mainz, Secretary-Treasurer of the ACS Division of the History of Chemistry, for solving some of our logistical problems. And kudos to our fine editor, Jeremy Lewis, who gave the final form to this book, for which we are most grateful.

Last in this list of acknowledgments, but no less important, we remember the collaboration of a marvellous citizen of the world who devoted her life equally to astrophysics and civil rights: Margherita Hack (1922–2013). She supplied us with all the information concerning the false discoveries of extraterrestrial elements, such as *etherium, newtonium, nebulium*, and *coronium*.

NOTE TO THE READER

This book is divided into seven sections, arranged largely chronologically.

Part I consists of announcements of discoveries that precede the formulation of the concept of "chemical element" in 1789, the conventional date that coincides with the publication of the "Traité Élémentaire de Chimie" by Antoine Lavoisier, which is considered the first modern treatise on chemistry.

Part II embraces the period from 1789 to 1869, the date of Dmitri Mendeleev's formulation of the periodic table of the elements.

Part III ends at the very beginning of World War I (1914), a period of relative elemental chaos, with research following the guiding logic of Mendeleev's empirical organizing principle but lacking a theoretical basis.

Part IV (1914–39) takes us through Moseley's revolution and Soddy's isotopic theory, to the advent of the synthesis of new nuclides with the aid of the first linear accelerators and cyclotrons.

Part V takes us up to the present day and consists largely of the syntheses of the transuranium elements, but also includes fanciful and imaginative stories of elusive elements whose atomic numbers were less than one ("heaps" of neutrons).

Part VI is devoted to those elements so bizarre that, if they had ever been discovered, they would never have found a place in the periodic table.

Part VII is dedicated to the most recent attempts at chemical transmutation, not in the alchemical sense, whose history would lead us far from the aim of this volume, but with physical instruments or cumbersome apparatus whose use was carried on in a "better way to obtain erroneous results."

Elemental names appearing in italics throughout the book are those of chemical elements that were false, spurious, not confirmed, or even correct but that have fallen into oblivion or whose use has been lost or changed over time. They occupy a separate index at the end of this volume.

A word about units: The units used in this book are the standard international units but, where appropriate, units mentioned in the original documents but no longer in use are also reported. A word about the persons who appear in this book: For those persons whose scientific discoveries are pertinent to the narration, a fairly extended biography is supplied. Birth and death dates (when known) are contained in the Name Index at the end of this volume and follow the person's name when first mentioned in the book.

INTRODUCTION

WHY COLLECT INTO ONE VOLUME THE DISCOVERIES OF ELEMENTS THAT HAVE BEEN SHOWN TO BE ERRONEOUS OR HAVE BEEN FORGOTTEN?

In 1961, Denis Duveen asserted that we cannot properly understand chemistry without a knowledge of its history.[1] This idea has subsequently been enlarged upon in the literature of chemical education. Its history "is not only a chronologically organized set of facts, but also a coherent picture of the origins of ideas, their development, and their influence and consequences for human civilization,"[2] and an aid to understanding how chemists have solved problems in the past, thus revealing the nature of the scientific process.[3] It is hoped that the contents of this volume will help readers understand that the pathway to the classification of the elements was fraught with obstacles and errors that actually, in the long run, helped to clarify the nature of these fundamental units of matter. One might even attribute the role of catalyst to some of these errors, in much the same way that the famous, but brilliant, blunders of Charles Darwin (1809–82), Linus Pauling (1901–94), and Albert Einstein (1879–1955) have become the stuff of legend.[4]

Although physicists have as their purview the birth of the universe, and biologists concentrate on the origin of life, chemists have a unique role to play in the ordering of the building blocks of nature, namely, the development of the periodic table of the elements. This single document embodies much of our knowledge of chemistry and, as such, has become emblematic of our discipline. However, the table as it has come down to us has undergone many changes over the two centuries of its evolution. Although certain relationships were initially discerned among the elements, how to order them was not always clear. An order based on atomic weight seemed to present the best approach in the mid-19th century, but many atomic weights had been determined erroneously, and, in addition, some anomalies in the properties of elements were observed. So it gradually became clear that there were missing pieces to the puzzle that had to be found. As well, the ordering attempt revealed some obvious gaps that led chemists to seek the missing elements—an enterprise that was notably successful.

And perhaps too successful—because once chemists realized that there were elements "out there" to be discovered, it was open season with no limits, particularly when it came to the higher atomic weight elements. It was only with the establishment of the atomic number as the primary ordering principle in 1913 that some "sense" could be made of the table. Yet, at the same time, the discovery of radioactivity and the seemingly endless "new" elements that made their appearance in research laboratories only served to create more confusion in what belonged and what did not belong in the periodic table.

Although today's periodic table presently comprises 118 elements, 114 of which bear definitive names, over the many decades of its creation this was not always so. In fact, there are many more elemental "discoveries" later shown to be false than there are entries in the present table. Some of these were good-faith errors, some were the result of personal wishful thinking, some were the fantasy children of pseudoscientists—and all have their fascinating stories that serve to illustrate the fact that our present knowledge came about in fits and starts, with many dead ends, regrettable personal and political battles, and sad retractions.

This fascinating journey, one that predates Dmitri Mendeleev (1834–1907) by several centuries, is what prompted us to write this book. Gathering the stories and the documentation of these erroneous, spurious, nonexistent "lost elements" into one place is our attempt to faithfully reconstruct the "scene of the crimes," so to speak. It should be borne in mind that many of the elements presently familiar to us did not have the same names that we use today: names that fell into oblivion were often the result of false claims and priority struggles. Other false elements actually occupied space in the periodic table as temporary "tenants" until they were proven false. The trail we have chosen to embark on is not exactly a beaten path, which is why it has taken the authors 14 years to collect and filter the material for this volume. It was first necessary to trace the history of the concept of the element, how elements were eventually defined, and then how scientists went about identifying them. The latter endeavors are documented mainly from primary sources.

With these premises, our book was born. We have written it as an informative and sometimes anecdotal compendium of the shadow side of the chemical elements, mirroring the tenacious dedication of Mary Elvira Weeks[5] (1892–1975) to the bright side of their discoveries. We feel that this effort is important because if we accept the premise that the history of science is not a collection of information but a tool to analyze that information and arrive at valuable conclusions,[6] then we offer this volume as an analytical tool to anyone who wishes to use it to develop research ideas and to draw helpful and valuable conclusions.

Finally, we would like to mention a popular website dedicated to the elements, http://elements.vanderkrogt.net/, an endeavor different from, but parallel to, our own. Both projects developed side by side over the past decade, and much of our own work was shared with and acknowledged by the developer of the website. On this site, you will also find the names of false elements, but only for those that eventually became attached to elements presently in the periodic table.

HOW "AN ELEMENT" BECAME A "CHEMICAL ELEMENT"

Before the modern model of the atom evolved, the concept of an element had been purely speculative. Aristotle (ca. 382–22 BCE), one of the greatest philosophers of antiquity, theorized on the nature of what he called principals, elements, substances, and numbers, frequently interweaving all four in a complex philosophical dance. Drawing on the writings of others, he said "Empedocles... was the first to speak of four material elements; yet he does not use four, but treats them as two only; he treats fire by itself, and its opposite—earth, air, and water—as one kind of thing," and again, "[Leucippus and Democritus] say the differences in the elements are the causes of all other qualities."[7] He enlarged on these ideas thus: "the elements out of which... units are said to be made are indivisible parts of plurality" and "everything that consists of elements is composite."[8] Aristotle's teaching was the theoretical foundation of alchemy and of various Western schools of natural philosophy for many centuries thereafter.

In contrast to the Aristotelian idea of earth, air, fire, and water comprising the four elements, early Chinese naturalists centered on five: earth, wood, metal, fire, and water—not so much as five types of fundamental matter, but as five ways in which matter was fundamentally related through process and only manifest when they were undergoing change.[9] This idea of change as a necessary property of matter presages, although not in so many words, the entire basis of the discipline of chemistry.

In the 16th century, Aureolus Philippus Theophrastus of Hohenheim, called Paracelsus (1493–1541), a famous physician and scientist, brought the elements "down to earth." Still believing in the four elements,[10] he introduced the idea that on another, spiritual level, all substances consist of three sources: mercury, salt, and sulphur, which are the carriers of three qualities—volatility, solidity, and inflammability.[11] Clues to a proper understanding of the nature of elements can be found in the teaching of Robert Boyle (1627–91), an outstanding 17th-century English chemist. In his book *The Sceptical Chymist*, Boyle criticized the view of elements as carriers of certain qualities. Elements, according to Boyle, must be material in their nature and constitute solid bodies. He contended that mere theoretical examination, without experiment, was quite insufficient, remarking with a stiff dose of sarcasm that "when I took the pains impartially to examine the bodies themselves that are said to result from the blended Elements, and to torture them into a confession of their constituent Principles, I was quickly induc'd to think that the number of the Elements has been contended about by Philosophers with more earnestness then [sic] success."[12] Boyle, although admitting that he had no effective system to offer in place of the philosophies that he attacked,[13] also spoke against the belief that the number of elements is limited, thus opening up the possibilities for the discovery of new elements.

Antoine-Laurent Lavoisier's (1743–94) views were a considerable step forward in this direction. Early on, he cast doubt on the idea of four basic elements as propounded by the Greeks:

> On ne manquera pas d'être surpris de ne point trouver dans un traité élémentaire de chimie un chapitre sur les parties constituantes et élémentaires des corps; mais je ferai remarquer ici que cette tendance que nous avons à vouloir que tous les corps de la nature ne soient composés que de trois ou quatre éléments tient à un préjugé qui nous vient originairement des philosophes grecs. L'admission de quatre éléments, qui, par la variété de leurs proportions, composent tous les corps que nous connaissons, est une pure hypothèse, imaginée longtemps avant qu'on eût les premières notions de la physique expérimentale et de la chimie.[14]

He clearly stated his concept of simple bodies: he believed that all substances that scientists had failed to decompose in any way were elements, and he divided all simple substances into four groups. The first group consisted of oxygen, nitrogen, and hydrogen, as well as light and *caloric* (which was, of course, an error). Lavoisier considered these simple substances to be real elements. In the second group Lavoisier included sulfur, phosphorus, coal, a radical of muriatic acid (later called chlorine), a radical of hydrofluoric acid (fluorine), and a radical of boric acid (boron). According to Lavoisier, these all were simple nonmetallic substances capable of being oxidized and of producing acids. The third group contained 17 simple metallic substances "oxydable and acidifiable," ranging from antimony to zinc. And, last, the fourth group included salt-forming compounds (earths), which, however, were known to be complex, including lime (calcium oxide), alumina (aluminum oxide), silica (silicon oxide), and magnesia (magnesium oxide).[15] In 1789, the idea that these substances were oxides of unknown elements was only a conjecture. This classification and the comments about it were still greatly confused and unclear, but, nevertheless, they served as a program for further research into the nature of the elements.

It has been argued that Lavoisier drew no distinction between the concepts of "an element" and "a simple body." In fact, in his preface to the *Traité*, he commented that the elements "were all the substances in which one is capable by any means of reducing... by decomposition, and if they may be compounded, we should not suppose them to be so unless this can be proven by experiment and observation." This so-called analytical approach[16] to the elements, one that concentrated on concrete laboratory substances as opposed to metaphysical speculation about the ultimate components of substances, was central to the chemical revolution. The caveat is that this idea is at best a criterion for when a substance should be recognized as an element rather than a definition of "element." Lavoisier's notion of element is thus compositional: he understands the behavior of composite substances to be a direct consequence of the basic substances that they contain. Essentially, Lavoisier assumed that elements survived in their compounds and that they could be recovered from their compounds by the process of decomposition. This is opposed to Aristotle's theory of chemical combination, in which the ultimate components do not persist unchanged in more complex bodies. Actually, Lavoisier did not consciously seek to demolish an abstract concept of elements, but he did seriously compromise the "Aristotelian four" by demonstrating that fire was not a weighable substance but a phenomenon. Seventeenth-century conceptions of the elements—in which they were not viewed as material components of laboratory substances but simply as contributing to the characteristics of composite substances—were much closer to the Aristotelian view.

Paralleling Lavoisier's ideas regarding the nature of elements was his concern with the state of chemical nomenclature at the time. Early on, he criticized the vagueness of chemical expression compared to the precision he found in mathematics and physics. In describing the results of some of his pioneering experiments, particularly with gases, he found it necessary to coin some terms and to find a way of expressing the difference between gases and their aqueous solutions. Meanwhile, Louis-Bernard Guyton de Morveau (1737–1816),[17] probably in early 1787, traveled to Paris to discuss the new antiphlogistic theory with Lavoisier. Guyton de Morveau, under the influence of Torbern Bergman (1735–84), had been ruminating about a new system of chemical nomenclature for many years, so their conversation quickly turned to that topic. Joining them to examine Lavoisier's experiments in support of the new oxygen theory were Antoine-François Fourcroy (1755–1809) and Claude-Louis Berthollet (1748–1822). From discussing the new theory, they went on to discuss the possibility of reforming chemical nomenclature. This historic meeting resulted in the collaborative publication of the *Méthode de Nomenclature chimique*[18] in the summer of that same year.[19] The new nomenclature was itself based on the principle that a body's name ought to correspond to its composition, thus consolidating one more brick in the structure of the chemical revolution.

John Dalton (1766–1844), in 1808, presented a theory of atomism that mirrored Lavoisier's compositional theory: each of Lavoisier's elements possessed a stable, substance-specific kind of atom that survived chemical change.[20] Whereas Lavoisier's very successful definition made no reference to atoms (thus making it acceptable to anti-atomistic chemists such as Wilhelm Ostwald and Marcellin Berthelot), Dalton connected his hypothetical atoms with the elements, proposing that the chemical elements were composed of atoms and that the atoms of a given chemical element were all identical, having the same mass.[21] These ideas were widely accepted and greatly clarified over the course of the 19th century owing to the development of atomic and molecular theory and to the work of Dmitri Ivanovich Mendeleev. The development of the concept of the chemical element at this time was as twisted a pathway as the discoveries of the individual elements, both true and false, and makes for very interesting reading.[22,23]

Amazing new discoveries and developments toward the end of the century heralded great changes in what, by now, was considered the classical concept of the element. When, in 1894, Lord Rayleigh (1842–1919) and William Ramsay (1852–1916) announced the discovery of a monatomic gas with an atomic weight of 39.8, it was thought that this event presaged the toppling of the periodic system. When, in 1896, Henri Becquerel (1852–1908) realized that the penetrating emanation coming from uranium ore was a property of the material itself and not the result of impinging radiation, he had to hypothesize that the uranium was spontaneously undergoing a change. But what was it changing into? How could a hitherto stable, substance-specific simple body be changing right before his eyes? In 1897, J. J. Thomson (1856–1940) discovered corpuscles (later called electrons) being ejected from the atoms of gases in his cathode ray tubes and concluded from further experiments that they were fundamental to all matter, he demonstrated that atoms were not indivisible. In 1898, Marie Skłodowska Curie (1867–1934) and Pierre Curie (1859–1906) discovered two new elements that were far more radioactive (a word coined by Marie) than the parent substances. And, in 1902, Ernest Rutherford (1871–1937) and Frederick Soddy (1877–1956) realized that radioactive substances were actually spontaneously transmuting into new chemical substances.[24, 25] It was clearly time to reexamine the classic idea of the nature of the element and the nature of the atom.

The most immediate problem centered around the idea that atoms were not the immutable building blocks of nature but actually possessed a composite nature consisting of at least electrons and other yet to be determined components. Radioactive transmutations were seen to undermine the very foundations of chemistry. Hence, Mendeleev, and with him many other chemists, was hostile to these new discoveries and to the conclusions drawn from them. Worst of all, these scientists envisioned the actual demise of chemistry by a descent back into alchemy, on the one hand, and a loss of autonomy to physics, on the other.

The year 1913 was crucial. In that year, Henry Moseley (1887–1915) demonstrated that one could identify an element and its numerical place in the periodic table purely by measuring the X-rays it emitted. Frederick Soddy proposed the notion of *isotopes*, wherein two atoms could be chemically identical, with the same atomic number (a consequence of Moseley's Law), but have different atomic weights. These notions were very difficult for chemists to accept: many denied that isotopes behaved in exactly the same way. Gradual acceptance followed, helped along by the discovery of hafnium (the first element to be discovered on the basis of atomic number) and impeded in other ways by the differing mindsets of traditional chemists and those trained to think in terms of physics and physical chemistry. Echoes of these differences resound throughout this book, and they will be easy to identify.

We can think of no better way of expressing the evolution of the concept of element than these words of Tenney L. Davis:[26]

> During the period preceding [Stanislao] Cannizzaro[27] an element was a substance whose combining weight was one particular number or some small multiple of that number. Not long thereafter an element became a substance which had one atomic weight and only one, but that state of affairs did not long prevail. Isotopes were discovered, and an element now is a substance which has an atomic number. The weight-test has disappeared. We resort to X-ray spectra. To the question, What sort of things are the elements? the answer was once given—hard impenetrable atoms which differ in shape, then atoms which differ in weight, compressible atoms, arrangements of electrons and protons, and now apparently arrangements of waves. Yet our abstract notion of element—the natural body or bodies, one or many, of which all things consist, from which they arise, into which they pass away—is the same today as it was in the time of Lavoisier or Boyle, Aristotle or Thales.

IS THERE ANY ORDER TO THE DISCOVERIES OF THE ELEMENTS?

Which element was discovered first?[28] For almost 10 elements, chemistry books report only the words "known from antiquity." The concept of antiquity is rather loose, and the words mean that these elements were known long before our time. Of course, we do not know who discovered them, although archaeologists can give us more or less reliable information on the time when an element was first used by humans in antiquity (without, of course, being perceived as an element). Elements known in antiquity were iron, carbon, gold, silver, mercury, tin, copper, lead, and sulfur. All these elements differ broadly in their properties. Are they the most abundant elements on earth? As regards abundance, only iron and carbon are among the 10 most abundant elements. Sulfur is also fairly abundant, but the other elements are quite rare on earth. The most abundant elements are oxygen, silicon, and aluminum. Oxygen, the most abundant, was not recognized as a chemical element until the end of the 18th century. Silicon, the main solid component of the earth's crust, was discovered only in the 19th century, as was aluminum, although clay (alumina) had been used for ages.

The natural abundance of the terrestrial chemical elements is by no means related to the dates of their discoveries. Hence, most of the elements known from antiquity occur in nature as simple substances. Gold, silver, and sulfur occur on earth in the free state (although sulfur is mainly a constituent of minerals); copper and mercury are encountered in the free state much less frequently. But the reason that these elements were among the first to be discovered is that their compounds are easily reduced in the presence of carbon (charcoal). Many scientists believe that our forebears first began to use iron in the free state as meteoritic iron.

The age of discovery of chemical elements began only in the second part of the 18th century.[29] Preceding millennia had seen the discovery of only five new elements: arsenic, antimony, bismuth, phosphorus, and zinc. They were discovered by alchemists who were vainly looking for the Philosopher's Stone but who were also engaged in metallurgy, medicine, and other material occupations that increased the frequency of their chance discoveries. As time moved on, discoveries became increasingly linked to the interpretation of observations and the incorporation of facts and their interpretations into some kind of theoretical framework, along with scientists' greater skill in handling the complex materials given them by Nature.[30]

The discovery of new chemical elements became a routine matter and not a stroke of good luck only after two main conditions had been fulfilled[31]: first, chemistry began to take shape as an independent science, and scientists learned how to determine the composition of minerals. Second, most scientists at last reached a consensus on the concept of chemical element. It was the beginning of the great analytical period in the history of chemistry, in the course of which many of the naturally occurring elements were discovered.

Various analytical methods, constantly being improved, were the key factors that led, step by step, to the discovery of new chemical elements. But chemical analysis by itself was not enough to fill all the boxes in the periodic table. Scientists divined the existence of many new elements not because they discovered them, figuratively speaking, lying on the bottom of a test tube.

Some elements do not form their own minerals but exist only as admixtures to all sorts of minerals containing other elements. They seem to be widely dispersed in the earth's crust and are called *trace elements*, often announcing their presence through a peculiar "visiting card": their spectrum. If a grain of a substance is introduced into the flame of a gas burner and the light is passed through a prism, the refracted light contains a number of differently arranged spectral lines of various colors. By studying the spectra of known elements, scientists came to the conclusion that each element had its own "spectral portrait." Spectral analysis was immediately recognized as a powerful research tool. If the spectrum of a certain substance contained unknown lines, it was logical to assume that this substance contained a new element. However, in such cases, it took courage for scientists to announce the existence of new elements because they did not have a single atom in their hands and did not know the unknown element's properties.

Naturally, the history of the discovery of chemical elements was to a certain extent affected by the abundance factor: those elements less abundant in nature were discovered later than many others. All of them were discovered within about a quarter of a century, from the very end of the 19th century into the beginning of the 20th century. These elements would have remained hidden for an even longer time if analytical techniques had not included the measurement of radioactivity.

Some substances spontaneously emit electromagnetic radiation and matter. At first, it was believed that this phenomenon was a property of certain minerals, but later it was realized that radioactivity was an atomic property, typical of heavy elements like uranium and thorium. When scientists noticed that the radioactive output of a given mineral was greater than its uranium and/or thorium content, they assumed the presence of another element: an unknown element. Polonium and radium were thus discovered. This led to another research method—the radiometric method— that, in the long run, led to the discovery of other naturally radioactive elements.[32]

Radiochemistry gave rise to the development of a new method of analysis, much more sensitive than those previously used. Through its use, by the end of the 1920s, all naturally occurring elements had been discovered. However, this was not the end of the discovery of elements.

In 1934, Irène (1897–1956) and Frédéric Joliot-Curie (1900–58) created the first synthetic isotopes of naturally occurring elements.[33] Thus the expression "discovery of a new element" acquired a new meaning. In 1937, with the aid of nuclear reactions, the chemist Carlo Perrier (1886–1948) and the physicist Emilio Gino Segrè (1905–89) identified the first "artificially synthesized element."[34] From that year on, the discoveries of artificial elements became the purview of physicists and nuclear chemists. This field of research uses complex techniques in which radiometric methods play an important role. All synthesized elements are radioactive, and some of these elements possess an extremely short half-life. Their synthesis and characterization were full of scientific and technical complexities, requiring massive government funding and the collaboration of scientists on an international scale.

This brief summary of the elements that now reside in the periodic table can also apply to those "elements" that have no place there. False discoveries of chemical elements are also the product of the methods used in discovery; their histories are intimately intertwined with the real elements, like the *basso continuo* that accompanies the melody in a baroque concerto. In this book, we bring back to life the history of these false discoveries, for the most part with respect to their chronology.

Although we have drawn on many sources, both primary and secondary, we would like to call attention to two papers, published 76 years apart, that summarized the "lost" elements then known: Charles Baskerville's (1870–1922) 1904 address to the American Association for the Advancement of Science (AAAS)[35] and Vladimir Karpenko's 1980 paper in *Ambix* that examines two outstanding cases of elemental error and presents a comprehensive table, in alphabetical order, of more than 175 erroneous discoveries, complete with references and remarks.[36] A more recent addition to this literature is a paper by J. A. Bustelo, J. Garcia, and P. Román that concentrates on the lost names of the true elements. The paper contains a comprehensive, fully referenced table of these names, proposed and not accepted.[37]

THE DEVELOPMENT OF THE PERIODIC TABLE

"The periodic table...is a map of the way in which electrons arrange themselves in the atoms of a particular element...Its constant use by chemists emphasizes the central role. . .[it]...plays in making sense of what otherwise might be a chaotic jumble of facts about the elements and their many molecular combinations."[38] Today, we can find works that emphasize some of its many other facets—it is no longer a map but a kingdom, with its own limits, rules, and alliances.[39] It has become a cultural icon that can make unlikely connections, such as that between Michelangelo's *Moses* and Cleopatra's ingestion of the ultimate calcium supplement[40]; a mine of colorful anecdotes and odd facts about the discovery of its elements[41]; a system that represents the elements as human figures with periodically changing hairstyles[42]; a thing of beauty and a joy forever[43]; or the source of a life-altering encounter.[44] It is also probably the greatest piece of chemical research accomplished in the process of writing a chemistry textbook!

Although it is not the purpose of this book to exhaustively document the development of the periodic system,[45] especially since this book is devoted to identifying "illegal tenants" who have occupied it from time to time, the topic deserves a few words to set it into context. As with every other scientific breakthrough, the compilation of the periodic table began modestly, with collections of facts about material substances assembled over the course of the centuries. As the concept of "element" became clearer, and as data about the known elements began to accumulate, scientists began to look at the phenomenonological relationships among the elements. But, as van Spronsen points out, "the periodic system was comparatively... late in coming... due not so much to technical imperfections in atomic weight determinations as to the... fact that the theory of chemical bonding, based on Avogadro's hypothesis, was not unanimously accepted."[46]

From his careful work on mineral analysis and composition, in 1817, Johann Wolfgang Döbereiner (1780–1849) identified a triad of elements with similar properties in the mineral celestine. Other triads were added over the following decades, indicating a growing awareness of possible "families" of elements that had in common a simple numerical relationship in their atomic weights. Others who expanded on this idea were Leopold Gmelin (1788–1853) in 1827, Oliver Gibbs (1822–1908) in 1845, Jean-Baptiste Dumas (1800–84) in 1851, and William Odling (1829–1921) in 1857. In 1860, Stanislao Cannizzaro delivered his fiery speech on the importance of atomic weights at the Karlsruhe Conference. In 1862, the French mineralogist and geologist at the Paris École des Mines, Alexandre Emile Beguyer de Chancourtois (1820–86), proposed a natural system of classification embodied in a graphical representation that he dubbed "Vis Tellurique." By plotting atomic weights along a helical curve whose base has a circumference of 16, similar elements tended to arrange themselves in vertical columns.[47] The actual diagram did not appear until the publication of his book a year later.[48] Two years later, the English chemist John Alexander Reina Newlands (1837–98) arranged the known elements in order of increasing atomic weight and noticed that this arrangement allowed one to attribute some order, at least partially, to the properties of the elements (although his idea was scorned when he presented it to the London Chemical Society). In 1869, the Russian chemist Dmitri Ivanovich Mendeleev presented his paper to the Russian Chemical Society "On the Relationship Between the Properties and the Atomic Weight of the Elements,"[49] and he considered this discovery "the direct consequence of all the deductions drawn from the accumulated experiments done towards the end of the decade 1860–1870." In 1870, in Liebig's *Annalen*,[50] Julius Lothar Meyer (1830–95) arrived at the same conclusions, publishing a periodic table of the elements similar to that of Mendeleev. In addition to Meyer's grouping of the elements according to their atomic weight, in many respects resembling our modern periodic table, he also prepared a graph plotting the atomic volumes of the elements against their known atomic weights—a plot that clearly shows the periodic variation of this property.[51] The periodic table was born; it allowed scientists to predict the existence of elements not yet known and also to attribute chemical properties to them.

From these developmental steps, it is quite clear that chemists were beginning to converge on the phenomenological concept of linking the elements by their basic properties. Many historians of science think that Cannizzaro's ideas propounded at Karlsruhe were the catalyst that precipitated the simultaneous discovery of the system a decade later. Both Mendeleev and Meyer had attended the Karlsruhe Conference. Both were influenced by Cannizzaro's paper. Both came up with uncannily similar periodic tables—but Mendeleev was a year earlier than Meyer and less tentative in his conclusions.[52] These ideas had percolated in their minds for a long time. I. S. Dmitriev, director of the Mendeleev Institute at the University of Saint Petersburg, writes: "Mendeleev's discovery of the Periodic Law did not follow a linear pathway, but rather one that was complicated, difficult, winding, one that utilized various criteria over a period of time."[53]

One could certainly say the same for Julius Lothar Meyer. The two chemists arrived at strikingly similar conclusions, and both are accorded equal credit for the discovery. So why is the periodic law associated exclusively with the name of Mendeleev in the popular mind and in much popular literature? Some might say that the discovery of a new element, as almost precisely predicted by the gaps left in Mendeleev's table, seemed to have clinched his claim.

But Mendeleev was not the first to correctly predict the existence of a "missing element." That honor goes to Newlands who, in 1864, correctly predicted an element with an estimated atomic weight of 73 that would lie between silicon and tin. He was very close to the accepted value of 72.59 for germanium, discovered by Clemens Alexander Winkler (1838–1904) in 1886.[54]

Be that as it may, an unfortunate priority dispute between Meyer and Mendeleev ran on throughout the 1870s, and it seems to have revolved around the mistranslation of a single word from Mendeleev's Russian into the German article published in the *Zeitschrift für Chemie: periodicheski* to *stufenweise*. Apparently, the translator did not think it was important to emphasize periodicity and believed that the word for "gradual" or "stepwise" would do quite nicely,[55] whereas Meyer took that word to mean that Mendeleev had not recognized the repeat pattern of properties implicit in the word "periodic." Mendeleev took the initiative in defending "his" system, insisting that it was not enough to simply organize the elements, but also to be able to have an instrument with predictive properties, an idea that he propounded until his death in 1905. With Meyer's death in 1895, there was no one left to take up his cause, so the balance shifted in Mendeleev's favor, helped along by Russia's growing economic importance.[56,57] Thus, Mendeleev is immortalized with a box in his own table, an honor accorded so far to only 14 other human beings.

As technology advanced, many elements were discovered that confirmed Mendeleev's initial predictions. Some bumps along the road were how to accommodate the plethora of rare earth elements and the unexpected discovery of the noble gases and of numerous radioactive species that seemed to be individual new elements until the existence of isotopes came to be understood. Moseley predicted that his X-ray method would "prove a powerful method of chemical analysis.... It may even lead to the discovery of missing elements, as it will be possible to predict the position of their characteristic lines."[58] Following on the results of this landmark paper, chemists realized that only seven of the naturally occurring elements remained to be discovered, thus cutting down drastically the number of reported false discoveries and setting in motion an element hunt full of controversial competing claims that lasted for decades.[59]

Once chemists realized that not only could the periodic system bring order out of chaos and that it had predictive possibilities, but also that it served as a theoretical tool—as a map of the way in which electrons arrange themselves in atoms—it quickly took its rightful place as the "chemist's Bible." It has gone through many revisions since it was first visualized by Mendeleev and Meyer.

The lanthanides, elements 57–71, resemble one another so much that it took the better part of a century and a half to separate and characterize them. Their signature characteristic is that, as the atomic number increases along the series, they are filling in inner f-orbitals with electrons while the chemical properties of the preferred +3 oxidation state remain relatively unchanged. The facts that there is little covalency, that the +3 oxidation

1	2	3	4	5	6	7	8	9	10	11	12	13	14	15	16	17	18
1																	
1A																	8A
1 **H** 1.00794	2A											3A	4A	5A	6A	7A	2 **He** 4.002602
3 **Li** 6.941	4 **Be** 9.012182											5 **B** 10.811	6 **C** 12.0107	7 **N** 14.0067	8 **O** 15.9994	9 **F** 18.9984032	10 **Ne** 20.1797
11 **Na** 22.989769	12 **Mg** 24.3050	3B	4B	5B	6B	7B		8B		1B	2B	13 **Al** 26.9815386	14 **Si** 28.0855	15 **P** 30.973762	16 **S** 32.065	17 **Cl** 35.453	18 **Ar** 39.948
19 **K** 39.0983	20 **Ca** 40.078	21 **Sc** 44.955912	22 **Ti** 47.867	23 **V** 50.9415	24 **Cr** 51.9961	25 **Mn** 54.938045	26 **Fe** 55.845	27 **Co** 58.933195	28 **Ni** 58.6934	29 **Cu** 63.546	30 **Zn** 65.38	31 **Ga** 69.723	32 **Ge** 72.64	33 **As** 74.92160	34 **Se** 78.96	35 **Br** 79.904	36 **Kr** 83.798
37 **Rb** 85.4678	38 **Sr** 87.62	39 **Y** 88.90585	40 **Zr** 91.224	41 **Nb** 92.90638	42 **Mo** 95.96	43 **Tc** [98]	44 **Ru** 101.07	45 **Rh** 102.90550	46 **Pd** 106.42	47 **Ag** 107.8682	48 **Cd** 112.411	49 **In** 114.818	50 **Sn** 118.710	51 **Sb** 121.760	52 **Te** 127.60	53 **I** 126.90447	54 **Xe** 131.293
55 **Cs** 132.9054519	56 **Ba** 137.327	57-71 Lanthanides	72 **Hf** 178.49	73 **Ta** 180.94788	74 **W** 183.84	75 **Re** 186.207	76 **Os** 190.23	77 **Ir** 192.217	78 **Pt** 195.084	79 **Au** 196.966569	80 **Hg** 200.59	81 **Tl** 204.3833	82 **Pb** 207.2	83 **Bi** 208.98040	84 **Po** [209]	85 **At** [210]	86 **Rn** [222]
87 **Fr** [223]	88 **Ra** [226]	89-103 Actinides	104 **Rf** [267]	105 **Db** [268]	106 **Sg** [271]	107 **Bh** [272]	108 **Hs** [270]	109 **Mt** [276]	110 **Ds** [281]	111 **Rg** [280]	112 **Cn** [285]	113 **Uut** [284]	114 **Fl** [289]	115 **Uup** [288]	116 **Lv** [293]	117 **Uus** [294]	118 **Uuo** [294]

Lanthanides	57 **La** 138.90547	58 **Ce** 140.116	59 **Pr** 140.90765	60 **Nd** 144.242	61 **Pm** [145]	62 **Sm** 150.36	63 **Eu** 151.964	64 **Gd** 157.25	65 **Tb** 158.92535	66 **Dy** 162.500	67 **Ho** 164.93032	68 **Er** 167.259	69 **Tm** 168.93421	70 **Yb** 173.054	71 **Lu** 174.9668
Actinides	89 **Ac** [227]	90 **Th** 232.03806	91 **Pa** 231.03588	92 **U** 238.02891	93 **Np** [237]	94 **Pu** [244]	95 **Am** [243]	96 **Cm** [247]	97 **Bk** [247]	98 **Cf** [251]	99 **Es** [252]	100 **Fm** [257]	101 **Md** [258]	102 **No** [259]	103 **Lr** [262]

FIGURE 0.02. Periodic Table. The grayscale boxes indicate the locations where the false discoveries of elements have been most prevalent. The reason that some elements have been a source of error while others have not is due substantially to two factors, which may be taken individually or combined: the extreme scarcity of some of these elements in nature (such as the elements of atomic numbers 43, 61, 85, 87, and 91) or the similarity of their chemical properties with other nearby elements (rare earths, rhenium, and hafnium). For the transuranium elements, as well as for nitrogen, vanadium, niobium, some noble gases, beryllium, and other alkaline-earth elements, the case is quite different: they are anomalous due to the priority disputes of two or three academic research groups. Finally, other false elements, because of their bizarre nature, have not found a well-defined place in the periodic table of the elements.

state is preferred under normal conditions, and that the atomic radii are not very different all contribute to their chemical similarity. To accommodate the lanthanides would make for a "super long form" table, and so they are often displayed separated from the rest of the table for the sake of convenience.

Alfred Werner (1866–1919) was the first to recognize that yet another intergroup accommodation might be necessary for the heavier elements beyond uranium,[60] a suggestion taken up by Glenn Seaborg (1912–99) in 1944 while his group was struggling with the placement of the transuranium elements in the periodic table. In his own words:

> I began to believe it was correct to propose a second lanthanide-style series of elements… [starting]… with element number 89, actinium, the element directly below lanthanum on the periodic table. Perhaps there was another inner electron shell being filled. This would make the series directly analogous to the lanthanides, which would make sense, but it would require a radical change in the periodic table… [Wendell] Latimer told me that such an outlandish proposal would ruin my scientific reputation. Fortunately, that was no deterrent because at the time I had no scientific reputation to lose.[61]

So today's most common form of the periodic table (Figure 0.02) consists of a main body that includes the s-block, the d-block, and the p-block, along with the lanthanides and actinides that ride along below to better indicate the difference in their inner-electron arrangement.

Notes

1. Duveen, D. I. Personalized bibliography: An approach to the history of chemistry. *J. Chem. Educ.* **1961**, *38*, 418–21.
2. Rodygin, M. Y.; Rodygina, I. V. A course in early chemistry for undergraduates. *J. Chem. Educ.* **1998**, *75*, 1320–22.
3. Fine, L. W. Discussion: History-assisted instruction and the uses of human issues in chemical education. *J. Chem. Educ.* **1975**, *52*, 20.
4. Livio, M. *Brilliant Blunders.* Simon and Schuster: New York, 2013.
5. Weeks, M. E. *Discovery of the Elements*: Reprinted from a series of articles published in the *Journal of Chemical Education*; Kessinger Publishing: Whitefish, Montana, 2003.
6. Giunta, C. J. Using history to teach scientific method: The case of argon. *J. Chem. Educ.* **1998**, *75*(10), 1322–25.
7. Aristotle. *Metaphysics*, Book I, Part 4. W. D. Ross, tr. ebooks@Adelaide, 2007.
8. Aristotle. *Metaphysics* Book XIV, Part 9. Aristotle's views on what constitutes an element have often been oversimplified; an antidote is Paul Needham's essay, Aristotle's theory of chemical reaction and chemical substances. In Baird, D.; Scerri, E.; McIntyre, L., Eds., *Philosophy of Chemistry: Synthesis of a New Discipline*. Springer: Dordrecht, The Netherlands, 2006; 43–67.
9. Ronan, C. A. *The Shorter Science and Civilisation in China: An Abridgement of Joseph Needham's Original Text*, vol. 1. Cambridge University Press: Cambridge, U.K., 1978, pp. 146–47.
10. *Das Buch Paragranum* (c.1529–30). In *Paracelsus: Selected Writings*, Jacobi, J., Ed.; Princeton University Press: Princeton, NJ, 1951, pp. 133–34.
11. Paracelsus. *The Hermetic and Alchemical Writings of Aureolus Philippus Theophrastus Bombast, of Hohenheim, called Paracelsus the Great*; J. Elliott and Company: London, 1894, p. 162.
12. Boyle, R. *The Sceptical Chymist*. Project Gutenberg ebook, 2007 (prepared from the 1661 first edition); Physiological Considerations, Part of the First Dialogue. http://www.gutenberg.org/files/22914/22914-h/22914-h.htm (accessed April 9, 2014).

13. Siegfried, R. *From Elements to Atoms: A History of Chemical Composition.* American Philosophical Society: Philadelphia, 2002, p. 47.
14. Lavoisier, A.-L. *Traité élémentaire de chimie*; Chez Cuchet: Paris, 1789, p. 5. "We would be surprised not to find in an elementary treatise of chemistry a chapter on the constituent and elementary parts of bodies; but I note here that our tendency to want that all bodies in nature are composed of only three or four elements is due to a bias that comes originally from the Greek philosophers. The introduction of four elements, which, by the variety of their proportions, comprise all the bodies that we know, is pure conjecture, imagined long before we had the first notions of experimental physics and chemistry."
15. Donovan, A. *Antoine Lavoisier: Science, Administration, and Revolution*; The University Press: Cambridge, England, 1993, p. 180.
16. Hendry, R. F. The elements and conceptual change. In Beebee, H.; Sabbarton-Leary, N., Eds., *The Semantics and Metaphysics of Natural Kinds.* Routledge: New York, 2010, pp. 137–58.
17. Siegfried, R. *From Elements to Atoms: A History of Chemical Composition.* American Philosophical Society: Philadelphia, 2002, p. 184. Guyton de Morveau's nomenclature of neutral salts in 1782 became the crucial beginning of the successful reform of chemical nomenclature in 1787. His fundamental principle was that "the denomination of a chemical compound is clear and precise only to the extent that it recalls its component parts by names conforming to their nature." Guyton de Morveau, L. -B. Sur les dénominations chymiques. *Observations physique* **1782**, *19*, 37–382 (at 374).
18. Chez Cuchet: Paris, France, 1787.
19. Crosland, M. P. *Historical Studies in the Language of Chemistry.* Harvard University Press: Cambridge, MA, 1962, pp. 168–77.
20. Klein, U. Origin of the concept of chemical compound. *Science in Context* **1994**, *7*, 163–204.
21. Kragh, H. Conceptual changes in chemistry: The notion of a chemical element, ca. 1900–25. *Stud. Hist. Phil. Mod. Phys.* **2000**, *31*(4), 435–50.
22. Ede, A. *The Chemical Element: A Historical Perspective.* Greenwood Publishing Group: Westport, CT, 2006.
23. Hendry, R. F. Substantial confusion. *Stud. Hist. Phil. Sci.* **2006**, *37*, 322–36. Hendry discusses the "core conception" of a chemical element as based on three assumptions: that they survive chemical change, that they are the components of chemical compounds, and that the elemental composition of a compound explains its behavior. He distinguishes between the pragmatic concrete concept of element of Lavoisier (free element or simple substance) as opposed to the abstract concept of element of Mendeleev (element or basic substance). See also the important writings of F. A. Paneth in this regard: The epistemological status of the chemical concept of element. *Found. Chem.* **2003**, *5*, 113 (reprinted from *Brit. J. Philos. Sci.* **1962**, *13*, 1 and 144); Chemical elements and primordial matter: Mendeleeff's view and the present position. In *Chemistry and Beyond*, Dingle, H.; Martin, G. R., Eds., John Wiley: New York, 1965, pp. 53–72.
24. Rutherford, E.; Soddy, F. The cause and nature of radioactivity. *Phil. Mag.* **1902**, *4*, 370–96.
25. Reeves, R. *A Force of Nature: The Frontier Genius of Ernest Rutherford.* W.W. Norton: New York, 2008, pp. 49–51.
26. Davis, T. L. Boyle's conception of element compared with that of Lavoisier. *Isis* **1931**, *16*, 82–91.
27. Cannizzaro, S. *Il Nuovo Cimento* **1858**, *7*, 321–366. This influential paper, originally published as a letter to a Professor S. DeLuca and delivered at the Royal University of Genoa, was reprinted and distributed at the Karlsruhe Conference (September 3–5, 1860), the first worldwide international chemistry conference, organized to discuss matters of chemical nomenclature, atomic weights, and chemical notation. This paper can be accessed in English at http://www.chemteam.info/Chem-History/Cannizzaro.html (accessed April 13, 2014).
28. Goldwhite, H.; Adams, R. C. *J. Chem. Educ.* **1970**, *47*, 808.
29. Stark, J. G.; Wallace, H.G. *Education in Chemistry* **1970**, 7(4), 150–153; Etzioni, S., *Sch. Sci. Rev.* **1980**, *61*(216), 518–20.

30. Kuhn, T. S. Historical structure of scientific discovery. *Science*, New Series **1962**, *136* (No., 3518), 760–64.
31. Calascibetta, F. *Atti del VII Convegno Nazionale di Storia e Fondamenti della Chimica, L'Aquila, Italy, Oct. 8–11 1997*, 259–72.
32. Dinga, G. P. *Chemistry* **1968**, *41*(2), 20.
33. Joliot, F.; Curie, I. *Nature* **1934**, *133*, 201.
34. Perrier, C.; Segrè, E. *Atti della Accademia Nazionale dei Lincei, Classe di Scienze Fisiche, Matematiche e Naturali, Rendiconti* **1937**, *25*, 723.
35. Baskerville, C. The elements: Verified and unverified. Address to the Chemistry Section at the 53rd Meeting of the American Association for the Advancement of Science, Saint Louis, MO, 1903–04; *Proc. AAAS* **1904**, pp. 387–442.
36. Karpenko, V. The discovery of supposed new elements: Two centuries of errors. *Ambix* **1980**, *27*(2), 77–102.
37. Bustelo Lutzardo, J. A.; García Martínez, J.; Román Polo, P. Los elementos perdidos de la tabla periódica: sus nombres y otras curiosidades. *An. Quim.* **2012**, *108*(1), 57–64.
38. Stwertka, A. *A Guide to the Elements*, 3rd ed. Oxford University Press: NewYork, 2012, p. 15
39. Atkins, P. W. *The Periodic Kingdom: A Journey into the Land of the Chemical Elements.* Basic Books: New York, 1995.
40. Aldersey-Williams, H. *Periodic Tales. The Curious Lives of the Elements.* Penguin Books: London, 2011, pp. 274–75.
41. Kean, S. *The Disappearing Spoon: And Other True Tales of Madness, Love and the History of the World from the Periodic Table of the Elements.* Little, Brown & Company: New York, 2010.
42. Yorifuji, B. *Wonderful Life with the Elements: The Periodic Table Personified.* No Starch Press: San Francisco, CA, 2012
43. Gray, T. *The Elements: A Visual Exploration of Every Known Atom in the Universe.* Black Dog & Leventhal Publishers: New York, 2009.
44. Sacks, O. *Uncle Tungsten.* Alfred A. Knopf: New York, 2001, p. 194.
45. This task is done very well by numerous books on the periodic table written over the course of many decades. Outstanding among them are van Spronsen, J. W. *The Periodic System of Chemical Elements: A History of the First Hundred Years*, Elsevier: New York, 1969; Mazurs, E. *Graphical Representations of the Periodic System During 100 Years*, University of Alabama Press: Tuscaloosa, AL; 1974; Scerri, E. *The Periodic Table: Its Story and Its Significance*, Oxford University Press: New York, 2007; Emsley, J. *Nature's Building Blocks, An A–Z Guide to the Elements, New Edition*, Oxford University Press: New York, 2011; and Jensen, W. B., Ed. *Mendeleev on the Periodic Law: Selected Writings, 1869–1905*, Dover Publications: Mineola, NY, 2005. An excellent website containing much more information is "The Internet Database of Periodic Tables" at http://www.meta-synthesis.com/webbook/35_pt/pt_database.php?Button=Reviews+and+Books (accessed April 12, 2014).
46. van Spronsen, J. W. *The Periodic System of Chemical Elements: A History of the First Hundred Years*; Elsevier: New York, 1969, p. 21.
47. de Chancourtois, A. E. B. *Compt. Rend. Chim.* **1862**, *54*, 757–61; 840–43; 967–71.
48. de Chancourtois, A. E. B. *Vis Tellurique: Classement des Corps Simples ou Radicaux, Obtenu au Moyen d'un Système de Classification Hélicoïdal et Numérique*; Mallet-Bachelier: Paris, France, 1863. See http://web.lemoyne.edu/~giunta/ea/mendelann.html (accessed April 12, 2014), Carmen Giunta's website page on Dmitri Mendeleev's Faraday Lecture, note 13, where an image of the *vis tellurique* can be accessed.
49. Mendelejeff, D. Ueber der Beziehungen der Eigenschaften zu den Atomgewichten der Elemente. *Zeitschrift für Chemie* **1869**, *5*, 405–06.
50. Meyer, L. Die Natur der chemischen Elemente als Function ihrer Atomgewichte [1869]. *Annal. Chem. Pharm.* **1870**, *7* (Suppl. Bd.), 354–64.
51. Newman, J. R. *The World of Mathematics*, vol. 2. Simon and Schuster: New York, 1956, p. 911.

52. Strathern, P. *Mendeleyev's Dream. The Quest for the Elements*. St. Martin's Press: New York. 2000, p. 289.
53. Dmitriev, I. S. Scientific discovery in statu nascendi: The case of Dmitrii Mendeleev's Periodic Law. *Historical Studies in the Physical and Biological Sciences* **2004**, *34*(2), 233–75.
54. Greenwood, N. N.; Earnshaw, A. *Chemistry of the Elements*. Pergamon Press: New York, 1984, p. 32.
55. From Ref. 49, p. 405: "Die nach der Grösse des Atomgewichts geordneten Elemente zeigen eine *stufenweise* Abänderung in den Eigenschaften." The elements ordered according to their atomic weights show a *gradual* change in their properties. Gradual does not mean periodic!
56. Meyer. M. An element of order. *Chemical Heritage* **2013**, *31*(2), 32–37.
57. Masanori Kaji offers an additional reason for Mendeleev's ascendancy over Meyer: "Whereas Mendeleev discarded the atom and relied solely on the refined concept of a chemical element, Meyer embraced the atom and even supported the speculation of Prout's hypothesis of a primordial matter (hydrogen) as the building block of the elements. This prompted Meyer to underestimate his findings and prevented his having full confidence in his discovery of 1869." Kaji, M. *Bull. Hist. Chem.* **2002**, *27*(1), 4–16.
58. Moseley, H. G. J. The high frequency spectra of the elements. *Phil. Mag.* **1913**, 1024.
59. Scerri, E. *A Tale of 7 Elements*. Oxford University Press: New York, 2013.
60. Thyssen, P: Binnemans, K. Accommodation of the rare earths in the periodic table: A historical analysis. In *Handbook on the Physics and Chemistry of Rare Earths*, vol. 41. Gschneidner, K. A., Jr.; Bünzli, J.-C. G.; Pecharsky, V. K., Eds.; North Holland (Elsevier): Amsterdam, The Netherlands, 2011; 1–94, p. 37.
61. Seaborg, G. T.; Seaborg, E. *Adventures in the Atomic Age: From Watts to Washington*. Farrar, Straus and Giroux: New York, 2001, p. 127.

The Lost Elements

PART I

Before 1789

Early Errors and Early Elements

Sola manent interceptis vestigia muris,
ruderibus latis tecta sepulta iacent.
Non indignemur mortalia corpora solvi:
cernimus exemplis oppida posse mori.
[Only traces are left in ruins and remains of walls,
Roofs lie buried in vast ruins.
Let us not be resentful that mortal bodies disintegrate:
Behold: Even cities can die.]
—Claudius Rutilius Namatianus

PROLOGUE TO PART I

In Part I, we examine, analyze, and discuss the errors in discovery that occurred some years prior to Lavoisier's masterful contribution that moved the chemical sciences toward a new understanding of the concept of a chemical element. Therefore, these errors cannot be judged by the same standards as those treated in other sections of the book. These cases are confined to a very short period of time leading up to 1789, the date we arbitrarily select as the dividing line between protochemistry and chemistry. Prior to 1750, there is no evidence of a false discovery among solids, liquids, or gases.

Due to both the brief period of time under investigation and the limited number of chemists and technologists working in the Western world in the second half of the 18th century, fewer than a dozen erroneous findings are discussed in this section. Some of the substances are called "earths" according to the prevailing custom of naming the oxides of an element. In other cases, element names are given in Latin, the official language of culture and science for more than a thousand years, and in yet others, names were given in local languages.

The cases examined in Part I deal exclusively with solids, for three major reasons:

- Very few elements are liquid at room temperature. Mercury was known since antiquity; the likelihood of coming across another, such as gallium or iodine, in that period was very poor;
- Regarding the gaseous elements, pneumatic chemistry had reached a sufficient degree of experimental sophistication, and gas analysis was reasonably reliable. However, the technology for liquefying and distilling the constituents of air (noble gases) was well beyond the reach of 18th-century chemists;
- There was an increased interest in the study of metals and minerals in Europe during this period, due in large part to the vital link between the prosperity of a nation and the productivity of its mines. Consequently, the degree of development of analytical chemistry and metallurgy was in direct proportion to the scientific exploitation of mining.

I.1

THE BEGINNING OF A LONG SERIES OF SCIENTIFIC BLUNDERS

I.1.1. TERRA NOBILIS

The enthusiasm that often characterizes researchers can at times distort certain preconceived convictions and deceive the scientist into believing that a controlled experiment has produced the correct result when, in fact, it is erroneous due to insufficient or incorrect data. This is the case for the discovery of a mysterious *terra nobilis* made by the chemist Torbern Olof Bergman.

Bergman was born on March 20, 1735, in Katrineberg,[1] Sweden. He was a chemist and mineralogist who became famous in 1775 for printing the most extensive tables of chemical affinity ever published at that time, and he was the first chemist to use letters of the alphabet as a notation system for chemical species. He took his doctorate at the University of Uppsala in 1758. After initially holding the professorship of physics and mathematics, he later took the chair in chemistry, which he retained for the rest of his life. Bergman made significant contributions to progress in quantitative analysis and metallurgy, and he developed a classification scheme of minerals based on their chemical characteristics.

In 1777, Bergman confidently announced[2] the result of an extremely expensive investigation. He studied the behavior of diamond with a blowpipe, and, aside from the presence of silicon, he seemed to have generated an unknown compound. He extracted the oxide of a metal from the diamonds, which, according to the custom of the time, he called *terra nobilis*. His discovery was quickly forgotten, not least because his life soon took a tragic turn.

After marrying Margareta Catharina Trast in 1771, he enthusiastically continued his activities as a synthetic and analytical chemist,[3] but on July 8, 1784, at the age of only 49, he died in Medevi, Sweden. It is believed that he fell victim to poisoning from the chemical substances he used in his research. At the time of his death, he had been a member of the Royal Society of London and the Swedish Royal Academy for many years, and he was certainly one of the most famous chemists of his time. In fact, his funeral eulogy was conducted by Marie Jean Antoine Nicolas de Caritat, Marquis de Condorcet (1743–94) and the anatomist Felix Vicq-d'Azyr (1748–94).

I.1.2. SIDERUM AND HYDROSIDERUM

At the end of the 18th century, there existed a particular type of iron called *fer cassant a froid* due to its tendency to crumble when cold; when hot, however, it was malleable like a common metal. In Uppsala during those years, Torbern Olof Bergman was investigating the origin of this strange property of iron, while on the opposite shore of the Baltic sea at Szczecin (Stettin), Poland, Johann Karl Friedrich Meyer (1733–1811) was also studying the same problem. Between 1777 and 1778, both scientists, working independently, discovered that by the action of sulfuric acid on *fer cassant a froid*, a white powder was formed. By a process of reduction (the details of which are not known), the white powder was converted into a fragile gray substance with an appearance similar to that of a metal but not very soluble in acids. The specific gravity of the substance was 6.700. It did not melt easily and, once combined with iron, it regenerated the substance *fer cassant a froid*. Both proposed that the unknown substance mixed with iron could be a new element. Bergman suggested that it be called *siderum,* whereas Meyer proposed the name *hydrosiderum*[4] or *wassereisen*. Martin Heinrich Klaproth (1743–1817), working in Berlin, immediately connected these observations with his own experiments in the field. He had noted that when iron was combined with phosphorus, a white product was formed with an appearance similar to that of the so-called *fer cassant a froid*. His initial suspicion that *siderum* and *hydrosiderum* were in fact an alloy of iron and phosphorus was confirmed by subsequent chemical analysis. To resolve the mystery, Klaproth heated phosphoric acid with iron and carbon and obtained a white powder very similar to the *hydrosiderum* of Meyer and the *siderum* of Bergman.[5] Meyer, when Klaproth informed him of the outcome of his experiments, replied that he, too, after a long analysis, had found that *hydrosiderum* contained phosphoric acid.[6] Shortly afterward, the chemist Carl Wilhelm Scheele (1742–86) became interested in the problem of the composition of *fer cassant a froid*. He approached the problem by decomposing the substance, thus identifying phosphoric acid and iron.[7] Although it contained the same elements as *hydrosiderum* (iron and phosphorus), in *siderum,* phosphorus was not present as phosphoric acid but as elemental phosphorus or as iron phosphide.

Finally, the Swedish chemist Sven Rinman (1720–92) managed to demonstrate that the fragility and other poor qualities (from a metallurgical point of view) of *fer cassant a froid* could be removed by heating the alloy in a reducing atmosphere.[8]

Johann Karl Friedrich Meyer worked as a pharmacist for many years in his native town and, in 1784, became member of the Academy of Sciences in Berlin. He died at Szczecin on February 20, 1811, at the age of 78.

I.1.3. SYDNEIUM OR AUSTRALIUM

In 1779, during a session of the House of Commons, the naturalist Joseph Banks (1743–1820) expressed the urgent need to establish a new crown colony in Australia. The government demonstrated an interest in creating new penal colonies, although to observers of the time it seemed almost insanely expensive. However, due to the loss of the North American colonies, there was terrible overcrowding in the English prisons and a solution needed to be found quickly. The first colonists and prisoners set sail a few years later and, in January 1788, the first settlement was established in Sydney Bay, in a region now called New South Wales. Two years later, an article was published in the Royal Society's *Philosophical Transactions*, of which Banks was president, with the title "On the Analysis of a Mineral Substance from New South Wales."[9] The author was the well-known producer of and expert on ceramics, Josiah Wedgwood (1730–95), who from 1783 had been an elected member of the Royal Society for his studies on pyrometry. Josiah Wedgwood was born at Burslem on July 12, 1730. He had an excellent sense of observation, but only limited schooling. His contact with chemistry and mineralogy came through meetings with members of the Lunar Society of Birmingham—the source of the scientific notions that were to lead him to become a master producer of ceramics and porcelain,[10] renowned throughout the world.

This informal but notable Society, so-called because its members used to meet every month on the Monday closest to the full moon, counted among its members many distinguished scientists, including James Keir (1735–80), Joseph Priestley (1733–1804), James Watt (1736–1819), William Withering (1741–99), and Erasmus Darwin (1731–1802), grandfather of the more famous Charles Robert Darwin.

The cover letter of Wedgwood's article to the *Philosophical Transactions* of the Royal Society, addressed to Banks, reveals the reasons for Wedgwood's commercial interest in certain new minerals originating in Australia: "I have the pleasure of acquainting you that the clay from Sydney Cove, which you did the honour of submitting to my examination, is an excellent material for pottery, and may certainly be made the basis of a valuable manufacture for our infant colony there." He then elucidated arguments of a more chemical nature: "The other mineral. . . seems to contain one substance hitherto unknown." Wedgwood reported the analysis that he and his assistant (Mr. Chisholm) had carried out on that material. The mineral was described as mixture of fine white sand, particles of mica, and a soft white soil. The analysis proved to be quite difficult due to the modest instrumentation available and because the material was soluble only in hot hydrochloric acid and precipitated as a white powder upon addition of small amounts of alkali. They succeeded in melting the white powder (perhaps an oxide) at high temperatures but could not obtain the free metal on reduction with charcoal. Wedgwood concluded that the substance was not a combination of an unknown earth with an "acid radical." Furthermore, he was unable to say whether the new substance was "an earth" (metallic oxide) or metallic, although his experience suggested the first hypothesis. The German anthropologist Johann Friedrich Blumenbach (1752–1840), at the University of Göttingen, confirmed some of Wedgwood's observations and thus enabled the presumed discovery to become known on the continent. Although Wedgwood had not given a name to the new earth, in the textbooks of the time the names *sydneia* (element = *sydneium*), *australa* (element = *australium*), and *terra australis* were attributed to it. In French texts, for a short period, it was known as *terre de sidnei*, whereas for the English it was *austral sand*. In Berlin, in 1797, the chemist Martin Klaproth undertook a careful and detailed investigation of the so-called *sydneia* or *austral sand*, finding only aluminum, silica, and traces of iron. William Nicholson (1753–1815), writer, editor, and officer of the East India Company, opposed Klaproth's findings. He asserted that because Klaproth had not analyzed the same material as Wedgwood there were no grounds for rejecting Wedgwood's discovery. Because it was well known that Nicholson had traveled around the world on behalf of the Wedgwood Company, his statement was confidently regarded as correct. The controversy was finally laid to rest the following year when Charles Hatchett (1765–1847) stated, contrary to Nicholson, that he had proof that the samples analyzed by Wedgwood and Klaproth had the same provenance. In fact, both derived from the original sample donated by the president of the Royal Society, Joseph Banks, to Wedgwood.

Shortly before his death, the Viennese chemist Karl Haidinger (1756–97) gave to Klaproth the sample of Australian sand that he himself had obtained from Hatchett. Hatchett had already analyzed a small amount of the mineral and found the same substances subsequently reported by Klaproth.[11] The two analyses were in accord, except that Hatchett also found traces of graphite.

Thus, the names *sydneia* and *austral sand* disappeared from the list of new substances. Today, Wedgwood's results can be explained by the simple hypothesis, already aired by Hatchett, that he had used poor-quality reagents that led his analysis into error. Wedgwood did not suffer the shame of a public retraction because he died a few years before his discovery was rejected.

Today, the name of Wedgwood is still well known, although it is no longer linked to chemistry but rather to the ceramic industry. He created a pottery industry indebted to his scientific experiments and was one of the greatest pioneering industrialists of his age. He experimented with several clay mixtures that were to contribute considerably to his success as a pottery producer (producing such colors as antique red, cane, drab, chocolate, and olive). Wedgwood died at the age of 64 on January 3, 1795.

Wedgwood's designs are still used today by the Wedgwood company. His original pieces are highly regarded antiques, and Wedgwood porcelain and ceramic products are used around the world, making the company one of the most famous producers of porcelain.

I.1.4. THE ELEMENT THAT BREATHES

The German physician Samuel Christian Friedrich Hahnemann (1755–1843), born in Meissen on April 10 or 11, 1755, is credited as the founder of the alternative form of medicine called homeopathy.[12] His publications, however, also covered topics in chemistry.[13] He studied medicine at the University of Leipzig and subsequently at Erlangen, where he graduated in 1779. During this period, he became a Freemason.[14] In 1782, he married Johanna Kuchler, with whom he had 11 children. The family moved continually from one town to another within Prussia. Hahnemann did not practice medicine, but followed with interest new discoveries in chemistry and dedicated his time to the study and translation of medical texts. In 1801, while in Hamburg, Hahnemann announced that he had discovered a new alkali metal[15] that displayed properties very different from those of the other first-group elements known at that time. The most unusual of these properties concerned the effect of temperature: upon heating the material, its volume increased by a factor of up to 20. The name that he chose for the new substance reflected this peculiar property, which was similar to "inorganic respiration"—*pneum-alkali*, an alkaline element possessing a lung. In the solid state, the element *pneum-alkali* was characterized by hexahedral crystals, lack of ability to absorb humidity from the air or to display efflorescence, and solubility in hot water. Hahnemann did not reveal how he had discovered the new metal, but he did present a long and detailed analysis of its most common derivatives. The sulfate of *pneum-alkali* was not soluble in "ardent spirits," but dissolved in nitric, phosphoric, and acetic acids. The salt derivatives of *pneum-alkali* were soluble in water, whereas the chlorides had a "feathered" appearance. It also displayed the characteristic strongly reducing property of all alkaline elements. On combining the elemental form of *pneum-alkali* with transition metal salts, they were reduced to the metallic state. The new metal seemed to possess a wide range of properties, such as the saponification of vegetable oils and the capacity to react with both the oxychloride and nitrate of mercury. It was able to change the color of certain natural pigment-based dyes from blue to green. These wide-ranging properties convinced Hahnemann to commercialize his discovery by opening a shop in Leipzig where he sold vials containing an ounce (0.03 kg) of the metal for the price of one gold coin (issued by the King of Prussia, Friedrich II). The news of

Samuel Hahnemann's discovery did not pass unnoticed. The fact that he was an atypical physician and a chemist hostile to established ideas led many in the scientific community to press for an investigation. The discovery of the new alkaline element, combined with Hahnemann's desire to gain personal benefit from it, provided an ideal occasion for this, offering a pretext for the major chemists of the time to discredit him.[16]

As soon as the discovery of *pneum-alkali* was announced, the Society of the Friends of the Natural Sciences of Berlin, which counted among its members many of the principal chemists of the time, obtained a sealed, intact vial containing an ounce (0.03 kg) of *pneum-alkali*. The outcome of the analysis carried out by three illustrious chemists, Martin Klaproth, Dietrich Ludwig Gustav Karsten (1768–1810), and Sigismund Friedrich Hermbstaedt (1760–1833), left no doubt: *pneum-alkali* was not a new metal but simply a borate. In their communication, the three chemists invited Hahnemann to publish a full retraction and offer compensation for the fraudulent sale of an ounce (0.03 kg) of *pneum-alkali* for a gold coin when the same quantity of borate could be bought in any pharmacy for a few pennies. A violent attack on Hahnemann's work was also reported by Johann Bartholomäeus Trommsdorff (1770–1837), professor at Erfurt (Germany). He found that the sealed vials sold for an exorbitant price contained only borate and natron.[17] Trommsdorff attacked Hahnemann thus: "A great deal of impudence is required to pull the leg of the worthy German chemical fraternity, and to defraud them of their money." Hahnemann replied to the accusations by publishing a letter proclaiming his innocent intent: "I am incapable of wilfully deceiving: I may however, like other men, be unintentionally mistaken. I am in the same boat with Klaproth and his diamond spar."[18] Hahnemann continued providing a detailed explanation of the causes of his errors. Professor Alexander Nicolaus Scherer (1771–1824), who had published the first results of the discovery of *pneum-alkali*, remained loyal to Hahnemann, counterattacking Trommsdorff and reminding him of the many mistakes that he had also committed during his career as professor of chemistry at Erfurt. However, it was now too late for such exchanges, and Hahnemann was banned from the scientific community. His exclusion was not so much the consequence of his mistake, but rather because he was considered different, an exponent of "heretical" ideas within the scientific establishment.

I.1.5. THE BIRTH OF HOMEOPATHY

After the controversy concerning the discovery of *pneum-alkali*, Hahnemann directed his interests toward the medical field. He believed that medicine at that time caused more harm than good, typified by common practices such as bloodletting (which remained widely used until the end of the 19th century) and purgative and emetic practices that were supposed to remove illness from the patient and restore the correct balance of the four "humors."[19] He refused to accept the concept that to cure an illness, the causative matter should be removed from the body. He advocated instead that to restore harmony and equilibrium within the body the patient needed fresh air, good food, and exercise. Hahnemann's proposal was certainly more humane and less dangerous than the most widely used medical practices of the time, and Hahnemann formulated the basis of homeotherapy while translating the volume *Materia Medica Pura* (Pure Medical Matter), by the Scottish physician William Cullen (1710–90). At that time, malaria was treated by use of an extract from the bark of cinchona: Cullen believed that the effectiveness of quinine was due to its "tonic effect on the stomach." Hahnemann dismissed this idea because many other substances more astringent than quinine did not relieve the fever; thus, some other property had to be the origin of quinine's therapeutic effects. Hahnemann decided to experiment with quinine on himself and, after taking it for several days, he believed that he had developed the symptoms of malaria. He hypothesized that a series of symptoms could be cured by the substance that in a healthy person produced the same effects. In 1806, Hahnemann published his first important work *The Medicine of Experience*, which already contained the fundamental principles of homeopathy (from the Greek *omeos*, "similar" and *pathos*, "illness"), but the basic aspects of his methods had been published 10 years earlier:[20]

- *Experimenta in homine sano*: the effect of medicines can only be discovered by experiments on healthy people because in ill people the symptoms of the illness are obscured by those caused by the medicine.
- *Similia similibus curentur*: the medicine must be chosen on the basis of the similarity between its effects and the symptoms of the patient, without reference to the presumed illness that caused the symptoms.
- *Doses minimae*: medicines must be administered in small doses.
- *Unitas remedii*: the treatment should be repeated only if the symptoms return.

In 1810, Hahnemann published the first edition of his principal theoretical work, *The Organon of Rational Healing*, later retitled as *The Organon of the Art of Healing*. This edition was followed by five others, the last published posthumously in 1921.

Upon his return to Leipzig for the fourth time, Hahnemann began lecturing on homeopathy at the university, where he encountered strong opposition from other physicians and pharmacists. During this period, he carried out many experiments with a small group of students to test the effects of numerous substances. The results were published in a text of six volumes called *Materia Medica Pura* (Pure Medical Matter). It should be recalled that the importance of self-suggestion (the *placebo effect*) was not well understood at that time, and Hahnemann's experiments took no account of it; his students knew which substances they were taking and what effects were expected.

In 1820, Prince Karl Philipp zu Schwarzenberg (1771–1820), an Austrian field marshal and hero of the battle of Leipzig, went to Hahnemann for a cure for his disabling stroke. Unfortunately, the prince died, his death was blamed on Hahnemann's incompetence, and the physicians and pharmacists of Leipzig managed to obtain an order to impede Hahnemann from distributing his medicine. Unable to practice his profession, Hahnemann left Leipzig. In 1821, he moved to Kothen, where he subjected his theory to a profound re-evaluation in order to reply to the many criticisms leveled against it. His growing support for the doctrine of vitalism is evident from this study. To avoid the collateral effects of his medicines, Hahnemann continually reduced the doses, reaching extremely low levels. To combat the objection that such low doses could not be effective, Hahnemann replied that the efficacy of his remedies was considerably increased by a process called "dynamization," which consisted of repeatedly shaking the product up to 100 times.

At the same time, he developed his theory on chronic disturbances. In 1827, he confided to his two most trusted students that he had discovered the causes of all chronic diseases and how to cure them, which he published in the treatise "Chronic Illnesses."[21] In Hahnemann's view, all chronic illnesses, except those caused by orthodox medicine

FIGURE I.01. Monument to Christian Friedrich Samuel Hahnemann (1755–1843). Founder of homeopathy and discoverer of the element that breathed, hence the name *pneum-alkali*. Hahnemann is buried in Paris's Père Lachaise Cemetery, Division 19. The plaque on the left is a partial catalog of his works on homeopathy; on the right are quotations from his major work of 1810, *The Organon of Medicine*. Photograph by Mary Virginia Orna.

or a bad lifestyle, were caused by four kinds of "miasma" or poisonous vapors: syphilis, psychosis, tuberculosis, and psoriasis. Contradicting his own principles, Hahnemann experimented with his products on chronic patients, leading him to attribute to his medicines a series of symptoms that were in fact caused by the illnesses themselves. Although Hahnemann's first criticisms of orthodox medicine were empirically based, this evolution of the theory was based primarily on the doctrine of "vitalism" and not on a correct application of scientific method. For this reason, he was increasingly criticized, even by some of his followers. The first controversies among homeopaths were fostered by Hahnemann himself, who attacked without reserve as "traitors" and "apostates" those who brought about even small changes to his "medical theory."

After the death of his wife in 1835, at the age of 80, he married Marie Melanie d'Hervilly (1800–78), who was little more than 30 years old.[22] Shortly afterward, they moved to Paris, where Hahnemann died in 1843. He was buried in the cemetery of Père Lachaise (Figure I.01). Hahnemann's controversial ideas continue to find followers even today, despite the warnings of the modern medical profession.

Notes

1. Apparently a hamlet presently reduced to a single mailbox in Låstads parish, according to element sleuths James and Jenny Marshall. Private Communication, January 21, 2013.
2. Hibben, J. G. *Inductive Logic*; Read Books: Alcester, UK, 2007, p. 272.

3. Among Bergman's many discoveries, one must certainly recall fixed air (CO_2), oxalic acid, and hepatic gas (H_2S).
4. Meyer, J. K. F. *Schriften der Gesellsch. naturf. Freunde* **1780**, *2*, 334; Meyer, J. K. F. *Schriften der Gesellsch. naturf. Freunde* **1780**, *2*; 380; Meyer, J. K. F. *Beschaft. der Gesellsch. naturf. Freunde* **1782**, *3*, 74.
5. Klaproth, M. H. *Crell's Annalen* **1784**, *1*, 390.
6. Meyer, J. K. F. *Crell's Annalen* **1784**, *1*, 195.
7. Scheele, C. W. *Crell's Annals*, English Translation, **1784**, *1*, 112.
8. Rinman, S. *Ann. de Chimie* **1829**, *42*, 831.
9. Wedgwood, J. *Phil. Trans.* **1790**, *80*, 306.
10. Although it digresses from the principal theme, it is worth spending a few words on Wedgwood. Wedgwood's fame was not due to his work as a chemist or mineralogist, but for his ceramics and porcelain inspired, in form and decoration, by classical antiquities. After a short partnership with John Harrison at Stoke-on-Trent in Staffordshire, in 1759, he founded a business with Thomas Whieldon. In 1765, Wedgwood began the production of a cream-colored durable ceramic known as creamware. This ceramic was much appreciated by Queen Charlotte (1744–1818); after receiving her patronage the name was changed to Queen's Ware. The publicity that followed led to its becoming the preferred choice for high society in England and overseas. Many manufacturers suffered from the appearance on the market of this new competitor, even those of international standing such as Sèvres and Meissen. Those that survived began to imitate the cream-colored Wedgwood ceramic, which on the continent was called Fine Faienceware or English Faience. From 1768 on, production was directed toward hard porcelain without enamel, decorated with subjects of the classical world. The most important being black basalts, which imitated Greek vases and jaspers, which were very fine-grained porcelain glasses obtained by the effect of very high temperatures on clay mixtures containing barium sulfate. The Jasperwares, porcelain products prepared from jaspers, were imitated by Sèvres and Meissen, where it was called Wedgwoodwork. In 1774, the Empress of Russia, Catherine the Great (1729–96), a noted collector of porcelain and ceramics, commissioned a service of 952 pieces from Wedgwood. In the same year, an opaque glass ceramic was invented that was to become important for the creation of engravings, cameos, and medallions. Many of the products from the Wedgwood factory were reminiscent of the ornaments originating from the most recent archaeological sites, although the colors of the decorative elements were modified to render them colder and more delicate, according to the tastes of the time.
11. Hatchett, C. *Phil. Trans.* **1798**, *88*, 110.
12. von Lippmann, E. O. *Chemiker Zeitung* **1926**, *50*(4), 25.
13. Kleiner, I. S. *Sci. Mon.* **1938**, *46*, 450.
14. Cook, T. M. *Samuel Hahnemann: The Founder of Homeopathic Medicine*; Thorsons: Wellingborough, U.K., 1981.
15. Anon., *A Journal of Natural Philosophy, Chemistry and the Arts*, **1801**, *4*, 523; Hahnemann, C. F. S. *Scherer's Journal of Chemistry* **1801**, *5*, 665.
16. Haehl, R. *Samuel Hahnemann: His Life and Work*; Jain B. Publisher: New Delhi, India, 1995.
17. The term "natron" was used to indicate hydrated sodium carbonate, chemical formula $Na_2CO_3{\cdot}10H_2O$. Its name derived from the Latin word for sodium, *natrium*. In Egypt, where there were extensive deposits, it was used in mummification due to its property as a drying agent.
18. Professor Klaproth, who at the time of the accusations against Hahnemann was the most prominent living German chemist, just a few years earlier in 1788 had also made a blunder, mistakenly identifying a substance as being newly discovered and calling it *diamanthspatherde* or *terra adamantina* (adamantine earth) (Klaproth, M. *Beschaft. Ges. Nat. Fr.* Berlin, **1788**, *8*, 4; *Crell's Chem. Annalen* **1789**, 5). He indicated *terra adamantina* as a new substance resistant to acids. *Diamanthspatherde* could be melted by addition of alkalis, and its properties were

similar to those of silicon. Today, we know that *diamanthspatherde* is constituted by one-third corundum, Al_2O_3, and two-thirds alluminite, $Al_2(SO_4)(OH)_4$ 7 H_2O.

19. The four humors, phlegm, black bile, yellow bile, and blood were part of the cosmology passed down from the ancient Greek philosophers and were allied with the cosmology of the four Aristotelian elements, earth, air, fire, and water. See https://www.nlm.nih.gov/exhibition/shakespeare/fourhumors.html (accessed April 9, 2014).
20. Pingel, S. *Der Hautarzt: Zeitschrift fuer Dermatologie, Venerologie, und verwandte Gebiete* **1992**, *43*(8), 475.
21. Bradford, T. L. *The Life and Letters of Dr. Samuel Hahnemann*; Boericke & Tafel: Philadelphia, PA, 1895.
22. Marie Melanie d'Hervilly belonged to one of the oldest noble families of Paris. A successful painter and poet, her illness due to cholera directed her attention to homeopathy and the treatment by Dr. Frederick Foster Hervey Quin (1799–1878) during the 1832 cholera epidemic in Paris. Quin was a prominent English homeopath who had been a student of Hahnemann. She obtained a translation of the 1829 fourth edition of the *Organon*, which led her to travel to Kothen to be treated by its author, Dr. Hahnemann. She arrived on October 7, 1834, and began her cure with Hahnemann, who not only treated her illness but also secretly courted and then married her on January 18, 1835.

I.2

THE ELEMENTS HIDDEN BY ALTERNATIVE NAMES

Some well-known elements could have had quite different names. The discoverers of these elements, after "baptizing" them one way later renamed them, more or less voluntarily, with names that are still in use today. This is the case for *ceresium*, a metal now known as palladium, and for *ochroite*, which, if it had been accepted, would have replaced the name cerium.

Tellurium might also have had another name, *metallum problematicum*, although the practicality of such a name in common usage is doubtful. The discovery and naming of *erythronium* or *panchromium* presents a complex issue because, although the substance—later known as vanadium—was indeed an element, it was not recognized as such at the time of its discovery.

I.2.1. METALLUM PROBLEMATICUM OR TELLURIUM

The discovery of tellurium was the unexpected result of an analysis by Hungarian chemist Leopold Anton Ruprecht (1748–1814). In 1782, his interest was focused on the analysis of a rock (nagyágite[23]) coming from Transylvania. This substance was a true chemical puzzle and was suspected of containing a large amount of gold. Ruprecht did not accept this idea, which a quick analysis might suggest, but claimed to have found antimony instead.

Ferenc Müller von Reichenstein (1740–1825), one of Ruprecht's teachers and an inspector of mines in Transylvania, analyzed the rock vein where nagyágite had been obtained and demonstrated the presence of bismuth. Ruprecht replied confidently that the constituent in question could not be bismuth. Müller von Reichenstein admitted his mistake but remained convinced that the mineral contained an unknown metal. Shortly afterward, Ruprecht also admitted that he was no longer convinced of the presence of antimony.

After this exchange of opinions, in 1783, Müller von Reichenstein published an article on the composition of nagyágite that reported the presence of an unknown semimetal that he called *metallum problematicum*.[24] He listed the characteristic reactions of this element and concluded by declaring that he would send a sample of the new substance to Torbern Bergman in Sweden, requesting that he should confirm the new discovery (Bergman was considered to be the greatest living mineral chemist). Bergman began his analyses but shortly afterward asked Müller von Reichenstein to send by ship a more abundant amount of the sample. It is very likely that Bergman never received the new samples since he died 2 months after sending the request.

At this point, when all seemed to be leading to inevitable success, Müller von Reichenstein interrupted his research on *metallum problematicum*. Ten years later, in 1793, when the whole episode was nearly forgotten, the chemist Martin Heinrich Klaproth asked

Müller von Reichenstein for samples of the presumed metal so that he could carry out a detailed analysis in the hope of confirming the discovery. The outcome of the analysis supported Müller von Reichenstein's stance: the sample contained an unknown element. Unfortunately, Müller von Reichenstein was unable to provide the new substance with a name to his liking, thus allowing Klaproth to take advantage of the situation—within the scientific community, the accepted name of the new substance became tellurium.

I.2.2. OCHROITE OR CERIUM

Ten years later, Martin Heinrich Klaproth[25] was again in the limelight for the discovery of a new metal. In 1803, Klaproth, working in Germany at the same time as the Swedish chemists Jöns Jacob Berzelius (1779–1848) and Wilhelm Hisinger (1766–1852), found a new substance that he called cerium. The metal was extracted in the form of an oxide from the minerals cerite and ochroite. The properties of cerium oxide were reminiscent of those of the recently discovered yttrium oxide. In fact, they seemed identical except that yttrium was insoluble in a solution of ammonium carbonate, and yttrium oxide acquired a brown color upon heating.

As in the case of the first lanthanide, also discovered simultaneously by Klaproth and Berzelius, each decided to propose a name of his own choice for the new metal. The name "cerium," still in use today, was given by Berzelius, who had been inspired by the name of the asteroid Ceres, discovered 2 years previously (1801). Klaproth instead suggested the name "ochroite" due to the yellow-brown color of the metal oxide (in English, *ochre* and in Greek, ὠχρός).

The attribution of the name was complicated by inappropriate behavior on both sides. Berzelius and Hisinger sent the results of their experiments to Adolph Ferdinand Gehlen (1775–1815), editor of the German *Neues Allgemeines Journal der Chemie*. To support their claim, they printed, at their own expense, a small pamphlet[26]; limited to only 50 copies, today, it is a true collectors' item. Independently, Klaproth, who had analyzed the same tungsten-rich mineral from Bastnäs in Sweden, presented his results, using the name "ochroite," to Gehlen's *Journal*. His article appeared in an issue preceding that of his Swedish colleagues, Hisinger and Berzelius.

It is not clear in what order Gehlen received the two manuscripts, but in a letter sent to Hisinger in May 1804, Gehlen credited Hisinger and Berzelius as the discovers of the metal and gave them the honor of naming it. Klaproth accepted the decision against him with good spirit, suggesting only a slight modification of the name to *cererium*, adding a syllable to emphasize that the etymology of the new metal's name derived from the Roman divinity Ceres and not from the Greek κηρα, which means wax.[27]

As in the case of the name *ochroite*, the modification of the name cerium into *cererium* was not accepted. This double failure undoubtedly represented a difficult (albeit brief) period for Klaproth. His fame as a chemist was growing by leaps and bounds, and not only in Germany. A few months later, he learned that J. F. John had decided to dedicate a new element to him, calling it *klaprothium*[28] (during the year following Klaproth's death, a similar suggestion was made). Martin Heinrich Klaproth died in Berlin on January 1, 1817, at the age of 73.

I.2.3. CERESIUM OR PALLADIUM

The bizarre story of the discovery of palladium involves Andreas (or Jedrzej) Śniadecki (1768–1838), known as a talented Polish scientist and an advocate of Lavoisier's innovative ideas. He became professor of chemistry and pharmacy in the city of Wilnius, at that time part of the Russian empire. This gave him full opportunity to begin the characterization of minerals from rich platiniferous deposits in the Urals. In fact, while analyzing platiniferous material, Śniadecki suspected that he had identified a new element that he called *vestium* (or *vestalium*), fascinated as he was by the recent discovery of the asteroid Vesta.[29,30] The Paris Academy of Sciences, in conjunction with the Academy of Saint Petersburg, never confirmed Śniadecki's results, although the Russian academy, after an initial decree to suppress publication, allowed publication of the presumed discovery. In the eyes of many academics, however, Śniadecki had simply rediscovered palladium. James and Virginia Marshall present a cogent argument for this conclusion in their 2011 paper.[31]

The actual discovery of palladium was made by William Hyde Wollaston (1766–1828) who, 5 years earlier, in 1803, had isolated the noble metal at the same time as rhodium. Wollaston found it in a platiniferous mineral from South America. He dissolved the rock in *aqua regia,* subsequently neutralized the solution with sodium hydroxide, and treated it with ammonium chloride to precipitate the platinum in the form of ammonium chloroplatinate. Upon subsequent addition of mercury cyanide to the remaining liquid, palladium cyanide formed that, when heated in a reducing atmosphere, produced metallic palladium. Initially (i.e., in 1802), he considered calling this metal *ceresium.*[32] However, 2 years later, perhaps because Hisinger and Berzelius had proposed a similar name, "cerium," for their element in 1803 (before Wollaston published his own findings), he decided to use the name of one of the first observed planetoids: palladium, in honor of the asteroid Pallas,[33] discovered 2 years previously.

Exactly 100 years after the death of Jedrzej Śniadecki, in 1938, an article was published in Poland with the intention of restoring credibility to the discovery of *vestium.*[34] However, due to the imminent war that was to overrun Poland, the claim proposed by his compatriots passed unnoticed. In 1967, the idea put forth in 1938 was again considered by other Polish chemists who hypothesized that the *vestium* isolated by Śniadecki in 1808 could have been ruthenium, a metal unknown at that time.[35] However, these nationalistic predispositions lasted only briefly. The following year, on the occasion of the bicentenary of Śniadecki's birth, Polish chemist Kazimierz Sarnecki announced after lengthy analyses and with some reluctance that, due to irreconcilable differences, *vestium* could not possibly have been ruthenium.[36]

I.2.4. ERYTHRONIUM, PANCHROMIUM, OR VANADIUM

The element with atomic number 23 that we know as vanadium was identified for the first time in 1801 by the Spanish chemist Andrés Manuel del Rio (1764–1849) while he analyzed minerals from Mexico. As with the other elements already mentioned, it is important to note several fundamental points: this discovery was made before John Dalton propounded his atomic theory between 1803 and 1808, before the formulation of the periodic table by Dmitri I. Mendeleev in 1869, and more than a century before the concept of the atomic number, elucidated largely by the work of Henry G. J. Moseley in 1913. Hence, the possible points of reference for a scientist in those far off days were fairly limited.

Del Rio was born in Madrid on November 10, 1764, and studied at the University of Alcalá de Henares, subsequently obtaining his doctorate at the Accademia Mineraria de

Almadén. He left for England in 1791, moved to France during the turbulent days of the revolution to study under Lavoisier, and finally traveled to Germany, to the renowned Royal Mining School of Freiburg in Saxony. In Germany, he established a solid and long-lasting friendship with naturalist Alexander, Baron von Humboldt (1769–1859).

In 1794, while the revolutionary winds from France were blowing strongly throughout Europe, the youthful Del Rio was named professor of mineralogy at the mining school of Mexico, which had been recently founded by the chemist Don Fausto d'Elhuyar y de Zubice (1755–1833).[37] Del Rio set sail[38] and established himself in the New World. Some years later, while examining lead-containing minerals from Zimapán, in the province of Hidalgo in central Mexico, he isolated several compounds of a substance that initially he called *panchromium*. The name quite appropriately described the multicolored salts of the new metal. Shortly afterward, the name was changed to *erythronium* (from Greek, ἐρυθρός, meaning red) due to the predominantly red color of the solutions obtained by treatment with acids. There are some doubts about the effective date of the discovery (1801 or 1802) because the original documents are no longer available, but it is clear that the brown lead-containing mineral from Zimapán was in fact vanadite: $3Pb_3(VO_4)_2PbCl_2$. The oldest document regarding *erythronium*[39] was published in the *Gazeta de México* on November 12, 1802. Del Rio[40] gave many mineral samples to his friend von Humboldt[41] when the latter visited Mexico in 1803. The German naturalist in turn sent some of these samples, together with several scientific considerations, to the Institut de France, but unfortunately they never arrived at their destination because the ship transporting them to Europe was lost.

Because the properties of *erythronium* were very similar to those of chromium, an element discovered in 1794 by the French chemist Louis Nicolas Vauquelin (1763–1829), Del Rio lost faith in his work and rejected his discovery. In 1805, the mineral suspected of containing *erythronium* was analyzed by mineralogist Hippolyte-Victor Collet-Descotils (1773–1815), a friend of Vauquelin. He erroneously concluded that Del Rio's new metal was actually basic lead chromate.[42]

A quarter of a century later, in 1830, Nils Gabriel Sefström (1787–1845) described a new element that he had found in iron deposits from Taberg, in the region of Småland, Sweden. He noted that the properties of the iron extracted from those deposits were marked by peculiar features possibly related to the presence of a new metal[43] that he immediately named vanadium after the Scandinavian divinity "gottin Freya Vanadin." A few months before the discovery of vanadium, Friedrich Wöhler (1800–82) had come close to its rediscovery,[44] but without explanation abandoned his samples and dedicated his time to other pursuits. This neglect of his experiments cost him dearly: when, in 1831, he realized that Sefström had discovered the same metal, he had no choice but to give the credit to his colleague—and immediately made his claim for Del Rio's *erythronium*. The missed opportunity must have been a cause of much frustration to him. In fact, Berzelius wrote personally to console him, emphasizing that the name Wöhler would be immortal due to his many other important discoveries.[45]

Berzelius also created an imaginative story for the public regarding the discovery of vanadium:[46] "In the distant North there lived a fascinating and gracious goddess, Vanadis. One day a person arrived at her house and knocked at the door. The goddess, who was not in a hurry, did not move and thought—they'll knock again if they want to see me—but she heard nothing. The surprised goddess asked herself, who could the mortal be that did not have the patience to knock again to meet her, and ran to the window. She recognized Wöhler who was slowly walking away. A few days later someone else knocked vigorously and repeatedly at the door; the goddess opened the door and so it was that Sefström discovered vanadium." Vanadium was obtained in the metallic state only many years later by the chemist Henry Enfield Roscoe (1833–1915).[47] Once the identity of vanadium and *erythronium* had been universally recognized, the controversy took the form of the priority of discovery. Who should be given recognition for the discovery and, consequently, the right to propose a name? Del Rio, who had been the first to discover it but had rejected his own discovery? Or Sefström, who had rediscovered it almost 30 years later? The answers to these important questions were not limited only to achieving a consensus among chemists.

In August 1947, two Mexicans, physicist Manuel Salvador Vallarta (1899–1977) and historian Arturo Arnaiz y Freg (1915–82) of the Universidad Nacional de México, asked the International Commission for Chemical Nomenclature, which at that time was examining names to adopt for the elements of atomic numbers 43, 61, 85, and 87, to consider the possibility of exhuming the name *erythronium*[48] for the element with atomic number 23. The reply of the eminent chemist Friedrich Paneth (1887–1958) appeared at the end of the Letter that the two Mexican physicists published in *Nature*.[49] Paneth, who had taken English citizenship, was firm in his intent that the name vanadium should not be changed. Many supported his arguments, but Paneth's success was mostly due to the fact that the Mexicans' attempt to change the name came more than a century after the element's discovery. Too much time had elapsed to expect that the name already in use by a decree of the international commission could be changed.

In 1834, Manuel Andrés Del Rio, after weathering the events linked to vanadium, took the chair of geology and mineralogy at the University of Mexico City, where he worked indefatigably into his 80s. In 1845, due to recurrent health problems and poor eyesight, he was excused from his teaching duties, and on March 23, 1849, at the age of 84 and still active in his work, he died of a cerebral thrombosis.[50]

Notes

23. The empirical formula of this mineral is $AuPbSb_{0.75}Bi_{0.25}Te_{2.5}S_6$.
24. Müller von Reichenstein, F. *Physik. Arbeiten der einträchtigen Freunde in Wien* **1783**, *1*(1), 57.
25. Martin Heinrich Klaproth was born in Wernigerode, on December 1, 1743. A noted German chemist, in 1789, he discovered uranium and zirconium; he was the first to characterize them as distinct elements, although he did not manage to obtain either in the pure metallic state. Furthermore, he clarified the composition of numerous poorly characterized substances, including compounds of elements that had been only recently discovered: tellurium, strontium, cerium, and chromium.
26. *Cerium, en ny metal* (Cerium, a new metal); Nordstrom: Stockholm, Sweden, 1804.
27. Anon. *Ofv. Ak. Stockh.* **1873**, *30*, 13.
28. Mellor, J. W. *A Comprehensive Treatise on Inorganic and Theoretical Chemistry*; Longmans Green: London and New York, 1946, p. 404.
29. The asteroid Vesta was discovered by the German astronomer Heinrich Wilhelm Olbers (1758–1840) on March 29, 1807. Olbers gave the honor of choosing a name for the new asteroid to the mathematician Carl Friedrich Gauss (1777–1855), and it was named in honor of the Roman goddess Vesta.
30. Śniadecki, A. *Memoires de l'Academie imperiale des sciences de St.-Petersbourg* **1810**, *2*, 30.
31. Marshall, J. L.; Marshall, V. R. "The curious case of 'Vestium.'" *The Hexagon* **2011**, Summer, 20–24.

32. The name *ceresium* derives from Ceres (Cerere), an asteroid observed in 1801 by the Theatine monk, mathematician, and astronomer Giuseppe Piazzi (1746–1826), who in turn took the name from the Roman goddess of the harvest, of growing plants, and motherly love. The name ceased to exist due to the wishes of its discoverer, an event unique in the literature of the discovery of the elements.
33. Pallas or Pallade was discovered on March 28, 1802, by Heinrich Wilhelm Olbers, while, from his private observatory installed at his house in Bremen, he was trying to find Ceres and verify its orbit, which had been predicted mathematically. Olbers himself named the asteroid Pallas after one of the three daughters of Triton in Greek mythology. An ambiguous figure, Pallas was the playmate of the young Athena who mortally wounded her. In honor of Pallas, Athena erected a statue in her likeness, the palladium. Moreover, to remember her friend forever she decided that she would call herself Pallas Athena. However, it has also been suggested that Pallas is the epithet given to the goddess Athena as protector and serious advisor in war.
34. Plesniewicz, S.; Sarnecki, K. *Przemysl Chemiczny* **1938**, *22*, 88.
35. Znachko-Yavorskii, I. L. *Kwartalnik Historii Nauki i Techniki* **1967**, *12*(1), 47.
36. Sarnecki, K. *Kwartalnik Historii Nauki i Techniki* **1968**, *13*(4), 799.
37. In 1783, Don Fausto together with his brother Juan José d'Elhuyar (1754–96) discovered tungsten, later called wolfram.
38. The trip from Cádiz to the New World required 11 weeks, which clearly describes how far Mexico of the 18th century was from the European intellectual centers. This isolation had a strong influence on the events regarding the discovery of the new element.
39. Greenwood, N. N. *Catal. Today* **2003**, *78*, 5.
40. Del Rio, M. A. letter addressed to Baron von Humboldt in 1817 and published in *Mercurio de España* **1819**, 173.
41. von Humboldt, A. *Ann. Mus. Nat.*,**1804**, *3*, 396; *Gehlen's Journal* **1804**, *2*, 695; *Gilbert's Ann.* **1804**, *18*, 118.
42. Collet-Descotils, H. -V. *Ann. Chim. Phys.* **1805**, *53*, 268.
43. Sefström, N. G. *Acad. Handl. Stockholm* **1830**, 255; *Schweigger's Journ.* **1831**, *62*, 316; *Pogg. Ann.* **1831**, *21*, 43; *Ann. Chim. Phys.* **1831**, *2*(46), 105; *Phil. Mag.* **1831**, *2*(10), 151.
44. Wöhler, F. *Schweigger's Journ.* **1831**, *62*, 124; *Pogg. Ann.* **1831**, *21*, 49; *Ann. Chim. Phys.* **1831**, *1*(46), 111.
45. Berzelius, J. J. *Phil. Mag.* **1831**, *2*(10), 151; *Phil. Mag.* **1831**, *2*(10), 157.
46. Mellor, J. W. *A Comprehensive Treatise on Inorganic and Theoretical Chemistry*; Longmans Green: London, 1939; vol. IX, p. 714.
47. Roscoe, H. E. *Phil. Trans.* **1868**, *158*, 1; *Proc. Roy Soc.* **1868**, *16*, 220.
48. Vallarta, M. S.; Arnaiz y Freg, A. *Nature* **1947**, *159*, 163.
49. Paneth, F. *Nature* **1947**, *159*, 164.
50. Weeks, M. E. *J. Chem. Educ.* **1935**, *12*, 161.

PART II

1789–1869

From Lavoisier to Mendeleev: The First Errors at the Dawn of the Concept of the Chemical Element

The power of teaching is rarely very efficient except for those happy situations where it is almost superfluous.
—Richard Feynman (1918–88)
Nobel Prize for Physics 1965

PROLOGUE TO PART II

In the 80 years covered by Part II of this volume, scientists confronted the concept of the chemical element for the first time. However, many of them were caught in a conflict between growing scientific realism and the visionary utopia of a protoscience still linked to alchemy. Those individuals (some of whom still believed in the phlogiston!) who engaged in the isolation of new elements through a reliance on their considerable technical skills rather than on the new theories expounded by Lavoisier sowed the seeds of failure in their works. The growth and development of analytical techniques played an important role not only in the discovery of new elements, but also in the more accurate determination of atomic weights. Part II lays before the reader a mix of published and unpublished texts relating to false discoveries of elements, as well as to the rediscovery of simple substances already known.

II.1

ANALYTICAL METHODOLOGY FROM LAVOISIER TO MENDELEEV

Within the period covered by Part II, 1789–1869, 37 true elements, almost all of them metals, were discovered. Prior to this time, about 14 metals had been discovered, excluding those that had been known from ancient times. The discovery of the elements during this period of interest is intimately related to the analytical methodologies available to chemists, as well as to a growing consciousness of just what an element is. Because these methods were also available to the less competent who may have lacked the skills to use them or the knowledge to interpret their results, their use also led to as many, if not more, erroneous discoveries in the same period. One can number among the major sources of error faulty interpretation of experimental data, the "rediscovery" of an already known element, sample impurities, very similar chemical properties (as in the case of the rare earths), the presence of an element in nature in very scarce or trace amounts, gross experimental errors, confusion of oxides and earths with their metals, and baseless dogmatic pronouncements by known "authorities" in the field.[1]

Antoine Laurent Lavoisier's conceptualization of what constitutes an element was a radical break from the principles of alchemy.[2] His stipulation that an element is a substance that cannot be further decomposed conferred an operational, pragmatic, concrete definition on what had previously been a more abstract concept. At the other end of the spectrum was the intuition of Dmitri Mendeleev who, contrary to the prevailing acceptance of Lavoisier's concept, stressed the importance of retaining a more abstract, more fundamental sense of an element—an idea that in the long run enabled the development of the periodic table. What both men had in common is that they defined and named individual elements as those components of substances that could survive chemical change and whose presence in compounds could explain their physical and chemical properties.[3] Mendeleev's table has been immortalized in every chemistry classroom—and also concretely in Saint Petersburg, the city that saw most of his professional activity, by a spectacular building-sized model (see Figure II.01)

The analytical chemist depends on both of these concepts and indeed, analytical practice preceded Lavoisier's concept by at least a century. In essence, chemists of the 18th century had already put into practice an idea that Lavoisier would later conceptualize.[4] And it was only through the development of a system of chemistry with logical interrelationships and accurate qualitative and quantitative analyses that a considerable number of pure substances could be identified and incorporated into a consistent whole.[5] The principal methodologies, in addition to the alchemical methods of cupellation, smelting, and distillation, were (1) blowpipe analysis, (2) qualitative and quantitative analysis, (3) electrolysis, and (4) spectroscopy (after 1860). The most important criteria for these methods were speed, selectivity, and sensitivity.

FIGURE II.01. Statue of Dmitri Mendeleev (1834–1907) on Moskovskii Prospekt next to the building of the Bureau of Weights and Measures where Mendeleev worked as director. During his tenure, he introduced new standards for the production of vodka. The building now houses the Mendeleev All-Russian Institute of Meteorological Research and sports a building-high version of his famous periodic table. Photograph by Mary Virginia Orna.

II.1.1. BLOWPIPE ANALYSIS

For at least a thousand years, jewelers and metal workers had used the blowpipe as a tool, but its first appearance in the chemical literature comes in the late 17th century, and references to its use rapidly multiplied thereafter.[6] Axel Fredrik Cronstedt (1722–65), who utilized this instrument to discover nickel in 1751, honed its use to a fine degree. In his many papers, he described its systematic application, and he also made many structural improvements that were then later improved upon by others.[7] The blowpipe method, in the right hands, enabled the analyst to detect very small differences in composition simply and speedily. Considered an indispensable tool in the chemistry laboratory for more than a century, it figured in the discovery of at least a dozen elements from Cronstedt's discovery of nickel to Ferdinand Reich's (1799–1882) and Theodor Hieronymus Richter's (1824–98) discovery of indium in 1863. Working in the second decade of the 19th century, Johan Gottlieb Gahn (1745–1818), discoverer of manganese in 1774, refined and extended the technique of blowpipe analysis. J. J. Berzelius summarized and verified Gahn's experimental work (Gahn published virtually nothing) in his own book on the subject, averring that Gahn always carried his blowpipe with him, even on his shortest journeys.[8]

II.1.2. QUALITATIVE AND QUANTITATIVE ANALYSIS

Generally considered the founder of the science of mineralogy, Axel Cronstedt was the first to make use of chemical criteria for the purpose of mineral classification, thereby bringing about radical changes in mineral taxonomy.[9] His system was based on chemical composition instead of the time-honored external characteristics of color, hardness, luster, and form. Qualitative identification of the metals and nonmetal radicals was carried out by Torbern Olof Bergman,[10] Martin Heinrich Klaproth, and Heinrich Rose (1795–1864) but became much more systematic in the 1840s, led by Carl Remigius Fresenius (1818–97). The systematic basis for making group separations thenceforward was developed gradually by numerous analysts.[11] For example, Klaproth isolated zirconia from zircon in 1789, and in that same year, he erroneously discovered uranium in pitchblende.[12] In 1795, he discovered titanium, and in 1797, he isolated chromium. Klaproth followed a well-worn path to success, using knowledge of chemical reactivity, melting points, and solubilities of known chemical elements and compounds to separate them out from new and unknown materials. Once additional tests showed that a new material had unique characteristics, the analyst was well on the way to discovery—and the possibility of many missteps caused by the presence of impurities, the removal of which was sometimes very difficult.[13]

Gravimetric and volumetric methods of analysis were also used extensively. Although gravimetric methods were favored because one had a weighable material on hand, volumetric methods allowed for speedy multiple determinations.

The principal gravimetric tool then, as now, was the analytical balance or one of its modern derivatives. In 1785, Lavoisier stressed that his published results were based on repeated weighing and measuring experiments that were the only criteria for admitting anything in physics and chemistry. It is well-known that he spent a fair amount of his fortune on the best scientific apparatus that money could buy, often designing the apparatus himself and then having it purpose-built by a specialist. (The enormous precision analytical balances that he used are now on display in Paris's Musée des Arts et Métiers and were made by the best manufacturer of the time).

Precise balances also played a role in pneumatic chemistry, the study of gases, and their isolation and identification. Several key elements (H, N, O, Cl) were discovered using gasometric methods—the determination of gas density and other physical properties of gases and how these relate to their chemical properties.

Lavoisier's insistence on the need for precision instruments was not lost on those who followed him, although some of his pieces of apparatus were so unique and so complex that scientists of lesser means were forced to improvise and therefore sacrifice both accuracy and precision. However, gravimetic analysis advanced and, with it, more correct atomic weights for the elements and, by extension, more exact stoichiometry and thus better knowledge of an element's chemical properties—two pieces of information that allowed Mendeleev and others to deduce the connections between atomic weight and chemical properties.

II.1.3. ELECTROLYSIS

Alessandro Volta's (1745–1827) invention in 1800 of the voltaic pile, made by multiplying the number of metal–metal junctions in series, soon was utilized by William Nicholson

(1753–1815) and others to electrolyze water. Humphry Davy (1778–1829) of the Royal Institution of Great Britain was swift to recognize the pile's potential and quickly constructed one consisting of more than 250 plates. By 1807, he had succeeded in isolating sodium and potassium, and soon thereafter he produced barium, magnesium, calcium, and strontium in the elemental state.[14] Trevor Levere remarks:

> Davy was the most successful of those who accepted the challenge implicit in Lavoisier's definition of elements as the last products of analysis. If he could decompose one or more of Lavoisier's elements, then he would have discovered new ones. As Lavoisier had observed, there was no telling where this process of discovery through decomposition or analysis would lead. Davy aimed to find out. He produced a series of ever more powerful electric piles.[15]

Ironically, Davy's success was no success at all—at least to him. He carried electrolysis to the extreme not to discover new elements, but to discover the ultimate particle from which all elements were made. He kept on making increasingly powerful batteries and discovering, reluctantly, more and more chemical elements, but never the one he was truly looking for. The end point of his quest lay far into the future and was accomplished not by the chemical means of a voltaic pile but by the instruments of high-energy physics and the logical consequence of Lavoisier's definition of element.[16]

Faced with the question of "elementality," Davy was not the only scientist to fall into conceptual error. Pérez-Bustamante has catalogued a number of misconceptions, including Klaproth's reluctance to classify potassium and sodium as real metals because of their strikingly low densities; Berzelius's assumption that nitrogen was an oxide of a hypothetical radical, *nitricium* (1803); and Christian Friedrich Schönbein's (1799–1868) idea that chlorine was a peroxide of *murium* (1865). Actually, when HCl, known to the alchemists, was dissolved in water, a typical acidic solution was formed, and it was named muriatic acid, from the Latin *muria*, meaning brine. In 1779, Lavoisier wrongly concluded that oxygen was present in muriatic acid and that this made it an acid. He predicted the existence of another element or *muriatic radical* and eventually, in 1810, Davy recognized it as an element and gave it its modern name—chlorine.

Once the free alkali and alkaline earth elements were widely available, they were often used to reduce less active metals, giving rise to additional elemental discoveries.

II.1.4. EMISSION SPECTROSCOPY

In 1859, Gustav Kirchhoff (1824–87) and Robert Wilhelm Bunsen (1811–99) were making studies to characterize the colors of heated elements. It was Kirchhoff who realized that the observed frequencies of the various elements' emission lines were characteristic of a given element. Kirchhoff suggested to Bunsen that they systematize their studies and try to develop a device that would form spectra of these colors by using a prism. By October of that year, they had invented an appropriate instrument, a prototype spectroscope, by which they were able to identify the characteristic spectra of sodium, lithium, and potassium.[17] After numerous laborious purifications, Bunsen proved that highly pure samples produced unique spectra. In the course of this work, Bunsen detected previously unknown new blue spectral emission lines in samples of brine water from some well-known German spas. He realized that these lines were the signature of a hitherto unknown element, which he named cesium, after the Latin for "deep blue." In the following year, he and one of his students discovered rubidium (after *rubidus*, meaning "deep red") by a similar process.[18,19] In the same year that rubidium was discovered via spectroscopy, Sir William Crookes (1832–1919) discovered thallium by the same method,[20] although credit is also given to Claude-Auguste Lamy (1820–78) who discovered it almost simultaneously, working independently. Indium, gallium, and some rare earths followed in quick succession.[21]

Spectroscopic identification became the method of choice in analytical chemistry, but its use over the years spanning the invention of the spectroscope to the discovery of the atomic number and beyond was fraught with error, as discussed in upcoming parts of this book.

This overview of analytical methods was necessarily brief, but some fine books and articles on the history of analytical chemistry are available. Perhaps the most comprehensive is the volume by Ferenc Szabadvary,[22] which discusses not only the discipline but includes many delightful mini-biographies of its practitioners.

Notes

1. Pérez-Bustamante, J. A. Analytical chemistry in the discovery of the elements. *Fresenius' Journal of Analytical Chemistry* **1997**, *357*(2), 162–72.
2. Scerri. E. What is an element? What is the periodic table? And what does quantum mechanics contribute to the question? *Found. Chem.* **2012**, *14*, 69–81.
3. Hendry, R. F. Lavoisier and Mendeleev on the elements. *Found. Chem.* **2005**, *7*, 31–48.
4. Brock, W. H. *The Norton History of Chemistry.* W. W. Norton & Co.: New York, 1993, p. 119.
5. Bartow, V. Axel Fredrik Cronstedt. *J. Chem. Educ.* **1953**, *30*, 247–52.
6. Jensen, W. B. The development of blowpipe analysis. In Stock, J. T.; Orna, M. V., Eds. *The History and Preservation of Chemical Instrumentation.* D. Reidel Publishing Company: Dordrecht, The Netherlands, 1986; pp. 123–49.
7. Griffin, J. J. *A Practical Treatise on the Use of the Blowpipe in Chemical and Mineral Analysis.* R. Griffin & Co.: Glasgow, Scotland, 1827, p. 14.
8. Berzelius, J. J. *The Use of the Blowpipe in Chemistry and Mineralogy,* 4th ed. William D. Ticknor: Boston, MA, 1845.
9. Oldroyd, D. R. A note on the status of A. F. Cronstedt's simple earths and his analytical methods. *Isis* **1975**, *65*(4), 506–12.
10. In his book on blowpipe analysis, *De tubo feruminatorio*, Bergman is the first chemist to make reference to the difference between qualitative and quantitative analysis. Bergman, T. O. *Opuscula physica et chimica* **1779**, *2*, 452–505.
11. Ihde, A. J. *The Development of Modern Chemistry.* Dover Publications: New York, 1984, p. 277.
12. Henri Victor Regnault (1810–78) pointed out that Klaproth's "metal" did not obey the Law of Dulong and Petit; Eugène Melchior Peligot (1811–90) then revised the discovery: realizing it was uranium oxide, he succeeded in obtaining the pure metal by reduction of uranium chloride with potassium in 1841.
13. Ede, A. *The Chemical Element: A Historical Perspective.* Greenwood Publishing Group: Westport, CT, 2006, p. 65.
14. Leicester, H. M. *The Historical Background of Chemistry.* Dover Publications: New York, 1956, p. 165.
15. Levere, T. *Transforming Matter: A History of Chemistry from Alchemy to the Buckyball.* Johns Hopkins University Press: Baltimore, MD, 2001, p. 88.

16. Bensaude-Vincent, B. Mendeleev's periodic system of chemical elements. *Br. J. Hist. Sci.*, **1986**, *19*, 3–17.
17. Orna, M. V. *The Chemical History of Color.* Springer: Heidelberg, Germany, 2013; pp. 99–105.
18. Kirchhoff, G; Bunsen, R. W. Chemische Analyse durch Spectralbeobachtungen (Chemical analysis by observation of spectra). *Pogg. Ann.* **1860**, *110*, 161–89.
19. Kirchhoff, G; Bunsen, R. W. Chemische Analyse durch Spectralbeobachtungen. *Ann. Phys. Chem.* **1861**, *189*(7), 337–81.
20. Crookes, W. On the existence of a new element, probably of the sulphur group. *Chem. News* **1861**, *3*, 193–94.
21. Lamy, C.-A. De l'existencè d'un nouveau métal, le thallium. *Comptes Rend. Acad. Sci. Paris* **1862**, *54*, 1255–62.
22. Szabadváry, F. *History of Analytical Chemistry.* Gordon & Breach Science Publishers: London, 1992.

II.2

THE ELEMENTS OF THE KINGDOM OF NAPLES

In 1786, Immanuel Kant (1724–1804) wrote "Chemistry can become nothing more than a systematic art or an experimental doctrine, but never a true science, because its principles are merely empirical and. . . are incapable of the application of mathematics."[23] Despite Kant's rigid view, chemists have managed to formulate a number of general laws, the first of which was the Law of Conservation of Mass in 1774, generally attributed to the French scientist Antoine Laurent Lavoisier.[24] Moreover, Lavoisier helped to define a clear concept of the chemical element. Based on this clarification and the refinement of analytical methods, new elements were discovered more frequently. Tungsten (or wolfram) was discovered in 1781 by Carl Wilhelm Scheele and was then isolated by Fausto de Elhuyar y de Zubice, a Spaniard, in 1783; tellurium, discovered by J. F. Müller von Reichenstein in 1782, was isolated only 16 years later by M. H. Klaproth, who in 1789 also discovered uranium[25] and zirconium. In the same year, William Gregor (1761–1817) discovered titanium.[26] These last three elements were only isolated in a very pure state more than a century later. In 1790, as many as six elements were "discovered": strontium by Adair Crawford (1748–95) and William Cruikshank (~1745–~1810); and *borbonium, apulium, austrium, parthenium,* and *bornium* by Anton Ruprecht and his student Matteo Tondi (1762–1835). As it turned out, only strontium found its way into the roll call of genuine elements.

II.2.1. RUPRECHT AND TONDI: TWO METALLURGISTS WITHOUT METALS

Antal (or Anton) Leopold Ruprecht (1748–1814) was born in Schollnitz (present-day Smolnik in Slovakia) in Hungary, in 1748. He obtained his diploma from the Mining Academy of Selmecbány under the supervision of metallurgist Ignaz Elder von Born (1742–91), renowned for introducing into Austria the method of amalgamation for the separation of gold and silver from gangue. He continued his studies in Freiburg and in Scandinavia (1777–79), where he analyzed the composition of many minerals while working with the well-known chemist Torbern Olof Bergman. Finally, in 1779, he was named professor of chemistry and metallurgy at the prestigious Bergakademie (Mining Academy) at Schemnitz (present-day Banská Štiavnica in central Slovakia), where he remained until 1792, when he was called to the Hofkammer of Vienna. In Vienna, at the age of 44, he took Born's place (Born had died the previous year).

Ruprecht was the first chemist to succeed in melting platinum and, according to some, he played a minor role in the discovery of tellurium. At the end of the 18th century, Austria was a scientific and cultural focal point for all the countries bordering the Danube and those of southern Europe. Although Berlin and Stockholm were rivals for supremacy in the field of metallurgy, in the sciences, the title was held by Paris, and the prestige of Lavoisier was felt throughout Europe.[27]

The other personality in this story is Matteo Tondi. He was born in 1762 at Sansevero in Apulia and studied medicine at the University of Naples. According to the custom of the time, he also followed courses in botany, physics, and chemistry. In 1789, he went to Austria to learn the art of mining. During his travels across the mining districts of Europe, Matteo Tondi collected, meticulously and with great passion, 35 chests of minerals, which were sent to the Mineralogical Museum in Naples. According to Poggendorff's *Biographisch-Literarisches Handwörterbuch*, the chemical fractionation of minerals to isolate new elements constituted for Tondi only a brief episode in his career, one that, after achieving fame and success, he may have tried to forget.[28]

II.2.2. PLAYING BINGO WITH FIVE ELEMENTS

In 1791, the newly published *Annales de Chimie*[29] reported an extract from a paper by Professor Anton Ruprecht and Dr. Matteo Tondi (the entire article was published in *Crell's Annalen*) in which the two scientists described a process for reducing a number of wolfram and molybdenum minerals.[30] The content of the article was quite modest, both scientifically and for its lack of originality, and it could have easily been forgotten were it not for a harsh reply by Andrea Savaresi (1762–1810).[31] He criticized the article so severely that there followed a regrettable diatribe. The resentment between Savaresi and Tondi probably existed before this event, fired by an intense rivalry between them when they were working in Born's laboratory. In a letter, now part of the Waller collection in the library of the University of Uppsala, Matteo Tondi vented his anger against Savaresi: "please tell Mr Born in confidence that perhaps I will ask to leave in order to free myself of Savaresi, who openly works against me with Lippi."

Savaresi and Carminantonio Lippi (1760–1823), together with Tondi, were among six young scientists[32] sent to Vienna by a high-ranking official of the king of Naples to receive training in metallurgy and the art of mining. The desire to excel must have led the three into intense competition. The real dispute, however, began following the publication of the discovery of five elements by two disciples of Ruprecht and Tondi: Microszewski and Bienkowski.

In a note published in *Annales de Chimie*, the two Polish students briefly described how Ruprecht and Tondi had isolated five new simple bodies.[33] According to the knowledge of the period, it was the norm to believe that "earths" (today we would use the term "oxides") not decomposed (today we would say "reduced") could conceal new elements. According to the two Polish students, Dr. Matteo Tondi and Professor Anton Ruprecht had extracted and immediately given names to the metals *borbonium, austrium, parthenium,*[34] *apulium*, and *bornium*. The paper described both the scientific and the more mundane aspects of the work. Tondi proposed three of the names of the new elements: *apulium* after the Italian region, Apulia; *parthenium* after the city of Naples; and *borbonium* after the reigning dynasty of the Kingdom of Naples. All the names were easily traceable to Tondi's native country. Conversely, *austrium* from Austria and *bornium* from Ignaz von Born, clearly reflect Ruprecht's desire to honor his mother country and his professor.[35]

Returning to the first important aspect of Tondi's discovery, the "earths" from which the new metals were extracted (alumina, barytes, lime, and magnesia) and the boric acid contained elements that, in 1791, were still unknown: boron, aluminum, barium, calcium, and magnesium. Before publishing their results, Ruprecht and Tondi decided to repeat their experiments (with a positive outcome) at an extraordinary meeting of the Academy in Vienna called expressly for this purpose. The results were published in *Annales de Chimie*. On the page following their note, two detailed letters were published, one[36] by Martin Klaproth addressed to Claude Berthollet (1748–1822) and the second[37] by Joseph Franz Freiherr von Jacquin (1766–1839) to Bertrand Pellettier (1761–97). They not only criticized the work of Tondi and Ruprecht, but tried to discredit their scientific status. Savaresi also attacked their work, publishing part of his correspondence with Berthollet and Klaproth. As if all of this were not enough, Savaresi sent a note to the *Annales de Chimie* highly critical of Tondi, discrediting him in the eyes of the academic world of the time.[38]

II.2.3. THE EXTRACTION PROCEDURE OF THE NEW METALS

The chemical process used by Tondi to reduce the five earths was simple. First, he prepared a paste of the oxide he wanted to reduce using linseed oil and powdered carbon. He then spread it on the internal walls of a Hesse crucible[39] up to two-thirds of its height; inside the cavity, he placed carbon powder reaching the same height as the paste. On top of this material, he placed ground deer bones and antlers and powdered lye,[40] completely filling the container. The crucible was then put in an oven and heated at a high temperature. For each earth examined, this process gave one or more metallic lens-shaped disks. Tondi believed that these metal-like spherules were new elements. Although the preparation process reported here seems crude and amateurish, it was standard practice at the time. Tondi took the time to describe aspects of his experiment that today are considered irrelevant (e.g., "a good flame") instead of mentioning the quantity of material used or at least the ratio of the weights of the reagents. Nevertheless, one should recall that as little as two centuries ago the need to weigh both reagents and products was not universally recognized. This fundamental requirement, necessary for the advancement of chemistry, was introduced and widely promulgated by Lavoisier precisely during the years Tondi worked.

II.2.4. RIGHT OR WRONG, WAS TONDI THE VICTIM OF A SWORN ENEMY?

On February 27, 1791, Klaproth communicated the results of his experiments to Berthollet in Paris. Klaproth had previously read the same memo in a very important assembly, the Academy of Sciences of Berlin. He had repeated Tondi's experiments, trying in vain to reduce the crude earths (oxides of Ca, Ba, Mg, Al) and the sedative salt boric acid (H_3BO_3). He concluded that Tondi and Ruprecht's work was an illusion and that they had fooled both themselves and the public. Klaproth believed that Tondi and Ruprecht had obtained *hydrosideron* (probably iron phosphide, FeP),[41] produced by the reduction of iron oxides present in the clay of which the crucible was made and by the phosphorus present in the deer antler powder. To verify his hypothesis, Klaproth, after following the procedure of Tondi and Ruprecht using a Hesse crucible, repeated the experiment with a porcelain mortar. He found no metallic grains among the reduction products.

Obviously, Ruprecht and Tondi had no intention of falsifying their results, but, at the same time, their work cannot be judged leniently. Although not superficial, their work was certainly ill-considered and their conclusions poorly defined. In fact, Klaproth's harsh judgment of Tondi's work was salutary, urging as it did all chemists to use appropriate care in their work. Tondi, highly criticized, portrayed himself as an unjustly persecuted victim (although, from the documents available to us, he appears to have been a restless person of very argumentative character), and some others agreed. Unable to reply openly to a professor of Klaproth's status, he vented his anger against his colleague Savaresi. However, if a piece of work is shown to be incorrect, the merit of those who note and report the error should be recognized. More than anyone else, Klaproth did justice to the role of Savaresi, commending him for his impartiality and scientific rigor.[42]

II.2.5. THE ELEMENTS THAT REPLACED THOSE OF TONDI

Calcium: Calcium oxide or lime had been known since ancient times and was long considered an element or undecomposable "earth." This concept was questioned in the 18th century, but only in 1808 was Sir Humphry Davy able to establish that lime is a combination of oxygen with a metal. In a communication to the Royal Society of London on June 30, 1808, Davy reported that he had obtained a new alkaline element by electrolysis.

Barium: In 1772, Carl Wilhelm Scheele noted that in pyrolusite (MnO_2) there were several small crystals that he recognized as a new earth (barium oxide, BaO). Two years later, Johan Gottlieb Gahn (1745–1818) found at Falun the same oxide in spar (in German *Schwerspat*, $BaSO_4$). Scheele called it *Schwerspatherde* (heavy earth from spar). This "heavy earth" was renamed *barote* by Louis Bernard Guyton de Morveau (1737–1816) and later changed again to baryte, from the Greek *barys* (βαρυς) that means "heavy." The name was changed for the last time to *baryta* by Lavoisier. The element was isolated by Sir Humphry Davy who communicated the discovery together with that of calcium.

Aluminum: Hans Christian Oersted (1777–1851) is credited with being the first to prepare metallic aluminum.[43] In 1825, he isolated a small quantity of impure material, but the discovery passed unnoticed as a consequence of its publication in an unknown Danish journal. Moreover, he was not completely convinced of his discovery. He spoke of his work with Friedrich Wöhler, who refined Oersted's procedure and, in 1827, obtained a reasonable amount of the metal.

Boron: In 1702, Wilhelm Homberg (1652–1715) used borax, a substance thought at that time to be made by synthesis, to prepare a white powder resembling snow that he called sedative salt (boric acid, H_3BO_3). In 1747–48, Théodore Baron de Hénouville (1715–68) discovered that the active substance in the preparation was another material, which he identified as Na_2O. Louis-Joseph Gay-Lussac (1778–1850) and Louis-Jacques Thénard (1777–1857) in France[44,45] and Humphry Davy[46] in England isolated the element in 1808. On June 21, 1808, the two French chemists announced that they had decomposed boric acid with potassium and, subsequently, had resynthesized it from its elements. At the end of their communication, they proposed a name for the element: "Nous désignerons par la suite ce radical sous le nome bore, qui est tiré de celui du borax."[47] Nine days later in England (June 30, 1808), Sir Humphry Davy presented a communication to the Royal Society in which he announced the discovery of metallic boron by heating boric acid and potassium in a copper tube. During the Bakerian Lecture, read the following December

15, Davy proposed calling the new substance *boracium*. In the Anglo-Saxon world, *bore* and *boracium* were modified to "boron" to make it sound similar to carbon.

Magnesium: In 1808, Sir Humphry Davy isolated a metal and called it *magnium*, in order not to confuse it with the name manganese, a metal that had been found in *magnesia nigra*:

> These new substances will demand names; and on the same principles as I have named the bases of the fixed alkalis, potassium and sodium, I shall venture to denominate the metals from the alkaline earths barium, strontium, calcium, and magnium; the last of these words is undoubtedly objectionable, but magnesium has been already applied to metallic manganese, and would consequently have been an equivocal term.[48]

Contrary to Davy's proposal, the term magnesium persisted to indicate the metal that had been discovered in *magnesia alba* (magnesium oxide). In Russian and other Slavic languages, the form *magnium* (МАГНИЙ) is still in use. Davy isolated the new element in a very impure state. The French chemist Antoine Alexandre-Brutus Bussy (1794–1882) is credited with obtaining it in a very high state of purity, although in powdered form, in 1828. In the same year, Johann Bartholomäus Trommsdorff (1770–1837) proposed calling this metal *talcinium* from the name of the mineral *talkerde* (magnesium oxide), but this suggestion was quickly forgotten.[49]

II.2.6. POSSIBLE PRESENT-DAY INTERPRETATIONS

The metals that Tondi and Ruprecht were looking for did not exist in the free state in nature. The two chemists tried to extract them from very common compounds in which they are found in nature, summarized in Table II.1.

Today, we know that the only ways to isolate calcium and barium are (1) by reduction from their compounds using other electropositive metals, such as aluminum; (2) by electrolysis of concentrated solutions with a mercury cathode; and (3) by electrolysis of the halogen derivatives of the alkaline-earth metal in the molten state. With regard to boric acid, other reduction procedures are used. The oxide can be prepared from boric acid, $B(OH)_3$, by heating. The reduction of the oxide takes place by reaction with magnesium:

$$B_2O_3 + 3Mg \rightarrow 2B + 3MgO \quad \text{(Eq. II.1)}$$

Table II.1 Summary of the Work of Ruprecht and Tondi

Proposed Name	*Source*	*Chemical Formula*	*Unknown Element Present*	*Date of Isolation*	*Authority*
Borbonium	Barite	$BaSO_4$	Ba	1808	Davy
Austrium	Magnesia	MgO	Mg	1808	Davy
Parthenium	Calcite	CaO	Ca	1808	Davy
Apulium	Alumina	Al_2O_3	Al	1827	Oersted/Wöhler
Bornium	Boric acid	H_3BO_3	B	1808	Gay-Lussac/Thénard

Small quantities of boron can also be prepared by thermal decomposition of BBr_3 in a stream of hydrogen in a tantalum tube heated to more than 1,000 °C. It is impossible to obtain the elements that Tondi was looking for (Ca, Ba, B) following his procedure because carbon, linseed oil, and powdered bones are not sufficiently good reducing agents.

Savaresi's and Klaproth's analyses[50] clearly revealed the nature of the product obtained by Tondi in his attempt to reduce the five earths: ferric oxide present in the Hesse crucible + powdered bones + carbon → *hydrosideron* + fixed air, or, in modern-day notation:

$$2Fe_2O_3 + P_4O_{10} + 8C \rightarrow 4FeP + 8CO_2 \quad \text{(Eq. II.2)}$$

Some doubts have been expressed regarding whether Tondi isolated aluminum or magnesium. In the vast literature that exists on aluminum, there is even a report attributed to the epoch of the emperor Tiberius (41 BCE–37 CE), recorded in 77 CE by Pliny the Elder (ca. 23–79 CE) in his *Naturalis Historiae.* According to this document, a metal similar to silver but much lighter and malleable as iron could be extracted from clay. The work of Tondi was noted by J. T. Kemp[51] and J. L. Howe in 1902 and published in *Chemical News* the same year.[52] The procedure described by Pliny the Elder, although vague, seems to resemble that of Ruprecht and Tondi. J. W. Mellor (1873–1938) explained how this might be possible. He believed that Kemp and Howe had interpreted too freely some parts of Pliny's report.[53] Mary Elvira Weeks, who did not mention Tondi,[54] was convinced that Ruprecht could have extracted a new element (*austrium*) from magnesia, which could have been magnesium. This case again met with some controversy in the chemistry literature. In 1821, shortly after Davy's discovery, Edward Daniel Clarke (1769–1822), an acclaimed professor at Cambridge, hypothesized that finely ground magnesia (MgO) mixed with oil and then placed in a blowpipe flame produced an inconsistent and crumbly material.[55] The proposed deoxygenation reaction was as follows:

$$nMgO + (n - \tfrac{1}{2}m)\,C \rightarrow nMg + (n - m)\,CO + \tfrac{1}{2}CO_2 \quad \text{(Eq. II.3)}$$

Clarke did not live long enough (he died the following year) to know that others were to demonstrate the inaccuracy of his hypothesis. Many years after Davy's discovery, in 1884, J. Walter tried again to reduce magnesium oxide with carbon, following the same procedure used for zinc. A temperature of white heat was reached, and the appropriate precautions taken in case the metal was formed in either the liquid or vapor state, but the result was negative.[56] At the turn of the 20th century, the well-known Henri Moissan (1852–1907), using an electric furnace of his own invention, repeated this experiment many times, observing that magnesia could be melted and maintained in the liquid state in a carbon crucible without reducing the oxide to magnesium.[57] Some years later, in 1907, Oliver Patterson Watts (1865–1953) observed that by passing carbon vapor over a magnesia bed, metallic magnesium was formed:[58]

$$MgO + C \rightarrow Mg + CO \quad \text{(Eq. II.4)}$$

Finally, in 1915, O. L. Kowalke and D. S. Grenfell[59] found that the reduction of magnesia with carbon began very slowly at around 1,950 °C.

Returning to the experiments of Tondi and Ruprecht (who did not leave a detailed description of their process), we can hypothesize that they may have worked with a common furnace used for forging iron. In these furnaces, temperatures of about 1,200–1,500 °C can be obtained by burning oak and pine wood. At the end of the process, which could last for several days, the furnace was destroyed and at the bottom a block of steel was collected. Because Tondi's procedure did not involve heating for several days in the crucible nor reaching a temperature of about 2,000 °C (necessary to trigger the reaction of magnesium oxide with carbon), one can be reasonably certain that he did not manage to obtain alkaline-earth elements in the metallic form.[60]

II.2.7. REVOLUTION OFFERS A SECOND CAREER POSSIBILITY

The events following the announcement of the discovery of the five elements could have created an insurmountable obstacle to Tondi's career. Shortly afterward, in 1799, he left the Habsburg empire and sought refuge in France for political reasons. He became assistant in Mineralogy at the Museum of Natural History in Paris, directed by the famous Deodat de Dolomieu (1750–1801) and later by the Abbé René Just Haüy (1743–1822), who was to revolutionize the old concepts of the mineralogical sciences.[61] In this environment, Tondi learned the new techniques of crystallography. A relationship of reciprocal esteem was to develop between Tondi and Haüy, who delegated Tondi to collect the essential samples for his research. In 1811, Tondi returned to Naples and subsequently took the chair of geology at the Royal University of Naples and became curator of the Royal Mineralogical Museum. Twenty years later, on November 17, 1835, at the age of 73, he died at Naples.

Notes

23. Kant, I. In *Sämmtliche Werke*; Rosenkranz, K.; Schubert, F. W., Eds.; Leopold Voss: Leipzig, Germany, 1840; vol. 5, p. 310.
24. Although recent literature has called into question Lavoisier's originality, particularly with respect to his being credited with the formulation of the Law of Conservation of Mass (see Bensaude-Vincent, B.; Simon, J. *Chemistry—The Impure Science*, 2nd ed. Imperial College Press: London, 2012; pp. 86–88), one must admit that he accomplished his revolution by a consistent application of the principle of conservation of weight as a way of determining and confirming the results of chemical experiments. He was the first to make it a systematic instrument of experimental investigation and confirmation. (See Siegfried, R. Lavoisier and the Conservation of Weight Principle. *Bull. Hist. Chem.* **1989**, *5*, 24–31). Justus von Liebig (1803–73) said of him "He discovered no new body, no new property, no natural phenomenon previously unknown, His immortal glory consists in this—he infused into the body of science a new spirit." As quoted by Jaffe, B. *Crucibles: The Story of Chemistry*, 4th revised edition. Dover Publications: New York, 1976, p. 72.
25. Uranium has had a long, colorful, and explosive history since its discovery. For an excellent biography, please see Zoellner, T. *Uranium*. Viking Press: New York, 2009.
26. Gregor, a British clergyman and mineralogist, discovered a new simple substance while investigating a magnetic sand he found in Menachan, Cornwall. He named the black sand menachanite and the new element *menachite* (see Ohly, J. *Analysis, Detection and Commercial Value of the Rare Metals*, 3rd ed.; Mining Science Publishing Co.: Denver, CO, 1907, p. 209.) In 1794, while engaged in the study of rutile, Klaproth announced the discovery of a new earth to which he gave the name titanium, alluding to the Titans, the fabled giants of ancient mythology. Only 3 years later, he found that titanium was identical to Gregor's *menachite*, so although Klaproth retains the honor of naming the element, credit for discovery goes to Gregor.

27. Seligardi, R. *Lavoisier in Italy*; Leo S. Olschki: Firenze, Italy, 2002.
28. Poggendorff's *Biographisch-Literarisches Handwörterbuch*; Leipzig, Germany, 1863; Zweiter Band M-Z, p. 1115 (Online: http://www.scholarly-societies.org/history/Poggendorff.html; accessed April 14, 2014).
29. *Les Annales de Chimie ou recueil de mémoires concernant la Chimie et les Arts qui en dépendant* were founded by A. Lavoisier, among others, in 1789. Lavoisier was also one of its first editors.
30. Ruprecht, A. Tondi, M. *Ann. Chim.* **1791**, *8*, 3.
31. Savaresi, A. *Ann. Chim.* **1791**, *8*, 9.
32. More correctly termed *pensionari* (salaried) of the king of Naples.
33. Microszewski, M.; Bienkowski, M. *Ann. Chim. Phys.* **1791**, *9*, 51.
34. The name of this element derives from the city of Naples, which in turn refers to the Kingdom of Naples, of which Tondi was a citizen. The story of Parthenope has more than 26 centuries of history. In 680 BC, the Cumaeans founded the city of Parthenope (at the place where the homonymous Siren is said to have died) that, with the passage of time, became the refuge for the unhappy inhabitants of Cumae, who quickly became so numerous so as to pose a threat to the mother country. In 470 BC, Parthenope was destroyed (according to the legend) by Neptune, the father of the Sirens, to punish its inhabitants. A little later in the same place, the city of Neapolis (properly *Nea-Poleis*; i.e., Newtown) was founded (present-day Naples).
35. Ruprecht, A. *Crell's Chem. Ann.*, **1792**, 195.
36. Klaproth, M. H. *Ann. Chim.* **1791**, *9*, 53.
37. Jacquin, J. F. *Ann. Chim.* **1791**, *9*, 54.
38. Savaresi, A. *Ann. Chim.* **1791**, *9*, 157; Savaresi, A. *Ann. Chim.* **1791**, *10*, 61; Savaresi, A. *Ann. Chim.* **1791**, *10*, 254.
39. The Hesse crucible takes its name from the city of Hesse in Germany, where it was made. It was composed of a mix of clay and fine sand. Such crucibles were highly regarded because they could sustain abrupt changes in temperature without damage.
40. Lye is a concentrated solution of sodium hydroxide, or historically, potassium hydroxide, or caustic potash.
41. Iron phosphide (FeP) has metallic characteristics. In fact, the metallic aspect of this compound, having a blue color with shades of gray, could have misled both Tondi and Ruprecht.
42. Klaproth, M. H. *Ann. Chim.* **1791**, *10*, 275.
43. Oersted, H. C. *Danske Vid. Selsk. Forh.* **1825**, 15.
44. Gay-Lussac, L. -J.; Thénard, L. -J. *Bull. Soc. Philomath.* **1808**, *1*, 10.
45. Gay-Lussac, L. -J.; Thénard, L. -J. *Ann. Chim. Phys.* **1808**, *68*, 169.
46. Davy, H. *Phil. Trans. Roy. Soc. London* **1808**, *98*, 333.
47. "Finally, we will designate this radical by the name boron, that is derived from borax."
48. Henry, Lord Brougham. *Contribution to the Edinburgh Review*; Richard Griffin & Co.: Glasgow, Scotland, 1856; vol. III, p. 239.
49. Trommsdorff, J. B. *Trommsdorff's Journ.* **1828**, *17*, 50.
50. Savaresi, A. *Ann. Chim.* **1791**, *11*, 38.
51. Kemp, J. T. *Chem. News* **1902**, *86*, 171.
52. Howe, J. L. *Chem. News* **1902**, *86*, 293.
53. Mellor, J. W. *A Comprehensive Treatise on Inorganic and Theoretical Chemistry*; Longmans Green: London and New York, 1937.
54. Weeks, M. E. *Discovery of the Elements*; Journal of Chemical Education: Easton, PA, 1935.
55. Clarke, E. D. *Ann. Phil.* **1821**, *17*, 421.
56. Walter, J. *Dinglers Polytechn. J.* **1884**, *252*, 337.
57. Moissan, H. *Einteilung der Elemente*; M. Krayn: Berlin, Germany, 1901; Moissan, H. *The Electric Furnace*; Edward Arnold: London, 1904.
58. Watts, O. P. *Trans. Amer. Electrochem. Soc.* **1907**, *11*, 279.
59. Kowalke, O. L.; Grenfell, D. S. *Trans. Electrochem. Soc.* **1915**, *27*, 221.

60. One might hypothesize, however, that since the alkaline-earth metals are volatile (boiling points for Mg, Ca, and Ba are 1,100, 767, and 1,640 °C, respectively), it might be possible to overcome the unfavorable equilibrium associated with carbon reduction by driving the reaction to the product side by distillation of the metal out of the furnace in the absence of oxygen and CO in the furnace atmosphere.
61. Macorini, E. *Enciclopedia della Scienza e della Tecnica*, 7th ed.; Mondadori-McGraw-Hill: Milan, Italy, 1980.

II.3

AUSTRIUM: ONE ELEMENT, TWO ELEMENTS, THREE ELEMENTS, AND FINALLY, ZERO ELEMENTS

II.3.1. THE FIRST FLEETING ATTEMPT TO NAME AN ELEMENT AUSTRIUM

One of the five elements claimed to have been discovered by Leopold Anton Ruprecht and Matteo Tondi carried the name *austrium*, a name suggested by Ruprecht in honor of his native country.[62] As we have seen, these scientists attempted to isolate this element by heating magnesia in an iron crucible with charcoal; they obtained by reduction an unknown substance in the form of a metallic button.

However, it is difficult to believe that they succeeded in carrying out such a complex reduction process using the experimental conditions described.[63] In fact, it was not until 1808 that the metal contained in magnesia in the elemental state and with a sufficiently high degree of purity was isolated by the English chemist Sir Humphry Davy. Davy managed to decompose magnesia using a more elegant process. He mixed it with a few drops of mercury and heated it to white heat, while at the same time passing potassium vapor through it. The reagents had been sealed in a platinum tube in the absence of air. The amalgam of magnesium and mercury distilled in the absence of air released the alkaline earth metal in the elementary, but relatively impure, state. Many years later, Antoine Alexandre-Brutus Bussy was the first to isolate magnesium with a reasonable level of purity. Initially, Davy wanted to call the new metal *magnium*[64] to avoid confusion with manganese, which at that time had the name *manganesium*, whereas several German chemists preferred the name *talcium*.[65] Ruprecht took no further interest in his discovery. Although still alive in 1808, we do not know whether he was aware of Davy's excellent work. Anton Leopold Ruprecht continued his studies in chemistry, in particular tellurium, in Vienna after moving there from Banská Štiavnica in 1792.

II.3.2. AUSTRIUM: A POSTHUMOUS ELEMENT

The second attempt to attach the name *austrium* to a chemical element was made a century after the events just discussed. On May 6, 1886, at the Viennese Academy of Sciences, a letter sent by Professor Eduard Linnemann (1841–86)[66] from the German University of Prague a few weeks earlier was read to the assembly of Austrian scientists. In a curious and tragic twist of fate, during the time between his sending the letter, in which he announced his discovery, and its reading to the assembly, Linnemann suddenly died. The announcement of his death was reported by many scientific journals; curiously, the journal *Nature* inserted his obituary beside a description of the preparations for the celebration of the 100th birthday of Michel Eugène Chevreul (1786–1889).[67] Professor Friedrich Lippich[68] recounted the discovery of the new metal by his recently deceased colleague to the Viennese Academy. Linnemann had extracted the metal from samples of orthite[69] originating from Arendal. He had been so sure of his discovery that in the first paragraph of the article he proposed a name for the new metal "hatte ich Gelegenheit, die Gegenwart eines neuen Metalles, welche, ich mit dem Namen 'Austrium' bezeichne, festzustellen."[70] Linnemann also wrote that he wanted the symbol for *austrium* to be the three letters Aus, a type of designation never given to any other element (until the IUPAC provisional three-letter designation of transuranium elements). After this brief initial digression, Linnemann described the chemical analysis. The mineral was subjected to attack by hydrochloric acid and after removing the lead, copper, arsenic, and tin with hydrogen sulfide, the acidity of the resulting solution was reduced with sodium acetate to be as close as possible to neutral. Subsequently, H_2S was again bubbled through the solution and many precipitation products were collected: Cu, Pb, Zn, Cd, Tl, Fe, Ca, Mg, and *austrium*. The precipitate was dissolved in hot hydrochloric acid, treated with excess sodium hydroxide, and filtered; the filtered material was then gently heated in a flame. Linnemann then separated most of the transition metals by addition of Na_2S. The *austrium* remained, according to the analysis of the author, in solution. However, upon prolonged exposure to air, Linnemann observed the incipient precipitation of *austrium* carbonate mixed with sulfur. He repeated the process of purification several times by adding acetic acid and H_2S, obtaining a precipitate composed of *austrium* slightly contaminated by traces of copper, zinc, and lead. Professor Lippich, who read of Linnemann's posthumous discovery, reported the observation of two spectral lines of the new element in the violet region of the electromagnetic spectrum at $\lambda = 4{,}165$ and 4,030 Å.

In the same year, another well-known chemist and spectroscopist, Paul Emile (*dit* François) Lecoq de Boisbaudran (1838–1912), examined Linnemann's orthite samples and concluded that he had not discovered a new element.[71] Not only did he contest Linnemann's work, but he also explained his errors. Lecoq de Boisbaudran recognized what had deceived Linnemann: gallium contaminated by traces of copper, lead, and zinc. In the final analysis, Linnemann had been very able in determining the impurities present in the gallium, but unable to recognize the most abundant metal present. Unfortunately, Eduard Linnemann could not reply to the accusations advanced after his death.

Linnemann was born at Frankfurt am Main (Germany) on February 2, 1841. He studied chemistry at Heidelberg and obtained his doctoral degree at the Polytechnic of Karlsruhe. He started his career as the assistant of Friedrich August Kekulé von Stradonitz (1829–96) at the University of Ghent. Then he worked with Leopold von Pebal (1826–87) at the University of Lemberg where, in 1869, he was appointed full professor. From 1872 to 1875, he was professor at Brno and, from 1875 until his death, professor at the German University of Prague. The chemists of that time had placed great hope in Linnemann's experimental skills. Unfortunately, he died at Prague at the age of 45 on April 24 (or April 27 according to some sources), 1886. However, in 1900, Richard Pribram (1847–1928) published an article in which he acknowledged the gross error committed by his old colleague. His work confirmed the conclusions of Lecoq de Boisbaudran, but from a more careful spectroscopic study of orthite he noted unknown bands presumably from a new metal that he again proposed to call by the name *austrium*,[72] even though he knew that this element could not be the same *austrium* as Linnemann's. The confusion that could have arisen concerning this argument was easily overcome by his colleagues who completely ignored the announcement of his discovery.

Richard Pribram was born in Prague on April 21, 1847. He studied chemistry in Prague and Munich under the guidance of Justus von Liebig, eventually becoming an assistant in organic chemistry at the University of Leipzig. In 1872, he earned his habilitation at Prague, where he worked as a lecturer for the next 2 years. He later became full professor of general and analytical chemistry at the University of Czernowitz, where he served first as dean of the faculty and subsequently as rector. He devoted his genius to analytical chemistry and is especially remembered for his analyses of the mineral springs in Bukovina (a historic region in central Europe currently straddling Romania and Ukraine). He had just turned 80 when, on January 7, 1928, he died in Berlin.

II.3.3. THE "AUSTRIAN ELEMENT" OF A CZECH CHEMIST

The first determinations of the atomic weight of tellurium were due to Jöns Jacob Berzelius who, by transforming tellurium into tellurium dioxide using nitric acid, obtained different values: 128.9 in 1818 and 128.3 in 1832. Some time later, Carl Auer Freiherr von Welsbach (1858–1929) found the weight to be 127.9. Dmitri Mendeleev, in his first note on the periodic law (1869), which was based upon the arrangement of the elements in order of increasing atomic weight, stated that the experimental values of the atomic weights of some elements might have to be corrected. The anomalous atomic weight of tellurium was very troubling to Mendeleev because it would place tellurium after iodine (atomic weight = 127), thus making it a "halogen" and making iodine an uncomfortable inhabitant of the sulfur-selenium group. Mendeleev resolved this untenable position by declaring that the determinations of Berzelius and von Welsbach, which assigned a value of 128 to tellurium, had to be wrong, and he assigned an atomic weight of 125 to this element.

In 1889, Mendeleev was invited by the Chemical Society of London to pay tribute to Faraday, and he gave an address on his periodic law of the elements. While speaking about the forecasts that could be made for the discovery of new elements, he indicated the properties of an element analogous to tellurium,[73] *dvi-tellurium* (Dte) that should be positioned after bismuth, which had an atomic weight of 212. The first experiments of the young Czech chemist Bohuslav Brauner (1855–1935) attributed an atomic weight of 125 to tellurium and appeared to endorse the forecast for *Dte*. Over the next 6 years, Brauner tried in various ways to determine the atomic weight of tellurium, obtaining values in the range of 125–140. The most convincing result was 127.64. Two factors cast doubt on Brauner's determination: the high atomic weight of Te and its position in the periodic table. After having verified that he had not made any experimental errors, he felt obliged to introduce a hypothesis in which tellurium was not a simple substance but a variety of intimately bound elements. The *Chemical News* reported Brauner's remarks read before the Chemical Society on June 6, 1889 as follows:

> By submitting tellurium solutions to a systematic fractional precipitation, he has, in fact, succeeded in obtaining a variety of substances, some of which are undoubtedly novel elements. One of these it is proposed to call Austriacum (Austrium). In all probability this is the dvi-tellurium. . . the probable existence of which was pointed out for the first time by Mendeleeff in his recent Faraday lecture. From analyses made with material the uniformity of which is not yet quite established, the author is satisfied that the atomic weight of the element in question approaches very closely to that indicated by Mendeleeff.[74]

Interestingly, in Brauner's published paper,[75] he merely states that "tellurium is not a simple substance" and that "if the periodic law is true, we may conclude by deduction that tellurium contains a foreign substance which renders its 'atomic weight' higher." He speculates that these substances could be Se, Sb, Bi, or unknown elements, including Mendeleev's *dvi-tellurium*. But he does not use the name *austriacum* in the published paper, and the evidence presented for the existence of other components in tellurium is pretty slim. In the discussion following Brauner's paper, John Alexander Reina Newlands (1837–98) "remarked that he had always placed tellurium below iodine; he had no doubt that the exceptional atomic weight would ultimately be rectified, and that true tellurium, when isolated, would be found to have an atomic weight near 125."[76] It is not difficult to imagine the great confusion among chemists regarding this topic.

Bohuslav Brauner (see Figure II.02) was born in Prague on May 8, 1855. His father was a famous lawyer, and his mother was the daughter of the well-known chemist Karel Augustin Neumann (1771–1866), first professor of chemistry at the University of Prague. The maternal grandfather of Bohuslav Brauner, Karel Augustin was in turn the nephew of Caspar Neumann (1648–1715), professor of pharmacy at Berlin. At 18, Bohuslav attended the German university of his native town, following the lectures in inorganic chemistry of Frantisek Stolba (1839–1910), in organic chemistry of Adolf Lieben (1836–1913) and

FIGURE II.02. Bohuslav Brauner (1855–1935). A renowned inorganic chemist chiefly concerned with the isolation and study of the rare earth elements. While a student of Bunsen's in Heidelberg, he recognized the complexity of didymium, but his prudence and reluctance to publish his results came at a price: Carl Auer von Welsbach came to the same conclusions and discovered neodymium and praseodymium. Years later, Brauner made a doubly-wrong announcement, claiming to have discovered *meta-cerium* and *austriacum*.

Eduard Linnemann, and in physics of Ernst Mach (1838–1916). After initially working with Linnemann in the field of organic chemistry, he started to take an interest in inorganic chemistry. The discovery of gallium by Lecoq de Boisbaudran in 1875 and the studies of Dmitri Mendeleev on the periodicity of the chemical properties of the elements had a major impact on Brauner. In 1880, he obtained his PhD in chemistry and, in autumn of the same year, left for England to work with Sir Henry Enfield Roscoe, renowned for his studies on vanadium. After spending a brief period in Bunsen's laboratory, in 1882, he returned to Prague as lecturer at the Univerzita Karlova (Charles University). He became assistant professor in 1890 and, 7 years later, obtained the chair in chemistry, which he maintained until his retirement in 1925. While still in London, Brauner began a correspondence with Mendeleev that continued until Mendeleev's death. Brauner was a tireless communicator and spoke many languages: Czech, German, English, and French, as well as Russian and a little Italian. In fact, he corresponded for a brief period with the Italian inorganic chemist Augusto Piccini (1854–1905) who, like Roscoe, was well known for his studies on vanadium.

It was not the first time that Bohuslav Brauner had proposed the type of "chemical splitting" discussed earlier. In 1882, while working in Bunsen's laboratory, he had proposed, correctly, that *didymium*, purified from all the known rare earths, was in reality a mixture of two elements. He did not follow up experimentally on his intuition, and the isolation of *didymium*'s two components was ultimately achieved in 1885 by Carl Auer von Welsbach, to Brauner's regret. In later life, Brauner confided to his friend and colleague Georges Urbain that Auer von Welsbach had plagiarized his work and ideas during the period when they were both working in Heidelberg.[77] A few years later, in 1895, the French chemists Paul Schützenberger (1829–97) and Octave Boudouard (1872–1923) asked if cerium might also be a mixture of elements,[78] as was the case of *didymium*. Bohuslav Brauner worked all his life on the isolation of the rare earths, seeking an opportunity to make a sensational discovery and thus repair his image after failing to separate the components of didymium and making the false announcement of the discovery of *austriacum*.

Brauner asserted that he had isolated a cerium sulfate similar to that described by Paul Schützenberger. He had isolated cerium hydroxide from the salt, and this new compound formed the basis of the announcement of his discovery. In fact, the hydroxide was present in two distinct forms: the first soluble in cold hydrochloric acid, whereas the second was soluble in the same acid but at lower temperatures. After many fractionations, Brauner obtained two oxides that, in his view, constituted two distinct elements. The first was white whereas the other was dark pink "and called by the author *metacerium*."[79] The arc spectrum of cerium and *metacerium* displayed both lines in common and lines that were characteristic only of one or other element. Brauner attributed the common lines to impurities of one element in the other. His work was marked by other inaccuracies, such as his conviction that the valence of the rare earth elements increased as the atomic number increased. The following year, the first timid denials of Brauner's work appeared. Paul Gerard Drossbach (1866–1903) was unable to confirm the existence of *metacerium*.[80] The work of Schützenberger and Boudouard, on the possibility of splitting cerium into two elements, aroused not only Brauner's interest, but others. In 1897, the Russian (naturalized French) chemist Gregoire Wyrouboff (1843–1913) and Auguste Victor Louis Verneuil (1856–1913) disproved Brauner's discovery.[81] Their work showed unequivocally that both Brauner and Schützenberger had made erroneous analyses and drawn like conclusions as a result of using impure cerium samples. Bohuslav Brauner never retracted his discovery, but he also never mentioned *metacerium* again in any of his scientific publications. Because he held other prestigious positions in the international scientific community, such as membership on the atomic weights commission, a retraction should have been a moral obligation. His silence on these unfortunate events had little impact on his reputation, yet throughout his life his attempts to split tellurium and cerium were tacitly ignored.[82] Even after his death on February 15, 1935, at almost 80 years of age, the various national chemical societies (Czech, English, French, and German) that were to pay tribute to him were careful not to mention *austrium, austriacum,* or *metacerium.*[83]

II.3.4. A THIRD "SPLIT" FOR BOHUSLAV BRAUNER

In 1901, while working with some thorium samples, Bohuslav Brauner once again claimed to have achieved a split, this time into two different fractions that he termed thorium-α and thorium-β.[84] Brauner processed a sample of a thorium salt with ammonia and oxalic acid. He collected a number of positive fractions that he called Th-α and a number of negative fractions that he called Th-β. He used the word "negative" because the atomic weight of the fractions he designated Th-β was far from ordinary thorium. For Th-α, he found A = 233.5 according to both the sulfate and oxalate methods of precipitation. For Th-β, he found a much lower atomic weight (A = 220) and density, as well as a very different spectrum from that of Th-α. Brauner never clarified the nature of this supposed new element.

Notes

62. Ruprecht, A. L. *Crell's Chem. Ann.* **1792**, 195.
63. Duché, M. T. *Elements of Chemical Philosophy*; Corey & Fairbank: Cincinnati, OH, 1832, p. 553.
64. LeNormand, Payen, *et al. New Universal Dictionary of Technology or Arts and Trades*, Tipografia Giuseppe Antonelle: Venezia, Italy, 1834; vol. 34, p. 168.
65. Mellor, J. W. *A Comprehensive Treatise on Inorganic and Theoretical Chemistry*; Longmans Green: New York and London, 1946, vol. 4, p. 251.
66. Linnemann, E. Austrium, ein neues metallisches Element. *Monatshefte für Chemie* **1886**, *7*(1), 121.
67. Anon. *Nature* **1886**, *34*(864) 59.
68. Friedrich Lippich was the uncle of the future Nobel Prize winner Ferdinand Cori (1896–1984).
69. Orthite is one of the synonyms of the mineral allanite-Y, the chemical formula is: $(Y,Ce,Ca)_2(Al,Fe^{3+})_3(SiO_4)_3(OH)$.
70. I had the opportunity to determine the presence of a new element to which I gave the name Austrium.
71. Lecoq de Boisbaudran, P. E. *Compt. Rend. Chim.* **1886**, *102*, 1436.
72. Pribram, R., *Monatshefte für Chemie*, **1900**, *21*, 148–55.
73. The discovery of tellurium is marked by many little-known aspects; one of these regards the two names that the metal could have been given. The new metal was first described in 1782–85 by Hungarian mineralogist Ferenc Müller von Reichenstein. In 1796, the Irish scientist Richard Kirwan (1733–1812) called the native mineral *sylvanite*, alluding to the locality where it had been found in Transylvania. In 1798, M. H. Klaproth changed the name to tellurium (from Tellus, the Roman goddess of the earth and fertility). However, before Klaproth's intervention (at the beginning of the last decade of the 18th century), the new metal had been rediscovered independently by Hungarian chemist Pál Kitaibel (1757–1817) in mineral samples originating from Deutschpilsen (now Nagyborzsony), Hungary. Kitaibel's work passed unnoticed for

two centuries; only recently have his manuscripts shown that not only had he discovered tellurium independently, but he had also proposed the name *pilsum*. See Papp, G. *Hung. Magyar Kemikusok Lapja* **2001**, *56*(5), 179.

74. Brauner, B. *J. Chem. Soc. Trans.* **1889**, *55*, 382–411.
75. Brauner, B. *Chem. News* **1889**, *29*, 295.
76. The homologous successor to iodine is astatine, which has an atomic weight of 210, not 125. It should be pointed out that in the 1880s, the periodic table was sometimes presented in a fashion rotated 90 degrees from the one that is standard now: the groups were listed in order from 1 to 8 from top to bottom. But even if we assume that Newlands had in mind such a table, his comment still doesn't make sense because, in such a table, Group 7 appears above Group 8, not below. A more likely explanation of Newlands's puzzling comment, in the discussion following the reading of Brauner's paper, is that he said that he always placed tellurium *before* iodine, but the reporter for *Chemical News* misheard and thought he said *below* instead of before.
77. Fontani, M.; Costa, M.; Cinquantini, A. *Memorie di Scienze Fisiche e Naturali.* Accademia delle Scienze detta dei XL, **2003**, 413.
78. Schützenberger, P. *Compt. Rend. Chim.* **1895**, *120*, 663.
79. Brauner, B. *Chem. News* **1895**, *71*, 283.
80. Drossbach, P. G. *Ber. Dtsch. Chem. Ges.* **1896**, *29*, 2452.
81. Wyrouboff, G.; Verneuil, A. V. L. *Compt. Rend. Chim.* **1897**, *125*, 950; *Compt. Rend. Chim.* **1897**, *125*, 1180.
82. Urbain, G. *Recueil des Travaux Chimiques des Pays-Bas et de la Belgique* **1925**, *44*, 281; Jorissen, W. *Chemisch. Weekblad* **1925**, *22*, 242.
83. Levy, S. I. *J. Chem. Soc.* **1935**, 1876; J. H. (Obituary written by Jaroslav Heyrovský, Nobel laureate and Brauner's student) *Nature*, **1935**, *135*(3413), 497; Ulrich, F. *Priroda*, **1935**, *28*, 121; Dorabialska, A. *Roczniki Chemii* **1935**, *15*, 415; Druce, G. *Nature* **1942**, *150*(3813), 623; Anon. *Nature* **1955**, *175*(4462), 796.
84. Brauner, B. *Proc. Chem. Soc.* **1900**, *17*, 67.

II.4

THE RETURN OF THE OLYMPIANS: SILENE, ARIDIUM, SATURNUM, PELOPIUM, DIANIUM, NEPTUNIUM, AND PLUTONIUM

The concept of the twilight of the gods—*Ragnarok* in Norse mythology, the *Kali-yuga* recounted by the sacred texts of ancient India—has its parallels in Greek mythology as well, indicating a time marking the decline of the divine and the transcendent with respect to the human horizon. In this case, the following account introduces a brief history of those elements—all false discoveries—that were called by the names of pagan gods, a practice first suggested by Martin Heinrich Klaproth and followed thereafter by a majority of chemists. A metaphor that unites the past with the present also seems to unite—as remarked by the great scientist and philosopher Blaise Pascal (1622–63)—"men who, in the end, are defeated by something that is greater than themselves."

II.4.1. SILENE

Louis Joseph Proust (1754–1826)[85] is commonly remembered as the chemist who stated the Law of Definite Proportions. Less well known is the fact that he passed a large part of his scientific career in Spain, first at the seminary of Vergara, then at the Royal Artillery College of Segovia, followed by the University of Salamanca, where he taught chemistry thanks to an agreement between Louis XVI of France (1754–93) and Charles III of Spain (1716–88). Finally, he was appointed director of the splendid Royal Laboratory of Madrid, where it was said that "even the apparatus in common use was made of platinum." He returned to France in 1806 for family reasons, but was then unable to leave the country because Napoleon Bonaparte (1769–1821) had declared war on Spain. In 1808, while Proust was still in France, his Spanish laboratory was ransacked by the people of Madrid during the siege of the town by the army of Napoleon I.

Louis Joseph Proust was primarily an analytical chemist. He published many papers on the composition of minerals but was above all the first chemist to lay the basis of chemical analysis by wet methods, developing the classical scheme of systematic analysis. In this regard, he introduced the use of hydrogen sulfide as a precipitant of heavy metals. In 1802, at the end of a long and careful analysis of several lead mineral samples from deposits in Hungary, Proust sent a detailed letter describing his experiments to Jean Claude Delamétherie (1743–1817), editor of the *Journal de Physique*; the letter was subsequently published.[86] Proust had obtained the oxide of an unknown metal. He was unable to isolate the metal, which he called *silene*, due to the difficulty of reducing the compound using the chemical means available to him. However, he managed to characterize two oxidation states of the metal: the higher oxidation state gave yellow solutions, whereas the lower oxidation state gave green solutions. His research on the unknown metal continued, and, at the end of the same year (1802), he sent another letter to Delamétherie in which he summarized the new results on *silene* in one sentence:[87] "The new metal can only be uranium." Proust wanted to publish his results because, he believed, they would enrich the chemical literature of this metal that had been discovered several years earlier by M. H. Klaproth. The article arising from the first letter of Proust was continued by Ludwig Wilhelm Gilbert (1769–1824) and published in *Annalen der Physik*, of which he was the editor. In the German version, the metal was called *silene*, but since 1826, the name *silenium* has been used in the collective index of *Annalen der Physik*. Although at the time Proust's discovery was known to be mistaken, the description of the properties of uranium were extremely accurate, if not prophetic. Today, Klaproth is accepted as the true discoverer of uranium, which he had extracted from samples of pitchblende in 1789. Klaproth gave the name *uranit* to the new element in honor of the planet Uranus, discovered 8 years earlier in 1781. In the years that followed, he changed the name into uranium for conformity with the names of the other metals. Klaproth reported that by reduction he had obtained the metal in the elemental state in the form of a black powder. Berzelius, who had also characterized the metal, agreed with Klaproth that uranium was in absolute terms one of the easiest elements to reduce. Johann Gottfried Leonhardi (1746–1823) was to suggest that the new metal should have the name of its discoverer and not that of a Greek god. Hence, he proposed the name *klaprothium* in the third edition of his translation of Macquer's *Chemical Dictionary*.[88,89] It goes without saying that the proposal was not accepted; in fact, modern chemists have continued to use names from mythology to name the transuranium elements. It is difficult to imagine how we would cope phonetically if the synthetic elements were indicated as *transklaprothic*. Proust, who rediscovered uranium, died at the age of 72 in Angers, on July 5, 1826. More than half a century after the "first" discovery of uranium, in 1841, the French chemist Eugène Melchior Peligot (1811–90) vindicated his compatriot Proust. Reducing UCl_4 with potassium, he obtained for the first time the metal with atomic number 92 and demonstrated that the material considered by both Klaproth and Berzelius to be elemental uranium was, in reality, UO_2.

II.4.2. ARIDIUM

At the moment of the announcement that could have assured his fame forever, the Swedish chemist Clemens Ullgren (1811–68) was 39 years old. Born in Stockholm, he began his studies in chemistry under the guidance of J. J. Berzelius. In 1850, Professor Wallmark reported to the Stockholm Academy of Sciences that Ullgren had discovered an unknown element[90] in iron deposits from the Norwegian mining town of Røros. Upon analyzing the solutions in which iron from Røros had been dissolved, Ullgren realized that they behaved differently from those containing iron from other sources. He then meticulously analyzed those minerals from which the iron had been extracted that showed anomalous properties. He reacted the iron- and chrome-containing minerals with hydrochloric acid, concentrated the solution, and separated the silicates, which he then saturated with hydrogen sulfide. Upon repeating the operation a number of times, he was able to extract the green chrome oxide. After many other separations, Ullgren obtained a dark brown powderlike precipitate similar to iron peroxide that was soluble in soda. When the mysterious precipitate mixed with borate was heated in a flame, a gray-yellow pearl formed. The oxides of the presumed new metal had chemical properties that were so similar to those of iron that Ullgren thought to call it (following a rather tortuous reasoning) *aridium*. This name has Greek roots (Αρης, Ares, the god of war, and είδος, appearance or shape).[91] In his reasoning, Ullgren associated the "chemical similarity" of the new metal with that of iron, a metal linked from the beginnings of alchemy to the god of war. The work of Ullgren was centered only on the properties of three presumed compounds of *aridium*: the oxide, peroxide, and protoxide. He indicated the latter as being the lightest oxide.[92] In 1853, perhaps in part due to his discovery, Ullgren became professor of chemistry and technological chemistry at the Royal Institute of Technology, Stockholm. Less than a year later, he was to learn that his research on *aridium* was largely in error. Johann Friedrich (Jön Fridrik) Bahr (1815–75) was surprised by the completely unexpected results of his compatriot and decided to carefully reexamine the mineral in question; he found that it did not contain a new metal. Bahr even offered an explanation for Ullgren's errors. *Aridium*, far from being a new element, was a complex mixture of oxides of iron, chrome, and phosphorus.[93] By the irony of fate, Bahr suffered the same fate as Ullgren. In 1862, he believed that he had discovered a new element, which he called *wasium*, but a year later other chemists were to prove that his work was incorrect. Clemens Ullgren died in Stockholm at 57 years of age on November 6, 1868.

II.4.3. SATURNUM

In 1784, the chemist and metallurgist Antoine Grimoald Monnet (1734–1817) held the position of general inspector of mines in Brittany. He had obtained the mineral galena with a high content of lead (up to 85% by weight) from the deposits in Poullaonen. During the melting process to extract the metal, Monnet obtained as a side product a supernatant above the molten lead that attracted his curiosity. He collected the metal, which he thought could be a new metallic substance.[94] The substance's color, weight, and reactivity with acids were similar to lead, but there were considerable differences between them. The new metal had a greater luster and was more easily melted and volatilized.[95] Shortly afterward, the Irish chemist Richard Kirwan,[96] who had heard about Monnet's work, proposed the name *saturnite* for the new mineral in a letter to Monnet.[97] In fact, the new metal took the name *saturnum* in England and France and *saturnit* in Germany. In the meantime, two other chemists, Hassenfranz and Giroud, contested Monnet's discovery.[98] In the publication of their analysis, they stated that *saturnum* was simply a mixture of the sulfides of copper and lead, with traces of silver and iron. The percentages of the elements present varied according to the provenance of the mineral. Monnet published a reply to these criticisms the following March, although the arguments in his defense were quite weak.[99] He insisted on the fact that the substance that he had analyzed was different from that studied by Giroud and Hassenfranz. He revisited the mines in Brittany to collect new mineral samples and repeated the analysis, with negative results. Nonetheless, even though he failed to isolate *saturnum*, he never changed his conviction that his first analysis was correct. Many years later, in 1815, Delamétherie[100] analyzed the mineral and obtained a result in complete agreement with Monnet's critics. Monnet was a prestigious chemist, an important public personality, and a fervent follower of what was called "*ancienne chimie*," a term used to distinguish the traditional chemistry of the time from the new or Lavoisierian chemistry. Not only did he reject the progress that chemistry had made due to the discoveries of Lavoisier, Berthollet, and Priestley, but he also demeaned himself by exploiting the power he retained as chief inspector of mines in France to further his cause.

The removal of *saturnum* from the catalog of elements apparently did not disturb him too much because a year later he became enthusiastically involved in politics. At the beginning of the French Revolution, he declared himself to be an ardent adversary of the abolition of the privileges of the ruling classes. He was deprived of all his possessions but was not declared fit for armed service. He was sentenced to almost complete isolation and spent the last days of his life in Paris, where he died on May 23, 1817, two years after the restoration of the Bourbons.

II.4.4. PELOPIUM

In 1801, the English chemist Charles Hatchett[101] decided to examine in depth a mineral that had arrived in England from Massachusetts 50 years earlier but had remained unexamined at the British Museum in London. It was a heavy black block sent by John Winthrop (1681–1747), grandson of the Governor of Connecticut, John Winthrop (1609–76), who was also an alchemist, doctor, and collector of rocks (although there is no evidence that the specimen was originally owned by him). On concluding his work, Hatchett established that the mineral contained “a new earth,” namely the oxide of a new element, which he proposed to call *columbite*, in honor of the famous navigator from Genoa, Christopher Columbus (1451–1506). Hatchett purified the oxide and separated from this a new element that he called *columbium*.[102] The element was subsequently given its own symbol, the letters Cb.

In 1802, Anders Gustaf Ekeberg (1767–1813) independently discovered a new metal and called it tantalum. Ekeberg was born in Sweden and obtained his degree from the University of Uppsala in 1788. He taught at Uppsala, where he came across the minerals collected at Ytterby, (in Sweden, and they particularly attracted his attention. He isolated a new element from one of them (now recognized as yttriotantalite) that he proposed to call tantalum, probably because of the considerable time he had to apply to the analysis given its extremely low reactivity with acids. (Tantalus, in Greek mythology, was a son of Zeus and the nymph Plouto; he was condemned to eternal torment for having revealed the secrets of the gods to mankind.) Neither Hatchett nor Ekeberg realized that they had identified the same element until 1809, when William Hyde Wollaston analyzed the samples of both tantalite and columbite and reached the conclusion that *columbium* and tantalum were the same. In 1844, the studies of Heinrich Rose (1795–1864) on tantalite and columbite seemed to provide an explanation able to clarify the whole story. He noted that both minerals contained not only tantalum but also two other elements.

Rose accepted Wollaston’s work and believed that the elements of Hatchett (*columbium*) and Ekeberg (tantalum) were the same.[103] (Rose followed Berzelius’s recommendation that the new element be named “tantalum,” even though Hatchett had discovered it first, because Ekeberg characterized it more fully.) When Rose thought he had discovered two new elements in the same source mineral, he named them niobium and *pelopium*.[104] Niobe and Pelops were two of the three children of Tantalus.[105]

However, instead of clarifying things, more confusion arose within the chemistry community. The supporters of *columbium* thought it appropriate to sow a seed of doubt: could *columbium*, discovered in 1801, not be tantalum but one of the other two elements discovered by Rose? Ten years later, Arthur Connell, professor of chemistry at

Saint Andrew's, Scotland, was the first to point out that Hatchett's *columbium* and Rose's niobium were the same element.[106]

Many years were to pass before the situation was clarified. The names *columbium* and niobium were both used to identify the same element for a century. In 1949, the International Union of Pure and Applied Chemistry (IUPAC) officially adopted niobium as the name of this element. Since old habits tend to die slowly,[107] and because the Anglo-Saxon chemists considered themselves defrauded of the discovery,[108] a number of experts in metallurgy still use the term *columbium*.

Returning to the announcement of Rose's double discovery, Marshall and Marshall[109] conclude that he was misled because, after extracting niobic acid ($Nb_2O_5 \cdot nH_2O$) from tantalite from Bavaria and mixing it with carbon and then heating it in the presence of chlorine gas,[110] he obtained three chlorides: $NbCl_5$, $NbOCl_3$, and $TaCl_5$. The situation was complicated by the fact that $NbCl_5$ and $TaCl_5$ had such similar physical properties that they were virtually impossible to separate from one another, leading to a fatal flaw in Rose's research: undoubtedly, all his preparations were mixtures. Hence, in 1846, Hans Rudolph Hermann (1805–79) disproved Rose's conclusion that he had discovered two new elements. Hermann's analysis showed that niobium was effectively an element, but not *pelopium*, the latter being niobium contaminated by impurities.[111]

The pitfalls of working with the tantalite system become obvious from the next two sagas in which Hermann himself, working in Moscow, got entangled: he announced the double discovery of *ilmenium* (1874) and *neptunium* (1877) more than a dozen years after Munich chemist Wolfgang Franz von Kobell (1803–82) announced his discovery of *dianium* in 1860. All of these "discoveries" were later shown to be mixtures of niobium and tantalum containing other impurities.

The "earth acids" that Rose was dealing with were quite different from the basic rare earths and metal oxides to which chemists had become accustomed. Rose had, in fact, experienced the first example of two elements that exhibited the same chemical behavior due to the lanthanide contraction. He recognized this in his lament, "Such behavior is so unusual that in the whole realm of chemistry I know of no analog." What was Rose's mindset? He was not devious or egocentric—he was methodical, complete, objective, and correct in all his other analytical work. Perhaps anyone faced with such confusing data would fall into the same trap. Many years later, just 1 year before his death, Rose finally realized that his niobium was actually Hatchett's *columbium*.

II.4.5. DIANIUM

On March 10, 1860, Wolfgang Franz von Kobell, a well-known professor of mineralogy at the University of Munich, presented a paper to the Bavarian Academy of Sciences concerning the discovery of a new metal.[112] He had analyzed columbite from Greenland, and tantalite, samarskite, and œschynite from Finland. After reacting the minerals with hydrochloric acid, Kobell was able to isolate a new oxide similar to those of the niobium and tantalum family. Furthermore, he had identified several characteristics of the new metal. Upon addition of hydrochloric acid and tin, the color of the solution became an intense blue. This behavior was not observed for the analogous oxides of niobium and tantalum.

Kobell proposed calling the metal that he had examined *dianium*.[113] In accord with the well-entrenched custom of the time, Kobell chose the name *dianium* in honor of the

Roman goddess of the hunt. The American journal *The Living Age* reported that Kobell's choice lay in his interest in alpine sports.[114]

Wolfgang Xavier Franz von Kobell was born in Munich on July 19, 1803. From 1820 to 1823, he studied mineralogy at Landshut, and in 1826 he became professor at the University of Munich. In 1855, he invented the "stauroscope," an optical instrument used to study the properties of minerals under polarized light. He published many scientific papers and described many new minerals. The mineral kobellite was named in his honor. In more recent years, Henri Sainte-Claire Deville (1818–81) analyzed with great care minerals containing niobium and had gained some familiarity with the problems linked to their analysis. Upon hearing of Kobell's work in 1861, he wrote to Augustin Alexis Damour (1808–1902). The two examined Kobell's publications and repeated his experiments. As a member of the Paris Academy of Sciences, Damour presented and subsequently published (together with Deville) an article in which it was clear that they had strong doubts concerning the existence of *dianium*, declaring that it should not be considered a distinct chemical species.[115] The final confirmation of the error committed by Kobell was obtained a few years later. In 1864, the Swedish chemist Christian Wilhelm Blomstrand (1826–97), and 2 years later the Swiss chemist Jean Charles Galissard de Marignac (1817–94), proved that *dianium, columbium,* and niobium were the same element.

Kobell was not only a chemist and mineralogist, but also a historian, and he wrote short essays in the dialect of Upper Bavaria. His best known work is perhaps the popular Bavarian comedy called *Der Brandner Kaspar und das ewig Leben,*[116] written in 1871. This piece was transformed by Joseph Maria Lutz (1893–1972) into a theatrical script in the 1930s. Subsequently adapted for the movies, *Der Brandner Kaspar und das ewig Leben* seems to live the eternal life of its title, being presented by Bavarian television every year on the eve of All Saints. Many years later, for his numerous scientific and literary works, Kobell was ennobled with the title of baron (Ritter, in German). Franz von Kobell died on November 11, 1882, in his native town at the age of 79.

II.4.6. NEPTUNIUM

In 1877, inorganic chemist Hans Rudolph Hermann carried out a careful and extensive investigation of the metals belonging to the tantalum group. After fractionating a sample of columbite coming from the deposits at Haddam, Connecticut, he hypothesized that columbite contained not only niobium and *ilmenium*, but also a new element that he called *neptunium*.[117] Hermann's story had begun in 1846, over 30 years earlier, when, after examining some samples of yttriotantalite, he announced his first discovery of a chemical element. He had called the presumed element *ilmenium*,[118] with the symbol Il, after the Ilmenian Mountains near Minsk. The metal had a black appearance, a calculated atomic weight of about 104.6, and a density of 5.94. Hermann determined several properties of *ilmenium*: the oxide was white, the chloride green, and the sulfide black. Although Frederick Augustus Genth (1820–93) credited Hermann's discovery,[119] many other chemists were skeptical. In 1867, Galissard de Marignac criticized Hermann's work, stating that *ilmenium* was a mixture of titanium, niobium, and tantalum.[120]

At the age of 72, tormented by the repeated allegations that *ilmenium* was a false discovery, Hermann reconfirmed his work of 30 years earlier and even discovered another element, which he called *neptunium* with the symbol Np. Upon melting the mineral with potassium hydrogen sulfate, he reported the proportions for the four metals as shown in Table II.2.

Table II.2 R. Hermann's Analysis of Yttriotantalite

Compound	*Weight Percent*
Ta_2O_5	32.39
Nb_4O_7	36.79
Il_4O_7	24.52
Np_4O_7	6.30

He separated the four elements by first dissolving the hydrated oxides with HF and KF. He removed the fluorides of potassium and tantalum with water. Subsequently, by numerous fractional crystallizations, he also removed the fluorides of niobium and *ilmenium*, collecting the fluorides of *neptunium* and potassium in the mother liquor. By addition of NaOH, he precipitated first amorphous sodium *neptuniate* ($NaNpO_3$) and then crystalline sodium niobate ($NaNbO_3$).

Because the solubility of the two salts was different in boiling water, Hermann was able to separate the two elements completely and determined the atomic weight of *neptunium* to be 118. Unfortunately, Hermann's satisfaction was short-lived: only 2 years after the discovery of the second metal, he died at the age of 74. The difference between tantalum and niobium was clarified years later by Henri Sainte-Claire Deville, Louis J. Troost (1825–1911), and A. Larsson.[121] They also demonstrated unequivocally the nonexistence of *neptunium* and determined the compositions of many compounds of the two transition metals.

II.4.7. PLUTONIUM

Edward Daniel Clarke was the first to hold the chair of mineralogy at the University of Cambridge, from 1808 until his death. He was born in the same year as Napoleon Bonaparte, 1769, and at the age of 17 had already obtained a position at Jesus College, Cambridge. In 1790, he became tutor to Henry Tufton (1775–1849), nephew of the Duke of Dorset. In 1792, he began his travels, visiting many countries including Rhodes, Egypt, and Palestine. After the capitulation of the French army at Alexandria, Egypt, Clarke sent to England the artistic heritage of statues, sarcophagi, manuscripts, and maps that had been collected by the French. In 1803, he was appointed to a position at Cambridge University, partly due to the donation of an enormous statue of the Eleusinian Ceres. Clarke followed his studies passionately, not only in mineralogy but also in chemistry, making many discoveries and technical innovations among which, without doubt, should be mentioned the blowpipe, which he refined to a high degree of perfection. In 1817, he was appointed librarian of the University, and 2 years later he was one of the founders of the Philosophical Society of Cambridge. In 1816, Clarke became involved in a dispute with the celebrated chemist Sir Humphry Davy. In 1808, Davy had isolated by electrochemistry the element found in baryta ($BaSO_4$) and called it barium.

Eight years later, Edward Daniel Clarke stated that he had decomposed baryta by exposing it to an oxyhydrogen flame of his invention. He isolated the metal and called it *plutonium*[122] instead of the name given to it by Davy, largely because he felt the name barium, meaning heavy, was inappropriate for what he considered a light element. In spite of Clarke's considerations, but also due to Davy's previous work, the name of the metal remained barium. Nevertheless, a careful examination of Clarke's reports reveals that he thought baryta was the oxide of barium and not its sulfate.

In April 1817, William Hyde Wollaston visited Clarke at Cambridge. Wollaston wanted to see the experiments that Clarke had used to isolate *plutonium*, but when Clarke tried to repeat the experiments he did not find the new metal. He explained this unexpected result by proposing that the samples of baryta had become hydrated. Wollaston left disappointed by the outcome of the experiments and was convinced that Clarke had altered the previous results. Clarke was invited shortly afterward to the Royal Institution in London (where Davy was director) to try to persuade them of the correctness of his work. Once again, he was not able to successfully repeat his first experiment to extract *plutonium* from the mineral. He would not admit the failure of his experiments, but rather took a quite personal view of the facts in contradiction to the reality.[123] Many hypotheses were advanced about Clarke's work with the oxyhydrogen flame, but it seems unlikely that he really isolated barium following this approach. The chemist Joshua Mantell from Lewes in southern England proposed that he may have accidentally isolated strontium or silicon by means of the blowpipe, but this suggestion was discounted by Thomas Thomson (1773–1852). In the end, Clarke was obliged to admit that the metallic-like substance that he had obtained in the first attempt to reduce baryta with the oxyhydrogen flame could not be a metal. Two years later, he had the opportunity to amend his past mistakes by observing lithium while analyzing the lithium-containing mineral petalite ($LiAlSi_4O_{10}$), but he was unable to exploit the occasion and was not credited as its discoverer.[124] As Clarke's biographers were to write, he could not have been an original or acute thinker—as were many of his colleagues of the time—and was not able to attract admiration to himself.[125] Furthermore, many of the improvements of his scientific instruments had been suggested by others, and his experiments were far from being innovative. His most original theoretical speculation regarded volcanism but unfortunately it was shown to be mistaken.[126]

Clarke was a man of his time; an optimistic scientist, devoted to family and religion, although he had a quite particular conception of Creation: "all the constituents of created nature are combustible." Edward Daniel Clarke died in London on March 9, 1822, just 5 years after the events described.

Notes

85. Joseph Louis Proust was born in Angers, on September 26, 1754. The son of a spice merchant, in 1775, he was pharmacist of the Hôpital de la Salpêtrière in Paris. He was a friend of Professor Jacques Alexandre César Charles (1746–1823) who involved him in his experiments on aerostats and, in 1784, he took part in a flight in the hot air balloon, the *Marie-Antoinette*. He became famous particularly for his Law of Definite Proportions (1799), based on extensive quantitative chemical experiments. He demonstrated that a compound in the pure state conserves the definite proportions of the elements. It declares that a given pure chemical compound always contains exactly the same proportion of elements by mass.
86. Proust, J.-L. *Journ. de Phys.* **1802**, *55*, 297.
87. Proust, J.-L. *Journ. de Phys.* **1802**, *55*, 457.
88. Macquer, P. J. *Chymisches Wörterbuch*, 3rd Ed.; Weidmann: Liepzig, Germany, 1809; vol. 6, p. 597.
89. De Ment, J.; Dake, H. C. *Uranium and Atomic Power*; Chemical Publishing Co.: Gloucester, MA, 1945, p. 107.
90. Ullgren, C. *Öfversigt af Kongl.vetenskaps- akademiens förhandlingar* **1850**, no. 3, 55.

91. Fownes, G. *Elementary Chemistry, Theoretical and Practical*; Blanchard and Loeb: Camp Hill, Pennsylvania, 1855, p. 266.
92. Ullgren, C. *The Chemical Gazette* **1850**, *3*, 289.
93. Bahr, J. F. *J. prakt. Chem.* **1853**, *9*, 27.
94. Monnet, A. G. *J. Physique* **1786**, *28*, 168.
95. Delamétherie, J. - C. *Théorie de la terre;* Chez Maradan Publisher: Paris, 1797, 314; available from Kessinger Publishers LLC: Whitefish, MT.
96. Kirwan was an eclectic scientist, active in chemistry and geology. Although today his name is not well-known, at the time, he enjoyed some fame and was scientifically respected. He was in close contact with the major chemists in Britain and on the continent: Lavoisier, Black, Priestley, and Cavendish. Kirwan was one of the last supporters of phlogiston in England, and his use of this theory in the publication "Essay on Phlogiston and the Constitution of Acids" (1787) led him to clash with Lavoisier. In 1791, after more relaxed discussions with Lavoisier, he accepted, unwillingly, Lavoisier's theories.
97. The name of the planet Saturn was associated with lead. The name *saturnum* was chosen because the presumed new metal was extracted from lead minerals.
98. Cronstedt, A. F. *An Essay Towards a System of Mineralogy*; E. and C. Dilly: London, 1788, p. 872.
99. Monnet, A. G. *J. Physique* **1787**, *29*, 169.
100. Delamétherie, J.-C. *J. Physique*, **1815**, *80*, 270.
101. Hatchett was born in London, on January 2, 1765. In 1801, while analyzing a mineral at the British Museum in London, he discovered *columbium* (Cb). On November 26 of the same year, he announced the discovery before the Royal Society. In later life, Hatchett abandoned his work as a full-time chemist, devoting all his energy to the family business of carriage building. He died in London where he had lived all his life on March 10, 1847.
102. Hatchett, C. *Phil. Trans.* **1802**, *92*, 49; *Ann. der Physik* **1802**, *11*(5), 120.
103. Rose, H. *Ann. Phys.* **1844**, *63*, 317–41.
104. Rose, H. *Compt. Rend. Chim.* **1844**, *19*, 1275; *Pogg. Ann.* **1846**, *69*, 115.
105. Tantalus had three children: Pelops, Niobe, and Broteas. Tantalus, who was envied for his riches, was invited to dine with the gods and, having heard their conversations, became immortal. To exchange the hospitality, he invited the gods to a banquet at which, to honor the gods, he dared to offer that which was most dear to him. He killed his son Pelops, who he then cut up and boiled to serve to the gods. The gesture was interpreted as a test of the omniscience of the gods, as well as their wickedness. The guests refused to eat the meal offered by Tantalus except for Demeter who, perhaps still upset over the loss of her daughter Persephone, ate Pelops's shoulder. At the end of the feast, Hermes went to Hades to bring back the shade of Pelops, and he was restored to life. His body was reassembled, the missing shoulder being replaced by one of ivory. The sacrifice of Pelops was not the only evil deed Tantalus did; invited as a guest of the gods, he stole ambrosia and nectar (the food and drink of the gods). He then served these divine treats to his mortal friends.
106. Connell, A. *Phil. Mag.* **1854**, *7*, 461.
107. A similar attitude can be found in chemists of other nations. The French have unwillingly renounced the names *celtium* and *glucinium*, elements claimed to be discovered by French chemists but subsequently renamed hafnium and beryllium.
108. Smith, E. F. *Proc. Am. Phil. Soc.* **1905**, *180*, 15.
109. Marshall, J. L.; Marshall, V. R. *The HEXAGON of Alpha Chi Sigma*, **2013**, *104*, xx.
110. Berzelius, J. J. *Jahresbericht* **1848**, *27*, 56.
111. Hermann, R. *J. prakt. Chem.* **1846**, *38*, 91; *J. prakt. Chem.* **1849**, *40*, 457.
112. von Kobell, F. *Bull. d. K. Bayr. Ak. d. Wissen. München* (II Classe), Sitzung (1860); *Ann. Ch. Pharm.* **1860**, *114*, 837.
113. von Kobell, F. *J. prakt. Chem.* **1860**, *79*, 291; *Pogg. Ann.*, **1861**, 283.

114. Anon. *The Living Age* **1860**, *65*, 728.
115. Sainte-Claire Deville, H; Damour, A. A. *Compt. Rend. Chim.* **1861**, *53*, 1044.
116. "*Der Brandner Kaspar und das ewig Leben*" or "Kaspar Brandner and eternal life" is a comedy recited entirely in the Bavarian dialect, a dialect that is not understood even in Berlin, which is perhaps just as well, as in "Brandner Kaspar" we discover that there is no place in Paradise for Berliners. Kaspar Brandner, a seraphic peasant, instead knows that he has a place reserved for him there, but has no wish to start the trip. The problem lies in the fact that it is not enough to have a pure spirit to be accepted into Paradise; one (normally) arrives there dead. And Kaspar is very fond of his life, convinced that he is already living in a paradise or, to be more precise, his native Bavaria.
117. Hermann, R. *J. prakt. Chem.* **1877**, *15*, 105.
118. Hermann, R. *J. prakt. Chem.* **1846**, *38*, 91.
119. Genth, F. A. *Am. J. Sci.* **1853**, *15*, 246.
120. Galissard de Marignac, J. C. *Ann. Chim. Phys.* **1867**, *9*, 249.
121. Larsson, A. *Z. Anorg. Allg. Chem.* **1896**, *12*, 189.
122. Thomson, T. *A System of Chemistry*, 5th ed.; Smith, Elder & Co.: London, 1817.
123. Clarke, E. D. *Ann. Phil.* **1817**, *10*, 310.
124. Clarke, E. D. *Ann. Phil.* **1819**, *14*, 142.
125. Oldroyd, D. R. *Ann. Sci.* **1972**, *29*(3), 213.
126. Clarke's hypothesis was that volcanic events were the result of the combustion of hydrogen and oxygen in the heart of the earth. It was certainly a bizarre concept, which in fact never achieved recognition. The theory of his rival Davy was quite daring. He believed that volcanic events were due to the interaction among water, sodium, and potassium under the Earth's crust.

II.5

THE UNFORTUNATE AFFAIR OF A STUDENT OF KANT: A CAREER SOLDIER, BUT A CHEMIST BY PASSION

Jeremias Benjamin Richter (1762–1807) was born in Hirschberg Germany (present-day Jelenia Góra in Poland). He began his studies at Koenigsberg in 1785, following the lectures of Immanuel Kant (1724–1804). Four years later, he obtained a doctorate in mathematics but, not having the possibility to follow an academic career, to make ends meet, he became the chemistry consultant of Baron von Lestwitz at Gross-Ober-Tschirnau. In 1795, he became secretary and analyst at the Office of Mines in Breslau (present-day Wrocław in Poland) and in 1800, he went to Berlin to become "second Arcanist" or chemist at the dye laboratory of the Royal Porcelain Works of Berlin. Richter became a member of many scientific societies at Berlin, Munich, Göttingen, Potsdam, and Saint Petersburg. He refused to accept the atomicity of matter and was therefore sidelined by more eminent chemists. He was also a convinced supporter of the phlogiston theory, a choice that contributed considerably to limiting his standing as a scientific figure. He tried to classify chemistry as a branch of applied mathematics, identifying distinct regularities in the constitution of matter and the combining proportions of elements. By studying the laws that govern the constitution of matter, he determined the law of neutrality and a table of chemical equivalents. Such regularity classified equivalents according to arithmetic and geometric progressions, and all of his quantitative work became part of the volume *Stoichiometry* (a word introduced by Richter). In 1803, Richter, following the example of Ernst Gottfried Fischer (1754–1831) (who himself had been inspired by Richter's work), published a table of equivalent weights for 30 bases and 18 acids, from which one could calculate the relative weights of the constituents of every neutral compound. He observed that some metals, among them iron, could combine with other elements in proportions different from that for oxygen. He also studied chemical affinity and became convinced that there was a proportionality between affinity and the combining proportions between the elements he had examined. Richter's interests covered a wide range of topics; some worthy of note were calorimetry, colloidal gold, and the discovery of a new metal.

II.5.1. NICCOLANUM

Two years prior to his premature death, Richter was involved in what was discovered only 20 years later to be a gross error of analysis. For several years, he had been analyzing cobalt deposits in the Kingdom of Saxony, increasingly convinced that they contained not only cobalt but also arsenic, copper, nickel, iron, and another metal with properties similar to those of nickel.[127]

He was surprised that nickel, after being purified from cobalt, arsenic, and iron and then reduced, never had the appearance of a compact mass, but rather the granular appearance of "vitrified copper." This substance was not magnetic and did not have a metallic aspect. Richter dedicated much of his remaining energy to isolating what seemed to be a new substance; after many weeks of work, the effort appeared to be crowned with success.

He added charcoal to the ground granular mass in a porcelain capsule and heated it for 18 hours until completely reduced. He obtained a metallic disk similar in size to a button. Before describing the properties of the new metal, Richter named it *niccolanum* because it was always found together with nickel in mineral deposits.[128] It had a silver color like steel but contained shades of red, was malleable when cold but not when hot, and was attracted by a magnet placed nearby. Its specific gravity was 8.55. It reacted with nitric acid to form a dark green solution, which when concentrated formed a gel. Richter observed the formation of a pale blue precipitate when adding potassium carbonate to a solution of *niccolanum*, whereas the addition of ammonia gave rise to a red solution without a precipitate. Finally, he noted the formation of two distinct oxides of *niccolanum*. In conclusion, Richter compared and contrasted the characteristics of niccolanum with those of cobalt and nickel.

Niccolanum's similarities to cobalt were its solubility with acids and its reducibility only in the presence of carbon; its differences were the colors of its solutions, carbonates, and oxides.

Niccolanum's similarities to nickel were its magnetism, malleability, and intense green solutions; the loss of the green color on dehydration; and the red solution color in excess ammonia. Its differences were its reducibility only in the presence of carbon, its easier oxidation by nitric acid, its high oxidation number, the red color of its dehydrated salts, and the blue color of its carbonate.

II.5.2. THE ROAD FROM OBLIVION

The discovery of the new metal was not accepted favorably by the chemists of the time. No one succeeded in repeating Richter's experiments in isolating *niccolanum*. Thus, its discovery remained shrouded in doubt until 1822, when J. J. Berzelius, commenting on a number of false discoveries (namely, *wodanium* and *vestium*), stated that Richter had been deceived by the presence of arsenic and iron together with nickel.[129] In particular, Berzelius noted that the arsenates of iron, which are often present in nickel minerals, have the characteristic of dissolving in acids as if they were weak bases. Under alkaline conditions, iron arsenates precipitate without alteration of their nature. When oxides of arsenic are reduced with charcoal to arsenides, they assume a metallic appearance.

Jeremias Benjamin Richter never became part of the academic community but believed, ahead of his time, that chemistry could not develop without a mathematical basis. Richter's work was taken up by Louis Joseph Proust who, while undertaking an accurate analysis of neutralization relationships, generalized the Law of Definite Proportions in which the component reagents combined only in well-defined ratios. Richter wrote several treatises that were not widely read. After his death in Berlin on April 4, 1807, at the age of only 45, many of his minor discoveries and publications were attributed erroneously to Carl Friedrich Wenzel (ca. 1740–93).

Notes

127. Richter, J. B. *Gilbert's Ann.* **1805**, *19*, 377; *The Repertory of Arts, Manufactures and Agriculture, Second Series*, London, **1806**, no. 18, p. 288; *Retrospect of Philosophical, Mechanical and Agricultural Discoveries*; J. Wyatt: London, 1806, p. 373.
128. Richter, J. B. *Gehlen's Jour.*, **1808**, *4*, 392; Thomson, T. *A System of Chemistry of Inorganic Bodies*, 7th ed.; Baldwin & Cradock: London, vol. 1, 1831, p. 258.
129. Berzelius, J. J. *The Annals of Philosophy*, new series **1822**, *3*, 206.

II.6

ANDRÉ-MARIE AMPÈRE BURSTS ONTO THE CHEMISTRY SCENE

Any chemist who has worked with fluorine and its simple compounds can attest to the energy and vigor with which this element reacts. The chemical literature is marked by stories of failed and often fatal attempts by chemists who, either through ignorance or lack of attention, were victims of this aggressive element. Although today the danger of this gas is well known, at the beginning of the 19th century, the situation was rather confused, and only André-Marie Ampère (1775–1836) seemed to foresee its destructive capacity. Fluorine was not isolated until relatively late because in its elemental state it immediately attacks surrounding materials. It was finally isolated by electrolysis in 1886, by Henri Moissan,[130] although gaseous hydrofluoric acid (HF) and its aqueous solutions had been known for some time. However, for many years, no one was able to understand and decompose the mysterious substance that generated HF. The mineral fluorite was described in 1529 by Georgius Agricola (1494–1555) as a flux in the smelting of ores. The name fluorine derives from the Latin verb *fluere*, "to flow."

In 1670, H. Schwandhard discovered that when treated with a mineral acid, fluorite could be used to etch glass. Carl Wilhelm Scheele and many other scientists, among them Humphry Davy, Joseph Louis Gay-Lussac, Antoine Lavoisier, and Louis Thénard, carried out experiments with hydrofluoric acid that, in a few cases, sometimes ended tragically.[131] Scheele described it as an acid characterized by peculiar properties. It soon became famous and was inserted by Lavoisier into his new system of nomenclature under the name "*acide fluorique*." According to his theory, all acids contained oxygen and an unknown element; in this case, the element was named *fluorium*.

II.6.1. "PHTORE"

André-Marie Ampère was born at Polémieux-le-Mont-d'Or, near Lyons, on January 22, 1775. He was a famous physicist, credited with being one of the greatest scientists in the field of electromagnetism: the unit of electric current, the ampère, was named in his honor. He demonstrated his inclination toward mathematics and science at an early age; although his father would have liked to teach him Latin, he stopped when he discovered the boy's passion for mathematics. The young Ampère learned Latin anyway, enabling him to master the works of Leonhard Euler (1707–83) and Daniel Bernoulli (1700–82). His interests embraced almost all learning: history, travel, poetry, philosophy, natural sciences, physics, and chemistry. Ampère was also interested in probability theory, particularly within the context of gambling. In his small treatise of 1802 "Considérations sur la théorie mathématique du jeu," he demonstrated that the chances of winning were against the bettor. The publication caught the attention of mathematician Jean Baptiste Joseph

Delambre (1749–1822) who managed to obtain a position for him as a teacher in Lyons and subsequently (1804) a position at the Ecole Polytechnique of Paris, where Ampère was named professor of mathematics in 1809. He carried out his scientific research and many studies with continued passion in this institution of well-known excellence and in 1814 was elected a member of the newly formed Institut Impériale, the parent organization of the state Academy of Science.

In two letters to Humphry Davy, dated November 1, 1810 and August 25, 1812, Ampère suggested that HF was similar to muriatic acid (HCl) and did not contain oxygen, as Lavoisier thought, but was a binary compound with hydrogen and an unknown element,[132] similar to chlorine. The replies from Davy, dated February 8, 1811 and March 6, 1813, respectively, were inserted as footnotes in a subsequent publication by Ampere.[133] In the first note, Davy expressed a cautious view and, in line with Lavoisier's hypothesis, proposed that HF contained oxygen in a manner analogous to that of silicon tetrafluoride (SiF_4). In the second, he had changed his opinion and wrote: "Your ingenious views respecting fluorine may be confirmed." It is noteworthy that, in the space of 2 years, between the first and second letter, Davy had completely changed his ideas within the field of nomenclature.

The chemists of the time agreed that muriatic acid and hydrochloric acid contained chlorine, but they could not find an agreement regarding if, and how much, oxygen was present in these compounds. In the meantime, Davy had demonstrated that muriatic acid was a binary compound without oxygen.[134] Consequently, Ampère was convinced that hydrofluoric acid had an analogous composition. Thus, if his intuition was correct, the name should be changed to conform with the nomenclature for chlorine. He proposed three different names for the element still to be isolated and left the choice to Davy: "*fluore,*" "*fluorure*" (which was dismissed immediately due to the difficult pronunciation), and "*phtore,*" which was Ampère's preferred choice.

The word *phtore* derives from Greek and can signify either "destroy" or "corrupt"; the choice of the name was made bearing fully in mind the properties of the element in question. The binary acid would take the name "acide *hydrophtorique,*" an eventual acid containing oxygen would be known as "acide *phtorique,*" and the corresponding salts would be named "*phtorates.*" Davy was not able to isolate the elemental gas,[135] but was willing to accept the first name advanced by Ampère, *fluore*, which was then corrupted to fluorine. The influential English chemist Thomas Thomson, who was not openly contrary to changing the names of the elements as long as the decisions were not ill-considered, accepted the name of fluorine and included it in his textbook,[136] but rejected Ampère's *pthore*, resisting the tendency to coin new names arbitrarily.

Ampère's fame lies principally in the service he provided to science by establishing the relationships between electricity and magnetism. Moreover, his fundamental contribution to the impressive development of this new science led his contemporaries to call him the "Newton of electromagnetism."

In 1796, Ampère met Julie Carron, whom he married 3 years later. Their son Jean-Jacques (1800–64) would become a famous traveler and historian. In 1803, his wife, who had become an invalid after the birth of their son, died. Ampère never recovered from the blow, even though 3 years later he married Jeanne-Françoise Potot (1778–1866), with whom he had a daughter, Albine (1807–42). Ampère had suffered from poor health since childhood, but from 1827 on his condition became progressively worse. On June 10, 1836, while in Marseille to carry out an inspection of the Lycée Thiers, his condition worsened. He was taken to the infirmary of the school, where he died suddenly. He was buried in the cemetery of Montmartre, in Paris.

Notes

130. Moissan, H. *Compt. Rend. Chim.* **1886**, *93*, 202.
131. Many common people as well as scientists lost their lives as a result of this gas. During the 1950s in England and Germany, many grain harvests were destroyed in the vicinity of factories producing fluoride compounds. In 1933, in the Meuse valley, 60 people suffered a horrendous death following the escape of fluorine from a local industry.
132. Joubert, J. *Ann. Chim. Phys.* **1885**, *4*(6), 5.
133. Ampère, A. M. *Ann. Chim. Phys.* **1816**, *2*(2), 5.
134. Davy, H. *Phil. Trans.* **1810**, *100*, 231.
135. Davy, H. *Phil. Trans.*, **1813**, *103*, 263; Davy, H. *Phil. Trans.* **1814**, *104*, 62.
136. Thomson, T., *A System of Chemistry of Inorganic Bodies*, 7th ed.; Baldwin & Cradock: London, vol. 1, 1831; footnote, page 89.

II.7

CADMIUM: "BONE OF CONTENTION" AMONG CHEMICAL ELEMENTS

The story of the discovery of cadmium is complicated, both with respect to disagreements about its name and because of a dispute regarding the attribution of the discovery involving a pharmaceutical inspector, J. C. H. Roloff; a chemist, Carl Samuel Hermann; and a professor, Friedrich Stromeyer.

At the beginning of the 19th century, zinc oxide was used in a number of popular pharmaceutical formulations, and German authorities of the day used physicians to monitor the quality of commercial pharmaceutical products such as zinc oxide. In September 1817, Inspector J. C. H. Roloff (or Rolow) of Magdeburg[137] found zinc oxide of dubious content in many German provincial pharmacies. All the material originated from the Chemische Fabrik zu Schönebeck, owned by chemist Carl Samuel Hermann (1765–1846), who in turn obtained the zinc from deposits in Silesia. The zinc oxide was confiscated and analyzed. The first unofficial tests, carried out by Roloff himself, led him to believe with some apprehension that the confiscated medicines contained arsenic. However, on further analysis he realized his mistake: the commercial zinc oxide did not contain arsenic, but did contain an unknown and possibly new element. In February 1818, Roloff sent the results of his research to Dr. Christoph W. Hufeland for publication in the *Journal für die praktischen Heilkunde*. Unfortunately, publication was delayed until the following April.

In the meantime, Roloff's samples had been sent to Berlin and subjected to a careful official analysis by two public analysts, Kluge and Staberoh, who reached the same conclusion as Roloff: the sample contained a new element. On April 25, 1818, Kluge and Staberoh proposed calling the new element *klaprothium*, in memory of chemist Martin Heinrich Klaproth who had died in Berlin on January 1, 1817. At the same time, Hermann, without informing Roloff, had extracted the new metal and sent a sample to Friedrich Stromeyer (1776–1835) so that the discovery could be confirmed. At that time, Stromeyer was a renowned professor of metallurgy at the University of Göttingen and a general inspector of pharmacies. He confirmed Roloff's hypothesis and, in the autumn of 1817, called the new element *kadmium*. Then, the story became even more complicated.

On April 14, 1818, Roloff sent a sample of the new metal to Stromeyer, attaching a flattering accompanying letter in which he asked that, in case his hypothesis was confirmed, he be allowed to name the new element. It is not clear how Roloff reacted when he received Stromeyer's reply from Göttingen claiming that he, Stromeyer, had already found the same element in the samples that Hermann had sent to him. To further enliven the already complicated discovery of cadmium, in May 1818, an article entitled "Discovery of Two New Metals in Germany" appeared in *Annalen der Physik*, edited by Ludwig Wilhelm Gilbert.[138]

II.7.1. A RELATED DISCOVERY INCREASES THE CONFUSION: VESTIUM

While these events were taking place, chemists throughout Europe were competing to name the new element that would eventually be called cadmium. Ludwig Wilhelm Gilbert (1769–1824), reporting on an article published in an Austrian newspaper, noted that Dr. Lorenz Chrysanth von Vest (1776–1840), professor of chemistry and botany at the "Johanneum"[139] of Graz (presently the University of Graz), had found in 1817 a new element in the nickel and cobalt-pyrites deposits of Schladming located in Upper Styria. Von Vest claimed that this material had chemical characteristics that were entirely different from anything yet known. As editor of *Annalen*, Gilbert suggested the name *junonium*, in honor of the discovery of the asteroid Juno in 1804. This discovery had not yet been confirmed when Gilbert sent a reply to Hermann's letter, in which he agreed to accept the name cadmium proposed by Stromeyer. In the following issue of *Annalen*, however, Gilbert published the report from von Vest in which the name of the metal was reported as *vestäium* or *vestium*.[140] On June 15, 1818, von Vest wrote a note to Gilbert stating that he would not accept the name *junonium* for his metal because this name had already been used by Thomas Thomson. Although it was known that the *junonium* of Thomson was in fact *cererium* (cerium), von Vest did not want his metal to carry the name of a "defunct element." Because no other heavenly body was readily available to provide a name for the new element (except Vesta, which could not be used due to its similarity with the name of the element's discoverer), von Vest reluctantly suggested calling the new metal *sirium*.[141] However, in the final report released to the press, the metal continued to be referred to as *vestäium*, apparently the decision of editor Gilbert. At the end of this intricate affair, on July 30, 1818, von Vest noted grudgingly that Sir Humphry Davy, while staying in Graz during one of his never-ending scientific journeys around Europe, had begun a preliminary analysis of the so-called *vestäium*. Davy was not convinced that the presumed metal was an element, and initially it was considered to be impure tantalum. The *vestäium* samples accompanied Davy on the rest of his trip until he returned to England the following year, where they were finally analyzed by his assistant Michael Faraday (1791–1867). Faraday found that *vestium* was, in fact, impure nickel.[142]

At the end of an article entitled "Ueber das Cadmium," published in October 1818, Stromeyer claimed the discovery as his own.[143] In the introduction, he made a blatantly false statement, saying that both Roloff and Hermann had separately asked him to resolve the controversy regarding the discovery of the new element. Because this was not true, Roloff sent a letter to Gilbert on November 18, 1818, which was published in the following issue of *Annalen* entitled "Regarding the Discovery of Cadmium."[144] This was followed by a firm reply from Carl Samuel Hermann, who reported his version of the events leading to the discovery of cadmium.[145] As written in Gilbert's note, Stromeyer managed, for better or worse, to impose on the scientific community the name that he had given to the new element, cadmium. The name derives from the Latin *cadmia fornacea* or *fornacum*, which is the old name for zinc carbonate deposits. This name is, in turn, derived from the Greek καδμια γη (kadmeia gè, or "cadmea earth").[146]

Shortly afterward, Wilhelm Meissner (1792–1853), owner of the Löwenapotheke (Lion Pharmacy) in Halle, confirmed Hermann's discovery. He had received cadmium samples from Hermann (who had obtained them from Roloff) in Schönebeck. At this point, another chemist, Karl Johann Bernhard Karsten (1782–1853), came on the scene.[147] Karsten was in Berlin to examine mineral samples from the zinc deposits in Silesia. He

Table II.3 Names of Presumed Elements Surrounding the Discovery of Cadmium

Name of Presumed Element	*Authority*	*Element's Actual Name*
Nameless (sample given to Stromeyer to analyze)	Hermann	Cadmium
Nameless (sample given to Stromeyer to analyze)	Roloff	Cadmium
Kadmium	Stromeyer	Cadmium
Melinum	Karsten	Cadmium
Junonium	Thomson	Cerium
Sirium (another name for *Vestium*)	Von Vest	Impure Nickel
Vestäium or *Vestium*	Von Vest/Gilbert	Impure Nickel
Junonium (another name for *Vestium*)	Von Vest/Gilbert	Impure Nickel
Sirium (another name for *Vestium*)	Von Vest	Impure Nickel

found an unknown substance among those that he examined, and he decided to call the substance *melinum,* from the Latin *melinus* or "quince," because he was struck by the sulfur yellow color of this probable new element.[148] It is likely that, in this case, too, the element isolated was cadmium. Finally, Roloff, like Hermann, in order to defend his version of the events, published his story of the discovery of cadmium in *Gilbert's Annalen.*

Table II.3 is an attempt to clarify the "chemical confusion" created by chemists regarding the names of unconfirmed elements.

Notes

137. Roloff was both a pharmacist and a physician.
138. Gilbert, L. W. *Ann. der Physik* **1818**, *29*, 95; Anon. *Ann. der Physik* **1818**, *29*, 113.
139. Anon. *Phil. Mag.* **1819**, *53*, 463.
140. Von Vest, L. *Ann. der Physik* **1818**, *29*, 387; *Phil. Mag.***1818**, 463; *Ann. Philos. or Magazine of Chemistry, Mineralogy, Mechanics*, **1819**, 344.
141. He did not want to be accused of childish vanity if Gilbert introduced the name *vestäium* or *vestium*.
142. Faraday, M. *Ann. der Physik* **1819**, *39*, 80.
143. Stromeyer, F. *Ann. der Physik* **1818**, *30*, 193.
144. Roloff, J. C. H. *Ann. der Physik* **1819**, *31*, 205.
145. Hermann, C. S. *Ann. der Physik* **1820**, *36*, 276.
146. "*Terra cadmea*" (zinc carbonate deposits) was found for the first time near Thebes, a city founded in 1450 BC by the Phoenician prince Cadmus who, according to legend, apart from introducing the alphabet into Greece provided the knowledge to extract gold from its ore.
147. Karsten was born on November 26, 1782, in Butzow. His career was atypical of a chemist. Until 1810, he was councilor at the local mining authority in Breslau and subsequently manager of the mines in Silesia. During this period, he also held many seminars at Breslau and, from 1819, became advisor at the Ministry of Internal Affairs in Berlin. Karsten died in Berlin on August 22, 1853, at the age of 70.
148. Karsten, K. J. B. *Ann. der Physik* **1818**, *29*, 104.

II.8

A FIREPROOF FAMILY OF CHEMISTS

Luigi Valentino Gasparo was first in the Brugnatelli family line. He was followed by Gaspare, Tullio, and, last, another Luigi Valentino. These four men represent a celebrated dynasty of chemists, perhaps the most well known in Italy, which from the second half of the 18th century to the first half of the 20th held the chair of chemistry at the University of Pavia. The four chemists were also naturalists, physicians, engineers, and mineralogists. The first was a contemporary of Lavoisier, the last that of Marie Curie.

II.8.1. CHEMISTRY AS THE COMMON DENOMINATOR

Luigi Valentino Gasparo Brugnatelli (1761–1818) (Figure II.03) was born in Pavia on Saint Valentine's Day, February 14, 1761, into a less than affluent family. After a period of work in commerce, he obtained a doctorate in medicine in 1784, defending a thesis on the digestive power of gastric juices. He practiced as a physician for a very short time before turning to chemistry. He began his academic career as a temporary substitute for various professors before being named professor of chemistry at Pavia in 1796. He was a friend of and corresponded extensively with Alessandro Volta (1745–1827), whom he accompanied to Paris and the Congress of Lyon.

An untiring researcher, his 130 publications are collected in memoirs and four texts entitled: *Elementary Treatise on General Chemistry, General Pharmacopoeia, Vegetable and Animal Medical Matter,* and *Human Lithology.* Brugnatelli discovered numerous chemical compounds and also prepared fulminates of noble metals (Ag and Cu). In 1815, he found uric acid to be present in the droppings of the silkworm, but his most valuable work was focused on electrolysis, which was a new field of study at that time. His friendship with Volta encouraged him to undertake studies of electrical phenomena. His research on the coating of metals with noble metals (e.g., gold, silver, and copper plating) was of a very high level, and the University of Pavia has preserved in its museum 26 electroplated specimens produced by Brugnatelli consisting of insects, flowers, and leaves covered with copper. Brugnatelli is also remembered for having begun many journals which, between 1788 and 1827, published some of the most important work of that period in the field of experimental science.

He also attempted to reform Lavoisier's new chemical nomenclature. At the end of the 18th century, the "Age of Enlightenment," Brugnatelli published a paper on "oxygen and *thermoxygen*" that today is difficult to classify[149] He distinguished between two types of combustion: one with gaseous oxygen and the other with the unlikely element that he called *thermoxygen*, a form of oxygen combined with *caloric*. In the obsolete *caloric* theory introduced by Lavoisier,[150,151] heat consists of a fluid called *caloric* that flows from hotter to colder bodies. It is likely that Brugnatelli was strongly influenced by this theory in his formulation of *thermoxygen*. In the years that followed, Brugnatelli never mentioned

FIGURE II.03. Luigi Valentino Brugnatelli (1761–1818). Founder of a long line of Italian chemists and mineralogists. His son Gaspare, succeeding him at the University of Pavia, claimed to have discovered a new element, *apyre*, in human gallstones. Courtesy of Galileo Galilei Museum, Florence, Italy.

thermoxygen again, perhaps realizing the error he had made. Luigi Valentino Brugnatelli died in his native city on October 24, 1818, leaving a son just 23 years of age but who was already a university professor.

II.8.2. THE MOST IMPROBABLE OF THE CHEMICAL ELEMENTS

Gaspare is perhaps the least known member of the Brugnatelli family. Born in Pavia on April 25, 1795, in 1813, at the age of 18, he obtained his doctorate in chemistry. He traveled across Germany, Poland, and Hungary, and, in 1819, he became a professor at the University of Pavia. The year 1820 was most fruitful for the young Gaspare Brugnatelli (1795–1852). Soon after he published his "Guide to the Study of Chemistry" in three volumes, in December of that year, he wrote a new scientific paper entitled "Una nuova base salificabile."[152] The discovery hidden in this publication, if it had been confirmed, would be among the most original for chemical elements.

Gaspare Brugnatelli claimed to have discovered a new chemical element in the human body! In this paper, one can trace the influence made on the son by his father's work on uric acid and analysis of the excrement of butterflies. Gaspare analyzed the composition of human urinary and bile calculi, discovering therein what he claimed was a new alkali metal. He obtained a new "salified" base with which concentrated sulfuric acid was able to form a white neutral sulfate, not very soluble in water but much more so in alcohol. By adding potassium carbonate, he obtained a white precipitate that looked like light flakes. The new substance, curiously enough, in spite of its organic origin, was not destroyed by fire.[153] The element easily combined with phosphorus and iodine, and acid solutions of the new substance produced a blue precipitate by adding potassium cyanide; on the other hand, those solutions, when acidified with nitric acid, produced a green precipitate. The article that appeared the following year in France reported in the last sentence the proposed name of the new element: "L'auteur nomme ce nouvel alkali *apyre*, en raison de son indestructibilité au feu."[154]

According to the author's idea, *apyre*[155] was a new alkaline element, detectable only in the human body. However, exhaustive work by Alexander Marcet (1770–1822) cast increasingly grave doubts about the existence of this fanciful element until it fell completely into oblivion.[156] The complete analysis of urinary calculi, performed by Marcet in 1819,[157] did not uncover any evidence of a new element but only the oxalate, urate and phosphates of calcium, and traces of magnesia.

Gaspare Brugnatelli died at the age of 57 on October 31, 1852. His son, Tullio Brugnatelli (1825–1906), became full professor of chemistry at the University of Pavia and occupied that chair—which had been his grandfather's and later his father's—for more than 42 years. His son, Luigi Valentino Brugnatelli (1859–1928), was the next and last member of this "dynasty" of university professors at Pavia since he died childless. His large house and garden became part of the University of Pavia campus and later was transformed into a college for women.

Notes

149. Brugnatelli, L. V. *Ann. Chim. Phys.* **1799**, *17*, 29.
150. The description of the element introduced by Lavoisier was based on experimental data. He clarified the concept, which was very vague until then, according to which an element is a substance that could not be decomposed. His list of the elements included light and *caloric*, which he believed to be material substances. The name *caloric* appeared in the first reform of chemical nomenclature in 1787.
151. Raddi, A. *Philatelia Chim. Phys.* **2011**, *33*(3), 112.
152. Brugnatelli, G. *Brugnatelli Giorn. Fis.* **1820**, 3, 2.
153. "It refuses to burn, to decompose, or to volatilize."
154. Brugnatelli, G. *Archives des Découvertes et Inventions Nouvelles Pendant l'Année 1821*, Paris, **1821**, 99. "The author called the new alkali apyre, because of its indestructibility by fire."
155. The name of the new element spread under different spellings in the European countries: *apyre* (French, English, and Italian), *apyrit* (German). The etymology came from Latin *apyrus*, literally "fireproof."
156. Anon. *Handwörterbuch der reinen und angewandten Chemie*; F. Vieweg und Sohn: Braunschweig, Germany, 1859, p. 157.
157. Marcet A. *An essay on the chemical history and medical treatment of calculous disorders, 2nd ed.* Longman, Hurst, Rees, Orme, and Brown: London. 1819; pp. 63-69; Rosenfeld, L. "The Chemical Work of Alexander and Jane Marcet." *Clin Chem.* **2001**, *47(4)*, 784-92.

II.9

A BRIDGE OF FALSE HOPES BETWEEN DIVINITY AND FALSE ELEMENTS

II.9.1. CRODONIUM

In 1800, the German pharmacist Johann Bartholomäus Trommsdorff burst onto the scene by announcing that he had found a new earth in some samples of beryl coming from Saxony. He quickly called it *agusterde*, a name derived half from Greek and half from German: αγευστος (Greek, "without taste") and *erde* (German: "earth"), literally an "earth without taste."[158]

The renowned French analytical chemist Louis Nicolas Vauquelin was very suspicious and swore to himself that he would clarify matters. Having obtained a sample of Saxony beryl, he set about analyzing it, and after 4 long years he was able to refute Trommsdorff's claim. He announced that *agusterde* was only a mixture of phosphates and lime.[159,160]

In 1820, Trommsdorff, not at all discouraged by his unsuccessful announcement of the discovery of the new metal *agusterde*[161] 20 years earlier, published sensational news of the discovery of an unknown element.[162] The substance was extracted from incrustations that he found in bottles of sulfuric acid imported from England. He had already begun his analyses by the winter of 1818, and his work was certainly quite far along when news of it leaked out to the scientific world prior to his own publication.[163] He wanted to give his new element a name more splendid and magniloquent than the "nothing" name of *agusterde*, and thus he chose the name *crodonium*, derived from Crodo or Seater, a Saxon divinity corresponding to the Roman Saturn, who was adored in ancient Thuringia. With long hair and a beard, Crodo was represented as standing on a fish, signifying the help in adversity that he offered to his worshipers.

Trommsdorff's first announcement, in 1800, was made when he was not yet 30 years old. He was born in Erfurt on May 8, 1770, to Wilhelm Bernhard Trommsdorff (1738–82), a professor of medicine at the local university. Because his father died when he was 12, his mother sent him to study pharmacy at Weimar. Six years later, he returned to Erfurt and reopened the family business. He took his doctorate in 1794, and a year later became a lecturer in chemistry, physics, and pharmacy at the University of Erfurt.[164] He was very interested in the industrial production of pharmaceuticals and early cosmetics.

Shortly after the announcement of his discovery, Trommsdorff himself reported that *crodonia* (the oxide of the hypothetical *crodonium*) did not contain any new metal, but only magnesium and traces of copper, both as oxides.[165] Many years later, Townsend and Adams, in their detailed list of false elements, introduced these substances as "constituents" of *crodonium* and, perhaps using additional data, included iron and lime.[166]

In the history of pharmacy, one can see a rare record of more than two centuries of uninterrupted activity by the Trommsdorff family. It all started in 1734, with Johann Bartholomäus's grandfather, Hieronymus Jacob Trommsdorff (1708–68), who began his career as a pharmacist at the age of 26. Wilhelm Bernhard followed in his father's footsteps, as did his son, the protagonist of this chapter. Three years before his death in 1837, Johann Bartholomäus acquired the pharmacy Schwanen Ring, in partnership with his son Christian Wilhelm Hermann Trommsdorff (1811–84). In 1834, at Erfurt, the 50th anniversary of the beginning of Johann's pharmaceutical career was celebrated: at least 800 people were present, among whom were ex-students and admirers. Johann Bartholomäus continued to work in his pharmacy until almost the day of his death, at age 66, on March 8, 1837.

In 1984, the pharmaceutical house of Trommsdorff, at Alsdorf, celebrated its 250th anniversary.[167] Among the participants was Johann Bartholomäus's great grandnephew, Ernst Trommsdorff (1904–96).

Johann Bartholomäus Trommsdorff is considered the father of scientific pharmacology to this day.[168] He worked, as did many pharmacists of that era, as a polymath, eager to advance scientific knowledge. He was an indefatigable researcher and scholar, an ingenious chemist—he discovered acids that did not contain any oxygen—and he was also a physician, an excellent teacher, a pioneer in pharmaceutical journalism, the author of numerous tracts and manuals, and, not least, the philanthropic founder of a pension fund for pharmacists.

II.9.2. WODANIUM

Almost simultaneously with the discovery of *crodonium*, the renowned chemist Wilhelm August Lampadius (1772–1842) announced the discovery of a new metal.

Lampadius taught at the prestigious Freiberg Bergakademie (Mining Academy of Freiberg)[169] where he achieved his fame through the discovery of carbon disulfide, but he was also an author of scientific memoirs and technical manuals. When he was 46 years old, in 1818, he announced the discovery of a new metal that he had found in some mineral samples coming from a cobalt deposit at Topschau, Hungary. In an English work[170] faithful to the original and translated later into French,[171] Lampadius conferred the provisional name of *wodanium* (after the Wodan, the god of sky and war of German mythology) on his new element.

Lampadius asserted that the metal was 20% of the entire weight of the sample that he had analyzed, an exceptional fact that impressed scientists of the time. He isolated *wodanium* in the metallic state: it had a bronzelike appearance and was malleable and paramagnetic. Its specific gravity was a little more than 11. When it was heated in the presence of air, it formed a black oxide; if it was dissolved in mineral acids, the solutions were colorless but tending to yellowish. On addition of ammonia, Lampadius observed the slow formation of a bluish precipitate. Within a few months, the news of his discovery of *wodanium* was published far and wide on the European continent. For a very different reason, Lampadius was honored in Germany with the stamps shown in Figure II.04.

On March 16, 1820, Friedrich Stromeyer, professor of chemistry and pharmacy at Göttingen, presented a very careful analysis of the samples of mineral coming from the mines at Topschau to the local Royal Academy; among these samples was the mineral containing *wodanium*.[172] The mineral contained arsenides of nickel, cobalt, and iron, mixed with the sulfides of manganese, copper, lead, and antimony. He could affirm that there was no new metal.

Because of Stromeyer's reputation as an analytical chemist, the discovery of *wodanium* could have and should have immediately been set aside, but that was not the case. Three

FIGURE II.04. Wilhelm August Lampadius (1772–1842), a German chemist and metallurgist best known for his contributions to the development of gas street lamps, is honored in this pair of stamps issued in 1991 on the occasion of the 18th World Gas Congress in Berlin. From the Collection of Daniel Rabinovich, with his kind permission.

years later, Lampadius wrote to Ludwig Wilhelm Gilbert, editor of *Annalen der Physik*. Gilbert published Lampadius's account, in which he reported some observations on the presence of ammonium alum in the Bohemian mine at Tschumig. Just before the conclusion of the article, Lampadius referred to his discovery of *wodanium*. He notified the editor and the public that his work to confirm said discovery would necessitate more time and more analyses, and he concluded with the promise: "I consider it my duty to chemists to submit the work I have done." This pledge was never realized; no article on *wodanium* bearing Lampadius's name ever appeared again.

Wilhelm August Lampadius died at Fribourg on April 13, 1842, at the age of almost 70. For his discoveries in many areas of applied chemistry, he is considered even today as one of the most famous chemists of his era.

II.9.3. FALSE ELEMENTS EXCHANGED FOR ANOTHER FALSE ELEMENT

In 1906, *wodanium* would have enjoyed a new, albeit fleeting, revival a good 83 years after its first announcement through the work of an English chemist, C. T. Owen.[173] Owen tried, in his article in *Chemical News*, to offer an "honorable" conclusion to four false discoveries by claiming that the announcements for the discoveries of *wodanium, vestium, gnomium,* and *aurum millium* were nothing less than independent discoveries of the same metal. The alleged metals had uniquely in common the composition of the rocks that were supposed to contain them. These were rocks with high levels of arsenic, sulfur, nickel, and cobalt that, taken together, made up more than 95% of the total weight of the mineral.

In addition to *vestium*, another element that Owen absorbed into *wodanium* was *aurum millium*. In 1820, a letter from London was delivered to a gentleman in Baltimore (Maryland) announcing the discovery of a new metal by a certain Mr. Mills. *Aurum millium* (gold of Mills), as the discoverer hastened to name it, reminded one of gold (as did, strangely enough, *wodanium*) in color, hardness, and malleability.[174] It was suitable for the coinage and gold industries. Almost immediately, the American chemist Benjamin Silliman (1779–1864) contradicted this claim, asserting that *aurum millium* was, at the very most, an alloy and not an element.

Nearly 70 years later, in 1889, the "birth" of yet another new element was announced with the greatest publicity: *gnomium*. Two Germans, Gerhard Krüss (1859–95), of the University of Munich in Bavaria, and F. W. Schmidt (1829–1903) asserted that both nickel and cobalt, thought for decades to be elemental substances, actually contained a hidden element hitherto unknown[175]: "We announce the discovery of a new metal, found in minerals of cobalt and nickel, to which we give the name gnomium. An English chemist has succeeded in forming, with this substance, a product that has all the appearances of gold, and it also has similar malleability and ductility."[176]

Their naïve reasoning was based on an anomaly that created quite a few problems for chemists at the end of the 19th century. Mendeleev at first believed that he had identified the organizing principle of the elements in terms of their increasing atomic weight, but then discovered exceptions and inverted the positions of nickel and cobalt, and of iodine and tellurium in the periodic table. The anomalous atomic weight order of nickel and cobalt was the basis of Krüss and Schmidt's erroneous conviction that a new metal was hidden between them. They hastened to patent the process of *gnomium* extraction from the other two metals.[177] (The name, *gnomium*, from "gnome," was chosen because it was analogous with the name given to cobalt, whose etymology was rooted in Nordic mythology. The *kobolds* were mischievous sprites that inhabited caves and, by extension, the mines where cobalt was extracted. Naming the element for these malicious and deceptive creatures reflected the fact that cobalt-bearing minerals, because of their chemical properties, were often mistaken for more precious minerals like nickel and gold.)

The existence of an unknown element similar to cobalt and nickel was immediately suspect. To justify its existence, the two chemists put forth the idea that the atomic weight would be very similar to those of cobalt and nickel and that *gnomium* was actually the cause for the necessary inversion of these two elements in the periodic table. But the biggest problem that Krüss and Schmidt could not resolve was the fact that none of their colleagues could isolate this metal. Consequently, its existence grew more doubtful, and one anonymous critic openly referred to their work as the fruit of fantasy.[178] Krüss sought for incontestable experimental evidence by recording the arc spectrum of the new metal, but this result was, unfortunately, ambiguous.

Then, in 1891, when the dispute about the existence of *gnomium* was at its height, the chemist Hugh Remmler came on the scene.[179] He fractionated 1,200 g of cobalt chloride in search of the elusive *gnomium* but could not confirm its existence. Four years later, Krüss died at the young age of 35, but the final chapter in the story of *gnomium* had not yet been written.

In 1906, when referring to *gnomium* and its uncertain presence in nature, C. T. Owen was quietly optimistic, and he sought to support Krüss's work. Owen even affirmed Krüss and Schmidt's bizarre hypothesis, according to which: "An analysis of these might lead to the re-discovery of vestium." Never did prophecy seem more like guesswork.

In 1938, S. Plesniewicz and K. Sarnecki summarized what was known about *vestium* at the time, believing that it had been discovered decades earlier following the work of Russian chemist Karl K. Klaus (1796–1864).[180] Unfortunately, the *vestium* that both Plesniewicz and Sarnecki referred to was not the *vestium* of Lorenz von Vest, but an alleged new metal that the Polish-Lithuanian chemist Jędrzej Śniadecki believed he had isolated between 1806 and 1808.

Śniadecki was born November 30, 1768, in the united kingdom of Poland and Lithuania. In 1803, he became professor of chemistry and medicine at the University of Wilna. In 1806, he presented both at the Academy of Sciences in Paris and of Saint Petersburg news of the discovery of a new element that he wanted to call *vestium*, named, in this case, for the planetoid Vesta. His communication to Saint Petersburg was shelved and only 2 years later did the Academy of Sciences at Saint Petersburg accept Śniadecki's request to publish his notes on *vestium*, although what was published was not his full text.

Later, Śniadecki abandoned his work on this metal but remained very involved in chemistry. He translated into Polish the first modern chemical treatise, and he is also credited with having created modern chemical terminology in the Polish language. He died at Vilnius on May 12, 1838, not quite 70 years of age.[181]

In 1967, an attempt to confirm Śniadecki's presumed discovery was published by I. L. Znaczko-Jaworski,[182] but pressure from the Soviet government, which feared even the least reawakening of anti-Russian sentiment,[183] forced the Polish scientific community to publish a retraction of this work.

In many civilized countries, anniversaries help to hand on discoveries to posterity and to witness to individual genius. In mid-20th century, Poland, confined by the Warsaw Pact, this was not so. On the 200th anniversary of Śniadecki's birth, he died a second death: Kazimierz Sarnecki published a harsh refutation of his work and denied categorically that Śniadecki could have discovered ruthenium in 1808. Sarnecki wanted to show by every possible means that *vestium* and ruthenium could not be the same element.[184] An excellent recent review of the dubious discoveries of *vestium* is found in Marshall and Marshall.[185]

II.9.4. PTENE

As is generally accepted, Smithson Tennant (1761–1815) discovered iridium along with osmium in the summer of 1803, in the black residue obtained by the dissolution of native platinum in aqua regia.[186] Antoine François de Fourcroy, working with Nicolas Louis Vauquelin, took over the research on this black residue.

On October 10, 1803, they presented a paper to the Institut in Paris (the paper was later published in 1804) in which they described their study of this black solid. They fused it with potash, extracted the cooled melt with water (to produce a solution that they believed contained chromium but which may also have contained rhodium—later to be isolated by Wollaston in 1804), and treated the residue with more aqua regia. Addition of ammonium chloride to the latter produced, depending on conditions, red or yellow crystals. They thought that the red crystals, in addition to compounds of titanium, chromium, iron and copper, contained a compound of a new metal. These crystals could well have been or contained iridium as $(NH_4)_2[IrCl_6]$, but they chose not to name their "new element."[187]

On the same day that their first memoir was presented to the Institut, Hippolyte Victor Collet-Descotils, who had been Vauquelin's student, reported essentially similar results and published a more concise paper.[188] Like the prudent Fourcroy and Vauquelin, he did not name the new metal that he believed to be present in his flask, but said that he would assign it a name after further research.

The memoirs of Fourcroy and Vauquelin and of Collet-Descotils were known to Tennant when he presented his paper on June 21, 1804.[189] In it, he speaks of an unknown metallic ingredient that remains as a black powder after platina is put into solution, inferring that it may be iridium, as observed by Fourcroy and his colleagues. But he also remarks on the presence of another metal—probably osmium—"different from any hitherto known."

There are references in the literature to *ptene* or *ptène* (from the Greek πτηoζ, *ptènos*, "winged") as a name for osmium; indeed, Tennant is said to have proposed this name for it,[190,191] whereas Partington says that Fourcroy and Vauquelin proposed it.[192] (We can find no trace of this ungainly name either in Tennant's paper or in those of the French authors.) The symbol proposed by Berzelius was the familiar Os; for iridium, he proposed at first I, later changing it to Ir.[193]

II.9.5. DONARIUM

Scandinavian minerals play a special role in the history of the chemical elements. In 1851, Carl Wilhelm Bergemann (1804–84), at age 47, was an established professor of chemistry, having taught for more than 20 years at the University of Bonn. From that prestigious post, he announced the results of his analyses of some of the rarest minerals coming from Norway. Among them was one completely new to science that Bergemann called "orangite" because of its orange color. Analysis of this mineral did not cause him any difficulty, and Bergemann published his results: the oxides of silicon and calcium with traces of iron, potassium, magnesium, and manganese did not make up as much as a third of the mineral content. A good 71% of the total was composed of an oxide of a metal unknown to science.[194] He proposed the name *donarium* (symbol Do), after the Germanic god Donar (or Thor).[195,196]

Donarium was isolated as a black, powdery metal by reduction of the oxide with potassium. The new metal tended to oxidize spontaneously in the presence of moisture, forming a yellow-gray compound. This substance was converted into the sulfate by sulfuric acid or into a red oxide by reaction with aqua regia. The systematics of this element counted among them some rather ambiguous reactions: adding base to solutions of *donarium* produced a white precipitate, but there did not seem to be any reaction on addition of either hydrogen sulfide or potassium ferrocyanide.

The rivalry that cropped up between Bergemann and Karl Gustav Bischof (1792–1870), one of his colleagues at the University of Bonn,[197] was caused by trivialities but resulted in devastating consequences. When Bischof accepted into his laboratory, which he shared with Bergemann, four students gearing up for new lab stations, he had a run-in with his rival. This unpleasantness continued for more than a dozen years, distracting both men from their research,[198] and it could have caused Bergemann to commit some grave experimental errors.

Less than a year after the discovery of *donarium*, a French chemist, Augustin Alexis Damour (1808–1902), threw some light on the entire *donarium* affair. Except for only two properties, density and color, orangite seemed to be identical to thorium oxide. Damour, born in Paris, worked first at the French foreign ministry and only in 1854 did he dedicate himself entirely to mineralogy. His first analyses were published in 1837, and he was an active and prolific popularizer of science up until the age of 85.[199]

The erroneous property measured by Bergemann was the mineral's density, which was a simple enough error. However, the yellow-gray color of the calcinated material (the oxide) was explained by Damour by the fact that Bergemann had not been able to remove all the uranium and lead from his sample. In support of these claims, Damour, at the May 3, 1852 session of the Academy of Sciences of Paris, reported the complete analysis of orangite:[200] a hydrated silicate of thorium with an appreciable presence of oxides of uranium, lead, and calcium, as well as traces of iron, manganese, aluminum, and potassium.

In summary, Damour reported his unequivocal experimental evidence, according to which *donarium* and thorium were the same element, and thus orangite was none other than thorium oxide. Damour concluded by proposing the removal of the names orangite and *donarium* from the scientific literature.[201]

In that same year, Nils Johannes Berlin (1821–91) published his conclusions, which corroborated Damour's results.[202] In a letter to Heinrich Rose, a renowned mineralogist and analytical chemist at the University of Berlin, he said that many years earlier he had analyzed the mineral that Bergemann called orangite without finding any unknown element. After Bergemann's announcement of the discovery of *donarium*, Berlin decided to analyze the mineral again, and again he found no unknown element.

Following these two announcements, Bergemann admitted his mistake and his retraction was published in the same year.[203] In Bergemann's defense, it must be recognized that analysis of thorium minerals had created many problems for his fellow chemists of the time. Years earlier, J. J. Berzelius himself announced the discovery of thorium twice, but only on his second try was he able to prove its existence.

Bergemann could console himself in a certain sense with the fact that Damour's scathing pronouncements were not accepted by the entire chemical community: in fact, the name orangite is sometimes used even today to indicate a yellow-orange variety of thorite.

After his false discovery and retraction, Bergemann lived another 32 years, passing away in 1884 at the age of 80. His nemesis, Damour, was to surpass him even in longevity, actually living into the new century and passing away in Paris at the venerable age of 94.

Notes

158. Trommsdorff, J. B. *Almanach der Fortschritte in Wissenschaften, Künsten, Manufakturen und Handwerken* **1800**, *5*, 65.
159. Vauquelin, L. N. *Ann. der Physik* **1804**, *16*, 126.
160. Anon. *Chem. News* **1870**, 21.
161. Trommsdorff, J. B. *Almanach der Fortschritte in Wissenschaften, Künsten, Manufakturen und Handwerken* **1800**, *5*, 65.
162. Trommsdorff, J. B. *Ann. der Physik* **1820**, *36*, 208.
163. Gilbert, L. W. *Neuen J. der Pharmacie* **1819**, *3*, 2.
164. Rocchietta, S. *Minerva Medica* **1985**, *76*(45–46), 2219.
165. Trommsdorff, J. B. *Ann. der Physik* **1820**, *36*, 290.
166. Bolton, H. C. *Chem. News American Supplement* **1870**, *6*, 368.
167. Goetz, W. *Beitr. Gesch. Pharm.* **1985**, *37*, No. 26–27, 12/232–15/235.
168. Anon. *Farmaci e Farmacie* **1957**, 5.
169. Some renowned alumni and faculty members of this school were Andreas Manuel del Rio (1764–1849), discoverer of vanadium; the brothers Elhuyar, Don Juan José (1754–96) and Don Fausto (1755–1833), discoverers of tungsten; and Ferdinand Reich (1799–1882) and Theodor Hieronymus Richter (1824–98), co-discoverers of indium. Last in chronological order is Clemens Winkler (1838–1904) who, in 1886, discovered and isolated germanium.
170. Anon. *Ann. Philos.* **1818**, *8*, 232.
171. Thénard, L. J. *An Essay on Chemical Analysis*; W. Phillips: London, 1819, p. 414.
172. Stromeyer, F. *Taschenbuch für die gesammte Mineralogie*; J. C. Hermann: Frankfurt am Main, Germany, 1822, p. 225; Anon. *Journal de Pharmacie et des Sciences Accessoires* **1820**, *6*, 397.
173. Owen, C. T. *Chem. News* **1906**, 158.
174. Anon. *Blackwood's Edinburgh Magazine*, **1820**, 331; Anon. *The Edinburgh Monthly Magazine* **1821**, 381; Silliman, B., Ed. *American Journal of Science and Arts*, vol. II, S. Converse: New Haven, CT, 1820, p. 363.
175. Schunck, E. *Memoirs of the Manchester Literary and Philosophical Society* **1890**, *4*(3), 170.

176. Anon. *La Nuova Antologia* **1891**, *36*(3), 587.
177. Anon. *Chem. News* **1889**, 59; Krüss, G.; Schmidt, F. W. Ein neues Element, welches neben Kobalt und Nickel vorkommt. *Z. Anal. Chemie* **1889**, *28*, 340.
178. Anon. *Journal of the Society of Chemical Industry* **1892**, 711.
179. Remmler, H. *Z. Anorg. Chem.* **1893**, *2*, 221.
180. Plesniewicz, S.; Sarnecki, K. *Przemysl Chemiczny* **1938**, *22*, 88.
181. Chrzanowski, I.; Krzemiński, S.; Galle, H. *Wiek XIX sto lat myśli polskiej: życiorysy, streszczenia, wyjątki* **1906**, 409.
182. Znachko-Yavorskii, I. L. *Kwartalnik Historii Nauki i Techniki* **1967**, *12*(1), 47.
183. If the discovery of *vestium* had been validated, the name ruthenium (Latin for "Russia") would have been removed from the periodic table by the hand of a Pole.
184. Sarnecki, K. *Kwartalnik Historii Nauki i Techniki* **1968**, *13*(4), 799; Sarnecki, K. *Problemy* **1969**, *24*(6), 342.
185. Marshall, J. L.; Marshall, V. R. *Bull. Hist. Chem.* **2010**, *35*(1), 33–39.
186. Griffith, W. P. *Platinum and Metal Reviews* **2004**, *48*(4), 182–89.
187. Fourcroy, A. F.; Vauquelin, N. L. *Ann. Chim.* **1803**, *48*, 177; Fourcroy, A. F.; Vauquelin, N. L. *Ann. Chim.* **1804**, *49*, 188–219; summarized in *Phil. Mag.* **1804**, *19*, 117.
188. Collet-Descotils, H. V. *Ann. Chim.* **1803**, *48*, 153; Collet-Descotils, H. V. *J. Nat. Philos. Chem. Arts* **1804**, *8*, 118.
189. Smithson, T. *Phil. Trans.* **1804**, *94*, 411; Smithson, T. *J. Nat. Philos., Chem. Arts* **1805**, *10*, 24; Smithson, T. *J. Nat. Philos., Chem. Arts* **1804**, *8*, 220.
190. Webb, K. R. *J. Roy. Inst. Chem.* **1961**, *85*, 432.
191. Newton Friend, J. *Man and the Chemical Elements*, Griffin: London, 1951, p. 303.
192. Partington, J. R. *A History of Chemistry*, Macmillan: London, 1962; vol. 3, p. 105.
193. Berzelius, J. J. *Pogg. Annalen* **1834**, *32*, 232.
194. Bergemann, C. W. *Pogg. Ann.* **1851**, *82*, 56.
195. Bergemann, C. W. *Ann. Chim. Phys.* **1852**, 235.
196. Bergemann, C. W. *Ann. Chem. Pharm.* **1851**, 267.
197. Karl Gustav Bischof was born in Nuremberg, Bavaria, in 1792, and died in Bonn, in 1870. He was a professor at Bonn and is remembered as an expert experimental chemist and for his research on the inflammability of gases.
198. Schrubing, G. *Osiris* **1989**, *5*(2), 57.
199. Voit, C. *Sitzungsberichte der mathematisch-physikalischen Klasse der K. B. Akademie der Wissenschaften zu München*, **1904**, 33, 536.
200. Damour, A. *Compt. Rend. Chim.* **1852**, *34*, 685.
201. Damour, A. *Compt. Rend. Chim.* **1852**, *34*, 615.
202. Berlin, N. J. *Pogg. Ann.* **1852**, *85*, 556; *Pogg. Ann.* **1852**, *87*, 608.
203. Bergemann, C. W. *Pogg. Ann.* **1852**, *85*, 558.

II.10

GAHNIUM, POLINIUM, AND PLURANIUM

II.10.1. GAHNIUM

Jöns Jacob Berzelius is generally considered to have been one of the greatest chemists of the 19th century. In addition to his electrochemical studies, which led him to formulate the dualistic theory, he was involved in chemical nomenclature and notation, as well as in mineral chemistry. His interest in the latter has earned him a special place among the major chemists of all time: Berzelius discovered three elements, the first, cerium, when he was not yet 24 years old; the second, selenium, in 1817; and the third, thorium, a few months short of his 50th birthday. In addition to these achievements, Berzelius also claimed two discoveries that soon proved to be insupportable.

The first of these erroneous discoveries followed closely on the facts related to the discovery of cerium. Enthusiasm and youth led him to publish results recklessly and incompletely, and he even proposed a name for his supposed new metal, *gahnium*, in honor of the inspector of mines of Falun, Johan Gottlieb Gahn. Unfortunately, it soon became clear that *gahnium* was nothing more than zinc oxide.[204]

In 1815, Berzelius was involved in another discovery of a chemical element later proven to be nonexistent. In presenting this second announcement, he was more cautious and merely indicated the presence of a "new earth" to which he would not immediately give even a provisional name. Soon he realized that his caution was well-placed because the discovery proved to be false. In 1825, he published a letter in which he stated that his earlier tentative conclusions were totally wrong.[205]

Although he retracted his supposed discoveries, the echoes of Berzelius's failure were not immediately forgotten. The mineral in which he believed he had found a new metal was called by French mineralogist François Sulpice Beudant (1787–1850) "kenotime," from the Greek κενός, "vain," and τιμή, "honor," to highlight the regrettable failure of the famous Swedish chemist.[206] However, the original intent of punishing the vainglorious Berzelius was disappointed: with time, the word "kenotime" became "xenotime," a rare earth phosphate mineral whose major component is yttrium orthophosphate.

II.10.2. POLINIUM AND PLURANIUM

One of the first chemists to work on the identification of the platiniferous metals was Andreas (Jedrzej) Śniadecki who, from the beginning of the 19th century, carried on systematic analyses and investigations on these substances. However, a real breakthrough in the study of the noble metals was only possible after 1819, when the scientific analysis and exploitation of gold deposits in the Ural Mountains began. As a result of these analyses, chemists completed the isolation of ruthenium and were able to foresee rhenium appearing in metallic form.[207]

In 1823, by imperial decree, all the mines of the Ural Mountains had to send samples of platinum to the imperial capital—St. Petersburg—to be analyzed. Two years later, a state monopoly for noble metals was established. The Russian Minister of Finance, Frantsevich Kankrin (1775–1845), became interested in noble metals coming from the Urals, and from 1828 until his death in 1845, the Russian Imperial Mint produced platinum coins of three, six, and twelve roubles, respectively.

To stimulate research on platiniferous metals, Kankrin sent samples of native platinum to the most renowned chemists of his time, either Russians or foreigners. Berzelius received half a pound (0.2 kg) of platinum and conducted classic analytical experiments on it without finding anything new. Kankrin also sent 4 lb (1.8 kg) of platinum to Gottfried Wilhelm Osann (1796–1866), professor of chemistry at Dorpat (now Tartu, Estonia) and a platinum expert.

Osann was born October 26, 1796, in Weimar, and belonged to a family who numbered several university professors among its ranks; he studied chemistry and physics, and he became *privatdozent* in 1819 in physics and chemistry at the University of Erlangen. Between 1821 and 1823, he occupied the same position at the University of Jena. From 1823 to 1828, Osann taught chemistry and medicine at the University of Tartu. At the age of 32, he took up a teaching position at the University of Würzburg, not far from his Weimar birthplace.

In 1827, Osann announced that he had discovered three new elements[208] in the platiniferous material supplied to him by Count Kankrin. The names he suggested were *polinium,*[209] ruthenium,[210] and *pluranium.*[211] Osann isolated the three metals after having dissolved all the platiniferous material in aqua regia. The first metal was found in the insoluble residue, its oxide crystallizing in long prisms. The oxide easily sublimed; with the addition of ammonium sulfide, Osann obtained *polinium* sulfide, a gray, low-melting-point compound that could be reduced with a blowpipe.

The second metal also produced white crystals and was reduced to the metallic state with hydrogen. Metallic ruthenium appeared gray with red tinges. It was easily dissolved in aqua regia and precipitated as the sulfide. These two metals were found in small quantities, whereas the presence of the third element, *pluranium*, was surprisingly far greater. *Pluranium* was also soluble in aqua regia and formed alloys with other metals; for example, when alloyed with iron, the resulting metal had the unique feature of being extraordinarily resistant to acids.

Osann sent the three samples to Berzelius in Sweden for clarification and possibly a confirmation of his discoveries. Berzelius's verdict was as fatal as brutal. Where Osann had seen three unknown metals, Berzelius saw no new element. He added that the white oxide of ruthenium was actually a mixture of silicon, zirconium, and titanium with minimal traces of iron. Osann, unsurprisingly upset, harbored some doubts about the care with which these experiments were conducted, but the verdict was categorical, issued by a luminary in mineral chemistry, and Osann not only had to accept the condemnation but also retract his discovery.[212]

In addition, the quantity of the samples sent to Berzelius was extremely small; analytical chemistry, still in its infancy, could not provide a comprehensive response. Clarification of this matter had to wait for 17 years, when Karl Ernst Klaus (1796–1864) took up the platiniferous residue analyzed by Osann in 1827. Klaus, born in Dorpat (Tartu), was Osann's contemporary; he was self-taught and, at around the age of 30, he became Osann's assistant. In this way Klaus knew Osann's work methods and teaching, and, above all, he was able to "see" the samples of the alleged *polinium*, ruthenium, and *pluranium*.

Unfortunately, the material at his disposal was extremely scanty, and the mystery metal was even less abundant. Klaus spent many years at Osann's laboratory; eventually, he was appointed to a position at the University of Kazan, Central Asia. There, he was given 2 lbs (0.9 kg) of impure platinum as a gift. Two years later, he met Russian Minister of Finance Kankrin, and the latter promised to send him much more. In 1842, the Minister sent him 18 lbs (8.2 kg) of platinum from which, 2 years later, Klaus extracted a new metal.[213] Klaus reused the name that Osann had given to one of the presumed newly discovered metals, partly to get credit for the discovery. Having to choose between three names, Klaus gave pride of place to the most patriotic one, ruthenium. This choice allowed Klaus to come to the attention of Tsarist authorities, and it was also the occasion for him to be promoted and assigned to a more prestigious university.

Klaus thought that the white oxide of ruthenium could actually conceal traces of a new element, so he sent a few good samples to Berzelius. He waited, with trepidation, for the verdict of the by-now aged and irascible Swedish inorganic chemist. Berzelius seemed to be upset by the whole affair: it was the second time he was asked to validate the discovery of ruthenium. At first, he refused to help: in his eyes, this was nothing more than an attempt to restore Osann's reputation and work. A few weeks later, reluctantly, he began to analyze the material Klaus sent him. The first results were not encouraging: the metal was merely impure iridium.

We do not know what induced Berzelius to repeat the analysis with greater accuracy. In a second letter, he admitted to having been too harsh with Klaus, but it was the end of the letter that pleased Klaus much more than Berzelius's regret: Berzelius affirmed that ruthenium was indeed a new element. If it was a great satisfaction to Klaus to receive an apology from Berzelius, his suggestion that the discovery be made immediately public in the major German-language periodicals was really surprising.

Klaus followed Berzelius's suggestion and promptly published his results. As soon as the news appeared in the journals, Osann read the article and was deeply hurt by its contents. In the pages of *Poggendorff's Annalen*, Osann responded to Klaus's veiled criticisms and declared that Klaus's "ruthenium" was identical to his *polinium*,[214] which he had isolated as far back as 1828. Klaus replied with great kindness, but firmly refused to share credit for the discovery of ruthenium with anyone. According to Klaus, *polinium* was impure iridium.[215] Osann had previously admitted that he had mistaken iridium for the new element that he called *polinium*; Klaus took advantage of Osann's admission in an attempt to dissociate himself as a co-discoverer with Osann because the chemical community had already passed judgment on this supposed discovery.[216] However, to this day, the real identity of *polinium* has never been ascertained.

On the one hand, the white oxide of *pluranium*, which had a tendency to sublimate at relatively low temperatures, produced inert colorless crystals, was soluble in hydrochloric acid, and produced a brown precipitate generated by adding sulfuric acid to the solution, was a unique metal. On the other hand, these inhomogeneous properties made its existence rather suspicious. Berzelius was only able to identify the presence of *pluranium* once, whereas Friedrich Wöhler was never able to ascertain its presence among his samples. Both chemists, however, claimed that the samples they analyzed, although impure, did not contain tellurium, antimony, or bismuth.

Klaus politely but resolutely told Osann not to meddle again in "his" discovery (ruthenium). A tacit agreement in this unpleasant dispute was eventually reached, and the whole matter was apparently settled. This agreement gave to posterity a single discoverer of ruthenium: Karl Klaus.

Karl Karlovich Klaus was born in Tartu (Dorpat) on January 23, 1796—the same year of Osann's birth—in Estonia, then a province of the Tsarist Empire. Klaus lived nearly 20 years after the discovery of ruthenium and died on March 24, 1864, at the age of 68. Gottfried Wilhelm Osann survived his former assistant and rival by a little over 2 years, dying in Würzburg on August 10, 1866, shortly before his 70th birthday.[217]

Notes

204. Jorpes, J. E. *Jac. Berzelius. His Life and Work*; Almqvist Wiksell: Stockholm, Sweden, 1966, p. 29.
205. Berzelius, J. J. *Ann. Chim. Phys.* **1825**, *2(29)*, 337.
206. Beudant, F. S. *Traité élémentaire de Minéralogie*; Carilian: Paris, 1837; vol. 2, p. 552.
207. Turley, T. J. The Discovery of Ruthenium. *Earth Science* **1963**, *16*, 185–86.
208. Osann, G. *Pogg. Ann.* **1828**, *13*, 283.
209. The name comes from the Greek word *polinium*, πολιοσ, "gray"; in fact, the residues containing the supposed new metal were gray.
210. Ruthenium comes from Latin *Ruthenia*, the eastern region that would roughly correspond to the western borders of Tsarist Russia, now Ukraine.
211. *Pluranium* may be derived from the combination of the words "platinum" and "Ural," although there is no absolute certainty about the origin of name. The name could have arisen from a simple experimental observation: *pluranium* was supposed to be much more abundant than *polinium* in the samples observed.
212. Osann, G. *Pogg. Ann.* **1829**, *14*, 329; Osann, G. *Pogg. Ann.* **1829**, *15*, 158.
213. Klaus, K. *Pogg. Ann.* **1845**, *64*, 192; Klaus, K. *Justus Liebigs Ann. Chem.* **1846**, *56*, 257; Klaus, K. *Phil. Mag.* **1845**, *27*, 208. Klaus's discovery of ruthenium was first reported in Russian, in the journal *Uchenyia zapiski Imperatorskago Kazanskago universiteta* **1844**, Bk. 3, 15–200 (separately printed as *Khimicheskoe izsliedovanie ostatkov ural'skoi platinovoi rudi i metalla ruthenia*, Kazan, 1845), and, in abbreviated form, in the journal *Bull. Cl. Phys. -Math., Acad. Sci. St. Petersbourg* **1845**, *3*, col. 311–16. The reports in the German literature are subsequent translations of the articles in the *Bulletin*. A good review of the discovery of ruthenium is Pitchkov, V. N. *Platinum Met. Rev.* **1996**, *40*, 181–88. A more recent account is Kauffman, G. B.; Marshall, J. L.; Marshall, V. R. *Chem. Educator* **2014**, *19*, 106–115.
214. Osann, G. *Pogg. Ann.* **1845**, *64*, 208.
215. Klaus, K. *J. prakt. Chem.* **1846**, *38*, 164.
216. Klaus, K. *J. prakt. Chem.* **1860**, *80*, 282; Klaus, K. *Chem. News* **1861**, *3*, 194.
217. Hödrejärv, H. Gottfried Wilhelm Osann and Ruthenium. *Proc. Estonian Academy of Sciences. Chemistry*; Estonian Academy Publishers: Tallinn, Estonia, 2004; pp. 125–44.

II.11

ABERDONIA AND THE "SWEET" MAP OF OBLIVION

Every desire has its own plan, and every plan its own point of departure and arrival. Committed to making some sense out of the incessant coupling of molecules that combine and dissociate, chemists have tended to name these fundamental substances of matter—the elements—after cities, regions, or peoples so that their research might be better remembered by posterity and confer on their deeds a kind of human immortality.

II.11.1. DONIUM

At the end of 1835, one Doctor Davidson, roaming around granite deposits in the Aberdeen, Scotland, area, discovered a mineral to which he gave the name davidsonite.[218] He sent the mineralogical sample to an acquaintance, the chemist Thomson, for analysis. Too busy to do it himself, Thomson turned over the examination to an apprentice. The young man arrived at the conclusion that the mineral was composed of silica (66.59%), alumina (32.12%), and water (1.30%). The unexpected result caused the apprentice to request the aid of chemist Thomas Richardson, so that he might repeat the analysis of the unknown mineral.

After having reduced the mineral to a very fine powder, Richardson added sodium carbonate and melted the entire mass. The product obtained was treated with dilute hydrochloric acid, and, by filtration, the silica was removed. The mother liquor was concentrated, then neutralized with ammonium carbonate. On adding concentrated ammonia, Richardson obtained a white precipitate that was collected and dissolved again in acid. He carried out this procedure several more times on the mother liquor for the purpose of removing as much alumina as possible. When he was convinced that he had extracted the white product from the mother liquor, he added ammonium oxalate, without obtaining any precipitate.

The mother liquor was brought to dryness and a white mass appeared in the crucible; this mass, on heating, turned brown after a short period of time. Richardson thought that the brown residue, after filtration and washing with water, had to be "iron peroxide." The precipitate containing iron was dissolved in dilute hydrochloric acid. A large amount of soda was added, and the solution thus obtained was heated. A dark precipitate, attributed to iron, formed very quickly. The basic solution was separated from it, acidified, and concentrated with heat, which caused the precipitation of a white mass insoluble in caustic soda.

Richardson arrived at the conclusion that "there were several circumstances in the analysis which appeared to indicate that the mineral contained some other base besides alumina."

For this reason, he repeated the analysis using a larger amount of davidsonite. The results from these more extensive investigations confirmed the presence of a new substance or element whose properties were completely different from those of the alkaline elements and also from the more common transition elements. He put a great deal of energy into wet separation analytical techniques to characterize this element, and, at the end of his exhausting work, he proposed the name *donium*, a contraction of Aberdonia, the Latin name of Aberdeen.[219]

Old Aberdeen stands approximately on the location of *Aberdon*, the first settlement on the site of the present city. The name in Celtic means "at the confluence (prefix *aber*) of the Don (the local river) with the sea."[220]

The last effort that Richardson made was to isolate the metal in the elemental state. He placed a finely divided portion of the white oxide of *donium* in a glass tube and heated it to incandescence with an open flame while passing a stream of hydrogen through the interior of the tube for about an hour. At the end of the operation, he got a shiny metallic-looking slate-blue powder, indicating that *donium* was metallic.

II.11.2. TREENIUM

At the extreme southwestern end of Britain, in Cornwall, 37-year-old Henry Samuel Boase (1799–1883), a chemist and geologist, published an article—in some respects ludicrous and in others naïve—in response to Richardson's work on *donium*.[221] In it, he considered his own recent unpublished work which, in his opinion, allowed him to draw the same conclusions that Richardson had a few months earlier. Boase asserted that Richardson's discovery was precisely the same substance that he had found and named *treenium*, after Treene, the place where it was found, but that he had postponed publication in order to complete further examinations.

As in numerous other cases relative to the discovery of an element, Boase sought to overturn Richardson's work and establish priority for his own, now pointing out gaps in his colleague's article, now indicating errors in the analyses. Both Boase and Richardson were subjects during the brief reign of William IV (1765–1837) of Hanover. They were convinced that they lived in the most civilized country on the planet. They had been educated to always maintain a formal bearing; they knew how to express the most absurd demands with a mixture of good manners and hypocrisy, and this is how Boase ended his article: "Should my oxide prove to be the same as Mr. Richardson's Donium, my name of Treenium must of course give place to his, as the first had the honour of making it public, and I trust that this brief note will insure to me, if not the honour, at least, the credit of also having discovered Donium."

Beyond the analysis of the two elements' behavior, Boase did not know—and it could not have been otherwise in 1836—the number of elements yet to be discovered and therefore thought that the two elements were the same.

A week later, Boase wrote, at the request of journal's editor, an additional note with a more detailed account of his analysis of the new element. The data that he had collected led him to believe that *treenium* was a metal similar to tungsten and titanium, but his speculation was shown to be entirely erroneous.

Henry Samuel Boase studied chemistry and medicine first at Dublin and then at Edinburgh. After practicing medicine for a short time, he settled in his native Cornwall, devoting himself entirely and with particular enthusiasm to geology, which gave him his greatest satisfaction. He became secretary of the Royal Geological Society of Cornwall and in 1834 published the work that handed his name down to posterity: *Treatise on Primary Geology*. The following year, he settled in London and shortly afterward, he was elected a Fellow of the Royal Society. Proprietor of a bleachery in Dundee, he retired from business in 1871, but lived to almost 84 years of age, passing away on May 5, 1883.

II.11.3. THE DISCOVERY OF AN ALREADY KNOWN ELEMENT?

For a certain period of time some scientists thought that both *treenium* and *donium* had to be the oxide of beryllium, or *glucinium*.[222] In fact the properties of the white solid observed by Boase and Richardson fit very well with those of beryllium oxide, especially its solubility in ammonium carbonate in the cold. In addition, the composition of davidsonite argues in favor of this position: this mineral is a green-yellow variety of beryl, whose chemical composition is $Be_3Al_2(SiO_3)_6$. The element beryllium was discovered about 40 years prior to the work of Richardson and Boase.

In the middle of the complicated series of events surrounding these nonexistent elements, the mistaken discovery of a French chemist, Louis Nicolas Vauquelin, was in the process of being resolved.[223]

By the end of the 18th century, some chemists began to focus their interest on the composition of two gemstones that were similar to each other: beryllium and emerald. Martin Heinrich Klaproth analyzed some Peruvian emeralds given him by Prince Demetrius Augustine Gallitzin (1770–1840),[224] whereas Johann Jacob Bindheim (1743–1822) and others analyzed samples of beryl. The mineralogist René-Just Haüy was struck by the geometries of the two gemstones, which were very similar to each other; in 1798, Haüy asked the renowned Vauquelin to analyze both emerald and beryl. At the conclusion of his analysis, Vauquelin reported that the two gemstones were identical, with the exception of some traces of chromium[225] in emerald, but that they contained a new element. He read his report before the Academy of Sciences of Paris on February 15, 1798.[226] At the suggestion of the editor of *Annales de Chimie et de Physique*, he called the new earth, present in both gemstones, *glucine*, because of the sweet taste of its salts.[227] The name was taken from the Greek γλυκυς, meaning "sweet." At first, Vauquelin seemed reluctant to use this name, but he yielded under pressure from colleagues and proposed the symbol Gl for the new element. However, Klaproth noted that the name was too similar to *glycine*, an amino acid, and that it might create confusion between the two terms. Still others observed that some salts of yttrium also had a sweet taste and therefore the name *glucine* would not be completely suitable. Vauquelin suffered both from the criticism of his colleagues and for the ambiguity of the name *glucine*, especially because he had not chosen it of his own volition.

Just 10 years after the discovery of *glucine*, Sir Humphry Davy tried to isolate it in the elemental state. At the same time he was working on isolating aluminum (from alumina), silicon (from silica), and zirconium (from zirconia). The experiments were not going well, as he himself reported during a session of the Royal Society of London on June 30, 1808, but, nevertheless, he suggested a name for these elements:[228] "Had I been so fortunate as to have obtained more certain evidences on this subject, and to have procured the metallic substances I was in search of, I should have proposed for them the names of silicium, alumium, zirconium, and glucium." With the term *glucium* (later *glucinium*, for euphony),

Davy had only changed the spelling of the name proposed by Vauquelin and fully recognized the latter's priority of discovery. Klaproth, however, was of a different opinion and openly supported the name of *beryllia* for the oxide of the new element. He was in a group of those who hotly contested the fact that the salts of *glucium* were not the only ones that had a sweet taste. He proposed the name *beryllia*, from the Greek βηρυλλος, which ultimately came from the name of the mineral, beryl.

In 1828, the year before Vauquelin died, Antoine-Alexandre-Brutus Bussy[229] and Friedrich Wöhler,[230] independently of each other, isolated the first samples of elemental beryllium, but the controversy over the two names, *glucinium* and beryllium, as well as the denial of the existence of *treenium* and *donium*, had not yet arrived at its final chapter.

II.11.4. THE SWEET EPILOGUE LEAVES A BITTER TASTE IN THE MOUTH

In 1829, two great figures involved with the discovery and first attempts to isolate *glucinium* died: Sir Humphry Davy, whose health was undermined by years of inhaling toxic chemical substances, died in Geneva following respiratory failure. Nicolas-Louis Vauquelin survived him by a little less than 6 months, dying at the age of 66 the following November.

We must skip ahead to 1870 to put an end to the question of *treenium* and *donium*. Henry Carrington Bolton, in the pages of *Chemical News*, asserted that without a shadow of a doubt *donium* had nothing to do with beryllium. Furthermore, he went on to state how Richardson had actually erred: he had mistaken a mixture of iron-bearing and aluminum oxides for an elementary substance.[231]

For its unfortunate last chapter, *glucinium* had to wait another 80 years. In September 1949, in Amsterdam, during the 15th conference of the International Union of Pure and Applied Chemistry (IUPAC), the Commission on Inorganic Nomenclature met and with brutal pragmatism decreed the end of *glucinium* in favor of the more widespread name beryllium.[232] Presently, "beryllium" is in universal use except in the French scientific literature where, with a bit of chauvinism, the word *glucinium* and the symbol Gl are still used.

Notes

218. Manutchehr-Danai, M. *Dictionary of Gems and Gemology*: 2nd ed.; Springer: Heidelberg, Germany, 2005.
219. Richardson, T. *Record of General Science* **1836**, *3*, 426.
220. Graesse, J. G. Th. *Orbis Latinus: Lexikon lateinischer geographischer Namen des Mittelalters und der Neuzeit;* Richard Carl Schmidt & Co.: Berlin, Germany, 1909; a standard reference to Latin place names, with their modern equivalents (re-edited and expanded in 1972).
221. Boase, S. H. *Record of General Science* **1836**, *4*, 20.
222. Turner, E. *Elements of Chemistry: Including the Recent Discoveries and Doctrines*, 5th ed.; Desilver, Thomas and Co.: Philadelphia, PA, 1828.
223. Louis Nicolas Vauquelin was born on May 16, 1763, in Saint-André-d'Hébertot in Normandy. His first contact with chemistry came about when he was a laboratory assistant to a pharmacist in Rouen. He came into contact with Antoine-Francois de Fourcroy (1755–1809), and worked as his assistant from 1783 until 1791. He carried out numerous research projects, and during

his entire career he published 376 papers and discovered two elements, chromium in 1797 and beryllium in 1798.

224. Prince Gallitzin was the son of the Russian ambassador to the Netherlands. At the age of 17, he converted to Catholicism and donated most of his substantial fortune to the poor. He was ordained to the priesthood and served as a missionary in North America, working mainly in western Pennsylvania and Maryland. In 2005, his cause for beatification was entered to the Holy See.
225. A totally extraordinary fact: chromium was discovered the year before by Vauquelin himself.
226. By the calendar adopted after the French Revolution, the date is 26th day of the month of *Pluviôse* in the sixth year
227. Vauquelin, L. N. *Ann. Chim. Phys.* **1798**, *26*(1), 155.
228. Davy, H. *Phil. Mag.* **1808**, 333.
229. Bussy, A. A. B. *J. Chim. Medicale* **1828**, *4*, 453; *Dinglers Polytechn. J.* **1828**, *29*, 466.
230. Wöhler, F. *Ann. Chim. Phys.* **1828**, *39*, 77.
231. Bolton, H. C. *Chem. News* **1870**, 369.
232. Olander, A. *Svensk Kem. Tid.* **1949**, *61*, 275.

II.12

THE BRIEF PARENTHESES OF FOUR MISLEADING ELEMENTS

II.12.1. THE FLEETING EXISTENCE OF *THALIUM*

David Dale Owen (1807–60) was a distinguished American geologist. He was the third son of Robert Owen (1771–1851), a Welsh reformer who moved to the United States to accomplish his "social experiment" by creating the community of New Harmony in the state of Indiana. Owen lived in his father's community for about 30 years, and it was during that time that his interest in geology was awakened by the visits of geologist William Maclure (1763–1840) to his father. In 1836, he completed his first work, a geological survey map of Tennessee. Over the next 20 years, he was successively state geologist for Indiana, Kentucky, and Arkansas, but passed the last year of his life back in his home state of Indiana.[233]

In 1852, while conducting a geological survey of the states of Wisconsin, Iowa, and Minnesota, Owen discovered an amorphous mineral with a whitish cast and a consistency like wax on the northern shore of Lake Superior. Owen decided to call it thalite because he thought he recognized the presence of a new "radical." He was able to extract the oxide of the new substance, which he called *thalia*, but it was impossible to isolate the element in the pure state. Nevertheless, Owen proposed to call the element *thalium*,[234] fishing the name out of classical mythology.[235] He believed that the new substance was an alkaline earth metal with properties intermediate between magnesium and manganese. *Thalium*'s oxide had a pale green color and dissolved easily in hydrochloric acid. However, no other tests were made in support of the existence of this hypothetical metal.

Parenthetically, shortly following Owen's death in 1860, Sir William Crookes discovered a metal that he named thallium. The two names were never superimposed on one another. The first—*thalium*—was shown to be false and disappeared from the list of elements before the discovery of the second. However, the extraordinary similarity of the two names, *thalium* and thallium, could fool the casual reader. Thallium's etymology is derived from the Greek, θαλλος, meaning "green shoot." Figure II.05 is an image of William Crookes, a figure we will encounter repeatedly in this volume.

Later, both the mineralogist Frederick August Genth[236] and the chemists J. Lawrence Smith (1818–83) and George Jarvis Brush (1831–1912)[237] proved the inconsistency of the discovery of *thalium* and, in the end, Owen was forced to admit his error publicly.[238] The oxide of *thalium* was shown to be a complex mixture of lime and magnesia with traces of silica.

David Dale Owen's interests were not exclusively in geology. He helped his brother Dale Robert Owen (1801–77) establish the Smithsonian Institution in Washington, DC. A few years after the *thalium* affair, he died on November 13, 1860, at 53 years of age. And, although the U.S. scientific community has recently loaded his name with honors,

FIGURE II.05. Sir William Crookes (1832–1919) discoverer of thallium in 1862. In his later years, he began to show clear indications of scientific heterodoxy. A spiritist, he spoke of the inorganic evolution of the elements, proposing the concept of the *meta-element*; he asserted that he had identified the rare earths *monium* (or *victorium*), *jonium*, and *incognitum*. Courtesy, Fisher Collection, Chemical Heritage Foundation Archives.

identifying him as a pioneer of the geological sciences in America, it should be remembered that, at the time of the alleged discovery of *thalium*, some of his countrymen considered this American geologist a subject of Her British Majesty.[239]

With the passing of the decades, scientists have come to the realization that *thalite* was a variety of saponite[240] with a high level of aluminum, already known at the time.

II.12.2. THE METEORIC APPEARANCE AND DISAPPEARANCE OF COMESIUM

Notice of the discovery of a new element—*comesium*—was reported for the first time on April 14, 1880, at the annual meeting of the Naturhistorische Gesellschaft (Natural History Society) of Nuremberg. A Professor Speiss presented a memo from Doctor Hermann Kämmerer, a professor at the local Industrial School, in which he told of the discovery of a new metal with marked magnetic properties.

In 1870, Thomas Leykauf had been hired as Royal Professor of Chemical Technology at the Industrial School of Nuremberg, which was later transformed into Nuremberg Polytechnic Institute. He took on Kämmerer as professor of chemistry and mineralogy because he came highly recommended by Justus von Liebig (1803–73). When Leykauf died a year later, Kämmerer became his natural successor as director of the Technico-Industrial Department.

Kämmerer was born on April 7, 1840, at Mutterstadt, Germany, and had begun to study under Leykauf at Nuremberg. He went on to Leipzig, then Heidelberg, where he worked under the guidance of the renowned Robert Bunsen (1811–99) and later was a collaborator of Georg Ludwig Carius (1829–75). He then became Liebig's assistant at Munich in Bavaria.

Kämmerer gave the name *comesium*[241] to this magnetic substance. Details concerning the discovery were given during his oral presentation, and the discovery was reported uncritically by some journals,[242] but this erroneous announcement was soon forgotten by the scientific community. Hermann Kämmerer served continuously as director of the Industrial School of Nuremberg until his death on April 10, 1898.

II.12.3. THE MYSTERIOUS NATURE OF OURALIUM

The platiniferous minerals have always confused even the most expert chemists. Such was the case with Parisian chemist Antony Guyard (d. 1884). In 1869, he was analyzing commercial platinum coming from deposits in the Ural Mountains of Russia when he stumbled onto something that for all intents and purposes could have been a new metal. Ten years passed, and finally Guyard decided to publish news of his discovery. Guyard seemed able to obtain the substance in the elemental state: he had a considerable quantity of platinum at his disposition—2 kg—and he managed to get from it almost 2 g of the unknown metal.

The properties of the new metal were astounding: for brightness, it was second only to silver; its ductility and malleability could not be compared to the other noble metals, and they reminded one of lead. The melting point was similar to that of platinum, its specific gravity was calculated as 20.25, and its "molecular volume" was similar to osmium, palladium, and platinum.

Guyard was able to determine the atomic weight of the metal accurately and precisely, obtaining a value of 187.25. Some scholars continue to advance the hypothesis that Guyard could have discovered rhenium.[243] In fact, rhenium is found in nature associated with platinum, but the biggest puzzle is the two atomic weights: the one determined by Guyard and that of rhenium, which is 186.207. The only point that serves to discredit the effective discovery of rhenium are its chemical properties that, according to Guyard, were very much similar to those of platinum.

In his 1879 article, Guyard decided to name his new metal *ouralium* to memorialize its origin in minerals from the Russian Urals.[244] He also proposed its symbol, Ou. Unfortunately, although Guyard was able to present some characteristic reactions that led to favoring the existence of the new metal, he was not able to produce a complete spectroscopic examination by which he would have been able to clarify its nature.

The 2 g of *ouralium* that Guyard had obtained by electrolytic deposition on a copper plate from a solution of $OuCl_2$ and caustic potash were lost and with them went Guyard's hope of the credit for discovering rhenium. Later, when the news was reported in the British and American scientific literature, the name *ouralium* was changed to *uralium*. In the second half of the 20th century, a second metamorphosis of the name occurred: it was for good and all mangled into *oudalium* due to an erroneous transliteration from Russian into English of the book *Chemical Elements: How They Were Discovered* by the Trifonov brothers.[245]

II.12.4. THE BRIEF HISTORY OF IDUNIUM

The story of *idunium* creates in one a certain sense of disappointment because it suggests a lack of clarity of ideas inappropriate for a man of science at the end of the 19th century, especially for a university professor at one of the most prestigious academies in Europe. This professor's name was Friedrich Martin Websky (1824–86).

Websky was born on July 17, 1824, at Nieder Wuestegiersdorf, presently Głuszyca, in Poland. After he obtained his *matura* at Berlin in 1843, he returned to the mountainous region of Silesia to study mineralogy. From 1846 on, he pursued his studies at the Mineralogical Academy of Berlin. After having received his degree in 1853, Websky accepted a position as teacher at the local school of mineralogy in the mountainous mining district of the Tarnuv. In the following years, he became famous for his research and discovery of new minerals. In 1861, he moved to the local school of mineralogy at Breslau (present-day Wrocław). In 1865, he left his applied work once and for all and devoted himself entirely to research. As a matter of fact, the University of Wrocław had conferred on him the title of professor, *honoris causa*; a little later, he began to hold regular lectures in mineralogy there. In the environs of the university, he gathered a large collection of minerals, in part made up by his own personal samples. In 1868, he was made associate professor, and 6 years later, he accepted the chair of full professor in Berlin, left vacant after the death of Gustav Rose (1798–1873).

In 1884, at the age of 60, Websky found that he had a mysterious mineral on his hands, a very special gift that Professor Ludwig Brackebusch (1849–1906) had brought to Europe with him after a visit to the mine of Aquadita in the area of the Plata, Argentina. Websky analyzed it and, with great astonishment, found himself making the announcement of his lifetime.[246] If verified, his discovery would have been his epitaph; unfortunately, his discovery would be shown to be false.

The mineral was principally composed of zirconium-bearing lead vanadate. After initial processes to separate the metals present, Websky obtained vanadic acid, which he treated with ammonium chloride to precipitate as ammonium vanadate all the vanadium present in the samples. At the end of this operation, he realized that there was an unknown substance present in the mother liquor, and he guessed that this might be the acid of a new element. He quickly gave it a name: "Der neuen Körper, dem ich den Namen Idunium beilegen möchte."[247] The symbol proposed for the new metal was Id, and the new material was shown to have properties strangely similar to vanadium.

These repetitive likenesses, with the already known transition metals, should have been a warning, but Websky did not seem to be worried about this and briefly outlined the procedure he used to extract the new substance. On addition of ammonium sulfide to the mother liquor—rich, he supposed, in "*idunic* acid"—Websky collected a red precipitate that he thought was *idunium* oxide.

The particulars of the discovery were reported in England[248] and in the following year in France[249] and Italy.[250] On November 27, 1886, Martin Websky died at Berlin at the age of 62. The shaky discovery of *idunium*, no longer supported by the charisma of its discoverer, was easily incorporated into the circle of nonexistent discoveries.

Notes

233. Kimberling, C. *Indiana Magazine of History* **1996**, *92*, 2.

234. Owen, D. D. *Silliman's Amer. Jour.* **1852**, *13*, 4.
235. Thalia (Good Cheer), together with Aglaia (Splendor) and Euphrosyne (Mirth) were the Three Graces of Greek mythology. Thalia was considered the Muse of Poetry and also of Comedy. She was also the Grace associated with prosperity.
236. Genth, F. A. *J. prakt. Chem.* **1853**, *61, 8*, 378.
237. Smith, J. L.; Brush, G. J. *Silliman's Amer. Journ.* **1853**, *16, 48*, 365.
238. Hendrickson, W. B. *David Dale Owen: Pioneer Geologist of the Middle West*; Indiana Historical Bureau: Indianapolis, IN, 1943.
239. Anon. *Saint Louis Medical and Surgical Journal* **1852**, 288.
240. The mineral saponite has the formula: $(Ca/2,Na)_{0.3}(Mg,Fe^{2+})_3[(Si,Al)_4O_{10}](OH)_2 \cdot nH_2O$.
241. Anon. *Chem. Ztg.* **1880**, *17*, 273.
242. Wagner, J. R.; Fischer, F.; Schmidt, P. F.; Rassow, B.; Gottschalk, F. *Jahresbericht über die Leistungen der chemischen Technologie* **1880**, *26*, 4.
243. Powell, A. R. *Platinum Met. Rev.* **1960**, *4*(4), 144.
244. Guyard, A. *Monit. Scientif.* **1879**, *21*, 795; Anon. *Chem. News* **1879**, *40*, 57.
245. Trifonov, D. N.; Trifonov, V. D. *Chemical Elements: How They Were Discovered*; MIR Publishers: Moscow, 1982.
246. Websky, M. *Sitzungsberichte Berliner Akademie* **1884**, 331; Websky, M. *Sitzungs-berichte der Koeniglich Preussischen Akademie der Wissenschaften zu Berlin,* **1884**, 661.
247. "The new body, to which I would like to attach the name Idunium."
248. Websky, M. *Phil. Mag.* **1884**, *18* (series V), 232.
249. Websky, M. *Bulletin de la Société de Minéralogie et de cristallographie, Société chimique de France* **1885**, *8*, 657.
250. Anon. *Rendiconti della Reale Accademia dei Lincei* **1885**, *vol. 1 parte 1*, 34.

II.13

TWO IMAGINARY ELEMENTS: SULPHURIUM AND SULFENIUM

II.13.1. SULPHURIUM

Sulfur was known from antiquity, so much so that this element has been mentioned in the book of Genesis. The word sulfur almost certainly comes from the Arabic, *sufra*, which means "yellow" (the Sanskrit is *sulvere* and the Latin *sulfur*). Alchemists gave to sulfur a peculiar symbol: a cross surmounted by a triangle. Through their experiments, they found that mercury could be combined with it. Sulfur was not considered a simple body until the very end of the 18th century, when Antoine Lavoisier convinced the scientific community that sulfur was an element and not a compound. As soon as sulfur was recognized as an elementary substance, it became the object of attention by the nascent sulfuric acid industry.

Sulfur extraction began in Sicily at the beginning of the 17th century and developed rapidly. In 1820, it reached an annual production rate of 378,000 tons of raw material, equivalent to four-fifths of the world market. With the development of industrial production, in 1834, a census estimated that more than 200 sulfur mines were in operation, the product of which was being shipped by sea to Europe and the United States. England, the superpower of that time, required increasing amounts of sulfuric acid for its industries. The demand for raw materials, among which was sulfur, led mining and chemical engineers to look for new deposits all over the world.

In 1857, Joseph Jones of Bolton-le-Moors, located in Lancashire in the northern part of England, believed he had discovered a new metallic body in residues of sulfuric acid manufacture. The characteristics of the metal resembled arsenic, silver, and aluminum.[251] The news spread like wildfire and was reported in many local periodicals.[252]

Jones's interests were immersed more in the practical than in the classical, which led to his naming the new substance *sulphurium* without using any fanciful name or dabbling in linguistic elements. According to his belief, *sulphurium* oxide was nothing more than the industrial waste of brimstone manufacturing,[253] and it was without commercial value.

Because Jones had marked his discovery as commercially worthless, with some regret, he ended up paying workers by the wheelbarrow load to discard it. Further inquiries led Jones to publish some properties of *sulphurium*: it had the density of iron and the ductility and malleability typical of metals.

Two years later, another English chemist, deeply interested in the peculiar properties of *sulphurium*, tried to obtain a modest amount of it. It was hard to get and, eventually, although he had followed the instructions published by Jones, he failed.[254] He stated that the material he analyzed was a mass of already known elements, combined variously with each other. In the series of letters that followed, the author claimed that perhaps he gave short shrift to Jones, who may indeed have discovered what we now call thallium, but had not the means, prestige, nor interest to pursue it further.

Thallium was discovered, as noted in section II.12.1, in connection with the false discovery of a similarly named element, *thalium*. In 1861, the discoverer of thallium, Sir William Crookes, was performing spectroscopic measurements to search for traces of tellurium among residues of a sulfuric acid manufacturing facility. His work uncovered strong evidence of the existence of a new substance: a new spectral emission line of an intense green color. However, in 1862, the French chemist Claude-Auguste Lamy (1820–78) isolated thallium metal in macroscopic quantities, provided an ingot, and claimed in this way the priority of his discovery.

Finally, in 1863, a brief article appeared in the pages of a lesser known specialty magazine[255] reporting that Joseph Jones had discovered thallium in June 1857, and he had called it *sulphurium*. This toxic metal was isolated from the lead chambers used for sulfuric acid production.

According to some contemporaries, the error attributable to Mr. Jones, and which would have compromised his credibility, was to have believed that *sulphurium* was a "metallic base of sulphur," that is, a constituent of sulfur, thus admitting by implication that it could not be an element.

This defensive note on Jones's discovery went unnoticed, and the controversy over priority of the discovery of thallium remained confined between Crookes and Lamy, both personalities of great authority and prestige, highly esteemed by their peers: the former would become president of the Royal Society of London, the latter would later lead the Société Française de Chimie.

The interlude of the *sulphurium* affair did not stop the spasmodic search for new sulfur deposits to satisfy the increasing demand of industrialized nations. In 1867, vast underground deposits were discovered in Louisiana and Texas. However, the superficial layer of soil formed by shifting sands prevented mining by traditional methods. An entirely new procedure, called the Frasch Process, was therefore developed. This method allowed the mineral to be extracted from deep layers by means of the injection of superheated water into the subsoil. With this high-yield method, American sulfur, purer than the Sicilian variety, soon conquered the world markets.

II.13.2. THE ANCIENT MODERNITY OF SULFENIUM

About a century after the announcement of the discovery of *sulphurium*, a brash French research scientist came into the limelight of the "chemical stage." He stated that he had isolated a new element, *sulfénium*, belonging to the sixth group in the periodic table. His name was P. J. Marcel Duchaine, a name that certainly doesn't mean a thing to the majority of scholars today. Nothing joins these two events, so different in time and content, but the name of Duchaine's alleged discovery, which recalls the story of *sulphurium*.

Marcel Duchaine's ambit was a special field of research: he was interested in manufacturing artificial diamonds. This area had a long history, and it had an extraordinary "godfather" at the end of the 19th century: the 1906 Nobel Prize winner for Chemistry,[256] Henri Moissan, who, in 1893, was among the first to claim the production of synthetic diamonds, although today there is some doubt about his success (he may have mistaken carbide grit for his crude diamond sand).[257]

In 1963, Duchaine began to produce blue diamonds by reacting a mixture of CoF_2, $CoCl_2$, and NH_4F with diamonds and heating all the compounds in a furnace at temperatures of nearly 1,200 °C. This diamond-staining procedure was patented.[258] Five years later, he was able to patent his method[259] for the manufacture of "carbide diamonds" produced from an excess of carbon mixed with 1 part V_2O_5 and 5–6 parts of CaF_2.

After these initial successes, feeling ready to try his wings, in 1972, Duchaine set out to produce true artificial diamonds.[260] He learned that this synthetic strategy would require, based on an isolated case in the literature, the use of sulfides.

He worked out many possible synthetic strategies in order to identify the best one: pure carbon was mixed with, alternatively boron, beryllium, zirconium, hafnium, and silicon sulfides. The material was heated for several hours at 1,200 °C in a crucible of graphite, surmounted by a very special borosilicate glass made of rhenium and hafnium. After exhausting and expensive research, Duchaine claimed that he was able to produce some small diamonds. That same year, Duchaine applied for a patent for the synthesis of diamonds through the use of metal sulfides as catalysts.[261]

Meanwhile, Duchaine asserted that he had run into the unexpected discovery of a new element. When he applied for the patent,[262] he noted a new metalloid element, hitherto unknown, belonging to the sulfur group. He named it *sulfénium*, claiming all of its uses and applications in chemistry, metallurgy, and therapeutics.

His text referred to the study of diamond composition, to the determination of what Duchaine thought to be the "véritables éléments constituants ce minérals"[263] and their use in scientific and industrial production. Moreover, he suggested further technical applications for his discovery, including therapeutic use.

Duchaine performed a detailed study of the specific characteristics of the "new" element, but he soon stopped because of lack of enough diamonds. With so little available material, Duchaine could only note that diamond combined with pure iron produced small crystalline "geodes," completely different from either cast iron, or iron, selenium, and tellurium sulfide. Pure iron associated to diamond could be dissolved by mineral acids, liberating a gaseous compound containing *sulfénium*. Its characteristic reactions were similar to those of hydrogen sulfide, although it did not react with a solution of sodium nitroprussiate.

Duchaine was worried about the spectrum of this mysterious gas, which remained unexplained, although some chemical tests confirmed the possibility of the presence of H_2Se and H_2Te. Analysis did not show, in the initial diamond material, either the presence of selenium or tellurium.

Regardless of the fact that he was able to provide only scant evidence for his discovery, Duchaine went on to list possible commercial applications of this alleged metalloid. At the end of his patent application, he pointed out a possible pharmacologic use of *sulfénium*, either in its elemental state or in combination with organic compounds. Without having the faintest experimental clue, he predicted the use of *sulfénium* for treatment of bacterial infections and even of cancer, saying only that the healing properties of the new metalloid would be the subject for his next patent, which never saw the light of day.

Duchaine very probably came to his conclusions in perfectly good faith, but he mistook a mixture of sulfur and tellurium for a new element.[264]

Notes

251. Jones, J. *Mining J.* **1857**, *27*, July 14.

252. Anon. *Dublin Hospital Gazette* **1858**, 208; Anon. *Annual of Scientific Discovery: Or, Year-book of Facts in Science and Art* **1858**, 271.

253. The name "brimstone" comes from the fact that sulfur was often found as a yellow deposit around the brims of volcanic craters. Many publications of the 1800s often exchanged one term for the other.
254. Answers to Correspondents *Chem. News* **1863**, *7,* 263.
255. Anon. *The Mining and Smelting Magazine* **1863**, *3*, 359; Anon. *The American Chemist* **1870**, *1*, 1.
256. Moissan, H. *Le four éléctrique;* G. Steinheil: Paris, 1897.
257. Marshall, J. L.; Marshall, V. R. *The Hexagon* **2006** (Fall), 42.
258. Duchaine, M. P. J. *French Demande,* (February 1, 1963), 9 pp., FR 1316489.
259. Duchaine, M. P. J. *French Demande* (September 20, 1968) 2 pp., CODEN: FRXXAK FR 1539715.
260. Duchaine, M. P. J. *French Demande* (March 10, 1972) 4 pp., CODEN: FRXXAK FR 2094869.
261. Duchaine, M. P. J. *French Demande* (January 23, 1976) 10 pp., CODEN: FRXXBL FR 2276264.
262. Duchaine, M. P. J. *French Demande* (May 4, 1973) 4 pp., CODEN: FRXXBL FR 2149300.
263. "True elements constituting minerals."
264. Fontani, M.; Orna, M. V. *J. Sulfur Chem.* **2013**, *34*, 711–15.

II.14

THE ASTRONOMER "LEFT IN THE DARK"

Those who think that the history of the false discovery of chemical elements can be disposed of in a few pages or treated like a simple list of blunders by incompetent or amateurish scientists might be mistaken themselves. Because false discoveries are quickly forgotten, the provisional names given to the alleged elements can be "recycled" for other discoveries, ones that, in their own turn, can be true or false. The most striking of these cases was that of the "multiple" discovery of neptunium, which, before receiving its official "seal of approval" from Edwin M. McMillan (1907–91) and Philip H. Abelson (1913–2004) in the 1940s,[265] had to undergo the humiliation of three false claims to its discovery: the first[266] in 1858, the second[267] in 1877, and the final around a decade later.[268]

II.14.1. "LIGHT" AS A MEANS OF CHEMICAL INVESTIGATION

Between the end of 1859 and the beginning of 1860, in Heidelberg, the chemist Robert Bunsen and the physicist Gustav Kirchhoff put into practice a method of qualitative spectrochemical analysis, at the same time practical and effective, for chemical element research.[269] They had arrived at the conclusion, after long and precise studies, that a spectral line was an unambiguous characteristic for a specific element.[270]

Thus it was that light, conveyed by an emission or absorption spectrum, became a valid and irreplaceable investigative instrument in the hands of the chemical community. Not only did spectral analysis greatly simplify laboratory work, but it was also more sensitive than any wet method of chemical analysis and required a smaller amount of matter to examine.

Sir William Crookes sensed that this invention would have a great future in mineral analysis or in the search for new elements. Referring explicitly to Bunsen's work, he wrote: "With so delicate a reaction as the one just described, of an almost infinite sensibility, and applicable to all metals, the presence of elements, existing in so small quantities as to entirely escape ordinary analysis, may be rendered visible."[271]

Crookes had already shown interest in searching for yet-unknown elements. In fact, a couple of years earlier, he had engaged in copious correspondence with the astronomer John Herschel (1792–1871) when the latter announced that he had discovered an entire family of "photochemical" elements.[272]

The part of the inaugural discourse that John Herschel held at Leeds before the British Association for the Advancement of Science having to do with the discovery of five new elements was reported in its entirety in the pages of *Photographic News*.[273] Crookes was so interested in Herschel's work that he wrote to him to make himself properly conversant regarding his research. Crookes was honored by Herschel's kindness and did not lose the opportunity to glorify the astronomer's discoveries. The entire series of correspondence was published in the pages of *Photographic News* (letters dated October 8, 13, 15, 23, 1858).

II.14.2. A NEW FAMILY OF ELEMENTS FROM AN OLD FAMILY OF ASTRONOMERS

The Englishman John Frederick William Herschel, born at Slough on March 7, 1792, was an astronomer, chemist, and mathematician. He was the first astronomer to use the Julian Calendar in astronomy, and he made important contributions to improving the photographic processing of the period (daguerreotyping and calotyping) by discovering the properties of sodium thiosulfate for the fixing of images. In addition, he coined the terms *photography, negative*, and *positive*.

John was the son of William Herschel (1738–1822) and nephew of Caroline Lucretia Herschel (1750–1848), both famous astronomers. Although he had initially begun a career in law, he later devoted himself to astronomy and, when his father retired, he took upon himself the direction of the Royal Astronomical Observatory. He discovered that the Magellanic Clouds are formed from stars, and he published various star catalogues. In 1831, he was raised to the title of Knight of the Royal Guelphic Order. In 1848, he was named president of the Royal Astronomical Society and, in 1850, director of Her Majesty's Mint.

On October 29, 1858, William Crookes published the entire correspondence that he had carried on with Herschel.[274] The part that fascinated him most was the announcement of the discovery of five new elements and, in particular, one that later took the fleeting name of *junonium*.

The reason Herschel thought he had discovered so many new elements was due to the fact that he had prepared a certain number of light-sensitive films that, if exposed to the rays of the sun, produced five distinct reactions never observed before. These unusual phenomena, joined with a certain amount of ingenuousness, made him think that there were five new elements deposited on the films. He went a bit too far in hypothesizing that these made up a new class of "photochemical" elements that he named *junonium, vestium, neptunium, astaeum*, and *hebeium*. But Herschel went far beyond even this: he sent Crookes a sample of paper impregnated with a solution of sodium *junoniate*, inviting him to compare the distinct behaviors of this compound, when exposed to light, with respect to a sheet of paper impregnated with potassium iodide, potassium bromide, silver nitrate, or silver arsenate. Both Crookes and Herschel confronted the chemical side of the problem—to isolate *junonium, vestium, neptunium, astaeum*, and *hebeium*—but they were not successful. Given the complexity of the material treated and the lack of chemical information obtainable from Herschel's writings, it is not easy to determine what already known element (or mixture of already known elements) could have misled him. However, a year after his sensational announcement, John Herschel no longer upheld his discovery, writing[275] "Junonium (if it be really a distinct body) equals bromine in [its spectrum]." Herschel must have realized very quickly that the entire family of "photochemical" elements was at risk. With the passing of the months, the new elements remained elusive and intangible. Perhaps he himself had ceased to believe in their existence, but he never openly asserted that the five might be mixtures of already known elements. If Herschel were not able to defend his entire discovery, he had at least decided to defend the existence of one of these hypothetical elements to the bitter end. Contrary to every expectation, over the coming years, he did not abandon his fruitless attempts to isolate *junonium*.

II.14.3. NEPTUNIUM IS TEMPTING TO A LOT OF PEOPLE

Shortly after the discovery of *neptunium* by R. Hermann in 1877, another German chemist, the renowned Clemens A. Winkler (1838–1904), found himself in the situation of having to choose a name for a new metal. In 1886, Winkler was working at the Freiberg University of Mining and Technology. After a careful examination of the newly discovered mineral argyrodite ($GeS_2 \cdot 4Ag_2S$), Winkler was able to isolate a new element in the elemental state.[276] Initially, he wanted to call it *neptunium*, after the planet discovered in 1846 by astronomer Johann Gottfried Galle (1812–1910), but he realized that this name had already been bestowed on another element that later was shown to be false. He immediately rejected this name for fear of creating confusion, and he called his element germanium in honor of his native country.[277] He had hardly announced his choice when people began to murmur. For some, germanium sounded more like the flower *geranium* rather than the name "Germany." The protracted discussion caused Dmitri Mendeleev to write to Winkler encouraging him to use the name "germanium." But there was also some confusion as to where to place germanium in the periodic table. At first, Winkler thought it should go between bismuth and antimony, whereas Mendeleev thought, erroneously, that it was eka-cadmium. Not much time passed before two well-known German chemists, Theodor Hieronymus Richter (1824–98) and Julius Lothar Meyer correctly identified the new substance with Mendeleev's eka-silicon.[278] In the end, Winkler realized his own error and correctly positioned germanium where it belonged in the periodic table.[279]

Clemens Winkler was professor of chemical technology and analytical chemistry at the Freiberg University of Mining and Technology, which had its origins as a school of mines, so his interests were more centered on the technical aspects of chemistry rather than on pure research. One of his articles that caused a stir among chemists at the time for its great accuracy and wealth of details was not the one on germanium, but on the industrial production of sulfuric acid. In it, Winkler determined the stoichiometric mixture of oxygen and sulfur dioxide that led to the highest yield. The patent that followed was very advantageous to German industry; it, and numerous other technical advances made at this time, allowed Germany to evolve from the handicraft level to that of a great industrial power right up until the outbreak of World War I. Winkler, at the height of his fame by the end of 1902, retired from teaching on account of poor health. He lived another 2 years and died at Dresden on October 8, 1904, of a carcinoma.

II.14.4. CONCLUSION

John W. F. Herschel did not limit his explorations to the burgeoning science of photochemistry. He made noteworthy contributions to the fields of mathematics and epistemology. If he were not the first, he was certainly among the first to distinguish in a clear and sensible way between natural laws and general theories—a set pattern that tied together the laws of physics and the laws of chemistry. In addition, having taken on an important role in the creation of the idea of hypotheses, he also spoke of false theories and the need of the scientist and the researcher to place on the table all possible objections and to record meticulously all facts that might disprove a theory.

The passion to name new objects or to rename other people's discoveries must have been very strong in John Herschel: in fact, in addition to introducing the word "photography" into the language, he gave names to the seven moons of Saturn and to the four moons of Uranus, the planet discovered by his illustrious father. In the last years of his long life, he became a father for the 12th time. Of all his sons only his third, William James Herschel, (1833–1917), followed in his footsteps by choosing the field of astronomy. At the time of his death, on May 11, 1871, Herschel, although the son of a German astronomer, was regarded in England as one of its most prestigious polymaths. He was given a state funeral and was buried in Westminster Abbey among other illustrious personages of the nation.

Notes

265. McMillan, E.; Abelson, P. *Phys. Rev.* **1940**, *57*, 1185.
266. James, F. A. *Notes Rec. R. Soc. Lond.* **1984**, *39*(1), 65.
267. Hermann, R. *Chem. News* **1877**, *35*, 197.
268. Voronkov, M. G.; Abzaeva, K. A. *The Chemistry of Organic Germanium, Tin and Lead Compounds*; Rappoport, Z., Ed.; John Wiley and Sons: London and New York, 2002; vol. 2, ch. 1 & 3.
269. Bunsen, R. *Chim. Pharm.* **1859**, *3*, 257; *Phil. Mag.* **1860**, *18*, 513.
270. For a brief account of the impact of spectroscopy on chemistry, please see Orna, M. V. *The Chemical History of Color*; Springer Verlag: New York and Heidelberg, Germany, 2013; pp. 99–105.
271. Crookes, W. *Chem. News* **1860**, *2*, 281.
272. Herschel, J. F. W. *British Association for the Advancement of Science Reprints. Part 2*, **1858**, 41.
273. Crookes, W. *Photographic News* **1858**, *1*(4), 49.
274. Crookes, W. *Photographic News* **1858**, *1*(8), 86.
275. Herschel, J. *Photographic News* **1859**, *2*(46), 230.
276. Winkler, C. A. *J. prakt. Chem.* **1886**, *34*, 182; Winkler, C. A. *Ber. Dtsch. Chem. Ges.* **1886**, *19*, 210.
277. Figurovskii, N. A. *Discovery of the Elements and the Origin of their Names*; Science Ed.: Moscow, Russia, 1970, p. 65; Trifonov, D. N.; Trifonov, V. D. *How the Chemical Elements Were Discovered*; MIR Publishers: Moscow, Russia, 1982, p. 134; Moore, F. J. *A History of Chemistry*; McGraw-Hill: New York, 1939, p. 447.
278. Mendeleev, D. I. *Zh. Rus. Fiz. Khim. Obshch.* **1886**, *18*, 179.
279. Winkler, C. A. *J. prakt. Chem.* **1887**, *36*, 177.

II.15

BYTHIUM AND δ: TWO ELEMENTS THAT AROSE (AND VANISHED) VIA ELECTROLYSIS

In 1897, a new element was announced in *Electrochemische Zeitschrift* by Theodor Gross.[280] A fused mixture of silver sulfide and silver chloride was electrolyzed in a nitrogen atmosphere, using iridium-free platinum electrodes at currents between 3 and 10 amperes. In the melt, a dark gray powder, insoluble in aqua regia and ammonia, was found. Fused with alkaline carbonate, the mix gave a melt soluble in hydrochloric acid, which produced a brown precipitate when treated with hydrogen sulfide. The yield of the new substance was 5% of the original sulfur used. From the fact of this corresponding loss of sulfur, Gross thought that this new elementary body was formed by the decomposition of sulfur. Soon after, he admitted that there was also a small loss of chlorine (3%) in the electrolytic reaction. He suggested that the newly discovered element, which he called *bythium*, could also be formed by the decomposition of chlorine.

In a second experiment, he fused together ferrous sulfate and potassium chlorate and claimed to have obtained a new substance, *δ*, which had many of the properties of silicon. Although insisting that he had found a new element, for some unexplained reason, Gross did not see fit to prepare this substance on a large scale. He could not decide if this substance *δ* was the same as the *bythium* he had previously isolated.[281]

Gross did not publish the atomic weight of *bythium* for priority reasons, but an impending verdict soon demolished any hope of his discovery. In the following year, Gross continued his studies by melting a mixture of 5 parts of silver sulfate and 1 part of silver chloride between platinum electrodes at 15 volts and 5 amperes.[282] He observed an incandescence of the mass, presumably due to *bythium*, at the anode, accompanied by heavy gas evolution and dense, white steam, the vapors of which—he asserted—were not SO_3.

The discovery of these new elements did not pass unnoticed. It was curious that on the eve of the 20th century electricity would be used to decompose sulfur or chlorine: the idea was new at the time of Davy or his contemporaries earlier in the century, but definitely not in 1897. So it was that, in 1898, Alexander Hans tried to repeat Gross's experiment to shed light on this thorny problem,[283] but he could find no *bythium*.

Today, it is difficult to write a biography of Theodor Gross, the chemist. The only biographical information available refers to Dr. Theodor Gross, a chemical engineer hired by Graf (Earl) Ferdinand von Zeppelin (1838–1917) to improve airship construction. Gross tested possible engine materials to assess both fuel efficiency and power-to-weight ratios. When Zeppelin urged him to develop more efficient engines so as not to fall behind the French, Gross was unable to help his employer, and Zeppelin shortly afterward dismissed him, citing his lack of support and declaring that he was "an obstacle in my path."[284]

We do not know if this man is "our man." Although clues are not evidence, we may consider a few: the almost coincident interests in chemistry, the period in which Theodor Gross (1860–1924) lived, and such regrettable incompetency at work.

Notes

280. Gross, T. *Elektrochem. Ztschr.* **1897**, 4, 1–8.
281. Gross, T. *Elektrochem. Ztschr.* **1897**, 4, 112–15.
282. Gross, T. *Zeitschrift fuer Elektrochemie und Angewandte Physikalische Chemie* **1898**, 5, 48–52.
283. Hans, A. *Zeitschrift fuer Elektrochemie und Angewandte Physikalische Chemie* **1898**, 5, 93–95.
284. Dooley, S. C. *The Development of Material-Adapted Structural Form—Part II: Appendices* (Thèse no. 2986), École Polytechnique Fédérale de Lausanne, 2004; pp. 187 ff.

II.16

THE GHOSTS OF UNNAMED ELEMENTS

The elements that for simplicity we call "true" firmly occupy their places in the periodic table of the elements. The so-called false elements, difficult to systematize, also occupied, although for more or less extended times, a place within the periodic classification. But there are also those elemental discoveries, presumably false, whose uncertain discoverers did not dare to call them anything at all: these are the elements that have neither a name nor a label. Reconstructing their history is a very difficult task because these discoveries can easily be confused with one another. The following sections provide a brief review, along with the date of each discovery and the name of the presumed discoverer, of these discoveries.

For some (e.g., *brillium*), their story lies on the borderline between popular science and a fanciful joke: as reported by Charles Baskerville,[285] this supposed new element probably originated in the fertile mind of an unknown authority and found its way into the *Washington Post*[286] in the form of correspondence from two gentlemen from Newark, Delaware. They claim to have discovered *brillium* in coal ashes and that it had the peculiar property of producing more heat when placed under a furnace than ordinary fuel.

II.16.1. 1799: THE ELEMENT OF FERNANDEZ

The first in the list of these indeterminate elements seems to have been an oxide discovered by the chemist Fernandez toward the end of the 18th century.[287] Unfortunately, there is no available information about his chemical research nor is there any biographical information on the discoverer because the pertinent bibliography is very incomplete. Furthermore, it seems that none of the authors who cite Fernandez's work ever read his original writings. Biographical information about Fernandez was no less difficult to track down because no one ever mentioned his first name. The only chemist who might correspond to this name and could coincide with the discoverer of this presumed metallic element is D. Dominique Garcia Fernandez who, in 1799, holding the position of inspector of the Spanish Supreme Council of Commerce and of Mines, worked on the purification of nitric acid and on the influence of light on it.

II.16.2. 1852: THE ELEMENT OF FRIEDRICH AUGUST GENTH

Another 50 years passed before another announcement of the discovery of an unnamed element was published.[288] Friedrich August Ludwig Karl Wilhelm Genth was born in Wächterbach bei Hanau, on May 16, 1820. After taking his doctorate in chemistry at the University of Marburg, he became Robert Bunsen's assistant from 1846 to 1848. In 1849, he moved to the United States to take the position of superintendent of mines at Silver Hill, North Carolina.

In studying a sample of platinum from California, he recovered 2 grains[289] of a metal with an intense white color.[290] It was malleable and melted immediately in the presence of charcoal and on treatment with the oxyhydrogen blowpipe; it could be attacked by either hot hydrochloric or hot nitric acid, and, with hydrogen sulfide, it yielded a brown precipitate.[291] Successive studies showed that the substance that fooled Genth was a mixture of platinum cyanide and oxalate and of the chlorides of palladium and iridium.[292]

In 1872, Genth became professor of chemistry and mineralogy at the University of Pennsylvania in Philadelphia, where he spent the rest of his life. He served as president of the American Chemical Society in 1880. Genth died at Philadelphia on February 2, 1893.

II.16.3. 1852: THE ELEMENT OF CARL ANTON HJALMAR SJÖGREN

Carl Anton Hjalmar Sjögren (1822–93) was born on November 25, 1822, at Lofta in the county of Kalmar (Sweden). A celebrated geologist and mineralogist, he was the father of Sten Anders Hjalmar Sjögren (1856–1922), likewise famous in the same two fields. Sjögren finished his studies in chemistry at the University of Lund in 1842 and received his doctorate in 1847. Becoming tutor in mineralogy in 1848, 2 years later, he held the position of *auskultant i Bergskollegium* (student teacher in the School of Mines).

In 1852, while examining a sample of catapleiite, a rare zirconium mineral peculiar to Norway with the formula $(Na,Ca)_2ZrSi_3O_9.2H_2O$, Sjögren discovered an oxide that, in his opinion, could be that of an unknown element. Notice of the discovery of a new metal was quickly published,[293] but not even a year passed before an official retraction of the discovery was published in the same journal[294] due to the work of a fellow Swede, Nils Johannes Berlin (1812–97).[295]

In 1859, Sjögren became professor at the Falun School of Mines and, 4 years later, inspector of the mines of Wermland. Finally, in 1876, he was elected a member of the Vetenskapsakademien (the Royal Swedish Academy). He died on June 19, 1893, at Nynäs, in the region of Södermanland.

II.16.4. 1861: THE ELEMENT OF THE BROTHERS AUGUST AND FRIEDRICH WILHELM DUPRÉ

During a spectroscopic analysis of the water of London, the brothers August and Friedrich Wilhelm Dupré declared that they had isolated a new element. This element, like those already mentioned in this section, was never named, but nonetheless had its designated place in the periodic table. It would have been the heaviest alkaline earth metal after calcium, strontium, and barium.

The Duprés hypothesized that they had discovered the element that today we call radium. Curiously enough, another British subject, Henry D. Richmond, also claimed discovery of the same element three decades later, calling it *masrium*.[296]

Flame analysis of the residue from London water samples showed a weak blue line. The Duprés were not very accurate in calculating its exact position, limiting themselves to reporting that such a line could be found between the γ line of strontium and the β line of potassium, but closer to the first than the second. After this gross spectroscopic examination, the Duprés described their procedure for isolating what they thought was an alkaline earth metal: the residue obtained from the water was attacked by an excess of hydrochloric acid to which had been added a little sulfuric acid in order to clarify the solution. By concentrating the solid, the sulfates of barium, strontium, and the unknown element were obtained. The precipitate was filtered, washed and dried; then sodium carbonate was added. After melting, the mass was dissolved in water, and the solution was brought to a boil to eliminate the insoluble carbonates. The soluble carbonates were precipitated by concentrating the solution, and then resolubilized with a large quantity of dilute hydrochloric and sulfuric acids. After having removed all of the barium sulfate by the addition of alcohol, the two chemists obtained a second precipitate that consisted of the sulfates of strontium and the new metal.

The brothers Dupré converted the sulfate of the new alkaline earth metal into its respective oxalates and carbonates, and they determined the solubilities of these two compounds. If these values were well in accord with the solubilities of the respective salts of strontium and calcium, the chloride of the unknown metal looked like that of barium.

The two chemists should have been skeptical of the existence of this material. In fact, at the conclusion of their work, they repeated their flame experiment using a burner not made of brass, suspecting that the appearance of the blue line may have been due to the presence of copper in the alloy. They recorded their emission spectrum using as a flame source a common oil lamp and even a candle flame.

August Dupré, born in Mainz on September 6, 1835, was the younger, but more well-known, brother of Friedrich Wilhelm. They both became professional chemists. The Dupré family was originally French, but because they were Huguenots, they emigrated to Protestant Germany immediately after the repeal of the Edict of Nantes. The family settled in the Palatinate where, after many generations, the two chemist brothers were born. August obtained his doctorate at Heidelberg under Robert Bunsen and, in autumn of that same year, moved to London where he remained for the rest of his life.

In 1864, August Dupré joined the Westminster Medical School as a chemist. Two years later, he became a naturalized British subject. After having worked uninterruptedly at the medical school for 34 years, August retired to his country home at Mount Edgcumbe, Sutton, in Surrey. He died on July 15, 1907, after a long and difficult illness,[297] at the age of 72. Friedrich Wilhelm survived his brother by only a few months, passing away in 1908.

II.16.5. 1862: THE ELEMENT OF CHARLES FREDERICK CHANDLER

In 1862, celebrated American chemist Charles Frederick Chandler (1836–1925) announced his own discovery in a big way: he published his findings simultaneously in three journals[298] on two different continents.

Chandler was born on December 6, 1836, in the town of Lancaster, Massachusetts. He studied at the Lawrence Scientific School at Harvard University, then transferred to Göttingen, where he concluded his *cursus honorum* in chemistry. In 1856, he received his PhD and, in the following year, upon returning to the United States, he became assistant professor of chemistry under Charles Arad Joy (1823–91) at Union College in Schenectady, New York. A few years later, he succeeded to the chair of the department.

At the time of the discovery of the new metal, Chandler was not yet 26 years of age. However, during the preceding year, he had analyzed a mineral found in the Rogue River in Oregon. He wanted to wait for further confirmation of his discovery before making the announcement, but a series of coincidences forced him to publish his data early. In the first place, Chandler had worked with only 2 g of native platinum; he requested more material, but waited for it in vain for almost a year. Not having received it, he treated the little material he had at his disposal with hydrochloric acid and conducted a series of qualitative analytical tests on the resulting solution. Upon addition of hydrosulfuric acid (H_2S), he obtained a brown precipitate. The metal sulfide dissolved in both acid and in potassium chlorate. The addition of metallic zinc to this solution led to the formation of a precipitate similar to metallic tin, soluble in hydrochloric acid. The solution obtained in this manner, treated with mercury (I) chloride, yielded small crystals upon cooling. To obviate possible experimental errors, Chandler repeated these analyses three times, and, for all three times, the results were identical.

Speaking of his work to a friend, Chandler became aware of the announcement of the discovery of a new metal proposed 10 years prior by Friedrich Genth, and he was convinced that the two discoveries could be taken as the same substance. Perhaps because of the lack of other samples of native platinum to analyze, or perhaps suspecting that it would be useless to bring an action of priority against Genth, Chandler never returned to this subject—a decision probably all to the good, in light of his future academic career.

Charles Frederick Chandler held the chair of chemistry at Union College until 1864 when he was called to cover the duties of the assistant to Professor Thomas Eglestone (1832–1900) at the Columbia School of Mines. Afterward, he was "elected" professor of analytical and applied chemistry there; in 1877, he reorganized the school and succeeded to the prestigious chair of chemistry. He remained at the Columbia School of Mines until his retirement in 1903, holding the office of dean for almost 33 years.

The range of Chandler's chemical interests was extraordinarily broad, consisting of subjects like sugar, petroleum, minerals, illuminating gas, photographic materials, cosmetics, aniline colors, and electrochemistry (applied to analysis of water and food). His interests clearly show his practical frame of mind and sketch out brilliantly the diversity found among American university professors in the second half of the 19th century when contrasted with their European colleagues. Chandler considered industry an opportunity, an exciting career pathway for his many students; in this, he differed even more from some of his academic colleagues whose attitudes were more conservative.

He was, moreover, a genius who succeeded in organizing the American chemical community on the basis of modern realities. He was president of the committee of American chemists that, in 1874, met at the former home of Joseph Priestley (now the Priestley-Forsyth Memorial Library) at Northumberland, Pennsylvania, to commemorate in a solemn ceremony the centenary of the discovery of oxygen. At this meeting, he got the idea of founding an American chemical society at the national level.

Chandler made use of the pages of *American Chemist*, a chemical journal that Charles and his brother, William Henry Chandler (1841–1906), at that time professor of chemistry at Lehigh University, ran from 1870 to 1877 to spread this innovative idea. Finally, in 1876, the American Chemical Society was born, as well as its official organ, the *Journal of the American Chemical Society*; a great deal of credit in this enterprise is owed to the untiring willpower of Charles Chandler. Recognizing this, his colleagues elected him president of the Society from 1881 to 1889. Furthermore, he was the organizer and first president of the Chemists' Club, a group that had the goal of promoting a social and professional identity in the chemical community connected to the growing American chemical industry.

After a brief illness, on August 25, 1925, Charles Frederick Chandler died at his home in New York City at the age of 89. The young chemist who, in 1862, had the impudence to announce the false discovery of an element similar to platinum had become a national icon, celebrated both at home and abroad. On the occasion of his passing, he was solemnly honored in a way seldom accorded to American scientists.[299]

II.16.6. 1864: THE ELEMENTS OF WILLIAM NYLANDER AND CARL BISCHOFF

In 1864, Scandinavian chemist William Nylander (1822–99) submitted a memoir to the University of Lund with the intention that it be published. Its title was "Bidrag til Kännedomen om Zirkonjord" [On the Knowledge of the Composition of Zircon]. He had analyzed Norwegian eucolite, a mineral discovered in 1847. From his investigations, it appeared unequivocally that two "earths" were present that contained hitherto unknown elements, differing from one another in the solubilities of their respective double sulfates. Because Nylander had arrived at a dead end in his research, he never went beyond this announcement and did not name the presumed elements.

At the time of the discovery, William Nylander was about 42 years old, having been born at Oulu on January 3, 1822. He was an eclectic and versatile scientist who threw himself enthusiastically into many different areas of research only to suddenly abandon them. After obtaining his degree in medicine, he became head professor of botany at the University of Helsinki. After only 6 years, he grew tired of university teaching, left his job, and settled in Paris, where he spent a long and fruitful period of research in the 1850s, managing to make his living as an independent scientist. He soon became famous all over Europe because of his enormous collection of lichens that presently comprise the largest section of the University of Helsinki's museum. An introvert by nature, Nylander spent the last years of his life in almost complete isolation from the outside world. At the age of 77, on March 29, 1899, he died at his workbench, alone and forgotten.

The announcement of the discovery of a new element, also nameless, appeared during the same year in the work of the German chemist Carl Bischoff (or Bischof) (1812–84), born at Bad Dürrenberg on June 4, 1812. An excellent researcher and an expert alpinist, he studied chemistry, physics, and geology at Berlin. At the age of 17, in 1829, he distinguished himself by having constructed a small steam automobile capable of moving about on the streets; it was most probably the first vehicle of its type to roll on German soil. He was actively interested in mechanical and technological problems and, in 1839, invented a gas kiln; this invention brought about a complete transformation of heating plants in many sectors of industry. For this and for many other labors in the field of metallurgy he quickly became famous: on February 22, 1844, he was appointed—by official decree of Duke Alexander Carl von Anhalt-Bernburg (1805–63)—to the position of director of the Mägdesprung iron works. In 1856, he joined the association of German engineers, and in 1858, he was made president. Due to very poor health, in March 1863, he resigned his position and retired, but his scientific interests did not stop. The following year, while analyzing some calcareous minerals, he found what he believed was a new element; not being completely certain, he refrained from proposing a name.[300] His analytical work on the calcareous rock was taken up, along with other chemists, by John Percy (1817–89), a member of the Royal School of Mines (England), for the purpose of drafting a series of lectures that were ultimately published.[301] Bischoff had analyzed some dolomite, extracting from it the alkaline earth metals present, but the presumed discovery was not mentioned in the second article.[302] Perhaps because he was too ill to pursue his research, perhaps because of the uncertainty of his discovery, Bischoff quickly dropped the subject. On June 23, 1884, he died at Dresden three weeks after his 72nd birthday.

II.16.7. 1869: THE ELEMENT OF OSCAR LOEW

During the same year in which D. I. Mendeleev put order into the list of the chemical elements, in the United States, chemist Carl Benedict Oscar Loew (1844–1941) published his discovery of the oxide of a new metal.[303] He had analyzed some samples of zircon coming from North Carolina, finding in them what seemed to be a new elemental substance.

Loew, born on April 2, 1844, at Marktredwitz in Bavaria, was educated at the University of Munich and later at Leipzig. After having received his doctorate in chemistry in 1867, he moved to the United States and took part in the U.S. Expedition West of the 100th Meridian, an undertaking that forced him to travel uninterruptedly for 4 years (1867–71). During this American experience, Loew discovered the "earth" (oxide) of a new element for which he did not propose a name.

In 1871, he returned to Germany and there he remained for 22 years until, in 1893, he was invited to teach at the Imperial University of Tokyo. After his first American experience, Loew did not work on minerals again, and his presumed discovery fell into total oblivion. At the beginning of 1907, he was hired by the U.S. Department of Agriculture as a chemist at the experimental station in Puerto Rico.

Loew was a talented teacher and an untiring traveler: among his many students may be counted the Japanese Umetaro Suzuki (1874–1943), discoverer of aberic acid, a compound later called thiamine (vitamin B_1). In his old age, Loew returned to Germany. He lived 73 years beyond the announcement of his false discovery; he died at Berlin on January 26, 1941,[304] just prior to his 97th birthday.

II.16.8. 1878: THE ELEMENTS OF WILLIAM BALTHASAR GERLAND

In 1878, it was the turn of an English chemist to announce the possible existence of two new metals.[305] William Balthasar Gerland had been analyzing some minerals containing vanadates of copper and lead, found in a sandy ore vein, when he separated from them two totally unknown oxides.

He noted that the addition of alkali produced some insoluble precipitates. On the contrary, the alkaline carbonates yielded precipitates soluble upon the addition of an excess of reagent. The solution of the unknown metals yielded a precipitate if treated with bubbling carbon dioxide; the precipitate thus produced could be dissolved by bringing the solution to a boil.

One of the unknown oxides seemed to be an alkaline earth; it melted, giving a red color, but did not typically react like alumina. The pure salts of the new metals did not yield precipitates on the addition of calcium or barium carbonate, but in the presence of iron and alumina it was possible to obtain precipitation. The oxalates of the unknown metals were insoluble in water, acetic acid, and in dilute mineral acids, but became soluble in sodium carbonate and in concentrated mineral acids. With potassium sulfate, they formed slightly soluble compounds; the most soluble fraction showed an absorption spectrum that was brighter than that of didymium, compared with the less soluble fraction. However, the more soluble fraction had a spectrum that did not resemble that of erbium. The oxides obtained from the less soluble fraction possessed a brown leathery color; those from the more soluble part, a light brown color. The latter did not react with alkali, but they dissolved readily in acids, releasing a great deal of heat. With hydrofluoric, hydrochloric, chloric, perchloric, hydriodic, and periodic acids, they formed deliquescent salts soluble in water and in alcohol: the sulfates were less soluble and crystallized as hydrates. The latter lost their water of crystallization at temperatures above 180 °C.

Gerland was born on May 2, 1831. He took his doctorate in 1852 at Marburg, and then moved to Munich, in Bavaria, where he became the assistant to Hermann von Fehling (1812–85). After the false announcement of the discovery of the two new earths, in 1891, Gerland became a chemical consultant at Accrington, in Great Britain. In 1904, at the age of 73, he published his last scientific work.

Gerland wanted to hand down to posterity his motive for deciding to publish his discovery of the presumed elements. On August 3, 1878, he made note in the pages of *Chemical News* of his own experiments (which were, by his own admission, highly incomplete and full of gaps) to clinch the priority of his discovery. In fact, in that same year, American chemist John Lawrence Smith had gotten the jump on Gerland, giving to the press a brief memorandum on the discovery of a new metal that he called *mosandrium*.[306] Gerland maintained that his own discovery was the same as Smith's *mosandrium*; however, in time, both discoveries were found to be in error.

II.16.9. 1883: THE ELEMENT OF THEODOR EDUARD WILM

In 1883, in the course of his studies to determine the content of platinum in some platinum-bearing rocks, Theodor Eduard Wilm (1845–93) encountered a substance with peculiar properties, but the nature of which remained obscure. Wilm was born in Saint Petersburg on January 15, 1845. After having studied at Marburg and at Leipzig in the laboratory of Friedrich Konrad Beilstein (1838–1906), he received his doctorate in 1882 and 2 years later returned to the city of his birth as docent in chemistry at the local Kaiserlichen Ingenieur-Academie.

Wilm's work concentrated mainly on the characterization of minerals coming from the rich mines of the Urals. A solution of platinum obtained following complete solubilization of the native material showed no trace of osmium and iridium. Then Wilm treated the cold solution with an excess of barium sulfate; he obtained a precipitate that he dissolved in hydrochloric acid. He brought the solution to a boil, then saturated it with hydrogen sulfide. He added first nitric acid, then aqua regia to the solution. The residue was collected by filtration and treated with sodium chloride and with chlorine. The solid portion that it was not possible to dissolve in this way was treated by addition of sodium carbonate and melted. Then he added water to the melt. A white microcrystalline powder, completely insoluble, was the result. All of his attempts to characterize this powder came to nothing. Although Wilm was convinced that the substance under examination could contain a new element, he decided to publish this discovery with a simple introductory note on his work with platinum-bearing minerals.[307]

A few weeks later, Wilm took up his work with renewed ardor:[308] he repeated his previous operations and obtained again the precipitate that melted with sodium carbonate. To this, he added hydrochloric and nitric acids with the double intention of oxidizing and obtaining the chloro complex of the unknown metal. After adding ammonium chloride, he collected a precipitate that he suspected contained the ammoniacal chloride of ruthenium and iridium. After successive crystallizations, Wilm obtained some violet crystals; further chemical analysis of this substance gave such abnormal results that Wilm was even more convinced that the metal contained therein was of an unknown substance.

His work did not describe in enough detail the procedure used for the characterization of this supposed new element. As a result, it is difficult to believe that he had discovered rhenium almost a half century in advance of the work of the husband-wife team of Walter Noddack (1893–1960) and Ida Tacke Noddack (1896–1978): the analogies between the two metals were too weak. Most probably, Wilm obtained a mixed salt of ruthenium, rhodium, and iron. However, he never lived to see his fantastic hypothesis crumble; in November, 1893, he passed away at Saint Petersburg at only 48 years of age.

II.16.10. 1897: THE ELEMENT OF GETHEN G. BOUCHER AND F. RUDDOCK

In 1897, Gethen George Boucher[309] and F. Ruddock[310] announced that they had extracted and isolated a new element from pig iron. Gethen George Boucher was born in England on June 17, 1869. He had already worked with great success in the metallurgical field and, a year earlier, had developed an analytical method for the determination of sulfur content in steel.[311] After having isolated the new metal in a state of sufficient purity (although with extreme difficulty, given the fact that the weight percent was only between 0.0019 and 0.0060 of the mass of the impure iron), the two chemists recorded a certain number of chemical reactions typical of the new substance. When the solution of the metal in hydrochloric acid was heated to dryness, it became intensely blue in color. The oxide of the new metal remained virtually insoluble in hydrochloric, sulfuric, and nitric acids. The addition of stannic chloride to the aqueous solution of the metal yielded a blue precipitate when cold; when it was acidified with hydrochloric acid and brought to a boil, the color changed to brown. If sodium sulfate were added to a solution of the metal and brought to a boil, a blue color appeared. Reading Boucher's publication, the chemist C. Jones hastened to respond by publishing a note in which he asserted that the reactions described indicated that the metal could be molybdenum.[312] He was not gentle in judging the work of his colleague: according to him, Boucher had done nothing more in his research but "discover" molybdenum. Offended by Jones's insulting aspersions on his work, Boucher not only refused to accept the judgment of his colleague but, in a brief article that appeared shortly after, he also rejected out of hand Jones's accusation of superficiality.[313] But, despite the fact that Boucher did everything possible in repeating the tests necessary to exclude the presence of molybdenum in his samples, the accusations made by Jones were enough to demolish his credibility for the presumed existence of a new element hidden in pig iron.

II.16.11. 1904: THE RADIUM FOIL OF GEORGE FREDERICK KUNZ

In 1903, an American mineralogist and mineral collector, George Frederick Kunz (1856–1932) became associated with the young chemist Charles Baskerville when they conducted a meticulous investigation on the exposure of certain inorganic oxides and minerals to samples of radium-barium chloride and radium carbonate. They reported that all of the samples tested—oxides of uranium, zirconium, thorium, and of the rare earths, as well as the mineral kunzite (named after Kunz himself)—exhibited phosphorescence, whereas only two of them, namely the oxides of zirconium and thorium, phosphoresced when submitted to the action of ultraviolet (UV) light.[314] Because one of the nonradioactive oxides (zirconium dioxide) responded to UV light, and one of the radioactive oxides (uranium oxide) did not respond to the same stimulus, Kunz drew the conclusion that the zirconium and thorium oxides had something in common—perhaps even a common constituent—that differentiated them from the other tested samples. The two researchers hypothesized the existence of a new elementary substance that acted as a radium foil.[315] Unfortunately, there is no further word as to how this research progressed.

George Frederick Kunz was born in New York City on September 29, 1856. He became interested in minerals at a very young age. After attending Cooper Union (but leaving without a degree), he became special agent for the U.S. Geological Survey (1883–1909), research curator at the American Museum of Natural History (AMNH), and the leading advocate in the establishment of the international carat as a unit of measure for precious gems. He also assembled the Morgan-Tiffany collection of gems at the AMNH. Kunz proposed the adoption of the decimal metric system of weights and measures in the United States and was president of the American Metric Association. He became famous for identifying a new gem variety of the mineral spodumene,[316] which was named "kunzite" in his honor. Kunzite is a pink to lilac-colored gemstone with its color arising from trace to minor quantities of manganese; it is frequently irradiated to enhance its color, a fact that probably gave Kunz his research idea.

In the same year that Kunz published his work on *radium foil*, a report came in from Germany that a radioactive substance was found accompanying mercury earths, such as cinnabar, from a variety of sources. The substance, not identical to radium, was called *radiomercurium* because it was assumed to be part of the Zn-Cd-Hg series in the periodic table.[317] Having heard nothing further, we assume that it was not the higher homologue of Hg that we now call copernicium.

II.16.12. 1908: THE ELEMENT OF CLARE DE BRERETON EVANS

As has already been noted, Sir William Ramsay, the discoverer of almost all of the noble gases, had a laboratory at University College London in which many students worked on the search for chemical elements that were still missing from the periodic table.

The case of chemist Masataka Ogawa (1865–1930) is famous. Ogawa, after having obtained his degree at the University of Tokyo, came to Europe to study under the guidance of William Ramsay. The maestro entrusted to his older student the analysis of the mineral *thorianite*, with the secret hope of finding a yet-unknown radioactive element. From this mineral, Ogawa extracted and isolated a tiny quantity of an apparently unknown substance, but it was not radioactive. Shortly afterward, he announced the discovery of an element[318] that he called *nipponium* in honor of his native country.[319] The discovery was without foundation but aroused a great ferment of Japanese public opinion; even to this day, many articles published in Japan are inclined to be favorable to Ogawa's discovery,[320] arriving in some cases at the hypothesis that the element he isolated could have been rhenium, hafnium, or even protactinium.[321] This story is treated in more detail in Part III.16.

Less well-known is the fact that, in the same laboratory and almost contemporaneous with Ogawa's announcement, a second false discovery was in the making, the result of the work of an unknown English research chemist named Clare de Brereton Evans (b. ca. 1865). De Brereton Evans had been educated at Cheltenham Ladies College and had earned her BSc in 1889. Following graduation, she undertook research at the Central Technical College with Henry E. Armstrong (1848–1937), from which, in 1897, she was to be the first woman granted a DSc, for her research on aromatic amines. De Brereton Evans undertook part-time research at University College London as part of William Ramsay's group.

During the course of an analysis of a quantity of thorianite brought from the island of Ceylon, Clare de Brereton Evans found that about 1% of the material was made up of sulfides of silver, copper, and an unknown element apparently in the tin group. On closer examination, she found that a large part of the unknown metal was lead, which was present in the mineral at a level of about 3%. To confirm this discovery, de Brereton Evans decided to examine a larger amount of the mineral. From about 80 kg of the raw material, she obtained about 1 kg of mixed sulfates. At the end of many laborious chemical separations, she obtained about 150 g of the sulfide of the unknown element.[322] This salt was characterized by an intense brown color. It was immediately clear that the color of the sulfide should not be the basis upon which to claim a discovery because this was probably due to the presence of arsenic sulfide. Having removed this element, de Brereton Evans dissolved the remaining sulfide in nitric acid and, upon slow evaporation of the solution, a hygroscopic oxide precipitated that, when reduced in a stream of hydrogen at a temperature of 250–300 °C, led to a black substance of uncertain stoichiometry. Reduction at a higher temperature freed the metal in its elemental state. Its color was dark gray, and it was not volatile.

Clare de Brereton Evans sought to determine the atomic weight of the metal by two different methods: the first by electrolysis and the second by the ratio of its weight with its molecular oxide. The second method yielded a result with a value greater than that of arsenic, whereas the electrolysis of several milligrams of the chloride of the mysterious element yielded a weight close to that of antimony. Finally, a spectroscopic analysis of the remaining material was carried out, but this did not succeed in establishing the presence of any new spectral lines. With some caution, one might say that Clare de Brereton Evans had exchanged a mixture of the sulfides of selenium, antimony, arsenic, bismuth, and molybdenum for that of a metal not yet discovered.

This event would have ended here if German chemists A. Skrobal and Paul Artmann had not published an analogous work[323] in the pages of the journal *Chemiker-Zeitung*. The two authors undertook to examine the recent discoveries of Ogawa's *nipponium*, asserted to have been extracted from molybdenite,[324] and of the nameless element that Clare de Brereton Evans had isolated from *thorianite*. According to them, the elements discovered by the two Ramsay students were identical. Their observations went well beyond this: they asserted that the two metals, in addition to the fact that they corresponded to each other, would have been a rediscovery of the element claimed by F. G. Ruddock 10 years earlier in some steel samples and by G. G. Boucher in pig iron, as well as in a vanadoferric alloy. Paul Artmann at first was convinced that he was not dealing with a new element, but with traces of molybdenum; only after having read the work of Ogawa and Brereton Evans did he realize that he had discovered the same element, beating them to the punch. For this reason, a bitter exchange arose among the Germans, Ramsay's students, Ruddock, and Boucher in a competition to try to establish priority for the discovery of this nameless element.

II.16.13. 1913: THE ELEMENT OF H. C. HOLTZ

In 1913, the chemist H. C. Holtz began to analyze platinum-bearing minerals found in the Ural mountains. He dissolved his samples in mineral acid and from the solutions obtained in this manner precipitated $(NH_4)_2PtCl_6$. After filtration, he reduced the metal with zinc dust. A second precipitate, black and powdery, supposedly platinum, was collected and treated with nitric acid to remove palladium and copper. Holtz became aware that there was a discrepancy between the observed values and the amounts of metals present in his samples. To set the accounting straight, he hypothesized the presence of an unknown metal that had escaped his examination[325] but declined to give it a name.

The Spanish chemists Angel del Campo y Cerdán and Santiago Piña de Rubies repeated Holtz' experiments, but without success. Angel del Campo y Cerdán was a young chemist born at Cuenca on May 11, 1881. He took his doctorate in 1906, and 2 years later, he received a study grant from the Junta de Ampliación de Estudios e Investigaciones Cientificas and moved to Paris, where he was able to pursue the course in spectrochemistry given by Georges Urbain, with whom he carried out an exhaustive spectral study of pitchblende.[326]

Angel del Campo y Cerdán returned to Spain with a rich store of knowledge in the field of spectroscopy. In his work relative to rebutting Holtz's discovery, he was able to report the detailed spectrum of the fractions containing copper and palladium. Together with Santiago Piña de Rubiés, in two separate analyses, he found 360 and 600 lines, respectively, none of which appeared attributable to an unknown element. Tin, lead, chromium, and magnesium were observed in addition to the copper and palladium observed by Holtz. In their opinion, Holtz was fooled precisely because of the presence of the mixture of chromium, magnesium, and lead.

In 1915, Angel del Campo y Cerdán was appointed to the chair of Análisis Químico General at Madrid and later became a member of the International Commission on Chemical Nomenclature. In 1927, he was elected to the Real Academia de Ciencias Exactas, Físicas y Naturales of Madrid. He is credited with the development of a new type of fuel for airliners and a vaccine against pellagra. Angel del Campo y Cerdán died at Madrid November 4, 1944, at the age of 63.

Notes

285. Baskerville, C. "The Elements: Verified and Unverified." Address to the Chemistry Section at the 53rd Meeting of the American Association for the Advancement of Science, Saint Louis, MO, 1903–04; *Proc. AAAS* **1904**, pp. 387–442.
286. November 18, 1903.
287. Fernandez, D. *Scherer Allg. J.* **1799**, 3.
288. Genth, F. A. *Proc. Acad. Set. Phil.* **1852**, 6, 209.
289. The grain (gr) is an English unit of measure equivalent to 0.06479891 g. The same name is used as a unit of mass for precious stones, equivalent to 0.05 g (1/4 carat).

290. Genth, F. A. *Am. J. Sci.* **1853**, *15*(2), 246.
291. *Annual Scientific Discovery* or *Year-Book Facts in Science and Art for 1863*; David A. Wells: Boston, MA, 1865, p. 194.
292. Anon. *Hütten Ztg.* **1853**, *12*, 751.
293. Sjögren, K. A. H. *Journ. prakt. Chem.* **1852**, *55*, 298.
294. Berlin, N. J. *Journ. prakt. Chem.* **1853**, *57*, 145.
295. Nils Johannes Berlin, a Swedish chemist and pharmacologist born in Hernösand, was associate professor of pharmacology and later professor at Lund from 1845 to 1883.
296. Richmond, H. D. *Science* **1892**, *10*(488), 329.
297. Anon. *J. Chem. Soc., Trans.* **1908**, *93*, 2214.
298. Chandler, C. F. *Am. J. Sci.* **1862**, *33(2)*, 351; Chandler, C. F. *Silliman's Am. J.* **1862**, *33*, 1; Chandler, C. F. *Chem. News* **1862**, 30.
299. Bogert, M. T. *Biographical Memoir of Charles Frederick Chandler* (1836–1925). National Academy of Sciences Biographical Memoirs, Fifth Memoir; National Academy of Sciences: Washington, D.C., 1931; vol. XIV, p. 125.
300. Bischoff, C. *Pogg. Ann.* **1864**, *122*, 646.
301. Percy, J. *Chem. News*, **1864**, 67.
302. Bischoff, C. *Am. J. Sci.* **1864**, *38*, 420.
303. Loew, O. *New York Lyceum Annals* **1870**, *9*, 211.
304. Klinkowski, M. *Ber. Dtsch. Chem. Ges. A*: **1941**, *74A*, 115.
305. Gerland, B. W. *Chem. News J. Ind. Sci.* **1878**, *38*, 136.
306. Smith, J. L. *Compt. Rend. Chim.* **1878**, *87*, 831.
307. Wilm, T. E. *Ber. Dtsch. Chem. Ges.* **1883**, *16*, 1298.
308. Wilm, T. E. *Ber. Dtsch. Chem. Ges.* **1883**, *16*, 1524.
309. Boucher, G. G. *Chem. News J. Ind. Sci.* **1897**, *76*, 99.
310. Ruddock, F. *Chem. News J. Ind. Sci.* **1897**, *76*, 118.
311. Boucher, G. G. *Chem. News J. Ind. Sci.* **1896**, *74*, 76.
312. Jones, C. *Chem. News J. Ind. Sci.* **1897**, *76*, 171.
313. Boucher, G. G. *Chem. News J. Ind. Sci.* **1897**, *76*, 182.
314. Kunz, G. F.; Baskerville, C. *Am. J. Sci.* **1904**, *17*(4), 79–80.
315. Baskerville, C.; Kunz, G. F. *Am. J. Sci.* **1904**, *18*(4), 25–28.
316. Spodumene is lithium aluminum silicate, $LiAl(SiO_3)_2$.
317. Losanitsch, S. M. *Ber. Dtsch. Chem. Ges.* **1904**, *37*, 2904.
318. For a complete reprise, see Part III.16: The Setting of the Element of the "Rising Sun."
319. Ogawa, M. *J. Coll. Sci., Imp. Univ. Tokyo* **1908**, *25*, Art. 15.
320. Yoshihara, K. H. *Kagaku to Kyoiku* **2007**, *55*(6), 270; Yoshihara, K. H. *Spectrochimica Acta, Part B: Atomic Spectroscopy* **2004**, *59*B(8), 1305; Yoshihara, K. H. *Gendai Kagaku* **2004**, *398*, 36.
321. Yoshihara, K. H. *Kagakushi Kenkyu* **2007**, *34*(3), 153.
322. de Brereton Evans, C. *Journ. Chem. Soc. Transaction*, **1908**, *93*, 666.
323. Skrobal, A.; Artmann, P. *Chem. Ztg.* **1909**, *33*, 143.
324. Later, Ogawa asserted that he had also extracted *nipponium* from *thorianite*.
325. Holtz, H. C. *Ann. Chim. Phys.* **1913**, *27*, 559.
326. Urbain, G.; del Campo, A.; Scal, C. *Rev. Real Acad. Cien., Madrid* **1910**, *8*, 49.

PART III

1869–1913

From the Periodic Table to Moseley's Law: Rips and Tears in Mendeleev's Net

THE GREATNESS OF A MAN CORRESPONDS
TO THE PAINFUL DISPROPORTION
BETWEEN THE GOAL THAT HE AIMS AT AND
THE STRENGTHS THAT NATURE
HAS GIVEN HIM IN PLACING HIM INTO THE WORLD.
—MARGUERITE YOURCENAR (1903–87)

PROLOGUE TO PART III

In the third part of this book, we meet scientists, academics, and amateurs who were active from the second half of the 19th century until the discovery of the atomic number and the isotope in the early part of the 20th century. In addition to wet chemical analysis and fractional crystallization, they had at their disposal new tools for scientific investigation, tools that were predominantly physical such as spectroscopy and, later, chromatography and radioactivity. These investigators carried out their research supported by knowledge of the periodic law, a great advance described in Part II. Their findings were based on atomic weight, and the atomic weight anomalies in the periodic table—anomalies that would not be resolved until Moseley's discovery of the atomic number—weighed heavily on their minds. They also had improved analytical techniques at their disposal, in particular, visible emission and absorption spectroscopy.

But these were also persons of a decidedly Victorian stamp, especially the amateurs. Jealous of their independence, they kept a low profile, and with rare exceptions, stayed away from public life. They were aware of the social prestige they enjoyed and were often solitary investigators of the mysteries of nature and technology. In fact, in many encyclopedias and in the current scientific literature, many of them are not even mentioned. But if one minutely examines old bibliographical indices and the works of authors long past, it turns out that they are almost always recorded by appropriate entries.

III.1

THE FORERUNNERS OF CELTIUM AND HAFNIUM

OSTRANIUM, NORIUM, JARGONIUM, NIGRIUM, EUXENIUM, ASIUM, AND OCEANIUM

Of the naturally occurring nonradioactive elements, hafnium was the next to last to be discovered, preceding the discovery of rhenium by 3 years. It can boast of holding a very strange record: the number of claims for its discovery over the years is unequaled by any other element. This record was the cause of frustration for many scientists who, over the years, took turns in attempts to isolate it.

The reason that hafnium remained undiscovered until 1922 lay not so much in that its presence in nature (long known to be quite scarce) wasn't looked for, but in its peculiar chemical properties that bound it up intimately with zirconium. Toward the end of the 18th century, Martin Heinrich Klaproth melted some forms of yellow-green and red zirconium with sodium hydroxide and then digested the residue several times with hydrochloric and sulfuric acids to eliminate the extraneous silicon. The solution, thought to contain a number of elements, produced, upon addition of potassium carbonate, a generous precipitate. The oxide that Klaproth collected did not seem to belong to any known substance, and he called it *terra zirconia*.

With the passing of the years, he and many other chemists, among them the renowned Jöns Jacob Berzelius, determined the elemental composition of zircon and of its correlative minerals. Far from being simply $ZrSiO_4$, zircon contained traces of iron, aluminum, nickel, cobalt, lead, bismuth, manganese, lithium, sodium, zinc, calcium, magnesium, and uranium and small amounts of the rare earths.[1]

Some impurities persistently resisted separation from zirconium oxide or zirconia and were taken erroneously for oxides of new elements (new earths). In 1825, Johann Friedrich August Breithaupt (1791–1873) reported the presence of a new element, *ostranium*, isolated from ostranite, a mineral similar to zircon.[2]

Twenty years later, the Swedish chemist, mineralogist, and metallurgist Lars Fredrik Svanberg (1805–78) announced the discovery of a new element.[3] In his publication of 1845, he asserted that the zirconium oxide obtained from a variety of Siberian, Norwegian, and Indian zircon samples was in reality composed of two earths: one, *zirconia*, already noted, and another unknown earth. In particular, he extracted an oxide from nordite[4] and, as was the custom, he called it *norderde* (i.e., *terra noria*). The element present in this mineral was called *norium*, from *Nore*, an ancient name for Norway.

In a subsequent article, published during the same year, Svanberg found *norium* also in eudialite from Greenland. Because all of his attempts to separate *norium* from zirconium failed, he supposed that the samples he possessed contained variable quantities of the new element and that this was the cause of the variable solubilities of the zirconium and *norium* salts.

Unfortunately, no confirmation of this metal with the exception of his own was forthcoming from his colleagues because no one was able to obtain similar results.[5]

Another 24 years passed and, in 1869, Henry Clifton Sorby (1826–1908) announced the presence of a new oxide,[6] *jargonia*, in jargonite[7] originating in Ceylon. Sorby was born at Woodbourne, Attercliffe, on May 10, 1826. His family wanted him to follow a career in commerce because his father had a factory. However, at 15, Sorby decided to become a scientist. His family was in favor of this youthful inclination and gave him a suitable education by hiring a private tutor. In 1847, John Sorby died and left his son Henry enough money to allow him to live off the annuity. Instead of wasting it on parties and entertainment in London, following in the footsteps of some of the sons and heirs of the wealthy of his time, Henry built a laboratory on the ground floor of his home, where he carried on research for the next 61 years.

Sorby's interests ranged through many branches of science, with his primary passion being geology. In 1849, he "invented" petrographic microscopy, that is, microscopic examination of thin sections of minerals. For his notable scientific achievements, he was elected to the Royal Society at only 31 years of age. From the microscopic study of minerals, he passed on to the study of steel and alloys, inventing for this purpose the spectrum microscope. By means of this instrument, in 1869, Sorby was convinced that he had discovered an element that he called *jargonium*. The symbol he proposed for this new, although short-lived, element was Jg.[8]

After the death of his mother, Sorby, by this time a bit beyond 50 years of age, bought a yacht. He equipped the *Glimpse* as a floating laboratory and took many cruises. His interests in his old age were marine biology, meteorology, botany, architecture, and archaeology. In 1882, he contributed to the founding of the university in his native city, Sheffield, and, in 1905, donated a remarkable sum of money destined for the creation of a chair of geology there. He died at the age of 82 on March 9, 1908, neglected for the most part by the academic establishment of his day.

In the same year as the *jargonium* announcement, 1869, Arthur Herbert Church (1834–1915) announced the discovery of a new element associated with zircon.[9] He had observed the presence of this element through a careful spectroscopic study and proposed calling it *nigrium*. Church believed that among his preparations, he had three oxides (of zirconium, uranium, and *nigrium*), whereas in reality he had only two, which when mixed together gave rise to new lines in the spectrum, the ones that made him announce the discovery of *nigrium*.

Church asserted that Sorby's element was actually the same as his own and insisted that his discovery of *nigrium* ought to overshadow Sorby's.[10] It was inevitable that Sorby and Church would disagree vehemently with one another. In fact, Church, after having read the article in which Sorby announced the discovery of *jargonium*, protested loudly, pointing out to his colleague that as far back as 3 years earlier (1866) he had reported experimental evidence of the existence of *nigrium* (but without giving it a name).

Sorby's response was speedy: in a footnote in an article on *jargonium* he made claim to priority for his discovery.[11] He invited Church to look elsewhere; his element, *nigrium*, could have easily been the *norium* that Svanberg was unable to isolate. And the spectroscopic evidence was insufficiently strong enough to link *jargonium* to *nigrium*.

Furthermore, Sorby noted that Church had not explicitly spoken of a new element in 1866, but was driven to do so only after he (Sorby) had spoken of the properties of *jargonium* during the Royal Society session of March 6, 1869.

The disagreement between Sorby and Church continued unabated[12] until the appearance of an article in the *American Journal of Mining*. This journal reported on a note from the Polytechnic Association of the American Institute in which Isidor Walz, taking as his own the words of Professor Loew of the Lyceum of Natural History of New York, said that the discovery of *jargonium* was to be attributed to the independent work of an American, Church, and of a member of the Royal Society, Sorby. The American scientists recognized with bad grace the priority of Sorby's discovery, and this article was the true and proper death knell for *nigrium*. The American scientific community made sure to emphasize that Church's work came prior to that of his English colleague. If the new metal had kept its name, by now accepted as *jargonium*, it would have had two discoverers, Church and Sorby, not Sorby and Church. The uncomfortable understanding, however, lasted only a short time.

Little more than a year had elapsed before Sorby became aware of the errors he had made in his spectroscopic analysis, and he quickly published his retraction. He had confused the spectrum of impure zircon with that of the new element he called *jargonium*.

As a "flash in the pan" just before the close of the 19th century, the Danish chemist Julius Thomsen (1826–1909) reported experimental evidence for the existence of an element whose atomic weight would have placed it in the periodic table just before the element tantalum.

The dawn of the new century saw two very young German chemists taken up with research on this elusive element. In 1901, 31-year-old Karl Andreas Hofmann (1870–1940) and 23-year-old Wilhelm Prandtl (1878–1956) treated a small quantity of lead sulfate, produced as waste from the extraction of zirconium from its ore. From the raw material, they extracted zirconium oxide as a residue. The properties of the oxide, zirconia, were peculiar, and the two chemists guessed at the possibility that an unknown element was hidden in it. They began a complex process to separate out the new metal. In the end, they announced that about half of the so-called zirconia extracted from the mineral euxenite[13] consisted of the oxide of a new element that they provisionally called *terra euxenica* or *euxenerde*.[14] Because the unknown element had been recovered together with other tetravalent elements whose oxides were PbO_2, TiO_2, SiO_2, and ZrO_2, Karl Hofmann hazarded a guess that the new element also had a valence of 4. The equivalent weight of the sulfate was determined to be between 44 and 45, leading therefore to an atomic weight of 177.8, very close to that of the present element hafnium (178.49). It did not take long for the error to be discovered: in 1909, Otto Hauser together with his colleague F. Wirth published an exhaustive treatise[15] in which they rebutted the discovery of *euxenium*.[16] The two authors arrived at the conclusion, after a lengthy laboratory investigation, that zirconium was not present in great quantities in the samples of euxenite they examined and that the existence of the new element *euxenium* ought to be considered very uncertain, if not downright inconsistent, if only due to the spotty analytical characterization and the absence of spectroscopic investigations.[17]

Two years after the discovery of *euxenium*, at the age of 33, Hofmann became professor of inorganic chemistry at the Technische Hochschule of Charlottenburg. He did not publish another thing on the subject of *euxenium*, and he died at the age of 70 in 1940. Wilhelm Antonin Alexander Prandtl, after a brief career in industry, rose to a professorship in inorganic chemistry in 1910. In 1937, he left his post as a result of the rise of Nazism. He was recalled to his post in 1946 and made professor emeritus at the University of Munich; he died on October 22, 1956, at the age of 78.

In the first decades of the 20th century, in boundless imperial Russia, the hunt for the missing element between *lutecium* and tantalum began in earnest. In the years preceding World War I, the geochemist Vladimir Ivanovich Vernadsky (1863–1945) organized a special commission for research on minerals rich in radioactive substances from the Russian regions of central Asia.

Vernadsky was a scientist despised by the police because he had dared to criticize the Tsar's reactionary politics. In 1911, he was forced to leave his post at the University of Moscow, but in the following year he was nevertheless elected a member of the Academy of Sciences and in 1914 director of the Geological and Mineralogical Museum of Saint Petersburg. Using this new authoritative position, he organized numerous expeditions to central Asia in which he took an active part in finding new radioactive minerals and later extracting the radioactive elements contained in them.[18] He was obsessed with making Russia self-sufficient with respect to its supply of radium used in the treatment of neoplastic malignancies.

His first samples came from some deposits situated on the peninsula of Svjatoj Nos. The most interesting mineral sent to the laboratory belonged to a variety of orthite. Vernadsky thought that this mineral could contain a certain quantity of thorium and entrusted his young assistant, Konstantin Avtonomovich Nenadkevich (1880–1963), with the analysis. Although Vernadsky's work yielded excellent results (in 1918, he succeeded in extracting milligram quantities of radium from these minerals), that of his apprentice proceeded with difficulty. From the beginning, Nenadkevich believed that he had isolated thorium, but this certainty quickly vanished. He determined the atomic weight of the element to be an even 178, very different from that of thorium, 232. According to Mendeleev's periodic table, this element would have to occupy the box between lutetium[19] and tantalum.[20]

Nenadkevich excitedly reported his discovery to Vernadsky who, after ascertaining that the sample had come from the Trans-Baikal region in Asia, caught his assistant's enthusiasm and proposed the name of *asium* for the presumed new element. Unfortunately, no confirmation of their work ever came. The outbreak of World War I completely changed Vernadsky's research interests, and investigations to identify the new element were interrupted until the end of the Russian Revolution. The economic difficulties following the war, the Revolution, and the Russian civil war brought an end to the vision of isolating a new element. Publication of his results was so long postponed that, in the meantime, the element was "rediscovered" by other research groups.[21]

After the October Revolution, Vernadsky went to the Ukraine and was then, for a brief time, professor at the Sorbonne in Paris. He returned to Russia toward the end of the 1920s, where he became president of the Academy of Sciences and of numerous other institutions. At almost 82 years of age, he died at Moscow, on January 6, 1945, at the height of his fame, a few months before the Russian Army occupied Berlin and brought World War II to an end. Vernadsky was considered, and rightly so, one of Russia's greatest scientists, and his fame is great to this day. Streets, universities, volcanoes, mountain ranges in Antarctica, and even a crater on the dark side of the moon bear his name.

Nenadkevich's career pales by comparison. He was elected a correspondent to the Academy of Sciences in 1946, seemingly a consolation prize for his discovery of various new minerals.[22] He prudently never again mentioned *asium*, the hypothetical element that he believed he had isolated from orthite.[23] It would be difficult to effectively establish if he had observed a new element in his samples: the chemical composition of *orthite*, also known as allanite,[24] does not contain what would eventually be called element 72 in even trace amounts. Nenadkevich died on June 19, 1963, just 17 days after celebrating his 83rd birthday.

In the years following the end of World War I, Alexander Scott (1853–1947) announced the discovery of the higher homolog of zirconium. Scott was near 70 at the time, having been born December 28, 1853.[25] This event was interpreted as the crowning achievement of his work and the fruit of his brilliant scientific intuition. As a matter of fact, Scott never distinguished himself in any field of endeavor throughout his long scientific career.

In 1884, as his doctoral thesis bears witness, he determined the atomic weight of manganese. In future years, he would determine the atomic weights of 44 additional elements. His work was regarded as quite routine in his day.

In June 1884, he was given a position supervising the scientific activity of the laboratories of the Durham School. In December 1896, he moved to the Davy-Faraday Research Laboratory, where his superiors were none other than Lord Rayleigh (1842–1919) and Sir James Dewar (1842–1923). In this laboratory, Scott conducted the major part of his work

FIGURE III.01. Alexander Scott (1853–1947). In two successive publications, Scott claimed to have isolated an unknown element (with either atomic number 43 or 75) to which he gave the name *oceanium*. He was also president of the Chemical Society, London, from 1915 to 1917, and was instrumental in setting up the scientific laboratories of the British Museum.

on determining the atomic weights of elements. In 1898, he was elected an associate of the Royal Society. At the end of World War I, the Department of Scientific and Industrial Research asked Scott to evaluate the state of exhibits looked after by the British Museum after having spent the war years wrapped up in the basement of the building for fear of nightly German air raids. The work would not take more than 3 years. The museum outfitted a chemical laboratory for his use and, in addition to his work on the exhibits, Scott (Figure III.01) continued his study of the elements.

On February 1, 1923, Scott reported some results from studies conducted on black iron and titanium-bearing sands. This work had been begun many years before. In 1913, samples of sands from Maketu, New Zealand, were sent to him with the purpose of determining the assay of iron and thus to establish the commercial feasibility of exploiting this deposit. From a quick analysis, Scott found that the iron oxide content, as Fe_3O_4, was around 75%, whereas the other 25% consisted of TiO_2. A more detailed investigation, however, convinced him of the presence of a very small quantity of an insoluble substance; in 1918, he subsequently managed to extract 1.4 g of an earth (oxide) of an unknown metal that he labeled for simplicity's sake *new oxide*.

The atomic weight of the resulting element turned out to be 144. To confirm his work, Scott sent his samples to George de Hevesy (1885–1966) and Dirk Coster (1889–1950), who had just been credited with the discovery of element 72, hafnium. Scott asked them if such an element were present in his samples, in which case, he would be able to reclaim his discovery. The name he proposed was *oceanium*, from Oceanus, one of the Titans. The name of the new metal, *oceanium*, would furthermore contain the place of the discovery, a beach in New Zealand, one of the countries comprising Oceania.

The response from Copenhagen was rapid, but gloomy. There was no trace of element 72. The Danish group looked at the X-ray spectrum of the sample and also at the characteristic lines of element 75, which was still missing from the elemental roll call, but this research also yielded no fruitful results. The samples sent by Scott contained nothing but titanium and traces of other elements; nonetheless, he continued his research. At this point, his work took an unpleasant turn: in repeating the measurement of the atomic weight of *oceanium*, 2 months after the first announcement, it rose from 144 to 175.

Scott suspected that the insoluble residue present together with the titanium oxide could be SiO_2, but he did not lose hope that the sands of Maketu could contain a new element. Meanwhile, he ordered a large quantity of this material with the intention of investigating it minutely in all of its components. After exhaustive experimentation, the new earth was none other than silicon dioxide, and successive publications regarding *oceanium* never saw the light.

In the winter spanning the years 1923 and 1924, Scott visited the celebrated archaeologist Howard Carter (1874–1939) at Luxor. There, they struck up a cordial professional relationship. Scott became chemical consultant to Carter and, in the meantime, visited the excavations and the treasures recently removed from the tomb of the Pharaoh Tutankhamen. Scott collaborated with the famous chemist Alfred Lucas (1867–1945)[26] of the Service des Antiquités who had worked with Carter for nine seasons on the analysis and conservation of materials taken from the tomb.[27] Scott's presence at the British Museum continued well beyond the 3 years initially requested of him in the stipulations of 1919, so much so that, in 1938, the director of the Museum asked him, by now an octogenarian, to retire. He did so at age 85, and he died a very old man on March 10, 1947, in his 94th year.[28]

The presumed discovery of element 72 would not wait for long, and, indeed, it preceded Scott's announcement. In the summer of 1922, Georges Urbain and Alexandre Dauvillier (1892–1979), at Paris, announced the discovery of *celtium*,[29] but in January 1923, George de Hevesy and Dirk Coster, at Copenhagen, also claimed credit for the discovery. If the discovery of element 72 could be said to be over and done with, the same could not be said of its name. The complete tale of the *celtium*-hafnium controversy is reserved for Part IV.1.

Meanwhile, the French called element 72 *celtium*. De Hevesy leaned toward hafnium, although Niels Bohr (1885–1962), his patron at the Institute of Physics at Copenhagen, had insisted on the name *danium*.[30] In the middle of this confusion, the International Commission on Atomic Weights was invited urgently by both parties to make a decision. It held off for a number of years until, after the death of Georges Urbain, it opted for the name hafnium.

Notes

1. Later, it was found that zircon contained traces of radium and up to 9×10^{-7} mg of helium per gram of rock. Only much later was found that it contained traces of hafnium.
2. Breithaupt, A. *Pogg. Ann.* **1825**, *5*, 377.
3. Svanberg, L. F. *Pogg. Ann.* **1845**, *65*, 317; Svanberg, L. F. *Öfversigt af Kongl.vetenskaps- akademiens förhandlingar;* P. A. Norstedt & Söner: Stockholm, Sweden, 1845, pp. 34–37.
4. Nordite has the formula $(Ce,La,Ca)(Sr,Ca)Na_2(Na,Mn)(Zn,Mg)Si_6O_{17}$; it was found for the first time along the course of the Motchusuai River that flows on the Kola Peninsula in northern Russia.
5. Marignac, C. *Ann. Chim. Phys.* **1860**, *60*, 257.
6. Sorby, H. C. *Chem. News* **1869**, *17*, 511.
7. A variety of zircon.
8. Sorby, H. C. *Chem. News* **1869**, *18*, 7.
9. Church, H. A. *Intellectual Observer* **1866**, *9*, 201.
10. Church, H. A. *Chem. News* **1869**, *19*, 121.
11. Sorby, H. C. *Chem. News* **1870**, *20*, 197.
12. Sorby, H. C. *Chem. News* **1870**, *21*, 73.
13. Euxenite is an oxide containing titanium, niobium, tantalum, calcium, and many rare earth elements whose formula is $(Y,Ca,Er,La,Ce,U,Th)(Ni,Ta,Ti)_2O_6$.
14. Hofmann, K. A.; Prandtl, W. *Ber. Dtsch. Chem. Ges.* **1901**, *34*, 1064–69.
15. Hauser, O., Wirth, F. *Ber. Dtsch. Chem. Ges.* **1909**, *42*, 4443.
16. They were the first to coin the name *euxenium* for the mysterious element contained in *euxenerde* or *terra euxenica*.
17. Hofmann, K. A., Prandtl, W. *Ber. Dtsch. Chem. Ges.* **1901**, *34*, 1064.
18. Vernadsky, V. I. *Bull. Acad. Sci. Petrograd.* **1914**, 1353.
19. *Lutecium*, discovered by Urbain in 1907, kept the name given it by its discoverer until 1949, when its spelling was changed to lutetium.
20. Nenadkevitch, K. A. *Bull. Acad. Sci. Petrograd.* **1917**, *7*, 447.
21. Mellor, J. W. *A Comprehensive Treatise on Inorganic and Theoretical Chemistry*; Longmans, Green: London, New York, 1937; vol. V, p. 98.
22. In 1955, during a survey of the Kola Peninsula, M. V. Kuz'menko and M. E. Kazakova discovered a new mineral to which they gave the name nenadkevichite in honor of the chemist and mineralogist Nenadkevitch; its chemical composition is $(Na,Ca,K)(Nb,Ti)Si_2O_6(O,OH)_2 2H_2O$.
23. Nenadkevitch, K. A. *Compt. Rend. Acad. Sci. Russie* **1926**, *A*, 149.

24. Allanite is a hydroxysilicate of aluminum, iron(III), calcium, yttrium, and cerium with the formula $(Y,Ce,Ca)_2(Al,Fe)_3(SiO_4)_3(OH)$.
25. Scott, A. *J. Chem. Soc.* **1923**, *38*, 311.
26. Lucas, A. "The Chemistry of the Tomb." In *The Tomb of Tut-Ankh-Amon*; Carter, H., Ed.; George H. Doran Co.: New York, 1927; vol. 2, pp. 162–88.
27. Scott, A. *J. Chem. Soc.* **1923**, *105*, 881.
28. Anon. *J. Chem. Soc.* **1950**, *1*, 762.
29. The name *celtium* is derived from the Celtic peoples who, in the Roman era, occupied the region corresponding to today's France.
30. De Hevesy and Coster were at the Institute of Physics directed by Bohr at the time of their discovery. The work had been done in Denmark, and Bohr exercised pressure to name the new element *danium*, after Denmark. The two discoverers were not of the same opinion. The first, a chemist, was Hungarian, and the second, a spectroscopist, was Dutch. Coster proposed the name hafnium from the Latin name of Copenhagen (Hafnia), the city that had hosted them during their work. In the end, Bohr gave up his claim.

III.2

THE DISCOVERIES OF THE RARE EARTHS APPROACH THEIR END

PHILIPPIUM, ELEMENT X, DECIPIUM, MOSANDRIUM, ROGERIUM, AND COLUMBIUM

This section outlines the careers of two chemists who worked in the United States, but who were tightly bound to the cultural circles of Europe. The American, John Lawrence Smith, spent many years of specialized study in France. Marc Delafontaine (1837–1911) was born and educated in Switzerland and, after having worked for a period of time at the University of Geneva, moved permanently to the United States. In this story, almost all of which takes place in the French-speaking parts of Europe, the Swiss physicist Jacques Louis Soret (1827–90) also appears. He succeeded, albeit involuntarily, where the two chemists failed: he discovered an element, but then was robbed of the great glory of giving it a definitive name.

III.2.1. PHILIPPIUM AND ELEMENT X

In the 1870s, at Geneva, an excellent school of chemistry formed around the celebrated figure of Jean-Charles Galissard de Marignac. Among his students, the figure of Marc Abraham Delafontaine (Figure III.02) stood out. He was born March 31, 1837 or 1838, [31] at Céligny in Switzerland.

After finishing his studies in 1860, he was named "private docent" and subsequently professor of mineralogy and organic chemistry at the University of Geneva. In 1870, following the suggestion of chemist Jean Louis Agassiz (1807–73), he moved to Chicago, where he was named professor of toxicological chemistry at the Medical College. Well-versed in spectroscopy, he worked also as a professional chemist, equipping an analytical laboratory for this purpose and with his own funds; his expertise was found to be, on more than one occasion, a great help to the Chicago Police Department.

In 1878, both Jean-Charles Galissard de Marignac and Delafontaine published simultaneously an in-depth study of some minerals rich in the rare earths: the former analyzed 300 g of gadolinite[32] and the latter an almost identical amount of samarskite.[33]

Following the fractionation method proposed by Robert Bunsen, Marignac was able to obtain 18 fractions of different purities: on the one side, he obtained oxides of yttrium and, on the other, those of erbium and terbium. In March 1878, Delafontaine realized that terbium could be extracted more conveniently from samarskite.[34] At the conclusion of his article, Delafontaine observed that numerous circumstances led him to believe

FIGURE III.02. Marc Abraham Delafontaine (ca. 1837–1911). A Swiss chemist who became a naturalized American citizen, Delafontaine studied under the renowned chemist Jean Charles Galissard de Marignac, discoverer of ytterbium and gadolinium, and whose research on atomic weights led him to hypothesize on the existence of isotopes many years ahead of his time. Delafontaine taught for a time at the University of Geneva and later became professor of chemistry at a women's college in Chicago. He discovered two rare earth elements: *decipium* (believed today to be a mixture of elements already known) and *philippium* (today known as holmium). He was also a well-known spectroscopist and licensed forensic chemist.

in the existence of a hitherto unknown earth in the samarskite that he had examined. A second publication followed in which the author indicated the probable presence of a fourth earth, intermediate between the oxides of yttrium and terbium.[35] At the end of the description of his experimental work, he announced the discovery of a new oxide he named philippium in honor of his benefactor, M. Philippe Plantamour of Geneva.[36]

Not knowing that Delafontaine had supplied a name for this element, Jacques Louis Soret published his own note in which he said he was convinced that his Genevan colleague had effectively discovered a new element and proposed in his stead the symbol X.[37]

Soret, who was born in Geneva, on June 30, 1827, had personally known the young Delafontaine before he had departed for the United States. Correspondence between the two continued regularly, although with difficulty due to the enormous distance. This was precisely the reason why Soret was unaware of the name given to this new rare earth metal. Soret, a very experienced physicist, also supplied a detailed spectroscopic examination of the new earth.

Delafontaine also made note of the wavelengths of the new metal: the first in the blue ($\lambda = 450$ nm), the second in the violet region ($\lambda = 400$ nm). Without each other knowing it, Delafontaine thanked his colleague Soret for having convinced him that among his preparations a new element lay hidden.

The destiny of *philippium* seemed to be very hopeful, but almost like a lightning bolt falling out of a peaceful sky came the arrogant claim of American chemist John Lawrence Smith.

III.2.2. MOSANDRIUM

John Lawrence Smith was born near Charleston, South Carolina, on December 17, 1818. He enrolled at the University of Virginia in 1836 and spent 2 years studying chemistry, natural philosophy, and civil engineering, after which, for a year, he took up civil engineering. Abandoning this, he studied medicine and, in 1840, received his diploma at the University Medical Institute of South Carolina. Possessed of a restless personality, once he reached the finish line in this endeavor, he passed on to the assiduous study of chemistry, then moved to Paris where he became a specialist in chemical toxicology with Mathieu (Mateu) Orfilia (1787–1853), in physics with Claude Servais Poulliet (1791–1868) and Edmond Becquerel (1788–1878), in mineralogy with Armand Dufrénoy (1792–1857)[38] and Jean-Baptiste Elie de Beaumont (1798–1874), and in chemistry with Jean-Baptiste Dumas (1800–84), Théophile-Jules Pelouze (1807–67), and Michel Eugène Chevreul. After having also visited the laboratory of Justus von Liebig, in 1844, he returned to Charleston. He was among the founders of the journal *Science*.

Working as a chemist, he became interested in the composition of the soil around his native city, with the intention of bettering the cultivation of cotton. This area of investigation made him very famous, so much so that, in 1846, the sultan of the Ottoman Empire, Abdul-Mejid I (1823–61), asked President James Buchanan (1791–1868) to send an "instructor" who could teach Turkish agronomists how to introduce the cultivation of cotton into Asia Minor. The president sent John Lawrence Smith to Turkey, but, on his arrival, he realized that the Turkish government had already begun its experimentation—soon to be seen as disastrous—without waiting for him. So, the sultan "recycled" Smith by naming him superintendent of mines. For 4 years, he put all his energy into this work: in addition to chromium, he discovered deposits of coal and ercinite.[39] In 1850, he left Asia Minor and returned to Europe: during a stay in Paris, he invented the inverse microscope.[40] In 1851, he returned to the United States for good. A year later, he became professor of chemistry at the University of Virginia, and his interest in examining American minerals began. In 1854, he accepted the chair of medicinal chemistry at the University of Louisville. He then became interested in analyzing meteorites, in which he quickly became a world expert. His studies on meteorites led him into investigations of minerals held to be similar to them: among these, he analyzed samarskite, a mineral rich in the rare earths.

After having read a communication by Soret relative to the ultraviolet (UV) absorption spectra of the rare earths extracted from samarskite and wishing to take a precautionary measure regarding priority, he announced the discovery of a new element but declined to give it a name.[41] A short time later, he sent a second communication, very much more excited in tone than his preceding one, in which he decided to name the new metal.[42] He gave vent to his feelings to his colleague Soret, claiming to have communicated very early on to the Academy of Natural Sciences in Philadelphia—in their session of May 8, 1877—about the existence of a new element and the note that he had published in the following month.[43] Following this, Smith sent a sample of the new metal to Chicago, where Marc Delafontaine was working; Delafontaine responded privately, informing him that his metal was not at all new and actually was a sample of terbium. More weeks of feverish investigations followed.

In the course of his later investigations, Smith rejected Delafontaine's response; in the meantime, he found the article on the discovery of *philippium*. Smith was very disturbed, to say the least. He responded immediately, asserting that *philippium* was identical to his own element, to which he gave the name *mosandrium* in honor of the rare earth chemist Carl Gustav Mosander (1797–1858).[44] Delafontaine replied to Lawrence Smith's accusations with a brusque note that he presented to the Academy of Sciences at Paris.[45] He reviewed the events leading to the discovery of *philippium* and of *mosandrium* from his point of view and learned that, on July 22, 1878, Lawrence Smith had asked the Academy of Sciences to remove the seals from a packet he had sent to them some time earlier and that contained a note relative to the discovery of *mosandrium*. However, the contents of this note were the same as the communication on the presumed discovery that Smith had sent to Delafontaine at Chicago and that Delafontaine had returned to the sender branding it as false.

Furthermore, Delafontaine had made note of the fact that the presumed *mosandrium* was composed of about 75–80% terbium and a 20–25% mixture of yttrium, erbium, *didymium*, and *philippium*.

It was true that Smith had in his hands a new element, *philippium*, but an extremely small quantity of it and not at all free from contamination by the other rare earths. Meanwhile Delafontaine concluded, "I propose that the name mosandrium be removed from the list of elements, and I keep for myself the right to name the element whose existence I noted first and made known its distinctive characteristics."

In response to Delafontaine's criticisms, Smith presented a strongly argumentative note.[46] In it, he reviewed the entire set of events and cited numerous letters that were part of the written exchange between them. He cited the greatest experts on the rare earths: Marignac, Soret, and Delafontaine. In conclusion, he claimed priority of discovery and stuck to his chosen name for it, *mosandrium*.

On August 3, 1878, an English chemist from Macclesfield, Doctor W. B. Gerland, claimed that a good 15 years earlier one of his colleagues had given him a mineral sample, and, upon analysis, he had found that about 1% of the material defied classification. Initially, he had thought that it was an analytical error, but the announcement of the discovery of *mosandrium* induced him to claim his own discovery.[47] There were not many points of similarity between the two elements, nor was Gerland able to come back into possession of the samples he had examined in 1863, for which reason his note of protest ended up completely ignored.

III.2.3. DECIPIUM AND THE COMPLEXITY OF DIDYMIUM

The diatribe with Smith did not discourage Delafontaine from undertaking new investigations or from moving forward with those that he had left hanging. Continuing his studies on the chemical composition of samarskite coming from North Carolina, on October 28, 1878, Marc Delafontaine could assert with pride that[48] "I have succeeded in discovering a new metal that I will call *decipium* (from *decipiens*, deceiver)." The symbol proposed was Dp. The oxide of *decipium*, DpO, was yellow but turned white on strong heating in a stream of hydrogen, as opposed to the oxide of *philippium*, PpO, that had a "permanent" bright yellow color. The atomic weight of *decipium* was between 90 and 95. Its salts were colorless and characterized by two absorption bands placed at $\lambda = 416$ nm and $\lambda = 478$ nm on the scale proposed by Lecoq de Boisbaudran.

Contemporaneous to these events, Delafontaine's former mentor, Jean-Charles Galissard de Marignac announced the isolation of a new metal[49] from gadolinite,[50] ytterbium,[51] whose atomic weight was determined to be approximately 131.

Marc Delafontaine found himself examining, during the characterization of the two new substances, some material rich in *didymium*. A visible-region spectroscopic examination of this element, thought to be sufficiently pure, led him to observe two very sharp bands distinguished by the colors blue ($\lambda = 482$ nm) and green ($\lambda = 569$ nm). This experimental evidence of the complex nature of *didymium* convinced him that it contained a new element characterized by blue spectral bands.[52] His line of reasoning was correct and in certain ways anticipated the discovery of neodymium and praeseodymium[53] accomplished by Carl Auer von Welsbach (1858–1929) in 1885.[54] If Delafontaine had had more time and if he were not involved in those empty polemics with Smith, he may have been the first to succeed in separating *didymium* into its elementary components.

In the meantime, discoveries of new elements were going forward in frighteningly rapid succession. On July 28, 1879, Lecoq de Boisbaudran, in analyzing samples of samarskite, noted that on addition of ammonium hydroxide, an unknown earth precipitated before *didymium* oxide. Spectral analysis showed two bands in the blue region, apparently different from those of *decipium*. Lecoq called the new earth samaria, after the mineral from which it had been extracted, and he named the element samarium.[55] The mineral's name was taken from that of a virtually unknown Russian mining engineer, Vasilij Evgrafovič Samarskij-Byhovec (1803–70), who suddenly rose to prominence as being the first person whose name was definitively given to an element.

On August 16 of the following year, Jacques Louis Soret repeated the spectral analysis of samarium and confirmed the identity of Delafontaine's *decipium*.[56] Today, we are more or less certain that the latter was an impure sample of samarium.

Then, in 1880, Delafontaine systematically tackled the problem of putting in order the discoveries that had taken place in the previous 2 years.[57] Although he spoke extensively of his elements, and in particular of his unreliable *decipium*, he said very little about the discovery of *philippium*, now almost taken for granted. He also wrote of the recent discoveries of ytterbium, samarium, and scandium [58] and spoke of the very recent discoveries[59] of Per Theodor Cleve (1840–1905): thulium and holmium.[60] Delafontaine was careful to protect *decipium* from the claims of Lecoq de Boisbaudran, who was not aware of the danger that the discovery of holmium represented: this element was identical to *philippium* and only much later, when the name of holmium had already entered into common usage, was the validity of Delafontaine's discovery recognized. If the name of *philippium* has unfortunately disappeared forever, it must be said that today's inclination to re-evaluate the work of Delafontaine and Soret has led justifiably to their being included as co-discoverers of holmium.

Surprisingly, the name of element X remained in the scientific literature for a very long time: Subsequent to Soret's publication, Gerhard Krüss and Lars Fredrik Nilson (1840–99) carried out an extensive spectroscopic examination of seven minerals rich in the rare earths and from various sources. In 1887, the two authors arrived at the unsurprising conclusion that Soret's element X was actually a mixture of seven distinct elements[61] that they called $X\alpha$, $X\beta$, $X\gamma$, $X\delta$, $X\varepsilon$, $X\zeta$, and $X\eta$. In addition, from the same study, they asserted that erbium was composed of two elements, $Er\alpha$ and $Er\beta$, and that thulium and samarium were not exceptions, conferring on each two elements, *Tmα* and *Tmβ* and *Smα* and *Smβ*, respectively. For *didymium*, they found an astonishing ten constituents: *Diα, Diβ, Diγ, Diδ, Diε, Diζ, Diη, Diθ, Diι*, and the 10th indicated simply as *Di* without a Greek letter designation. Thus, one published article has the dubious honor of having brought together the false discoveries of 23 elements! Because this abundant crop of errors seemed highly unlikely, the authors' conclusions were not taken seriously.

III.2.4. ROGERIUM AND COLUMBIUM

Toward the end of 1879, John Lawrence Smith informally announced the results of some of his research on new elements. As we have seen, some years earlier, he had happily found a very promising field of scientific endeavor in the investigation of the ceric and yttric earths, and, in a short time, he had announced with great satisfaction that he had isolated a new substance he called *mosandrium*. From then on he dedicated himself to an assiduous study of samarskite, a mineral rich in the rare earths, and thus he claimed—or at least believed—that he had discovered two more new elements.[62] His great desire was to name the first one *columbium* and the second *rogerium,* in honor of his friend and instructor William Barton Rogers[63] (1804–82), just as Delafontaine sought to honor his teacher, Philippe Plantamour.

Because many other duties laid claim to his attention, Smith dedicated very little time to this line of investigation. With the exception of the purification of a small portion of *mosandrium,* he accomplished no other scientific research. Beginning in 1880, his health began to deteriorate, and he developed chronic liver inflammation. He often had to remain at home and confined to bed for long periods of time, thus interrupting his numerous and eclectic research investigations. On August 1, 1883, a violent attack of his illness left him bedridden for the remainder of his life. On October 22, after prolonged suffering, Lawrence Smith died at Louisville, Kentucky.[64]

During his lifetime, Smith collected many honors. He was made *chevalier de la légion d'honneur* in France; he was decorated with the order of Nichan Iftabar, the order of Medjidich from the Turkish government, and the order of Saint Stanislaus in Russia. In 1874, he became president of the American Association for the Advancement of Science and for a year (1877) he served as president of the newly formed American Chemical Society.[65]

III.2.5. CONCLUSION

On May 13, 1890, the chemist and spectroscopist Jacques Louis Soret died at the age of 63.[66] He had occupied the chair of chemistry at the University of Geneva since 1873, which he had left 3 years earlier to take up the chair of medical physics. He became famous not only for having determined the formula for ozone, but for his studies in electrolysis, spectroscopy, and fractionation of the rare earths. He climbed Mont Blanc and determined its altitude. He was interested in optics and in many other fields, in both physics and chemistry. On March 17, 1890, the French Academy of Sciences, by a vote of 41 to 4, gave joy to his last days by electing him foreign associate to take the place of James Prescott Joule (1818–89).

After the death of his colleague, Marc Delafontaine devoted himself once again to the extraction of *philippium*. On April 24, 1897, he asserted that he had extracted the metal by digesting 500 g of fergusonite[67] with 1,500 g of concentrated hydrofluoric acid.[68] In consecutive fractional crystallizations, with the method based on the differential solubilities of the nitrates of the rare earths, he extracted a sufficient quantity of *philippium* to carry out a certain number of chemical tests: he prepared the *philippous* and *philippic* salts. He obtained the sulfates, nitrates, oxalates, chloride, formates, and other salts of *philippium*, and he succeeded in determining their solubilities. Yet all his work fell into oblivion. No chemist who worked on the isolation and characterization of the lanthanides ever mentioned his *philippium* again.

The last word with respect to *element X* and *philippium* appears to have been made by the young Georges Urbain who wrote his doctoral thesis under Charles Friedel (1832–99). In 1898, he published a work on a new method of fractionation of the rare earths based on ethyl sulfate.[69] In it, he hypothesized that some bands of *element X*, erbium, and *didymium* would be close to one another and therefore highly deceptive. He also asserted that *element X* was actually impure holmium mixed with traces of dysprosium. Urbain offered the opinion that Delafontaine had fallen into a similar error in announcing the discovery of *philippium*.

The last scientific work of Marc Abraham Delafontaine was given to the publisher on June 3, 1905, and consisted of an impressive listing of the spectral lines of many rare earth elements. He used the most sophisticated state-of-the-art equipment, among which was the Rowland diffraction grating; he cited the work of important inorganic chemists of the caliber of Robert Tobias Thalén (1827–1905) and Robert Bunsen. He described the spectrum of terbium, yttrium, and other metals, but this time he made no mention of the second of his "creatures," *decipium*.

With respect to *philippium*, Delafontaine published a semi-retraction in which he stated that his sample was impure and that many of the spectral lines attributed to it actually could be attributed to other rare earths. After this partial "admission," he made note that at the moment of his announcement yttrium was not a metal, but a complex mixture of elements among which appeared to be scandium. This retraction, even though partial, did nothing to help rehabilitate his actions; however, recent analyses tend to confirm more than Delafontaine could ever have hoped for: the spectral lines of holmium and *philippium* in the visible region coincided, thus making Delafontaine the first discoverer of the element with atomic number 67.

By now very old and embittered by a lack of recognition, Delafontaine withdrew from public life and prophesied a gloomy future for the study of the rare earths.[70] For the most part, Delafontaine's lack of wider public recognition was due to the fact that he worked outside of the narrow and elite circle of university professors and also to the fact that the better part of his work was lost in a laboratory fire in Chicago. His fame as an expert analytical chemist was based on his ability to resolve many complicated homicide cases. He lived his last years with his son Jules and his daughter-in-law, Elizabeth Farson. Delafontaine had a lively temperament and was jovial by nature; in his late old age it was not unusual, passing by the window of his home, to hear the sounds of his happy voice or to experience a spirited laugh. No documents exist that report his death; it is assumed that he passed away in 1911.

Notes

31. There are two versions relative to Delafontaine's date of birth. *Poggendorff's Biographisch-Literarisches Handwörterbuch*, Verlag von Johan Ambrosius Barth, Leipzig, 1898; p. 344 speaks of 1838; on the other hand, M. E. Weeks's *Discovery of the Elements*; 7th Ed. Journal of Chemical Education: Easton, Pennsylvania, 1968, p. 677, gives the date as 1837.

32. Marignac, J. -C. Galissard de *Ann. Chim. Phys.* **1878**, *5*(14), 247.
33. Delafontaine, M. *Ann. Chim. Phys.* **1878**, *5*(14), 238.
34. Samarskite is a radioactive rare earth mineral series which includes samarskite-(Y) with formula: $(YFe^{3+}Fe^{2+}U,Th,Ca)_2(Nb,Ta)_2O_8$ and samarskite-(Yb) with formula $(YbFe^{3+})_2(Nb,Ta)_2O_8$. The formula for samarskite-(Y) is also given as: $(Y,Fe^{3+},U)(Nb,Ta)O_4$
35. Delafontaine, M. *Compt. Rend. Chim.* **1878**, *87*, 559.
36. Philippe Plantamour (1816–98) was a descendant of a family that took refuge in Switzerland in the 18th century for religious reasons; he was a chemist, a physicist, and the inventor of the limnograph by means of which he made regular observations on the variation of the level of Lake Geneva.
37. Soret, J. L. *Compt. Rend. Chim,* **1878**, *87*, 1062.
38. He was the son of the well-known painter, writer, and composer of erotic poetry Adélaïde-Gillette Dufrénoy (1765–1825), as well as the nephew of the goldsmith to the last kings of Poland.
39. Ercinite is a mineral with the formula $FeAl_2O_4$.
40. In the inverted microscope, the light source and the condenser are placed above the stage, whereas the objectives and the turret are under the stage. The inverted microscope is used to observe living cells and organisms at the bottom of a large container under natural conditions rather than on a glass slide.
41. Smith, J. L. *Compt. Rend. Chim.* **1878**, *86*, 146.
42. Smith, J. L. *Compt. Rend. Chim.* **1878**, *86*, 148.
43. Smith, J. L. J. *Am. J. Science* **1877**, *13*, 359; Smith, J. L. *Am. J. Science* **1877**, *13*, 509.
44. The chemist referred to is the Swede, Carl Gustav Mosander (1797–1858), discoverer of lanthanum.
45. Delafontaine, M. *Compt. Rend. Chim.* **1878**, *87*, 600.
46. Smith, J. L. *Compt. Rend. Chim.* **1878**, *87*, 831.
47. Gerland, W. B. *Chem. News* **1878**, *39*, 136.
48. Delafontaine, M. *Compt. Rend. Chim.* **1878**, *87*, 632.
49. Marignac, J. -C. Galissard de *Compt. Rend. Chim.* **1878**, *87*, 578; Marignac, J. -C. Galissard de *Ann. Chim. Phys.* **1878**, *5*(14), 247.
50. Gadolinite has the chemical formula $Y_2FeBe_2Si_2O_{10}$.
51. In 1907, Georges Urbain disproved the elementary nature of this element and from it he extracted *neo-ytterbium* and *lutecium*.
52. Delafontaine, M. *Compt. Rend. Chim.* **1878**, *87*, 634.
53. Welsbach von, C. A. *Monats. f. Chemie* **1885**, *6*, 477.
54. Adunka, R. *Carl Auer von Welsbach: Entdecker—Erfinder—Firmengründer*; Kärntner Landesarchivs: Klagenfurt, Austria, 2013, p. 17.
55. Lecoq de Boisbaudran, P. E. *Compt. Rend. Chim.* **1879**, *89*, 212.
56. Soret, J. L. *Compt. Rend. Chim.* **1880**, *91*, 378.
57. Delafontaine, M. *Compt. Rend. Chim.* **1880**, *90*, 221.
58. Cleve, P. T. *Compt. Rend. Chim.* **1880**, *89*, 419.
59. Cleve, P. T. *Compt. Rend. Chim.***1879**, *89*, 478; Cleve, P. T. *Compt. Rend. Chim.* **1879**, *89*, 708.
60. The names thulium and holmium derive respectively from Thule, the ancient name of Scandinavia, and Holmia, the latinized name of Stockholm.
61. Krüss, G.; Nilson, L. F. *Ber. Dtsch. Chem. Ges.* **1887**, *20*, 2134.
62. Wyckoff, W. *Nature* **1879**, *21*, 146; Crookes, W. *Phil. Trans.* **188**3, *174*, 891; Crookes, W. *Chem. News* **1883**, *49*, 182; Smith, J. L. *Am. Chem. Journ.* **1883**, *5*, 73.
63. Professor Rogers was, among other things, president of the National Academy of Sciences.
64. Silliman, B. *Journal of the National Academy of Sciences* **1884**, 4, April 17, 217.
65. Hirsch, R. F. *Anal. Chem.* **2002**, *74*, 153A.
66. Anon. *Chem. News* **1890**, *63*, 289.

67. Fergusonite is a mineral that owes its name to Robert Ferguson (1767–1840); its chemical formula is $(Ce,La,Y)NbO_4$.
68. Delafontaine, M. *Chem. News* **1897**, *75*, 229.
69. Urbain, G. *Bull. Soc. Chim.* **1898**, *19*, 376.
70. Delafontaine, M. *Chem. News* **1905**, *92*, 5.

III.3

LAVŒSIUM AND DAVYUM: THE RISE AND FALL OF TWO METALS WITH ILLUSTRIOUS NAMES

The inside stories of how the elements were accepted into the periodic table are complex and fascinating, especially when they bring to light lost facts and false discoveries separated from us by 130 years of history, as is the case with *lavœsium* and *davyum*.

III.3.1. THE DISCOVERY OF LAVŒSIUM

In 1861, Jean-Pierre Prat,[71] returning from an excursion to Ariége in the French Pyrénées, carried with him some specimens of minerals. Among these was one that had struck him particularly: a stone, black, compact, with a metallic reflection similar to graphite. Chemical analysis of this rock was difficult because the mineral was so complex: a mixture of the sulfides, oxides, selenides, tellurides, carbonates, numerous sulfates, and silicates of the alkali and alkaline earth metals. It also contained, in the elemental state, manganese, iron, nickel, and an unknown metal.

Prat was a member of the Société des Sciences Physiques et Naturelles de Bordeaux and, in a presentation read on February 5, 1862, at a meeting of the Society, he expressed his desire to call the new metal *lavœsium* (symbol Ls) in memory of the great chemist Antoine Laurent Lavoisier.[72] The new metal had a silvery color, was malleable, and had a melting point above 600 °C; it had a density around 7 g/cm^3 and demonstrated some peculiar chemical and physical properties.

Lavœsium reacted with chlorine, bromine, and iodine in the elementary state to form white, insoluble salts. With hydrosulfuric acid, it produced a tawny yellow precipitate. None of these characteristic reactions seemed to be similar to those of elements already known. From existing documentation, it does not seem that Prat was about to make an attempt—nor did he have the necessary instruments—to determine the atomic weight of the new element. Later investigations[73] suggested to the author the idea that the element might be more widespread in nature than he had previously thought: in fact, he found it again in hydroxysilicates of nickel and manganese coming from New Caledonia[74] and more abundantly in pyrites.[75] *Lavœsium* was, in all probability, being confused with one or more of the metals listed in Table III.1.

For 15 years, no one spoke of *lavœsium*. Only in 1877 did notice of the new metal leap the boundaries of provincial France: first in the magazine *Le monde pharmaceutique et médicale*,[76] and then, a few months later, in the journal *La Nature*[77] through the work of the young Gaston Bonnier (1853–1922), a future naturalist, at the time chemist *préparateur* at the École Normale Supérieure at Paris. The French review was translated in the pages

Table III.1 List of Some Elements and Their Analytical Characteristics (Lavœsium (Ls)) in all probability was confused with one or more of these)

Metal	*Cl_2 (color)*	*Br_2 (color)*	*I_2 (color)*	*H_2S (color)*
Ls	$LsCl_2$ (white)	$LsBr_2$ (white)	LsI_2 (white)	Ls_2S (tawny-yellow)
Zn	$ZnCl_2$ (white)	$ZnBr_2$ (white)	ZnI_2 (white)	ZnS (white)
Cd	$CdCl_2$ (white)	$CdBr_2$ (white)	CdI_2 (white)	CdS (yellow-orange)
Cu	CuCl (white)	CuBr (pale yellow)	CuI (white)	Cu_2S (blue-black)
Fe(II)	$FeCl_2$ (white)	$FeBr_2$ (yellow)	FeI_2 (gray violet)	FeS (white; black if impure)
Fe(III)	$FeCl_3$ (green)	$FeBr_3$ (red)	–	–
Mn	$MnCl_2$ (pink)	$MnBr_2$ (pink)	MnI_2 (pink)	MnS (pink)
Ni	$NiCl_2$ (yellow)	$NiBr_2$ (yellow)	NiI_2 (black)	NiS (black or yellow)
Tl	TlCl (white)	TlBr (yellow)	TlI (yellow)	Tl_2S (black)

of Sir William Crookes's *Chemical News*[78] and in *Chemischer Jahresbericht.*[79] In his work, Bonnier did not cite the original 1862 work of J.-P. Prat and, consequently, the scientific community of the time believed the year of the publication of the Bonnier article, 1877, was the date of the discovery of *lavœsium*, thus postdating the presumed discovery of the metal by 15 years. In 1883, a table of elements printed in Italy reported the name of this metal followed by numerous questions in the correspondence regarding its atomic weight, density, and specific heat. The existence of this metal remained uncertain for such a long time that, in the end, without any retraction, it disappeared into the world of complete indifference.

III.3.2. A RESIDUE OF WORK ON PLATINUM: DAVYUM

On June 28, 1877, a few months after the second announcement of the discovery of *lavœsium*, a Russian chemist, Sergius Kern, from the iron works of Obouchoff in Saint Petersburg, wrote in the pages of *Comptes Rendus* that he had isolated a new metal belonging to the platinum group. He named it *davyum* in honor of Sir Humphry Davy.[80]

Kern found the metal[81] by subjecting to various treatments a small quantity of a dark red residue obtained from platinum-bearing gangue coming from Russian deposits in the Urals. Upon heating, the dark red material gave a spongy mass that melted under an oxyhydrogen flame and yielded a small ingot of *davyum* (symbol Da). The metal in his possession was very scarce (0.27 g) but sufficient to determine some physical properties: hardness and ductility. The measured density was 9.385 g/cm^3. Kern observed that the metal was attacked by aqua regia, but not by hot sulfuric acid. A short time later, he published other properties of the metal[82] and asserted that *davyum* could occupy the empty box lying between molybdenum and ruthenium in D. I. Mendeleev's periodic table. The atomic weight was estimated at around 100, corresponding to present-day technetium. He gave his sample of *davyum* to the engineer Alexeieff[83] so that he could determine the exact atomic weight. The results were different from those hoped for: 154 (an atomic weight that we now know is intermediate between that of europium, 151.96, and gadolinium, 157.25). The first chemist to become extremely dubious about Kern's work was A. H. Allen who sarcastically remarked that if Kern often blundered in arithmetic,

Sistema periodico degli elementi chimici.

H =1	R^2O 1	RO 2	R^2O^3 3	RO^2 RH^4 4	R^2O^5 RH^3 5	RO^3 RH^2 6	R^2O^7 RH 7	RO^4 8	9	10	11	
I.	Li 7	Be 9,1	Bo 11	C 12	N 14	O 16	Fl 19					He 4 Ne 20
II.	Na 23	Mg 24,4	Al 27,1	Si 28	P 31	S 32	Cl 35,5					Ar 39,9
III.	K 39,1	Ca 40	Sc 44	Ti 48,1	V 51,2	Cr 52,4	Mn 55	Fe 56	Co 58,6	Ni 58,9	Cu 63,3	
IV.	Cu 63,3	Zn 65,3	Ga 69,9	Gm 72	As 75	Se 79	Br 80					
V.	Rb 85,4	Sr 87,5	Y 89	Zr 90,7	Nb 94	Mo 96	— 99	Ru 101,5	Ro 104	Pd 106	Ag 107,9	Kr 81,8
VI.	Ag 107,9	Cd 112	In 113,7	Sn 118	Sb 120	Te 125	J 126,9					
VII.	Cs 132,8	Ba 137	La 138	Ce 140	Di 142,5	— 147	— 149	Sm 150	— 150	— 151	— 152	Xe 128
VIII.	— 152	— 155	— 156	— 160	— 164	— 165	— 166					
IX.	— 169	— 171	Yb 173	— 178	Ta 182,8	W 184	— 187	Os 191	Jr 193	Pt 194,8	Au 196,6	
X.	Au 196	Hg 200	Tl 204	Pb 206,9	Bi 208	— 212	— 215					
XI.	— 220	— 224	— 228	To 233	— 236	U 240	— 244	— 246	— 247	— 248	— 250	

FIGURE III.03. Historic Periodic Table of Hugo Schiff (1834–1915). A periodic table of the elements designed personally by Hugo Schiff (1834–1915), about 10 years after Mendeleev's. Note the zero group of noble gases that has been added and some errors: the symbols of Gm for germanium, Ro for rhodium, To for thorium, Fl for fluorine, Bo for boron, J for iodine, Jr for iridium, and the presence of a nonexistent element, *didymium*, with the symbol Di. Courtesy, Chemical Heritage of the Department of Chemistry of the University of Florence, Italy.

then his chemical analyses would also be likewise imprecise. Allen concluded his article: "unfortunately, that gentleman's contributions contain little that is novel, and that little is mostly incorrect."[84]

The following year, Kern[85] reported other observations on *davyum*, among which was its extreme rarity in nature, but no one seemed to be any longer interested in this metal, whose existence was so uncertain. In 1895, the Swedish chemist P. J. F. Rang published an updated Mendeleev periodic table in which he placed *davyum* in the seventh group under manganese. Like *lavœsium*, the table of elements printed in Italy reported a new metal with the symbol Da. This list was updated personally by Hugo Schiff (1834–1915), who followed with attention the tumultuous series of announcements of new element discoveries, but he declined to mention both *davyum* and *lavœsium*. Schiff's historic table is shown in Figure III.03. In 1898, Professor John William Mallet (1832–1912) of the University of Virginia decided to repeat Kern's experiments.[86] He obtained a mass of 15 g from an original 35 kg of platinum-bearing mineral that he had ordered from the same deposit in which *davyum* had been found, and he confirmed the reactions described by the Russian chemist 20 years earlier.

III.3.3. LAVŒSIUM FALLS INTO OBLIVION

In the enthusiastic review of the discovery of *lavœsium*, Gaston Bonnier had already inadvertently explained the cause of Prat's error: there was no spectroscopic evidence.

Prat's first communication reported only the chemical tests that he had carried out to identify the new metal. As Gaston Bonnier in fact explained, the pharmaceutical chemist of Bordeaux only later recorded the spectrum of *lavœsium*, which curiously resembled the spectrum of copper.

Bonnier did not seem to doubt the truth of Prat's discovery, and he did not take care to analyze the curious resemblance between the spectral lines of *lavœsium* and copper. A more in-depth spectroscopic investigation would have immediately excluded the presence of a new element, but it was not done.

On April 13, 1877, a week after *Chemical News* announced the discovery of the new metal, a reader, W. H. Walenn,[87] sent a letter to the editor in which he cautiously observed the strong resemblance between *lavœsium* and zinc.[88] He worked in the field of electrodeposition of brass in alkaline baths; by subjecting to electrolysis a solution of Zn^{2+}, Cu^{2+}, potassium cyanide, and ammonium tartrate, slightly above the freezing point, he obtained extremely pure zinc at the negative electrode. When he raised the temperature, he noticed that the color of the metal deposited changed from silvery to yellowish. Walenn hypothesized that Prat's metal could have been made with a solution that he had placed under stress.

Walenn's hypothesis was correct: the presumed *lavœsium* could be none other than a type of brass containing a greater amount of zinc.[89] This hypothesis also explained the spectrum of *lavœsium* (coinciding with that of copper) and its chemical properties, virtually identical to those of zinc. We are tempted to hypothesize that Prat's sample also contained cadmium because the systematic chemistry of this element does not contradict the observations he made, but not having the original sample in hand precludes our going any further with this speculation. It's possible to believe that Prat had rediscovered thallium a year after Sir William Crookes and Claude-Auguste Lamy[90] and that he had called it *lavœsium*. In fact, this third-group metal can be found in pyrites and in zinc-containing rocks and can be extracted by roasting the raw mineral, materials and processes that Prat described in his work. But, unfortunately, even this hypothesis has to be discarded: thallium's spectrum is unequivocal, presenting only one green line in the visible spectrum, in contrast to that of *lavœsium* which is much richer, according to Prat's analysis, being composed of three distinct groups totalling 23 lines.

III.3.4. DAVYUM'S LONG AGONY

Meanwhile, the existence of Kern's metal seemed to be increasingly in doubt.[91] In 1910, Julius Ohly dedicated an entire paragraph to *davyum* in his text on the rare metals but did not introduce anything new except a spelling error in the name and the symbol of the presumed metal: *davyium* and Dm. Even if no one any longer believed in the existence of this element, some authors still felt the need to include it in their treatises on inorganic chemistry (further mangling its name to *devium*).[92]

On May 20, 1950, one month prior to his death, John Gerald Frederick Druce (1894–1950) wrote his last article just in time to read it in the pages of *Nature*.[93] In his youth, he had been involved in the false discovery of a missing element.[94] Before he died, he wanted to restore *davyum* to its place in the periodic table, but he did not succeed. He hypothesized that rhenium was none other than the element discovered by Kern back in 1877. This was not the first time that Druce[95] tried to "rehabilitate" a discovery, as one can see in his 1936 article in which he proposed that *davyum* was actually the precursor of *masurium* (presently technetium).

Druce's second hypothesis (1950) is in error because we now know that technetium is not present (in any appreciable amount) in nature and that, therefore, Kern would not have been able to melt a small ingot of this metal and call it *davyum*. With respect to rhenium, there are too many and too great incongruities between Kern's work and the results obtained by the actual discoverers of rhenium (Otto Berg [1874–1939] and the husband-wife team of Walter Noddack and Ida Tacke Noddack): the atomic weight calculated for *davyum*, 154, is nowhere near that of eka-manganese (Re), which is 186.21. The densities of the two metals are clearly not the same: rhenium's is 21.04 g/cm^3, while that calculated for *davyum* is little more than a third of that amount. Finally, it is difficult to believe that some chemical tests with potassium thiocyanide could possibly confirm the identities of these two elements. In our view, a likely explanation of the errors committed by Kern are that the reaction of the chloride of *davyum* with potassium thiocyanide gave a red color like that produced with ferric salts; the brownish-black precipitate of presumed DaS could very well have been FeS; and the density of Da, 9.385 g/cm^3, could have been the combined weighted densities of some lighter elements like iron and copper (known to be present in the sample at a level of about 7%) with traces of much heavier elements like platinum, iridium, palladium, and ruthenium, known to be present in the sample in a total amount of more than 90%.

We cannot be certain that Kern did not actually find some traces of rhenium, which is almost always present in platinum-bearing materials, but it is not possible to believe that he could have isolated it using the normal investigative techniques of his time. In all likelihood, Kern had obtained a mixture of iron, platinum, and iridium with variable traces of copper, palladium, and ruthenium.

III.3.5. CONCLUSION

In 1888, Henry Carrington Bolton (1843–1903)[96] published a detailed list regarding the discovery of the elements[97] in the decade 1877–87. The discovery of lavœsium first appeared in the reports of the Société des Sciences Physiques et Naturelles de Bordeaux in 1862. However, Gaston Bonnier gave an erroneous date of 1877 in his publications, and this error was picked up in later reports. Apparently no one had read the original reports of the discovery. Bolton, too, fell into this unfortunate error when he reported the discovery of *lavœsium* as being in 1877, along with *davyum*, *mosandrium*, and *neptunium*. Table III.2 is a historical summary of these four presumed elements.

Neither *lavœsium* nor the other three discoveries were ever proved to be true elements, nor does the name *lavœsium* occupy a place in the periodic table. At present, the last five elements to receive official names are darmstadtium[98,99] (Z = 110; symbol Ds),[100] roentgenium (Z = 111; symbol Rg), copernicium (Z = 112; symbol Cn), flerovium (Z = 114; symbol Fl), and livermorium (Z = 116; symbol Lv). However, four additional elements, with the atomic numbers of 113, 115, 117, and 118, remain unnamed. As the authorities work to decide an appropriate name for one of these transuranium elements, it would be appropriate to remember that Lavoisier's work, along with that of Mendeleev, to classify the elements is one of the pillars on which our present knowledge of chemistry is based.

Table III.2 List of the Presumed Elements Discovered in 1877, the Minerals from Which They Were Extracted, the Person Claiming the Discovery, and the Bibliographic Citation

Name	*Material*	*Discoverer*	*Bibliographic reference*
Neptunium[a]	Columbite $(Fe,Mn)(Nb,Ta)_2O_5$	R. Hermann	*J. prakt. Chem.* **1877**, *2*, 15; 105; *Chem. News* **1877**, *35*, 197
Lavœsium	Pyrites FeS_2	J.-P. Prat	*Le monde pharmaceutique* **1877**, *8*, 4
Mosandrium	Samarskite $(Y,Fe,U)(Nb,Ta)_2O_5$	J. L. Smith	*Compt. Rend. Chim.* **1877**, *87*, 148 *Chem. News* **1877**, *35*, 100
Davyum	Platinum-bearing deposits	S. Kern	*Compt. Rend. Chim.* **1877**, *87*, 72

[a] This "*neptunium*" is not to be confused with element 93 discovered in 1940 by Edwin McMillan and Philip H. Abelson.

Notes

71. Prat, J.-P. *Memoires de la Société des Sciences Physiques et Naturelles de Bordeaux*, **1862**, Séance du 5 février.
72. Lavoisier was considered the first modern chemist, raising the new discipline, chemistry, from the hazy foundations of alchemy. His work remained unsurpassed for many years. He made essential contributions to the definition of "element," so much so that in the 1930s, the chemist Georges Urbain asserted that he was closer to Lavoisier's concept than to that provided by nuclear physicists.
73. Prat, J.-P. *Memoires de la Société des Sciences Physiques et Naturelles de Bordeaux* **1876**, des procès verbaux; Séance du 21 décembre.
74. Prat, J.-P. *Memoires de la Société des Sciences Physiques et Naturelles de Bordeaux* **1877**, des procès verbaux; Séance du 14 juin.
75. Prat, J.-P. *Memoires de la Société des Sciences Physiques et Naturelles de Bordeaux* **1877**, Séance du 8 février; Prat, J.-P. *Memoires de la Société des Sciences Physiques et Naturelles de Bordeaux* **1877**, Séance du 22 mars.
76. *Le Monde Pharmaceutique et Médicale* **1877**, *8*, no. 1, 4.
77. Bonnier, G. *La Nature* **1877**, Première Semestre, 363.
78. *Chem. News* **1877**, *35*, 137.
79. *Chemischer Jahresbericht* **1877**, 275.
80. Kern, S. *Compt. Rend. Chim.* **1877**, *87*, 72.
81. Kern, S. *Chem. News* **1877**, *36*, 4; *The Manufacturer and Builder* **1877**, *9* (September 9), 206.
82. Kern, S. *Chem. News* **1877**, *36*, 114; Kern, S. *Chem. News* **1877**, *36*, 155; Kern, S. *Chem. News* **1877**, *36*, 164; Kern, S. *Compt. Rend. Chim.* **1877**, *87*, 667.
83. Kern, S. *Compt. Rend. Chim.* **1877**, *87*, 623.
84. Allen, A. H. *Chem. News* **1877**, *36*, 33.
85. Kern, S. *Chem. News* **1878**, *37*, 65.
86. Waggoner, J. *Chemistry* **1975**, *48*(10), 25.
87. Walenn, a member of the Physical Society, was an expert in thermoelectric batteries.
88. Walenn, W. H. *Chem. News* **1877**, *35*, 154.

89. Cotton, F. A.; Wilkinson, G.; Murillo, C.; Bochmann, M. *Advanced Inorganic Chemistry*; John Wiley & Sons: New York, 1999.
90. Lamy, C.-A. *Compt. Rend. Chim.* **1862**, *54*, 1255.
91. Thorpe, E. *History of Chemistry*, vol. 1; Watts & Co.: London, 1914.
92. Loring, F. H. *Chem. News* **1930**, *140*, March 21, 178.
93. Newton Friend, J.; Druce, J. G. *Nature* **1950**, *165*, May, 819.
94. John Gerald Druce, chemist and student of Jaroslav Heyrovský, Nobel Laureate in Chemistry, 1959, was first Englishman to receive a doctorate at the Univerzita Karlova (Charles University), Prague. In the 1920s, together with the chemist F. H. Loring, he did researched the missing element with atomic number 87. Later, Loring believed that he had found it and wanted to call it *alkalinium* (Ak).
95. Druce, J. G. *Chem. Ind.* (London, U.K.) **1936**, *2* (July), 577.
96. Henry Carrington Bolton, chemist, historian and bibliophile. In 1999, the Chemical Heritage Foundation established the Bolton Society in his honor for the collection of material applicable to chemistry and its related technologies.
97. Bolton, H. C. *Chem. News* **1888**, *58*, 188.
98. Karol, P. J.; Nakahara, H.; Petley, B. W.; Vogt, E. *Pure and Appl. Chem.* **2003**, *75*, 1601.
99. Recommendations, *Pure and Appl. Chem.* **2012**, *84*(7), 1669–1672.
100. By curious coincidence, darmstadtium had for a certain period the symbol Da, like that proposed for *davyum*. Later, the IUPAC changed the symbol so that it would not be confused with the unit dalton.

III.4

THE COMPLEX EVENTS SURROUNDING TWO "SCANDINAVIAN" METALS: NORWEGIUM AND WASIUM

In the 2-year period between 1877 and 1879, the Mendelevian revolution arrived at its 10th anniversary and seemed destined to continue to upset the peaceful landscape of inorganic chemistry. The search for new elements to fill up the empty spaces of the periodic table seemed to be giving rise to an unstoppable harvest of new discoveries.

The time seemed ripe for chemists and mineralogists to discover, in a very short time, all of the missing elements. The announcements of such discoveries became so profuse that it seemed that Western science was taking over the whole world. The continents, explored for centuries and exploited for their agricultural resources, were now furnishing new minerals rich in metals, both base and precious, and indispensable for the ever-hungry metallurgical industry of the Old World. Between the second half of 1877 and the first half of 1879, the pages of *Comptes Rendus de l'Académie des Sciences de Paris* and of *Chemical News* reported the discoveries of 20 elements: *neptunium, lavœsium, mosandrium, davyum*, an element observed as its oxide and generically called *new earth*, ytterbium, scandium, *ouralium*, samarium, terbium, holmium, thulium, *philippium, decipium, element X, barcenium,*[101] *columbium, rogerium, vesbium*, and *norwegium.*[102] If these discoveries had been confirmed as correct, one could speak of these as two wonderful years in inorganic chemistry. But the reality was very different: of these discoveries, 14 were false.

Although very few of these discoveries retained the dignity of being truly elements, some, such as *philippium, decipium*, element X, and perhaps *mosandrium*, came to be seen as true discoveries or rediscoveries of other rare earth elements. Unfortunately, the initial announcements of their discovery were passed over virtually unobserved, and later their respective discoverers engaged in long academic diatribes to establish priority. This poisoned the academic atmosphere both in Europe and in the Americas, and national rivalries were heightened. The European scientific community even arrived at the less than edifying assumption of according privilege to the more recognized and "consecrated" schools of scientific thought: the French, English, and German-Scandinavian schools.

One of the first controversies of this era arose when, from Norway, an announcement came of the discovery of a new metal that, in the end, was shown to be nonexistent. Even though this announcement was made in the pages of a minor journal, the presumed discoverer and his colleagues hastened to spread the notice abroad by publishing long excerpts in European scientific journals and sending letters to members of prestigious academies.

III.4.1. THE ANNOUNCEMENT OF THE DISCOVERY OF NORWEGIUM

Tellef Dahll (1825–93) was born in the Norwegian village of Kragerø[103] on April 10, 1825. In 1846, after having completed his regular university studies in mineralogy, he was hired as a mining official and later as inspector of the iron works at Fossum at Skien. Dahll confirmed the presence of coal on the island of And, and of gold in some areas of Finnmarken. From 1858, together with T. Kierulf, he was head of the Norwegian Geological Association. With Kierulf, he oversaw the publication of geological maps of Norway,[104] Scandinavia, and northern Finland.[105]

In 1872, he became superintendent of all the mines in the district of Sunnanfjällen.[106] On the occasion of celebrating the jubilee of the University of Uppsala, in 1877, he was granted a degree *honoris causa*.

In 1879, Dahll collected some samples of nickel arsenide and gersdorfftite[107] on the small island of Oterö, very close to the place where he was born. In his laboratory, he sought to accomplish a detailed chemical investigation on these samples. To his great surprise, he found that these rocks might contain a hitherto unknown element. Dahll was 54 years old and beginning to embark on a project that was beyond his strengths and talents, yet he published the discovery of a new metal in a monograph with the title "On Norwegium, a New Heavy Metal."[108]

The international scientific community became aware of this notice when Torstein Hiortdhal (1839–1925), professor of chemistry at the University of Christiania,[109] sent a letter to the Sorbonne, where the renowned Henri Etienne Saint-Claire Deville[110] was established, begging him to publish Dahll's findings in *Comptes Rendus de l'Académie des Sciences de Paris.* Almost immediately, Saint-Claire Deville sent the article to press and thus the discovery became widely known on the European continent.[111] Dahll, for his part, did not delay in publishing his discovery in the *Zeitschrift der Geologischen Gesellschaft*.[112] At the same time, announcements of the Norwegian geologist's claim became known in the United States.[113]

III.4.2. NORWEGIUM

Tellef Dahll subdivided his research into two parts: first, he isolated the unknown metal; later, he carried out some analytical tests to establish what group it belonged to. The mineral was roasted to remove the arsenic present; the residual material was then dissolved in aqua regia and precipitated with hydrogen sulfide. The precipitate was washed with water, then roasted again to remove sulfur and some remaining traces of arsenic. The final product, in Dahll's opinion, was the oxide of *norwegium*, which was redissolved in aqua regia and precipitated with a stoichiometric quantity of caustic potash. The free metal was recovered by heating the oxide in a carbon crucible, placing it in contact with a stream of hydrogen. The metal that remained in the crucible had a bright white appearance, was malleable, and had a specific gravity of around 9.5. *Norwegium* was easily attacked by oxidants such as nitric and sulfuric acids but not by complexing acids such as hydrochloric. The acidic solution was blue and turned green when diluted with water. On addition of excess caustic soda, ammonia, and sodium carbonate, this was replaced by an insoluble emerald-green precipitate that could only be dissolved with acid. The acidic solution of *norwegium*, upon addition of H_2S, was replaced with a sulfide with a dark brown color. These and other properties were studied again and then modified: at first, Dahll reported NgO as the formula for the oxide, to which he had assigned an elemental atomic weight of 145.95; later, he suggested that the oxide had the formula Ng_2O_3, causing the atomic weight to rise to 218.9.

The European scientific community, immediately rather skeptical, realized that an unfounded discovery had been published abroad only when, in October 1880, chemist George A. Prochazka (1855–1933) of the Tartar Chemical Company in Jersey City, New Jersey, reported that he had observed *norwegium* in American lead.[114] From then on, no announcement regarding this metal was published in any journals in the Old World; on the contrary, in the United States, the scientific community was very attentive to the study of the new metal and many new articles appeared as a result.[115]

About a year after the publication of the discovery of *norwegium*, George Prochazka advanced the curious hypothesis that American lead could conceal a mysterious unknown metal. He arrived at this conclusion after having examined some materials produced in the refinement of lead in several American factories. Prochazka suspected that an unknown metal, present in some American minerals and not completely separable using the established industrial processes, might be responsible for the gray or red tinges seen in refined lead.

III.4.3. A SECOND CLAIMANT

Professor Koenig of the University of Pennsylvania joined in the competition created by the announcement and propagation of the discovery of *norwegium* in the United States.[116] George Augustus Koenig (1844–1913), a naturalized American citizen, was born in Willstedt in the Grand Duchy of Baden in 1844. After having attended the Polytechnic Institute of Karlsruhe, he studied chemistry and mineralogy at Heidelberg and Berlin. Before coming to the United States in 1868, he spent a year in specialized study at the famous school of mines at Freiberg in the Kingdom of Saxony.

In the New World, he changed occupations several times, and he worked for the American mining company Tacony that had vast properties in the Chihuahua district of Mexico. Finally, in 1874, he became associate professor of chemistry and mineralogy at the University of Pennsylvania. Five years later, he occupied the chair of geology at the same university, but then left it because he preferred mineralogy and metallurgy. Shortly after becoming a full professor, in 1879, Koenig found himself involved in the discovery of a new metal. He made his announcement but with little scientific evidence to shore up the "vague suspicions" that his work led to. Koenig was examining in his chemical laboratory an unknown mineral found in Magnet Cove, Arkansas, when he became convinced that a white powder extracted from the rock and at first thought to be titanic acid was actually an unknown oxide. Thinking that he had discovered a new metal, he carried out a second chemical examination of the mineral and, for the second time, obtained the same results. Unfortunately, he did not succeed in isolating the metal in the elemental state, and his chemical tests were conducted on the oxide.

American journals, eager to show their public that the young North American republic was not dominated by European science, immediately embraced with ill-concealed parochialism the discoveries of *norwegium* and of the metal reported by Koenig.[117] Koenig, although respected as the most accomplished American mineralogist, was not very fortunate even in his own field. He described 12 new mineral species, but today we know that only two of them are still valid: bementite and paramelaconite.[118] In 1912, he published an exhaustive compendium of his chemical and mineralogical observations.[119] He died the following year at the age of 69.

III.4.4. THE "LAUNCHING" OF WASIUM

Some years before Mendeleev published his famous periodic table of the elements, the Swedish chemist Johann Friedrich (Jön Fridrik) Bahr reported that he had isolated a new earth, an oxide, from a sample of orthite. His work was similar to the research done on *norwegium*, although the two were displaced chronologically by about 14 years; in fact, at the moment that Dahll was publishing his results, Bahr had been dead 4 years. He was born at Visby on the island of Gotland, Sweden, on July 17, 1815. After finishing his studies at the Technological Institute of Stockholm, he became interested in analyzing Swedish minerals. Although he was an analytical chemist gifted with a very sharp mind, his work went almost unnoted by his contemporaries; very few of his articles were translated from his mother tongue.

At the age of 47, in 1862, Bahr published his discovery, "On a New Metal Oxide," which, for better or worse, was the reason for his ephemeral fame.[120] The article describing his discovery of a new metal was quickly published on the European continent thanks to the numerous translations of his original work, the first in German,[121] followed by French,[122] and finally English.[123]

A short time earlier, Bahr had received a sample of a mineral from an unidentified individual who worked in the mines at Rönsholm, a small island in the Gulf of Stockholm. The mineral, with a dark brown color, showed conchoidal fracturing properties. At first, he thought the mineral was only orthite, but quantitative analysis yielded difficulties and lots of surprises. The elements contained in it (silicon, aluminum, yttrium, manganese, magnesium, some alkali metals, calcium, iron, cerium, and *didymium*, as well as traces of lanthanum, uranium, and thorium) were found for the most part as oxides. At the end of a long and complex analysis of the mineral, about 1% of the oxide of an unknown metal remained. This was a yellow powder that, on heating, gave off dense red fumes and left a residue that looked like a white oxide. Bahr interpreted these results as the unequivocal presence of a new metal, and he gave much thought to an appropriate name for this new element. Wishing to give it a name that would honor both his country and the House of Vasa,[124] he decided on the name *wasium* (symbol Ws).

In all, he had not obtained more than a gram of the new element. Bahr determined the specific gravity of the metal to be about 3.726 g/cm^3. He also tried to record the spectrum of the unknown metal; unfortunately, *wasium* did not give a flame-test spectrum. On the contrary, the arc spectrum showed a series of very intricate lines; some of them belonged to the elements contained in the arc electrodes themselves; others, according to him, could have been the "fingerprints" of *wasium* itself.

Not long before the publication of the English version of his article, Bahr also analyzed the contents of two other minerals, among them gadolinite coming from Ytterby, in which he found traces of the new metal.

III.4.5. THE "SHIPWRECK" OF WASIUM

The name of the celebrated Vasa (or Wasa) dynasty,[125] sovereigns of Sweden, Poland, and Lithuania, appears to be synonymous with fame followed by sudden disaster. As with the celebrated galleon *Wasa*, which sank in 1863, so it was that less than a year after its birth *wasium* was taken into the ranks of "defunct elements" by an American chemist who was something of a jester.[126]

Almost a year having passed since the 1862 announcement, the first denial of the discovery of *wasium* came from Paris: the French chemist Jérôme F. J. Nicklès (1820–69), analyzing Norwegian orthite in search of Bahr's metal arrived at rebutting—not without a certain degree of arrogance—the discovery of his Swedish colleague, asserting that wasium was nothing more than a mixture of already-known elements, perhaps yttrium mixed with a little didymium or terbium.[127,128]

Nicklès satisfaction did not last very long; in 1869, he was the victim of lethal poisoning by fluorine in the fruitless attempt to isolate it in its elemental state. Also in 1864, Swiss chemist Marc Delafontaine published an article in which he demolished the discovery of *wasium*,[129] which was, in his opinion, nothing more than a mixture of rare earth elements. Then it was the turn of O. Popp, who arrived at the same conclusion as Delafontaine.[130]

As insistent voices on the nonexistence of *wasium* increased in the literature, at great risk of compromising his own scientific reputation, Bahr decided to drop out of the struggle and to clarify once more his own position on the subject. Thus, in July 1864, two articles in German appeared almost simultaneously.[131] In these, Bahr admitted his error relative to the hypothesis of the existence of *wasium*, but immediately afterward attacked the work of Delafontaine and Nicklès, criticizing both of them roundly. *Wasium* did not exist—that was clear to all—but in reversing his own position, passing from accused to accuser, Bahr stirred up a pernicious but sterile polemic. He who had originally committed an analytical error accused his colleagues—who had refuted the discovery of *wasium*—of grave shortcomings in chemical analysis. In fact, according to Bahr, *wasium* was not a mixture of yttrium, terbium or other rare earth elements, but was identical to thorium. The bitterness of this clash was interrupted by Nicklès's premature death. Bahr did not survive him by very long: in the spring of 1875, he died at Uppsala, not yet 60 years old.

III.4.6. THE EPILOGUE TO NORWEGIUM

Granted that *norwegium* could not find a place in the periodic table, we examined the systematics of this element. We assumed for simplicity that the qualitative analysis done by George Prochazka was the same as that done by Tellef Dahll. We then assumed that the sulfurous minerals examined by Dahll contained only the elements mentioned by him (Fe, Co, Ni, As, and S), as well as those that usually accompany arsenic, namely antimony and bismuth. Given these hypotheses, we can try to discover which element might have fooled the two investigators. Dahll, and later Prochazka, could have easily ended up by not separating out any elements at all, neither new nor already known. They simply took for a new element an alloy with a large amount of bismuth with nickel or with cobalt. In the first place, in favor of this hypothesis, it was necessary to report the molecular weight of *norwegium*, which was very close to that of bismuth. Furthermore, it is commonly known now that many alloys of bismuth have a melting point lower than that of the free metal, as did the alloy (Bi, Co, Ni) taken for Ng. The valence of 3 of the supposed new metal matches that of bismuth, the probably predominant component of the alloy mistaken by Dahll for *norwegium*. Finally one can take into account the numerous points of

Table III.3 Comparison of the Properties of Norwegium with Those of Possible Associated Elements with Which It May Have Been Confused

	Ng	*Fe*	*Co*	*Ni*	*As*	*Sb*	*Bi*
Atomic Weight	145.95; then 218.9	55.845	58.933	58.6934	74.921	121.76	208.98
T_{fus} (°C)	254	1,535	1,495	1,453	613	273	271.2
D (g/cm³)	9.44	7.86	8.9	8.9	5.7	6.69	9.8
Color of oxide	Emerald green (for NgO, or Ng_2O_3)	FeO Black Fe_2O_3 Brown	Olive green (CoO)	Dark green (NiO or Ni_2O_3)	White As_2O_3	Yellow Sb_2O_3	Yellow Bi_2O_3
Color of sulfide	Dark brown	Black	Black	Black	Gray	Orange red	Dark brown
Color of metal	Silvery white	Gray	Silvery white	Silvery white	Gray	Bluish white	Pinkish white

similarity at the qualitative level, such as the specific gravity, the colors of the metals, and also the colors of the oxides and sulfides.

The qualitative reactions of the presumed *norwegium* agree well with those of bismuth and of its oxide and hydroxide. In fact, Bi_2O_3, which one obtains by roasting its sulfide, is a polymeric covalent oxide with no acidic characteristics. From Bi_2O_3, one can recover metallic bismuth by reduction in a stream of hydrogen gas. On addition of hydroxides of the alkaline metals or of ammonia to solutions of bismuth, a hydroxide precipitates, $Bi(OH)_3$, which, like the oxide, has basic properties. Table III.3 compares the properties of other elements with which Dahll may have confused *norwegium*.

Although Dahll's discovery was not confirmed, neither was it completely refuted. The years passed and no one spoke of *norwegium* again. About a decade later, in 1888, a pleasant gratification greeted the elderly Tellef Dahll: his brother Johann, also a geologist, discovered a new mineral and called it dahllite[132] in his honor.

Tellef Dahll loved the outdoors and the possibilities offered by his profession in this regard were not lacking: he took many trips and compiled many details relative to the Scandinavian Peninsula. No longer young, in 1893, he undertook, in northern Norway, what turned out to be his last voyage. At first reluctant, because the state of his health seemed to preclude a strenuous expedition, an unexpected improvement induced him to put aside his hesitancies and to make the journey. On June 17, while Dahll was in the city of Telemark, the burdens of his work and the hardships of constant travel got the better of his hitherto strong constitution.

On the other side of the Atlantic, George A. Prochazka seemed to live through the events of *norwegium* with apparent unconcern. Having left his job at the Tartar Chemical Company, the following year he was hired by the Heller and Mertz Company.[133] Restless, but with an infinite capacity for renewal in the field of research, he passed from one industrial job to another, until his retirement in 1924. From then on, George Prochazka dedicated himself to his many other eclectic interests: from the color industry to political economy, from European voyages to musical comedies to Biblical exegesis.

Notes

101. The presumed metal was discovered in a Mexican mineral by John W. Mallet. The name *barcenium* seems to have originated through a simple misunderstanding on the part of the editor of the journal *Jahresbericht*. Taken from: Anon. *Transactions of the New York Academy of Sciences* **1887**, 200.
102. Anon. *Chem. News* **1879**, *41*, 267.
103. Some authors give the name Krager, others Kragerö.
104. Dahll, T. *Skand. Naturf. Foerlandl.* **1860**, *8*, 6.
105. Dahll, T. *Vid. Selsk. Forth.* **1868**, *10*, 9.
106. Norway belonged to the Swedish crown until it obtained its independence peacefully in 1905.
107. This mineral, discovered in 1843, takes its name from Johann von Gersdorff. It is a sulfo-arsenide of nickel and has the chemical formula NiAsS.
108. Dahll, T. *Vid. Selsk. Forth.* **1879**, *21*, 4.
109. In the period between 1878 and 1924, the city of Oslo was named Christiana or Kristiania, in honor of King Christian IV of Denmark (1577–1648).
110. There is a curious detail regarding the double citizenship of Saint-Claire Deville. French by birth, he was in fact the son of the French Consul to the Virgin Islands and hence born in a Danish colony.
111. *Comptes Rendus de l'Académie des Sciences de Paris* **1879**, *89* (I), 47; Excerpt of a letter of M. Hjortdahl to M. H. Saint-Claire Deville.
112. Notice reported by C. Rammelsberg in *Ber. Dtsch. Chem. Ges.* **1879**, *13*, 250.
113. Anon. *J. Am. Chem. Soc.* **1879**, *1*(9), 398; Anon. *The Manufacturer and Builder* **1879**, *11*(10), 231; Anon. *The Manufacturer and Builder* **1879**, *11*(11), 254; Anon. *The Manufacturer and Builder* **1879**, *12*(5), 110.
114. Prochazka, G. A. *J. Am. Chem. Soc.* **1880**, *2*(5), 213.
115. Anon. *The Manufacturer and Builder* **1880**, *12*(10), 230.
116. Anon. *The Manufacturer and Builder* **1879**, *11*(8), 183.
117. Anon. *The Manufacturer and Builder* **1879**, *11*(10), 231.
118. Bementite and paramelaconite have the respective chemical formulas: $Mn_8Si_6O_{15}(OH)_{10}$ and $Cu(I)_2Cu(II)_2O_3$.
119. Koenig, G. A. *Journal of the Academy of Natural Sciences* **1912**, *15*, 405.
120. Bahr, J. F. *Stockholm Ak. Handl.* **1862**, *19*, 8.
121. Bahr, J. F. *Ann. Phys. Chem.* **1863**, *119*, 572; Bahr, J.F., *J. prakt. Chem.* **1863**, *91*, 316.
122. Bahr, J. F. *Compt. Rend. Chim.* **1863**, *57*, 740.
123. Bahr, J. F. *Chem. News* **1863**, *8*, 175; Bahr, J.F. *Chem. News* **1863**, *8*, 185.
124. The House of Vasa (or Wasa) was the royal dynasty of Sweden (from 1523 to 1654) and of Poland (from 1587 to 1668). The two branches of the royal family embraced two different confessions: Protestant in Sweden and Catholic in Poland. This situation led to not a few wars between the two states. Even though the direct line of the Swedish royal family ended with Queen Christina, who converted to Catholicism and was forced to abdicate, the successive royal Houses of Palatinato-Zweibrücken and the present of Bernadotte have emphasized the fact that they are related through the female branch of the House of Vasa.
125. The launching took place during the afternoon of August 10, 1628. Under the admiring gaze of the whole population of Stockholm, the *Vasa* weighed anchor and set out on its first voyage on the open seas. The captain of the ship, Serin Hannson, betook himself to supper, satisfied at having culminated his career at the command of the admiralty of the fleet of King Gustavus Adolphus II while the *Vasa* coasted along the Sodermalm. Navigation was proceeding peacefully when, all of a sudden, a blast of very strong wind made the ship list heavily to port. The first mate raced below and ordered the crew to move the portside cannons to starboard and to

close the lower portholes. But it was already too late. The water came in in such quantities that the *Vasa* sank quickly, together with its 64 bronze cannons and all its crew.

126. Bolton, H. C. *Chem. News (American Supplement)* **1870**, *6*, 367; Bolton, H. C., *The American Chemist—New Series* **1870**, *1*, 1.
127. Nicklès, J. F. J. *J. Pharm. Chim.* **1864**, *45*, 25.
128. Nicklès, J. F. J. *Compt. Rend. Chim.* **1863**, *57*, 740.
129. Delafontaine, M. *Arch. Science Genève* **1864**, *18*, 369.
130. Popp, O. *Liebig's Annalen* **1864**, *131*, 364.
131. Bahr, J. F. *Ann. der Chemie* **1864**, *132*, 227; Bahr, J. F. *J. prakt. Chem.* **1864**, *91*, 316.
132. Dahllite is a synonym for hydroxyapatite. This mineral belongs to the phosphate group and has the formula $Ca_5PO_4(CO_3)_3(OH)$.
133. Olsen, J. C. *Ind. Eng. Chem.* **1933**, *25*, 711.

III.5

VESBIUM: AN ELEMENT FROM THE CENTER OF THE EARTH

"On December 16, 1631, Mount Somma, otherwise known as Vesuvius, six miles distant from Naples, produced a very strong eruption, as at other times in past centuries, spitting fire, smoking rocks and ashes for thirty miles around with thundering noise and with frequent and destructive earthquakes. The material hurled from the mountain destroyed castles and villas, and killed people and animals alike, with damages amounting to many millions, and the fire lasted many days."[134] This eruption, on account of its unusual violence, remained long in the collective memory not only of Neapolitans, but of others as well: the following story, one that took place two centuries later, makes clear reference to the 1631 catastrophe.

In 1879, the mineralogist Arcangelo Scacchi (1810–93) was a professor of mineralogy with a past career rich with numerous awards, both domestic and international. He was 69 years old, the director of the Museum of Mineralogy at the University of Naples, and an associate of at least 18 Academies, among which those of the Sciences at Paris and at Saint Petersburg stood out for their prestige. He was likewise an associate of several Italian Academies: the Linceo, the Georgophile, and the Pontifical. When he began his work, he undertook something that no other Italian had ever tried to do: isolate an unknown chemical element.

"The immense torrents of lava that issued forth from Vesuvius. . . in 1631 very often have the walls of their fissures carpeted with very thin green crusts, to which more rarely others with a yellow color are joined and which are all mixed up with the former." With these words, Scacchi began a description of his analytical work on the crusts arising from the eruption of 1631. As he emphasized in his long essay, for at least 3 years, he was intent on discovering the chemical composition of the mysterious incrustations. He described the processes carried out on the material, and he began his article by giving notice of the date of his claimed discovery of a new element he called *vesbium* (from the yellow crusted material, vesbine) before explaining his investigative work.[135] *Vesbium* was derived from the ancient Latin name for Vesuvius, recorded by Galen (ca. 130–201) in *De morbis curandis.*[136] The test for recognizing and isolating the substances contained in vesbine was based on an attack of the raw material with dilute hydrochloric acid, followed by selective precipitations that were needed to eliminate the presence of other metals in the original material. In fact, Scacchi complained about the fact that the incrustations were so stuck to the rock that he could not separate them by mechanical means. The solution produced by HCl attack was bluish, and into this would have gone the *vesbium*, copper, and silica. Following reprecipitation, reheating with concentrated hydrochloric acid, and filtering, he was left with a solid with a dull green color that he called a *vesbiate*, from which he derived an oxide with an uncertain stoichiometry: AgO,VbO_3 or AgO,VbO_5. From these

FIGURE III.04. Arcangelo Scacchi (1810–93), discoverer of the hypothetical metal called *vesbium*, which had been found in some volcanic deposits on the slopes of Vesuvius in 1879. Courtesy, Museum Center of Natural Sciences, University of Naples Frederick II.

molecular formulas arose the atomic weight of *vesbium*, which would have been 81.29 g/mol if the first formula were correct and 65.29 g/mol if the second were correct. The latter atomic weight was almost identical to that of zinc (65.38 g/mol), whereas that of the complex AgO,VbO_3 did not correspond to the atomic weight of any existing element, being intermediate between that of bromine (79.904 g/mol) and krypton (83.80 g/mol). However, Scacchi nursed serious doubts about the true stoichiometry of *vesbiate* and, in fact, added a shrewdly expressed note: "if on the other hand the formula of the vesbiate of silver were, as that of the phosphate of silver, $3AgO,VbO_5$, the equivalent weight of the vesbiate would be found to be much larger."

Scacchi, in decomposing potassium *vesbiate*, found what he considered two forms of *vesbic* acid: one red and scale-like and the other white and powdery. He ought to have certainly spoken of these two discoveries to some colleagues since, after having formulated two hypotheses regarding them, he defended the second with great tenacity, hypothesizing different allotropic forms.

Scacchi quickly became aware that he was not sure what he had on his hands.[137] He acknowledged that many gaps were certainly present in his work; however, the chemical knowledge of the time and the continued discoveries of new elements[138] contributed to make the existence of *vesbium* somewhat probable for several years.

In 1880, Arcangelo Scacchi (Figure III.04) received a pleasant letter from the Ministry of Public Instruction of the Kingdom of Italy that conferred on him a grant of 2,000 lire for "the study of a substance recently discovered at Vesuvius and that has been named Vesbium."[139]

The work had not yet been taken up again when one of his colleagues, Professor Alfonso Cossa (1833–1902), expressed the idea that *vesbium* could be vanadium. It thus

FIGURE III.05. Ferruccio Zambonini (1880–1932), Italian Chemist and Mineralogist. In 1910, Zambonini carried out a complete analysis of *vesbium* from which he deduced that it was identical to vanadium. Courtesy, Chemical Heritage of the Department of Chemistry of the University of Florence, Italy.

became necessary, in 1880, for Scacchi to publish a brief note defending his previous memorandum,[140] and he became much more cautious, asserting that "Recent research demonstrates that there is a great resemblance between vesbium and vanadium, as I did not neglect to mention in that same memorandum." However, to Arcangelo Scacchi, the idea that *vesbium* might not exist was not at all a pleasant one, and, at the conclusion of his letter written in response to his colleague Cossa he states "and up to now it is given to me to conclude that if vesbium is not the same thing as vanadium, at least it is similar."

Not much time passed before Scacchi realized that many of the chemical tests he had conducted on the original material were not decisive, whereas other analytical results were explainable if one admitted to the presence in the samples of various elements in trace amounts. He gradually became convinced that his new element was not necessarily new; he accepted the idea of having rediscovered vanadium but with a small gesture of pride he wrote: "there would always remain a fact of some importance: the presence of vanadium in volcanic lavas demonstrated by the yellow incrustations that with the name of vesbine I reported among the mineralogical species."

For many years, no further studies were done on vesbine because of its rarity. In 1910, the chemist and mineralogist Ferruccio Zambonini (1880–1932) (Figure III.05) succeeded in doing a complete analysis of vesbine,[141] thus demonstrating that it did not contain traces of any unknown element but many metals of the first transition series and the rare earths. Finally, in 1927, Zambonini and Guido Carobbi (1900–83) conducted an exhaustive spectroscopic and chemical study[142] on vesbine from which emerged two significant pieces of data: (1) the identification of *vesbium* with vanadium was amply demonstrated by physical and chemical methods; and (2) the chemical composition of vesbine, a hydrated form of the mineral species that goes under the name of cuprodescloizite, can be represented by the formula $(Pb,Cu)_3V_2O_8 \cdot (Pb,Cu)(OH)_2 \cdot 5H_2O$.

Arcangelo Scacchi was born at Gravina, in Puglia, on February 8, 1810; he studied at the Seminary of Bari till the age of 18, after which he transferred to Naples to study medicine. Receiving his degree in 1831, he was called 10 years later to be an assistant to the chair of mineralogy at the Royal University of Naples. Three years later, he became a permanent professor and director of the Museum of Mineralogy. Between 1879 and 1891, he participated, although in a peripheral way, in the drafting of the geological map of Italy. He kept his chair for a surprisingly long period: 50 years, retiring ill and infirm in 1891, the day of his 81st birthday. He survived, but in a precarious state of health, until October 11, 1893.

Notes

134. Zibaldone di Padre Matteo Pinelli (1577–1669) priore di Cerliano (Notebook of Father Matteo Pinelli, prior of Cerliano); Pacciani, S., Ed.; Gianpiero Pagnini Editore: Bagno a Ripoli, 1997, p. 114.
135. Scacchi, A. *Atti Acad. Sci. Napoli* **1879**, 13 Dicembre, 1.
136. Galen, *De Moribis Curandis*, bk. 5, 12.
137. A handwritten correction of a typographical error by Scacchi; the words in the text "separated out" were changed to "determined."
138. Gallium was discovered in 1875, ytterbium in 1878. Scandium, thulium, and samarium were discovered in 1879; and gadolinium in 1880.
139. Letter preserved in the Scacchi Letter Book, at the Royal Mineralogical Museum, presently the University of Naples.
140. Scacchi, A. *Rendiconti Reale Accademia delle Scienze fisiche e matematiche di Napoli* **1880**, Fascicolo 3°–4°, March–April, 1.
141. Zambonini, F. *Mineralogia Vesuviana* **1910**, 315.
142. Zambonini, F.; Carobbi, G. *The American Mineralogist* **1927**, *12*, no. 1, 1.

III.6

THE CURIOUS CASE OF THE TRIPLE DISCOVERY OF ACTINIUM

Nowadays, when chemists refer to actinium, their thoughts go to the radioactive element isolated in 1899 by André Debierne (1874–1949)[143] in the residues left over from work with pitchblende. However, almost two decades earlier, another chemist had announced the discovery of a different element to which, curiously enough, he gave the same name. The latter, far from being extracted from exotic and radioactive minerals, was found in very ordinary material.

In chemistry, this case of two elements with the same name is not unique: in 1812, Edward Daniel Clarke asserted that he had reduced barium oxide with an oxygen-hydrogen blowpipe. He proposed that the new metal be called *plutonium* since "all the tests showing its existence belonged to the realm of fire," but H. Davy, J. J. Berzelius, and other chemists preferred the name barium.[144]

A century later, Glenn T. Seaborg (1912–99), Edwin M. McMillan, Joseph W. Kennedy (1916–57), and Arthur C. Wahl (1917–2006) synthesized element 93 by bombarding uranium with deuterons. When they were requested to give a name to the new element, Seaborg said "we decided to name the element plutonium after the planet Pluto, just like uranium is named after Uranus and neptunium after Neptune."[145]

However, the case of actinium proved to be much more intriguing than that of plutonium. Thirty-four years after the discovery of the "true" actinium, André Debierne, by this time almost 60, believed that he had found a second element in it: *neoactinium*.

III.6.1. THE FIRST ANNOUNCEMENT OF THE DISCOVERY OF ACTINIUM

The first person to announce the discovery of *actinium* was the Englishman Thomas Lambe Phipson (1833–1908). He came from a well-to-do family from Ladywood, Birmingham. Born on May 5, 1833, he was the son of Samuel Rayland Phipson (1803–87) and Ellen Emma Elizabeth Lambe (1813–99). Phipson's father, because of some bad investments, was forced to move his family to Belgium where the cost of living was much lower. In 1855, Thomas Lambe Phipson received the title of *doctéur-ès-science* at the University of Brussels, where he had studied natural sciences. Later, he spent some time in Paris, returning to Belgium in 1859. In 1860, he was named adjunct professor of analytical chemistry, an office he kept until his permanent transfer to England, where he assumed the office of director of the Putney-London Chemical Laboratory.

On September 30, 1865, he married Catherine Julia Taylor (1837–1920). A man of multiple interests, he spent much of his free time on music and on diversified scientific lines

FIGURE III.06. Thomas Lambe Phipson (1833–1908). In 1881, Phipson exerted a great deal of energy describing, in at least eight publications, the presence of a photochemical element (called *actinium*) in the paint he used to decorate his own mailbox.

of research.[146] In 1881, Phipson published a brief article in which he defined "A Curious Actinic Phenomenon." This was the first in a series of 11 scientific communications on the same subject: eight articles out of the eleven were written by himself; of these, two were published in French journals and six in English journals. The photochemical phenomenon described by him was observed on the door of a mailbox painted with a new white pigment based on zinc. The door looked black during the daytime and white at night, and then became black again at sunrise. The effect was apparently due to a component in the paint that was sensitive to sunlight. According to Phipson, the pigment consisted of about 1.80% of a mixture of $BaSO_4$ and ZnS. Phipson discovered that the darkening of the product could be reproduced by exposing the pigment to direct sunlight for about 20 minutes. The original white coloration could be recovered by maintaining the object in the dark for 2–3 hours. The color-changing capacity from white to black was lost over a period of a few days, although some samples of the paint kept this property for months. Phipson did not observe any phosphorescence, but he realized that a piece of glass from an ordinary window, placed above the white paint, prevented darkening. Being an expert analytical chemist, he believed that he had discovered an unknown element within the paint to which he attributed the unusual properties already described. Phipson was cautious in his statements, but nevertheless proposed a name for the unknown metal: *actinium*, from the Greek ακτις, "ray." He published a second note[147] in French, which was picked up almost simultaneously by various French journals.[148] Figure III.06 is an image of this Renaissance-type polymath.

Meanwhile, the producer of the pigment felt the need to respond to Phipson's assertions.[149] He emphasized that the phenomenon described by Phipson was far from new, although he did not know the physical principles giving rise to it. Moreover, he criticized Phipson's results. Some weeks later, a new article appeared on the subject.[150] The author was an American paint manufacturer. More caustic than his English colleague, J. Cawley[151] even came to doubt the credibility of Phipson's chemical analyses and added the observation that a sheet of glass could not possibly prevent the darkening. Phipson's response[152] added little or nothing to the subject, except for the fact that he maintained that the process of darkening was due to a reversible oxidation of a new unknown metal mixed together with zinc in the pigment.[153]

Two months later, Phipson reported[154] with considerable pride to having "isolated the oxide and sulfide of the new metal in a state of tolerable purity." Although he never succeeded in obtaining *actinium* in the metallic state, he was convinced that he had discovered an important phenomenon: the phosphorescence of *actinium* would not be an atomic property, but the result of the combination of *actinium* with sulfur. As proof of this, he asserted that the oxide of *actinium* did not change color when exposed to sunlight. Cawley responded again, asserting that part of the zinc oxide, formed during calcination, would have reacted with zinc sulfide to produce zinc and sulfur dioxide, resulting in a blackening of the entire mass. This explanation agreed with facts that (1) the sensitivity to light was greater when the sulfide was in aqueous suspension as opposed to when it was in an anhydrous state, and (2) the presence of magnesium prevented darkening. J. Cawley also described the process of manufacture. His article ended with a clear rejection of Phipson's alleged discoveries, remarking that Phipson had the peculiar talent of deducing a great deal from scanty data.[155]

Phipson wasted no time in responding to the paint manufacturer, and his very caustic reply[156] asserted that his critics did not have the slightest idea of the cause of the phenomenon under discussion.[157] A short time later, Phipson informed the readers of *Chemical News* that he had succeeded in isolating *actinium*[158] by precipitation of an ammoniacal solution of magnesium. According to the author, the metal had formed a light gray deposit that, on compression, became white like silver. Phipson also described other chemical properties of *actinium* and, in the following year, published his last article on the subject.[159] From then on, the existence of *actinium* was held to be extremely dubious and no other mention of this metal was made.

III.6.2. CONFESSIONS OF A VIOLINIST

The curious diatribe concerning *actinium* was concluded by the hand of Thomas L. Phipson himself. Perhaps becoming aware of having committed an error, he abandoned the research and characterization of the presumed new metal and turned his attention to the drafting and editing of the latest number of his own journal, the *Journal of Medicine*. The chemical and physical process that was the basis of the phosphorescence of zinc sulfide was fully understood only many years later,[160] after new materials were discovered that illustrated the phenomenon of luminescence induced by exposure to electromagnetic radiation with very short wavelengths (X-ray or ultraviolet). The phenomenon of induced luminescence was seen when the source of electromagnetic radiation was removed. Luminescent pigments are polycrystalline inorganic substances containing zinc sulfide or alkaline-earth sulfides. The Phipson phenomenon could be attributed to some impurities that, illuminated by blue or ultraviolet light, gave rise to phosphorescence in samples of ZnS.[161]

Phipson was a man of vast interests that ranged from the sciences to music: he had a brilliant and eclectic mind. It is possible to say the same thing about his family, whose members brought together outstanding talents in science, music, and art. Samuel Phipson's son, Wilson Weatherley Phipson, was a versatile and innovative engineer, not to mention a very able pianist, whereas his brother, Thomas Lambe, united his passion for science with an excellent talent for music in general and for the violin in particular. Wilson Weatherley Phipson (1838–91), furthermore, had a beautiful tenor voice. The Phipson family often entertained friends and acquaintances with entirely "domestic" concerts: the cast of musicians was made up of all the members of the family playing various roles. In fact, Thomas's mother was an excellent lyric singer. Furthermore, he did not limit himself to just playing the violin, but wrote a number of musical pieces and pamphlets.[162] Shortly before he died, Phipson published his delightful autobiography full of anecdotes, episodes, and personal recollections.[163] The versatility and eclecticism of Thomas Lambe Phipson—characteristics rather common among men of the well-to-do class in the Victorian era—permitted him to dedicate himself to the most diverse scientific disciplines but inevitably ended by doing him harm, limiting his stature and condemning him to remain a dilettante in all the fields in which he engaged. Thomas Phipson died on February 22, 1908, at the age of almost 75.

III.6.3. DID THE SEARCH FOR NEOACTINIUM REALLY DELAY THE DISCOVERY OF FRANCIUM?

The announcement of the discovery of actinium was repeated in 1899 by the famous chemist André-Louis Debierne.[164] Born at Paris on Bastille Day, 1874, Debierne was a precocious student: at only 16 years of age he was admitted to the Ecole de Physique et de Chimie following with great profit the lessons of Alsatian chemist Charles Friedel. Having just completed his university studies, Debierne began research with little enthusiasm in the field of organic chemistry, in particular on the racemization of camphor with aluminum trichloride.[165] After Friedel's death, he began to work on mineral chemistry. He was assisted in this by Georges Urbain,[166] who was his elder by 2 years. In the laboratories of the Ecole de Physique et de Chimie, he became acquainted with Pierre Curie who later, together with his wife Marie Skłodowska Curie, welcomed him into their laboratory. For the rest of his life, Debierne was associated with the husband-wife team in deep friendship.

His career rise was very rapid: at the age of 25, after treating enormous amounts of pitchblende furnished by the Curies,[167] he found a radioactive element that he called actinium. He was certainly one of the youngest chemists to discover an element. This discovery—although it brought a certain amount of fame to him and a secure position at the university—happened when he was too young, having the effect of forcing him to never separate himself from the Curies or to undertake independent research.

Debierne studied the phenomenon of induced radioactivity with Pierre Curie and continued the work of William Ramsay and Frederick Soddy on the production of helium on the part of radioactive elements, developing an apparatus to manipulate small quantities of gas with which he was able to directly determine Avogadro's number.

After the tragic death of Pierre Curie in April 1906, Debierne helped Marie Curie to isolate and characterize radium. Debierne collaborated patiently, for more than 35 years, in the shadow of the overbearing figure of Madame Curie. When, in July 1934, the discoverer of radium and polonium—consumed by the radiation to which she had been exposed for so long—died, he became the director of the Institut du Radium and professor of general physics and radioactivity.[168] André Louis Debierne was by that time 60 years old. The preceding year, he had noticed some anomalies in his radioactive preparations. He took up research on these mysterious substances, thinking to find them in some minerals that he subjected to fractionation. At the conclusion of numerous tests, he was convinced that, along with the radium-bearing products present, there might be some new radioactive substances not yet characterized.

This phenomenon was interpreted by hypothesizing that part of the radium did not decay into radon but into another radioactive substance having a half-life of a few hours. He observed that the properties of this substance were virtually analogous to those of barium and radium. Debierne always thought in terms of "new radioelements," and he believed that the anomalies he observed could explain the fine structure of the spectrum of α particles caused by the disintegration of radium.[169] He remarked that the new radioactive substances seemed to have chemical properties very similar to those of radium but were not isotopes, and since it was impossible to assign them a place in the periodic table, he proposed to name them *neo-radium* (NeRa), distinguishing each by Roman numerals, I, II, and so on.

Debierne continued in the work of fractionating minerals rich in actinium, discovering within them new substances that he generically called *néo-elements*. The first of these to have a name was *néo-actinium*, followed closely by *néo-radium*. Debierne held that the new radioactive substances (*néo-radium* and *néo-actinium*) could be two distinct excited states of radium, but with chemical properties appreciably different between them. Following his speculations, these substances would have had their origins by nuclear rearrangement after the emission of γ radiation by a nucleus of radium or actinium.

To confirm Debierne's hypothesis, two young researchers from the Institut du Radium, one of whom was Bertrand Goldschmidt (1912–2002),[170] were asked to reproduce the experiment, but both failed in this undertaking. Some years later, other radiochemists showed that Debierne had fallen into a deplorable error, taking isotopic impurities for new radionuclides.[171] Debierne rejected the experimental evidence that dismantled his research, remaining strongly convinced that he had discovered a new nuclear phenomenon and to have isolated *néo-actinium*.

Continuing to follow the mirage represented by *néo-actinium*, Debierne slowly began to lose his authority in the laboratory he directed as well as his credibility at the international level. Many of his colleagues, among them the husband-wife team of Joliot-Curie, placed themselves in open conflict with him. In the middle of such confusion, however—in the same Curie Pavilion of the Institut du Radium directed by Debierne—a young laboratory technician brought to conclusion the discovery of the last naturally occurring element in the earth's crust. This element, known today by the name of francium, was one of the products of the radioactive decay of uranium-235. The discovery of the 87th element was done by Marguerite Perey (1909–75) in January 1939, when she was not yet 30 years old, and it marked the greatest scientific success to occur within the walls of the Curie Institute since the death of its founder in July 1934.

III.6.4. A COLD SHOWER AT THE END OF A CAREER

Although André Debierne and Marie Curie's daughter worked elbow to elbow at the Institut du Radium, things did not go well. In 1938, unaware of Debierne's goals for his research, Irène Joliot-Curie asked Marguerite Perey independently to prepare a sample of very high purity actinium. Joliot-Curie intended to measure the half-life of actinium with extreme precision, whereas Debierne wanted to search for the elusive element that he called *néo-actinium*. Marguerite Perey had accumulated copious experience in the chemistry of actinium thanks to the work done under the personal guidance of Marie Curie, and she was perfectly suited for this task. It was during the preparation and purification of some samples of actinium that the 29-year-old Perey discovered the presence of an unknown element with a very short half-life; after the necessary characterizations, she decided to name it francium.[172] She had to wait for months—and a painful compromise between André Debierne and Irène Joliot-Curie—before she could be credited with the discovery and allowed to suggest a name for the new element.[173]

In his memoirs, Bertrand Goldschmidt, the last French chemist remaining in the Curie laboratory, didn't spare a certain amount of sarcasm directed at Debierne:

> Debierne's personality was not well suited to running a lab. He was an introvert who was gradually becoming more and more reclusive. Months of unopened mail piled up on his desk. Never married, and with few friends, his relationships might be said to last as long as he had someone in sight. He could finish a conversation with a person, shake hands, and turn out the lights as he exited the room—completely forgetting the individual left behind in the dark![174]

On the other hand, Gaston Dupuy, *chef de travaux* at the Ecole Supérieure de Physique et de Chimique at Paris, had a completely different opinion and described Debierne with almost reverential affection:

> He was a man of extreme reserve, [Debierne] never spoke of his discoveries, not even during his lessons; he shunned honors, publicity,. . . for. . . his entire life he demonstrated the noblest performance; he is the exemplar of complete dedication to science. . .. Everyone who met André Debierne recognized that just under the surface of a slightly cold appearance, which intimidated one initially, was hidden great kindness and generosity. Very sensitive with respect to justice, he was always ready to intervene for others even when others had not done so for him.

In the last years of his life, Debierne seemed obsessed with the desire to find new areas of research, as if he wanted in a certain sense to break away from the overpowering figure of Marie Curie, with whom he had been associated and remembered as her assistant. Perhaps driven by the understandable desire to do something uniquely his own, Debierne began to study the transformations of materials at low temperature. When helium or hydrogen (and in some measure also neon) were put in contact with carbon at the temperature of liquid nitrogen, he observed the emission of a great deal of heat. He discarded the possibility that the heat emitted might be due to an allotropic modification of the carbon or a chemical change of some impurities present in it, determining similar processes highly improbable. Debierne asserted that the release of heat might be due to a nuclear reaction of an imprecise nature, although he never succeeded in putting this hypothesis to the test. The ex-co-worker and colleague of Madame Curie decided nonetheless to give a name to the phenomenon: *frigadréaction*.[175]

Unfortunately, it was shown that this hypothesis was also erroneous. André Debierne wrote his last contribution to science in 1947, when he sent to press a curious article: he studied the color of the clouds that were formed following the explosion of an atomic bomb on the Bikini atoll in the Marshall Islands. Debierne hypothesized that following a nuclear explosion, atmospheric nitrogen could be oxidized to nitric acid (theoretically 50 tons).[176] He emphasized that, from the meteorological point of view, the nitric acid could bring about increasing damage to marine flora and fauna.

While at his vacation home at Arcouest in Brittany, Debierne began complaining of the first symptoms of an illness that would kill him just a few days later. Coming back to Paris, he died on August 31, 1949, one month after he had celebrated his 75th birthday. Because he did not have any family or close relatives, his funeral was attended by a few friends, some surviving colleagues, and a meager group of his students.

Notes

143. Debierne, A. *Compt. Rend. Chim.* **1899**, *129*, 593; Debierne, A. *Compt. Rend. Chim.* **1900**, *130*, 906.
144. Webb, K. R. *Nature* **1947**, *160*, 164.
145. First Report (March 21, 1942); subsequently published in *JACS*, 1948.
146. Phipson's research was expansive, encompassing meteorological problems, the chemical composition of the atmosphere, the determination of fluorides and other elements in fossils, paleontology, lithography, photography, the properties of phosphorescent materials, desert plant life, the chemical composition of meteorites, organic chemistry, and entomology.
147. Phipson, T. L. *Compt. Rend. Chim.* **1881**, *93*, 387.
148. Phipson, T. L. *La Nature* **1881**, *IX année*, deuxieme semestre, 243.
149. Orr, J. B. *Chem. News* **1881**, *44*, 12.
150. Cawley, J. *Chem. News* **1881**, *44*, 51.
151. No personal data on J. Cawley is known except that he was an English merchant; in his response to Cawley's criticisms, Phipson transcribed—willfully or not?—the name of his colleague into "Crawley," mangling the spelling.
152. Phipson, T. L. *Chem. News* **1881**, *44*, 73.
153. Phipson, T. L. *Brit. Assn. Reps.* **1881**, 51, 603.
154. Phipson, T. L., *Chem. News* **1881**, *44*, 138.
155. Cawley, J. *Chem. News* **1881**, *44*, 167.
156. Phipson, T. L. *Chem. News* **1881**, *43*, 283.
157. Phipson, T. L. *Nature* **1882**, *25*, 394.
158. Phipson, T. L. *Chem. News* **1881**, *44*, 191.
159. Phipson, T. L. *Chem. News* **1882**, *45*, 61.
160. Ives, H. E.; Luckiesh, M. *Proc. Am. Phys. Soc.* **1911**, *32*, 240; Garlick, G. F. J.; Mason, D. E. *Proc. Phys. Soc. A*, **1949**, *62*, 817.
161. At the dawn of nuclear physics, zinc sulfide (ZnS) was used by Ernest Rutherford and other scientists to construct scintillation detectors. This material emits visible light when excited (bombarded) with X-rays or by an electron beam. For this reason, cathode ray tubes and X-ray screens were coated with ZnS.
162. Phipson, T. L. *Famous Violinists and Famous Violins*; Chatto & Windus: London, 1903; Phipson, T. L. *Voice and Violin.* Chatto & Windus: London, 1898; Phipson, T.L. *Biographical Sketches and Anecdotes of Celebrated Violinists Since Lulli*; R. Bentley & Son: London, 1877.
163. Phipson, T. L. *Confessions of a Violinist: Realities and Romance.* Chatto & Windus: London, 1902.

164. Debierne, A. *Compt. Rend. Chim.* **1899**, *129*, 593; Debierne, A. *Compt. Rend. Chim.* **1900**, *130*, 906.
165. Debierne, A. *Compt. Rend. Chim.* **1899**, *128*, 1110.
166. Urbain, G.; Debierne, A. *Compt. Rend. Chim.* **1899**, *129*, 302.
167. The Curies' work with tons of pitchblende slag heaps from the St. Joachimsthal mines in Bohemia was legendary.
168. Dupuy, G. *Bull. Soc. Chim. Fr.* **1950**, 1023.
169. Debierne, A. *Compt. Rend. Chim.* **1933**, *196*, 770.
170. Adloff, J.-P. *Quimica Nova* **1979**, October, 137.
171. Haïssinsky, M. *XIII International Congress of History of Science—Colloque du 75e anniversaire de la radioactivité*, Moscow, 1971.
172. Adloff, J.-P.; Kauffman, G. B. *The Chemical Educator* **2005**, *10*(5), 387.
173. Grinstein, L. S.; Rose, R. K.; Rafailovich, M. H. *Women in Chemistry and Physics: A Biobibliographic Sourcebook*, Greenwood Press: London, 1993.
174. Goldschmidt, B. *Atomic rivals*, Rutgers University Press: New Brunswick, NJ, 1990.
175. Debierne, A. *Compt. Rend. Chim.* **1937**, *205*, 141; Debierne, A. *Compt. Rend. Chim.* **1937**, *205*, 321.
176. Debierne, A. *Compt. Rend. Chim.* **1947**, *224*, 1220.

III.7

THE IMPROBABLE ELEMENTS OF A COUNTRY GENTLEMAN

In 1886, while a good part of upper-class Victorian England followed with bated breath the seesaw change of government between Robert Arthur Talbot Gascoyne-Cecil, Third Marquess of Salisbury (1830–1903) and William Ewart Gladstone (1809–98), a country gentleman of the county of Selkirk was occupied with a more earthly problem.

Alexander Pringle was a wealthy landowner whose property extended along the banks of the River Tweed in southern Scotland. A good part of his territory was barren and rocky; the very ancient mountainous elevations had been smoothed and softened by the perpetual action of the weather.

Pringle, in his prose, far from academic but anecdotal and colorful, told of having found some unknown metals in rock samples from the Paleozoic Era. Pringle's research, evidenced by an immense expenditure of his own means, had as its only desire that of "getting of the rarer ones [elements] in a small quantity such as might suffice to please a chemist if he found them upon his own estate."

Alexander Pringle began with the collection of a large quantity of quartz developed in large veins and easily visible at the edges of the rocks exposed at the foot of a glacier that, by his reckoning, would have diverted, like a funnel, all the material indispensable to him for his research. According to Pringle, the quartz crystals would have acted like a filter holding back, during the erosion due to rain and snow, traces of the overhanging rock. Although he regretted that the amount of material to collect, crush, and process was exceeding every expectation and that the unknown substances enclosed in the quartz were less in amount than his worst expectations, the result obtained was equal to the enormity and difficulty of his effort: no less than four new elements emerged!

Pringle described with unique accuracy one of the four presumed new elements, "the one that has given me the most trouble." He called it *polymnestum* (Pm), "because its compounds combined with those of several other elements all at one time."[177] *Polymnestum* had the appearance of a dark metal not easily melted and with an equivalent weight that hovered around 74. Pringle maintained that he had isolated two sulfides and four distinct oxides: PmS, PmS_2, PmO, PmO_2, PmO_3, and PmO_5.

Pringle found it hard to believe that one equivalent of *polymnestum* could combine with five equivalents of oxygen to form the pentoxide. For this reason, he repeated the experiments again; all his tests confirmed his hypothesis, and, in the end, he himself had to admit to the existence of PmO_5 from its delicate pink shades. Thanks to these measurements, repeated many times, Pringle was able to determine the atomic weight of the metal (74.01) with more accuracy than he could have done with other presumed elements. Of all the oxides that he had prepared, the more characteristic one was found to be the trioxide, obtained easily by attacking the metal in acid medium, whereas the monoxide had a very particular green coloration. He realized that *polymnestum* had properties similar to iron, inasmuch as it was attracted to a magnet placed in the immediate vicinity. Many other proofs and tests seemed to indicate and intimate a resemblance of this metal to iron, except the most decisive one: the test with potassium ferrocyanide.

Another element isolated by Pringle was *erebodium*, for which he proposed the symbol Eb. This metal appeared black like coal, and its atomic weight was shown to be 95.4. The only oxide that he succeeded in characterizing was the dioxide, EbO_2, which had an unexpected resemblance to bismuth oxide. The etymology of the name of this element came almost certainly from the black color ("like night") of the metal. In fact, Erebus (or Tenebrae), the son of Chaos, was a Greek mythological figure, brother and spouse of the Night and father of the Day and of Heaven (Ether). According to Hesiod, Erebus was the name of primordial darkness.[178]

Pringle called his third element *gadenium*. He succeeded in determining the atomic weight of this new metal with extreme accuracy (43.547), but he did not succeed in melting it. The powdery metal had an appearance tending toward gray. The monoxide of *gadenium* (curiously, Pringle did not give a chemical symbol for this metal) was red, whereas the dioxide had a creamlike color. The fourth and last element seemed to be a semimetal with an atomic weight of approximately 45.2. According to Pringle, this metalloid, which he called *hesperisium* (Hs) due to its "sunset sky" coloration, was red and had a bright metallic look: The Hesperides, from which this element draws its name, were mythological Greek figures, the daughters of the Night. They dwelt in the remote lands of the west and were the guardians of the garden of the golden apples of Hera.[179]

Hesperisium formed a monoxide (HsO), a dioxide (HsO_2), and a sesquioxide (Hs_2O_3); the last two oxides formed their respective acids; H_2HsO_3 and $HHsO_2$. Furthermore, both the hydride (H_2Hs) and the fluoride (F_2Hs) were gaseous. Because of these and other similarities, Pringle saw a certain resemblance between *hesperisium* and selenium.

Pringle concluded his article by asserting that, beyond the above-named four elements, he had glimpsed another, similar to lead with respect to ductility and color. This would be easily volatilizable and would have a low melting point. Unfortunately, he did not succeed in determining its atomic weight and therefore decided, without much regret, to abstain from proposing a name for it. Perhaps it was just as well; by this decision he avoided bringing to five the number of false elements present in his only publication!

The information left by Pringle contained many gaps and, in many cases, inaccuracies; the only quantitative data were the atomic weights. Based on this one objective datum, unfortunately, we can affirm that he did not discover any new elements: no value corresponds to that of a metal or semimetal known or unknown at the time of his analysis (1886). Table III.4 is a summary of Pringle's presumed elemental discoveries.

The data that best agree with the atomic weights reported are found in Table III.4. The oxides of *polymnestum*, however, do not possess the coloration of those of germanium or arsenic. The fact that this supposed metal possessed ferromagnetic properties probably gives greater credibility to the second hypothesis: that Pringle had isolated a metal of the first transition series (Fe, Co, or Ni) in a very impure state, contaminated with one or more elements with greater atomic weight.

In the case of *erebodium*, the only clue furnished by the author, excepting its atomic weight, was the qualitative description of the oxide: curiously, it agrees with that of molybdenum. But, unfortunately, molybdenum in the elemental state, far from being black like coal, has a coloration that is silvery-white. An analogous case can be seen with

Table III.4 Pringle's Presumed Discoveries

Element presumed to be discovered	*Element suspected of being already in existence*	*Atomic weight of the presumed new element*	*Atomic weight of the element already in existence*
Polymnestum (Pm)	Germanium (Ge)	74.01	72.64
	Arsenic (As)		74.92
Erebodium (Eb)	Niobium (Nb)	95.4	92.91
	Molybdenum (Mo)		95.94
Gadenium	Scandium (Sc)	43.547	44.96
Hesperisium (Hs)	Selenium (Se)	45.2	78.96

gadenium: the red oxide of the presumed new metal fits badly with chemical elements having an atomic weight similar to the presumed *gadenium*. In fact, the trioxide of scandium is white, whereas the monoxide of calcium is pale yellow. Also, in this case, one has to assume the presence of traces of another metal (perhaps an oxide of iron) capable of elevating the atomic weight of *gadenium* and at the same time conferring on it a dark red color.

The case of *hesperisium* clearly brings to light Pringle's double failure to step into the shoes of both the analytical and theoretical chemist: the error in the determination of the atomic weight of this element and the positioning in the periodic table of a metalloid like *hesperisium*. However, despite the vagueness of his prose, he actually suggested the nature of the real element that misled him: selenium. Again, in this case, his determination of the atomic weight is very unreliable.

Instead of moving the science of chemistry forward, Alexander Pringle's essay turned out to be a genuine impediment. And, aside from the harm done to Pringle's reputation, the real damage was sustained by *Chemical News*, which, in reporting this news, discredited itself as a scientific journal.[180]

Notes

177. Pringle A. *Chem. News* **1886**, *54*, 167.
178. Besides Heaven and Day, Erebus begot other children by Night: these were not true and proper divinities but personifications of human behaviors and fears. Among these there were Thanatos and Hypnos, held by the Greeks to be twins, the one the god of death and the other the god of sleep. Momus, the "black sheep," was exiled from Mount Olympus for having harshly criticized Zeus and Aphrodite, to name but a few.
179. The point at which the sun set had often been associated with the kingdom of Death and certain myths place the kingdom of Hades there. It is not out of the question that the three Hesperides would be associated with the kingdom of darkness as keepers of the fruits that conferred immortality.
180. Compared with prestigious scientific journals, *Chemical News*, published for the first time in 1859, occupied a relatively marginal position in the chemical literature. Directed at times by unscrupulous amateurishness, as can be witnessed by the fact that many false discoveries were reported in its pages, thus giving them great prominence, publication was definitively suspended in 1932.

III.8

A BRIDGE BETWEEN THE PROTOCHEMISTRY OF THE PHARAOHS AND THE ARAB WORLD: MASRIUM

Eleven decades ago, when the immense country of Egypt was, in its own unique way, part of the British Empire, an efficient chemical laboratory managed by the London government existed in Cairo. The main interests of this establishment, as well as the efforts of its researchers, were aimed at the service of archaeology—but not only archaeology.

In 1890, 23-year-old Henry Droop Richmond (1867–1931), after having studied at University College, Finsbury Technical College, and having passed a period of specialization in an analytical laboratory under the guidance of Otto Hehner (1853–1924), was hired as a second chemist at the Khedivial Laboratory of Cairo. In 1890–91, Johnson Pasha, English Viceroy of Egypt, gave to the laboratory where Richmond worked together with Dr. Hussein Off some specimens of fibrous alums that he had found during geological studies in the most remote parts of Egypt. In the middle of 1891, Johnson Pasha made a gift of another 100 kg of this type of alum so that the two chemists could examine it with greater accuracy than they had been able to do the previous year with the more modest samples at their disposal. The aim of this research was directed toward the commercial exploitation of metals that might be present in the mineral. In these samples, the two chemists found a quantity of cobalt. Johnson Pasha and his financial backer, who had received the rights of extraction from the Egyptian government, thought that they could evaluate and then commercially exploit the deposits.

During their analytical investigations to establish the composition of these alums, Off and Richmond observed that their samples contained a percentage, variable between 1% and 4%, of an unknown element.[181] After a long series of chemical processes done to separate the new element from the mineral matrix, Richmond and Off obtained a solution of the presumed new element. The chemical tests suggested that they were dealing with a divalent metal. Although the results were not totally convincing, and definite confirmation of the presumed new element was late in arriving, Richmond and Off concluded that they were dealing with an alkaline earth metal. The two chemists proposed the name *masrium* in honor of the country in which the alums had been found and where the two chemists had carried out their research (*Masr* is the Latin spelling of the Arabic name "Egypt"). At the same time, they proposed the name masrite for the fibrous alums.

The precipitate obtained on adding oxalate to the solution gave information relative to the atomic weight of *masrium*. On titration with permanganate, Richmond and Off determined the amount of oxalic acid ($H_2C_2O_4$) in the oxalate of *masrium*, MsC_2O_4,[182] and thus, after calcination of this compound, they determined the formula of the oxide, MsO. The method and the calculation of the atomic weight were marked by inaccuracies and errors. Richmond reported a value of 228 for the new element, basing this on the fact that his samples contained an appreciable amount of impurities.

Today, only the name masrite remains in use; it is a variety of *halotrichite* containing manganese and cobalt. *Masrium* is an element that does not exist. The atomic weight of this element would make it identical to radium, unknown at that time.[183] Today, we know that radium is not present in the alums that Richmond and Off were examining. A reexamination of the chemical processes of separation of the metals in the mineral, using today's more advanced knowledge of analytical chemistry, shows that Richmond and Off separated out aluminum and perhaps traces of manganese. If the latter had been present in any appreciable quantities, it would have been oxidized and separated out as the dioxide: however, manganese should have been extracted as the sulfide in the very first processes.[184]

The only element present in the silicate in appreciable quantities that could have survived until the last process was thus aluminum.[185] The presence of this element definitively broke down Richmond's hypothesis, according to which Ms would have been a divalent element, but it explains the disagreement between the value observed experimentally and the theoretical value for the atomic weight of the hypothetical *masrium*.

The chemical composition of masrite did not turn out to be exactly what Richmond proposed: $(Al,Fe)_2O_3(Ms,Mn,Co,Fe)O_4SO_320H_2O$. Today, we suggest $(Al,Fe)_2(Ms,Mn,Co,Fe)(SO_4)_424H_2O$. Richmond was aware of the difficulty of determining the amount of water of crystallization in the molecule.

The presence of cobalt in this mineral caused Richmond to propose a unique hypothesis: that the element could have been used by ancient Egyptian artisans to decorate the temples and tombs found along the course of the upper and lower Nile. Driven by curiosity, Richmond went to the director of the Cairo Museum (which, after 1891, had been moved to Giza), a Frenchman named M. Grébaut (1846–1915), and asked him for some samples to examine. Unfortunately, no manufactured color or smalt-containing traces of cobalt were found upon analysis.

The lifetime of *masrium* was quite short: the isolation of radium wrecked any possible reconciliation between Richmond's discovery and the experimental data collected by the Curies. It is ironic that it fell to a person of the same age as Richmond—Madame Curie—to get the credit for the discovery of the last remaining naturally occurring alkaline earth element. In J. W. Mellor's monumental treatise, the discoveries of the erroneous elements are reported meticulously, but the data relative to *masrium* are full of gaps; next to the *masrium* box is laconically written "discovery not confirmed."[186] The ephemeral existence of *masrium* represented nothing more than the results of an analytical error, and it was thus understood by the majority of chemists in its day.

At the end of 1892, Henry Droop Richmond returned to England. He did not take up mineralogy again, but became a chemist at the Aylesbury Dairy Company until 1915. In those years, he published numerous and original analytical works on the chemistry of food, in particular, milk. From 1915 to 1931, the year of his death, he was chief analytical chemist at Boots Pure Drug Company, Ltd. [187] During his career as a chemist, he was able to draft numerous monographs that appeared in the journal *Analyst*.[188] Fellow of the Royal Society of Chemistry since the age of 20, Richmond published only one article in the pages of the prestigious *Journal of the Royal Society of Chemistry*—the 1892 article in which he and Off announced their discovery of *masrium*.

Notes

181. Richmond, H. D.; Off, H. *J. Chem. Soc. Trans.* **1892**, *61*, 491; *The Manufacturer and Builder* **1892**, *24*(6), 127.
182. Ms was the symbol chosen by Richmond and Off for *masrium*, the element they hypothetically discovered.
183. In 1892, in the periodic table to which Richmond made reference, there was an element from the group *glucinium*, calcium, magnesium, strontium, and barium, the not-yet-discovered *eka-barium*, to use Mendeleev's terminology; that is, radium, which would be discovered by Pierre and Marie Curie in 1898. The atomic weight predicted for eka-barium by Mendeleev's and Newlands's theories was 225, which was in good agreement with the values determined by Richmond and Off. *Glucinium*, reported in the paper of Richmond and Off, was the name of element number 4 until 1921, when the International Commission on Atomic Weights decided to change it to its present name, beryllium.
184. Treadwell, F. P.; Hall, W. T. *Analytical Chemistry, vol. 1: Qualitative Analysis*, 7th ed. D. Van Nostrand Co.: New York, 1930.
185. Moeller, T. *Qualitative Analysis;* McGraw-Hill Book Company: New York, 1958.
186. Mellor, J. W. *A Comprehensive Treatise on Inorganic and Theoretical Chemistry*, vol. XVI. Longmans, Green: London and New York, 1937.
187. F. L. P. *J. Chem. Soc. Trans.* **1931**, *2*, 3382.
188. Richmond, H. D. *Analyst* **1908**, *33*, 179; Richmond, H. D. *Analyst* **1908**, *33*, 209; Richmond, H. D. *Analyst* **1908**, *33*, 305; Richmond, H. D. *Analyst* **1910**, *35*, 516; Richmond, H. D. *Analyst***1911**, *36*, 9; Richmond, H. D.; Huish, H. C. *Analyst* **1912**, *37*, 168; Richmond, H. D. *Analyst* **1918**, *43*, 167; Richmond, H. D. *Analyst* **1920**, *45*, 260; Richmond, H. D. *Analyst* **1923**, *48*, 67; Richmond, H. D. *Analyst* **1925**, *50*, 260.

III.9

THE DEMON HIDDEN IN THE RARE EARTHS

While testifying as an expert witness in a trial one day, Henry Rowland was asked during cross-examination what qualified him to serve as such a witness. "I am," the professor replied, "the greatest living expert on the subject under discussion." Some time later a friend, well aware of Rowland's usual modest and unassuming manner, expressed his surprise at this uncharacteristically grandiose remark. "Well, what did you expect me to do?" Rowland asked. "I was under oath.[189]

Much was said and even more was written about the American Henry Augustus Rowland III (1848–1901), as famous for his talents in experimental physics as for his reserve and modesty. The episode described in the preceding quote was somewhat unique in his brilliant career: it was the incident in which he was found in the guise of an "improvised" chemist.

III.9.1. PROVINCIAL AMERICA SUITS THE GREAT PHYSICIST JUST FINE

When Rowland was born on November 27, 1848, the United States was not yet the technologically advanced country that we know today: both its research laboratories and many of its university professors were guilty of provincialism and, unlike today, scientists could only specialize after taking their degrees by crossing the Atlantic and doing a residency in the famous English, French, or German laboratories. With a name lacking in originality, but blessed with a sharp intuition, Henry Augustus III was the son of the Reverend Henry Augustus II (1804–59) and nephew of Henry Augustus I, a theologian and son of a clergyman. The latter joined to his fervent faith an anti-English political fanaticism beyond the ordinary; a supporter of American independence from the British Crown, he used the pulpit to spread his ideas.

Although the young Rowland was expected to follow a normal course of studies, he could not tolerate the study of the classics. He was a very good electrochemical experimenter and wanted to study engineering; his parents, who at first thought of enrolling him at Yale, found themselves obliged to enroll him at Rensselaer Technical Institute (later Rensselaer Polytechnic Institute, RPI) where he took his degree in civil engineering in 1870. He passed a year in Europe and stayed for a long time in the laboratory of Hermann L. von Helmholtz (1821–94) at Berlin.[190] The anecdotes about his life and scientific career are numerous, beginning when he tried to publish his first article on physics in an American journal and it was rejected. It was immediately clear to him that the American scientific community was still very narrow-minded, but Rowland had such a high opinion of himself that he would not give in. He sent his work to the greatest physicist in the world, James Clerk Maxwell (1831–79), who immediately sent it to press. In 1876, Rowland became a professor at The Johns Hopkins University, perhaps the most prestigious U.S. university at the time, where he remained until his death.[191] In 1883, in recognition of the diffraction grating that he had invented and was named after him, he was elected a member of the Society for the Advancement of the Sciences and received the Rumford Prize. But even in his privileged position, Rowland felt the distress of not being able to move freely, yoked to a science incapable of fully appreciating his genius. His colleagues considered him a hostile and intolerant character.

Although he was the author of more than 100 patents, Rowland's major contribution to science was the invention of the concave diffraction grating,[192] capable of greatly improving the resolution of the spectrographs in use at the end of the 19th century. These instruments were of fundamental importance to the spectroscopists of his day and also for generations to come. At the beginning of the 1930s, Emilio G. Segrè (1905–89) is said to have observed that the Rowland grating was the most precious instrument[193] in the laboratory of Nobel Laureate Pieter Zeeman (1865–1943).

III.9.2. THE SON OF A PROTESTANT PASTOR DISCOVERS A DEMON

Rowland was a skilled engineer and inventor, as well as a versatile physicist and astrophysicist.[194] However, much less well-known was the chemical side of this "polyhedric" figure. In 1894, at the end of a years-long ambitious and systematic project in the separation and spectroscopic study of the rare earths, he published his results.[195] The rare earth group of 14 elements, with chemical properties so similar among them, was a real headache first for chemists and later for physicists: their complete isolation and the organization of their characteristics and properties had required over 113 years of work. Starting with confirmed discoveries, Rowland proposed to study the spectra of all the rare earth elements with his diffraction grating. By doing this, he believed that he could have the last word in the question of the rare earths, a true *terra incognita* for understanding the periodic table. Unfortunately, although he used an investigational instrument far superior to what had come earlier, just like his other famous colleagues, he ran into the snares represented by the separation of these elements and inevitably found himself involved in the announcement of a false discovery.

To study and characterize the rare earths, Rowland availed himself of the materials furnished to him by the chemist Oliver Wolcott Gibbs (1822–1908) and by the mineralogist Frank Wigglesworth Clark (1847–1931),[196] whereas for the identification of the new element, he had recourse to a sample of impure yttrium given him by Professor Gerhard Krüss of Munich.

Rowland, like a minority of his contemporary scientists, believed that some of the rare earths were not elementary substances. Following on this thought, Rowland believed that erbium, yttrium, and cerium were in reality mixtures of elementary substances not yet isolated. To use his own terminology, he divided erbium into its presumed constituents and did the same thing with yttrium and cerium: the "constitutive substances"[197] were indicated by the letters *a, b, i, d, h, n, c, k*.

Henry Rowland, like many other investigators before him, began the fractionation of the rare earths starting from the following minerals: samarskite $(Y,Fe^{3+},U)(Nb,Ta)_5O_4$;

cerite $(La,Ce,Ca)_9(Mg,Fe^{3+})(SiO_4)_6[SiO_3(OH)](OH)_3$; gadolinite, $Y_2Fe^{2+}Be_2Si_2O_{10}$; and yttrialite $(Y,Th)_2Si_2O_7$. Using acid attack, he dissolved these four mineralogical samples to obtain a mixture of oxides of La, Ce, Pr, Nd, and Th, as well as eight new substances (indicated by the letters *a, b, i, d, h, n, c, k*). Rowland sought to separate these last elements following the method of fractional crystallization commonly employed for the separation of the yttric and ceric earths. The mixture, consisting predominantly of oxides of the rare earths and indicated generically as L_2O_3 (where L = La, Ce, Pr, Nd, *a, b, i, d, h, n, c, k*), was dissolved in a solution of nitric acid and then diluted with water. After having heated the solution, some sodium sulfate was added in successive amounts with constant agitation until the neodymium lines disappeared from the spectrum. The precipitate was separated from the mother liquor and treated with potassium hydroxide, and the mixture of oxides coming from this operation (L_2O_3) was subjected to the same cycle of fractional crystallizations dozens of times. In this way, Rowland thought that he had separated the elements *a, b, i, d* in the first fractions, while succeeding fractions were enriched in elements *d, n, c, k*. Finally, the last were rich in component *h*.

Through techniques of fractional crystallization, Rowland succeeded in isolating element *a*, about which he reported some properties of the oxide and of the oxalate, but the element to which he decided to give a name was *d* because of its persistence and ubiquitousness in the preparations that he examined.

He first observed *d* by spectroscopic means in the yttric sample furnished by Krüss because it was much more abundant than elsewhere. However, he was not able to separate it from components *b, i, h, n, c*. Because of the chemical difficulties he had in trying to isolate it, and because it seemed to be present everywhere, Rowland suggested that it be called *demonium*: "On account of the trouble caused by it and its universal presence, I propose the name demonium for it. Its principal spectrum line is at wave-length 4000.6 nearly."[198]

The life of *demonium* was, fortunately, brief. With like irony and apparently with lack of consistency, William Crookes, editor of the journal that had accepted Rowland's article, published—almost by return mail and in the pages of the same journal—an unpleasant denial of his discoveries.[199] "Rowland's substances are already known as accepted elements; the white yttrium oxalate and oxide are—[for chemists]—far from novelties." After the lull caused by the false announcement of the discovery of *demonium* and of the other six "substances," Rowland did not completely abandon his study of the rare earths and obtained excellent arc spectra for the lanthanides, zirconium, vanadium, and many other elements.[200]

III.9.3. THE TRAGIC CONCLUSION

On June 4, 1890, Henry Augustus III married Henrietta Harrison. The joy of this event was of short duration: Rowland was diagnosed with a serious form of diabetes, at that time an incurable illness. Knowing that he would soon die (beyond all expectations, he lived for another decade), Rowland wanted to assure his family of a comfortable economic future. He spent the last decade of his life in a fruitless attempt to commercialize some of his patents, for example, the multiple telegraph,[201] which, although technically sound, made a fortune only after his death. The more his health deteriorated, the more his fame as a physicist became widespread beyond the borders of the United States: in 1890, he received the Grand Prix of the universal exposition at Paris; he was the first American to receive (1895) the Matteucci medal of the Italian Society of Sciences; and, in 1899, he was elected a foreign Fellow of the Royal Society of London.

Henry Augustus Rowland III died on April 16, 1901, in Baltimore. By his express desire, he was cremated and his ashes immured in a wall of the basement of his house where he had outfitted his personal laboratory; only later was a permanent resting place found in a suitable niche at The Johns Hopkins University.

Notes

189. Quoted from *Today in the Word*, August 5, 1993. http://www.sermonillustrations.com/a-z/h/humility.htm (accessed April 14, 2014).
190. Anon. *Nature* **1901**, *64*, 16.
191. Anon. *Science* **1901**, *36*, 681.
192. Rowland, H. A. *Phil. Mag.* **1884**, *17*, 25.
193. Segrè, E. G. *A Mind Always in Motion: The Autobiography of Emilio Segrè*. The University of California Press, Berkeley, CA; 1993.
194. Rowland, H. A. *Chicago Astr. Journal* **1895**, *2*, 117.
195. Rowland, H. A., *Chem. News* **1894**, *70*, 68.
196. In 1902, F. W. Clark wrote a heartfelt commemoration of Rowland, which is preserved in the archives of the Smithsonian Institution, Record Unit 7320, National Museum of Natural History, Division of Mammals, Biographical File, 1860–1973 and undated, box 14, folder 24.
197. Using a poor choice of a word, he called them "ingredients."
198. The unit most likely being referred to is the ångström.
199. Crookes, W. *Chem. News* **1894**, *70*, 81.
200. Rowland, H. A.; Tatnall, R. R. *Chicago Astr. Journal* **1895**, *2*, 3; Rowland, H. A.; Harrison, C. N. *Chicago Astr. Journal* **1897**, *7*, 17; Rowland, H. A.; Harrison, C. N. *Chicago Astr. Journal* **1897**, *7*, 22.
201. In 1906, during the eruption of Vesuvius, Rowland's multiple telegraph was used by the Italian government to transmit messages from Naples to the rest of the peninsula. For a certain period, these types of messages were the only means of communication between Naples and Italy.

III.10

DIM LIGHTS AND DARK SHADOWS AROUND "LUCIUM"

III.10.1. PREVIEW OF THE DISCOVERY

On September 25, 1896, the eclectic British inventor and scientist Sir William Crookes published in the monthly journal of which he was editor a brief summary of what the French chemist Prosper Barrière had announced: the discovery of a new metal by the name of *lucium*.[202] The report mentioned explicitly a commercial use for the new element as an incandescent filament for illumination, an alternative to the already well-known gauze filaments of Auer von Welsbach. Curiously, the article cited the names of four chemists of international fame: the professors Paul Schützenberger of Paris, Per Theodor Cleve (1840–1905) of Stockholm, Carl Remigius Fresenius (1818–97) of Wiesbaden, and finally the celebrated rare earth chemist Paul Emile Lecoq de Boisbaudran. These, as was reported in the brief communication, had characterized the new metal chemically and spectroscopically, recognizing its elemental nature. Strangely, the spectrum of the metal had many similarities with little known erbium, but in the article this aspect was cleverly underplayed.

In all probability, *Chemical News* had taken the notice from the essential details of Barrière's patent, which he had filed the preceding year, requesting commercial use for this element.[203] Notice of the presumed discovery was reported rather acritically by many French[204] and German[205] journals, spreading the news rapidly in the major centers of European research.

The illuminating properties possessed by the various metal oxides, among which were zirconium, lanthanum, yttrium, thorium, and magnesium, suggested to Prosper Barrière the idea of utilizing these substances as components for illumination along the lines of the Auer gauze.[206] His numerous experiments allowed him to discover the presence of a new simple substance within the mineral monazite, which had a somewhat variable composition. In the American patent, which came about a year later than the French one, we read: "This body, to which I have given the name "lucium," has properties different from those possessed by the substances used hitherto, and as to the constituency of which new body I am unable to state definitely at present whether it consists in a new element, a mixture of old elements, or a mixture of a new element and old elements."[207]

III.10.2. THE DISCOVERY OF THE FIRST "PATENTED" ELEMENT

The interests of Prosper Barrière were concentrated almost exclusively on the protection of his patent rather than on the discovery of a new metal. As one can see in the article that

Table III.5 Composition of Monazite Sands Reported by Barrière

Substance	*Percentage*
SiO_2	69.70
P_2O_5	6.00
Fe_2O_3	1.92
Al_2O_3	15.00
Ce, La, Di	2.13
Wet component	2.05
Lime, Magnesia	2.00
Element A	1.80

appeared on October 30, 1896, in *Chemical News*, Barrière gave a lengthy description of the chemical isolation of the metal, whereas the latter part, relative to the employment of the new substance, was treated in a clearer and almost abrupt manner.[208]

In Table III.5, the composition of the monazite sand is reported as found by Barrière; the values are the average of the results of many analyses. The gangue was fused with sodium carbonate in a suitable oven after which, once cooled, the mass obtained was leached to remove the silicates and phosphates. The carbonates were treated with sulfuric acid, and the sulfates obtained were dissolved in the cold with water and reprecipitated with ammonia. Finally, the precipitate thus obtained was dissolved in a solution of hydrochloric acid. The aluminum and iron were then removed, precipitating them as oxalates. Successive treatments with potassium sulfate, sodium sulfate, and sodium hyposulfate had the effect of separating out substance A (later called *lucium*) from the other components. Barrière realized that some traces of zinc oxide in the solution gave, if heated, a phosphorescence to the entire solution. He, in contrast to Thomas L. Phipson,[209] recognized and described accurately the photochemical properties of this metal.

The procedure for the production of incandescent gas for illumination was accurately described by Barrière: the solution of metal A or *lucium* was mixed together with a small amount of zinc oxide or other oxide able to increase its illuminating power, after which it was repeatedly absorbed on fibers of linen, muslin, or tulle, then fixed with two final washes, one acidic and one alkaline. Finally, the solution was evaporated and the textile cut in strips of 20 cm and formed into a wick. Every wick contained up to 6 cm^3 of solution and was attached to a nickel hook. The flame of a Bunsen burner was passed rapidly around and above the wick while a second burner was passed around the bottom. The textile burning away left an oxide skeleton of zinc and *lucium*. Calcined for a half hour thus, the gauze of *lucium* and zinc was ready to be sold: it emitted an intense and brilliant light when heated.

Particularly curious is the conclusion of the article. After having cited in a rather brusque way the desire to name the new metal, "the novel illuminative body which I have referred to as A, I have named lucium,"[210] Barrière passed on to list the results of his research in four points. Three of these refer to the practical use of *lucium* as an "instrument" for illumination. There is no reference made to the properties of the new body, for example, its atomic weight. The value for an atomic weight equal to 104 was reported in the preceding article of September 25, and this number seemed to arise from the information supplied by the four international chemists who, as reported by Crookes, had analyzed the material.

III.10.3. THE INTERVENTIONS OF CROOKES, FRESENIUS, AND SHAPLEIGH

Another month passed and Sir William Crookes felt it necessary to intervene directly in the case of *lucium*, publishing a long article in the pages of his journal.[211] For not very clear reasons, Prosper Barrière had given William Crookes a solution of the nitrate of *lucium*, and Crookes lost no time in analyzing it. The experiments that Crookes conducted convinced him of the error in the discovery of *lucium*. Preliminary spectroscopic examination of the solution had shown the presence of erbium and *didymium*. Nevertheless, Crookes evaporated the solution of the supposed nitrate of *lucium*, placed the residue in an empty tube, and recorded the phosphorescence spectrum. The lines obtained coincided with those of the well-known yttrium.

Crookes also took photographs of the ultraviolet spectrum of the supposed new metal but, once again, these indicated the presence of yttrium. At this point, he decided to examine the arc spectrum of the sample of *lucium* and of 3 samples of yttrium. The first sample, which he used as a reference sample, was ultrapure and had been furnished by Cleve; the second, relatively pure, had been prepared years earlier by J. Galissard de Marignac; the last had been prepared by Crookes himself and was, as he said, "as pure as I could make it."

The spectral analysis of the four samples showed unequivocally that *lucium* was nothing other than impure yttrium. At this point, Crookes brought up a certain Professor Schützenberger of Paris, thanking him for having furnished those precious chemical details relative to the extraction of the oxide of *lucium* from the monazite sands. In this way, Crookes implicitly shifted the blame onto Schützenberger's shoulders of having confirmed the existence of *lucium*, even though there was no confirmation in the literature except that made by Crookes. Thus was the only mention made of the celebrated Paul Schützenberger, chemist and expert on the rare earths, living at that time in Paris and professor at the Sorbonne. According to the French professor—and therefore also to Barrière, who seemed to have used Schützenberger's material—sodium thiosulfate could quantitatively precipitate yttrium, when in reality this did not happen. Consequently Schützenberger—or Barrière—had continued to work on relatively pure yttrium, confusing it with the new metal.

With a final cutting remark, Crookes sought to clarify how the atomic weight of 104 of the presumed *lucium* could refer back to that of the much lighter yttrium (89). The impurities that he had found in the samples of nitrate of *lucium* were of elements with higher atomic weights than that of yttrium: *didymium*, samarium, ytterbium, and erbium. An average value for the atomic weight of yttrium ($A = 89$) with, for example, traces of erbium ($A = 166$) would have raised the weight to 104, precisely as had been reported for *lucium*.

So far, the evidence for endorsement of the false discovery fell on Professor Schützenberger, although it appears that he had never published anything about it. A very curious fact was that an atomic weight of *lucium* equal to 104 was first cited and later denied only in the work of Crookes and never in the article or in the patents of Barrière.

Among those who had confirmed the existence of *lucium*, the first person mentioned in the anonymous article, which appeared in *Chemical News* on September 25, 1896, was the German professor Carl Remigius Fresenius. Wishing to emphasize that he had had no part in the entire story, he sent a letter to the editor of *Chemical News*[212] dated November 17, 1896, in which Fresenius said he was saddened to have been cited inappropriately in the work on the recognition of *lucium*, work that he had never done.[213] Because he had been engaged for the entire preceding year on the study of the fluorescence of the rare earths with his nephew, Dr. E. Heinz, he would never have been interested in the possibility of research on this element in his samples of thorite. Six months later, at the age of 79, Carl Remigius Fresenius died at Wiesbaden, leaving the mystery of *lucium* unresolved.

The last article on *lucium* was published in *Chemical News* in the July 23, 1897 issue,[214] a month after the death of Paul Schützenberger. The person who wrote it was Waldron Shapleigh, an American chemist who was one of the founders of the American Chemical Society in 1876. He asserted that he had received in May 1896 various samples of *lucium* coming from Paris and isolated by Schützenberger, but not by Paul Schützenberger, professor of chemistry at the Sorbonne as had been cited in the anonymous article of September 25, 1896, in *Chemical News*, but by his son, Léon Schützenberger.[215]

Shapleigh fractionated a small quantity of monazite sands coming from deposits in North Carolina and arrived at the same results[216] as Crookes. Furthermore, he examined three other minerals in which he suspected the presence of *lucium*: samarskite, xenotime, and euxenite. He concluded his article: "In order to obtain a larger sample of "lucium" to work with,. . . I took several hundred samples of North Carolina monazite sands, carefully following Barrière's method and failed to obtain any earth answering to the reaction of "lucium".. . . "Lucium" is not entitled to a place in the list of elements."

III.10.4. WHO WAS MANIPULATING LUCIUM'S STRINGS FROM BEHIND THE SCENES?

In conclusion, we can point out some peculiar aspects of this story: we know nothing about Prosper Barrière except that he was French. Aside from the one article published in *Chemical News*, publications by him do not exist in any relevant international scientific journals. Furthermore, although Barrière was named as discoverer of the metal first and pointed out as the person responsible for the terrible error later, he never defended his work. At this point, two hypotheses might be raised. The first sees Sir William Crookes (not new to publicizing false discoveries: he himself made five, all of which appeared in his journal) publishing fragmentary notices about an unknown French chemist. Perhaps he enriched them with somewhat inexact details such as the initial confirmation of Barrière's work by four famous, elderly, foreign chemists, of whom two died within a year of the discovery. One of these managed to write to Crookes denying completely any involvement in the discovery of *lucium*. Another from far-away Stockholm never responded, whereas the third, Paul Schützenberger, died while a polemic was raging about his name and his work in the pages of *Chemical News*. Perhaps Crookes, who was in contact with Schützenberger because of their mutual interest in isolating the rare earths, linked the name of the French luminary with that of Barrière, never imagining that Barrière could advance without some support from the academic community.

The second hypothesis involves Paul Schützenberger's son, Léon. Granting that what Shapleigh said was true, Léon would have sent him the samples of *lucium*. Could he not have also sent Crookes the results that confirmed Barrière's data, without the knowledge of his father and the other chemists? This, however, does not square with Crookes's assertion, according to which he would have received the sample of the nitrate of *lucium* directly from Prosper Barrière, returning the latter to the role of major suspect.

The complex figure of Paul Emile Lecoq de Boisbaudran must also be analyzed. No reference seems to draw him into the struggle following his involuntary involvement in the story of *lucium*, yet a certain cloud seems to thicken over this personage. Not associated with academe, as was the case with Prosper Barrière, Lecoq de Boisbaudran was never a university professor, but came from a wealthy wine-making family from Cognac. As an amateur, he accomplished spectroscopic studies with the purpose of characterizing the chemical composition of many minerals, studies that led him to discover three elements: samarium, dysprosium, and gallium. It was precisely this last discovery that allowed Lecoq de Boisbaudran to be the author of a kind of practical joke at the expense of the scientific world. The name "gallium" comes from the Latin Gallia, a Roman province that corresponds to present-day France, mother country of Lecoq de Boisbaudran. Yet others immediately saw in this name a left-handed trick on the part of the chemical amateur. It was said that Lecoq had named the new element for himself; in Latin, the word *gallus* when translated into French means *le coq*. Lecoq denied this in 1877.

Because 2012 marked the centenary of his death, Lecoq de Boisbaudran was commemorated by a retrospective article in the *Chemical Educator*.[217]

To conclude, perhaps it is worthwhile to return to the figure of William Crookes: in his old age, the English scientist was involved in highly controversial discussions in support of spiritualism. Inadvertently, he built up the case for *lucium* and soon thereafter he demolished it by demonstrating its nonexistence. Perhaps on account of prudence or maybe for lack of interest, he never returned to this subject. Sir William outlived all four of the chemists involved in the *lucium* affair, passing away at the age of 87 on April 4, 1919. With the death of Crookes, the person who could have provided the clearest, simplest, and perhaps the most "enlightening" explanation of this mystery also disappeared.

Notes

202. Anon. *Chem. News* **1896**, *74*, 159.
203. Barrière, P. French patent no. 246,163, dated March 28, 1895; on September 8, 1896, the patent was also extended to the United States: patent number US 567571, application number US 95-568836.
204. Urbain, G. *Bull. Soc. Chim.* **1896**, *18*, 540.
205. Anon. *Chemisches Repertorium (Supplement zu "Chemiker-Zeitung" no. 92)* **1896**, *26*, 265.
206. Auer von Welsbach studied the rare earths at length and succeeded in finding practical applications for these elusive elements: the Welsbach gauzes were made of thorium nitrate and small amounts of cerium nitrate. Thanks to this gauze, gas lighting remained competitive with electrical lighting for some time. By the time of Auer's death in 1929, his factory had produced more than 5 million of these gauzes.
207. US Patent Number 567571, dated September 8, 1896.
208. Barrière, P. *Chem. News* **1896**, *74*, 213.
209. In 1881, Phipson had fallen into the unfortunate error of attributing the luminescence of some of his preparations to the presence of a new element that he wanted to call *actinium* (whereas it was in actuality zinc). See Phipson, T. L. *Compt. Rend. Chim.* **1881**, *93*, 387. Many years passed and Phipson, far from admitting his own error, wished to explain the nature of X-rays as proving his own false discovery: roentgen, or X-rays were nothing more than an "electromagnetic manifestation" of his *actinium*. See Phipson, T. L. *Chem. News* **1896**, *74*, 260.
210. "The new luminous substance that up to now I have designated as A, I wish to call '*lucium*.'" The name *lucium* (from the Latin, *lux*) was certainly not chosen by chance.
211. Crookes, W. *Chem. News* **1896**, *74*, 259.
212. Fresenius, R. *Chem. News* **1896**, *74*, 269.
213. It is well to remember that the editor of the *Chem. News* was Sir William Crookes.

214. Shapleigh, W. *Chem. News* **1897**, *75*, 41.
215. The Schützenberger family was quite famous in France. It put down its roots at the beginning of the 18th century when the head of the family, Jean-Daniel, married the widow of the proprietor of the Royal Brewery, and the entire factory became a family business. The sons and nephews of this couple were French or German, depending on how the Rhineland border was moved. Léon Schützenberger's nephew, Marcel-Léon Schützenberger (1920–96), the last descendant of the dynasty, was a renowned mathematician.
216. Anon. *J. Franklin Inst.* **1897**, *144*(July), 75.
217. Kauffman, G. B.; Adloff, J.-P. *Chem. Educator* **2012,** *17,* 213–19.

III.11

IN THE BEGINNING THERE WAS DIDYMIUM. . . AND THEN CHAOS AMONG THE RARE EARTHS

Ever since the first element was isolated, the group consisting of the 14 rare earth elements bode no good for chemists. The chemical properties of these elements were so similar to one another that to separate them was a great challenge for many generations of chemists. Every new metal hid traces, more or less abundant, of the nearby elements. Until systems of separation were developed based on the fractional crystallization of immense quantities of material—and these were not perfect—it was not possible to isolate the 14 metals in the group. Fractional crystallization was developed through the work of Jean Charles Galissard de Marignac, Paul Schützenberger, Marc Abraham Delafontaine, Bohuslav Brauner, Carl Auer von Welsbach, Georges Urbain, Charles James (1880–1928), and many others.

It was a difficult, monotonous, time-consuming work, full of traps and snares that chronologically overlapped three centuries: from 1794 when Johan Gadolin (1760–1852) discovered yttrium,[218] to 1907 when Urbain isolated lutetium.[219] It was an enterprise perhaps unique in its duration, and it was a source of many failures in the field of the chemical sciences. As the irony of fate would have it, just 5 years after the discovery of lutetium, the physicist Henry G. Moseley found via experiments in the field of X-ray spectroscopy that the frequencies of the rays emitted by each element vary proportionally to the square of the number of the order (atomic number) of the element. This law could have resolved in a very short time the dilemma that worried generations of chemists at the moment of announcing the discovery of an element: had they isolated a new substance or, more frequently, did they have in their hands a complex mixture of substances?

III.11.1. DIDYMIUM: AN AWKWARD LODGER IN THE *f*-FAMILY

In 1839, Carl Gustav Mosander showed to chemists that *ceria* was in reality a complex earth. He was able to separate from it a white oxide that he called *lanthana*.[220] *Didymium*[221] was discovered in the same way in 1842. Following this, Mosander, a renowned Swedish chemist[222] born at Kalmar, on September 10, 1797, discovered two other elements: terbium and erbium. He attended elementary school at Kalmar until he was 12 years old, when he moved with his mother to Stockholm. There he became an apprentice in Ugglan's pharmacy. After passing the examination to become a pharmacist in 1817, he became interested in medicine, and, in 1820, he matriculated at the Karolinska Institute where he took his degree in surgery 4 years later. Following this, he taught chemistry at the same institute and soon became assistant curator of the mineralogical collection at the Swedish Museum of Natural History. In his youth, during his medical studies, Mosander had as his chemistry teacher Jöns Jakob Berzelius, whom he replaced in 1836 as professor of that same discipline at the Karolinska Institute.

Didymium was discovered in 1842, during one of his chemistry experiments devoted to decomposing a sample containing cerium nitrate with dilute nitric acid. He was motivated by the conviction that *ceria* (the oxide of cerium), discovered and isolated by Berzelius in 1803, could be in reality a mixture of cerium, lanthanum, and a new metal that he called *didymium*. The name *didymium* came from the Greek διδυμοι, meaning "twin," because it accompanied cerium and lanthanum in all the cerium-bearing minerals (allanite, cerite, yttriocerite, cryptolite). Immediately after the announcement of the discovery, Friedrich Wöhler, although a very close friend of both Berzelius and Mosander, raised harsh objections about the selection of the name. In fact, in German, *didymium* became *didym* and sounded somewhat foolish and infantile. Furthermore, he added with a touch of malice, Mosander had chosen the name "*didymium*-twin" because he had four children who were two pairs of twins!

Mosander was immovable and replied angrily that he would not hear of changing the name of his element. His intention, he claimed, was to use a name that began with the letter "d" in order to have an atomic symbol totally different from any that existed up to that point.

Although Mosander asserted erroneously that *didymium* was a new element, he was correct in his hypothesis that the Berzelius's *cerium* contained other elements, among which were lanthanum and cerium. However, the approximation that he made, that is, that the rest of the material would consist of a single element, *didymium*, was later shown to be false. At that time, spectroscopy had not yet been invented, and chemists could not avail themselves of this ancillary technique for the analysis of minerals. However, the three elements (cerium, lanthanum, and *didymium*) added together constituted only 95% of the content of the rare earths present in the original cerite coming from Bastnäs, thus creating more problems for analytical chemists.

The success of *didymium*, which lasted for more than 40 years, was due in part to its easy availability and identification: its salts had a pinkish color. The most difficult undertaking for chemists remained the extraction of lanthanum by crystallization because its salts usually were colorless. For the entire time that *didymium* was thought to be a distinct element, it had the symbol Di. Mosander died at Lovö, near Stockholm, on October 15, 1858, at the age of 71 and convinced in his heart of the good outcome of his discovery. However, the first cracks that would undermine the existence of *didymium* appeared very quickly, and they were quite visible to the expert eye. In 1848, Jean Charles Galissard de Marignac found that the atomic weight of the element seemed to be 496 (taking oxygen as equal to 100). The value therefore would have been too low for a rare earth element $(496 \times 16)/100 = 73.36$.

Six years later, Galissard de Marignac[223] repeated the determination of the atomic weight and found a value of 96. In 1853, Hermann calcined the oxide of *didymium* and extrapolated the atomic weight to 95.84. However, the values observed up to this point showed them incompatible with the correct positioning of the new metal among the rare earths; its atomic weight could be compared to that of yttrium, but certainly not to the "lightest" of the rare earths, cerium, with an atomic weight of 140.116. Certainly, to Mosander's error one can add the analytical errors of colleagues that only increased the confusion. In this way, on the one hand, the lifetime of *didymium* was lengthened, and on the other, a resolution to its mystery was also prolonged.

Years later, many other chemists,[224] this time by means of spectroscopy, raised ever greater doubts about the nature of *didymium*. In 1874, Per Teodor Cleve asserted that *didymium* was composed of only two elements. But the most significant criticism was made in 1878, by the chemist Marc Abraham Delafontaine.[225] He showed that *didymium*, extracted from a mineral coming from a different location, gave different absorption spectra. According to him, this strange phenomenon could be explained by admitting that the sample examined was not composed of a single element but of a mixture. From this mixture, Delafontaine thought, with good reason, that he had isolated a new element that he called *decipium*. At virtually the same time, Lecoq de Boisbaudran made a similar announcement, adopting the name samarium.[226] Other scientists also arrived at the same conclusions, but their work came to light too late when, on the international scene, an article appeared written by a young student of Robert Bunsen, Carl Auer von Welsbach.[227]

III.11.2. THE SPLITTING OF DIDYMIUM: PRAESEODIDYMIUM AND NEODIDYMIUM

In 1885, Carl Auer von Welsbach, working in Robert Bunsen's laboratory at Heidelberg, succeeded in separating *didymium* into two substances that today are known as praeseodymium and neodymium[228] using fractional crystallization of the double nitrates of ammonium in acid medium. He did not want to give the two elements the names that we presently use, but rather *praseodidymium* and *neodidymium* (i.e., green-*didymium* and new-*didymium*). Unfortunately, his hopes vanished almost immediately when the syllable "di" was lost in both names. 1885 did not signal the end of *didymium*, however: the name survived in the glass industry and in mineralogical tests. During World War I, mirrors made of *didymium* were used to send and receive signals by naval units. *Didymium* also survived in the mining industry associated with the exploitation of the rare earths.

Even up to the end of the 1920s, the name "salts of commercial *didymium*" was used to indicate the mixture of the elements of the rare earths that, after a crude removal of cerium, were present in monazite sands. By starting with these "*didymium* earths," in 1922, Luigi Rolla (1882–1960) and Lorenzo Fernandes (1902–77) undertook an immense investigation in a search directed at the isolation of element 61, which at that moment took the name *florentium*. This research, carried out at the Istituto Superiore di Studi Pratici e di Perfezionamento di Firenze, was historic on account of the enormous quantity of commercial *didymium* used for the fractional crystallizations: almost 2 tons of raw material.

The typical composition of commercial *didymium* was about 46% La, 34% Nd, and 11% Pr; the remaining 9% was composed principally of Sm and Gd. The percentages reported for Pr and Nd varied according to the region where the monazite sands were extracted. The two elements in *didymium* were not isolated in their pure forms until 1925, when the American chemist Edward Kremers (1865–1941),[229] using electrochemistry, first reduced a mixture of Nd_2O_3, NdF_3, and KF and successively of anhydrous $PrCl_3$.

III.11.3. A "COLORFUL" WAR: GLAUCODIDYMIUM OR GLAUCODYMIUM

The announcement of the discovery of neodymium and praseodymium was not a joyful event for most chemists. Two years after Auer von Welsbach's announcement, Gerhard

Krüss and Lars Fredrik Nilson[230] spectroscopically analyzed the oxides of praseodymium and neodymium and asserted that they were not in reality two metals but a mixture of eight elements. A short time later, Claude Metford Thompson (1855–1933) arrived at the same conclusion and confirmed the complexity of neodymium and praeseodymium; in all, he believed there were probably five unknown elements. Sir William Crookes[231] was not slow in affirming that "Neodymium and praeseodymium are not to be considered to be the names of actual elements, but rather the names of complex groups of molecules with which the complex molecule didymium splits up by one particular method of fractionation; other methods of fractionation would probably split didymium into different products."

With the passing of time, many chemists focused their skepticism on Auer von Welsbach's praeseodymium. The crystallographer Friedrich W. Muthmann (1861–1913), together with his assistants, chemists L. Stützel and C. R. Böhm, categorically denied its elemental nature.[232] However, only one article on the subject was published, by a Russo-German chemist.

Konstantin Dimitrievic von Chrustchoff (1852–1912),[233] was born in 1852[234] in Lipino, near the city of Charkow in the Tsarist empire. After having obtained his doctorate in chemistry at Tübingen, he returned to his home country and was appointed to the chair of mineralogy at the University of Saint Petersburg. In 1897, he announced that he had split didymium into three components. He gave recognition to Auer von Welsbach with respect to the isolation of neodymium; as for praeseodymium, he said it was composed of two elements. To one of these he gave the name chosen by Auer, and for the other he proposed the name of *glaucodidymium*[235] because of the blue color of its oxides. This name also lost the syllable "di" and became *glaucodymium*. Because of the peripheral location of the Russian empire, little notice was taken of the work of this chemist, and still less to the demise of his *glaucodydium*. Joseph William Mellor, in his encyclopedic treatise "Inorganic and Theoretical Chemistry," dismissed *glaucodymium* with the words "mixture of known elements."

Ten years before the facts just noted, von Chrustchoff also fell into a similar error by announcing the isolation of a new rare earth element; he had spectroscopically analyzed some monazite sands and had found a substance with an unknown spectrum. Von Chrustchoff became enamored of the idea of calling this new substance *russium* in honor of his native country.[236] The chemists of the time did not give much weight to this discovery, and they put *russium* into the same doubtful category as *lucium*, recently discovered by Prosper Barrière and whose existence was never proven.

Von Chrustchoff continued his interest in mineralogy, and particularly in crystallography, up until the time of his death in Saint Petersburg, on April 6, 1912, at almost 60 years of age.

III.11.4. CLAUDE-HENRI GORCEIX AND BOHUSLAV BRAUNER INTERVENE IN THE CHAOS

The first chemist who with good reason could be considered a co-discoverer of neodymium and praseodymium was the Czech Bohuslav Brauner. Born at Prague, on May 8, 1855, he studied first at the University of Heidelberg under the mentorship of Robert Bunsen and later set sail for Manchester where he worked in the laboratory of Henry Roscoe. When he returned to his native country in 1883, Brauner was named lecturer in chemistry at Charles University, Prague, and finally, in 1890, professor.

He dedicated his entire career to the study of the lanthanides. A profound authority on the periodic table and personal friend of Dmitri Mendeleev, solely on chemical grounds, in 1902, he advanced the hypothesis that a missing element could exist intermediate between neodymium and samarium. Later on, this rare earth element took the provisional names of *florentium, illinium,* and *cyclonium* before definitively arriving at promethium.

When Brauner was still in Bunsen's laboratory, he had the opportunity to meet Auer von Welsbach. Some years earlier, he had discussed the complex nature of *didymium* and had advanced the hypothesis that it was a mixture of three elements:[237] *didymium* true and proper, *didymium-β*, and samarium, discovered by Lecoq de Boisbaudran in samarskite. In the same year, he followed up with a more detailed article on the same subject; in it, Brauner did not claim priority for his discovery, as did the Swedish chemist Cleve,[238] but instead limited himself to the notification that his investigations had arrived independently at the same results.[239]

For many years, Cleve had suspected that *didymium* might be a mixture of rare earth elements, but this was never expressed clearly in affirming the presence of a new element. In 1882, supported by having succeeded in recording an unknown spectral line of wavelength 4333.5 Å and at the same time having succeeded in dividing *didymium* into two fractions having atomic weights of 146 and 142, Cleve called the two substances by the provisional names of *didymium-α* and *didymium-β*. The author admitted that his work, begun in 1874, was far from complete 10 years later.

The young Brauner, while at the University of Manchester, confirmed the work of his Swedish colleague and introduced some minor corrections to the values of the atomic weights. The work, which came to light in 1883, was the conclusion of his doctoral research with Roscoe. When, about 2 years later, Auer von Welsbach[240] published the work that was to mark the date of the birth of neodymium and praseodymium, Brauner, who had discovered the same elements but had chivalrously recognized Cleve's priority, seeing that his work had not been mentioned, was exceedingly offended. The conflict with Carl Auer von Welsbach broke out into open warfare in 1908, when Welsbach began stirring up the German scientific community against Georges Urbain, the French discoverer of lutetium. Many details of this diatribe remain unknown even today, although the Austrian version has always been directed at sweetening the facts and covering up this regrettable event.[241] According to Dr. Soňa Štrbáňová of the Institute of Contemporary History of the Academy of Sciences of the Czech Republic, this bitter controversy almost ended up in a courtroom.

In 1925, Charles University solemnly celebrated the 70th birthday of its most famous professor, Bohuslav Brauner. For the occasion, Georges Urbain,[242] of the Sorbonne in Paris and an intimate friend, also participated. Brauner lived another 10 years; he died following a brief illness on February 15, 1935.

Less noted is the work of French chemist Henri Gorceix, occurring simultaneously with the events just narrated, but geographically at the opposite end of the Earth. Gorceix succeeded in splitting *didymium* into two elements in Brazil 6 weeks earlier than Auer von Welsbach. Chemist and geologist Claude-Henri Gorceix (1842–1919) had been recommended to the Emperor of Brazil, Pedro II (1825–91) by the then-director of the School of Mines of Paris to create a similar institution in Brazil.

Gorceix (Figure III.07) was born October 19, 1842, at Saint-Denis des Murs (Haute-Vienne). After having finished at the Ecole normale supérieure in 1863, at the

FIGURE III.07. Claude-Henri Gorceix (1842–1919). Chemist and geologist, Gorceix was recommended by the then-director of the School of Mines, Paris, to the Emperor of Brazil, Pedro II, for the purpose of creating an analogous institution there. Simultaneously with Auer von Welsbach, he split *didymium* into its two constituents. Gift of Professor Juergen H. Maar, University of Florianópolis, Brazil.

age of 31, he embarked for Brazil to assume the post of director of the School of Mines at Ouro-Preto. In 1885, he reported a complete spectroscopic analysis of the monazite sands coming from Bahia.[243] His work preceded that of Auer von Welsbach by 6 weeks, and that would have served as irrefutable proof of the existence of neodymium and praeseodymium.[244]

Because Brazil was not the center of the scientific world, Gorceix sent his manuscript to Paris so that it would be published in the prestigious *Comptes Rendus,* giving his ideas and discoveries worldwide reach. But what impeded his work was his own uncertainty about his findings. He asserted that monazite was composed of phosphates and oxides of didymium, cerium, and perhaps of lanthanum, but also of an unknown substance accompanying lanthanum. He lamented the fact that his analyses were not reproducible and thought that a fraction of didymium unexplainably was contaminating his fraction of cerium.

After the Brazilian revolution in 1889, which cost Emperor Pedro II his throne, Gorceix returned to France. His scientific work, almost exclusively focused on the composition of minerals, was abruptly interrupted, perhaps due to problems related to his professional reintegration.

One year after the end of World War I, Henri Gorceix died at Limoges on September 6, 1919, at the age of 77.

Notes

218. Gadolin, J. *Svenska Akad. Handl.* **1794**, *15*, 137; Gadolin, J. *Crell's Annalen* **1796**, *I*, 313.
219. Urbain, G. *Compt. Rend. Chim.* **1906**, *142*, 785.
220. Mosander, C. G. *Pogg. Ann.* **1839**, *46*, 648; Mosander, C. G. *Compt. Rend. Chim.* **1839**, *8*, 356.
221. Mosander, C. G. *Pogg. Ann.* **1842**, *56*, 503; Mosander, C. G. *Liebig's Annalen* **1842**, *44*, 125; Mosander, C. G. *Skand. Naturfor. Forh.* **1842**, *3*, 387; Mosander, C. G. *Phil. Mag.* **1842**, *3*(23), 241.
222. Schofield, M. *Ind. Chem. Chem. Manuf.* **1958**, *34*, 420; Jorpes, J. E. *Acta Chem. Scand.* **1960**, *14*, 1681; Tansjoe, L. In *Episodes from the History of the Rare Earth Elements*; Evans, C. H., Ed.; Springer: Heidelberg, Germany, 1996, p. 37.
223. Marignac, J. C. G. de *Ann. Chim. Phys.* **1853**, *3*(38), 148.
224. Among these appear the names of A. Damour, J. H. Gladstone, O. L. Erdman, O. N. Rood, J. F. Bahr, and R. Bunsen.
225. Delafontaine, M. A. *Compt. Rend. Chim.* **1878**, *87*, 632.
226. Lecoq de Boisbaudran, P. E. *Compt. Rend. Chim.* **1879**, *89*, 212; Lecoq de Boisbaudran, P. E. *Chem. News* **1879**, *40*, 99.
227. Baumgartner, E. In *Episodes from the History of the Rare Earth Elements*; Evans, C. H., Ed.; Springer: Heidelberg, Germany, 1996, p. 113.
228. Auer von Welsbach, C. *Sitzber. Akad. Wien* **1885**, *92*, 318.
229. Kremers, E. *Trans. Am. Electrochem. Soc.* **1925**, 47; Wierda, J.; Kremers, H. C. *Trans. Am. Electrochem. Soc.* **1925**, 48.
230. Krüss, G.; Nilson, L. F. *Ber. Dtsch. Chem. Ges.* **1887**, *20*, 2134.
231. Crookes, W. *Proc. Roy. Soc.* **1886**, *40*, 236; Crookes, W. *Proc. Roy. Soc.* **1886**, *42*, 111.
232. Muthmann, F. W.; Stützel, L. *Ber. Dtsch. Chem. Ges.* **1899**, *32*, 2654; Muthmann, F. W.; Böhm, C. R. *Ber. Dtsch. Chem. Ges.* **1900**, *33*, 48.
233. *Jahreshefte der Gesellschaft für Naturkunde in Württemberg* **1975**, *130*, 64.
234. In the preceding reference, his birth date is given as "31 June."
235. von Chrustchoff, K. D. *Journ. Russ. Phys. Chem. Soc.* **1897**, *29*, 206.
236. von Chrustchoff, K. D. *Berg. Hutt. Ztg.* **1887**, *46*, 329.
237. Brauner, B. *J. Am. Chem. Soc.*, **1882**, *4*(10), 240.
238. Cleve, P. T. *Compt. Rend. Chim.* **1882**, *94*, 1528.
239. Brauner, B. *Compt. Rend. Chim.* **1882**, *94*, 1718.
240. Brauner and Auer were fellow countrymen in the sense that they were part of the heterogeneous union that made up the Austro-Hungarian Empire, but although Auer felt that he was Austrian, Brauner considered himself for all intents and purposes Czech.
241. Soukup, R. W. "*Die wissenschaftliche Welt des Carl Auer von Welsbach: seine bedeutendsten Lehrer, Kollegen und Mitarbeiter.*" Forum Naturwissenschaftlicher Unterricht. http://pluslucis.univie.ac.at/FBW0/FBW2008/Material/Soukup/Soukup.pdf (accessed April 12, 2014).
242. Urbain, G. *Recl. Trav. Chim. Pays-Bas Belg.* **1925**, *44*, 281.
243. Dutra, C. V. *R. Esc. Minas* **2002**, *55*(3), 185.
244. Gorceix, H. *Compt. Rend. Chim.* **1885**, *100*, 356.

III.12

SIR WILLIAM RAMSAY: THE MOST "NOBLE" OF CHEMISTS

Ramsay was the only scientist to have discovered or to have contributed to the discovery of all the elements in a group: Group 0, or the noble gases. Even if the recent discovery of the noble gas[245] with atomic number 118 has been definitely proven,[246] to Ramsay remains the great merit of having isolated all six of the noble gases present in nature. These discoveries make the Scottish scientist one of the greatest chemists of his time.[247] Today, his name has virtually disappeared from textbooks, and his figure is unknown to students. He did not discover laws or reactions that bear his name, but his masterful experimental work and his discoveries are and will remain a milestone in the continual progress of science and a testimony to his greatness.

Sir William Ramsay received the Nobel Prize for discovering an entire group of elements: the noble gases. He isolated helium,[248] an element already discovered in 1868 by the French astronomer Pierre Janssen (1824–1907) in the solar spectrum and believed at first to be present only in the sun.[249] In 1894, Ramsay,[250] independently of John W. Strutt, third Baron Rayleigh, discovered and isolated a rare gas with the atomic weight of 40. In the same year, at the meeting of the British Association, the two scientists by mutual agreement called this element argon (i.e., "inactive"). One of the first pieces of research that Ramsay conducted on this new gas was aimed at determining its chemical nature. By measurements conducted on the propagation of a sound wave in the new gas and in air, Ramsay discovered that argon was a monatomic gas. He obtained the same result some years later for neon and krypton. For Ramsay, the most surprising thing about argon was that it had no tendency whatsoever to react with other elements.

Between 1894 and 1898, William Ramsay, together with his assistant Morris William Travers (1872–1961), discovered krypton, neon, and xenon. Later, when radioactivity was discovered, he recorded the spectrum of radon, the noble gas with the highest atomic weight.[251]

III.12.1. THE FIRST DISCOVERIES

In 1895, while Lord Rayleigh and William Ramsay were discovering the first two noble gases, helium and argon, Julius Thomsen, in Germany, proposed an updated periodic system of the elements in which a new group appeared for the first time. According to Thomsen's reasoning, to render more gradual the passage from the doubly negative-valent oxygen to the singly negative-valent fluorine to the singly positive-valent sodium, it was necessary to introduce a group with a valence of zero.

The rare gases would thus have constituted the intermediate group for the passage of the elements monovalently electronegative of the seventh group to those electropositive, always monovalent, of the first group. Julius Thomsen indicated also the presumed atomic weights of all the elements that Ramsay soon was to discover: 4, 20, 36, 84, 132, and 212.[252] In 1887, Paul Emile Lecoq de Boisbaudran and F. Flawitzky, speculating on Medeleev's periodic table, had already predicted the existence of new gases in the atmosphere.

In 1897, Ramsay, in his inaugural address to the British Association for the Advancement of Science, reported in his capacity as president of the chemistry section, the following:

> The discovery of argon at once raised the curiosity of Lord Rayleigh and myself as to its position in this table. With a density of nearly 20, if a diatomic gas, like oxygen and nitrogen, it would follow fluorine in the periodic table; and our first idea was that argon was probably a mixture of three gases, all of which possessed nearly the same atomic weights, like iron, cobalt, and nickel. Indeed, their names were suggested, on this supposition, with patriotic bias, as Anglium, Scotium, and Hibernium.[253]

Anglia, Scotia and Hibernia are Latin names for England, Scotland, and Ireland, the three principal subdivisions of the United Kingdom. In the late 19th century, the names of many elements paid homage to their discoverers' homelands, despite the fact that Lavoisier had suggested that names given to new elements should reflect information about their properties.

The year 1898 opened with the announcements of the discovery of three new noble gases. At that time, Ramsay and his young assistant decided to analyze a large quantity of air that had been liquefied on a large scale by Kamerlingh Onnes (1853–1926), Sir James Dewar, and William Hampson (1859–1926). Hampson gave a deciliter of liquid air to Ramsay, who conducted his investigations on the residue after having evaporated a large part of it. The residue contained a gas that showed two brilliant lines, one yellow and one green. Its density was greater than that of argon. Ramsay and Travers called it krypton (i.e., "hidden").

In the meantime, Travers had prepared 15 liters of crude argon by removing from the air the oxygen and nitrogen and then forcing the rare gas into a bulb immersed in liquid air. Under these conditions, argon formed a colorless liquid, mobile, similar to water. Ramsay slowly removed the liquid air, and the argon began to boil. Ramsay suspected that, by distilling the raw argon, he could separate out other gases with higher or lower boiling points. If this had contained other "liquids" with lower boiling points, they would have first been distilled and collected separately. The heaviest gas would have been the last to distill off. Ramsay's hopes were not in vain because the first part of the gas evaporated was considerably lighter than argon and had a much lower boiling point. However, after a few distillations, Travers and Ramsay found that the liquid air was not sufficient to condense this gas until it liquefied. William Travers knew the way to go. He constructed an apparatus with which hydrogen could be liquefied. With liquid hydrogen, the two chemists cooled the gas mixture separated from the distillation of argon. Two-thirds of the volumes of these gases were condensed, while the rest remained gaseous. The gaseous portion, they discovered by spectroscopic investigation, was helium. The first portions of this new gas shone brightly with a "brilliant color of fire if you passed through it an electric discharge."[254]

The bright spectrum of this element showed many red and orange lines. That evening at dinner, Sir William reported the discovery to his family: Willie, Ramsay's 13-year-old son, asked his father what he called the new element. Ramsay replied that he had not thought of it yet. Young Willie (1886–1927) jumped the gun on his father by proposing: "I should like to call the new gas novum."[255]

William Ramsay slept on the idea. The following morning, he told his son he had accepted his proposition—but under one condition: he wanted the new element's name to be derived from the Greek, as were the names of his other three discoveries, helium, argon, and krypton. Willie changed the name from *novum* to neon.

Ramsay separated krypton from argon by fractionation. He observed that at the bottom of the container a tiny bubble of liquid remained. This residue also showed a spectrum characteristic of a new element, one that they called xenon (i.e., "foreign"). Ramsay and Travers published many of their papers in the prestigious pages of *Comptes Rendus de l'Académie des Sciences de Paris*, and it was there that the discoveries of the new elements krypton and neon appeared. In an appendix to their article on the discovery of krypton and neon, Marcellin Berthelot (1827–1907),[256] a long-time friend of Ramsay, added the following note: "The intense green line of krypton at 5566.3 Å coincides appreciably with bright line number 4 (5567 Å) in the aurora borealis. Therefore one could call this gas by the more euphonious name of eosium, a name that I take the liberty of suggesting to Mr. Ramsay." (The name *eosium* is derived from the Greek Ĕōs, meaning "dawn.")

III.12.2. A WRONG TRACK

In the spring of 1898, Ramsay and Travers announced virtually simultaneously in the journals *Comptes Rendus* and *Nature* the discovery of "a new gas in the atmosphere."[257] From the fractionation of a large quantity of liquid air, about 100 barrels, they obtained 10 cm^3 of a new gas that they sent to Lord Rayleigh so that he could determine the density. The density was 19.87; that of argon was 19.94. However, the spectra of the two gases were very different. In particular, two lines stood out, one green and one violet, that did not match any of the lines known for neon, krypton, and argon. Ramsay thought he had discovered a new noble gas, probably with an atomic weight of 175. In 1910, in follow-up work to this discovery, Ramsay reported the list of elements in Group 0. Between xenon and radon, he placed a box in which he indicated the atomic weight and, in the place of a name, a "?". Ramsay was convinced of this discovery and decided in good faith to call this new gas *metargon* or *metaargon*.[258] On June 30 of the same year, contrary voices were already raised against this discovery. Professor Schuster sent a letter to *Nature* in which he strongly criticized this last discovery of Ramsay and Travers. According to him, *metargon* could not exist as an element, but it had to be the result of an experimental error or of contamination in the preparation. Ramsay (Figure III.08) replied immediately (July 14) to defend his position. He denied that he could have made an error of that sort, and he reconfirmed the results he had obtained.

Ramsay was a scrupulous man, and he repeated the experiments that Schuster had rejected, doing them with much greater care. The results were not encouraging. He realized that the discovery of *metargon* was the result of an error, and he immediately

FIGURE III.08. Sir William Ramsay (1852–1916). Caricature of Ramsay in his classroom at University College, London, where he isolated neon, argon, krypton, and xenon. An extremely able experimentalist, he was a giant among chemists and physicists, although he, too, was involved in the discovery of the clearly false element, *metargon*. His friend Marcellin Bertholet, in telling the press about the discovery of krypton, suggested the more harmoniously sounding name of *eosium*. Courtesy, Fisher Collection, Chemical Heritage Foundation Archives.

published a retraction. He clearly explained the cause of the error, behaving like a great scientist, capable of managing the unpleasant task like a true and proper gentleman. In his retraction he did not skimp on humor, totally British, aiming a sarcastic stab at his colleague Schuster: "We are not infallible; and in this case there is always a large number of good friends who correct our inexactitudes with maximum care."

III.12.3. ANOMALOUS ARGON: THE ELEMENT THAT WOULD NOT FIT

Argon's discovery in 1895 caused a flurry of speculation and activity among scientists worldwide. First of all, it simply did not fit into the periodic system, even admitting to a possible new group as proposed by Thomsen. Ramsay found that argon was monatomic, its atomic weight was about 40, and it was an entirely unreactive chemically. He believed that there simply was no room for this species in the periodic table, and he actually ended up apologizing for it to his scientific colleagues.[259] Because of this anomaly, numerous attempts were made to try to somehow accommodate it. Dewar thought it might be N_3, as did Bohuslav Brauner[260] and Mendeleev, but the latter went further and suggested that if it were not triatomic nitrogen, it might be an entity he called X_6, where X was a hitherto unknown element between hydrogen and lithium.[261]

One of Mendeleev's greatest difficulties in accepting the monotomicity of argon was its atomic weight, which was greater than potassium. The other known pair inversions at the time, nickel-cobalt and tellurium-iodine, were viewed with the suspicion that more refined atomic weight measurements would resolve the issue to put them in line with Mendeleev's orderly arrangement of increasing atomic weight.[262] Other speculations included various configurations of argon, but our interest is in the proposals for the existence of unknown elements that the discovery of argon spawned. In addition to the three proposed by Ramsay himself, *anglium, scotium*, and *hibernium*, there were several others.

Rang proposed that A (argon) had an atomic weight of 13 and a valence of 4 and that it formed a triatomic doubly bonded molecule, A_3, greatly resistant to chemical action. A violated the law of Dulong and Petit, as well as Avogadro's Law; apart from H, it had the highest specific heat of all known elements, and its atomic volume corresponded exactly to its place in the atomic volume series. Rang placed it in Group IV of his own table, under He, above an unknown element, +, and followed by Ge, Sn, and Pb. He claimed, furthermore, that "my period-table is the truest and best tabular arrangement of the elements yet produced; that the table has place for all elements, and fulfills every proper requirement of to-day."[263]

George Johnstone Stoney (1826–1911), famous for having introduced the term electron ("the fixed charge of electricity, the same in all cases, which is associated with each chemical 'bond'") stated that all possible alternatives regarding the nature of argon be set down. He suggested that its discovery may have placed in our grasp the possibility of a much greater discovery—the six other elements between hydrogen and lithium that may have escaped gravitational attraction during the formative stages of the earth. He offered as the most probable of these *infra-carbon*, with an atomic weight of 2.5 or 3, as part of a series of *infra-elements* lying between *infra-beryllium* and *infra-fluorine*.[264]

Lancelot Winchester Andrews (b. 1856) of Mount Allison University in New Brunswick, Canada, ambitiously projected the curves of atomic volume, melting point, and acid and alkaline power into what he termed the "vacant space between hydrogen and lithium" to see if the Periodic Law could be made an "instrument of prophecy," as it had already served for the discoveries of Sc, Ge, and Ga. He discerned a family of *supra-elements* lying above the main group elements such as beryllium and boron and proposed that argon might be *supra-beryllium* or *supra-boron*. Because the quotient of the seeming atomic weight of argon (40) divided by the atomic weight of the hypothetical *supra-Be* (1.5) is approximately 28, then a "polymerization" of 28 of these single atoms could explain the observed properties of the species. He ended his paper by claiming that "argon and helium will drop into their places and open up new vistas of analogy and suggestion."[265]

Those new vistas, as we have seen, quickly were realized with the discovery of an entire new family of noble gases, putting an end to these forms of speculation, but opening up vast fields of elemental research. And they were indeed accommodated in Mendeleev's existing scheme in what he himself referred to as a magnificent survival of a critical test.[266]

III.12.4. A PAUSE IN RESEARCH

At the dawn of the new century, Ramsay cut down on his scientific contributions, but his active mind never abandoned research. In November 1900, William Ramsay departed with his wife for what was then Bombay, a major port of India. A wealthy Indian, J. N. Tata, had left £400,000 (a dizzying sum) for the construction of an entire university, and the new university's foundation had asked Ramsay to travel the subcontinent for the purpose of finding a suitable place for it. From Bombay, he went to Calcutta, Madras, Delhi, and Lucknow, finally selecting Bangalore as the site for the construction of the university buildings.

The direction of the fledgling university was entrusted to Morris William Travers, then only 29.[267] He returned to England at the beginning of World War I in 1914 to work in the war industry. His career after having left his mentor Ramsay was rather disappointing in comparison to the discoveries made in his youth. At the age of 55, in 1927, he returned to university teaching, as honorary professor of applied chemistry at Bristol. He left teaching when he reached the age limit in 1937. Between 1953 and 1955, Travers wrote a monumental biography of his mentor,[268] published in 1956. He died August 25, 1961, at the age of 89.

Ramsay, on his return from India, collaborated with Frederick Soddy, a person with innovative ideas and a lively genius. The latter had worked at Montréal under the guidance of Sir Ernest Rutherford in the field of radioactive elements. Soddy remained at Ramsay's side until the year that he won the Nobel Prize in chemistry.

By a curious succession of events, the 24-year-old Soddy influenced the new line of work for the 50-year-old Ramsay and thus opened the last chapter in Ramsay's research: radioactivity. Soddy later was distinguished for the discovery of the law of chemical displacement[269] relative to α-decay and was covered with glory for having defined the concept of the isotope (1913). The idea of isotopes revolutionized the definition of atomic weight and changed how scientists look at atomic structure. In 1915, he got into a battle with Sir William Crookes who claimed, rather unjustly, to have discovered the road to the concept of isotopes as early as 1883, when he had published some bizarre articles on *inorganic evolution* and *metaelements.*[270] In 1919, he gained the chair of organic chemistry at Oxford but lost the creative inspiration of youth that, in addition to the concept of isotopes, had led him to hypothesize the creation of an atomic bomb 20 years before its time. In 1921, at the age of 44, Soddy was honored with the Nobel Prize in chemistry. He retired from teaching in 1937 on the occasion of the death of his wife; he lived until he was almost 80.

In the summer of 1904, with the departure of Soddy, Ramsay had a Dr. Collie as a co-worker. But 1904–05 were full of enjoyable events that kept Sir William away from his laboratory. After having toured the length and breadth of the United States, he attended the annual meeting of the Society of Chemical Industry, of which he was president, at New York, and afterward he went to the World Exposition at Saint Louis. Finally, at the conclusion of this *annus mirabilis*, he went to Stockholm to receive the Nobel Prize in Chemistry for 1904. Yet, despite the fact that Ramsay was often away from London, he followed with lively interest the tumultuous developments in the chemistry of radioactive substances.

III.12.5. RADIOACTIVITY AND THE DISCOVERY OF NITON

Radon, the last noble gas, existed in three isotopic forms that originated from the radioactive decay of the families of ^{238}U, ^{235}U, and ^{238}Th. In 1898, the Curies and Gerhard Schmidt (1865–1949) discovered independently that thorium was radioactive. Pierre and Marie Curie isolated two new elements in the radioactive material: polonium and radium. They used the term "radioactivity" for the first time and described the spontaneous procedure (i.e., the decay) of the emission of α and β rays. In 1899, Ernest Rutherford and Robert Bowie Owens (1870–1940) discovered the isotope ^{220}Rn, *thoron*. Often the gas generated by the α decay of thorium had been called *emanation of thorium* or more simply, *emanation*. In fact, Rutherford realized that passing air over a salt of thorium or making it bubble in a solution of the salt dragged with it an *emanation* that had the power to discharge an electroscope for a short time. Later, the same properties were observed for radium as well, although this property lasted for many days. For thorium it was a question of minutes. In the periodic tables of the 1910s and of the first years of the 1920s, the box corresponding to the element with atomic number 86 contains the symbol Em for *emanation*.[271]

In 1904, André Debierne and Friedrich Oskar Giesel (1852–1927) isolated the isotope ^{219}Rn, *actinon*, coming from the transmutation of ^{235}U.

The isotope ^{222}Rn was discovered by the German chemist Friedrich E. Dorn (1848–1916)[272] in 1900. This originated from radium that, in turn, belonged to the radioactive family of ^{238}U. This isotope had a longer half-life, 3.8 days, compared with less than a minute for *thoron* and 4 seconds for *actinon*. It is for this reason that element 86 was given the name radon. The isotope discovered by Dorn was the most stable and therefore the easiest to study.[273] As long as Dorn lived, the name *emanation* was widespread, especially in the Anglo-Saxon, French, and Italian scientific worlds.

Ramsay had completed his work on the inert gases when, in 1904, with Robert Whytlaw-Gray (1877–1959), he determined by spectroscopy the presence of the last rare gas, radon. In his attempt to eliminate the use of "*emanation of radium*," he suggested the term *exradium* if radium were the source of the gas, and likewise *exthorium* and *exactinium* if the sources were thorium and actinium, respectively.[274]

The rising discipline of radiochemistry had discarded the basis for an understanding of the atom, but at the same time the concept of the isotope, not yet clear, meant that each radioactive isotope of the same element would be considered a new simple substance. Radon was not an exception: initially, the different names *radon, actinon, thoron* were left alone, but already before World War I the first efforts were made to give symbols to these three gaseous elements generated by three different decay "families": ^{222}Rn, ^{220}Rn, and ^{219}Rn. Table III.6 summarizes the situation.

Table III.6 Proposed Names for the Isotopes of Element 86

Authority proposing the name of the isotopes of element 86	*Radium Emanation*	*Thorium Emanation*	*Actinium Emanation*
Ramsay, Collie[a]	Exradium	Exthorium	Exactinium
Perrin[b]	Radeon	Thoreon	Actineon
Dorn[c]	Radon	Thoron	Akton
Adams[d]	Radon	Thoron	Actinon
Subsequent identity	^{222}Rn	^{220}Rn	^{219}Rn

[a]Ramsay, W.; Collie, J. N. *Proc. R. Soc. London* **1904**, *73*, 470.

[b]Perrin, J. *Ann. Physique* **1919**, *11*, 5.

[c]Dorn, F. E. *Abh. Naturf. Ges.* (Halle) **1900**, *22*, 155.

[d]Adams, E. Q. *J. Am. Chem. Soc.* **1920**, *42*, 2205.

Ramsay may have crossed swords with some opponents of his previous findings, but he was by now well established through the authority of other researchers. In 1906, Rudolf Schmidt, by means of fractional distillation, arrived at the conclusion that xenon was not an element but a mixture of some gases.[275] He harshly criticized the spectrum of this element that had been recorded years earlier by Edward Charles Cyril Baly (1871–1948), to whom Ramsay had given samples of the gas in order that the spectrum be recorded.

Two years later, Ramsay published a work in which he reported the complete fractionation of 120 tons of liquid air. Although he and Richard B. Moore, the spectroscopist, reported the same work in two distinct articles, the results were rather different. Although he never succeeded in separating out any new element heavier than xenon from liquid air, Ramsay remained much more of a "possibilist" than his colleague. He had, in fact, entertained the possibility that two inactive elements in the eighth group still existed and were waiting to be discovered, one with an atomic weight of 172–175 and the other with one of greater than 200. According to Ramsay, two of the three gases produced by the radioactive decay of thorium, actinium, and radium could be those sought. The third element was inappropriately inserted into the periodic system. Ramsay's two errors were that he did not recognize that the element with atomic weight 174 was lutetium, a rare earth, recently discovered by Urbain, and that the three gases *thoron*, *radon*, and *actinon* were isotopes of the same element. He did not correctly position the rare earths and, because of this, the periodic table that he constructed left space for a further gaseous element with a mass identical to that of lutetium. A more minor error was that of assuming that the three gases radon, *actinon*, and *thoron* were three distinct elements, but this error, given the knowledge of the time, was difficult to avoid. If he had been able to compare the spectra of the three isotopes he may have recognized that, in reality, he was not dealing with three distinct elements. However, the difficulty of recording the spectra of the isotopes of *actinon* and *thoron*, with their extremely short half-lives, hindered this comparison. Moore was much more skeptical; in his article, he rejected the existence of the other noble gases in the 120 tons of liquid air that he had examined.[276]

In the years between 1904 and 1910, Ramsay first succeeded in recording the spectrum of the *emanation*. Together with Whytlaw-Gray, he discovered that helium had formed in the ampoule where he had collected the *emanation*. This phenomenon was observed in different spectra recorded in succession. Slowly, as the amount of *emanation* declined, that of helium rose from zero to a maximum value. This was the last contribution of great relevance Ramsay made to science. His work consisted of collecting five different samples of *emanation* of radium in extremely thin capillaries, from which he succeeded in determining both the volume and weight.

In prior work, André Debierne had found the atomic weight of the *emanation* by indirect measurements, comparing measurements of the velocities of various gases made to pass through a slit. This was the first experiment to give an approximate value for the atomic weight of the gas. The value found by Debierne was 220 ± 6.

In calculating the same atomic weight, Ramsay used a balance with a sensitivity of a half-millionth of a milligram.This balance was constructed by Steeb, a former student, who dealt with similar issues and knew Ramsay's exacting specifications. The balance was based on the principle that when imperceptible variations in pressure are made, a small ampoule of silica, containing a known weight of air, changes weight. In this way, in 1910, Ramsay published a work in *Comptes Rendus* in which, from the average of five weighings, he found the atomic weight of the *emanation*: 222.5. From what was written by

Ramsay in 1910, one can deduce that he continued to believe in the existence of another noble gas between xenon and the *emanation*.

Ramsay successfully made important measurements on the last of the noble gases. He determined its atomic weight more accurately than any recorded up to this time. He also succeeded in recording the spectrum of the gas.[277] His spectroscopic work was indeed difficult considering that the quantities he had were very modest: about sixty-thousandths of a cubic millimeter of gas for every measurement. Bearing in mind his unique reputation in terms of the noble gases, Ramsay was conscious of having done his best work. He had discovered almost all of the noble gases, and those he had not discovered he had isolated. Since the suggestion of Ramsay and Whytlaw-Gray of 1904 relative to the nomenclature of the last noble gas had not been accepted, in 1910, he proposed another name for it: niton (symbol Ni), meaning "brilliant," by reason of its phosphorescent properties.[278,279]

At the conclusion of Ramsay's request to call the last of the noble gases *niton* is an ironic printing error: the symbol proposed was shown as Ni (i.e., nickel). It didn't matter: the name *niton* and its correct symbol (Nt) soon disappeared: World War I overshadowed both the name and the author of the presumed discovery. In 1912, the International Commission for Atomic Weights accepted the name *niton*, although until 1923 its three isotopes were called emanation of radium, emanation of thorium, and emanation of actinium. In 1923, the International Committee for Chemical Elements and the Union Internationale de la chimie pure et appliquée selected for these isotopes the names proposed by Schmidt and Adams; radon, *thoron*, and *actinon* (Rn, Tn, and An, respectively). A little later, however, this decision fell into disuse, with the isotopes coming to be called by their mass number and not by name. The only survivor by name was the isotope with the longest half life, radon.

III.12.6. A HARVEST OF LAURELS AT THE CONCLUSION OF HIS CAREER

After Ramsay left teaching at the age of 70, he retired to live in the country. He loved to travel and learned languages with great facility: he spoke and wrote fluent German, French, and Italian. While still very young, in 1872, he had worked in the German laboratory of Wilhelm Rudolph Fittig (1835–1910) at Tübingen, where he had specialized in the study of organic chemistry. From 1880 to 1887, he occupied the chair of chemistry at University College Bristol, and from 1887 until he retired in 1913, he was a teacher of inorganic chemistry at University College London. For his 50th birthday, Edward VII (1841–1910) made him Knight Commander of the Order of the Bath, bestowing on him the title of Sir. A little later, Kaiser Wilhelm II (1859–1941) awarded him the highest chivalric honor in Prussia: *Pour la merité*. The King of Italy, Victor Emmanuel III (1869–1947) made him *Commendatore*. In France, Ramsay was created Official of the Legion of Honor.

The years following his retirement from teaching were dramatic and in a certain way probably contributed to the deterioration of his health. When in March 1913 he left his post to Professor Frederick G. Donnan (1870–1956), he was allowed to move with all of his apparatus to Hazelmere, but he preferred to settle into his own laboratory, which he had outfitted in his large home. Of the work of the last years of his life we know very little and that little is known from his laboratory notebook. Among the interests that occupied Ramsay from 1913 to 1916 was the pockets of helium present in English coal mines.

He also conducted extensive studies on behalf of the government on the permeability of helium through the casings used in airships. As his last work, Ramsay prepared radium bromide and zinc sulfide for phosphorescent screens.[280] The last entry in his laboratory notebook is dated December 1, 1915, a date after which Ramsay never returned to his laboratory because of constant pain from a terminal malignant tumor.

The outbreak of World War I disconcerted Ramsay because he was very much associated with the German scientific world. Nevertheless, he signed the manifesto that British intellectuals[281] composed in response to that published by his German colleagues, in support of the entrance of Britain into the Great War. The war sadly ruptured a long friendship with Emil Fischer (1852–1919) as Ramsay entered the enterprise of national defense and worked actively almost until his death. As many articles in the *Times* of London testify, in the last months of 1914 and the beginning of 1915, Ramsay showed himself to be increasingly disdainful toward Teutonic science in general.[282] In the last years of his life, frustrated by the knowledge that he would not see the end of the war because of his illness, his ideas became extreme, and he became radicalized with respect to his scorn for the entire German population: "The greatest advances in scientific thought have not been made by members of the German race; nor have the earlier applications of science had Germany for their origin. . . Much of their previous reputation has been due to the Hebrews resident among them."[283] Many historians hold that his open hostility and excessive anti-German sentiment could well have been due to a change in his mental state caused by the pain of the tumor in his nose. Sir William died on July 23, 1916, at High Wycombe in Buckinghamshire.[284]

Ramsay's life was his own personal tribute to science. Perhaps nothing better than his own words, now distant in time, can sum up his activities and his life: "Being a son of parents like my father and my mother, and having a collaborator like my wife, have given me a happiness that I must reward with the greatest gratitude; and both my birth and my career correspond so to my inclinations and to my intentions, that if I were allowed to choose, I would hardly have changed the rules of God."[285]

III.12.7. POSTSCRIPT: KRYPTON II

Shortly after the discovery of the noble gases by Ramsay and his colleagues, Albert Ladenburg (1842–1911), professor of chemistry at the University of Breslau (presently Wrocław, Poland) became interested in their isolation in order to determine their correct place in the periodic table. With the help of his student Curt Kruegel, he examined the least volatile portion of a large quantity (850 L) of liquid air that had previously been purified of oxygen and nitrogen. The 3.5 L of gas that remained were condensed in a bath of liquid air under increased pressure and then subjected to a kind of fractional distillation. The first fraction, with a boiling point of –181.2 °C, exhibited a complete argon spectrum, while the fraction obtained from the crystalline residue exhibited a bright krypton spectrum. Ladenburg determined the atomic weight of krypton to be 58.81 (we now know that it is 83.8), causing him to suggest that these new atmospheric elements be placed before Group I in the periodic system: argon with atomic weight 39 before potassium, and krypton, with atomic weight of 59, before copper (copper was in Group I in the "short form" of the periodic table).[286] Later that same year, Ladenburg reconfirmed his original atomic weight of krypton and reaffirmed his hypothesis regarding its position in the periodic system. Although his idea was never accepted by the scientific community, he never retracted it. One can surmise today that his gas samples were contaminated by other volatile substances or that he made gross errors in his experimental measurements.

Albert Ladenburg was born in Mannheim on July 2, 1842. At the age of 15, he enrolled in the Karlsruhe Technische Hochschule where he studied mathematics and modern languages. He then moved on to the University of Heidelberg where he studied chemistry and physics with Robert Bunsen, under whose guidance he received his PhD. He then worked for 6 months with Kekulé, who introduced him to his structural theory. Ladenburg theorized that the structure of benzene was prismatic, which turned out to be wrong with respect to benzene but prescient with respect to the form: the compound prismane was synthesized in 1973. He visited England and then went to Paris to work for 18 months with Charles-Adolphe Wurtz (1817–84) on organosilicon and tin compounds. He moved to Kiel in 1873 as professor of chemistry and, in 1889, was appointed to the chemistry chair at Breslau. His later research focused on organic chemistry, a field in which he was extraordinarily successful. He isolated hyoscine, also known as scopolamine. In 1905, he was awarded the prestigious Davy Medal "for his researches in organic chemistry, especially in connection with the synthesis of natural alkaloids." He died on August 15, 1911, at the age of 69 in Breslau.

Notes

245. Ninov, V. et al. *Phys. Rev. Lett.*, **1999**, *83*, 1104; Ninov, V. et al. *Editorial Note, Phys. Rev. Lett.* **2002**, *89*, 039901(E); Oganessian, Y., et al. *JINR Preprints and Communication* **2002**, D7-2002-287.
246. After the announcement (1999) and the retraction (2001) of the discovery of element 118, it seems that, in 2003, Yuri Oganessian achieved the synthesis of the last noble gas by means of a nuclear fusion reaction of californium and calcium.
247. Kauffman, G. B.; Priebe, P. M. *J. Chem. Educ.* **1990**, *67*, 93; McKie, D. *Proc. Roy. Soc.* **1962**, 377; Kauffman, G. B. *J. Chem. Educ.* **1982**, *59*, 3.
248. This gas was discovered in *cleveite*, a radioactive mineral with the formula UO_2 (containing up to 10% of the rare earths) that, by α decay, emits He. Only later, Ramsay found traces of helium in liquid air that he subjected to fractional distillation; having removed argon, oxygen, and nitrogen, Ramsay collected a fraction composed of a mixture of neon and helium.
249. Only later this element, of which the only evidence of its existence was its spectrum, was called helium by the astronomer Norman Lockyer (1836–1920).
250. In those years, Ramsay was studying the physicochemical properties of liquid air; it was thus that in distilling it to separate the oxygen from the nitrogen that he discovered argon.
251. In 1871, Ramsay perfected his knowledge of spectroscopy in Heidelberg, Germany, under no less a figure than Robert Bunsen (1811–99), the father of this science.
252. Presently, the atomic weights are: He (4.00); Ne (20.18); Ar (39.95); Kr (83.80); Xe (131.29); Rn (222.0).
253. Ramsay, W. *Nature* **1897**, *56*, 378.
254. Initially, no commercial use was found for this new element. French chemist and inventor Georges Claude (1870–1960) was the first to apply an electric discharge to a sealed tube containing neon (1902). Under these conditions, neon gave out a soft red light. Claude had the idea of producing artificial illumination different from that known at the time (thermoluminescence). On December 11, 1911, in Paris, he displayed his first neon lamp, and it was a failure. In 1923, he sold seven discharge tubes in the form of a luminous sign—"Packard"—to a Packard auto dealer for $24,000. Georges Claude founded the company Claude Neon Lights and quickly amassed an immense fortune with this discovery. His invention fascinated his contemporaries;

because neon light shone vividly even in broad daylight, in English, his invention took on the name "liquid fire."

255. His father liked the idea, but suggested using the Greek word for "new," *neos*. Thus, the element was named neon.
256. Marcellin Berthelot was one of the fathers of organic chemistry: he threw new light onto the nature of alcohols and sugars. He conducted important experiments on calorimetry, isomerism, and the chemistry of fermentation and foods. Senator for life since 1881, few people know that, in 1885, he was foreign minister of the French Republic and in that role signed the Anglo-French Treaty for the management of Siam and Cochin-China. Although he was older than Ramsay by 25 years, a profound friendship grew between them. In his memoirs, Ramsay mentions him as one of the greatest chemists of his time: "He [Berthelot] was one of the greatest among the illustrious men of whom France can be proud."
257. Ramsay, W.; Travers, M. W. *Compt. Rend. Chim.* **1898**, *126*, 1610; Ramsay, W.; Travers, M. W. *Nature* **1898**, *58*, 245.
258. Ramsay expected that there would be a noble gas with an atomic weight near 175 because he did not accept the existence of the inner transition (lanthanide) series which, by 1910, almost everyone else had acknowledged.
259. Giunta, C. Argon and the periodic system: The piece that would not fit. *Found. Chem.* **2001**, *3*, 105–28
260. Brauner, B. Some remarks on "argon." *Chem. News.* **1895**, *71*, 79.
261. Mendeleev, D. *The Principles of Chemistry* (translated from 6th Russian edition by George Kamensky); Longmans, Green, London, 1897, vol. II, pp. 491 ff.
262. One might consider it unfortunate that of all the noble gases, argon, with its anomalous atomic weight, was discovered first. This fact certainly generated a lot of thinking and perhaps was a catalyst toward realizing that atomic weight might not be the principal ordering system for the elements.
263. Rang, F. The period table. *Chem. News.* **1895**, *72*, 200–201.
264. Stoney, G. J. Argon: A suggestion. *Chem. News.* **1895**, *71*, 67–68.
265. Andrews, L. W. The position of argon in the periodic system. *Chem. News.* **1895**, *71*, 235.
266. Scerri, E.; Worrall, J. Prediction and the periodic table. *Stud. Hist. Phil. Sci.* **2001**, *32*, 407–52, at p. 445.
267. Travers was suggested for the directorship in 1901, but was not actually appointed to it until 1906.
268. Travers, M. W. *A Life of Sir William Ramsay*. Edward Arnold: London, 1956.
269. The Law of Soddy, or of Chemical Displacement, says that every transformation of an element by emission of an α particle lowers the atomic weight by 4 and the atomic number by 2.
270. Crookes, W. *Proc. Roy. Soc.* **1883**, *35*, 262; Crookes, W. *Report of British Association* **1886**, 558.
271. In the 1920s, the scientific literature was full of a mélange of names for element 86: *niton* (Nt), which was the name adopted by *Chemical Abstracts; emanation* (Em); radon (Rn); *thoron* (Tn); *actinon* (At); and finally, emanation of radium. In 1923, the International Committee on the Chemical Elements, comprising F. W. Aston, G. P. Baxter (1876–1953), B. Brauner, A. Debierne, A. Leduc, T. W. Richards (1868–1928), F. Soddy, and G. Urbain, adopted the name in use today, radon. The true elemental nature of radon was established by the Curies and Rutherford. Around the time of Dorn's death (1916), Rutherford was no longer conducting experiments on radon, but both he and Marie Curie were consulted by the International Committee and approved the names of the three isotopes selected by the Committee. The International Committee had sought the opinions of the two great scientists because it was held that they were the two discoverers of radon. A curious episode took place later: Curie tried to influence the decision of the Commission for its nomenclature, also involving Rutherford, proposing the names: *radioneon*, or *radion*, for element 86. Rutherford declined the honor of giving a name to the last noble gas. Unlike Curie, there is indeed no evidence that Rutherford laid claim to the discovery of radon.

272. Dorn, F. E. *Abh. Naturf. Ges.* Halle, **1901**, *23*, 1–15.
273. Gessel T. F. Background atmospheric ^{222}Rn concentration outdoors and indoors: A review. *Health Physics* **1983**, *45:2*, 289.
274. Ramsay, W. *Z. Phys. Chem.* **1903**, *74,* 44.
275. Schimdt, R. *Ber. Dtsch. Chem. Ges.* **1906**, *4*, 277.
276. Moore, R. B. *Proc. R. Soc. London* **1908**, *81*, 195.
277. Ramsay, W. *Proc. R. Soc. London* **1908**, *81*, 178; Ramsay, W. *Nature* **1907**, *76*, 269.
278. Ramsay, W.; Gray, R.W. *Compt. Rend. Chim.* **1910**, *151*, 126.
279. John Norman Collie (1859–1942), chemist, mountaineer, and explorer, worked with Ramsay on inert gases until he became professor of organic chemistry at University College of London. His grandfather was an explorer, too. He served on the warship that annexed Western Australia to the British Crown, and he also discovered the Collie River; the town of Collie was named for him.
280. Ramsay, W. *Nature* **1914**, *94*, 137.
281. Among the signers appear the names of J. Norman Collie, Sir William Crookes, Lord Rayleigh, William H. Perkin Jr. (1860–1929), William H. Bragg (1862–1942), Joseph John Thomson (1856–1940), and many others.
282. Ramsay, W. *Times (London)*, October 24, 1914; *Times (London)*, January 21, 23, 29, 30; March 6; April 6; July 2, 15, 19, 26; August 12, 13; December 15, 1915.
283. Badash, L. British and American views of the German menace in World War I. *Notes and Records of the Royal Society of London* **1979**, *34*, 91–121.
284. W.A.T. *J. Chem. Soc.* **1917**, *112*, 369.
285. Ramsay, W. *Chimica e chimici; saggi storici e critici*; Remo Sandron: Firenze, Italy, 1913.
286. Ladenburg, A.; Kruegel, C. *Sitzungsber. K. Preuss. Akad. Wiss.* **1900**, 212–17.

III.13

CONFEDERATE AND UNION STARS IN THE PERIODIC TABLE

III.13.1. INTRODUCTION

Over the course of years, American chemists have engaged in an intensive search for new elements. At first, the route was very difficult, the terrain rugged, and the setbacks many. From the beginning of the 20th century until after World War II, the discoveries of *carolinium* (1901), *illinium* (1926), *virginium* (1930), *alabamine* (1931), and californium (1950) were announced. (For three of these elemental discoveries, scientists chose the names of U.S. states that had been part of the Confederacy during the Civil War: North Carolina, Virginia, and Alabama.) The last to make its appearance in the periodic table, californium, was the only discovery that turned out to be correct.[287]

When the discovery of *carolinium* was announced in 1901, the United States was still something of a frontier country, and the research done there was in certain respects marginal: the driving force of ideas and the source of new and great discoveries were still coming from the Old World. For example, when Ludwig Boltzmann (1844–1906) came to the United States from Austria at the beginning of the century for a series of conferences, he was deeply shocked by the raw state of American culture. He was a refined man, coming from an empire that was undergoing slow decline and a golden decadence with the alternating inevitability and light-heartedness of the waltzes played in the cafes of Vienna. Boltzmann joined in his scientific prose an extraordinary refinement of style to a crystalline scientific clarity. The young republic's *naïf* culture and rough and unrefined attitudes, where his colleagues walked around the universities of the Midwest and California with bandoliers and pistols in their belts, horrified him. Even Ernest Rutherford turned down an invitation to take a position at Yale (in New England, not in the Wild West) because of the reputation American universities had of being "places more adapted for students rather than for researchers."

However, with time, the American university scene improved: early in the 20th century many of those same students completed their education in Europe and, in the process, learned new theories and state-of-the-art research methods in the more lively and stimulating laboratories of Manchester, Paris, Heidelberg, and Copenhagen. As the new century progressed, American science was becoming increasingly competitive with that of the Old World. However, although great strides were made, scientific advances were always behind technological progress.

The 19th century gave rise to the genius of J. Willard Gibbs (1839–1903), a theorist who long remained an isolated "parenthesis" in the annals of American scientific literature.

At the beginning of the 20th century, the American scientific community was more attached to applied rather than fundamental research. In the era of Thomas Alva Edison (1847–1931), America also produced the clever and skillful Ernest Orlando Lawrence

(1901–58), by education more an engineer than a physicist: he read patent manuals as his textbooks while his European colleagues (Paul Dirac, Enrico Fermi [1901–54], Werner Heisenberg [1901–76], Wolfgang Pauli [1900–58], and the Joliot-Curie team) had read the works of Rutherford, Arnold Sommerfeld (1868–1951), and Niels Bohr. Later in the century, in the 1930s and 1940s, American research received an important supply of fresh energy from European scientists who were refugees fleeing from the Nazis.

As with other eras and other nations, it was inevitable that among U.S. practitioners of the official chemical and physical sciences, the first half of the 20th century would also produce those who interpreted the results of their scientific research a bit too hastily.

III.13.2. CAROLINIUM (AND BERZELIUM)

In 1901, the discovery of a new element, *carolinium*, was announced. Its discoverer, Charles Baskerville, was an active member of the American Chemical Society, the Society of Chemical Industry, the American Electrochemical Society, and the New York Academy of Sciences. Born in Mississippi, on January 18, 1870, he began his chemical studies at the University of Mississippi, then moved to the University of Virginia to study under John W. Mallet. His academic career matured at the University of North Carolina, where, from 1891, he moved up the professorial ranks from assistant, to assistant professor, to professor and then chair (from 1901 to 1904) of the department of chemistry. During this time, he traveled to Europe to study with August W. Hofmann (1818–92) at the University of Berlin. In the 14 years that he was associated with the University of North Carolina, his commitments were divided between his passion for teaching and his interest in studying the rare earths, as different articles in the *Journal of the American Chemical Society* attest and in which he presented, among others, the discovery of two new elements associated with thorium:[288] *carolinium* and *berzelium*. It is curious to note that when Baskerville was not yet 34 years old and at the height of his scientific activity, his research interests turned suddenly toward more technical and practical areas of chemistry and he abruptly abandoned the field in which he may have received the greatest recognition. His contemporaries interpreted this choice as an example of intellectual versatility; as a matter of fact, Baskerville is more famous for his contributions in the fields of anesthetics and the food and textile industries than for his studies on radioactivity and the properties of the rare earths.[289] Charles Baskerville died at the age of 52 on January 28, 1922, following a bout of pneumonia (Figure III.09).

The discovery of *carolinium* should be seen as only one of a number of false discoveries common when dealing with elements of the rare earth group. The discovery of these elements covers a period of about 120 years, reaching back to the various trial-and-error methods used before chemists had available more reliable investigative instruments and adequate supporting theories. The character of a new chemical element, in fact, was determined based on properties like atomic weight, separability, the color of its compounds, its crystal form, and its reactions. The close resemblance among the properties of the rare earths, however, made it much more difficult to isolate these elements and led to a situation in which mixtures of several elements were taken for elemental species. Furthermore, the values of the atomic weights were found to be unreliable, given that poor separation of the elements from one another could mean that more than one element was still present in an oxide, and this influenced the weight. Purer samples of the rare earths became a reality only in the second half of the 20th century; up until then, fractional crystallization

FIGURE III.09. Charles Baskerville (1870–1922). Professor and active member of the American Chemical Society, in 1901, Baskerville announced the discovery of two elements associated with thorium, *carolinium* and *berzelium*, named in honor of the state of North Carolina and of the chemist Berzelius, respectively. After the let-down he suffered upon the retraction of this double discovery, he redirected his interests to food chemistry, textiles, and anesthetics. Courtesy, William Haynes Portrait Collection, Chemical Heritage Foundation.

was the only method of purification and, in many cases, could require hundreds of fractionations and many months of work.

In 1904, an article appeared in *Chemical News* that made this point in trying to acquire information about thorium from the time of its discovery until the time the article was published. Its author, Charles Baskerville, emphasized the complexity of this element whose salts with organic bases (e.g., phenylhydrazine) were consistent with atomic weights that varied between 212 and 252 (actual atomic weight is 232.5). Studies on its radioactivity were just beginning, and there were conflicting versions describing the radioactive properties of this element; Marie Curie was of the opinion that: "the property of emitted rays—which act on photographic plates is a specific property of uranium and thorium." Whereas Baskerville asserted that "thorium is not a primary radioactive body."[290] His version was in perfect accord with the discovery of similar compounds derived from fractions of thorium that differed in their radioactivity and that were, according to him, attributable to new elements associated with this metal.[291] These elements assumed the names of *carolinium* and *berzelium*: the first in honor of the state of North Carolina; the second in memory of the scientist who first encountered thorium: Jöns Jacob Berzelius.

Baskerville's words regarding the choice of names for the two new elements were reported in the pages of the *Journal of the American Chemical Society* after his announcement at the meeting of the American Chemical Society at Denver in 1901: "on account of the extensive occurrence in this state (North Carolina) of the monazite sands from which the original material was obtained, if the investigation give successful issue, I should like to have the element known as *carolinium*, with the symbol Cn" and "on account of his [Berzelius] beautiful pioneer researches in the difficult field, as the discoverer of thorium from which it comes, it is only proper that it should bear his name, so I have designated the element *berzelium*, with the symbol Bz."

Atomic weights were found experimentally (respectively 255.6 and 212) for the two elements.[292] The properties of the new elements effectively differed from the compounds of thorium from which they were originally extracted. Furthermore, the dioxide of thorium showed phosphorescence when exposed to ultraviolet radiation, while the analogous oxides of *carolinium* and *berzelium* did not respond to this stimulus. The thorium obtained after the extraction of *carolinium* and *berzelium* from the original sample, in addition, emitted a more marked phosphorescence in accordance with the decrease in the amounts of the new substances. All of the new oxides were found to be radioactive, especially *carolinium*'s; in fact, some ammoniacal washes, obtained in the process of extraction and purification of the thorium from the monazite sands resulted in, after evaporation of the mother liquor, residues that produced a marked effect on a photographic plate and that had a level of radioactivity three times higher than that of thorium, using the apparatus of Dolezalek.[293] In support of the hypothesis of the discovery there were the differences in chemical behavior observed in the new compounds: for example, the oxide of *carolinium* was shown to be soluble in concentrated HCl, which was not the case with its analogs of thorium and *berzelium*. The arc and spark spectra were shown, however, to be identical, causing the author to suppose that the material examined was not completely pure or that the spectral data were not sufficiently complete. But the discovery was never confirmed, and it remains difficult even today to reconstruct the factors that could have caused experimental errors and consequently erroneous hypotheses, given that the original publications and calculations relative to the atomic weights were not reported with sufficient accuracy. Certainly we know that, then as now, the extraction of thorium was done starting with the mineral monazite which, beyond the 6–7% thorium, contains many other rare earths with variable composition. The other source of thorium on the other hand, thorite, contains thorium and uranium in a matrix of silicates and remains today the most common mineral of thorium, despite the fact that monazite commands the major part of the thorium mineral market. The isolation and purification of this element, from the others also present in the minerals being extracted, was accomplished only in 1904 through the work of D. Lely Jr and L. Hamburger;[294] even up to the present day, the so-called pure samples of thorium are sometimes contaminated by other elements. An error because of a flaw in the atomic weight estimate, consistent with the presumed discovery of *berzelium*, could have been due to the presence of traces of cerium (Ce_2O_3, yellow-green; CeO_2, yellow-white) in residues of ThO_2 (white, turning gray in air) obtained from the various crystallizations. Cerium (atomic weight = 140), in fact, is the lower analog of thorium and as such presents quite similar physico-chemical properties; this does not make them easily separable from one another. Certainly, we can exclude the possibility that Baskerville succeeded in isolating the isotope ^{212}Th, given that he had carried out only chemical reactions, and it is not possible to separate two isotopes of the same element by chemical means. The error most likely responsible for the discovery of *carolinium* is perhaps consistent with the presence of mixed oxides of uranium (UO_2, brown; UO_3, yellow-orange) that, with the other, were found to be soluble in concentrated HCl, as opposed to ThO_2 and CeO_2, which are not. The presence of thorium oxide in all three of the samples guarantees that they were radioactive, and the fact that the radioactivity was greater in the mixture attributed to *carolinium* could be attributed to a higher percentage of uranium present and without nonradioactive contamination.

III.13.3. CONCLUSION

It's common for some to think that the periodic table constructed by Mendeleev back in 1869 is the same as the original: this couldn't be further from the truth. The periodic table of the elements has evolved with the passing of the decades and the centuries, slowly growing in dimensions as new elements were added to the list of those whose discovery was already confirmed. The arrangement of the elements in periods was known by chemists in the 19th century, but it was the physicists who, utilizing the concept of atomic orbitals, subdivided the table into blocks of elements designated by the letters *s, p, d, f, g,* and so on according to the electronic shells being filled.

Thus, over the years, false announcements piled up even as true discoveries of missing elements went on to fill the empty boxes in Mendeleev's original table.[295] One hundred eighteen elements are known today, from hydrogen to ununoctium[296] (Z = 118), but the discoveries of elements shown to be false are almost equal in number.

In the period examined, 1901–50, that is, from the announcement of the discovery of *carolinium* to that of californium, science has made giant strides forward, followed by similar strides in technology.[297] Science has passed from the discovery of X-rays to the atomic bomb.[298] The chemical laboratories of the beginning of the last century were more similar to the laboratories of medieval alchemists than to those of the present day: some were without electricity, all lacked gas lines. Refrigerators did not exist, nor did apparatus that today we take for granted: magnetic or mechanical stirrers, fume hoods, rotary evaporators, ice machines, pH meters, sensitive analytical balances, and more. Also, the spread of ideas was much slower. The radio did not exist (not commercially until the end of the 1920s); European scientific journals arrived in the New World by steamer just like the immigrants who boarded the *Mauretania* or the *Lusitania*. It is therefore not surprising that in the discovery of *carolinium* (1901) only arc and spark spectra were used, not X-rays.

As an example of the limitations of research in the first half of the 20th century, Baskerville—a good teacher and an excellent disseminator of information—was not, in a certain sense, a very good experimentalist. The experimental techniques he used were not much different from those of J. J. Berzelius who, in 1828, had discovered thorium. He did not take advantage of the more recent discoveries of radioactivity (as did the Curies, A. Debierne, and F. Dorn) to isolate new radioactive elements. With respect to his discoveries, he was very young: 31 at the moment of his first announcement and 34 at the time of the second. He was a full professor, free to develop his research as he pleased in an American university in all regards rich in funds and furnished with the best laboratories. He started out with great advantages, yet he failed while the Curies triumphed using the highly inadequate means at their disposal. One of the reasons for his debacle was the provincialism of research in the United States in those years. After his double failure,

FIGURE III.10. Robert Bunsen (1811–99). Renowned German chemist and founder of analytical spectroscopy, Bunsen was responsible for educating generations of chemists and physicists in his laboratory. Using a spectroscope that he invented himself, he discovered rubidium and cesium. Courtesy, Chemical Heritage of the Department of Chemistry of the University of Florence, Italy.

Baskerville abandoned his research in the area of radioactivity (in a period still rich with discoveries) and turned to organic chemistry and the chemistry of anesthetics.

Regarding the discoveries of *alabamine, virginium*, and others, as we have emphasized previously, scientists resorted to new investigative techniquess during this era. Chemists like Bunsen (Figure III.10) and Kirchhoff, inventors of the spectroscope and spectral analysis, identified cesium and rubidium using instruments of their own invention, just as Humphry Davy had extracted by electrolysis many alkali and alkaline-earth metals.

Notes

287. *Illinium*, as well as *alamabine* and *virginium*, continued to appear on periodic tables in textbooks and wall charts until well into the 1940s. See Trimble, R. F. "What happened to alabamine, virginium and illinium? *J. Chem. Educ.* **1975**, *52*, 585.
288. Baskerville, C. *J. Am. Chem. Soc.* **1901**, *23*, 761.
289. Venable, F. P. *J. Ind. Eng. Chem. (Washington, D.C.)* **1922**, *14*, 247; Moody, H. R. *Chem. Metall. Eng.* **1922**, *26*, 280; Smith, E. F. *J. Chem. Soc.* **1923**, *124*, 3421.
290. Baskerville, C. *Chem. News* **1904**, 87, 151.
291. There are 25 isotopes of thorium known today with masses between 212 and 236, all unstable, with decay rates that make it impossible to isolate them in any appreciable quantity except for

the isotope ^{232}Th which is found in nature and has a half-life of 1.4 x 10^{10} years. It is an α emitter, and undergoes a series of six α decays and four β before becoming the stable isotope ^{208}Pb. ^{232}Th is radioactive enough to expose a photographic plate in a few hours. Thorium decays with the eventual formation of *thoron* (^{220}Rn), an α-emitter. Following bombardment with slow neutrons, ^{232}Th is transformed into the nuclide ^{233}U.

292. Baskerville, C. *J. Am. Chem. Soc.* **1904**, *26*, 922; Baskerville, C. *Science [N.S.]*, **1904**, *19*, 88.
293. The electrometer that Friedrich Dolezalak (1873–1920) devised was an instrument to measure potential difference by utilizing electric repulsion and attraction. The experiments on radioactivity using the electroscopic method instead of measuring the effect of particle emission on a photographic plate was based on the determination of the intensity of the radiation, measuring the conductivity of the air exposed to the rays' activity.
294. One of the industrial methods for obtaining metallic thorium from monazite involved attacking the mineral with sulfuric acid; after removing the excess acid from the solution, as well as the phosphates and most of the other cations, a thorium salt was obtained containing various impurities that were subsequently removed by selective extraction. By this method, the chloride could be obtained, from which, by electrolysis in a solution of NaCl and KCl, powdered thorium resulted, which could be purified from the salts by aqueous washings and then drying and desiccating to form the solid metal. Alternatively, thorium could be produced by the reduction of ThO_2 at 1,400 °C. However, ThO_2 was commonly reduced with Ca to give Th with a high degree of purity.
295. Ridgway, D. W. *J. Chem. Educ.* **1975**, *52*, 70.
296. Literally *one-one-eight.*
297. Henahan, J. F. *Chemistry* **1978**, *51*, 26; Seaborg, G. T. *The Science Teacher* **1983**, *50*, 29.
298. Seaborg, G. T. *Elements of the Universe*; E. P. Dutton: New York, 1958.

III.14

TWO ELEMENTS FROM THE DEPTHS OF PROVINCIAL AMERICANA

During the early 20th century, new discoveries were accomplished almost every day: those relative to radioactivity, to X-rays, and to the special theory of relativity in the area of physics; and to the isolation of polonium, radium, and actinium in chemistry. Yet, to all of these great works, produced like the links of a long and uninterrupted chain that stretches to our own day, William M. Courtis paid not the least attention.

William M. Courtis (1842–1920s?) was born at Marblehead, Massachusetts, in January 1842, son of Mehitabel and William Courtis, merchant by profession. In 1871, he began regular studies in mining engineering and found work in a foundry in Stonewall, Virginia. In 1870, he met and later married Lizzie E. Folger. In 1880, he changed jobs, moving to Silver City, New Mexico, but maintaining a home in Detroit, Michigan from 1874 to 1920. William M. Courtis performed many analyses on the waters coming from Gila Hot Springs in Grant County, New Mexico.[299]

According to some sources, Courtis may have discovered a new element[300] in deposits coming from the mining district of San Pedro when he worked in New Mexico as assistant in the geographic investigations done by the governor of the territory. The date was not specified, but presumably he would have been posted there in the last quarter of the 19th century. According to other sources, Courtis might have made the discovery of the metal that in 1901 he named *amarillium* when he was analyzing rocks coming from a copper deposit in Similakameen, British Columbia.[301]

Notice of the discovery of *amarillium* spread in the following years and, in 1903, reached the English-language periodicals. *Amarillium* was considered a metal of the platinum family; it had a bronze-like look, and its unique chemical property was that it was confirmed to be soluble in aqua regia.

In 1912, the chemist T. A. Eastick again brought *amarillium* to the attention of the scientific community. He was looking for experimental evidence in support of a metal discovered the year earlier by A. G. French,[302] called *canadium*. The more Eastick tried to find it, the less he succeeded. At the end of his work, he advanced the hypothesis that *canadium* was *amarillium*, if not the downright elusive element named *josephinium*.[303] The existence of *josephinium* seems to have also been advanced by Courtis in a 1903 article that appeared in *Transactions of the American Institute of Mining*.

Today, the name *josephinium* (synonym of *awaruite*) is understood as the mineral constituting the natural alloy of Ni and Fe whose composition varies from Ni_2Fe to Ni_3Fe. It is highly probable that Courtis confused the natural iron-nickel alloy with a new element that he had extracted under the form of nuggets from the detrital deposit along a creek feeding the Josephine River (hence the name of the element). Around 1912, another curious hypothesis was advanced according to which, observing amorphous silver under an ultramicroscope, this same material could have been confused for *amarillium*.

The discoveries of both *amarillium* and *josephinium* passed completely unnoticed, so much so that, a few years later, when some chemists were trying to reconstruct the entire affair, they attributed to William's son, Stuart A. Courtis, the credit for having isolated the first of the two metals.[304]

Shortly after World War II, long after its first appearance, an article appeared in *The New International Yearbook* describing the destiny reserved for *amarillium* and, unfortunately, for its discoverer; both were "consigned to the haven of lost elements."[305]

In 1921, William M. Courtis left his residence in Detroit and disappeared from history. The reason for this disappearance is probably attributable to his death, with some sources affirming that he died sometime during the 1920s.

Notes

299. Jones, F. A. *New Mexico Mines and Minerals*; The New Mexican Printing Company: Santa Fe, 1904, p. 355.
300. Courtis, M. W. *Trans. Am. Inst. Min., Metall., Pet. Eng., Soc. Min. Eng. AIME* **1903**, *33*, 1120.
301. Courtis, M. W. *Trans. Am. Inst. Min., Metall., Pet. Eng., Soc. Min. Eng. AIME* **1901**, *31*, 1080.
302. French, A. G. *Chem. News* **1911**, *104*, 283.
303. Eastick, T. A. *Chem. News* **1912**, *105*, 36.
304. *Detroit Educational Bulletin* **1921/1922**, *5*, 18.
305. *The New International Year Book*, **1966**, 134.

III.15

THE EARLY SUCCESSES OF THE YOUNG URBAIN

Georges Urbain was virtually the only chemist of the first half of the 20th century who knew how to improve his own working methodology by exploiting the progress made in physics. He dedicated his whole existence to the identification and isolation of the rare earth elements and to hafnium, a constant commitment that stretched over a period of more than 40 years.

When he first started to do this research, he used classical methods: fractional crystallization, arc and incandescence spectroscopy, cathodic phosphorescence, and magnetism; during the second part of his scientific career, he was among the first chemists to use X-ray spectroscopy with great success. The watershed of his research activity occurred during World War I, which coincided with his maturation and successive abandonment of the laboratory bench in favor of academe and theoretical chemistry.

From his youth, Urbain was primarily an experimental chemist: he raised the rank of thermogravimetry to an analytical discipline, studied the magnetic properties of the rare earth elements in depth, and discovered the ferromagnetism of gadolinium,[306] as well as isolating three new elements. During his professional maturity, he tried to formulate a unifying theory for the chemical sciences. Always retaining an interest in chemistry, with the years, he developed a passion for the arts, becoming a musician, sculptor, and talented painter. Hand-in-hand with his youthful investigations on the missing elements, he also proved the groundlessness of some presumed discoveries: if his discovery of three elements brought him fame and honor, unmasking false discoveries gained him many enemies.

His experiments with X-ray diffraction on various elements showed that, for every element on the continuous radiation spectrum, a characteristic spectrum consisting of a certain number of lines was superimposed. Following up on this direction of research, Henry G. J. Moseley found a simple formula that correlated the frequency of the lines with the atomic number of the emitting element. This relationship allowed for the determination of the atomic number of every element with extreme precision.

With this methodology, it was finally and definitively possible to position the elements in the periodic table. Furthermore, from the point of view of the internal structure of the atom, it turned out that the atomic numbers were more important than atomic weights. This decisive step was verified just at the beginning of World War I; before that time, chemists were engaged in tedious fractional crystallizations in attempts to isolate the last elements that escaped their sieve.

Before the war, Urbain had discovered *neo-ytterbium* and lutetium (1906) and after the war, *celtium* (1922). He himself recorded how the work to identify lutetium had occupied him for many years in extensive manipulations: concentrations, purifications, and transformations of the corresponding mineral in the earth (or oxide) and finally to the volatile halides for the purpose of determining the atomic weight of the new metal. Some years later, in June 1914, Urbain visited Moseley at Oxford to subject some samples of rare earths to a check. Urbain recorded this episode in a letter to Ernest Rutherford:[307] the rapidity and reliability of Moseley's technique left him speechless. He left eight samples with Moseley for further analysis and when he returned to Paris, he had with him the solution to the dilemma[308] that had worried six generations of chemists.

If we consider the discovery of a single element a great success, Georges Urbain was a giant—not only was he an experimental chemist, he also sought to formulate a unifying theory for the chemical sciences (Figure III.11). The chemical scene in those years was in great ferment; almost every year, discoveries of new elements were announced. Urbain, with his enormous store of knowledge on the fractionation of the lanthanides and with suitable spectroscopic expertise, was virtually the only chemist able to prove the veracity of a discovery. We can say that he "wrote the book" on the study of the rare earths, even discrediting some of the discoveries of the famous chemist Sir William Crookes. The young Urbain was 32 years old and not yet a professor when he demolished the 73-year-old Crookes's discoveries of *monium* or *victorium*, *incognitum*, and *ionium*. Hard on this, Urbain then proved the groundlessness of Crookes's meta-elements, of the elements $Z\alpha$, $Z\beta$, $Z\gamma$, $Z\delta$, $Z\varepsilon$, and $Z\zeta$ of the renowned Paul Emile Lecoq de Boisbaudran,[309] and finally of the elements Σ, Γ, Δ, Ω, and Θ proposed by Eugène Demarçay (1852–1903).[310] In 1907, he tackled the subject of *bauxium*, a hypothetical element present in bauxite, and he eliminated it from the periodic table. During World War I, he challenged the entire work of the Austrian chemist[311] Josef Maria Eder (1855–1944) and part of the work of Eder's colleague, Hofrat Eduard Valenta (1857–1937). As a result, the discoveries of five elements, *denebium*, *dubhium*, *neo-thulium*, *euro-samarium*, and *welsium*, were retracted.[312] Finally, in 1910, he pronounced the final word on the experiment of X_2, an element whose existence was put forward only a few years earlier.

Here, we concentrate our attention on the first part of Georges Urbain's career, from his debut as a research chemist in the field of the rare earths (1899) to the outbreak of World War I (1914).

III.15.1. BAUXIUM

The production of aluminum metal on a vast scale came about in 1886. Paul Louis Toussaint Héroult (1863–1914) and Charles Martin Hall (1863–1914) had developed independently an electrochemical method of obtaining the metal starting with the oxide. Although it was an immediate success, the Hall-Héroult industrial process had a defect: the cost of purifying alumina was surprisingly high and could compromise the economic initiative of the two chemists. The following year, Karl Josef Bayer (1847–1904) resolved this problem: he developed and later patented a process of purifying alumina starting from bauxite.[313]

Bayer was born in Bielitz (Bielsko in present-day Poland) on March 4, 1847. His education was somewhat disorganized. After having abandoned his studies in architecture, he studied chemistry at Wiesbaden and obtained his doctorate with Robert Bunsen at Heidelberg. After having taught for a brief time at the University of Brünn, he built his career in the industrial sector, first in Bohemia and later at Saint Petersburg. In 1889, he discovered the method, still famous, for extracting aluminum from bauxite. The "Bayer

Process" consisted of four steps: the first was digestion, which consisted of mixing bauxite powder with a solution that contained bicarbonate, followed by heating the mixture to a temperature of 250 °C at a pressure that reached 30 atmospheres. The second step, clarification, removed the insoluble impurities from the bauxite. Left to settle, the solution precipitated or crystallized to form aluminum hydroxide, $Al(OH)_3$. The last step was calcination, which consisted of heating the $Al(OH)_3$ to about 1,000 °C. This process produced a white powder of alumina with a purity higher than 99%. Bayer tenacity and intuition repaid his efforts; his process is still used today, more than a hundred years after its development.

In May 1894, a German newspaper carried a brief article by Bayer regarding the possible discovery of a new element. The new substance—without a name—had been found in the mother liquors coming from the reaction of bauxite with soda. Bayer described different qualitative reactions of the new metal and reported the principal properties of some of its compounds. Although he had succeeded in obtaining 2 g[314] of the oxide, for unspecified experimental reasons, he was not able to determine its atomic weight. Completely intimidated by this, although he had chemically treated 1,000 tons of bauxite, he postponed the solution of the problem to a later investigation.[315]

Virtually in the same period, a "M(onsieur) Bayer" presented an abstract on his work on French bauxite to the Société chimique de Paris. In December of the same year, the complete report of Bayer's work came to light in the French journal of the Chemical Society. Translations and abstracts in other languages soon appeared. If, on the one hand, the material examined was always the same, on the other hand, the author's name seemed to change with each version: R. S. Bayer in the German and English versions and Dr. Beyer in the American version. However, none of these journals made mention of the name of the presumed element. It is reasonable to think that both Dr. Beyer and R. S. Bayer were none other than Karl Josef Bayer, arrived in Paris in May 1894 to discuss his patent on the chemical treatment of bauxite.

In 1907, Georges Griner and Georges Urbain felt the need to shed some light on this mysterious element.[316] Although they correctly attributed the discovery of the element to Karl Josef Bayer and likewise correctly verified, via spectroscopy, the groundlessness of his discovery (the mysterious substance was a mixture of abundant amounts of vanadium and tungsten, with traces of molybdenum, copper, bismuth, lead, calcium, and sodium), they were not altogether correct in the way they did it. Bayer's nonexistent element had remained nameless for 13 years; now, at the moment of discrediting it, Urbain and Griner, for no apparent reason, referred to it by the name *bauxium*: "M. Bayer obtained crystals of ammonium molybdate and some green crystals which he mechanically separated from the former. These green crystals contain molybdenum and another substance that the author considers a new element: bauxium."

This curious example of nomenclature bestowed in the death throes of an element did not disturb Karl Josef Bayer: he died suddenly on October 4, 1904, in Rietzdorf.

III.15.2. FROM MONIUM TO VICTORIUM AND IN PURSUIT OF IONIUM AND INCOGNITUM

The British Society for the Advancement of Science, founded in 1831, brought together every year the most famous English scientists and men of culture. From the speaker's platform of these congresses, discoveries were announced, controversies arose, and

FIGURE III.11. Georges Urbain (1872–1938). Professor of chemistry at the Sorbonne, president of the French Chemical Society, member of the Academy of Sciences, talented musician, composer, sculptor, and painter, in 1907, Urbain separated *neo-ytterbium* and *lutecium* from ytterbium. In 1911, he announced the discovery of *celtium*, mistaken for element number 72, but in reality lutetium. Eleven years later, he revisited a similar but improved work; however, his previous error had irremediably undermined his credibility so much so that very few recognized the merit of his discovery.

prophecies were made. William Crookes, in his 1898 inaugural address as president of the society, described the great scientific events of the past year since the last meeting of the society. After a wandering discourse on the discovery of polonium, the theory of radioactivity, and the fixation of nitrogen, Crooke announced his own latest discovery, a new element he called *monium* (meaning "alone" because its spectral lines stood apart at the end of the ultraviolet spectrum).[317]

Although this discovery was shown later to be false because it was not a new rare earth but a mixture of gadolinium and terbium, applause filled the great hall at Bristol. Strengthened by this encouragement, Crookes then touched on a theme that held great interest for him—spiritualism—and he appealed to the scientists present to consider conducting experiments on the phenomena associated with this belief, citing the work of Pierre Janet (1859–1947) in France, Sigmund Freud (1856–1939) in Austria, and William James (1842–1910) in the United States.

Crookes was a controversial Victorian scientist, a man emblematic of his era. He embodied the best virtues of English society at the end of the 19th century, but he was also determined by the limits that society imposed.[318] At the moment of his announcement of the discovery of *monium*, the 19th century was about to end, William Crookes was nearly 70 and was about to receive a knighthood from Queen Victoria. He was a rich man, famous, and, above all, respected. We will examine certain points in his tumultuous, polycentric 1898 inaugural address in greater detail later.

William Crookes was born June 17, 1832. He did not have an orthodox education. His father Joseph (1791–1882), beginning in very modest circumstances as an apprentice tailor, became rich cutting trousers for the London bourgeoisie and planned a solid career in architecture for his first-born son. But William, not yet 16, instead enrolled at the Royal College of Chemistry in 1848, under the guidance of the renowned Augustus Wilhelm von Hofmann. At 19, he published his first work, in Germany, on some compounds of selenium. In 1855, he was a chemistry teacher at Chester Training College; 6 years later, he had a stroke of fortune: Robert Bunsen announced in the spring of 1860 the discovery of two new elements, rubidium and cesium, detected by the new spectroscopic techniques developed together with Gustav Robert Kirchhoff. Crookes, who had already worked between 1853 and 1857 on photographic problems, turned his attention to the spectra of certain selenium-bearing materials that Hofmann had given him at the time of their collaboration. In their spectra, Crookes observed an unpublished green line. Following this line of research, he discovered a new element, thallium. The notice of this discovery appeared on March 30, 1861, in *Chemical News*, a journal that Crookes founded in 1859 and of which he was director and proprietor. If it were due to chance that Crookes had in his possession these particular samples, it was certainly not by chance that he had the ability to determine the chemical and physical characteristics of the new element; the accuracy with which he established the atomic weight remained exemplary for decades. On the wave of his discovery of thallium, Crookes became, at the age of 31, a member of the Royal Society, obtaining the highest honor that could be given to an English scientist. During his very long life, he became interested in various branches of chemistry and physics: his innumerable research projects and scientific forays were spread out over 68 years, from 1851 to 1919.

Crookes conducted useful applied research (e.g., on the health of workers), but he also published some "offbeat" pieces as well. In the July 1870 *Quarterly Journal of Science*, the honorable member of the most prestigious scientific society in the world published an article with the ambiguous title "Experimental Research on a New Force." This new force was essentially that of the spiritualist medium Daniel Douglas Home (1833–86), a "spiritist" well-known to the public of the time. In the eyes of many of his colleagues, these investigations were a betrayal of true scientific merit. Crookes, however, continued his investigations unperturbed and never retreated from his heterodox positions.

In 1898, Crookes had arrived at the conclusion that he had discovered a new element in the course of his research on the phosphorescence spectra of the rare earth elements. The fractions of rare earths, sealed in discharge tubes under vacuum, were induced to emit an ultraviolet phosphorescence. This discovery came to light after some bizarre work that he had done about 10 years earlier. On that occasion, he thought he had separated the oxide of yttrium into nine new earths that he called generically *meta-elements*; these, he demonstrated, derived one from the other as from a mother earth.

The results of his 1898 discovery were exhibited at a soirée of the Royal Society[319] on May 3, 1899. Unlike in his September 1898 address to the British Society for the Advancement of Science, where Crookes had proposed the name *monium* for the new element, in this second announcement he chose to call the new rare earth *victorium*, probably in recognition of Queen Victoria (1819–1901) who had, a short time earlier, created him a Knight of the British Empire and who was celebrating her 80th birthday that year.

Crookes's starting material was the oxide of impure yttrium taken from minerals like samarskite, gadolinite, and cerite. After some years of work, Crookes obtained a fraction that he said was the oxide of *victorium*; from the formula Vc_2O_3, which he thought to be reliable, he was able to determine the atomic weight as 117.

Crookes's prudence did not increase along with his advancing years. Indeed, on December 15, 1905, Sir William asserted that, over the course of the years, he had discovered different groups of lines isolated in the phosphorescence spectra of fractions of yttrium- and samarium-bearing earths; these, he said, were unequivocal signs of the presence of new elements: not one element, but two, perhaps three new simple substances.[320] Between the green lines of samarium, he spied a new group that he attributed to a new element that he called *Gβ*.[321] Beyond the brilliant blue lines of ytterbium, Crookes observed the presence of an entire new group of lines, ascribed to the presence of a new element, *ionium*. At the extreme opposite end of the spectrum, a "rather cloudy" line that he said could not belong to any element already known confirmed for him the existence of yet another a new element, in whose name was inherent all of the uncertainty of Sir William Crookes held at the moment of announcing the discovery: *incognitum*.[322] The spectrum ended with the unequivocal lines of *victorium*, which were all exceptionally strong and perfectly separated from one another.

Crookes's joy over his new discoveries was shadowed; at the conclusion of his article, among the notes, he wrote: "Since writing the above, I see that M. Urbain. . ., in a paper to the Académie des Sciences, contends that the substance that I have called victorium may be a compound containing gadolinium."[323]

In fact, Georges Urbain's response was swift. Four articles published in the prestigious pages of the *Comptes Rendus de l'Académie des Sciences de Paris* refuted the existence of *victorium, ionium, incognitum*, and, finally, of the bizarre hypothesis of the *meta-elements*.

In 1905, Georges Urbain first announced[324] that he saw no difference between the cathodic ultraviolet phosphorescence spectrum of gadolinium and that of the presumed *victorium*. Urbain did not categorically condemn Crookes's work, but said that he hoped to do additional investigations, which he quickly did. In 1881, Crookes had discovered that a great number of substances, particularly the rare earths, emitted a bright light (phosphorescence) if they were exposed to cathode rays in a discharge tube.[325] What Crookes had not understood, and moreover had badly interpreted, was the continual appearance of new spectral lines. Basing his ideas on their existence, he hypothesized the existence of *meta-elements* or of downright new elements. Urbain was a more attentive observer. He realized that the lines appeared and then disappeared from the addition of impurities[326] in the rare earth elements, but also from the effect of the chemical nature of the ligands (sulfates, ethylsulfates, oxides, etc.).[327] It was precisely this last piece of data that demolished Crookes's bold hypothesis. Urbain established that the lines of *Gβ* were nothing other than those of the well-known terbium.[328]

Urbain did not limit himself to criticizing the work of colleagues older than himself—to disprove part of Crookes's work, an attentive study of the phenomenon allowed him to formulate the law of "the optimum of cathodic phosphorescence in binary systems":

> I have determined in a great many cases that phosphorescence is observed in mixtures in which certain trace substances act as exciters while the large mass of matter acts as a diluent. In general, the exciter element (like Mn, Sm, Eu, Tb, etc.) is not or

is minimally phosphorescent in the pure state; the same may be said for the diluent elements (like Ca, Al, Gl, Y, etc.). Phosphorescence therefore passes necessarily through a relative maximum that always corresponds to small amounts of the exciter element.[329]

Urbain's demonstration possessed all of the elegance of a mathematical proof; he started by assuming as absurd that his samples of gadolinium were polluted by traces of the hypothetical *victorium* and thus raised an objection to the existence of Crookes's element. At the conclusion of his work, Urbain was able to assert that Crookes's samples were composed of Gd_2O_3 and CaO in amounts of 2.8% and 92.2%, respectively. Finally, he observed that the lines of the hypothetical *victorium* recorded by Crookes were still visible after adding 200 parts of Gd per million of Ca.

Another 2 years passed before the existence of *incognitum* and *ionium* would be openly placed in discussion. Two days before Christmas 1907, Georges Urbain published a memoir with the title "Sur la nature de quelques éléments et meta-éléments phosphorescent de Sir W. Crookes."[330] In 1906, he partially refuted Crookes's results by accurately recording the spectra of terbium and gadolinium.[331] The following year, Urbain, using the law of optimum of the cathodic phosphorescence of binary systems, felt ready to assert that Crookes's *ionium* and *incognitum* were nothing more than mixtures of terbium and gadolinium in the following proportions:

- *Ionium* = Tb: 0.5–1.0%; Gd: 99.5–99.0%
- *Incognitum* = Tb: 2.00%; Gd: 98.00%

Fortunately for Crookes, the *annus horribilis* for his elements was lightened by the celebration of his silver wedding anniversary. Ten years later, in 1916, Crookes's wife, to whom he was profoundly attached, passed away. From then on, his health declined rapidly but he remained active until the end.[332] Sir William died April 4, 1919, 6 months after the conclusion of World War I.

III.15.3. THE ELEMENT E OR X_2

In 1910, the 38-year-old Georges Urbain had been professor of mineralogy at the Sorbonne for 4 years; at virtually the same time, he had been elected a member of the International Commission on Atomic Weights.[333] Three years earlier, he had discovered *neo-ytterbium* and lutetium.[334] His was a career full of hard work and success. Always up to date on the latest discoveries, Urbain was usually a participant in the international conventions and thus had a way of publicizing his most recent successes. His only limitation, it seems, was language. He spoke only French and for this reason gladly participated in international conventions where this language was the one most used by his audiences: aside from conventions at home, he frequently visited Belgium, Spain, Portugal, Czechoslovakia, and even Romania. All of these countries (with the exception of Czechoslovakia) spoke romance languages and, in these countries between the two wars, France's cultural-economic influence was very strong by virtue of strong anti-German military ties.[335] In 1910, Urbain sent to press his account of the communication held on May 28, 1909 during the 10th Congress of Chemistry in London.[336] Also appearing among his

FIGURE III.12. Franz Serafin Exner (1849–1926). Famed Austrian spectroscopist and physicist, at the beginning of the 20th century, with his colleague Eduard Haschek (1875–1947), Exner found five unknown lines in the spectra of *cassiopeium* and *aldebaranium*. He believed that these were due to traces of a new element and didn't waste any time in calling it "Element X." Courtesy, Archiv der Universität Wien.

most recent works was a detailed list of the presumed elements that he had occasion to study and then disprove. One of these was the element provisionally called X_2.[337]

Between 1895 and the beginning of World War I, two Austrian physicists, Franz Exner (1849–1926) and his student Eduard Haschek (1875–1947) measured more than 100,000 lines of all of the known elements.[338] They then constructed a three-volume spectral atlas.[339] Franz Exner (Figure III.12) was born in Vienna on March 24, 1849. He studied at the University of Vienna and at Strasbourg where, in 1873, he obtained his doctorate. Exner was interested in electrochemistry, atmospheric electricity,[340] and spectrographic analysis. In 1907, he was elected rector of the University of Vienna.

In 1910, Exner and Haschek were immersed in a period of intense spectrographic study when they recorded the existence of some unknown spectral lines; they attributed these lines to an unknown element.[341] The notice was not overlooked by Georges Urbain who rapidly refuted the discovery: "[The element]. . . X_2 of MM Exner and Haschek are identical to the dysprosium of Lecoq de Boisbaudran."

The name X_2 can be explained for two reasons: the first and most obvious is tied to the fact that with the letter X one commonly indicates an unknown object or concept, in this case an element; the second hypothesis on which they built their presumed discovery was tied to some of Urbain's earlier work[342] in which he refuted the presumed discovery of element X[343] by the Swiss chemist Jacques Louis Soret. In this case, the subscript 2 was specified to distinguish the provisional name given by Urbain (X_2)[344] from that proposed many years earlier by Soret (X).

Exner was deeply involved in spectroscopic studies even into his old age: his spectral atlas, *Die Spektren der Elemente bei normalem Druck*,[345] was reprinted until 1924. Advancing years did not diminish Exner's passion for symposia and seminars: he gave many public discourses in which he explained his vision of the world. Beginning with the empiricism of Ernst Mach, to the atomism of Ludwig Boltzmann and the interpretation of probability, Exner deduced, long before the advent of quantum mechanics, that the basic laws of nature were in and of themselves indeterminate, whereas deterministic and rigorous laws could only be applied in a very limited way on the macroscopic scale.

His vision of "ethical evolution" was more complex: according to him, both science and humanistic studies could be united by means of the law of large numbers. The entire world was governed by the global tendency toward the most probable states. This unifying view placed Exner among the "reductionists." Foremost in Exner's mind was that even culture could be a natural outcome of human growth and decline, which, via history, showed continual ethical progress that would arrive at its fullness in a scientific vision of the world.[346]

One can easily imagine that, to Franz Exner, taken up as he was by these highly idealistic matters, Georges Urbain's unmasking of the erroneous discovery of the phantom element X_2 would matter little. And, in fact, Exner never responded to Urbain's challenges. Franz Exner died October 15, 1926, in the city of his birth, at the age of 77. This episode is expanded upon in Part IV.6.

III.15.4. THE META-ELEMENTS

Many years after he undertook the study of the rare earths, Georges Urbain wrote of the situation besetting mineral chemistry at the beginning of his career:

> In 1898 the subject of the rare earths was in great confusion. There was an abundance of documents of very uneven value. In these truth and error were closely associated, and there was no way to distinguish with certainty one from the other. It was even quite difficult to separate facts from the hypotheses and interpretations which, more often than not, needlessly encumbered them.[347]

Spectroscopic study to understand atomic complexity began in the second half of the 19th century, but only with the turn of the new century was significant progress in this technology able to expand scientific knowledge instead of becoming a source of error.

In 1862, Crookes discovered thallium via spectroscopy, and for the remainder of his life, he remained devoted to spectral investigations and greatly contributed to the development of this discipline, although his own discoveries were at times incorrect. At the beginning of the 1880s, his research focused on the analysis of spectral phosphorescence. Crookes aimed at discovering a method for researching trace elements in order to apply this technique to the discovery of new chemical elements.[348] The discontinuous spectra of the rare earths seemed to be very complicated, and their details seemed to vary in a way that puzzled him. Sir William, in addition to being a practical-minded scientist, also was endowed with a whimsical imagination and wrote inventive prose: "It was impossible for me to get rid of the conviction that I was observing a group of autographs in the molecular world. . . I needed a Rosetta Stone."[349]

Working on the fractionation of yttrium oxide with dilute ammonia, Crookes succeeded in separating some fractions of the rare earths with different basicities that he then examined.[350] After 2 years of work, Crookes obtained a series of earths that he placed in a vacuum tube in which he struck an electric spark and obtained the emission of phosphorescent light. Crookes observed the presence of many different spectra, both as lines and bands, and with varying relative intensities. Sir William interpreted this result as evidence of a separation of yttrium into its major components. The final result at which he arrived was the discovery of five constituents into which yttrium could be separated and that he indicated with letters: G_α, G_β, G_γ, G_δ, and G_ζ. (See Table IV.2 in Section IV.9 for Crookes's spectroscopic data.) He later hypothesized that yttrium could be split into as many as eight components. To explain the experimental evidence of his spectra, Crookes proposed going beyond the traditional idea of a chemical element and introduced the concept of the *meta-element*. Every chemical element—derived from the progressive cooling of a primordial material—would be nothing other than the sum of different *meta-elements* characterized by small variations in atomic weight. Some chemists, especially those in England, were in favor of this hypothesis because it allowed one to salvage (William) Prout's hypothesis (1785–1850), according to which the atomic weights were whole numbers. The atomic weight of an element, in this new framework of ideas, was none other than the weighted average of the individual *meta-elements*. Crookes noted that there was no way of separating the *meta-elements* from one another, and he did not succeed from a chemical point of view in splitting up what he called the yttrium molecular group. The idea of *meta-elements* was wrong, and the evidence from chemical fractionation was shown to be due to the presence of impurities in the phosphorogenic samples used by Crookes.

The *meta-elements* concept did not disclose any new facts,[351] and it was eliminated from the history of rational science by Georges Urbain who demonstrated how alterations in the phosphorescence spectra happened by adding other earths in trace amounts.[352] To prove this, Urbain studied the spectrum of a rare earth element in the pure state, then added to the sample traces of other elements belonging to the same family, a little at a time. The spectrum at first showed no appreciable changes; adding an amount of the phosphorogenic element beyond a certain threshold, however, changed the spectrum radically. With this method, Urbain discredited the idea of the *meta-elements*, but it also allowed him to formulate his law of optimum of phosphorescence of binary systems.[353] Urbain gave a detailed description of the impurities present in the samples of yttrium examined by Crookes, as well as their quantities.

Crookes took the blow with apparent indifference; however, some years later, in 1915, he claimed to have anticipated the concept of the isotope. Frederick Soddy, author of the revolutionary concept of isotopes, which refuted Prout's hypothesis, cited Crookes's work as historical support for his own intuition.[354,355]

Recently, Christian K. Jørgensen (1931–2001), looking for clues showing how our modern knowledge came about, published an article with the engaging title[356] "Lanthanides Since 1839: From Crowded Elements to a Quantum-Chemical Rosetta Stone." In it, Jørgensen re-proposed some of Crookes's ideas about *meta-elements* in an attempt to rehabilitate the Victorian chemist's idea: in Jørgensen's reconstruction, the *meta-elements* seemed adaptable to the concept of isotopes and their formation under extreme astrophysical conditions. This hypothesis is debatable, but perhaps acceptable—if the *meta-elements* had been a mere hypothesis. But Crookes's announcement of their existence was based on an egregious error, albeit one committed in good faith.

III.15.5. THE ELEMENTS OF PAUL EMILE (FRANÇOIS) LECOQ DE BOISBAUDRAN AND OF EUGÈNE-ANATOLE DEMARÇAY

Lecoq de Boisbaudran held that the phosphorescent bands observed by Crookes in the visible region of the spectrum of some samples of yttrium oxide belonged to two rare earths that he had already observed 1885 and that he had provisionally named[357] Z_α and Z_β.

But were Z_α and Z_β distinct elements, or did they have a more complex nature? Lecoq de Boisbaudran himself reported how Z_β became concentrated in the darker fractions of terbium while Z_α went into the clearer fractions. He believed he was in a position to discover, by studying the absorption bands of their spectra, new elementary substances: Z_γ, Z_δ, and a third element that he called dysprosium.[358] Some years later, between 1892 and 1893, Lecoq de Boisbaudran believed that he had seen two new elements while fractionating samarium oxide with ammonia.[359] In the "reverse spectrum,"[360] he observed a new line that he attributed to an element that he provisionally called Z_ε and a band that he thought belonged to another unknown element: Z_ζ. In 1893, Lecoq de Boisbaudran noted a marked resemblance between his two elements, Z_ε and Z_ζ, and Crookes's element S_δ, but he did not express an opinion on the identities of the three elements. Only in 1896 was Eugène Anatole Demarçay able to establish the identity of the three substances as europium.

At the turn of the 19th century, French chemist and spectroscopist Eugène-Anatole Demarçay (1852–1903) was seen as Lecoq de Boisbaudran's natural heir. Because of his ability to read a complex spectrum like an open book, he was frequently asked to verify new elements. Even the Curies went to him in 1898 to confirm via spectroscopy the nature of the element radium.[361]

Two years later, in examining the oxides of gadolinium, terbium, erbium, and yttrium, Demarçay, via spectroscopy, observed the presence of four unknown elements.[362] These substances were provisionally designated with the Greek letters: Γ, Δ, Ω, and Θ. The spark spectrum from the brown oxide obtained from the more soluble fraction of the nitrates of gadolinium and magnesium "*déjà assez pur*" (deemed already sufficiently pure), showed the presence of some lines that could be attributed to pure terbium, but, uncertain of the purity of the material he used, Demarçay preferred to attribute them provisionally to an element he designated Γ. In other oxides with lighter shades, other unknown lines were observed that had been assigned by Lecoq de Boisbaudran to the element Z_γ. In this case, Demarçay named the provisional new element Δ. The same fate befell certain spectral lines found in intermediately soluble fractions of yttrium, purified from holmium and terbium, that Demarçay attributed to the hypothetical element Ω. Finally, in the spectrum recorded for some fractions of the basic earths intermediate between erbium and terbium, unknown lines were observed that he attributed to the element Θ. The work of Demarçay in this complicated field of the rare earths had, in fact, an antecedent. By 1892, this element, discovered by Lecoq de Boisbaudran in 1886, was thought by its own discoverer to be a mixture of two elements. The hypothesis was correct, but Lecoq de Boisbaudran did not succeed in presenting experimental evidence in its favor. In 1896, Demarçay began to fractionate a sample of samarium. He obtained from the nitrate of samarium a new earth[363] that he, again in a provisional way, decided to call Σ.[364] Demarçay worked for another 5 years with the technique of fractional crystallization with magnesium nitrate before announcing that his Σ was in reality a new element.

In 1900, Demarçay still had not succeeded in completely isolating his new element, but he was nonetheless certain of its presence, saying that[365] "I previously announced that the impure oxide of samarium contained a newly characterized element with different lines and an atomic weight greater than that of samarium and less than that of gadolinium; I have designated it with Σ waiting until I have actually isolated it."

Eugène-Anatole Demarçay was born in Paris on January 1, 1852. After attending the Lycée Condorcet, he visited England and then concluded his studies at the École Polytechnique. His early interest in terpene and ether research made valuable contributions to the perfume industry. Later, he moved into the field of inorganic chemistry; he accidentally lost his right eye in an explosion while studying nitrogen sulfides. Nevertheless, he continued to carry out studies in vacuum on volatility and on the temperatures of spark spectra. In carrying out these analyses, he noticed that the sparks generated by platinum electrodes produced luminous lines that were very useful in studying the rare earths.

In 1901, his elaborate crystallizations with magnesium and samarium nitrates led him to observe new spectral lines that he attributed to the presence of a new "earth" or oxide that he called *europia* (by which the element contained in it, europium, was called).[366] Eugène-Anatole Demarçay dedicated a good part of his brief life to study and research, exposing himself without precautions to radiation, harmful substances, and toxic vapors. With serene resignation, he saw his health deteriorate rapidly: "Demarçay was seen to be slowly dying, a lover of life who yet was abandoning it, but conscious of accomplishing his duty, and happy with the years he had lived."[367] He spent his last months in the consoling and positive faith that the progress of the human race would be made by science and its laboratories. Demarçay (Figure III.13) died at the age of 51 in 1903, and the figure of this great scientist unjustifiably fell quickly into oblivion.[368]

Some years later, when Georges Urbain succeeded in obtaining many elements of the rare earths in a high state of purity, he could assert that two of presumed elements, *Γ* and *Δ*, were in reality terbium[369] and dysprosium,[370] respectively, highly contaminated with traces of other rare earths. Until today, the remaining of Demarçay's elements, *Ω* and *Θ*, have not been confirmed. It is almost certain that in their case unrecognized impurities contaminated Demarçay's samples of yttrium and ytterbium. Because the ytterbium available at the end of the 19th century would have been shown to be a mixture of two elements, it is possible to conjecture that Demarçay could have observed lutetium some years prior to Urbain's discovery. However, such is not the case: the ultraviolet spectrum of Demarçay's element Θ does not correspond to that observed by Georges Urbain.[371]

Paul-Émile (dit François) Lecoq de Boisbaudran was born in Cognac in the Hôtel Templéreau de Beauché, his ancestral family home, on April 18, 1838. He was the eldest child and only son of Paul-Aimé Boisbaudran Lecoq (1799–1870) and Anne-Louise-Alexandrine Joubert (1814–91).[372]

Lecoq de Boisbaudran was attracted by chance to the study of the lanthanides, a group of elements whose very existence was doubted by some and for which information at that time was extremely scanty. At the very beginning of his scientific career, while walking

FIGURE III.13. Eugène Anatole Demarçay (1852–1903), Chemist and Spectroscopist. In 1896, Demarçay discovered europium and, a few years later, he recorded the spectra of radium and polonium. He observed the presence of four unknown elements that were provisionally designated with Greek letters Γ, Δ, Ω, and Θ He died at only 52 years of age in 1903.

along the corridors near the University laboratories, he saw a series of jars filled with rocks containing rare earth elements. Like St. Paul on the road to Damascus, he was instantaneously "converted" to inorganic chemistry and wanted immediately to commence research. However, being still young and lacking a diploma, his professors refused to allow him to handle such high-priced materials, fearing that he might lose these precious samples in botched experiments. He set aside his desire for a time. It was only after his discovery of gallium (1875), when his financial situation was a little better, that he could acquire the valuable minerals necessary to implement his collection and carry on research on rare earth elements. When he moved to Paris, he lived in a two-room apartment: one room was used as his bedroom, the other as a laboratory where he installed his glassware and his famous spectroscope. There he spent many hours precipitating, washing, extracting, and dissolving his many mineral samples. Because of their high prices, he was still constrained to work on small samples, but this dearth of rare earth elements did not prevent him from discovering three elements: gallium, samarium, and dysprosium.

Lecoq de Boisbaudran (Figure III.14) contributed a prescient idea to the development of the periodic classification of elements by proposing, soon after its discovery, that argon was a member of a new, previously unsuspected chemical series of elements, later to become known as the noble gases.

FIGURE III.14. Paul Emile (François) Lecoq de Boisbaudran (1838–1912). A descendant of the Lecoq family, lords of Boisbaudran, in his youth Lecoq de Boisbaudran was engaged in the family wine business but soon started to study chemistry in earnest, although without a formal education. A stranger to academic circles, he worked privately in his home laboratory, where he discovered gallium, samarium, and dysprosium. He suggested to his friend Ramsay that a new group should be added to the periodic table in order to include the noble gases.

The last descendant of the lords of Boisbaudrant, on December 27, 1897, Paul-Emile Lecoq de Boisbaudran married a young widow, Jeannette Nadault–Valette (1852-1926). The marriage was immensely happy, but of short duration. Beginning in 1900, his health began to decline rapidly: he was struck by a long and painful arthritis that made it very difficult for him to remain physically active, rendering him vulnerable to other diseases. Lecoq died childless at his home on May 28, 1912, at the age of 74.

With the exception of his friend Sir William Ramsay, Lecoq de Boisbaudran was not much appreciated as a scientist beyond French borders. Georges Urbain paid tribute to his colleague thus:

> One day, face to face with him, I lamented the fact that his work was so little known and I reproached him for not having done enough to make it better known; he replied with a peaceful smile that science certainly was not lacking in impartial historians. I cannot recall this without being moved. . . I then realized that I had before me not just a great scientist, but also a great person.[373]

III.15.6. THE TERBIUM-I, TERBIUM-II, AND TERBIUM-III OF WELSBACH

The friction between Georges Urbain and Auer von Welsbach following the discovery of the elements called *neo-ytterbium* and lutetium by Urbain and *cassiopeium* and *aldebaranium* by von Welsbach never resolved. The two chemists remained bitter rivals for

FIGURE III.15. Carl Auer von Welsbach (1858–1929), Austrian Chemist, Inventor, and Entrepreneur. In 1885, he split didymium into *neodidymium* (neodymium) and *praeseodidymium* (praeseodymium). In 1907, independently of Georges Urbain at Paris, he split ytterbium into two elements that he called *aldebaranium* and *cassiopeium*. Many of his experiments were carried out in his private laboratory located at his castle in the Carinthian Alps. Courtesy, Auer von Welsbach Museum, Althofen, Carinthia, Austria.

their entire lives. To make matters worse, 5 years after the verdict of the International Commission fell in favor of Urbain, a new conflict arose between them.

Within the walls of his castle in the Carinthian Alps, the 56-year-old baron Auer von Welsbach kept at the tedious work of fractional crystallization with inflexible determination. He was working on an impure sample of terbium using the method of the double oxalate of ammonium (a method introduced some time before by his rival). He first separated gadolinium, which produced the insoluble double salts of ammonia; among the intermediate products he found the oxide of terbium, whereas the oxide of dysprosium was isolated last because of its marked difficulty of crystallization. The three fractions had different colors, and the first optical analyses confirmed the absolute purity of the three samples. With understandable amazement, Auer von Welsbach ascertained that the presumed earths of gadolinium and dysprosium isolated by him in the samples of terbium were not actually gadolinium and dysprosium but new elements. He arrived at the conclusion that the terbium discovered years earlier was in reality a mixture of three elements. He analyzed via spectroscopy samples of Gd, Tb, and Dy and found that each one of these substances had spectral lines in common. These lines became strongly intensified in the fractions where he had concentrated the new elements, leaving no room for uncertainty or hesitation.

With incredible speed, Auer von Welsbach (Figure III.15) collected his data and sent them to the Viennese Academy of Sciences so that they could be published as soon as possible. Chemists long suspected that the terbium discovered by Mosander back in 1843 was highly impure, but everybody was convinced that the impurities that contaminated this element were due to traces of elements already known. Auer von Welsbach, by contrast, had for the first time discovered that the contaminants were in reality two distinct, absolutely new elements. After his disappointment concerning the discovery of *neo-ytterbium* and lutetium, Auer decided to take his deserved revenge on the young and arrogant Urbain. Not wishing to be upstaged again, he announced his discovery, despite the fact that the data at his disposition were still quite scarce, certain he had split an element thought to be "pure" into three new simple bodies that he designated[374] *terbium-I, terbium-II,* and *terbium-III.*

Gross incongruities soon came to light, and, unfortunately for Auer, they did not escape the notice of many chemists, among whom was his bitter rival, Georges Urbain: the spectroscopic properties of *terbium-I* and *terbium-III* were very similar to those of gadolinium, whereas *terbium-II* resembled dysprosium. Auer ignored his critics, announcing that he had prepared the new metals in large quantities and at high levels of purity without much difficulty.

Urbain didn't let a second chance slip by to deal a death blow to the shaky discovery of his Austrian colleague.[375] The presumed discovery of *terbium-I, terbium-II* and *terbium-III* was quickly interpreted correctly as the contamination of gadolinium, dysprosium, and terbium with very small traces of rare earths. In explaining Auer's error, Urbain was pitiless, referring once again to his famous law of optimum of cathodic phosphorescence in binary systems.[376] The second conflict between the two scientists also seemed to go in favor of the Frenchman, but a few years later the unpleasant verdict regarding *celtium* in some way avenged the Austrian for the frustrations he suffered.

On September 1, 1928, Auer traveled to Berlin where the Deutsche Chemische Gesellschaft was toasting his 70th birthday. An untiring worker, he continued working in his laboratory until the first days of August 1929, when piercing abdominal pains signaled an imminent end. Then he passed among the great crystallization dishes placed on his laboratory benches, lovingly caressed the spectroscopes, and covered them with white sheets. He went up to his room and awaited the end, which came on August 4, 1929.

Despite his vigilance, it seems strange that the scrupulous, almost punctilious Urbain failed to correct two more false discoveries—*neo-holmium* and *neo-erbium*—when he had refuted similar discoveries of supposed new rare earth elements. Eder and Valenta, in their long careers as spectroscopists, recorded many arc and spark spectra of the rare earth elements. They prepared holmium with a high level of purity that they called *neo-holmium*[377] in order not to confuse it with commercial holmium, at that time very impure. At the end of the 19th century, Gerhard Krüss (1859–95) did the same thing in purifying impure erbium, suggesting the name *neo-erbium* for the purified element.[378] Perhaps out of devotion for his deceased teacher, Karl Andreas Hofmann, expert in the study of the rare earths, confirmed Krüss's hypothesis as late as 1908.[379]

Georges Urbain noted many years later that only X-ray spectroscopy brought order to the rare earths; visible, arc, and scintillation spectroscopies did not have the same usefulness as Moseley's method and often led scientists into error. He also emphasized the fact that this technique was not at all helpful in the isolation of elements and that fractional crystallization remained the only method of merit in the search for new elements.

Notes

306. Urbain, G.; Weiss, P.; Trombé, F. *Compt. Rend. Chim.* **1935**, *200*, 2132.
307. Heilbron, J. L. *H. G. J. Moseley: The Life and Letters of an English Physicist*; University of California Press: Berkeley, 1974.
308. This dilemma consisted in establishing the "exact" number of elements that made up the so-called rare earth group.
309. It was in fact Georges Urbain who, in 1912, published two obituaries on his friend and colleague Paul Emile Lecoq de Boisbaudran: see Urbain, G. *Mon. sci.* [5] **1912**, *2*, 591; Urbain, G. *Chem. -Ztg.* **1912**, *36*, 929.
310. Marshall, J. L.; Marshall, V. R. *The Hexagon* **2003**, Summer, 19. This reference documents the 1903 death of Eugène Demarçay, verified by the authors who actually visited the family and saw the original death certificate. Every other reference we have found gives a date of 1904, apparently from an incorrect redaction of the professional death certificate.
311. Friend and colleague of Eder, Baron Carl Auer von Welsbach discovered 4 elements, but in his old age, he announced the false discovery of a fifth between erbium and thulium.
312. Fontani, M.; Costa, M. *La Chimica e l'Industria* **2003**, *85*(2), 59.
313. The mineral, rich in aluminum, known by the name of bauxite, was discovered in 1821 by Pierre Berthier (1782–1861), professor at the Ecole des Mines de Paris, while he was prospecting iron-bearing deposits in the south of France. He called this oxide "Terre d'alumine des Beaux" in honor of the village of Les Beaux near Marseilles, where he made the discovery. The red color of the deposit struck him as a possible source of iron. Instead, he found not iron, but an abundant amount of Al_2O_3. The name was subsequently changed to "beauxite" and later to "bauxite." The initial analyses reported incorrectly that bauxite contained alumina dihydrate, $Al_2O_3(2H_2O)$; later, by means of a thermogravimetric technique introduced by G. Urbain, it was shown to be composed of a mixture of the hydroxides $Al(OH)_3$ and AlO(OH).
314. Bayer, M. *Bull. Soc. Chim. Fr.* **1894**, *11* [3], 534.
315. Bayer, M. *Bull. Soc. Chim. Fr.* **1894**, *11* [3], 1155.
316. Griner, G.; Urbain, G. *Bull. Soc. Chim. Fr.* **1907**, *1* [1], 1158.
317. Crookes, W. *Proc. R. Soc. London: Report of the Meeting of the British Association for the Advancement of Science* **1898**, 3–38.
318. For more information on Crookes as a "character," see Brock, W. H. Chemical characters: Sir William Crookes (1832–1919). In Patterson, G. D.; Rasmussen, S. C., Eds., *Characters in Chemistry: A Celebration of the Humanity of Chemistry*; American Chemical Society: Washington, DC, 2013, pp. 73–100.
319. Crookes, W. *Proc. R. Soc. London* **1900**, *65*, 237.
320. I am in possession of good evidence pointing to the existence of two, if not three, new bodies.
321. According to Crookes's idea, *Gβ (terbium)*, would be the second meta-element of terbium (after Gα) and not a true and proper element. The concept of *meta-elements* was introduced by Crookes within a general vision of the inorganic evolution of the elements. Both these ideas were shown to be a scientific blind alley, not at first, but over the years they contributed to making research in this area ever more complex and intricate.
322. Crookes W. *Proc. Roy. Soc., London* **1886**, *40*, 7.
323. Crookes, W. *Chem. News* **1905**, 273.
324. Urbain, G. *Compt. Rend. Chim.* **1905**, *140*, 1233.
325. Crookes, W. *Compt. Rend. Chim.***1881**, *92*, 1281.
326. Urbain used the word "diluent."
327. Urbain, G. *Compt. Rend. Chim.***1905**, *141*, 954.

328. Urbain, G. *Compt. Rend. Chim.***1905**, *141*, 521.
329. Urbain, G. *Compt. Rend. Chim.* **1905**, *141*, 955.
330. Urbain, G. *Compt. Rend. Chim.***1907**, *145*, 1335. On the nature of some phosphorescent elements and meta-elements of Sir W. Crookes.
331. Urbain, G. *J. Chim. Phys.*, **1906**, *4*, 321.
332. Witness to the fact that Crookes worked assiduously up to the time of his death, his last article appeared 2 years after his passing: Crookes, W. *Chem. News and Journal of Industrial Science*, **1921**, *123*, 81.
333. Landolt, H.; Ostwald, W.; Wallach, O. *Siebenter Berichte der Commission für die Festsetzung der Atomgewichte. Ber. Dtsch. Chem. Ges.* **1906**, *39*, 2176.
334. Urbain, G. *Compt. Rend. Chim.***1907**, *145*, 759.
335. Urbain represented the last gasp of the culture of the "continental school" that had spread throughout Europe.
336. Urbain, G. *Rev. R. Acad. Cienc. Exactas, Fis. Nat. Madrid* **1910**, *7*, 974.
337. Mellor, J. W. *A Comprehensive Treatise on Inorganic and Theoretical Chemistry.* Longmans, Green: New York, 1956; vol. V, p. 496, refers to this element, indicating it with the letter "E."
338. Exner, F.; Haschek, E. *Sitzungsber. Akad. Wiss. Wien* **1896**, *105*(IIA), 989; Exner, F.; Haschek, E. *J. Phys. Chem.* **1898**, *2*(3), 207; Exner, F.; Haschek, E. *Sitzungsber. Akad. Wiss. Wien* **1895**, 104 (IIA) 909; Exner, F.; Haschek, E. *J. Phys. Chem.* **1897**, *1*(9), 624–625; Exner, F.; Haschek, E. *Sitzungsber. Akad. Wiss. Wien* **1897**, 105 (IIA) 389, 503, 707; Exner, F.; Haschek, E. *J. Phys. Chem.* **1897**, *1*(9), 624.
339. Exner, F.; Haschek, E. *Die Spektren der Elemente bei normalem Druck*; F. Deuticke: Wien, 1911.
340. This research area aroused the interest and was taken up and extended by his student, Viktor F. Hess (1883–1964), whose discovery of cosmic rays won him the Nobel Prize in Physics for 1936.
341. Exner, F.; Haschek, E. *Sitzungsber. Akad. Wiss. Wien* **1910**, *119*, 771.
342. Urbain, G. *Bull. Soc. Chim. Fr.* **1898**, *19*, 376.
343. Soret, J. -L. *Compt. Rend. Chim.* **1878**, *86*, 1062.
344. We could not find any reference to "X_2" (the name found in the bibliography cited by Urbain) but only the name "E" (found in Mellor, cited earlier). It is possible that Urbain had not read the original article and had himself, as in the case of *bauxium*, coined a name for the element that, not long afterward, he would have refuted.
345. The spectra of the elements at atmospheric pressure.
346. Stoltzner, M. *Phys. Perspect.* **2002** *4*, 267.
347. Urbain, G. *Chem. Rev.* **1925**, *1*, 143.
348. Crookes, W. *Proc. R. Soc. London* **1883**, *35*, 262.
349. Crookes, W. *Report of British Association* **1886**, 558.
350. Crookes, W. *Report of British Association* **1886**, 583.
351. In retrospect, "Crookes' viewpoints threw considerable light upon the outlandish nature of the transition metals, and the concept of meta-elements greatly facilitated the development of alternative accommodation methodologies for the rare-earth elements." Thyssen, P: Binnemans, K. Accommodation of the rare earths in the periodic table: A historical analysis. In *Handbook on the Physics and Chemistry of Rare Earths*, vol. 41. Gschneidner, K. A., Jr.; Bünzli, J.-C. G.; Pecharsky, V. K., Eds.; North Holland (Elsevier): Amsterdam, The Netherlands, 2011; 1–94; p. 44.
352. Urbain, G. *Compt. Rend. Chim.* **1907**, *145*, 1335.
353. Urbain, G. *Compt. Rend. Chim.* **1908**, *147*, 1286.
354. Dekosky, R. K. *Br. J. Hist. Sci.* **1973**, *6*(4), 400.
355. Dekosky, R. K. *Isis* **1976**, *67*(236), 36.
356. Jørgensen, C. K. *Inorg. Chim. Acta* **1987**, *139*(1–2), 1.
357. Lecoq de Boisbaudran, P. E. *Compt. Rend. Chim.* **1885**, 100, 1437; Lecoq de Boisbaudran, P. E. *Compt. Rend. Chim.* **1885**, *101*, 552; Lecoq de Boisbaudran, P. E. *Compt. Rend. Chim.* **1885**, *101*, 588.

358. Lecoq de Boisbaudran, P. E. *Compt. Rend. Chim.* **1886**, *102*, 153; Lecoq de Boisbaudran, P. E. *Compt. Rend. Chim.* **1886**, *102*, 395; Lecoq de Boisbaudran, P. E. *Compt. Rend. Chim.* **1886**, *102*, 647; Lecoq de Boisbaudran, P. E. *Compt. Rend. Chim.* **1886**, *102*, 899; Lecoq de Boisbaudran, P. E. *Compt. Rend. Chim.* **1886**, *102*, 1003; Lecoq de Boisbaudran, P. E. *Compt. Rend. Chim.* **1886**, *102*, 1005; Lecoq de Boisbaudran, P. E. *Compt. Rend. Chim.* **1886**, *102*, 1436; Lecoq de Boisbaudran, P. E. *Compt. Rend. Chim.* **1886**, *103*, 113; Lecoq de Boisbaudran, P. E. *Compt. Rend. Chim.* **1886**, *103*, 627.
359. Lecoq de Boisbaudran, P. E. *Compt. Rend. Chim.* **1892**, *114*, 575; Lecoq de Boisbaudran, P. E. *Compt. Rend. Chim.* **1893**, *116*, 611; Lecoq de Boisbaudran, P. E. *Compt. Rend. Chim.* **1893**, *116*, 674.
360. In a reverse spectrum, a spark strikes between the surface of a concentrated hydrochloric acid solution of the substance under examination and a platinum wire placed a few millimeters away from it. The wire carries a negative charge and the solution a positive charge. A weak fluorescence appears at the surface of the liquid, giving a band spectrum that is rather "cloudy." L. de Boisbaudran called these spectra *les spectres de renversement* because the direction of the scintillation is the reverse of the spectra in common use.
361. Demarçay, E. -A. *Compt. Rend. Chim.* **1898**, *126*, 1039.
362. Demarçay, E. -A. *Compt. Rend. Chim.* **1900**, *130*, 1019; Demarçay, E. -A. *Compt. Rend. Chim.* **1900**, *130*, 1185; Demarçay, E.-A. *Compt. Rend. Chim.* **1900**, *131*, 387; Demarçay, E. -A. *Compt. Rend. Chim.* **1900**, *131*, 389.
363. Demarçay, E. -A. *Compt. Rend. Chim.* **1896**, *122*, 728.
364. Demarçay, E. -A. *Chem. News* **1896**, 170.
365. Demarçay, E. -A. *Compt. Rend. Chim.* **1900**, *130*, 1469.
366. Demarçay, E. -A. *Compt. Rend. Chim.* **1901**, *132*, 1484.
367. Etard, A. *Chem, News.* **1904**, 137.
368. Etard, A. *Bull. Soc. Chim. Fr.* **1904**, *31*(3), i; Weeks, M. E. *J. Chem. Educ.* **1932**, *9*, 1767.
369. Urbain, G. *Compt. Rend. Chim.* **1904**, *139*, 736; Urbain, G. *Compt. Rend. Chim.* **1905**, *141*, 521; Urbain, G. *Compt. Rend. Chim.* **1906**, *142*, 957.
370. Urbain, G. *Compt. Rend. Chim.* **1906**, *143*, 229; Urbain, G.; Demenitroux, M. *Compt. Rend. Chim.* **1906**, *143*, 598; Urbain, G. *Compt. Rend. Chim.* **1908**, *146*, 922.
371. Urbain, G. *Compt. Rend. Chim.* **1907**, *145*, 759.
372. Marché, M. Souvenir d'un Grand Chimiste Cognaçais. Brève notice sur la vie et les travaux de Paul-Émile (dit François) Lecoq de Boisbaudran. *Bulletin de l'Institut d'Histoire et d'Archéologie de Cognac et du cognaçais.* **1958**, *1*(3), 1–19.
373. Urbain, G. L'œuvre de Lecoq de Boisbaudran. *Revue Générale des Sciences.* **1912**, *17*, 15.
374. Auer von Welsbach, C. *Chem. -Ztg.* **1912**, *35*, 658.
375. Urbain, G. *Chem. Rev.* **1925**, *1*(2), 143.
376. Urbain, G. *Compt. Rend. Chim.* **1908**, *147*, 1286.
377. Eder, J. M.; Valenta, E. *Sitz. Akad. Wiss. Wien.* **1911**, *119*, 32.
378. Krüss, G. *Ber. Dtsch. Chem. Ges.* **1887**, *30*, 2143.
379. Hofmann, K. A.; Burger, O. *Acad. Sci. Munich Ber.* **1908**, *41*, 308; Hofmann, K. A.; Bugge, G. *Munich Ber.* **1909**, *41*, 3783.

III.16

THE SETTING OF THE ELEMENT OF THE "RISING SUN"

Among the British scientists who worked at identifying and isolating new chemical elements, William Ramsay stands out as a giant and deserves to be remembered for his great contribution to the completion of the periodic system. His discoveries of argon, neon, krypton, and xenon, together with his experimental confirmation of the existence of helium on Earth, furnished proof of the existence of a whole new family of elements. This work was made possible by the development of vacuum techniques at low temperatures and of optical spectroscopy. His role in inorganic chemistry in those years was very great and comparable to the work of Marie Curie, pioneer of the new discipline of radiochemistry.

Ramsay hosted in his laboratory many researchers and simple visitors. Some of these names still resound on the altars of science, men like Frederick Soddy and Otto Hahn (1879–1968). But another guest, less illustrious, was attracted also by Ramsay's activity in the field of inorganic chemistry: Masataka Ogawa.

Ogawa was born in Edo (present-day Tokyo) on February 21, 1865, three years before the Meiji reform that exposed Japanese society to rapid change. Under that reform, in the course of only a few decades, Japan emerged from an old, rigid feudal system thousands of years old into competition with more evolved Western nations. Masataka was the son of a samurai, a member of a warrior class that, with the coming of the reform, lost salary and social privilege. Because of this, the family retired to a more modest lifestyle in the country. His father died when Masataka was still very young, but he had the good fortune to receive a study grant to complete his education. He studied at the Imperial University of Tokyo, where he obtained his degree at 22 years of age. After a period of teaching chemistry in a high school, in 1896 he returned to the Imperial University of Tokyo to study inorganic chemistry under the guidance of Edward Divers (1837–1912).[380] Ogawa obtained a permanent position 3 years later and finally, in 1904, received funds to go to London to study chemistry under the protective wing of the renowned Ramsay.

Sir William welcomed the not-so-young Japanese student, giving him a gift of a sample of thorianite[381] that had been sent to him from the island of Ceylon. The English scientist cherished the idea that the mineral could hide one or more unknown elements in trace amounts.

Ogawa's work started immediately, only briefly interrupted when he went to Montréal to specialize under the guidance of Ernest Rutherford. Despite the fact that the teacher was actually 6 years younger than his 40-year-old student, a cordial and reciprocal sentiment of respect and collaboration developed between the two. In the laboratory, Ogawa worked for long hours fractionating thorianite, and his constancy was repaid: he found a substance that seemed to differ in its properties from all others known. Spectroscopic examination of the sample showed the presence of a new line at 4882 Å. As soon as Ramsay heard the news, he was convinced, and soon convinced Ogawa, that they were in the presence of a new element. Ramsay suggested that Ogawa name it *nipponium*, in honor of Ogawa's mother country, Japan.

The atomic weight was calculated a little later, and it seemed to fit perfectly into a vacant position in the periodic table, between molybdenum and ruthenium. In 1906, Ogawa left England and returned to Japan, continuing to occupy himself with the new element. The funds to obtain suitable minerals dried up, but fortunately he discovered that the new metal was abundant in Japanese molybdenite. After a period of time spent organizing a new laboratory, he published two monographs[382]—later translated into English with the help of Ramsay—on *nipponium*, extracted from thorianite[383] and from molybdenite.[384] A year later, the periodic tables of Great Britain, thanks to Ramsay's influence, were already showing the symbol of the new element, Np. That same year, Ogawa received a prize from the newly formed Japanese Chemical Society, and in 1910, at 45 years of age, he presented his doctoral thesis on the recently discovered element. His fame grew markedly in his own country, where, in a society in search of ways to emulate the West and eager for successes, this discovery was held up as a wonderful opportunity that ought not be allowed to fall into a vacuum.

In 1911, Ogawa was appointed professor at the Imperial University at Sendai, the third university in the country, and later he became director of the Faculty of Science. In 1919, he was elected to the post of rector. His status allowed him to have many students and to place himself at the head of a well-funded group of researchers. The numerous scientists around him were all involved in the concentration and extraction of *nipponium*.

His students described Masataka Ogawa (Figure III.16), in the act of crossing the threshold into his laboratory, as "a monk who enters church, in a mystical state of ecstasy, full of hope and faith in his work."

However, all efforts to isolate the metal were in vain: *nipponium* seemed to have vanished. Ogawa's discouragement was very great, and it was thus that he, having left his high academic post, took up the role of researcher and set himself personally to manipulate the raw mineral. After many attempts, he was able to extract a metal sample; full of joy, he invited his students into the Great Hall, gathered them around a table, and, with great ceremony, raised the cloth that covered a mass of metal saying: "The new metal, *nipponium*, is here! I will send this material to be analyzed by X-rays." Ogawa was certain of his discovery; he had no presentiment of what would happen soon after.

World War I had just ended. In Japan at that time, suitable apparatus for a complete X-ray examination did not exist, nor were there scientific personnel trained to a high enough level to confirm or deny the results of decades of Ogawa's research.

It so happened that, in 1924, his colleague in organic chemistry, Toshiyuki Majima (1874–1963) visited the Niels Bohr Institute at Copenhagen for some spectroscopic tests. There he found Hungarian George de Hevesy studying a mysterious sample of *nipponium*. When he returned to Japan, Majima asked about the source of the sample that was being analyzed at Copenhagen. Ogawa asserted that he had never sent his original sample to de Hevesy, or anyone. The idea that de Hevesy was analyzing *nipponium* must have alarmed Ogawa: in those years, de Hevesy occupied the delicate position of "censor" over the work of chemists who claimed the discoveries of new chemical elements. It is unknown if he ran his office with greater or less scrupulousness or partisanship than others, but the fact remains that he operated under the unquestioned figure of the great physicist Niels Bohr.

FIGURE III.16. Masataka Ogawa (1865–1930), Distinguished Japanese Scientist. Student of Sir William Ramsay at London and Lord Rutherford in Canada, Ogawa searched long and hard for element number 43, for which he proposed the name *nipponium*. Gift of Masanori Kaji.

In 1925, George de Hevesy published the results of his *nipponium* analysis[385]: "Mr. R. B. Moore, Senior Chemist at the Bureau of Mines in Washington, had the great kindness to send us a few crystals of the silicate of nipponium obtained by Ogawa. These crystals were composed of zirconium silicate having a content of 2% hafnium."

Ogawa's resentment toward de Hevesy was based on the fact that the Japanese chemist had never sent a sample to a Mr. Moore in America; furthermore, he had quantitatively removed all the silicates from the mineral examined. Finally, hafnium did not show at all the line at 4882 Å typical of *nipponium*. However de Hevesy's results were accepted by Western science, and Ogawa was given no possibility to respond. To further bury Ogawa's discovery, a few months later, chemists Walter Noddack and Ida Tacke Noddack and spectroscopic specialist Otto Berg announced the discovery of *masurium* and rhenium.[386]

Masurium, positioned in the periodic table right under manganese, seemed to correspond to *nipponium*, but the German chemists did not take the trouble to mention Ogawa's work, a reflection of the lack of consideration given to the Japanese chemist by European scientists after the death of his mentor, William Ramsay.

No one speaks of *nipponium* any longer, and the once-celebrated Ogawa fell into obscurity. He had sought a phantom element for decades, one that subsequent researchers realized was not present in nature. More recently, Professor H. Kenji Yoshihara of Tohoku University in Sendai advanced the hypothesis that Ogawa had not discovered eka-manganese (technetium), but dwi-managanese[387] (rhenium). In fact, the wavelength observed for the new element, 4882 Å, would seem to coincide, within experimental error, with that of rhenium (4889 Å). Furthermore, one can reasonably suppose that Ogawa had erroneously calculated the atomic weight of the mysterious element. He had hypothesized that *nipponium* had a valence of 2, but there are no certain proofs that verify this. In all probability, Ogawa obtained $ReOCl_4$, in which the metal has a valence of 6. On these assumptions, recalculating the atomic weight of *nipponium* using Ogawa's data gives a value of 185.2, compared to that of rhenium, which is 186.2. Finally, we know that the molybdenite analyzed by Ogawa contained a very large percentage of rhenium.

In 1927, another chemist returned from Copenhagen to Tokyo, Kenjiro Kimura (1896–1988). With him, he carried a precious Siegbahn X-ray spectrometer. It took about 2 years before the instrument was in full regular use. The metal Ogawa showed to his students in 1919 was analyzed in 1930; when Toshi Inoue (1894–1967), a friend of Kimura, read the results he exclaimed: "Truly a most beautiful sample of rhenium!"[388]

At that time, the spirit of *bushido* (chivalrousness) was blowing gently among Japanese scientists, and this *forma mentis* prevented Ogawa from engaging in even a minimal action of revenge against de Hevesy or the Noddacks. Furthermore, only a few more weeks of life remained to him. On July 3, 1930, he suffered a fatal gallbladder attack, and he died eight days later in Sendai hospital, after unspeakable suffering. His remains were interred in the temple of Shozanji, in Mita, Tokyo. After his death, donations were collected to build a small Japanese garden in his memory. This garden— called, because of its form, *sankaku koen* (i.e. "triangular park")—stands at the Katahira campus of Tohoku University.

Ogawa's death was tragic, not least because he was unable to respond to his critics. However, almost immediately after his death, many scientists began to rediscover his work on *nipponium* and re-evaluate its content. Among them, three of his eight children—Shintaro (1902–79), Eijiro (1904–44), and Shiro (1912–99)—deserve to be remembered for continuing, as chemists and physicists, Ogawa's legacy both as a scientist and loving father.

Geochemist Victor Moritz Goldschmidt (1888–1947) in his geochemistry textbook, mentioned *nipponium*:[389] "A supposed new element *nipponium* reported many years ago may have been a mixture of oxides of rhenium and molybdenum, as it has been isolated from Japanese molybdenite." Goldschmidt's was the first in a long series of posthumous tributes paid to this neglected and unfortunate Japanese chemist.[390]

Notes

380. Divers had gone to teach chemistry in Japan on the recommendation of the chemist Alexander William Williamson (1824–1904), Ramsay's predecessor at University College London.
381. Thorianite is a mineral with the chemical formula ThO_2.
382. Ogawa, M. *J. Coll. Sci. Imp. Univ. Tokyo* **1908**, *25*, Art. 15; Ogawa, M. *J. Coll. Sci. Imp. Univ. Tokyo* **1908**, *25*, Art. 16.
383. Ogawa, M. *Chem. News* **1908**, *98*, 249.
384. Ogawa, M. *Chem. News* **1908**, *98*, 261.
385. de Hevesy, G. *Kgl. Danske Videnskab. Selskab. Mat-fis. Medd.* **1925**, *VI*, no. 7.
386. Noddack, W.; Tacke, I.; Berg, O. *Naturw.* **1925**, *13*, 567.
387. Yoshihara, K. H. *Radiochim. Acta* **1997**, *77*, 9; Yoshihara, K. H. *Kagakushi* **1997,** *24*, 295.

388. Yoshihara, H. K. *Historia Scientiarum: International Journal of the History of Science Society of Japan* **2000**, 9(3), 258–269.
389. Goldschmidt, V. M.; Muir, A. (Ed.) *Geochemistry*; Clarendon Press: Oxford, 1954. This book was published posthumously; Goldschmidt, a geologist, died in 1947 as a result of the severities of being a prisoner of war under the Nazis in World War II. His manuscript, written in hospitals and nursing homes, was unfinished at the time. It was the work of years to finish the text, add the notes, and collect the scattered materials left by the author.
390. As a postscript, and realizing that what one finds on the internet is not peer-reviewed, we must note that the Wikipedia entry for rhenium at http://en.wikipedia.org/wiki/Rhenium (accessed April 12, 2014) lists Ogawa as the discoverer and the person to first isolate rhenium in 1908 and I. Tacke, W. Noddack, and O. Berg as the persons who conferred a name on this element in 1922.

III.17

THE TIMES HAVE CHANGED: FROM CANADIUM TO QUEBECIUM

Her Majesty's mail has always been delivered promptly: on November 23, 1911, Mr. Thomas French received a letter from his father, Mr. Andrew Gordon French, dated November 12. The sender was in the province of British Columbia, Western Canada; the addressee was thousands of miles away, in Glasgow, Scotland. The letter claimed that the elder French had discovered a new metal of the platinum group, a specimen of which he proposed to send to the Royal Society. French remarked that the metal, which he called *canadium*, was brighter and easier to work with than palladium.[391] The colorful particulars of this discovery were reported in the local daily newspaper,[392] the *Glasgow Herald*, on December 5, 1911.

Mr. Andrew Gordon French was a renowned metallurgist and native of Glasgow. Before leaving Scotland more than 20 years before the facts narrated in his letter, he had worked as a gold- and silversmith in many foundries. Later, he acquired a certain expertise in the extraction and working of metals of the platinum group.

The first peculiarity of his discovery is in regard to the fact that, in other deposits of a similar nature, he had not found *canadium*. This is incongruous when you consider the fact that this element's rarity (as reported by the same author) was not really so impressive: from 1 ton (909 kg) of the ore, French extracted about 3 ounces (0.09 kg)[393] of *canadium*. More worrying is the fact that he said had found the metal in the elemental state in semicrystalline grains or elongated prisms (0.5 mm long and 0.1 mm thick), white in color. Also strange is French's assertion that he found, in platinum-bearing deposits, an alloy of *canadium* with a scale-like form. The metallic particles had a color intermediate between blue and white. Placing a sheet of the alloy in a flame, a volatile metal (which the author identified as osmium) was removed, leaving a brilliant white pearl. French did not identify *canadium* with any other element. The analyses to which French subjected the material he discovered were inaccurate even for his time: *canadium* was too ductile to be confused with another element; furthermore, its melting point was incredibly low. French conducted other wet tests and found that it did not oxidize even after prolonged exposure to moisture. Furthermore, the oxidizing flame of a blowpipe did not corrode the metal.

The metal dissolved in aqua regia and in concentrated nitric acid. These latter solutions did not give a precipitate even after adding sodium chloride or potassium iodide. The metal did not darken either in the presence of hydrogen sulfide or alkali metal sulfides, nor by the action of iodine. Quoting the author: "Its [canadium's] melting point is somewhat lower than that of fine gold and silver, and very much lower than that of palladium."[394]

It is interesting to note the absence of information, even qualitative, in this sentence; the melting points of the three metals referred to are 1,065, 961, and 1,552 °C, respectively, presenting such great differences among them that cannot hope to resolve this puzzle.

Too many clues are missing to make it possible to formulate any hypothesis of error whatsoever. If we look at a periodic table of the time (1911), we know that many elements were still missing from the roll call: hafnium, technetium, rhenium, promethium, and astatine. We can eliminate the idea that French was looking for radioactive elements; he was predominantly a mining expert who was looking for platinum-bearing deposits. Radioactivity and the elements associated with it were beyond his horizon.

We can exclude the fact that he was looking for a rare earth element, not because the minerals containing the rare earths were different from platinum-bearing ores, but because the lanthanides were found in the form of earths (oxides) and not in the elemental state (because these metals reacted with water more or less energetically). We would be tempted to say that French found rhenium, but this element has a melting point that is much higher—in fact, double—than that of palladium. One reference suggests that perhaps he mistook an alloy, for example osmiridium, for the pure metal.[395]

Upon the recommendation of the editor of *Chemical News*, French reported that *canadium* could be an element not yet discovered and hazarded a guess that it was probably the missing element between molybdenum and ruthenium in the periodic table; namely, eka-manganese. However, the element we now call technetium is a radioactive element that has none of the chemical or physical characteristics cited by French; in addition, technetium is not present in nature in the quantities that he reported.

Missing from French's data are the presumed atomic weight of *canadium* as well as the valence of the metal, data indispensable in 1911 (along with the wet tests) to determine a material's nature. Lacking an analytical scheme, we can hypothesize that French analyzed an alloy of metals already known and confused cadmium or zinc for what he called *canadium*. In fact, the melting points of cadmium and zinc are much lower than those of gold, silver, or platinum. Furthermore, cadmium easily forms alloys with zinc, with which it is found associated in nature.[396] Andrew Gordon French was not an academic and, with the exception of the one work on *canadium*, had not published any research. However, between 1908 and 1916, he filed seven patents (in the United States, Britain, New Zealand, and Norway) that embraced most of the interests of chemistry of the time: the production of sulfuric acid, ammonium chloride, and manganese sulfate, and the refining of zinc and lead. The illusory discovery of *canadium* occurred the same year as that of Urbain's *celtium* but, unlike the latter, it was soon forgotten.

Some years later, a curious case revived memories of old the *canadium* discovery. Around 1935, the French-Canadian Léon Lortie (1902–85)[397] gave a lecture related to his work on cerium that he had done under the direction of Georges Urbain. In those years, Urbain's laboratory was the Mecca of inorganic chemists: Urbain, as head of the Institute, was the last element hunter still alive and he had discovered *neo-ytterbium* (hereinafter called ytterbium) and lutetium. The conference was held at the Ecole Polytechnique of Montréal, in Saint Denis. Léo Parizeau,[398] who sat in the front row of the vast auditorium, posed a question to the speaker: "If you were to find a new element, what would you name it?" With one accord, Léon Lortie's whole research group responded "*Canadium*!"

Times have changed, and the claims of the Francophone population of Canada have increased. On October 19, 1996, at the 31st Congress of the Association des professeurs de sciences du Québec, held in Hull, Canada, at the Ecole de l'Île, Québécois chemist Pierre Demers announced his discovery of a new element, one he chose to call not *canadium* but *quebecium*.[399] Its symbol is Qb; the atomic number is 118. (For the record, the discovery of element 118 was announced at the Lawrence Berkeley National Laboratory [LBNL] by

Victor Ninov's American team [together with element 116], but retracted by the same author in 2001.)[400]

Pierre Demers, in a speech characterized by conspicuous Gallic pride noted that all civilized countries have an element representing them in the periodic table, except Canada. (Although it must be said in passing that Italy also is missing in this roll call, as is Japan, although many scientists have been busy in this regard: *florentium, ausonium, hesperium, littorium*, and *nipponium* are some examples of fruitless searches.) According to Demers, the list contains countries such as Russia, America, France, Germany, Poland, India, and Samaria; regions such as Scandinavia and Asia; cities such as Paris, Copenhagen, Stockholm, Berkeley, and Dubna; and continents such as Europe and America. (He may have been ignorant of the etymology of two of the elements: samarium[401] and indium were not named for Samaria and India, but for a Russian mining engineer—Colonel Vasilij Evgrafovič Samarskij-Byhovec (1803–70)—for samarium and for the colors of its spectral lines for indium.)

III.17.1. WHO IS PIERRE DEMERS?

The first author of this volume knows professor Demers through close correspondence. Pierre Demers was born November 8, 1914; he began his primary schooling in 1922, in Paris. He matriculated at the University of Montréal in 1933 and obtained in 1936 a licentiate in physics and one in mathematics the following year. In 1937, he went to Cornell University, in Ithaca, New York, and the following year to the Ecole Normale Supérieure at Paris. Up until 1940, he worked as an adjunct at the Collège de France at Ivry, in the laboratory of atomic synthesis, under the direction of Frédéric Joliot, having as colleagues Hans von Halban (1908–64) and Lew Kowarski (1907–79). At the outbreak of World War II, Demers returned to America and briefly worked at the Massachusetts Institute of Technology, where he became interested in spectroscopy. From 1947 to 1980, he was professor at the University of Montréal. In 1950, he worked at the Institute of Physics of the University of Milan as adjunct professor.

In 1975, he founded the Centre Québécois de la Couleur, of which he was twice president, from 1975 to 1982 and from 1991 to the present. In addition to being a man of science, Demers was a powerful supporter of the rights of the Francophone population of Québec. A member from 1977 of the Parti Québécois, from 1985 he became a radical and joined the Parti Indépendentiste.

In 1995, Pierre Demers became interested in more eclectic areas of science. The deeply rooted stereotypes and rigidity of 20th-century science had partially marginalized him. Demers resembled much more a well-rounded Renaissance genius like Leonardo da Vinci (1452–1519) than he did a tireless, methodical scientist in the mold of Augustin Jean Fresnel (1788–1827).

The "official science" of the large research universities and multinational conglomerates gives short shrift to the informal and imaginative approaches that are sometimes condemned as "deviant science." At times, however, such forays into the unknown are ideas ahead of their time and there are examples of some of them that eventually enter mainstream science many years after they were propounded.

As for Demers, perhaps "official" science should take a greater interest in its so-called deviant colleagues. It should spend less time in ostracizing them and demonizing them (removing their subsidies because they move against the orthodox ideas of the establishment), and devote itself more to the spread of knowledge. The most well-known case of this sort in Italy is that of Giorgio Piccardi (1895–1972), whose research on fluctuating phenomena was repeatedly opposed by the Consiglio Nazionale delle Ricerche (Italy's national science foundation). If he had been listened to, maybe today his name would be associated with that of Ilya Prigogine (1917–2003) and remembered every time one spoke of irreproducible phenomena.[402] The Middle Ages is not that far behind us, at least with respect to certain attitudes and in certain mindsets.

After retiring from teaching, Pierre Demers entered a second season of intellectual and scientific productivity. Even beyond the age of 70, he was busy studying the magic numbers[403] for the masses of elementary particles.[404] In these years, he was always revising the concept of time as a tri-vector in space and in biological life.[405] One cannot fail to be impressed by his historic essays on Charles De Gaulle and by his studies on Louis Pasteur.[406]

Overall, Demers authored nearly 900 publications; his interests were a dizzying kaleidoscope.[407] Some of his writings were rejected outright, but the grit and curiosity that moved Pierre Demers are typical of a "noble" researcher, not simply one who kowtows to science. His publications also reach out into the political and sociological fields, as in the case of his study on the "Future of Québec,"[408] in which he developed biomathematical models based upon the data available from historic immigration patterns. Some titles of his publications are certainly cryptic to the eyes of a chemist and tend to make even more well-disposed persons turn up their nose: "Need for a Musicodynamic Quantum," or "On the New Analysis of the Muscial Scale of Elementary Particles."[409] Equally strange, but certainly not lacking in fascination, is Demers's "The Study of a Biomathematical Model of the Periodicity of Our Perceptions of Space, Color, Music, and Mass."[410]

In 1995, three years before the presumed discovery of element 118 appeared in the pages of *Physical Review*, Demers called his virtual element *quebecium*. This proved the key to creating a new periodic table. Demers was also concerned about safeguarding the name of his element because, in the meantime, Victor Ninov had announced the discovery of element 118, calling it temporarily ununoctium[411] or *ninovium*.[412] According to Demers, he had the right to name *quebecium* even though the Canadian physicist had not contributed in any way to its discovery. Things became simpler when, in August 2001, Ninov withdrew the discovery of this element as a result of refutations made by the French scientist Jean Péter who, in the laboratories at Ganil, was unable to obtain the same results that Ninov had collected 2 years earlier at the LBNL in California.

Demers, in response to a letter listing objective difficulties to the universal acceptance of the name *quebecium* by the International Union of Pure and Applied Chemistry (IUPAC) Commission for Nomenclature, responded thus:

> I make note at this point, regarding the official nomenclature, that we ought not follow Anglo-American decisions. My idea is to totally ignore the authority of IUPAC, whose importance derives from a fetishistic worship of the English language and American Imperialism. When the name of quebecium becomes well known, it will take such a natural priority over all other names that the whole world will come to respect it because it comes from a little country with great aspirations.

Demers had a particular interest in periodicity; browsing the list of his publications, the repetition of the word "periodicity" is impressive. (It should be noted that, since 1869,

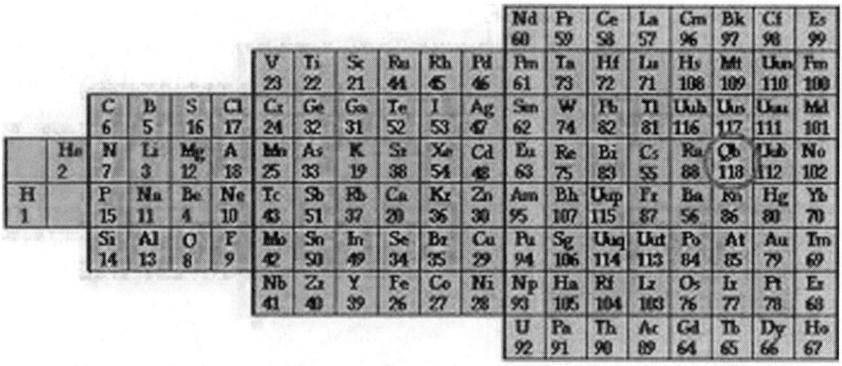

FIGURE III.17. A Representation of the "Québecium" System Proposed by Pierre Demers.

many "tables" have been proposed. In addition to the best-known version, which shows the elements in a two-dimensional "stacked block" format, new forms include elliptical, three-dimensional, and [among the most curious] helical tables.)

In the early 1990s, in an attempt to classify atomic particles, Demers created the "Periodic Table of the Elementary Particles,"[413] The "*quebecium* system" was propounded as a new periodic system of the 118 elements, represented in two dimensions by four square grids arranged one next to the side of the following one that had respectively 2, 4, 6, and 8 boxes on each side.

In three dimensions the system consisted of a compact stacking of 120 spheres on a triangular base having eight spheres to a side, representing a regular tetrahedron. Although his proposal was correct from the mathematical point of view, it was complicated and less intuitive than Mendeleev's table.

The filling in of the elements of the periodic table is neither that simple nor intuitive, but it is also not random. It does not rely on atomic orbitals or shells which were skillfully interpreted by the theory of Niels Bohr, but on three principles, which Demers himself developed:

- To get the atom with atomic number Z it is necessary to remove 118-Z electrons from an atom of quebecium and maintain Z constant.
- You remove the electrons in a descending mode starting from 118.
- In the case of the exceptional elements, the boxes of atoms with Z gradually increasing are not consecutive in the table of the elements. The sequential order of the occupied boxes goes from 1 to Z+b, with b gaps.

The *quebecium* table appears, in our opinion, to be a predominantly geometric construction. Demers succeeded in putting 118 boxes in order but, unlike Mendeleev's table, chemical properties are not taken into account. The groups and above all the periods are missing, so that the new table may be called a table of the elements, but it loses the adjective most important for chemists, "periodic."

Moreover, according to the author, his creature opens up the possibility of rather risky "marriages," such as the periodicity of the elements and the symmetry of order 4 of electromagnetic forces. Demers also made some analogies between his tetrahedral construction of the table of the elements and the genetic code, arguing that the *quebecium* system is nothing short of a profile of the system of the living world and vice versa! Finally, in Demers's opinion, his new "tables" of two or three dimensions should appear at the side of the Mendeleev's table, waiting to replace it.

We respect Demers's work, but we do not think that the name *quebecium* will survive. The struggle is unequal—pitting the powerful IUPAC on one side and an isolated retired professor on the other. His discovery does not bring new life to chemistry or physics, nor to the interpretation of the periodic properties of the elements. It only changes the graphics.

Pierre Demers has been accused of naiveté. We do not believe it. He was a versatile and eclectic physicist who sometimes exceeded the invisible boundaries of science. The Canadian press was divided between "innocent" and "guilty," as shown on the first page of the *Journal de Saint-Laurent* of May 4, 1996. From 1911 to the present, a century has passed, without any evidence of *canadium*,[414] and *quebecium* is in the mind of God. Demers is a committed Québécois, such that the love for his land, for his language, and for French customs is apparent in all of his writings. It will certainly be a disappointment for him to discover over the years that the name of element 118 will be different from what he had proposed.

ADDENDUM

In 2003, Professor Pierre Demers sent us a letter in perfect Italian. It contained an impassioned defense of his work. It seems correct to present this short passage to our readers:

> The truth has rights that error does not possess. The introduction of an innovation is more laborious. We need a great deal of understanding to accept a new system. There are established habits that can be overcome only with careful consideration.

Notes

391. Letter dated November 12, 1911, from Nelson, British Columbia.
392. *Glasgow Herald*, December 5, 1911.
393. An ounce corresponds to a little less than 30 g (i.e., exactly 28.3495 g).
394. French, A. G. *Chem. News* **1911**, *104*, 283.
395. Rayner-Canham, G. W. The curious case of canadium. *Canadian Chem. Educ.* **1973**, *8*(3), 10–11.
396. We can hypothesize with a degree of certainty that perhaps it was the element cadmium that French confused for his "new" one. The metal's blue-white color corresponds to the description given by French for his *canadium*, although cadmium is never found free in nature.
397. Léon Lortie received his degree in 1928, and, in the same year, he went to the University of Paris with a grant from the Rockefeller Foundation, where he earned his doctorate in physics in 1930. In 1931, he was appointed to a teaching post at the University of Montréal, where he remained until he was 60 years old. A prolific author, he specialized both in his science and in its popularization among the Francophone population of Canada. Among his publications were numerous articles on the history of science and science education. He received many honorary doctorates during his lifetime.
398. Léo-Errol Parizeau was born in Grenville, Québec, on May 24, 1882. He received his degree in medicine (1904) and during World War I he was a military radiologist in France. He was

a professor of radiology beginning in 1919. As a result of long exposure to X-rays, he lost his sight in one eye and retired to private life in 1938. He died on January 10, 1944, in Outremont.

399. Demers, P. *Le Nouveau Système des Elements: Le Système du Quebecium*; Presses universitaires: Montreal, Canada, 1997.
400. Acronym of Lawrence Berkeley National Laboratory.
401. The well-known science fiction writer and popularizer of science, Isaac Asimov (1920–92), wrote with respect to samarium: "Goodness knows, that for most of my life I had the impression that the name of this element derived from the 'Good Samaritan,' but obviously I was wrong."
402. Manzelli, P.; Costa, M.; Fontani, M. *Le Scienze* **2003** 420, 8; Costa, M.; Fontani, M. *Cifa News* **2003** 33, 8; Costa, M.; Fontani, M. *Cifa News, Supplemento di Ricerca Aerospaziale*, **2005**, *21, 2*, 4.
403. A magic number is the number of nuclear particles (either protons or neutrons) that confer unusual stability on a given nucleus. Such nuclei have higher than average binding energies per nucleon than would be predicted from mass-loss calculations.
404. Demers, P. sent October 9, 1990, to *Interface de l'ACFAS*, no reply.
405. Demers, P. *Il Nuovo Cimento* **1981**, *31*(7), 258.
406. Demers, P. *Science et Francophonie* **1990**, *29*, 3.
407. Demers, P. *Rev. intern. Biomathématique* **1997** (34), *137*, 4; Demers, P. *Rev. intern. Biomathématique* **1997** (35), *138*, 5.
408. Demers, P. *XVe congrès international de biomathématique*, Paris, September, 7–9, 1995.
409. Demers, P. *Revue de Biomathématique*, 2e trimestre **1994**, *35*, 126.
410. Demers, P. Proposed communication (December 23, 1990) to the ACFAS on the occasion of its 59th Conference, University of Sherbrooke, May 21–24, 1991. Rejected February 15, 1991.
411. According to the IUPAC recommendation.
412. According to what was reported in Demers's writings.
413. Demers, P. *Il nuovo Cimento, Brief note*, sent June 10, 1990. Rejected October 10, 1990.
414. It is strange to observe how the science fiction writer H. G. Wells (1866–1946) had anticipated the discovery of the nonexistent *canadium* in his tale "Tono-Bungay," published in 1909. It may be that Andrew French had read this story, and it left an impression on him. Or maybe not, thus allowing us to suspect that the two "discoveries" were just an odd coincidence.

PART IV

1914–1939

From Nuclear Classification to the First Accelerators: Chemists' Paradise Lost. . . (and Physicists' Paradise Regained)

CHEMISTRY HAS BEEN TERMED BY THE PHYSICIST
AS THE MESSY PART OF PHYSICS, BUT THAT IS NO REASON
WHY THE PHYSICISTS SHOULD BE PERMITTED TO MAKE A MESS OF CHEMISTRY
WHEN THEY INVADE IT.
—FREDERICK SODDY (AS QUOTED IN *AMERICAN JOURNAL OF PHYSICS, 1946, 14*, 248)

PROLOGUE TO PART IV

Moseley's discovery of the atomic number just before World War I gave research scientists the advantage of being able to exactly position supposed new elements in a designated box in the periodic table, definitely a great step forward. Furthermore, they had the added advantage of being able to pinpoint exactly which elements were missing from the roll call initiated by Mendeleev.

The theme of the missing elements, and of their identification, is circumscribed by a relatively brief period of time, spanning the formulation of the periodic table by Mendeleev in 1869 to the outbreak of World War II. Yet during these few decades, chemical scientists seemed to revive a Dark Age that might be called the "atomic myth." Ernest Rutherford's research on nuclear structure at the beginning of the 20th century signaled the sunset of the domination of chemists over the atom. The work of physicists to throw light on the structure of the nucleus, founded as it was on the concept of the atomic number, can be seen as the expulsion of chemists from "Atomic Paradise." From that moment, the atom disappeared from the chemists' horizon, never to return. To chemists remained only the dominion of those few elements that eluded identification and classification by the chemists of the 19th century. The time period encompassed by Part IV includes the sagas of the search for the last, and most elusive, of the naturally occurring elements: numbers 43, 61, 72, 75, 85, 87, and 91.

Once it was determined that elements could be placed in their proper boxes in the periodic table, another problem arose: some "elements" being discovered, whether stable or radioactive, were laying claim to the same box! It was not until Frederick Soddy's hypothesis about the existence of isotopes (*iso*, "same"; *topos*, "place") in 1912, confirmed experimentally the following year by J. J. Thomson, that these claimants could be classified and accepted as true members of the great elemental family. The concept of the isotope also cleared up the confusion about the vast number of radioactive species being discovered: many of them were variants on the same element, with different half-lives and mass numbers, but with the same atomic number.

Thus, Moseley's Law and Soddy's hypothesis dominated progress in atomic physics from the outbreak of World War I to the beginning of World War II.

IV.1

FROM THE ECLIPSE OF ALDEBARANIUM AND CASSIOPEIUM TO THE PRIORITY CONFLICT BETWEEN CELTIUM AND HAFNIUM

IV.1.1. A COLLECTIVE HISTORY: THE RARE EARTHS

In 1794, Finnish scientist Johan Gadolin discovered the first of the rare earth elements in some ore deposits at Ytterby, Sweden. He called the oxide of the new element that he had isolated ytterbia and ytterbite the ore from which he had extracted it. Three years later, Anders Gustaf Ekeberg verified Gadolin's discoveries and proposed the name of yttria (or yttric earths) for the oxide and gadolinite for the ore. For many years, chemists, among them L. N. Vauquelin, J. J. Berzelius, and M. H. Klaproth, wrestled with the problem that perhaps Gadolin's yttrium was not a simple body but in reality contained other elements.

In 1842, the Swedish chemist C. G. Mosander described how, by means of the fractional precipitations of the oxalates from dilute solutions of oxalic acid and by treatment of the hydroxides with dilute ammoniacal solutions, he seemed to have succeeded in extracting three new elements. The first was yttrium, the most basic; the second was erbium, the least basic; and the intermediate fraction he called terbium. The names terbium, erbium, and ytterbium derive from the name of the town, Ytterby. The names that Mosander gave to the three elements derived from the sequence in which they were separated: the name yttrium was not changed out of respect for Gadolin. The first element that he extracted, Mosander called terbium, and the following one he called erbium. He removed a letter from the word terbium because he had isolated it later.

In the following years, it was discovered that both erbium and terbium were not single elements but mixtures of elements yet unknown. A practice developed that we might call an *entente cordiale*: when a discoverer split a presumed element into its constituents, one element retained the name already given by its preceding discoverer. This usage was respected by everyone, including Urbain, who, in 1907, presented his discoveries with the names *neo-ytterbium* and *lutecium*.[1] Only Auer von Welsbach, a renowned Austrian chemist, did not respect this tacit "gentlemen's agreement" and called the elements with atomic numbers 70 and 71 *aldebaranium* and *cassiopeium*.[2]

IV.1.2. THE LIGHTS OF PARIS HIDE THE STARS

A few weeks after the announcement of the identification of *neo-ytterbium* and *lutecium*, Auer repeated these discoveries. Baron Carl Auer von Welsbach,[3] born in Vienna September 1, 1858, studied there and subsequently went to Heidelberg to specialize under the guidance of Robert Bunsen. At the beginning of the 1880s, Auer published his first works on the rare earths. In 1885, after an extremely careful work of separation, he realized that *didymium* was not an element, but a mixture of two simple substances. He crystallized solutions of *didymium* with ammonium nitrate and got a yield of green crystals. The metal contained in the salt he called *praseodidymium*. From the solution of the mother liquor, he isolated a pink salt and from that a new element that he called *neodidymium*. Later, the names of these elements were changed to praeseodymium and neodymium. Toward the end of that decade, Auer von Welsbach became interested in practical problems such as the synthesis of new alloys for incandescent lamp filaments and cigarette lighter flints, and he filed numerous patents in these years. He also established his own business, one that netted him a considerable fortune. He received the title of Baron (Freiherr, in German) from Emperor Franz Josef (1830–1916); the motto that he chose for the occasion was inherently linked to the work that brought him fame and fortune: *plus lucis* (more light)!

In 1905, he sent a brief communication to the *Akademie der Wissenschaften* of Vienna. In it, he acknowledged the complexity of the ytterbium discovered by Marignac. But, in 1904, in Paris, Urbain had already begun to investigate the presumed elemental nature of ytterbium. Many of Auer's supporters over the years have told of the episode when the problem of awarding priority of discovery for the new elements arose.[4,5,6] Auer reported his convictions, according to which the ytterbium isolated by Marignac in 1878 was not a simple substance but a mixture of more than one element.[7,8] However, he did not publish his results. Urbain also had the same intuition, but was quicker to draw his conclusions and separate and characterize the two new elements.[9] He substantially kept the name of the first unchanged by proposing only the prefix "neo" as well as a new chemical symbol, Ny, in honor of the work carried out by Marignac. The other element he called *lutecium*, with the symbol Lu, in honor of Paris, from the Latin name *Lutetia parisorum*.

A short time later, still in 1907, Auer von Welsbach discovered the same two elements spectroscopically. He separated them by the classical method of fractional crystallization, and he called them *aldebaranium* (Ad, atomic number 70, corresponding to present-day ytterbium) and *cassiopeium*[10] (Cp, atomic number 71, corresponding to present-day lutetium) after the star Aldebaran and the constellation Cassiopeia. The names of these two elements never came into common usage except for the fact that *cassiopeium* sometimes appeared in German-language journals. Auer's work on *cassiopeium* and *aldebaranium* was later, although by very little, than that of Urbain, and because of this an inevitable and hardly edifying controversy arose regarding the priority of the discoveries.

Urbain resolutely placed on the discussion table the data from his two publications, and von Welsbach's prestige availed him nothing. In 1909, the International Commission on Atomic Weights resolved the argument, pronouncing in favor of Urbain.[11] Some of his detractors asserted that he had influenced the decision of the Commission by the fact that he himself had been a member since February 1907, when he succeeded to the post of Henri Moissan. The presence of Urbain on the Commission did not in any way taint the priority of the discoveries of *neo-ytterbium* and *lutecium*, and, none of the names proposed by Urbain had the good fortune to remain unchanged. The first changed into ytterbium very shortly after the discovery, whereas the name *lutecium*, although it survived its discoverer, was changed to lutetium in 1949 by the International Union of Pure and Applied Chemistry (IUPAC) Commission.

Between the beginning of the 20th century and the outbreak of World War I, Auer von Welsbach was interested in searching for the other missing rare earth elements and, shortly after the end of the war, he searched actively for element number 61. He was also busy with the extraction of the radioactive elements from the deposits of pitchblende at Joachimsthal.

In 1911, when Urbain made the first announcement of the discovery of *celtium* from samples of gadolinite, Carl Auer von Welsbach was busy about the fractionation of thulium.[12] Auer was convinced that this element, like the *didymium* of Mosander or the ytterbium of Marignac, could contain three new elements that he called provisionally *thulium I, thulium II*, and *thulium III*. (Previously he had used a similar notation for doubtful discoveries for terbium.) His research was published in the same year and did not go unobserved, at least in the German world, although in translation in the Anglo-Saxon world thulium was confused with terbium. Auer reported a detailed spectroscopic study of both the arc and incandescent spectra of the three new elements, but he was not able to chemically separate them; he regretted that the three bodies—TmI, TmII, and TmIII—had properties so similar that they could not be separated with the analytical techniques that existed at the time. They seemed to be so similar because the three elements did not exist at all!

Every year, from the years immediately following World War I until his death, Auer was a candidate for the Nobel Prize in Chemistry. Like Urbain, he came very, very close to this goal, receiving many votes from the Swedish academics. Later, at almost 71 years of age, Baron Carl Auer von Welsbach passed away at his castle in Carinthia on August 4, 1929.

IV.1.3. CELTIUM

Georges Urbain[13,14,15] already enjoyed great international renown when he began his research on *celtium*. Born in Paris on April 12, 1872, he left Charles Friedel's laboratory in 1899 upon the death of his mentor and devoted himself with great success to the study of mineralogical chemistry. This work absorbed him for more than 25 years and led to the isolation of two new elements: *neo-ytterbium* (Ny; later ytterbium) and *lutecium* (Lu;[16] later lutetium), and finally he announced the discovery of *celtium* (Ct).

Prior to these discoveries, Urbain was already an expert in the separation and characterization of the rare earth elements. He succeeded in isolating for the first time in their elementary state samarium, europium, gadolinium, terbium, dysprosium, and holmium;[17] in 1906, he succeeded in splitting the presumed ytterbium of Galissard de Marignac into the two elements that he called *neo-ytterbium* and *lutecium*.[18] Then, in 1907, Urbain announced the discovery of the element he called *celtium*, choosing this name in honor of the Celtic population that lived in modern-day France in pre-Roman times. Four years later, in 1911, he published his results, asserting that it was a new element found mixed with lutetium and scandium in the mineral gadolinite.[19] His work at this point was preliminary and in large part inaccurate.

At first, Urbain thought that *celtium* had chemical properties similar to the rare earth elements. In June 1914, he made a journey to Oxford, where Sir William Ramsay introduced Urbain to Moseley.[20] Moseley's work failed to confirm the presence of element 72 in Urbain's rare earth fractions.[21] In August 1914, World War I broke out, and Moseley enlisted enthusiastically. He died the following year on the beaches of the Dardanelles. Urbain interrupted his own work and enlisted in the national defense. He was discharged in 1919.

IV.1.4. NEO-CELTIUM

Shortly after the end of World War I, Georges Urbain again took up his studies of the rare earths. Although his academic duties would increase markedly after he was elected to the Académie des Sciences in 1921, he would also have the satisfaction of accomplishing what he felt was his last truly important discovery: *neo-celtium*. On May 22, 1922, Georges Urbain announced that he had arrived at the definitive identification of element number 72.

This new announcement was not looked on favorably by the international scientific community, possibly because it was tainted by the shadow of the first announcement of the discovery of *celtium* in 1907 and the subsequent publication of the results in 1911 that had been in error: the data were distorted by the presence of large amounts of lutetium. Furthermore, the discovery could not be verified by Moseley in 1914.

In 1922, the law of Moseley was seen as the necessary and sufficient condition to identify a new simple substance. Urbain's claim relied on the spectroscopic work of Alexandre Henri Georges Dauvillier, who published the spectroscopic lines of the new element.[22,23] Some people spoke of rediscovery because the name changed back to *celtium* once again.

Eight months later, in January 1923, George de Hevesy and Dirk Coster (Figure IV.01) announced in their turn the discovery of element 72, which they called hafnium.[24] When these colleagues of Niels Bohr at the University of Copenhagen announced their discovery, they set in motion one of the most storied (and heated) scientific controversies of the first half of the 20th century.

FIGURE IV.01. Dirk Coster (1889–1950), Dutch Chemist and Physicist. With George de Hevesy, 6 months after Urbain announced the discovery of *celtium*, he discovered hafnium. Courtesy Niels Bohr Archive, Copenhagen.

The controversy could possibly have been settled quickly if the solution were based strictly on scientific evidence. However, rampant nationalism in the wake of a world war, the ascendancy of a new physics based on atomic theory, and the decline of the old methods of chemistry based on endless fractional crystallizations and physical separations blinded people on both sides. For one thing, Urbain positioned the new element as the last of the rare earths, and he looked for it among the rare earth ores. On the other hand, de Hevesy and Coster postulated correctly that it should be the higher homologue of zirconium, in the next periodic group beyond the rare earths. Thus it was that de Hevesy and Coster started to examine zirconium-bearing minerals.[25,26] And indeed, that is where they found it.[27]

A phalanx of prestigious physicists was convinced of the nonexistence of *celtium*. Hence, they engaged in some activities that, by today's standards, would point to a lack of objectivity. For example, Niels Bohr, the director of the center where hafnium had been discovered, pressured the editor of *Nature* to let him know in advance what Urbain was publishing. The Anglo-Saxon press was divided: on the one hand, the journal *Chemistry and Industry* favored *celtium*; on the other, the editor of *Nature* was excessively partial to hafnium.[28,29,30,31,32,33,34]

On June 17, Lord Rutherford transmitted, in his own hand, an article to the journal *Nature* in which he praised the work of his French colleagues and reported some parts of their work in translation:[35]

> In two recent communications to the Paris Academy of Science by M. A. Dauvillier and Prof. G. Urbain respectively, very definitive conclusions have been reached as to the identity of celtium with the missing element of number 72.

At first, Rutherford was one of Urbain's earliest supporters, possibly because *celtium* was the fruit of Moseley's law and possibly because he did not fully accept Bohr's new ideas, even though Bohr had been one of his most outstanding students and they had an ongoing close personal and professional relationship.[36]

Bohr barraged Rutherford with letters accusing Urbain of incompetence,[37] and another prestigious physicist, Friedrich Adolf Paneth, strongly convinced that Urbain did not have any *celtium*,[38] tried to persuade many of his colleagues that Urbain's work was on the amateurish side. It was he who urged Swedish physicist Karl Manne Georg Siegbahn (1886–1978) to go to Paris expressly to investigate. Siegbahn went to Dauvillier's laboratory to look at the photographic plates and the characteristic lines of *celtium*. Upon his return to Stockholm, he issued a hardly edifying comment: "I did not see any celtium lines on the photographic plate that Dauvillier showed me. I think that they are probably only visible to Frenchmen."[39]

The letter, full of sarcasm, with terms couched in such a way as to discredit all of French science in general, arrived in Paris. Up until that moment, French resentment had been very hesitant and for the most part limited to the academic scene. Urbain always tried to be very fair and honest in claiming priority for his discoveries: "My efforts have led to imperfect separations, but by themselves they were sufficient to enable high-frequency spectra to assign atomic numbers unambiguously to the three components that I discovered: (neo) ytterbium 70, lutetium: 71; celtium: 72."[40]

After Siegbahn's sarcastic pronouncement, it was not long before inflammatory headlines directed at him appeared in the French press, which only served to fan the flames. In rapid succession, Bohr accused Urbain of exploiting the memory of the late Moseley to influence Rutherford and the Anglo-Saxon public. (As indicated earlier, Bohr was a friend of the editor of *Nature* and, in one of his articles, he did not forget to thank him for having let him see Urbain's galley sheets that would be published in that very same journal!) The attack on Urbain's scientific credibility produced many echoes. Simultaneously, Coster and de Hevesy accused him of plagiarism against von Welsbach.

However, the work of Urbain and Dauvillier was favorably received beyond French borders. In addition to Rutherford's opinion, already cited, one can add the appreciation of Bohuslav Brauner, Blas Cabrera (1878–1945), and others.[41] None of this was enough to save *celtium* from oblivion. The disappearance of Dauvillier[42] from the scene, as well as the de Broglie brothers, who turned their attention to other fields of physics, facilitated the acceptance of hafnium. In an essay by the Danish savant, Helge Kragh, one can find the solemnly pronounced sentence: "When in the summer of 1923, the controversies were de facto settled in favour of hafnium, the priority conflicts had worked as a fine propaganda for Bohr's ideas on atomic structure."[43]

IV.1.5. CELTIUM DOESN'T HAVE A LEG TO STAND ON[44]

There are many more ramifications to the demise of *celtium*, as the following details will show. In their first publication, in January 1923, de Hevesy and Coster proposed the name of hafnium for element 72. *Hafnia* is the Latinized name of Copenhagen, the Danish capital, where the Institute of Theoretical Physics was located and where all of their work took place. The director of the institute was the young and already famous Niels Bohr. He wanted the new element to have the name of *danium*, after Denmark, but one author asserts that de Hevesy and Coster categorically rejected this idea since neither of them was Danish.[45,46] Eventually, *danium* became the name of choice, but due to a printer's error, the name never made it into the first announcement of the discovery, caused some initial confusion, and eventually was dropped in favor of hafnium. Two years before hafnium was discovered, the Bonzenfrei group, shown in Figure IV.02, met in Dahlem, Berlin.

The publication of their results had hardly appeared in *Nature* when a violent conflict blazed up pitting Georges Urbain, the presumed discoverer of *celtium* on one side, and, on the other, de Hevesy and Coster, who had isolated it. Initially Rutherford and later Brauner lined up on Urbain's side; on the other side was Bohr, as well as the greater part of the German and Scandinavian scientific world. Unfortunately, *celtium* was not isolated in appreciable quantities nor was it possible to determine its chemical properties.[47] These were indispensable requirements to stand up to the expert Copenhagen researchers.

In the meantime, the two distinct research groups did their utmost to rapidly extract the element from its minerals. Urbain's research group was more numerous than the group working at Copenhagen, but he was making efforts to find *celtium* in yttric minerals, which had a very low concentration of the element. He could have swept aside all of hafnium's aspirations if only he had analyzed the zircon-bearing minerals that he had in his laboratory. These were samples from Madagascar; they were without doubt the minerals richest in *celtium*. But Urbain was obliged to publicly affirm the element's presence in the rare earth group. From them he had first observed it, and from them he wished to extract it. The presence of *celtium* in these minerals was very rare: the order of magnitude was about 1.0%, whereas in zircon, it could be found in quantities of as much as 17%.[48]

FIGURE IV.02. Das Bonzenfreie Kolloquium.(1920). Chemists and physicists gathered at Dahlem (Berlin) in 1920 at a colloquium organized for Niels Bohr by Lise Meitner ("bonzenfrei" literally means "without bigwigs"). Left to right: Otto Stern, Wilhelm Lenz, James Franck, Rudolf Walter Ladenburg, Paul Knipping, Niels Bohr, E. Wagner, Otto von Baeyer, Otto Hahn, George de Hevesy, Lise Meitner, Wilhelm Westphal, Hans Wilhelm Geiger, Gustav Ludwig Hertz, and Peter Pringsheim. George de Hevesy, a Hungarian refugee, worked with Niels Bohr at Copenhagen. Two years later, in January 1922, together with his colleague Dirk Coster, he discovered element number 72. For a very brief period, the name *danium* was considered for this element, but finally the name hafnium was decided on. Lise Meitner and Otto Hahn, in 1917, had discovered element number 91, protactinium. Among the more fanciful names proposed for this element were lisonium and lisottonium, with the aim of immortalizing their names. In more recent years, for the transuranium element number 105, the name *hahnium* was proposed, but later rejected. Hahn's colleague, Lise Meitner, although less fortunate in life, was amply honored in death: having fled from Nazi persecution, she missed recognition for the discovery of nuclear fission and the subsequent Nobel prize, but in 1997, the IUPAC Commission adopted the name meitnerium for element number 109. Courtesy Niels Bohr Archive, Copenhagen.

Thus, he lost precious time employed in a fruitless search, and he eventually paid bitterly for his stubbornness. Coster and de Hevesy, working in the manner alluded to earlier, succeeded in extracting hafnium very quickly and of surprisingly high purity. At this point, Urbain admitted, although very late, that he had believed *celtium* was a rare earth element, as he had written in 1911; he recognized the priority of Coster and de Hevesy in extracting the metal from zirconium-bearing minerals. However, he emphasized that the discovery was the result of his work in collaboration with Dauvillier. The date was May 1922. The concession he made was characteristically Urbain: mild and gentlemanly. It was instead interpreted as a sign of failure and weakness, and the scientific community reacted accordingly.

The French press had received the claim from Denmark by rallying around Urbain. In a dizzying spate of nationalism, French newspapers reported the discovery of hafnium and, at the same time, vilified the Danes. One of them carried the headline "The Stink of Kraut." The reference was obvious: De Hevesy had served in the Austrian army against the Entente, and, in 1923, France was still mourning its dead, exhausted after the long struggle of the war. Similar sentiments were common throughout the Western world. On the other side of the Channel, the editor of *Chemical News* and president of the Royal Society, W. P. Wynne (1861–1950) said: "We adhere to the original name celtium given to it by Urbain as a representative of the great French nation which was loyal to us throughout the war. We do not accept the name which was given by the Danes who only pocketed the spoil after the War."[49]

Urbain, however, never strayed from the professional integrity that distinguished him and never took profit from such incidents. He had dedicated his entire life to clarifying the dilemma of the rare earths: it must have seemed almost inevitable, although ironic, that the first element beyond the rare earths had escaped from his hands at the moment he had thought to reap the laurels of discovery.

The immense labor that led to the extraction of lutetium in 1906 was accomplished by more than 20,000 fractional crystallizations, in large part done by Urbain himself. If the technical efforts and the vigor of the preceding 10 years had been maintained, it is possible that Urbain would have obtained a demonstrable quantity of *celtium*. However, Dauvillier merely reported that one of Urbain's samples enriched in Lu and Yb gave X-ray spectra that contained two "extremely feeble" emission lines that could be assigned to element 72.

Coster and de Hevesy, having embarked on the correct path in searching for element 72 in the zirconium-bearing minerals, in 1926 were able to report the chemical, physical, and magnetic properties of hafnium. In his long treatise on element 72, however, de Hevesy reported the story of the discovery of hafnium in which he spared no criticism of his French colleague by mentioning only the episodes of Urbain's less precise work done in 1911.

In addition, he also raised serious doubts about Urbain's actual discoveries in 1906: *neoytterbium* and *lutecium*. Urbain never yielded an inch to these accusations and, for the rest of his life, he continued to ignore hafnium, although the international community, even before his death, had turned its back on him.

Meanwhile, Coster's ill will toward Urbain can be read in this citation: "In [1936] Coster. . . objected strongly to the proposal of Urbain as member of the Dutch Royal Academy and tried to make Bohr intervene against Urbain. . . 'who has shown a lack a reliability which is intolerable in a scientific man.'"[50] Later, Coster went to Austria to marshall another force, the renowned Auer von Welsbach, against Urbain. Auer had built a laboratory in his own castle and there he tried to extract element 61 from monazite rocks. Dirk Coster offered to carry some purified samples to Copenhagen and to study them spectroscopically. The condition he placed on von Welsbach was that he reject the 1909 decision of the International Commission on Atomic Weights in which Urbain's proposal to call elements 70 and 71 *neo-ytterbium* and *lutecium*. Auer agreed, and with him the entire German scientific community. They renamed element 71 with the old name of *cassiopeium* given it by Auer. In this, science gave way to a more sinister nationalism. For element number 70, it was decided to keep the name of ytterbium and not to change it into *aldebaranium*. The decision was motivated not so much out of respect for Urbain's discovery but in honor of the 1878 work of Marignac, who was a Swiss national, not French. Urbain found himself having to defend his own scientific credibility from the joint attacks of Auer von Welsbach, de Hevesy, and Coster.

Unexpected help came to Urbain from the 68-year-old Czech chemist Bohuslav Brauner. One of the reasons that led Brauner to enter the field spontaneously on Urbain's side went back a long way. The *didymium* discovered by Mosander had been spectroscopically examined by Brauner in 1882, and he observed two absorption bands, one in the blue and the other in the yellow region. At the time, he did not think seriously that he had observed the bands of two distinct elements making up *didymium*. It was only in 1885 that Auer succeeded in splitting *didymium* into its constituent parts, for which he proposed the names *neo-didymium* and praeseodymium.

Although Brauner had never (and would not during his lifetime) officially claim the discovery of these elements, he considered himself the true discoverer and felt that Auer von Welsbach had plagiarized his work. In his old age, he confided to Urbain his sentiments with respect to von Welsbach.[51]

The way in which Coster and de Hevesy involved Auer von Welsbach against Urbain reminded Brauner of his own conflict. Brauner informed Urbain of the risk to which the survival of *lutecium* was exposed. Georges Urbain returned Brauner's courtesy and found objective, and at the same time unwavering, words as much to thank him as to condemn the attempt of Auer, de Coster, and Hevesy to discredit him internationally: "It is well known that Lutecium failed to repeat the history of praeseodymium's, and you tremble more than me for the fate of celtium. I hope that in the future this argument will consider the balance in your favor, and that justice is finally and universally done to your credit."[52]

Urbain showed himself to be extraordinarily cautious and polite. He thanked Brauner, but he did not take advantage of his help to fuel the international controversy. He limited himself, and legitimately so, to wish that someday, like Auer, he could be pointed out as one who had discovered two elements.

After an endless dispute, carried on almost entirely within the pages of *Nature* and *Chemistry and Industry*, the International Commission on Weights and Measures came to a painful and disappointing conclusion: element 72 would have two names.[53] It would be correct to call it either hafnium or *celtium*. Both symbols, Hf and Ct, would be correct. This decision did not satisfy either party. In the space of a few years, and even before the International Commission erased the name of *celtium* from the list of elements, it had already been replaced.

This could seem paradoxical, but some attribute the final decision on the part of the International Commission to the waning power of the French Chemical Society, which was undergoing a period of decline within the international community.[54] However, others might argue that the problem lay with the fact that the chemical properties of celtium described by Urbain did not agree with theoretical considerations of atomic structure.[55] And still others, as we have seen, did not think that Urbain had any *celtium* at all.

Professionally, Urbain stood out from his colleagues then as now. His vast knowledge ranged from chemistry, to art, and to music. For more than 30 years, his lectures at the Sorbonne were memorable. Students competed to attend them, eager to hear the warm and persuasive voice of their teacher; those who wished to take part in his lectures (not all of them chemists!) ran between the arches and down the corridors of the University to reach the amphitheater and take their places in one of the 350 seats still free. With nostalgia, Urbain's last student, Georges Champetier (1905–80), destined later to succeed to his chair, recalled how the course was very difficult and demanding. Nevertheless, students were proud to be Urbain's disciples. When he entered the lecture hall, students would instinctively rise to their feet, not out of fear of their professor, with his seraphic face and gentle manner, but out of a desire to transmit their own affection and respect to this exceptional person: "The attention of the students bordered on perfection, but it was seldom that perfect silence was maintained for a full hour. It often happened that at least once in every hour Monsieur Urbain did so beautifully demonstrate some important point or let drop some so irresistibly cheerful, but always appropriate, remark that a burst of applause arose spontaneously from the students."[56]

In addition to his renowned career as a chemist, Urbain was also an amateur painter and sculptor: the bust of his friend Jean Perrin (1870–1942) at the Sorbonne is his own work. The same is true of the bust of his teacher Charles Friedel, which is kept in the secretariat of the Faculty of Science at Paris.

Urbain was also concerned with theoretical chemistry; he formulated the law of homeomerism and perfected the law of optimums in cathodic phosphorescence of binary systems. He made notable contributions to the concept of isomorphism and to the extension of the coordination theory of Alfred Werner (1866–1919). His most important theoretical work was almost certainly *Les notions fondamentales d'éléments chimiques et d'atome* (Gauthier-Villars: Paris, 1925) in which is found the most elegant, and perhaps also the most complete, definition, at least from the epistemological point of view, of the element.

Urbain was also an accomplished musician. As such, in 1921, he set to music his first composition: *A la veillée* and two melodies on the poetry of Paul Verlaine (1844–96): *Chanson d'automne* and *Sur l'herbe*. In 1922, *Magagnose et Dyonisos*, an opera in six brilliant variations, followed.

Georges Urbain was nominated for the Nobel Prize in Chemistry for the first time in 1912. In that year, winners in that discipline were his countrymen Victor Grignard (1871–1935) and Paul Sabatier (1854–1941). He received the candidacy 20 times: he was the natural candidate and flag bearer for a whole school of chemists who saw in him the person who had put the rare earths in order.

In 1925, Urbain was next in line to receive the coveted award. He had good chances of success also in 1927, receiving a number of votes equal to de Hevesy's. The same thing happened again in 1928 and 1933. (In 1933, the Nobel Prize in Chemistry was not awarded.) Urbain was nominated by many Nobel Laureates: Jules Bordet (1870–1961), Jean Perrin, Jaroslav Heyrovský (1890–1967), Charles-Edouard Guillaume (1861–1938), Hans Karl Euler von Chelpin (1873–1964), Frédéric Joliot, and Irène Joliot-Curie. Victor Grignard nominated his friend Urbain almost every year up until 1934. Urbain had his last nomination in 1936. The following year, Grignard suffered a sudden illness that, in 6 short weeks, caused his premature death.

In poor health, in the late spring of 1938, Urbain underwent surgery and seemed likely to fully recover. He spent the summer in Provence, where he lived the few months that remained to him immersed in a peaceful and familiar atmosphere before his unexpected end.

A lesser known aspect of this versatile chemist regards his ideas on international politics; in fact, he showed his hostility to the Nazi party from the beginning of Adolf Hitler's rise to power. In 1933, he sent a telegram to the Führer protesting the burning of the Reichstag, and he bluntly pointed to this incident as an intentional act of aggression on the part of the Nazis. For this reason, in 1937, he forbade his son, Pierre Urbain (1895–1968), to attend a conference in Germany, fearing Nazi retaliation.

Surrounded by family members, colleagues, and friends, to their dismay and astonishment and without any evident sign, Urbain unexpectedly expired on November 5, 1938, of an aggressive bladder infection.

Table IV.1 Nobel Nominations for Urbain (U), Auer von Welsbach (A), de Hevesy (H), and Coster (C)

Year of Nomination	*Year of Nomination*	*Year of Nomination*	*Year of Nomination*
1912—U	1919	1926—U, A, C	1933—U[a], H, C
1913	1920—U	1927—U, H, C	1934—U, H
1914—U	1921—U, A	1928—U[a], H	1935—H
1915	1922—U, A	1929—U, A, H, C	1936—U, H
1916—U	1923—U, A	1930—U	1937—H[b]
1917	1924—U, A, H, C	1931—U	
1918—U, A	1925—U[a], A[a]	1932—U	

[a]Came very close to winning.

[b]De Hevesy won the Nobel Prize in 1943.

Urbain's last Nobel nomination destined for failure took place in 1939. The event was both curious and dramatic in that the nomination was presented posthumously. Because of the Indian subcontinent's isolation from a good part of the academic world, Dahr Nil Ratan (1892–1987) of Allahabad, ignorant of Urbain's death, sent a letter to the Nobel Committee in which he proposed the French chemist for the prize for his excellent work in the field of the rare earths and for his contributions to theoretical chemistry, as well as for the fact that he represented, as the virtual dean of French chemists, the best of science in that country. In his nomination, Ratan wrote not merely of his scientific stature as a man of science, but described Urbain as a cultured, refined, and generous man. In fact, few persons during their lifetimes would be so appreciated and admired on the one hand and so slandered on the other. With Urbain's passing, not only did a scientist expire, but also a painter, a sculptor, a historian, an encyclopedist, and a man gifted with a rare mildness of manner.

Table IV.1 lists the Nobel nominations for the major players in the drama documented in this section. Of the four contenders, only de Hevesy's continued nomination eventually met with success.

Urbain was the last of the great classical chemists. Among the consequences that World War II, which loomed on the horizon, would bring were new discoveries. The creation of the first artificial elements was among them. The work of chemists like Urbain, who had searched out the last elements in nature, suddenly seemed anachronistic. In closing the eyes of Urbain, destiny would also close an era.

Notes

1. Urbain, G. *Compt. Rend. Chim.* **1907**, *145*, 759.
2. Auer von Welsbach, C. *Kaiserisch Akademie Wissenschaft Wiener* (Math. Natur. Klasse), **1907**, 468.
3. Gumtann, V. *J. Chem. Educ.* **1970**, *47*, 209.
4. Urbain, G. *Chem. Ztg.* **1908**, *32*, 730.
5. Auer von Welsbach, C. *Monatshefte für Chemie und Verwandte Teile Anderer Wissenschaften* **1908**, *29*, 181.
6. Urbain, G. *Z. anorg. Chem.* **1910**, *68*, 236.

7. Auer von Welsbach, C. *Monatshefte für Chemie und Verwandte Teile Anderer Wissenschaften* **1911**, *32*, 373.
8. Auer von Welsbach, C. *Monatshefte für Chemie und Verwandte Teile Anderer Wissenschaften* **1913**, *34*, 1713.
9. Urbain, G.; Bourion, F.; Maillard, J. *Compt. Rend. Chim.* **1909**, *149*, 127.
10. Auer von Welsbach, C. *Kaiserisch Akademie Wissenschaft Wiener* **1907**, 468.
11. Clark, F. W.; Ostwald, W.; Thorpe, T. E.; Urbain, G. *J. Am. Chem. Soc.* **1909**, *31*, 1–6.
12. Auer von Welsbach, C. *Monatshefte für Chemie und Verwandte Teile Anderer Wissenschaften* **1911**, *32*, 373.
13. Job, P. *Bull. Soc. Chim. France* **1939** 5e Ser. 6, 49.
14. Champetier, G.; Boatner, C. H. *J. Chem. Educ.* **1940**, *17*, 103.
15. Mémoires et Communication, *Compt. Rend. Chim.* **1938**, *207*, 66.
16. Urbain, G. *Compt. Rend. Chim.* **1908**, *146*, 406.
17. Urbain, G. *Chem. Rev.* **1924-1925**, *1*, 143.
18. Urbain, G. *Compt. Rend. Chim.* **1907**, *145*, 759.
19. Urbain, G. *Compt. Rend. Chim.* **1911**, *152*, 141.
20. Heilbron, J. L. *H. G. J. Moseley—The Life and Letters of an English Physicist—1887–1915*; University of California Press: Berkeley, 1974.
21. Mel'nikov, V. P. *Centaurus* **1982**, *26*, 317.
22. Dauvillier, A. *Compt. Rend. Chim.* **1922**, *174*, 1347.
23. Dauvillier, A. *Chem. Ind. (London, U.K.)* **1923**, *42*, 1182.
24. Coster, D.; de Hevesy, G. *Nature* **1923**, *111*, 79.
25. de Hevesy, G. *Ber. Dtsch. Chem. Ges.* **1923**, *56*, 1503.
26. Coster, D. *Phil. Mag.* **1923**, *46*, 956.
27. There is a popular account for this discovery that supposedly credits the quantum theoretical approach of Niels Bohr with correctly predicting that element 72 would be found among the transition elements rather than, as commonly thought, among the rare earths. However, the Danish chemist, Julius Thomsen, had predicted the transition metal nature of element 72 as early as 1895 and the English chemist, C. R. Bury, actually published its correct electronic configuration in 1921, preceding Bohr's publication by 2 years. For a detailed correction of this popular misconception, see Scerri, E. *Ann. Sci.* **1994**, *51*, 137.
28. Coster, D.; de Hevesy, G. *Nature* **1923**, *111*, 182.
29. Urbain, G.; Dauvillier, A. *Nature* **1923**, *111*, 218.
30. Coster, D.; de Hevesy, G. *Nature* **1923**, *111*, 252.
31. Coster, D.; de Hevesy, G. *Nature* **1923**, *111*, 462.
32. Urbain, G. *Chem. Ind. (London, U.K.)* **1923**, *42*, 764.
33. Brauner, B. Hafnium or Celtium. *Chem. Ind. (London, U.K.)* **1923**, *42*, 884.
34. Editorial, *Chem. Ind. (London, U.K.)* **1923**, *42*, 782.
35. Rutherford, E. *Nature* **1922**, *109*, 781.
36. Kragh, H. *Centaurus* **1980**, *23*, no. 4, 275.
37. Letter from Bohr to Rutherford, February 9, 1923 (Bohr's Scientific Correspondence).
38. Paneth, F. A. *Ergebnisse der Exakten Naturwissenschaften* **1923**, *2*, 160.
39. Heilbron, J. L. *H. G. J. Moseley: The Life and Letters of an English Physicist*; University of California Press: Berkeley, 1974.
40. Urbain, G. *Compt. Rend. Chim.* **1922**, *174*, 1349.
41. Brauner, B. *Chem. Ind. (London, U.K.)* **1923**, *42*, 884.
42. Dauvillier was born May 5, 1892, at Saint-Lubin-des-Joncherets. He had been a student of Urbain before the war. When he returned from the front, he began to study radiology. At the end of the war, he received his degree in physics and, in 1920, enrolled at the Ecole des Haute-Etudes under the mentorship of Maurice de Broglie (1875–1960), specializing in X-ray

diffractometry. He also made many valuable contributions in the field of television, especially in cathode-ray tube technology. He died in 1979.

43. Kragh, H. *Centaurus* **1980**, *23*(4), 275.
44. An extension of a later remark made by Ernest Rutherford in a letter to Niels Bohr (Letter of February 15, 1923, Bohr's Scientific Correspondence) and cited in R. Peierls, "Rutherford Memorial Lecture," November 10, 1987; reproduced in *Current Science* **1997**, *73*, 707–712: "I quite agree. . . Urbain has not a leg to stand on."
45. Heilbron, J. L. *H. G. J. Moseley: The Life and Letters of an English Physicist*; University of California Press: Berkeley, 1974.
46. Kragh, H.; Robertson, P. *J. Chem. Educ.* **1979**, *56*, 456.
47. Urbain, G. *Compt. Rend. Chim.* **1923**, *176*, 469.
48. de Hevesy, G.; Madsen, E. *Angew. Chem.* **1925**, *38*, 228.
49. Kragh, H. *Centaurus* **1980**, *23*(4), 275. This remark is referring to the fact that, after Germany had lost World War I, in which Denmark had remained neutral, the victors redrew the border between Denmark and Germany, returning northern Schleswig, that was originally Danish territory, to Denmark, whereas Alsace and Lorraine only officially returned to French rule at the price of France's participation in 4 years of war and millions of casualties.
50. Kragh, H. *Centaurus* **1980**, *23*(4), 292.
51. Weeks, M. E. *Discovery of the Elements;* Journal of Chemical Education: Easton, PA, 1968, p. 689.
52. Urbain, G. *Recueil des Travaux chimiques des Pays-Bas* **1925**, *44*, 281.
53. Aston, F. W.; Baxter, G. P.; Brauner, B.; Debierne, A.; Leduc, A.; Richards, T. W.; Soddy, F.; Urbain, G. Second Report of the International Committee of the International Union of Pure and Applied Chemistry on Chemical Elements, International Atomic Weights. *J. Chem. Soc.* **1925**, *127*, 913.
54. Claro Gomes, J. M. *Georges Urbain (1872–1938) Chimie et Philosophie*, Thesis to obtain the doctoral degree, University of Paris X Nanterre, 2003.
55. King, H. S. *Nature* **1923**, *112*(2801), 9.
56. Champetier, G.; Boatner, C. H. *J. Chem. Educ.* **1940**, *17*, 103–109.

IV.2

FROM THE PRESUMED INERT ELEMENTS TO THOSE LOST IN THE DEAD SEA

The following narrative is inspired by two alternative atomic theories developed simultaneously in England by Frederick Henry Loring[57] and in South Africa by James R. Moir (1874–1929). Surprisingly similar, these two events combine science, amateurism, and pseudoscience to produce some fascinating literary history.

IV.2.1. THE ATOMIC THEORY OF JAMES MOIR AND THE SUBELEMENTS X AND ZOÏKON

In 1909, South African James Moir was a young chemist with more than 10 publications to his credit. He was interested in deepening his knowledge of organic chemistry and working on a solution to some of his country's practical problems, as for example, the ventilation of mines. However, in the 2-year period of 1909–10, Moir earned public attention with three distinct works: the first was a suggestion for a new atomic theory,[58] and the other two were in regard to a method for "harmonizing" the atomic weights of the chemical elements.[59]

In suggesting a new and improbable atomic structure for the already known elements, Moir had recourse to the experimental data that both chemists and physicists had collected in the course of their more recent research. The atoms, as Moir understood them, consisted of arrangements or dispositions of four or five constituent principals that he called *primary materials*. The composite or secondary atom, like that of carbon for example, would be composed of four identical subatoms with atomic weights of 3, arranged in space in a tetrahedral structure. James Moir called the element with an atomic weight of 3 *zoïkon* and indicated it with the symbol Z. This element was one of his primary materials. The second subatom was hydrogen, to which Moir gave an intrinsic repulsive force. The third constituent of the ordinary atoms was a hypothetical element to which he did not give a name but only a symbol—*X*—and an atomic weight of 2. Like hydrogen, this would be monovalent, although unlike hydrogen it would not be completely capable of saturating another element by combining with it. The last two elements that completed the list of primary bodies were recently discovered by Sir William Ramsay: helium (atomic weight = 4.09) and neon (atomic weight = 19.7).

James Moir did not concentrate his interests in characterizing *zoïkon* or element *X*, but sought to explain the composition of the already known elements on the basis of his heterodox theory. For example, the metals would contain inside them some hydrogen by virtue of their electropositivity, whereas the halogens would contain element *X*, the possessor of electronegativity.

Moir furthered his explanation of how the other elements were constituted, beginning with the five primary substances, and he also illustrated the formulas of many of these.

For example, nitrogen would have the nuclear formula z_4X; the disposition of the subelement X in the tetrahedral structure of the nucleus (a species of nuclear isomerism *ante litteram*) would have allowed in this case a bivalent character; in the other, a tetravalent character. Proceeding with this idea, Moir illustrated the composition of many other elements: oxygen z_4X_2; lithium H_2Xz, and the like.

Moir continued to develop his ideas in his laboratory at the Department of Mineralogy at Johannesburg in South Africa. Within a month's time (December 1909 to the beginning of the New Year), he published two works in which he aired his unconventional ideas on valence. In his opinion, the typical valence of every element was caused by the presence of a subelement of an atomic weight that was 1/112 that of hydrogen, which he designated with the Greek letter μ. Therefore the monovalent elements would contain 1 μ, the bivalent 2 μ, and so forth. He also developed a paradoxical hypothesis to explain the atomic weights of the chemical elements: to arrive at their atomic weights, the major part of the atomic mass was due to the product of polymerization of an entity consisting of atoms of H minus a particle μ. For example, hydrogen would be the result of the nuclear reaction: $H = H- + \mu$; by the same token, silver would be $Ag = 108H- + \mu$. With the symbol "H–," Moir indicated the monomer from which all of the elements took their origin. Moir quickly realized that his system had a flaw: the atomic weights of some of the known elements did not fall within his system of "nuclear polymerization," and without any hesitation he created a new subelement with an atomic weight of 1/10 that of hydrogen to salvage his hypothesis. Both the hypothesis of the elements *zoïkon* and X, as well as that of the subelements and the tetrahedral aggregation of the atomic nuclei, were completely ignored, although in 1921 Moir reproposed these latest hypotheses in the light of the recent discoveries of Sir Ernest Rutherford.[60]

In subsequent years, James Moir returned to his interests in organic chemistry and in particular the constitution of natural pigments.[61] Moir died in 1929, but, as his publications testify, he was active right up until the end.[62] At the international level, his ideas went unnoticed, and he never received any kind of recognition, whereas in South Africa he was well-known as a pioneer in the chemical sciences and was twice elected to the post of president of the South African Association for the Advancement of Science. When James Moir died, a subscription was opened to institute a foundation that would carry his name, and when, a year later, on May 31, the subscription was closed, it had collected 800 pounds sterling. With such a sum, the South African Chemical Society established the "James Moir Medal" that is awarded to this day to those university students who have completed their course of study with distinction.

IV.2.2. THE HARMONIZATION OF THE ELEMENTS AND THE INERT ELEMENTS

The time was ripe for science to produce some explanations to certain phenomena if progress was to be made. And, in two opposite places on the globe, almost simultaneously and unknown to each other, two scientists responded to the appeal. In the southern hemisphere, James Moir expounded his concept of the harmonization of the atomic weights of the elements, whereas Frederick Henry Loring, in England, elaborated on a theory of the mathematical harmonization of the elements.

In 1909, an English chemist with many diverse interests, Frederick Henry Loring became aware of some regularities in properties of the known elements; he set about classifying them as a function of periodic recurrences, like specific heat and some chemical properties.[63] The discovery of the noble gases had in some way rendered it necessary to modify the periodic table postulated by Mendeleev, and Loring felt the need to put order into the classification of the elements. Loring did not deny the undoubted utility of Mendeleev's creature, but he was of the opinion that only by means of a mathematical equation could one arrive at the elegance of form and that power of concept that still were missing from chemistry. His empirical method was based on two simple operations. First, Loring proposed to order the elements—or better, their atomic weights—in a numerical series according to this equation:

$$W = \pm(4P) + K \qquad \text{(Eq. IV.1)}$$

where W was the atomic weight; P a number in the series 0, 1, 2, 3,. . .; and K an arbitrary constant whose value would be between 0 and 4.

On December 10, 1909, from his house in Doughty Street, London, Loring sent his third and last manuscript to press.[64] In it, he changed the form of his empirical equation: $W = 3.1 \pm 4n$. Moreover, he had substituted the experimental value 3.1 for the constant K. If the necessity of correcting and broadening the bases on which the concept of the periodic table rested was a real problem that many chemists faced at the beginning of the 20th century, Loring, like his other colleagues, erred in his approach: he had worked on the more insidious concept of atomic weight instead of concentrating on the more significant concept of atomic number, of which the potential had not yet been clarified.

The disposition of the atomic weights proposed by Loring had, in his opinion, two apparent incongruities: they created some gaps among some elements, and they did not take into account the existence of nitrogen or *glucinium* (beryllium). Loring did not pay much attention to these unexplainable consequences but rather utilized them as a basis for his theory. He proposed that these two presumed elements, N and Be, were actually a combination of a lighter nucleus with a gaseous element not yet discovered that he called *satellite* and to which he gave the symbol St. *Satellite* would have an atomic weight equal to 0.2684. In fact, subtracting this value from the atomic weights of nitrogen and *glucinium*, he found that the resulting values were in agreement with those predicted by his equation. If, in the first two publications, Loring had expressed a certain caution for the innovative hypotheses he presented, in his last one[65] he completely abandoned any semblance of prudence and hypothesized the existence of three new inactive gases.

Returning to the composition of nitrogen, whose atomic weight in 1909 was established as 14.007, Loring expressed the conviction that nitrogen would be the combination product of the three inert gaseous elements: "one of the most striking pieces of evidence is that the three component elements of nitrogen [are] *satellite, nitron* (Nt), the hypothetical inactive element,. . . and helium."

To make the sum of the atomic weights of these three gases match that of nitrogen, he fixed the value of *nitron* at 9.75. In fact, $0.27 + 3.98 + 9.75 = 14.00$. The author, aware that he was forcing the results into place by admitting that he knew, a priori, the value that he was trying to calculate, sought to stave off critics by introducing a note in which he reported on his own testimony a citation torn out of context by a true authority in matter: "the value for helium '3.98' was given to me by Sir William Ramsay."

Loring was criticized for not being able to isolate either *nitron* or *satellite*, but he defended himself by asserting that his work was merely theoretical, having collected the experimental data from renowned chemists of the time. Loring possessed an approach that was professionally objectionable: in fact, he did not spurn the use of data of dubious authenticity if they were in agreement with his theory. In his works, nonexistent elements like *nipponium* and *decipium* appeared, but they did not contradict his equation. However, Loring was a very intelligent person, and he knew how to distinguish good from mediocre work. The experimental data of Georges Urbain[66] were treated with much respect: the atomic weights of the lanthanides were the most accurate known, whereas the study of the magnetic properties of the rare earths was a field in which Urbain had no rivals. However, even exploiting data of the first order, the theory proposed by Loring was quickly shown to be false: it rested on a gross artifact, and he did not succeed in explaining the nature of the new matter—*satellite*—that he said was composed of many elements, among which were beryllium, nitrogen, many rare earths, and tellurium.

Always referring to his equation, which according to Loring "would have harmonized the atomic weights," he predicted the existence of another two inert gaseous elements whose atomic weights were fixed at 216 and 251, respectively. Loring expatiated both on *satellite* and on *nitron*, but to these two other elements he gave little notice, indicating them simply with the letters Z^1 and Z^2.

Although Loring's hypotheses were clearly inadmissible, he nevertheless made a contribution that turned out to be right: he discovered that atomic weights were exact mathematical functions. And, indeed, according to quantum mechanics, the mathematical basis for understanding and schematizing the atomic edifice, the atomic number supersedes the atomic weight: a classification and a mathematical theory that, unfortunately, were beyond Loring's knowledge and ability.

Loring's theories were faithfully reported in the pages of Crookes's *Chemical News* but because that journal had a broad readership on the Continent, his ideas encountered criticism from experts in the field. Other persons in Loring's position would have reflected on their own past errors, but not Loring. After a hiatus of more than two decades and with his full acceptance of the new nuclear model, Loring returned to the limelight with a new discovery. But before we discuss this discovery, it is necessary to introduce another and much younger English chemist: the name of John Gerald Druce became associated with Loring's later enterprise.

IV.2.3. FROM ENGLAND TO PRAGUE ON THE TRAIL OF ELEMENT NUMBER 75

If Druce and Loring had not known one another, Druce's position as director of *Chemical News* and the interests that they shared would certainly have ended up making them close friends. John Gerald F. Druce was born at Leamington Spa in 1894; he was educated at University College London, where he concluded his studies in 1921. In 1923, he obtained his doctorate at Charles University, Prague, in Czechoslovakia, a country to which he always felt close. He took up the post of director of *Chemical News* on the death of James H. Gardiner in 1924 and held this position for 6 years. During this period, Druce was also employed as chemistry master at the Grammar School of Battersea. The journal that he directed had lost much of its original prestige from the time when Crookes had been its director; then, in 1930, Druce passed the post on to a new editor, H. C. Blood Ryan, who, 2 years into this "disastrous appointment" managed to bankrupt the company.[67]

During Druce's editorial direction, Loring, first as sole author and later in partnership with Druce, published numerous monographs on the search for the missing elements.

In inadequate laboratories housed in the basement of the St. John Hill School, Druce started his hunt for the three missing elements with atomic numbers of 43, 75, and 87. In 1925, while the husband-wife team of Ida Tacke Noddack and Walter Noddack, together with Otto Berg, were isolating rhenium,[68] Druce prepared potassium perrhenate, starting with a sample of pyrolusite (manganese sulfide). His work appeared simultaneously with that of the German couple[69] so that, for a certain period, the English press encouraged by Druce, claimed for him credit for the discovery of element number 75. Later that same year, the Czech chemist and newly appointed university professor Jaroslav Heyrovský, using a polarographic technique that he had discovered and developed, dedicated himself passionately to the search for the same element, eka-manganese. Together with Vaclav Dolejšek (1895–1945), an expert spectroscopist, Heyrovský was aware that some samples of "crude" ores gave results that could be interpreted as admitting to the presence of an element analogous to Mn that was reduced potentiometrically together with manganese. At the end of their chemical and spectral analyses, Heyrovský and Dolejšek concluded that "this chemical behavior coincides with that mentioned by Dr. Gerald Druce."[70] Druce had probably known Heyrovský in London, where the latter had come to study under the guidance of Sir William Ramsay, or he had met him during one of his numerous stays in Bohemia, a land beloved by the English chemist as a second homeland. The two became friends and maintained a close correspondence over a period of time in a relationship consolidated by reciprocal esteem and loyalty.[71] In a private communication, Druce even proposed the name *pragium*,[72] after the city of Prague, for element number 75. This was most likely Druce's attempt to include Heyrovský in his discovery and form a common front against the claims of the Noddacks and Otto Berg. Although Heyrovský considered the spectroscopic work of Noddack, Tacke Noddack, and Berg inconclusive, he rejected Druce's proposal, not wishing to openly challenge the discovery of his three German colleagues. Nobel laureate Jaroslav Heyrovský is shown in Figure IV.03.

Thus, at the end of 1925, the question of who might be the real discoverer of element 75—the Germans, Heyrovský and Dolejšek, or Druce and Loring with their dvi-manganese remained fully open. A. N. Campbell made the objection that the polarographic technique was not sufficient and made note that the maximum potential observed at –1.00 V by Heyrovský did not correspond to that of dvi-manganese, but rather to that of hydrogen.[73] A year later, Zvjagintsev, Korsunski, and Seljakov[74] openly supported the work of the Czech chemists who, in their opinion, "seem to have chosen a more trustworthy way, assuming that the dvi-manganese is associated with manganese and not with platinum." Their intent was in fact to discredit the work of Noddack, Tacke Noddack, and Berg, who had announced that they had discovered element 75 by analyzing platinum-bearing rocks.

The role of arbiter in this controversy was assumed by Wilhelm Prandtl who, in a long and detailed work,[75] critically analyzed all the research conducted in the preceding years and ending with the discovery of the higher homologs of manganese (technetium and rhenium). His conclusions were not positive for any of the three teams involved in the controversy, but the heaviest verdict fell on his fellow Germans, whom he found guilty of falsifying the discovery of eka-manganese (element 43, technetium). However, Prandtl's severe criticisms were quickly overcome: a short time later, isolating macroscopic

FIGURE IV.03. Jaroslav Heyrovský (1890–1967), Czech Chemist, Inventor of the Polarograph and Nobel Laureate in Chemistry, 1959. By means of the polarographic technique, Heyrovský, together with his colleague Vaclav Dolejšek (1895–1945), presumably identified element number 75 and called it *pragium*. With the kind permission of the Heyrovský family.

quantities of element 75, the Noddacks were awarded its discovery, as well as the right to name it rhenium.

The judgment of the renowned German inorganic chemist on the work of the English chemists, on the other hand, left no room for appeal: "Obviously the presence of tungsten and lead simulated the presence of element 75 for Druce and Loring." At the conclusion of his article, Prandtl demolished Heyrovský and Dolejšek's claim by asserting that "even the purest of platinum exhibits traces of tungstic acid, zinc and cobalt, but no trace of eka-manganese."

Heyrovský accepted Prandtl's verdict and, after a long series of polarographic investigations, was able to affirm that there was no trace of element 75 in the samples of manganese coming from the Czech deposits, thus removing the obstacles to the Germans taking full credit for the discovery of rhenium.

IV.2.4. ON THE BANKS OF THE DEAD SEA: THE FIRST INVESTIGATIONS FOR THE IDENTIFICATION OF ELEMENT 87

On May 1, 1926, John Newton Friend (1881–1966) put pen to paper and decided to make public[76] the results of some of his odd researches that were started soon after the end of World War I. Newton Friend was convinced that the Dead Sea basin, because of its peculiar geoclimactic conditions, was the only place in the world where it would be possible to find the heaviest and rarest of the alkali metals, the element with atomic number 87.

Because the Dead Sea lacks an outlet for its waters, the accumulation of minerals in the water is very high and is further increased by the high rate of summer evaporation in this torrid region; the density of the water in the basin is around 1.25 g/cm^3. Under these conditions, thought Newton Friend, element 87 ought to be concentrated in the waters of the Dead Sea as in no other place on earth; therefore, even if it were present in microscopic quantities, the continuous action of accumulation would render it measurable. In June 1925, Newton Friend set out for Palestine, not as a pilgrim, but in the outfit of a chemist, burdened with numerous glass collection containers and a big bag of analytical instruments.

Maybe it was the mysticism of the place, maybe it was the idea of being able to resolve the enigma of element 87 with such simplicity and elegance that inspired in Newton Friend an indomitable perseverance; he traveled up and down the Dead Sea coast taking samples of the water. He rightly believed that *eka-cesium* had the properties of the alkali metals and, trusting that this would be the case, he undertook appropriate tests. After having removed all of the elements except the alkali metals, he performed gravimetric analysis. The results were not encouraging. So, Newton Friend sent a fraction suspected of containing traces of element 87 to the Hilger Adam Ltd. Company so that the experts who worked there could subject the salts to an accurate X-ray analysis.

Some photographic plates showed traces of a mysterious line that could coincide with the Lα line calculated for *eka-cesium*, but this remained the only proof of the presence of this element in the waters of the Dead Sea. Newton Friend eventually realized that the Dead Sea did not hold the element that he tenaciously sought, but he went on to hypothesize that the element with atomic number 87, being found in the periodic table between radium and radon, could be radioactive and have a very short half-life. It was a fortuitously correct guess at the conclusion of research that was in large part erroneous.

To his contemporaries, it seemed that Newton Friend was more disappointed in the failure of his elegant argument—which would have ensured not only fame and international recognition, but would also have allowed him to demonstrate that human ingenuity could overcome the lack of experimental equipment—than in his failure to discover element 87. However, in later years, this distinguished British chemist committed his vast knowledge to the preparation of a monumental, accurate, and elegant volume on the discovery of the chemical elements and the different uses that society had made of them over a period that exceeded 40 centuries.[77]

IV.2.5. ALKALINIUM

After Newton Friend's bizarre chemical expedition to the Holy Land failed, the search for element 87 passed into the hands first of Loring and later Druce. Frederick H. Loring was born in England; and although his date of birth is not known, it was some time in the last quarter of the 19th century, making him Druce's senior by about 20 years. Loring's broad and varied scientific career began with his first publications appearing in 1906 and, after several more or less productive periods, concluded in 1945.

After an initial pause marked by his attempt to propagate his theory of the harmonization of atomic weights, Loring changed his scientific interests with great frequency: from speculations on the theory of "associations"[78] through explaining the structure of the atom, and finally to taking up the ideas of Sir William Crookes on cyclic inorganic evolution.[79]

Soon after the end of World War I, Loring began, initially with marginal interest and then with increasing passion and tenacity, searching for the last chemical elements.[80] Simultaneously, he launched himself into an enterprise of secondary importance that was destined to have an unhappy outcome: to interpret and comment on the recent discoveries in atomic physics, on quantum theory and the structure of matter,[81] on the concept of isotopes,[82] and finally, on the hypothesis of an element with atomic number zero.[83]

On January 30, 1926, Loring published a brief article in *Nature* in which he clearly stated his view of the state of the search for the missing elements.[84] This article served as a sort of manifesto and, at the same time, as a turning point in Loring's scientific career: in it, one can recognize his determination to concentrate his efforts on the identification of only two of the missing elements—numbers 75 and 87—availing himself of the chemical work of Gerald Druce and of the X-ray information furnished by the Adam Hilger Ltd. Company. In fact, after some work initially done along parallel paths, Druce and Loring united forces and, between the end of 1925 and the beginning of 1926, numerous works appeared that bore both their names.

On November 6, 1925, Loring and Druce published in *Chemical News* (Druce at that time was director of the journal) a succinct article of scarcely a page relative to the identification of the 87th element.[85] Although the two chemists, with the help of measurements made by technicians at the omnipresent Adam Hilger Ltd. Company, were able to affirm that they had recorded a line at 1032 Å assumed to be a poorly resolved doublet of the $L_{\alpha1}$ and $L_{\alpha2}$ emissions of element 87, many uncertainties still remained. The two authors had not succeeded in obtaining samples of the element they were looking for that were free of traces of silver bromide, whose lines were interfering heavily, including in the region of the spectrum where the secondary lines of *eka-cesium* should have appeared. The position of the two British scientists was not easily sustainable: they were not holders of prestigious university chairs; on the contrary, they were little more than amateurs. Chemists or physicists who occupied academic positions much more solid than theirs, in much more prestigious schools, and who could put forth their ideas to the academic establishment with much more weight and self-confidence would be motivated to publicize some of their ideas on a particular phenomenon only in the rarest of cases.

Druce and Loring were so certain of their discovery that, even lacking irrefutable experimental evidence, prematurely and with inappropriate arrogance, asserted: "We have, for the present, designated the element eka-cesium in accordance with the nomenclature adopted by Mendeléeff." They concluded their article with the assertion that "Further work is being done to obtain this element, as free as possible, from other elements."

Not even a week after the publication of this article, a second work by Druce and Loring appeared in the pages of *Chemical News*.[86] In it, the two chemists revealed for the first time that the presumed samples of the oxide of *eka-cesium* were extracted from pyrolusite, a mineral rich in manganese. Their work was based on the analysis of this mineral done 30 years earlier by Hartely and Ramage, who had found in pyrolusite a certain quantity of the alkali metals.[87] Sadly, Loring and Druce contributed nothing toward identifying or isolating element 87. In the middle of all this approximation and uncertainty, the two chemists then stated that they had identified some characteristic lines in the X-ray spectrum of *eka-iodine*, but then, barely 2 weeks later, they published a retraction.[88] This retraction referred only to an error in the attribution of particular lines that did not belong to *eka-iodine* but to another missing element, number 93.

Thus, in the midst of their chaotic research to identify and isolate element 87, and during the week in which their correction appeared in *Chemical News*, Loring and Druce reported on their research aimed at identifying and isolating the illusory transuranium element[89] with atomic number 93. In this case, according the authors, while looking for element number 93 they found themselves *involuntarily* observing the higher homologs of iodine and cesium, although they regretted that the images of their characteristic spectra were not completely clear. The transuranium element was observed, but the possibility of proving its existence with absolute certainty was found difficult. This time, it was the technicians at Adam Hilger Ltd. Company who threw a monkey wrench into the British chemists' wheels; the spectroscopists there were unsure of the reliability of their measurements because the line being looked for was too close to the critical limits of the instrument.

Despite his many publications on the search for eka-cesium, Loring failed to offer anything more than vain hypotheses regarding its existence.[90] After 1925, for unpublicized reasons, the collaboration between Loring and Druce ended and their paths diverged, although the interests of both remained focused on the search for the missing elements.[91]

In January and February 1926, Loring took up with renewed ardor the subject closest to his heart,[92] the identification of element 87. In this last publication, after having reexamined the prior work done by Druce, he reported other measurements received from the technicians of Adam Hilger's research laboratory.[93] He was able to reconfirm with absolute certainty that he had recorded the $L_{\alpha 1}$ line of *eka-cesium*; as for the weak $L_{\beta 1}$ line, after numberless failed attempts and with the employment of a spectrum comparator, he was able to extrapolate it, although with difficulty, between the signals of silver bromide. Loring realized that the lone $L_{\alpha 1}$ line would not be enough to prove the existence of *eka-cesium* and therefore he tried to intensify the signal on the photographic emulsion, obtaining as a result both the intensification of the looked-for lines but also the appearance of new undesirable lines. He could not prove his discovery with only one line of element 87: it was too little for the scientific community to be able to award him credit for the discovery.

Between the months of March and June, Loring gradually detached himself from the search for elements 85 and 87 in nature and began to promote the hypothesis that they were members of the radioactive families. In the first of three articles published in that period,[94] Loring determined that much work remained to be done to identify element 87. He went on to explain that, after the first positive attempts done by Druce to isolate *eka-cesium*, he did not succeed in obtaining samples rich in this element. Perhaps in this we can deduce the reason for the end to their collaboration, and perhaps in these words, Loring was conveying a veiled indictment of young Druce for having tampered with the results. The matter remains a mystery; what is certain is that Loring was blinded by the illusionary mirage represented by the identification of the ephemeral element 87. After a long introduction, Loring expounded his hypothesis, according to which the lower homologs of *eka-cesium* (rubidium and cesium) might be emitting or absorbing electrons by way of nuclear disintegration—spontaneously or by exogenous induction—and transforming themselves into the element so tenaciously sought. Although referring to the experiments of famous physicists of the time, such as Ernest Rutherford and Patrick Maynard Stuart Blackett (1897–1974), Loring's theoretical reasoning was inaccurate and the conclusions he reached false.

Loring had noticed that the lines of bromine that appeared in the X-ray spectrum with increasing intensity with increasing exposure time of the samples being tested could be explained in two ways. The first explanation—the presence of bromine in the photographic emulsion—was discarded. The second route was more than a little bizarre, taking into consideration the possibility of atomic fusion but, contrary to every logical hypothesis, it was accepted by the author as the more probable. According to Loring, the bromine he observed in his spectra had been generated in the anticathode of the X-ray tube according to the following nuclear reaction:[95]

$$^{19}K + ^{20}Ca \rightarrow ^{35}Br \quad \text{(Eq. IV.2)}$$

Loring hypothesized, and later said that he had verified, that nuclear synthesis in the X-ray tubes led to the formation of other elements, but only those of Group 7. This bizarre assertion, with no demonstrable justification, was also found to be erroneous:

$$^{19}K + ^{28}Ni \rightarrow ^{43}Ma \quad \text{(Eq. IV.3)}$$ [96]

$$^{37}Rb + ^{20}Ca \rightarrow ^{53}I \quad \text{(Eq. IV.4)}$$

$$^{55}Cs + ^{24}Cr \rightarrow ^{75}Re \quad \text{(Eq. IV.5)}$$

$$^{37}Rb + ^{52}Te \rightarrow ^{85}85 \quad \text{(Eq. IV.6)}$$

It is immediately apparent that there is a lack of balance of the atomic numbers in these nuclear reactions: this fact did not escape Loring, who did not lose heart and hypothesized that the loss of four protons in every reaction with the forced insertion of four electrons during the X-irradiation could explain it. This expedient would have reduced the atomic number by four units, thereby restoring the count (the four protons would have been transformed into as many neutrons). The following month, Loring published a long, five-part article[97] in *Chemical News*. The first three parts offered a rather partisan view of the discovery of element 75. The author pointed out that, as in the case of the discovery of *celtium* (atomic number 72), a bitter dispute was in progress among the discoverers: on one side, the Frenchmen Georges Urbain and Alexandre Dauvillier, and on the other, the contenders from the Institute of Physics at Copenhagen, George de Hevesy and Dirk Coster. A similar dispute, Loring reminded the public, was in progress to award credit for the discovery of element 75: a contest between the Noddacks on the one hand and himself and Gerald Druce on the other.[98] The difference was that the first dispute was of international proportions; the Loring-Druce claim was far more modest in scope, raised almost exclusively by Loring and Druce in the pages of the Druce-directed *Chemical News*, such that the international scientific community and even the Noddacks appeared unaware of it. Loring furnished some details on the chemical separation done by Druce when he worked at Charles University in Prague and of the analyses done by Dolejšek on the enriched material, and then had these findings confirmed by a luminary in the field of X-rays, Manne Siegbahn. To Loring's great regret, the subject of the dispute was not yet *sub judice* of the International Commission for the Nomenclature of the Chemical Elements.

In the concluding paragraphs of his article, Loring dealt with two subjects substantially different from his search for element 75. After having spoken in the next to the last paragraph of the possibility of the existence in nature of element 93, Loring reported two new nuclear reactions for the synthesis of two chemical elements in vacuum tubes irradiated by X-rays:

$$^{55}\mathrm{Cs} + {}^{42}\mathrm{Mo} \rightarrow {}^{93}93 \qquad \text{(Eq. IV.7)}$$

$$^{37}\mathrm{Rb} + {}^{28}\mathrm{Ni} \rightarrow {}^{61}61 \qquad \text{(Eq. IV.8)}$$

At the conclusion of the article, Loring pointed out the following: "It is not yet proved that any of these elements can be formed in the X-ray tube as suggested, as a possibility, in the [previous article]."

With the years, Loring showed increasing signs of unease, and this restlessness was apparent in his writings. Ever since he put himself on the track of the elements not yet identified, others had succeeded while he had either failed or arrived on the scene too late. In 1930, when *celtium*, rhenium, and hafnium appeared to be solidly confirmed discoveries, and the elements 61 and 93 were too far away from his interests—excluding from the count the elusive element 85—nothing was left but to discover element number 87. Loring decided not to allow his last chance escape his grasp.

After his sensational series of articles in 1926, Loring withdrew into a long silence until he suddenly reappeared on the international scene with a startling announcement: the discovery of element 87, which he called *alkalinium*. Loring asserted that element 87 was not radioactive, using bismuth as an analogous nonradioactive element among short-lived products with both an odd atomic number and odd atomic weight.[99]

But if the assignment of the name "alkali metals" is universally accepted, then potential confusion might develop with the equivocal name of *alkalinium*; the name's root would stand both for the group of elements and for a single element. The following year, six monographs that carried Loring's name appeared, all of them dealing with element 87. The first[100] served not only to fix the melting point of *alkalinium* at 616 °C, but asserted that strong experimental evidence presupposed the existence of this element in the solar corona. Another publication justified the method of attributing the spectral lines to the element,[101] and the next[102] fixed the specific heat of *alkalinium* at 0.0338 at 0 °C. In the following articles,[103] Loring returned to the concept dear to his heart, the "harmonization" of the atomic weights; to justify the irregularities present among the isotopes of the alkali metals, he predicted that element 87 would possess a single isotope with a mass of 223, a hypothesis later shown to be effectively correct.[104]

During these same years, other scientists also embarked on the fruitless search for traces of naturally occurring element 87, among them Fred Allison (1882–1974) in the United States and Horia Hulubei (1896–1972) in France. They respectively named this element *virginium*[105] and *moldavium*.[106] In both cases, they affirmed the presence of the element in extremely small trace amounts, as did Loring. A stable isotope of eka-cesium was found in 1932, by Professor Gustaf Alfred Aartovaara (b. 1863) of Helsinki. He asserted that he had found element 87 in some Finnish feldspars in macroscopic quantities.[107]

IV.2.6. ALKALINIUM'S EPILOGUE

When on March 21, 1930 Loring announced his discovery and coined its name, he embarked on an undertaking without imagining what the consequences might be. Some years later, the verdict of science brought an end to the strange and uncertain existence of *alkalinium*: ironically, the etymological root of the word "alkali" is from the Arabic *qalaa*, "to roast," and the failed attempts to have *alkalinium* recognized ended up "incinerating" Loring's hopes.

The ups and downs of Druce and Loring concluded shortly after World War II. Not yet 56 years old, Druce died June 21, 1950, in a London hospital after a long and painful incurable illness,[108] thus removing himself from a trying situation and relieving the scientific community from the embarrassment of having to condemn him for having associated his name with that of Loring, responsible for the false discovery of *alkalinium*.

After the conclusion of World War II, Frederick H. Loring disappears from history. In 1940, his home address was London, but it is likely that he became an American citizen. In 1942, for commercial reasons tied to a patent related to the treatment of wheat flour,[109] Loring appeared on American soil. The previous year with his last publication,[110] reviewed in 1945, Frederick Henry Loring accomplished his last theoretical acrobatics in the field of the physical sciences. He reworked Lord Kelvin's theory of the atomic vortex,[111] connecting it to the atomic numbers of the inert gases and to a certain number of properties of atomic orbitals.[112]

With the passing of time, the memory of these men has assumed the semblance of an abandoned cemetery where they and their discoveries, true and presumed, lie forgotten.

Notes

57. Loring, F. H. *Chem. News* **1909**, *100*, 281.
58. Moir, J. *J. Chem. Metal. Min. S. Africa* **1909**, *9*, 334.
59. Moir, J. *J. Chem. Soc. Trans.* **1909**, *95*, 1752; Moir, J. *Proc. R. Soc. London* **1910**, *25*, 213.
60. Moir, J. *Chem. News J. Ind. Sci.* **1922**, *124*, 105; Moir, J. *Chem. News J. Ind. Sci.* **1922**, *124*, 118; Moir, J. *Chem. News J. Ind. Sci.* **1922**, *124*, 133; Moir, J. *Chem. News J. Ind. Sci.* **1922**, *124*, 149.
61. Moir, J. *J. Chem. Soc. Trans.* **1924**, *125*, 1134.
62. Moir, J. *Trans. R. Soc. S. Africa* **1929**, *18*(Pt. 3), 183; Moir, J. *Trans. R. Soc. S. Africa*, **1929**, *18*(Pt. 2), 137; Moir, J. *J. South African Chem. Inst.* **1929**, *12*, 16.
63. Loring, F. H. *Chem. News* **1909**, *99*, 148.
64. Loring, F. H. *Chem. News* **1909**, *99*, 241.
65. Loring, F. H. *Chem. News* **1909**, *100*, 281.
66. Urbain, G.; Jantsch, G. *Compt. Rend. Chim.* **1908**, *147*, 1286.
67. H. C. Blood Ryan was also vice-president of the *European Branch of the Muslim Association for the Advancement of Science* and of the *Institute of Criminology*.
68. Noddack, W.; Tacke, I. *Oesterreichische Chem. -Ztg.* **1925**, *28*, 127; Noddack, W.; Tacke, I. *Naturwissenschaften* **1925**, *13*, 567; Noddack, W.; Tacke, I. *Sitzb. Preuss. Akad. Wissenschaften* **1925**, 400; Berg, O.; Tacke I. *Naturwissenschaften* **1925**, *13*, 571.
69. Druce, G. *Chem. News* **1925**, *131*, 273.
70. Dolejšek, J.; Heyrovský, J. *Nature* **1925**, *116*, 782.
71. Druce, G. *Nature* **1942**, *150*, 623.

72. Karpenko, V. *Ambix* **1980**, *27*, 77; Ref. 44a.
73. Campbell, A. N. *Nature* **1925**, *116*, 866.
74. Zvjagintsev, O.; Korsunski, M.; Seljakov, M. *Nature* **1926**, *118*, 262.
75. Prandtl, W. *Z. Angew. Chem.* **1926**, *39*, 1049.
76. Newton Friend, J. *Nature* **1926**, *117*, 789.
77. Newton Friend, J. *Man and the Chemical Elements*. Scribners: New York, 1961.
78. Loring, F. H. *Chem. News J. Ind. Sci.* **1914**, *110*, 25.
79. Loring, F. H. *Chem. News J. Ind. Sci.* **1915**, *111*, 157; Loring, F. H. *Chem. News J. Ind. Sci.* **1915**, *111*, 181.
80. Loring, F. H. *Chem. News J. Ind. Sci.* **1922**, *125*, 309; Loring, F. H. *Chem. News J. Ind. Sci.* **1922**, *125*, 386; Loring, F. H. *Chem. News J. Ind. Sci.* **1923**, *126*, 1.
81. Loring, F. H. *Chem. News J. Ind. Sci.* **1920**, *120*, 105.
82. Loring, F. H. *Chem. News J. Ind. Sci.* **1920**, *120*, 181; Loring, F. H. *Chem. News J. Ind. Sci.* **1920**, *120*, 193; Loring, F. H. *Chem. News J. Ind. Sci.* **1920**, *120*, 205; Loring, F. H. *Chem. News J. Ind. Sci.* **1920**, *120*, 217.
83. Loring, F. H. *Chem. News J. Ind. Sci.* **1923**, *126*, 307; Loring, F. H. *Chem. News J. Ind. Sci.* **1923**, *126*, 325; Loring, F. H. *Chem. News J. Ind. Sci.* **1923**, *126*, 371; Loring, F. H. *Chem. News J. Ind. Sci.* **1923**, *127*, 225.
84. Loring, F. H. *Nature* **1926**, *117*, 153.
85. Loring, F. H.; Druce, J. G. F. *Chem. News J. Ind. Sci.* **1925**, *131*, 289.
86. Loring, F. H.; Druce, J. G. F. *Chem. News J. Ind. Sci.* **1925**, *131*, 305.
87. Hartely, R. *Trans. Chem. Soc.* **1897**, *71*, 533.
88. Loring, F. H.; Druce, J. G. F. *Chem. News J. Ind. Sci.* **1925**, *131*, 321.
89. Loring, F. H.; Druce, J. G. F. *Chem. News J. Ind. Sci.* **1925**, *131*, 337.
90. Loring, F. H. *Chem. News J. Ind. Sci.* **1925**, *131*, 371.
91. Druce, J. G. F. *Continental Met. & Chem. Eng.* **1926**, *1*, 111; Druce, J. G. F. *Science Progress* (St. Albans, U.K.) **1927**, *21*, 479; Druce, J. G. F. *Chemisch Weekblad* **1926**, *23*, 497; Druce, J. G. F. *Chemisch Weekblad* **1926**, *23*, 318; Druce, J. G. F. *Science Progress* (St. Albans, U.K.) **1926**, *20*, 690.
92. Loring, F. H. *Chem. News J. Ind. Sci.* **1925**, *131*, 371.
93. Actually, Adam Hilger, Ltd. was one of the leading English manufacturers, at a level a little higher than artisanal, of spectrochemical analytical instruments. It contributed to the development of technical apparatus that had many metallurgical applications.
94. Loring, F. H. *Nature* **1926**, *117*, 448.
95. Loring, F. H. *Chem. News J. Ind. Sci.* **1926**, *132*, 311. Loring annotated the atomic numbers as superscripts, whereas conventional practice places them as subscripts in nuclear reactions.
96. The symbol Ma refers to the hypothetical element with atomic number 43 *masurium* (or *eka-manganese*), today known as technetium.
97. Loring, F. H. *Chem. News J. Ind. Sci.* **1926**, *133*, 407.
98. Loring, F. H. *Chem. News J. Ind. Sci.* **1926**, *133*, 276.
99. Loring F. H. *Chem. News J. Ind. Sci.* **1930**, *140*, 178.
100. Loring, F. H. *Chem. News J. Ind. Sci.* **1931**, *143*, 18; Loring, F. H. *Chem. News J. Ind. Sci.* **1931**, *143*, 39.
101. Loring, F. H. *Chem. News J. Ind. Sci.* **1931**, *143*, 149.
102. Loring, F. H. *Chem. News J. Ind. Sci.* **1931**, *143*, 359.
103. Loring, F. H. *Chem. News J. Ind. Sci.* **1931**, *143*, 360; Loring, F. H., *Chem. News J. Ind. Sci.* **1931**, *143*, 408.
104. Asimov, I. *J. Chem. Educ.* **1959**, *30*, 616.
105. Allison, F.; Murphy, E. J. *Phys. Rev.* **1930**, *35*, 285.
106. Hulubei, H. *Compt. Rend.Chim.* **1937**, *205*, 854.
107. Aartovaara, G. A. *Tekniska Foereningens i Finland Foerhandlingar* **1932**, *52*, 157.

108. Orten, G. *Nature* **1950,** *166*, 134; Briscoe, H. V. A. *J. Chem. Soc. London* **1950**, 3358; Anon. *J. R. Inst. Chem.* **1950**, 348.
109. Loring, F. H. *Patent number GB 540687*, 1941.
110. Loring, F. H. *Chem. Prod. Chem. News* **1945**, *8*, 54.
111. Thomson, W. (later Lord Kelvin) *Proc. R. Soc. Edinburgh* **1867**, *VI*, 94; Thomson, W. (later Lord Kelvin) *Phil. Mag.* **1867**, *34*, 15.
112. Loring, F. H. *Chem. Prod. Chem. News* **1944**, *7*, 40.

IV.3

A SUCCESS "TRANSMUTED" INTO FAILURE

IV.3.1. BREVIUM

"Don't call it transmutation. They'll cut off our heads as if we were alchemists!"[113] This was the recommendation that 30-year-old Ernest Rutherford gave to his young student Frederick Soddy, but transmutation it was and with transmutation Soddy would deal for the rest of his life.

Soddy developed the revolutionary concept of the isotope and thus was able to predict that identical elements, with the same chemical properties, could differ in their atomic mass. As a result of this discovery, in 1921, he was awarded the Nobel Prize in Chemistry. Soddy was also interested in radioactivity: he hypothesized and demonstrated that α and β radiation accompanied a chemical transmutation of the element of interest via this physical process. In what was a real race to identify radioactive elements, he was considered the discoverer of protactinium (1918). Unfortunately, this was not the case.

After the discovery of radium, polonium, actinium, and radon, some chemists believed that uranium ores could contain other as yet undiscovered radioactive elements. This idea was proposed independently by Alexander S. Russell (1888–1972)[114] and Kasimir Fajans (1887–1975)[115] in 1912.

The chemist Kasimir Fajans was born in Warsaw on May 27, 1887. The year before World War I broke out looked very promising for him. Barely 26 years of age, he worked in Karlsruhe, Germany, where had been named *privatdozent* and had undertaken a ticklish project in the field of radiochemistry that culminated in the publication of six monographs and the discovery of an element. Fajans's work on UX, that at first chemists thought to be a single radioactive element, showed that it was actually a mixture of two radioactive elements: UX_1 and UX_2. Following the law of chemical displacement recently formulated by Soddy, Fajans succeeded in writing the first radioactive cascades in the uranium-238 decay chain:

$$U_1 \xrightarrow{\alpha} UX_1 \xrightarrow{\beta} UX_2 \xrightarrow{\beta} U_{II} \xrightarrow{\alpha} Io \qquad \text{(Eq. IV.9)}$$

that corresponds today to the following series:

$$^{238}U \xrightarrow{\alpha} {}^{234}Th \xrightarrow{\beta} {}^{234}Pa^{m} \xrightarrow{\beta} {}^{234}U \xrightarrow{\alpha} {}^{230}Th \qquad \text{(Eq. IV.10)}$$

In 1913, there was no known radioactive element in Group 5 of the periodic table between thorium and uranium. The periodic table at the time was written in compact form, and only the rare earth elements were placed outside the main body of the table itself.

Fajans and the 24-year-old Ostwald Helmuth Göhring (1889–1915?) realized that the substance known as UX gave rise to the product UX_2 through radioactive decay (from then on, in fact, chemists took to indicating UX as UX_1 and the product as UX_2) that was collected on a lead plate. UX_2 showed β activity and a very short half-life that could not be assigned to any radioactive element already known. Its chemical nature was confirmed by co-precipitating it with a solution of hydrated tantalum(V) oxide. They realized that UX_2 should occupy a vacant box in the periodic table and decided to name the new element.[116] The name they chose was *brevium*,[117] whose etymology was easy to interpret: the isotope of this element, discovered by Fajans and his assistant, had a half-life of little more than a minute. Figure IV.04 pictures Göhring and Fajans with another of their colleagues.

Between the discovery of *brevium* and the outbreak of World War I, Göhring[118] and Fajans looked for other isotopes[119] of element 91 and tried to publicize their discovery as much as possible. In 1914, Göhring was called to the front and probably perished in the dreadful slaughter: no publications carry his name after 1915.

FIGURE IV.04. Oswald Helmuth Göhring (1889–?), Kasimir Fajans (1887–1975, seated), and Max Ernst Lembert (1891–1925), Pictured in 1915 at the Technische Hochschule of Karlsruhe. Fajans had, 2 years earlier, discovered element number 91, protactinium, and called it *brevium*, but the discovery, possibly due to the imminent threat of war, was not recognized.

IV.3.2. LISONIUM AND LISOTTONIUM

Soon after the announcement of the discovery of *brevium*, Otto Hahn, together with his Austrian collaborator Lise Meitner (1878–1968), began the search for the other isotopes of this radioactive element. Their research was based on the possibility, later confirmed, that isotopes of element 91 with half-lives greater than that of *brevium* might exist. At the outbreak of World War I, the 35-year-old Hahn was conscripted but never went to the front: he entered the ranks of the chemists who, directed by Fritz Haber (1868–1934), made the first poisonous war gas.

Lise Meitner went back to Austria and lent her aid as a volunteer, like Marie Curie, in medical radiology. The experience of this work traumatized her and, in October 1916, she left her military hospital job and returned to Berlin. This return to the familiar chemistry laboratory in the elegant section of Berlin-Dahlem was beneficial to Lise Meitner, who strove to continue the research that had been interrupted for more than 2 years. Alone and with the pitiful imperial government subvention not destined for the war effort, she sought to advance research on the aforementioned isotopes. Finally, in January 1917, Hahn received a lengthy leave and was able to return to his laboratory. In the meantime, Meitner had developed a working method more accurate than that used by Fajans and Göhring for the discovery of *brevium*.

From a small quantity of pitchblende, she isolated 2 g of SiO_2. To that she added some potassium fluorotantalate and dissolved the mixture in HF, brought it to a boil in concentrated sulfuric acid, and obtained a precipitate of tantalum and the presumed parent of actinium.

For an entire year, Meitner and Hahn developed radiochemical tests to identify the possible radioelements present in their samples. At the end of the year, Lise Meitner went to Braunschweig to visit Friedrich Oskar Giesel, the famous industrial chemist who, shortly after André Debierne and independently of the latter, had discovered actinium, initially calling it *emanium*.[120,121] Giesel was intent on producing radioactive metals for therapeutic use. In addition to a comparison of her results with her eminent colleague, Lise Meitner obtained a promise from the industrialist of a kilogram of radioactive salts, the precious products of discarded material from the purification of radium. Giesel kept his word and, in December, with new samples of material at her disposal, Meitner accomplished the last step in the isolation of element 91. On March 16, 1918, the two researchers sent an article to the editor of *Physikalische Zeitschrift* carrying the title: "Die Muttersubstanz des Aktiniums; ein neues radioaktives Element der langen Halbwertzeit." With evident satisfaction, the two scientists reported that[122] "We have been able to discover a new element [from pitchblende] and we have shown that it is the mother substance of actinium. We propose consequently the name protoactinium."

The isotope of *protoactinium*[123] that Meitner and Hahn discovered in the winter of 1917–18, ^{231}Pa, has a very long half-life: about 32,700 years. Following the discovery, Meitner passed through a period in which she was burdened with commitments but, as she herself said, very pleasant ones, and both the conversations and the exchange of letters with Viennese physicist Stefan Meyer (1872–1949) testify to that.

Responding to a letter from Meitner, Meyer emphasized how he would have preferred that the name of element 91 be either *lisonium* or *lisottonium*, with the symbol Lo. His proposal reflected the names of the discoverers, Lise and Otto, and was an indication of how much of the work the Viennese physicist credited to Lise Meitner. (A few lines later, in the same letter, he admitted that the names *lisonium* and *lisottonium*, although pleasing, would never be accorded a favorable reception on the part of the whole scientific community.) Meyer then turned his attention to the symbol for the new element. He would have preferred Pn, taking the letter "n" from the last half of the name protoactinium analogously to the letter "d" in the case of palladium (Pd). The suggestion was not favorably received by Meitner.

In June 1918, Frederick Soddy and his young student John Arnold Cranston (1891–1972) published two articles[124] on the same subject. Their results were published 5 years later than those of Fajans and 3 months later than those of Hahn and Meitner. Cranston, after having started work in Soddy's laboratory at the University of Aberdeen, left for the French front in 1914, and only near the conclusion of the conflict did he return to his country and resume the experiments suspended for such a long time.

Soddy and Cranston had treated a certain quantity of pitchblende and then sublimed a radioactive substance whose properties matched those described by Hahn and Meitner. The British chemists, because of the meager amount of material isolated, were unable to completely describe the radioactive decay of element 91. For this reason, as well as for the fact that their publication came out 3 months after that of their German colleagues, they very chivalrously[125] recognized the priority of the work of Meitner and Hahn.

Hahn and Meitner had obtained an easy success on the English front, but the internal one was a bit trickier. The delicate question of the name came down to *brevium*. The discoverer of the latter element, Kasimir Fajans, was born in Tsarist Poland but had studied in Germany, and it was there that he was carrying out his research activities. Beginning his university studies first at Leipzig and then at Heidelberg, where he took his doctorate, at the end of the war he was working at the University of Munich in Bavaria. In the summer of 1918, Hahn visited Fajans, seeking to validate his right to name element 91. By virtue of the usage in force at the time, the naming of a new radioactive element belonged to the person who discovered the isotope with the longest half-life. Formally, Hahn was in the right: his isotope had a half-life about 10 billion times longer than that of the one discovered by Fajans. However, the latter, with an aggressive character and imperious temperament, did not immediately back down, maintaining that he had discovered the new radioactive element, that he had understood its elemental nature, and that he had published the results of his research 5 years ahead of Hahn and Meitner. However, Hahn and Meitner succeeded in preventing him from bringing an action against their priority, and Fajans's obstinacy and stubbornness, fortunately, did not degenerate into open hostility but instead left a humorous remembrance among his colleagues who, through the years, began to refer to him by the nickname "Kasimir the Great."[126]

By an irony of fate, although the discovery of protoactinium was accepted universally, later research on the origin of the decay chain of actinium became more complicated.

In 1921, Otto Hahn[127] discovered the third and last natural isotope of *protoactinium*, ^{234}Pa, that has a half-life of 6.7 hours. *Brevium* has the same mass (234), but in fact it is a metastable form of the latter.

Finally, in 1927, the 22-year-old Aristid Victor Grosse (1905–85) succeeded in preparing a very modest quantity of Pa_2O_5 in the form of a white powder.[128] Only in 1934 did he succeed in converting the oxide into the iodide from which, under vacuum and by the Joule effect, he obtained the elemental form deposited on a metallic filament:[129]

$$2PaI_5 \longrightarrow 2Pa + 5I_2 \qquad \text{(Eq. IV.11)}$$

IV.3.3. RADIO-BREVIUM AND THE MISSED DISCOVERY OF NUCLEAR FISSION

Aristid Victor Grosse, born in Russia, on January 4, 1905, passed his youth between Japan and Shanghai; he did his university studies at the University of Berlin from 1922 through 1926. Later, he worked in the laboratory of Otto Hahn and Lise Meitner, where he remained until 1927. He was hired for a brief period by Universal Oil Products Corporation, but before World War II started, he joined the Columbia University faculty in New York City. He also visited the best laboratories of Europe (the Cavendish Laboratories at Cambridge, the Institut du Radium at Paris, and the Institut für Chemie der Kaiser-Wilhelm Gesellschaft at Berlin).

In the mid-1930s, he studied using radiochemical techniques the presumed transuranium elements ($Z = 93$ and $Z = 94$) discovered by Enrico Fermi and his colleagues,[130] for which the names *ausonium* and *hesperium* had been proposed.[131] Grosse deserves the credit for having recognized the analogy in the chemistry of elements 90, 91, and 92 (thorium, *protoactinium*, and uranium) with the rare earths.[132] Following this discovery, he was convinced that the elements following actinium would comprise a family similar to the lanthanides. This was quite different from what Hahn and other radiochemists had proposed—that is, that thorium, *protoactinium*, uranium, *ausonium*, and *hesperium* would be the higher homologues of *celtium*, tantalum, wolfram, rhenium, and osmium, respectively.

In 1934, Grosse repeated Fermi's experiments. He bombarded uranium with slow neutrons and realized that the radioactive isotope produced, with a 13-minute half-life, thought by his colleagues at Rome to be an isotope of *ausonium*, was simply element 91.[133] The reaction he proposed was:

$$^{238}U + {}^{1}n \longrightarrow {}^{238}Rm + {}^{1}H \qquad \text{(Eq. IV.12)}$$

He wanted to call the new isotope of element 91 *radio-brevium*, with the symbol Rm. As a matter of fact, he followed the agreed-upon sequence introduced by Frédéric Joliot and Irène Curie to name every new artificial isotope by placing the prefix "radio" before the name of the element generated.[134] Strangely, in his articles, Grosse never called element 91 by the name given it by its discoverers, Hahn and Meitner (who were also his teachers and with whom he had collaborated in the 1920s): *protoactinium*. In his publications, he alternated between the names *eka*-tantalum (indicated by him curiously enough with the letters Et) and *radio-brevium* (Rm). In this way, he implicitly recognized the name *brevium* as the authentic one for element 91.

The results that Grosse arrived at, based essentially on chemical data, were erroneous. He recognized the similarity of the product formed by him (*radio-brevium*) with that of element 91. A study of the phenomenon from the physicist's point of view may have prevented him from arriving at this conclusion and would have allowed him to discover, 4 years in advance of Lise Meitner, the fission of uranium, perhaps even earning for him the Nobel Prize in Chemistry in 1944, which went instead to Otto Hahn.[135]

In 1939, Grosse left his post at Columbia University to become part of the Manhattan Project. Ironically, he found that Enrico Fermi, to whose neutron bombardment experiments on uranium he sought to give an explanation, was his superior. Grosse died in 1985 at the age of 80.

IV.3.4. BREVIUM'S LAST GASP

By the end of 1973, the ongoing argument tied to the controversy between *brevium* and *protoactinium* seemed to have been laid aside for good. The names of the discoverers and the priority of the discovery of element 91 seemed to have been well established and accepted by everyone. But a latent unpleasantness and a profound malaise, harbored by Fajans for 60 years, exploded on the occasion of the publication of the obituary of John Arnold Cranston.[136]

In 1972, Cranston died peacefully at the age of 81, comforted by the presence of his five children and numerous grandchildren. He was a man with a reserved and unruffled temperament, but also affectionate and brilliant; he was a man of versatile talents associated with profound culture. One could never imagine that the article dedicated to him at the time of his death could be "transmuted" into great unpleasantness for others. However, the article infuriated the 86-year-old "Kasimir the Great" who, in the pages of *Nature*,[137] responded obstinately to the person who, in praising the life and work of Cranston, also erroneously attributed to him the discovery of element 91.

Cranston did not claim for himself the discovery of element 91 for as long as his mentor Soddy was alive. Later, however, his attitude changed. In a colloquium held at the Department of Chemistry at the University of Glasgow on January 26, 1967, and published in the following year,[138] he made it clear: "Dr. Cranston was the co-discoverer with Soddy of Protactinium in 1917."

If Fajans had known about this definite assertion, he would have been far more furious than he was when he read Cranston's obituary. The article that appeared in *Nature* was Kasimir Fajans's last publication, and it is sad to think that his scientific efforts would be tinged with a strong sentiment of regret and distress. He lived for 2 more years; on May 18, 1975, he died in Ann Arbor, Michigan, where he had taken refuge in 1936 from the Nazi regime.[139]

Fajans's personal history,[140] tragic in certain respects, should not cause us sorrow. Today, as we tentatively reevaluate his work, we realize that his greatest misfortune was not the lack of recognition for the discovery of *brevium* but the fact that, for the greater part of his life, he was an exile, first in Germany and later in the United States.

Notes

113. Campbell, J. *Scientist Supreme*. AAS Publications: Christchurch, New Zealand, 1999.
114. Russell, A. S. *Chem. News* **1913**, *107*, 49.
115. Fajans, K. *Z. Physik* **1913**, *14*, 136.
116. Fajans, K.; Göhring, O. *Naturwissenschaften* **1913**, *1*, 399; Fajans, K.; Göhring, O. *Z. Physik.* **1913**, *14*, 877.
117. Fajans, K. *Ber. Dtsch. Chem. Ges.* **1913**, *46*, 3492.
118. Göhring, O. *Z. Physik.* **1914**, *15*, 642; Göhring, O. *Mitt. Chem. -Tech. Inst. Tech. Hochsch. Karlsruhe* **1914**, *8*, 297; Göhring, O. *Neues Jahrb. Min. Geol.* **1915**, *II*, ref. 15.
119. The two German researchers came very close to the concept of the isotope and arrived at actually proposing the term "pleiade" instead, as a clear reference to the spectacular galactic collection of the same name composed of a multitude of stars.
120. Giesel, F. O. *Ber. Dtsch. Chem. Ges.* **1902**, *35*, 3608.
121. Giesel, F. O. *Ber. Dtsch. Chem. Ges.* **1904**. *37*, 1696.
122. Hahn, O.; Meitner, L. *Phys. Z.* **1918**, *19*, 208.
123. In 1949, the IUPAC officially changed the name to protactinium.

124. Soddy, F.; Cranston, J. A. *Nature* **1918**, *100*, 498; *Proc. Roy. Soc.* **1918**, *94A*, 384.
125. It should be remembered that in June 1918, England (Cranston and Soddy's country) was at war with the Central Empires, birthplace of Hahn and Meitner.
126. Sime, R. L. *J. Chem. Educ.* **1986**, *63*, 653.
127. Hahn, O. *Ber. Dtsch. Chem. Ges.* **1921**, *54*, 1131.
128. Grosse, A. V. *Ber. Dtsch. Chem. Ges.* **1928**, *61*, 233; Grosse, A. V., *J. Am. Chem. Soc.* **1930**, *52*, 1742.
129. Grosse, A. V. *J. Am. Chem. Soc.* **1934**, *56*, 2200; Grosse, A. V. *Science* **1934**, *80*, 512.
130. Fermi, E. et al. *Nature* **1933**, *133*, 898.
131. Fontani, M.; Costa, M.; Manzelli, P. *Memorie di Scienze Fisiche e Naturali. Accademia delle Scienze detta dei XL*, 2001, *XXV*, serie V, parte II, tomo II, 433.
132. Grosse, A. V. Agruss, M. S. *J. Am. Chem. Soc.* **1935**, *57*, 438; Grosse, A. V. *J. Am. Chem. Soc.* **1935**, *57*, 440.
133. Grosse, A. V. *Phys. Rev.* **1934** *46*, 241.
134. Curie, I.; Joliot, F. *Compt. Rend. Chim.* **1934**, *133*, 898.
135. Sime, R. L., *Le Scienze* **1998**,April (*4*), 356; Crawford, E.; Sime, R. L.; Walker, M. *Physics Today* **1997**, *50*(9), 26.
136. Cumming, W. M. *Chemistry in Britain* **1972** *8*, 388.
137. Fajans, K.; Morris, D. F. C. *Nature* **1973**, *244*, 137.
138. Cranston, J. A. *Chemistry in Britain* **1968**, *4*(2), 66.
139. Schwab, G. M. *Bayerische Akademie der Wissenschaften, Jahrbuch* **1976**, 227.
140. Hurwic, J. *Actualité Chimique* **1976**, *1*, 28; Hurwic, J. *J. Chem. Educ.* **1987**, *64*(2), 122.

IV.4

FROM PLEOCHROIC HALOES TO THE BIRTH OF THE EARTH

The Nobel Laureate Emilio Segrè in his biography[141] reported that he had deposited some milligrams of technetium, the element discovered in 1937, on his mother's tomb. Technetium is a radioactive element with a rather long half-life. Because Segrè had emigrated to the United States, he rarely returned to his native Italy to visit his mother's tomb. For this reason, he wrote, technetium would last longer than an ordinary bunch of flowers! (It is worth mentioning that Segrè, like many Jews, was acquainted with the custom of placing stones, not flowers, on tombs.) The half-lives of the radioactive elements have also served excellently for other (more scientific) purposes. Almost a century ago, via his diligent study of radioactive substances, John Joly, an Irish physicist, was in a position to date the age of our planet.

IV.4.1. THE ORIGINS OF THE IRISH PHYSICIST

Right in the middle of World War I, two small independent Irish groups, the Irish Republican Brotherhood and the Irish Citizen Army, organized a rebellion that caught the armed forces of the United Kingdom by surprise. On Easter Monday of 1916, they marched on Dublin and took possession of some key points of the city. Their headquarters were set up in the central post office on O'Connell Street: from there, the revolutionaries read to perplexed passers-by a declaration that proclaimed the Republic of Ireland. After less than a week of furious combat, the rebels surrendered in the face of superior British forces.

While the city was in flames at the height of the rebellion, a lone figure ventured into the locality of Trinity College in a desperate attempt to prevent his laboratory from being destroyed and to save his documents from falling into the hands of the English. With a little bit of luck, John Joly survived these tragic days and much later became the grand and venerated "old man" of science on the island newly independent after more than three centuries of English domination.

John Joly (1857–1933) was born at Holywood House, at Bracknagh, in County Offaly. His date of birth merits discussion. Although the official certificate says November 1, 1858, his father, the Reverend John Plunket Joly (1826–58) noted in his diary that the day was November 1, 1857. Since the Reverend Joly died on March 3, 1858, it seems reasonable that his son was born on November 1 of the preceding year. His mother, the Countess Julia Anna Maria Georgina Lusi had Venetian roots that went far back; in fact, her maternal great-grandfather had been governor of Cephalonia on behalf of the Doge in 1772, before being accused of espionage, then pardoned, and finally recruited, in a rather amazing way, by Frederick of Prussia. From his father, Joly had French blood in his veins.

This kaleidoscopic genealogy, as he himself said, was responsible for his intellectual precociousness and his versatility in all the sciences.

In 1876, Joly entered Trinity College, and he took his degree in engineering in 1882. A short time later, he moved to the physics department. Then, in 1897, changing his research interests drastically, he obtained a chair in geology, a post that he held for 36 years. When he moved with the qualifications of assistant in the department of engineering to the department of physics, the mind of the young engineer adapted easily to this new area of research, testified to by his numerous inventions and patents on physics instruments: the meldometer, the constant-volume gas thermometer, the apophorometer,[142] the hydrostatic balance, the differential steam calorimeter, and the well-known photometer.

The most promising of his inventions was related to color photography. Work was done earlier in this area by J. C. Maxwell in England and by Gabriel Lippmann (1845–1921) in France. In 1894, Joly patented the first method for making a color photograph. He arranged three filters on a glass plate tracing a series of very thin lines—200 to the inch—in red-orange, yellow-green, and blue-violet. A photograph made through a similar filter reproduced the colors and gave the image a reasonable effect of depth. The method was commercialized as the Joly Process, but problems soon arose. In Chicago, a local inventor claimed this invention and took Joly to court. In the end, Joly won his case technically, but soon other speculators appropriated his discovery. Embittered by the experience, Joly retired and undertook a long voyage on the European continent with his half-brother Charles Jasper (1864–1906), an Alpinist and Astronomer Royal at Trinity College Dublin. The Jolys visited the Alps, finding them attractive and fascinating. As they traveled, John stopped everywhere to collect minerals, having developed a passionate interest in their study.

IV.4.2. RADIOACTIVITY MAKES DATING OF THE EARTH POSSIBLE

With the discovery of the radioactivity of the uranium-bearing minerals on the part of Henri Becquerel, another line of investigation opened up for Joly. In 1897, he had just changed over to the chair of geology. Having worked with the physicist Ernest Rutherford (1871–1937) before the outbreak of World War I, he had the advantage of understanding radioactive phenomena. On the basis of data collected from the radioactive decay of some minerals, he was in a position to fix the beginning of the Devonian period at 400 million years before the present. From there, he hypothesized that the earth would have been formed more than a billion years ago. We know today that his estimate was incorrect, but surprisingly insightful in that he had grasped the order of magnitude involved. Today estimated to be about 4 or 5 billion years old, earlier estimates of the age of the earth had been far from accurate. One of the most curious involved Archbishop James Ussher, also of Trinity College Dublin. James Ussher (1580–1655) taught theology at the same College and by simple data taken from the sacred texts, he estimated the divine creation of the world to have occurred on Sunday, October 23, 4004 years before Christ. Some centuries later, Lord Kelvin (1824–1907) and others, on the basis of the scientific data available to them at the time, sought to estimate the age of the universe, but without success. In 1908, the British Association for the Advancement of Science met at Dublin. As president of the Geology Section, physicist and geologist John Joly spoke of uranium and geology and described the role that radioactive substances had in the generation of the internal heat of the earth's crust. Joly's hypothesis on radioactivity was perceived as the only correct one to allow for the dating of our planet. Through his studies on radioactive substances, he persuaded the Royal Dublin Society to create the Institute of Radium, analogous to the one in Paris. From his preparations of radium bromide, he extracted a gas, an emanation of radon, that he sealed up in long capillary tubes. Together with Dr. Stevenson, a young physician who was his assistant, he used these sealed vials to cure facial tumors that were otherwise untreatable.

At the end of World War I, John Joly took up the geological studies that the conflict had interrupted. He was the first to introduce into this discipline contributions coming from the infant disciplines of nuclear physics and radioactivity, and he was occupied with these subjects for the last two decades of his life. His observations are collected in more than 200 articles and in numerous books that he published. In particular, Joly was the first to observe and correctly interpret the *pleochroic haloes*, the curious circular forms present in minerals like mica. The form of the pleochroic haloes is due to the fact that these bodies are spherical; when seen in thin sections of a mineral, they look like circles. They are caused by the radioactive decay of α particles, whose energy determines the radius of each sphere. α particles have a fairly large ionizing effect (in air, they have a "radius of action" of up to 7 cm); thus, they cause fluorescence in some materials; they expose photographic plates; they make many minerals change color when they are bombarded with them; they make glass and quartz fragile; and they darken mica, giving rise to the pleochroic haloes.

On April 8, 1922, Joly sent a letter to the journal *Nature* in which he announced the discovery of a new radioactive element.[143] He had discovered that the radii of the haloes were a function of the radioactive isotope present in traces in the mica and responsible for the emission of the α particles. In 1916, before the Easter Rebellion, Joly had already initiated a study of these curious formations in samples of black mica coming from the rich mineral deposit at Ytterby in Sweden. Unfortunately, the war and Ireland's struggle for independence distracted him from this study for 6 years.

The pleochroic haloes that Joly saw under the microscope were perfectly spherical and with a diameter of 0.01 mm. They could be counted in the thousands. As he himself stated, looking through the ocular, he could not believe his own eyes: a minuscule starry sky was revealed. Inside the haloes, he observed an opaque part caused by the daughter isotope of the element emitting the α particles. With a Swedish colleague by the name of Prior, he passed many samples of mica through a sieve: red mica from the Devonian coming from County Carlow, mica from Arendal dating back to the Archaeozoic and rich in traces of uranium, and many others. Just as for the mica from Ytterby, he found pleochroic haloes in the mica from Arendal. Their presence in this sample of a radioactive element like uranium made him suppose that the Ytterby mica also contained a radioactive element, but different, in that the radii of the haloes had different dimensions. The radii of the pleochroic haloes in the Arendal mica were 0.015–0.016 mm, about 50% larger than those of the haloes of Ytterby mica.

Joly tried to reproduce the conditions of pressure and temperature in which the minerals he had in hand were formed. At first he thought that it was the formation temperature of the mica that was responsible for the spherical forms, but then he arrived at the correct hypothesis of their radioactive origin. With incredible patience and painstaking thoroughness, he measured the distance between the pleochroic haloes and their nuclei—0.0045 mm—and used these measurements to support his hypothesis concerning the α particles. Joly was a skillful physicist. By means of complex calculations, he obtained the pathway of the α particles in the air from the data of their mean free path of the same in the mica. The conversion factor was fixed at 0.005 mm in the mica and about 1 cm in air. No daughter α particle of a radioactive element known at the time had an equal mean free path.

IV.4.3. HIBERNIUM: AN ELUSIVE ELEMENT

To Joly, it seemed reasonable that the traces observed were due to a new element.[144] He made his deductions through indirect proofs, anticipating the times in certain aspects. Today, the verification of a superheavy element's discovery happens by indirect means. Joly's prudence and scientific rigor are evident in the announcement of his discovery: "I wonder am I justified in naming an element for such evidence as I have found—the range of an α-ray?. . . If ever it is isolated I would ask the finder to call it Hibernium after this beautiful but most unhappy country."[145] Joly was certainly not interested in naming the new element; he was more involved in solving the puzzle of the pleochroic haloes. His hypotheses on the nature of the haloes was correct, but his hypothesis on the existence of a new element, although plausible, was incorrect. Perhaps Joly suspected this: he was not a chemist; he did not know how to treat the material at his disposition and isolate the presumed new simple substance to subject it to an accurate spectroscopic analysis. He was a talented physicist and, with the means at his disposal, he obtained truly remarkable results. To gain clarity with respect to his research and possible discovery, Joly spent the entire spring in close correspondence with a young physicist, Svein Rosseland (1894–1985) of the Institute of Theoretical Physics at Copenhagen.

Rosseland wrote a detailed account of his research to Joly. The conclusions at which he arrived were clear. *Hibernium*,[146] understood as a new element, did not exist. There could be no appeal to his researches, and Joly immediately published these results on June 3 of the same year.

Rosseland began his studies on the samples Joly sent to him by looking for traces of radioactivity, but unsuccessfully. The traces of *hibernium* were probably too small; from them, he would have been able to calculate the energy of the α particles and, consequently, he would have had a kind of fingerprint of the element in question. He then abandoned the study of the pleochroic haloes and concentrated on the central spots. Rosseland estimated a lower nuclear radius of the product of transformation following the emission of the α particle, and he arrived at the presumed atomic number of the radioactive element: $Z = 40$. At that time, scientists knew that a radioactive isotope of rubidium ($Z = 37$) was a β emitter; the Swedish physicist was led to hypothesize that the parent element was yttrium, with atomic number 39. Rosseland's hypotheses were just as uncertain as Joly's. The basis on which Rosseland, a student of Bohr, placed his arguments were shaky, and he himself emphasized that he was hazarding nothing more than a guess—proofs would have come if Joly or others had found traces of yttrium in the samples of mica under examination.

Joly found a work by Ivar Nordenskjöld (1877–1947) who had previously chemically analyzed two samples of black Ytterby mica, finding in them only traces of yttrium.[147] Joly arrived at the conclusion that the sample that gave a positive outcome had been changed during chemical manipulations. As a second operation, he set out to analyze the samples of mica in his possession, but the traces of rubidium that Rosseland had predicted, were not found.[148] Joly broadened his investigations even more: he looked for, without success, traces of strontium, the element that could have been the origin of the β-decay of rubidium.[149] The hypotheses advanced by Svein Rosseland appeared very improbable to Joly and, instead of demolishing his certainties, they actually helped boost his conviction for the existence of *hibernium*.

About a decade after these events, working independently, George de Hevesy[150] and Luigi Rolla[151] (who worked on the fractionation of the rare earths in an attempt to isolate the elusive *florentium*) made an unexpected discovery. Without knowing it, they identified the only natural radioactive isotope of an element already known for some time: samarium. Later, Joly's discovery (of *hibernium*) was rejected, and the radioactive element present at the center of the pleochroic haloes was found to be samarium.[152]

In the last years of his life, John Joly turned his interests to botany. While he was still young, with his colleague Henry Horatio Dixon (1869–1953), he had succeeded in a tentative explanation of why the sap in plants flowed in a direction contrary to the force of gravity (1893). He found that the force, actively capable of opposing the force of weight, arose principally from the evaporation of water from the leaves, increasing the effect of capillary action. The biophysicists of the time attacked him bitterly, but toward the end of his life, Joly had the satisfaction of seeing his hypothesis universally accepted. In his old age, Joly— the "grand old man of science"—became the image of the elegant Victorian scientist-gentleman: the black tie, the vest from whose pocket dangled a watch chain, the shirt with the starched collar, the obligatory pince-nez, and a pair of bushy white mustaches. For many years, he personified the icon of a world slowly disappearing.[153] John Joly died on December 8, 1933, at the age of 76.

Notes

141. Segrè, E. *Autobiografia di un Fisico*; Il Mulino: Bologna, 1995.
142. The *meldometer* was a kind of instrument that could determine the melting point of a mineral. In the *apophorometer,* the sample to be examined was placed on a tape *meldometer*; the sublimation product was collected on a silica disk and then subjected to chemical tests.
143. Joly, J. *Nature* **1922**, *109*, 517.
144. Joly, J. *Nature* **1922**, *109*, 578; Joly, J. *Nature* **1922**, *109*, 711.
145. Joly, J. *Proc. Roy Soc.* A **1922**, *102*, 682.
146. *Hibernium* is derived from Hibernia, the Roman name for what is presently Ireland, conquered by Julius Gnaius Agricola (40–93 CE), governor of Brittania, during the reign of Domitian.
147. Nordenskjold, I. *Bull. Geol. Inst. Upsala* **1923**, 9, 5.
148. Joly, J. *The Surface History of the Earth;* Clarendon Press: Oxford, 1925, p. 192.
149. Joly, J. *Nature* **1928**, *121*, 207.
150. de Hevesy, G. *Nature* **1932**, *130*, 846.
151. Rolla, L. *Atti dei Lincei* **1933**, *131*, 472.
152. Curie, M.; Tackvorian, S. *Compt. Rend. Chim.* **1933**, *196*, 933; Weaire, D.; Coonan, S. *Europhysics* **2001**, *32*, no. 2.
153. *Obituary Notices of the Royal Society of London*, no. 3, December 1934, 259.

IV.5

IF ANYONE HAS A SHEEP, WOLFRAM WILL EAT IT

Around the middle of the 16th century, Georgius Agricola[154] referred to a mineral called *wolf froth* (or foam),[155] which today we know by the name of wolframite[156] and in which was found a new element. Yet, with the discovery of wolfram (tungsten), some chemists speculated that its cunning oxides would conceal a new element, *neo-tungsten*. An analogous but fruitless investigation was carried on searching for so-called *neo-molybdenum*. The exhausting hunt for the two neo-elements started in the middle of the 19th century and had its unhappy conclusion in 1919.

In 1761, Johann Gottlieb Lehmann (1719–67)[157] melted wolframite with sodium nitrate and found that the fusion product dissolved in water, coloring the solution green, and that it later turned red because of the manganates and permanganates present. Lehmann added sulfuric acid and obtained a white spongy precipitate that turned yellow on long exposure to air. Eighteen years later, in 1779, the chemist Peter Woulfe (1727–ca. 1805) roasted some samples of wolframite with hydrochloric acid and obtained a product with a bright yellow color that made him hypothesize on the existence of some new elements within the mineral. In 1781, the renowned Swedish chemist Carl Wilhelm Scheele[158] analyzed a white mineral called tungsten (later called scheelite for obvious reasons). He perceived that this mineral was the calcium salt of a mysterious new acid that he wanted to call tungstic acid.[159] Of a contrary opinion was his countryman Torbern Olof Bergman,[160] who thought that the cause of the high density of scheelite was due to the presence of barium oxide and not to a new element. Since he was a very good chemist, he conducted his analyses scrupulously, and, when he found that the content of the mineral was siliceous rather than alkaline, as it would have to be if he were dealing with barium oxide, he was somewhat puzzled, but quick to recognize that his hypothesis was erroneous. Later, he realized that tungstic acid was the oxide of a new element,[161] one that he called *lapis ponderosus* or "heavy stone." The Latin name never took hold; on the contrary, people always seemed to use the Swedish translation that we use today: tungsten. The credit for having isolated the metal goes to two young Spanish noblemen: in 1783, Juan José Elhuyar y de Zubice,[162] a student of Bergman at Uppsala, together with his younger brother Fausto de Elhuyar y de Zubice, analyzed wolframite and found tungstic oxide.[163] Heating the tungstic acid (oxide of W) with charcoal powder at very high temperatures, they succeeded in reducing the element to the metallic state. They presented their discovery at the Academy of Sciences in Toulouse[164] on March 4, 1784: "we would like to call this new element volfram, borrowing this name from the matter from which it was extracted. . . this name is more suitable than tungust or tungsten because wolframite is a mineral that was known much earlier than tungsten." Curiously, in modern Swedish, the element is normally called *volfram*

FIGURE IV.05. Carl Wilhelm Scheele (1742–86). One of the greatest chemists of the 18th century, in his pharmacy his famous experiments allowed him to isolate oxygen and study its behavior in combustion. In addition, he discovered tungsten, molybdenum, nitrogen, chlorine (through the reaction of hydrochloric acid with manganese dioxide), and manganese. After his untimely death at the age of 43, it was proposed (by Martin Heinrich Klaproth) that a new element, *scheelium*, be named after him. Courtesy, Chemical Heritage of the Department of Chemistry of the University of Florence, Italy.

as the de Elhuyar brothers had suggested[165] and not *wolfram* as is the usage in the other Germanic languages.[166]

In 1811, Martin Heinrich Klaproth proposed the name *scheelium*,[167] in honor of the metal's discoverer. The renowned Jöns Jakob Berzelius, who initially supported this name, rethought his position a little later and changed his opinion, openly opposing the proposal to honor the memory of Scheele (Figure IV.05) in this way[168] and justifying his ill-concealed jealousy with the following words:[169] "C. W. Scheele had already immortalized his name by other great discoveries to such an extent as to preclude the necessity of its being handed down to posterity by the denomination of a substance."

IV.5.1. THE NEIGHBORS OF MOLYBDENUM AND TUNGSTEN

On November 22, 1915, a Monsieur Gerber[170] deposited a sealed packet at the Académie des Sciences. Almost 2 years later, he requested that the Académie break the seals and that the document be made public. Thus it was that between April and October of 1917, a long monograph appeared carrying the title "A la recherche de deux métaux inconnus," subdivided for convenience into four articles.[171] M. Gerber was not a university professor but an able amateur, with his own chemical laboratory in the city of Clermont-Ferrand. However, not having the spectroscopic apparatus necessary to analyze purified samples, Gerber turned to a renowned physicist, most probably Antoine Arnaud Alfred Xavier Louis de Gramont (1861–1923) for help.[172]

Gerber began with a study of the periodic classification proposed by Dmitri Mendeleev and observed that the manganese triad had remained incomplete for too long, despite European chemists searching for the two elements—called provisionally *eka-manganese* and *dvi-manganese* (indicated by Gerber by the symbols Km and Dm)—for almost half a century. Gerber himself furnished the reasons that drove him to dedicate himself with passion to this fascinating but unfortunate hunt for the missing elements: "The idea of my work came to me after long reflection on the gap, never explained, present in the Mendelevian series of elements, a problem still open [and vigorously debated] in the most recent studies on matter."

The fact then that both eka-manganese and dvi-manganese were not discovered for such a long time suggested to Gerber that the hypothesis based on Mendeleev's periodic law, despite the fact that it had predicted the existence of 10 elements, needed to be corrected. Gerber suggested that the elements with atomic numbers 43 and 75 ought to be looked for not among the minerals rich in manganese but among the elements in the preceding triad: Cr, Mo, and Tu (W). Initially, Gerber looked for the missing elements by carefully examining a great number of mineral samples coming from Germany, and he did not skimp on the efforts he used to also analyze commercial alloys based on manganese in the secret hope of finding traces of the two elusive elements. For this reason, he bought 50 kg of ferromanganic discards from the high ovens of the Société d'Outreau. From this sample, he extracted 27 g of metallic sulfides that, on successive chemical treatments, released about 9 g of molybdic acid that he very carefully compared with pure commercial samples; the chemical reactions, both qualitative and quantitative, that he conducted on his two samples did not fit together at all. Gerber, in his elegant and discursive prose, reported the following assertion:

> From this moment, the following questions were ones that demanded much of my attention: is molybdenum a simple substance? Can the same be true of tungsten? These two substances can't be anything other than mixtures, in various proportions, of two respective metals, very close to each other not only with respect to atomic weight and density, as laid down by the law of Mendeleev, but also with respect to many other similar chemical properties.

It was not the first time that chemists noted the complexity of the compounds of tungsten—and especially the tungstates—and also observed how incomplete and full of gaps their knowledge of their composition was. In 1847, Auguste Laurent (1807–53) asserted that he had prepared an ammonium iso-tungstate whose existence had always been held in doubt.[173] In 1860, perhaps the last scion of the Bernoulli dynasty, the chemist Friedrich-Adolph Bernoulli (1835–1915),[174] earned his PhD publishing as his thesis *De wolframio nonnullique ejus conjunctionibus*.[175] The following year, C. Scheibler was the first to hypothesize that tungsten might be in reality a combination of two or more elements.[176] According to the conjectures advanced by Gerber, tungstic acid[177] could contain *neo-molybdenum*.

At about the end of 1913, Gerber went on to analyze the major minerals rich in tungsten and molybdenum in his hunt for *eka-manganese* and *dvi-manganese*: molybdenite and thorianite. In particular, from a variety of Australian molybdenite coming from Glenn Innes in New South Wales, by successive fractional crystallizations, he realized that the metal, present in the most soluble sample of ammonium molybdate, showed an atomic weight higher (99.9) than that of molybdenum (96). The other perceptible difference between Mo and *neo-molybdenum* lay in the volatility of the oxides obtained by the decomposition of the ammonium salts. These properties were greater in the supposed compounds of the new metal. Gerber, after having thoroughly analyzed and characterized his samples, felt ready to announce that "the last sediments contained another material, probably eka-manganese, for which Mendeleev's classification predicted an atomic weight of around 100, and which I provisionally will call neomolybdenum."

Likewise unexplainable were both the chemical properties and spectroscopic properties of the two elements. To get to the bottom of this puzzle, Gerber ended up nurturing the hypothesis that *neo-molybdenum* and molybdenum could be two "metal-isotopes." This bizarre idea arose in Gerber's mind after he had sent his samples of isotungstic anhydride and polytungstic acid[178] to M. de Gramont for a spectroscopic examination. He was, in fact, so convinced that the isotungstic anhydride would contain the much sought-after *dvi-manganese* that he labeled "*sel nouveau de Dm*" on the test-tube that he sent to the spectroscopist. Gerber was likewise convinced that *neo-molybdenum* was not identical to *nipponium*, discovered shortly before by Masataka Ogawa,[179] asserting that "the metal that I isolated is completely different from Ogawa's." Unfortunately, de Gramont's report was not positive. In a few words, he summarized the concept that Gerber had expanded upon in four articles of more than 30 pages in length: "no new metal: only tungsten." The spectroscopic results would have discouraged any other scientist, but not Gerber. He simply adopted a new atomic theory that was more in accord with his needs: "And here's the question. How does one reconcile this result with the following determinations that point to a different metal both with respect to atomic weight and 'crystallinity' that does not exist in any other degree except in pure tungsten? One knows that I have dealt with this blind alley appealing to a new fact, as yet little studied, isotopy."

Gerber arrived at some erroneous conclusions because, fundamentally, he was an amateur: a good amateur, but only and always an amateur. In his articles, he never cited original work except in French. In addition, he was one of the last chemists to use the obsolete apical (superscripted) stoichiometric notation common in France until the end of the 19th century. Whether he freely twisted the concept of the isotope for his own gains or simply because he did not understand it is not clear from his writings. In any case, Gerber committed his errors in good faith. He asserted that de Gramont's discordant results could be explained thus: "in accordance with the most recent discoveries that have produced among physicists the idea of metal-isotopes, with the concept that there exist in nature spectroscopic doubles that can conceal the simple substances."

In the same way, Gerber believed that tungsten was not a unique element but concealed another inside itself, with chemical and physical properties so similar that they were indissolubly bound to one another. Because of the metal-isotope, the two metals produced, when analyzed spectroscopically, the same "image." Gerber also coined a name for this element and later determined its atomic weight: "I have not been able to determine up till now the heavy metal that I isolated from Tungsten of the type R_2O_7, that represents Dvi-manganese predicted by Mendeleev. I will provisionally call it Neotungsten."

He believed that metatungstic acid would have produced isotungstate, as described by Laurent. The isotungstic acid extracted from the minerals and subjected to predetermined chemical processes would contain the new metal, one with an atomic weight of about 187, as opposed to 184, the commonly accepted value of tungsten. Curiously, the fourth and last article relative to the presumed discovery of *neo-molybdenum* and *neo-tungsten* emphasized that his reasoning had been honestly guided by a purely intuitive hypothesis. His work closed by recommending his discovery to further research by his successors.

Sadly, there were no successors to Gerber, and the two new elements were never isolated. In fact, the results he arrived at were never confirmed by anyone else; on the contrary, P. Barbe, who only casually pursued the subject, in attempting to repeat Gerber's experiments found no evidence of the existence of the isotungstates nor did he record abnormal values for the atomic weight of the samples of tungsten analyzed.[180] Inadvertently, Barbe proved Gerber's error: his ammonium isotungstate was nothing more than sodium ammonium tungstate.[181]

Notes

154. A German scholar and a man of learning, known as the "Father of Mineralogy." His true name was Georg Bauer; Agricola is the Latinized version of his name; *agricola*, "farmer," is the Latin translation of the name Bauer. His most famous work was *De Re Metallica*, Books I–XII, a complete and systematic treatise on metallurgy and techniques for extracting metals from their ores.
155. The word "wolfram" is composed of the roots "wolf" and "rahm" (froth or slobber). The bizarre name of this mineral makes reference to its chemical properties: wolframite "eats" tin; that is, it reacts with it, forming a froth-like product like a hungry wolf (with slobbering mouth) eats sheep.
156. Wolframite: mineral with the empirical formula (Fe, Mn)WO_4; another name for wolfram, the element extractable from wolframite, is tungsten, taken from the mineral tungstenite and having the formula $CaWO_4$; in Swedish, its name means heavy (tung) stone (sten).
157. German geologist, director of the Imperial Museum of St. Petersburg, Lehmann was the first chemist to study wolframite in depth. His reports were published in 1761 in "*Physikalisch-chimische Schriften*": 82 pages of extraordinary scientific quality. He synthesized tungstic acid (later named by Scheele) but was unsuccessful in identifying the new chemical element it contained.
158. Scheele, one of the greatest experimental chemists of all time, was born in Pomerania during the era that it was under Swedish rule. He moved to Sweden, where he worked in a small pharmacy. Later, he became interested in chemistry and became a member of the Stockholm Academy of Sciences.
159. Scheele, C. W. *Akad. Handl. Stockholm* **1781**, *2*, 89.
160. Professor at the University of Uppsala. Bergman was one of the most outstanding analytical chemists of the 18th century.

161. In his article, Bergman did not forget to give Scheele the credit for the discovery of the new metal.
162. Later, Fausto de Elhuyar was made part of the Royal Spanish Commission for the creation and organization of the School of Mines of Mexico City and, in this position, planned the building, an architectural jewel known as the Palacio de Mineria. Elhuyar left Mexico at the end of the War of Independence in 1821, when the majority of the Spanish residents were expelled.
163. Whitaker, A. P. *JSTOR* **1951**, *31* (4), 557.
164. De Elhuyar, J. J.; De Elhuyar, F. *Mem. Acad. Toulouse* **1784**, *2*, 141.
165. The brothers de Elhuyar used "v" and not "w" because the latter did not exist in the Spanish alphabet until 1914.
166. Between the end of the 19th and the beginning of the 20th centuries, the symbol Tu was used for this element in Italy and France. Recently, the IUPAC Commission removed the name "wolfram" and other terms derived from it from its Red Book, which has engendered some controversy. See, for example, http://www.iupac.org/publications/ci/2005/2704/ud_goya.html (accessed April 13, 2014).
167. According to some, the name proposed was *Scheelerz*.
168. Berzelius, J. J. *Schweigger's Journ.* **1816**, *16*, 476.
169. Mellor, J. W. *Comprehensive Treatise on Inorganic and Theoretical Chemistry*; Longmans Green: London, 1956; vol. XL, p. 674.
170. There is practically no biographical information about Gerber except for the *Poggendorff Biographishces Wörterbuch* of 1904, which speaks of a certain Carl Ludwig Gerber born in Germany on January 1, 1854. Also, supposing that the "M." in front of Gerber's name means "Monsieur," it seems that Gerber published his articles for the most part in a French journal during World War I.
171. Gerber, M. *Le Moniteur Scientifique Quesneville* **1917**, *7*, 73; Gerber, M. *Le Moniteur Scientifique Quesneville* **1917**, *7*, 121; Gerber. M. *Le Moniteur Scientifique Quesneville* **1917**, *7*, 169; Gerber, M. *Le Moniteur Scientifique Quesneville* **1917**, *7*, 219.
172. Antoine Arnaud Alfred Xavier Louis de Gramont, Duke of Cavadal, was a direct descendant of the King of Navarre. Born on September 21, 1861, after studying organic chemistry with Charles Friedel, he later specialized in the new discipline of spectroscopy under the authoritative guidance of Paul Emile Lecoq de Boisbaudran. In 1913, he was elected a member of the Académie des Sciences. He died on October 31, 1923. It is not certain if the physicist mentioned by Gerber was indeed him or his son, Armand A. Agenor, Count of Gramont (1879–1962), who, like his father, had completed spectroscopic studies and founded the Institute of Applied and Theoretical Optics.
173. Laurent, A. *Ann. Chim. Phys.* **1847**, *21*, 54.
174. Friedrich-Adolph Bernoulli was born on September 16, 1835, and probably died shortly after the outbreak of World War I. He was still living in 1901, as reported by Ugo Schiff, who updated in his own hand his personal copy of the *Poggendorff Biographishces Wörterbuch*. Since Schiff himself died on September 8, 1915, we surmise that Bernoulli's death occurred some time later than Schiff's.
175. Bernoulli, F.-A. Of wolfram and some of its compounds. *Ann. Phys. Chem.* **1860**, *111*, 573.
176. Scheibler, C. *J. prakt. Chem.* **1861**, *83*, 273.
177. According to Gerber, isotungstic acid had the empirical formula $H_2W_2O_7$; it also could be found in the hydrated form: $H_2W_2O_7 \cdot H_2O$.
178. According to Gerber, respectively, W_2O_6 and $H_6W_6O_{21}$.
179. Ogawa, M. *J. Coll. Sci., Imp. Univ. Tokyo* **1908**, *25*, Art. 15; Ogawa, M. *J. Coll. Sci., Imp. Univ. Tokyo*, **1908**, *25*, Art. 16.
180. Barbe, P. *Le Moniteur Scientifique Quesneville* **1919**, *9*, 73.
181. $(Na,NH_4)WO_4\ 2H_2O$.

IV.6.

WHEN IT COMES TO NEW DISCOVERIES, THE MORE YOU ERR, YOU END UP ERRING MORE

"If you love research and if you cultivate her without pride, she will never be stingy with results."[182] With this sentence, Professor Giorgio Piccardi concluded his last lecture and took his leave of his students, exhorting them to uncover the secrets that Nature still held in reserve.

Making discoveries is part and parcel of a scientist's career. Some of these discoveries will be proved incorrect, others blatantly false; this is the common fate of the scientific life. You advance by trial and error until you arrive at the truth. And it's only natural that young and zealous researchers may commit more errors than their hoary colleagues. And then we find the elderly Josef Maria Eder, who was at the same time both executioner and victim in his own strange case.

At the height of his career and his fame, and already more than 60 years of age, Eder entered a field of research relatively new to him: the isolation of the last of the rare earth elements. Unfortunately, this was a field also full of traps and snares. Between 1916 and 1923, he was dragged along by the euphoria of ever rasher claims and in this way managed to collect at least five counterfeit discoveries.

The embarrassment of being forced to make a retraction after a discovery shown to be false costs the researcher a great deal, both emotionally and on an academic level. Eder proved an exception: although he had announced the discovery of five new elements, he never took the trouble to rectify his position and lived in apparent bliss until he was almost 90.

Josef Maria Eder was born in Austria, at Krems on the Danube, on March 16, 1855. After attending the local high school, he moved to Vienna. In the Habsburg capital, he took courses both at the university and at the polytechnic institute. He took an early interest in the chemical basis of the rising discipline of photography. With Viktor von Tóth, he developed methods of coloring photographs with ferricyanide. Around 1879, along with G. Pizzighelli (1849–1912), he introduced important modifications to photographic plates, using gelatin impregnated with silver chloride. Some time later, Eder perfected impressionable gelatin, thus introducing the use of silver bromide. These two discoveries won him immediate success and allowed him to produce photosensitive photographic film on a grand scale. In 1882, Eder was named professor of chemistry and physics at the most prestigious professional school in Vienna. On March 1, 1888, in a renovated building that he owned, he founded the Höhere Graphische Bundes Lehr (the Federal Advanced School of Graphic Arts).[183] One year later, he was elected its director.

In these years, he developed the concept of photochemistry, although his interests would always remain linked to the more practical aspects of the discipline. With a staff consisting of 10 technicians and 108 students, Eder was able to accomplish important and innovative experiments in photometry and X-ray photography.

In 1892, he was invited to teach at the Technische Hochschule (the Polytechnic Institute of Vienna).[184] In 1885, Eder had married Anna Valenta. With his brother-in-law, Hofrat Eduard Valenta, he began a fine scientific collaboration, with the work done entirely in the laboratory of the Federal Advanced School of Graphic Arts. Under their direction, this institute became an important center both as a resource and for technical and scientific instruction.

Just before the beginning of World War I, Eder, by now in his 60s, felt ready to make his great leap forward: from applied to pure research. Using high-resolution spectroscopes and the best state-of-the-art photographic equipment, he engaged in the identification of some of the missing elements.[185] Unfortunately, Eder was not current with the advances of science. His great limitation, and one that conditioned all of his research, was his lack of updating.

To characterize the presumed new elements, Eder availed himself of visible absorption and emission spectra. Before him, scientists like Robert Bunsen, Sir William Crookes, Paul Emile Lecoq de Boisbaudran, and others had already used the same investigative techniques. But all of these scientists belonged to the preceding generation. Eder's own generation, represented by scientists like Bohuslav Brauner and Carl Auer von Welsbach, worked in a completely different way. Brauner abandoned the search for new elements at the beginning of the century, after realizing that researchers better-trained than he were superseding his work; Auer, on the other hand, remained active in rare earth research all of his life. The constant contributions of the young scientists in his laboratory constituted an important resource in Auer's work, and the freshness of new investigative techniques and the zeal brought to them by his young researchers kept his own research current with the times.

In the 1910s, when Eder, by now no longer young, undertook his research, all the research groups were already using X-ray spectrometers to find the lines characteristic of new elements in their X-ray spectra. The ranks of new "element hunters," both distinguished or simply engaged in this work during the first quarter of the 20th century, included Georges Urbain, B Smith Hopkins, Charles King James, Luigi Rolla, George de Hevesy, Walter Noddack, and Ida Tacke Noddack. All of them made their progress based on the more recent discoveries in physics. If Eder had truly known of Moseley's discovery in 1913, he would not have imagined that the rare earths could contain as enormous a number of elements as he proposed.

In 1916, Eder presented to the Academy of Sciences of Vienna some measurements of wavelengths of many rare earth elements:[186] *cassiopeium* (Cp[187]), *aldebaranium* (Ad), erbium (Er), and thulium (Tm). He had obtained the spectral lines of all of these elements using a concave diffraction grating, of the type built by the renowned American physicist Henry Augustus Rowland III. Eder had compared the results he obtained with the arc spectrum of iron. Together with this work, he reported the carbon arc spectrum of *cassiopeium* (lutetium). In the region between 7,237 and 2,392 Å, the 260 emission lines obtained by exciting the metal appeared. In addition to the classic line spectrum, lutetium also presented a characteristic band. Eder observed many more lines (630 of them) in the arc spectrum of *aldebaranium* (ytterbium). Of these, he believed that only 422 belonged to ytterbium: the others were those of *cassiopeium* and thulium.

A few years earlier, Franz Exner and Eduard Haschek had found five lines in common in the spectra of extremely pure *cassiopeium* and *aldebaranium*.[188] They believed that these lines were due to a trace of a new element present in the two samples. They called it, provisionally and without much imagination, *element X*. Eder repeated Exner and Haschek's experiments and recognized the inconsistency of their work in that, of the five lines of the presumed *element X*, four belonged to a trace of *aldebaranium* in the sample of $Cp_2(SO_4)_3{\cdot}8H_2O$ and the other to thulium, also present in their samples of $Ad_2(SO_4)_3{\cdot}8H_2O$.

Eder's account continued with the study of a third preparation that had attracted his attention because of its complexity. The samples that he analyzed had been prepared with great care by his friend and countryman, Auer von Welsbach. Auer had separated the sample of thulium into three fractions. The two chemists were not actually convinced that thulium was an element, but thought that it might be a mixture of simpler substances. In the first preparation, which Welsbach called *aldebaranium-thulium* I, Eder perceived some new lines that he felt were due to the presence of a new metal, one he called *denebium*. Welsbach's second (and very impure) fraction, *aldebaranium-thulium* II, showed mostly thulium lines. In honor of the work of Per T. Cleve, the discoverer of thulium, Eder proposed that this be called *neothulium*, slightly modifying the name proposed by Cleve in 1879. The third fraction prepared by Welsbach with the name *aldebaranium-thulium* III gave a series of new spectral lines that Eder immediately attributed to the presence of a third element. He called the new element *dubhium*. The presentation he made before the Academy of Sciences ended with the analysis and interpretation of the spectra of samples of erbium. In this case as well, Eder was certain that erbium was not a simple body but a mixture of several elements. For the moment, he could not say how many, so he limited himself to calling it a complex. In giving names to the elements found in the fractions of *aldebaranium-thulium* I, II and III, he wanted to imitate his great friend Auer von Welsbach. The latter had given the names *aldebaranium* and *cassiopeium* to the two new metals with atomic numbers 70 and 71, names derived from the constellation Cassiopeia and the star Aldebaran. Unfortunately, Auer had been preceded by a few months by the untiring work of Georges Urbain and the only names now remaining were those proposed by Urbain: *neo-ytterbium*[189] and lutetium.[190] Eder was so convinced of his results that, together with the names, he also proposed symbols for the new elements. *Denebium*, whose name derived from the star Deneb, in the constellation *Cygnus* (the Swan), had the symbol De. The second element had the privilege of not having its name, *neothulium*, changed drastically, but only its symbol, Nt. Finally, the third element took the name *dubhium* from the star Dubhe.[191] (The name Dubhe, "the bear," is derived from the Arabic phrase "thahr al dubb al akbar," which means "the back of the great bear"; the star is found on the back of the constellation Ursa Major.) The symbol proposed for this element was Du, although in some later publications one finds it replaced by Db.[192]

In 1917, the year following the announcement of the discovery of the elements De, Nt, and Du, Eder presented a second communication to the Viennese Academy of Sciences.[193] The idea for this long dissertation coincided with the study left suspended the year earlier on the complex the nature of erbium. He skipped the study of erbium, although the year earlier he had suggested that it was a mixture.

In the second communication, Eder said that he had examined samples of europium in depth. Europium's spectrum was easy to recognize because its characteristic lines were particularly bright and very obvious. Eder was sent some preparations of europium by

Georges Urbain in order to compare them with those he had received from Auer von Welsbach. According to Urbain, his samples contained very pure europium. He also examined two preparations by Auer von Welsbach, one coming from a fractionation of gadolinium and the other of samarium. Using a Rowland-type spectroscope, he measured 1,171 lines of europium. According to Eder, some of the lines resulting from samples not polluted with samarium belonged to a new element that he did not hesitate to call *eurosamarium*.

Toward the end of World War I, Eder studied the spectrum of dysprosium. This element has atomic number 66 and follows terbium in the periodic table. He was deceived by some impure fractions of terbium and dysprosium. To have full certainty and to verify his intuition, Eder was sent a sample of very pure $Dy_2(SO_4)_3 \cdot 8H_2O$ by his friend Auer. Eder changed the sulfate into the more volatile chloride and studied the entire spectrum of dysprosium. In fact, the arc spectra of the rare earth chlorides give better resolution than the sulfates. Eder counted 4,385 lines, some of which were unknown, in the red and yellow regions and also near the ultraviolet region. Dysprosium's lines are predominantly in the green region. Eder tried in various ways to separate out the element that he believed was present in the dysprosium sample but did not succeed, not so much because of his fractionation techniques (very good in themselves) or purity-checking spectroscopic investigations (also the best), but for the simple fact that the dysprosium Auer von Welsbach gave him was a simple body and could not be split into more elements. Because Josef Maria Eder did not feel very secure about the results of his research, he declined to name the presumed new element.[194]

The last stop on Eder's express train to oblivion began in 1920 but had its roots decades earlier. In 1909, Eder and his brother-in-law Valenta had spectroscopically studied the elemental nature of terbium. Unlike the major spectroscopists of his time, Eugène A. Demarçay, Marc A. Delafontaine, and Georges Urbain, Eder did not believe that terbium was an element but a mixture of simpler substances. In the same work, Eder rejected the research of his two colleagues, Exner and Haschek, who, in studying some of Auer von Welsbach's preparations, believed that they had discovered a new element between Tb and Gd, which they called element "*E*." Eder was clever at reconstructing the spectrum of the hypothetical element "*E*" showing that it was the superposition of some lines of gadolinium with those of terbium. However, he was blind in the face of the same error that he himself was committing: he announced a discovery based on the same erroneous presuppositions that he had demolished. According to Eder, the last elements to be discovered would not lie between gadolinium and terbium, but between dysprosium and terbium. Thus he made his fifth erroneous announcement, asserting that he had found 300 new lines of an unknown element in the fraction of terbium and dysprosium furnished by Auer von Welsbach. Eder wanted to call this new element *welsium* in honor of his "private source" of rare earth samples, the baron Carl Auer von Welsbach.[195]

Later, Eder compared Auer's samples with those of two other colleagues, Eberhard and Urbain. The samples coming from France did not match the others. At first, this could have been an unpleasant event for Urbain. But with the passage of time, Eder's studies turned out to be a dangerous weapon against his friend Auer instead. Auer's samples turned out to be very impure. Urbain discovered, in fact, that the spectrum of a rare earth element could be changed perceptibly by the addition of traces of other elements. In a masterful piece of analytical work, Urbain refuted the existence of the *meta-elements* of Crookes, who 10 years before Eder had made the same errors. The only victor in this struggle between intricate spectroscopic lines was Urbain, who received double satisfaction. He discovered the cause of the errors, proving at the same time the greater purity of his own samples compared to those of his rival, Auer von Welsbach.

Despite all these events, in 1922, Eder reproposed the discovery of a new element between terbium and dysprosium. Unlike the announcement in 1920, he did not limit himself to reporting a list of 300 spectral lines but instead mapped out the entire new rare earth element. The complete count of the lines of *welsium* amounted to several thousand.[196]

In his "other" life, Eder was among the cleverest of photographic technicians. He was among the first to obtain X-ray photographs. He created a personal photographic collection that, in 1922, he sold to the Kodak Company. In 1949, one of the greatest photographic collections in the world belonged to the Eastman Kodak Research Laboratory, created over the years by the inventor and pioneer of photography George Eastman (1854–1932). A considerable part of this vast collection still comprises the Cromer Fund and the collection of Josef Maria Eder.

In 1923, Josef Maria Eder left the directorship of the Federal Advanced School of Graphic Arts to his brother-in-law Hofrat Eduard Valenta and, in the next year, he retired from university teaching. He continued, on and off, to publish works on spectroscopy. For the rest of his life, he was an exemplar of the times: in 1918, he saw the fall of the Habsburg Empire and the flight of the Emperor Charles I (1887–1922) from Vienna; in March 1938, by now quite elderly, he witnessed the *Anschluß*.[197] Eder published his last work in March 1938, a few days before the annexation of Austria to Nazi Germany. Embittered, he left Vienna and retired to the Tirol. Josef Maria Eder died at Kitzbühel, on October 18, 1944, during the most dramatic part of World War II, when the "relief map" of his native land was literally flattened by heavy Allied bombing.[198]

Notes

182. Personal communication to Mariagrazia Costa (1965).
183. Federal School of Advanced Graphic Arts.
184. Vienna Polytechnic Institute.
185. Eder, J. M. *Sitzungsber. K. K. Akad. Wiss. Vienna IIa* **1915**, *123*, 2290.
186. Eder, J. M. *Sitzungsber. K. K. Akad. Wiss. Vienna IIa* **1916**, *124*, 242; Eder, J. M. *J. Chem. Soc.* **1916**, *110 II*, 277.
187. For more than 40 years from the date of their isolation in 1907, until 1949, the elements with atomic number 70 and 71 had double names: respectively, *aldebaranium* and *cassiopeium*, in the German-speaking world; ytterbium and lutetium in the English- and French-speaking countries. In 1949, IUPAC sought to bring these names into line. The International Commission for Inorganic Chemical Nomenclature officially declared ytterbium and lutetium as the names of elements 70 and 71, recognizing the priority of Urbain's discovery, as well as the names that he proposed.
188. Exner F.; Haschek, E. *Sitzungsber. K. K. Akad. Wiss. Vienna IIa* **1913**, *121*, 1075.
189. The name *neoytterbium*, proposed by Urbain in 1907, was unexpectedly changed to the present-day ytterbium.
190. Urbain, G. *Compt. Rend. Chim.* **1907**, *145*, 1335.
191. Eder, J. M. *Sitzungsber. K. K. Akad. Wiss. Vienna IIa* **1916**, *125*, 1467.
192. By sheer coincidence, in 1996, the IUPAC Commission for Inorganic Chemical Nomenclature gave the same symbol, Db, to dubnium, element 105.
193. Eder, J. M. *Sitzungsber. K. K. Akad. Wiss. Vienna IIa* **1917**, *126*, 473.

194. Eder, J. M. *Sitzungsber. K. K. Akad. Wiss. Vienna IIa* **1918**, *127*, 1100.
195. Eder, J. M. *Sitzungsber. K. K. Akad. Wiss. Vienna IIa* **1920**, *129*, 421.
196. Eder, J. M. *Sitzungsber. K. K. Akad. Wiss. Vienna IIa* **1922**, *131*, 199.
197. "Union" with Germany. The term *Anschluß* means the end of the Austrian Republic, which arose in 1918 from the ashes of the thousand-year old Holy Roman Empire and its successor, the Austro-Hungarian Empire.
198. *Photographische Korrespondenz, Chem. Soc. Transactions*,1955, Nr. 1091 der ganzen Folke, Nr, 6/1955, 91.

IV.7

THE RADIOACTIVE ELEMENT OF THE HOT SPRINGS

During the November 21, 1921, session of the *Académie des Sciences de Paris*, physical chemist Paul Alfred Daniel Berthelot (1865–1927), son of world-renowned Marcellin Pierre Eugène Berthelot,[199] presented—in the name of the chemist Pierre Loisel—a rather curious communication.

Immediately after the end of World War I, Loisel began to investigate the composition of the waters in the hot springs[200] scattered around the region of Bagnoles-de-l'Orne.[201] He realized that these waters contained volatile radioactive substances dissolved within them,[202] substances whose activity underwent seasonal variations.[203] Loisel accomplished his research on many springs in a relatively widespread area—about 100 square kilometers[204]—and accurately measured the content of "emanation of radium"[205] dissolved in every sample of the water.

In 1921, Loisel succeeded in developing a technique to quantitatively measure the content of the "emanation of radium" present in the water examined. He made use of a gold-leaf electroscope, deemed necessary to correct the value of the ionization current of the air carried by the penetrating radiation. The method described by Loisel allowed him to calculate the current due to the ionization in part produced by the characteristic radiation of the "emanation of radium" and in part to the radiation of radium A, B, and C. The value of the current produced by the "emanation of radium" was calculated in a relatively simple way: he repeated the measurements, introducing fresh air into the electroscope; in other words, air free of the "emanation." Then he recorded the value of the current due to the radiation from radium A, B, and C and, by subtraction, obtained the data he was looking for.

In 1922, Loisel, together with his colleague M. Michailesco, observed the presence of the "emanation of radium" dissolved in the waters of the thermal locale of Băile Herculane[206] in Romania.[207] In a later work,[208] Loisel observed that the gas dissolved in a certain number of the hot springs gave an activity curve that could not be explained solely by the presence of the known "emanation"; that is, of radium. From the curves, he calculated that the half-life of the new radioactive product was 22 minutes. The variations in current on the curves that he recorded, however, were of the order of magnitude of the natural loss of electroscopic charge, and this ought to have alarmed the scientist. At first, this was so: Loisel admitted that additional investigations would be needed to clarify the presence or absence of a new radioactive body in his samples. Nevertheless, as a precautionary measure, although the data necessary for characterization were full of holes, Loisel concluded his article by proposing that he had determined the presence of a new element he called *emilium*.

In 1924, after some years of silence interrupted only by the writing of a book on the radioactive waters in the region of Bagnoles-de-l'Orne,[209] Loisel returned to the subject of *emilium*, dusting off the data from his investigations of the radioactive springs.[210] Loisel was convinced that the waters' radioactivity was due to the continuous erosion of neighboring granite rocks by aqueous action. The rocks must be radioactive, and the water nothing more than a means of transporting the radioactivity. Therefore, he collected some of this granite and dissolved it with heated mineral acids; later, he collected the "emanation of radium" that had accumulated for various periods from 2 to 65 days, during a period of time that stretched from December 1921 to May 1922. Instead of the characteristic curve of radon, he observed a progress from a minimum of ionization current after 15 minutes and a maximum at 28 minutes. Loisel thought that this behavior was caused by an unknown radioactive gas having a half-life of about 22 minutes. This gas would have been generated by an unknown radioactive substance having a much longer half-life and belonging to a new radioactive series; this was the substance Loisel dubbed *emilium*.

After this last article, Loisel disappeared completely from the scene, never publishing another scientific work. The scientist Pierre Loisel, like his presumed *emilium*, had an extremely short half-life.

Notes

199. Marcellin Berthelot was a chemist and a French politician. In 1856, he synthesized methane, the lightest hydrocarbon, using its constituent elements as his starting materials. His success in synthesizing organic compounds in the laboratory threw into doubt the vitalistic theory then dominant, according to which organic compounds could only be created by living organisms. Berthelot, who contributed to the body of knowledge of almost every class of organic chemical compounds, carried on important research on explosives, colorants, and heat released in chemical reactions. He died on March 18, 1907, a few moments after his wife, Sophie Niaudet (1837–1907) breathed her last. He was buried in the Panthéon together with the great men and women of France.
200. Loisel, P. *Compt. Rend. Chim.* **1919**, *169*, 791.
201. The town of Bagnoles-de-l'Orne, located in lower Normandy, has taken on a sad reputation in Italy: June 9, 1937, the deaths of the antifascists Carlo Rosselli (1899–1937) and Nello Rosselli (1900–37) took place at the hands of French assassins hired by the Italian Organization for Vigilance and Repression of Anti-Fascism (OVRA). OVRA was the secret police of the Kingdom of Italy, founded in 1927 under the regime of Fascist dictator Benito Mussolini (1883–1945).
202. Loisel, P. *Compt. Rend. Chim.* **1920**, *172*, 1484.
203. Loisel, P. *Compt. Rend. Chim.* **1920**, *171*, 858; Loisel, P. *Compt. Rend. Chim.* **1921**, *173*, 920; Loisel, P.; Castelnau, R. *Compt. Rend. Chim.* **1921**, *173*, 1390.
204. Loisel, P. *Compt. Rend. Chim.* **1922**, *175*, 890. This is an area of about 39 square miles.
205. Emanation of radium, or simply "emanation" had for a certain period of time the symbol Em; presently, this element is called radon (Rn).
206. Originally, the place was known to the Romans by the name of *Aqua Herculis*, in the province of Dacia. The locality was known at the beginning of the 20th century not only for its very warm thermal waters in which large amounts of sulfur, chlorine, calcium, magnesium, and sodium were dissolved, but also for the unusual property of its air, the result of negative ionization.
207. Loisel, P.; Michailesco, M. *Compt. Rend. Chim.* **1922**, *175*, 1054.
208. Loisel, P. *Compt. Rend. Chim.* **1921**, *173*, 1098.
209. Loisel, P. *Recherches sur la radioactivité des eaux de Bagnoles-de-l'Orne*; L'Expansion Scientifique Française: Paris, 1922.
210. Loisel, P. *Compt. Rend. Chim.* **1924**, *179*, 533.

IV.8

MOSELEYUM: THE TWOFOLD ATTEMPT TO HONOR A HERO

Filling the gap between manganese and rhenium by discovering the last element of the second transition series was long a Holy Grail of science. It was reasonable to believe that this element, with atomic number 43, should concentrate during the fractionation process because its properties were intermediate between those of Mn and Re. The empirical law stating that element 43 (like element 61) should not exist was not proposed until 1934 and improved 6 years later, thus coming well after the most famous attempts to isolate this element were made.[211]

A tentative identification of element 43, much less known among the more famous blunders of *lucium, nipponium,* and *masurium*, was carried out by physicists Claude H. Bosanquet (b. 1896) and T. C. Keeley (b. 1894)[212] in July 1924. About a year earlier, the two researchers examined a large assortment of minerals rich in manganese in the secret hope of finding traces of the elusive higher homolog. The minerals used for analysis were varied (psilomelane, torbenite, rhodonite, rhodochrosite, polyanite, pyrolusite, wad, manganite, and franklinite) and came from the four corners of the globe. Claude H. Bosanquet had worked with the 1915 Nobel laureate in physics, William H. Bragg (1890–1971) and with him had published, between 1921 and 1924, three articles on X-ray diffraction. Using this investigative technique, the two physicists set off on the hunt for element 43, convinced that the more traditional chemical methods would not be sensitive enough to record the traces present in their samples. The two spectroscopists photographed the X-ray spectrum with wavelengths between 620 and 720 Xu (Siegbahn units) of each mineral.

The spectrum was accurately calibrated with the lines of the K-series of molybdenum. The two young researchers availed themselves of the help of Professor Frederick Alexander Lindemann (1886–1957), a member of the Royal Society, scientific advisor to Winston Churchill (1874–1965) during World War II, and supporter of the carpet bombing of German cities. In spite of all the precautions taken to increase the sensitivity of the roentgenographic instrument and reduce the dispersion to 10 Xu, they could not confirm the presence of element 43. A weak line coincident with the theoretical Kα line of the sought-after element was perceived (intermediate between the Kα and the Kβ lines of molybdenum), but the attribution was so uncertain that the authors preferred to simply say: "The results have, so far, been negative."

Claude Bosanquet did not publish any article on element 43 but, unlike Keeley, he continued to work in physics, remaining active until 1959, when his last scientific monograph appeared.

One year later, in 1925, Professor Richard Hamer of the University of Pittsburgh proposed that element 43—about which rumors were flying that its discovery was imminent—be called *moseleyum*, with the symbol Ms, a name that he said: "would be better and more international in character, like true science itself, than a Latinized name of the discoverer's own country." He already had in mind the work of Bosanquet and Keeley.

His proposal was ignored by the entire scientific community if one excepts the editor of *Nature*, who observed that it would indeed honor the memory of the famous physicist Henry G. J. Moseley[213,214] despite being "rather suggestive of certain sepulchral monuments."

Hamer's proposal preceded by only a few weeks the announcement of *masurium*'s discovery by chemists Walter Noddack and Ida Tacke Noddack and spectroscopic specialist Otto Berg; their discovery was named after the eastern German (Prussian) province of Masuria.[215]

In 1947, Austrian radiochemist and naturalized Englishman Friedrich Adolf Paneth, at the conclusion of the entire series of events leading to the discovery of element 43, said: "We may rejoice that the nationalistic discoverer, Walter Noddack, had no wish to celebrate Moseley, since masurium does not and did not exist."[216]

Although Hamer's unusual proposal quickly disappeared from the minds of most chemists and physicists, many decades later, in 2005, the celebrated writer/physician Oliver Sacks (b. 1933)[217] wrote to the editor of the journal *Chemistry International*.[218] After a lengthy panegyric directed to members of the IUPAC commission, who had recently given element 111 the name roentgenium, he asked which name would be chosen for the next element, discovered some years earlier, yet still provisionally "labeled" *ununbium*. Using his scientific prestige and vast popularity as a writer, he advanced the following proposal:

> Two names immediately come to mind—names of great pioneers from the heroic early years of the twentieth century. . . One such pioneer was Frederick Soddy, who worked with Rutherford in their crucial years in Montreal, defining new radioactive isotopes and their pathways of decay. [It was Soddy who coined the word isotope and in 1921 he was awarded the Nobel Prize in chemistry.] The other is Henry Gwyn Jeffreys (Harry) Moseley, the dazzling young theoretical physicist who worked out the real meaning of atomic numbers, and then, in principle, completed the Periodic Table by predicting the existence of elements 43, 61, 72, 75, 85, 87, and 91, stressing that these, and these alone, remained to be discovered. He thus, in Soddy's phrase, called the roll of the elements. Moseley was killed, tragically, at Gallipoli, in 1915—he was only 27, and there is no saying what he might have achieved had he lived.[219,220]

Perhaps the greatest monument to Moseley's work was not to immortalize his name with a new chemical element—an honor that he certainly deserves for having discovered the correlation between the atomic number and the frequency of emitted X-rays—but a scholarship endowed in his name. The first students to take advantage of the Moseley scholarship knew how to bring their own knowledge to fruition. Both received the Nobel Prize: the chemist Robert H. Robinson (1886–1975), who was Moseley's student, and the physicist Patrick Maynard Stuart Blackett, a dominant figure among British physicists following World War II, as well as president of the Royal Society.

Notes

211. Mattauch, J. *Z. Phys.* **1934**, *91*, 361; Mattauch, J., *Ergebnisse der exakten Naturwissenschaften*, **1940**, *19*, 206.

212. Bosanquet, C. H.; Keeley, T. C. *Phil. Mag.* **1924**, *48*, S. 6, no. 283, 145.
213. Hamer, R. *Nature* **1925**, *115*, 545; Hamer, R. *Science* **1925**, *61*, 208.
214. Rayner-Canham, G.; Zheng, Z. Naming elements after scientists: An account of a controversy. *Found. Chem.* **2008**, *10*, 13–18. This paper also documents the controversies surrounding the naming of curium and seaborgium.
215. Noddack, W.; Tacke, I. *Oesterreichische Chem. -Ztg.* **1925**, *28*, 127; Noddack, W.; Tacke, I. *Sitzb. Preuss. Akad Wissenschaften* **1925**, 400; Berg, O.; Tacke, I. *Sitzb. Preuss. Akad Wissenschaften* **1925**, 405.
216. Paneth, F. A. *Nature* **1947**, *159*, 8.
217. Sacks is an English neurologist and writer who lives and works in the United States; author of various best-sellers, his *Uncle Tungsten—Memories of a Chemical Boyhood* (Vintage Books, 2001) is certainly a masterpiece of "chemical literature."
218. Sacks, O. *Chemistry International* **2005**, *27*(4).
219. This unlikely name, followed by the elemental name *soddyum*, caused confusion with the element with atomic number 11, sodium.
220. Although many people think that it was Soddy who coined the word *isotope*, it was actually a suggestion made to him by a distant relative, Dr. Margaret Todd, a classical scholar. Soddy did not coin the name; he discovered, or rather intuited, the phenomenon. For his discovery of the isotope, in 1921, he was awarded the Nobel Prize in chemistry.

IV.9

THE INORGANIC EVOLUTION OF ELEMENT 61: FLORENTIUM, ILLINIUM, CYCLONIUM, AND FINALLY, PROMETHIUM

The story of element number 61 is so unusual that it deserves a special chapter. The study of the rare earths came to a climax during the years when chemists sought to organize the chemical elements according to a rule. Despite the clarity offered by the periodic system, the rare earths continued to be a source of consternation and confusion because their valences did not correspond to expected periodic trends. However, great progress continued to be made despite the lack of a viable theory. *Didymium* ceased being referred to as a single element because from it the French chemist Paul E. Lecoq de Boisbaudran had extracted samarium, while 6 years later Carl Auer von Welsbach separated *didymium* into two other elements: *neo-didymium* (neodymium) and *praeseo-didymium* (praeseodymium). In 1886, William Crookes[221] asserted, erroneously, that Nd and Pr were a mixture of several elements, among which lay at least two yet unknown:

> It is obvious. . . that the element giving the band at 475 cannot be the same as the one causing the band at 451.5 and if the body giving the strongest of these is called dysprosium another name must be chosen for the element which gives rise to the absorption-band 475. And now comes the question: What is the origin of band 475? In remarks made on the band 443 I mentioned that it is accompanied by other fainter lines. One of these occurs at 475, and therefore I was prepared to connect these bands as being due to one and the same element; but M. de Boisbaudran, in his description of the spectrum of dysprosium, has shown that band 475 can be obtained strong in the absence of band 443. The bands 443 and 475 therefore are not caused either by didymium, dysprosium, or any hitherto identified element; consequently each must be regarded as characteristic of a new body.

Crookes went on to say that he identified additional lines indicating hitherto unknown elements but gave them only provisional names. Table IV.2 shows Crooke's data.

This hypothesis was also stated by Eugène-Anatole Demarçay,[222] who declared that at least two simple bodies were present in samarium, whereas Henri Becquerel[223] was a bit more cautious, simply stating that the use of optical analysis on crystalline substances made it possible to recognize not only the presence of different bodies, but also that of various chemical groups for the same body. He also remarked that the observation of uneven displacement of spectral bands under the experimental conditions he described provided a method for characterizing chemically different substances. In the late 1880s, other chemists, among them C. M. Thompson,[224] P. Kiesewetter, and G. Krüss,[225] were on the hunt for what they suspected were one or more yet undiscovered elements hidden

Table IV.2 W. Crookes's Data Table Identifying Nine "New" Elements

Position of Lines in the Spectrum	*Scale of Spectroscope*	*Mean Wavelength of Band or Line*	*1/λ²*	*Provisional Name*	*Probability*
Absorption bands in	8·270°	443	5096	Dα	New
Violet and blue. . .	8·828	475	4432	Sβ	New
Bright lines in–					
Violet. . .	8·515	456	4809	Sγ	Ytterbium
Deep blue. . .	8·931	482	4304	Gα	New
Greenish-blue (mean of a close pair)	9·650	545	3367	Gβ	Gadolinium or Zβ
Green. . .	9·812	564	3144	Gγ	New
Citron. . .	9·890	574	3035	Gδ	New or Zα
Yellow. . .	10·050	597	2806	Gε	New
Orange. . .	10·129	609	2693	Sδ	New
Red. . .	10·185	619	2611	Gζ	New
Deep red. . .	10·338	647	2389	Gη	New

among the rare earths. A year earlier, Krüss and Lars Fredrik Nilson concluded, for example, that didymium extracted from eight different mineral sources did not contain only three metals, but was possibly a mixture of as many as nine new elements.[226] At the beginning of the 20th century, and as reiterated in a letter to *Nature* years later on the occasion of the announcement of the discovery of *illinium*,[227] Bohuslav Brauner stated that:

> I arrived at the conviction that the gap between the neodymium and samarium was abnormally large. In my paper read. . . in St. Petersburg in 1902, I came to the conclusion—not reached by any chemist before—that the following seven elements, possessing now the atomic numbers 43, 61, 72, 75, 85, 87, and 89,[228] remained to be discovered. As regards element No. 61, the difference between the atomic weights of Sm—Nd = 6.1, and it is greater than that between any other two neighbouring elements.

As far back as 1902,[229] using terms of atomic weight rather than atomic number, Brauner wrote:

> Apart from the 10 elements already listed. . . and more or less accurately studied by me, about seven to ten additional elements could be placed in this group. . . It is not impossible that one would be able to split neodymium, Nd = 143.8, into at least one element with a smaller atomic weight, and into another element with a higher atomic weight of about 145 and, similarly, some more gaps lying in the area between Ce and Ta could be filled.

From 1913 on, using Moseley's law, scientists could speak in terms of atomic number and determine that there was really only one missing element between Nd and Sm. This discovery should have made the work easier; instead, the series of presumed discoveries of element

61 grew greater. In 1917, Josef Maria Eder, photographing the arc spectra of preparations of samarium, perceived unknown lines and attributed them to a new element.[230] In 1921, Charles James, investigating the solubilities of the carbonates of the rare earths, perceived, using only chemical means, the possibility that a new element existed between Nd and Sm.[231] The following year, A. Hadding obtained an X-ray spectrum from a sample of fluocerite[232] in which he observed some unknown lines.[233] In 1924, Wilhelm Prandtl and A. Grimm[234] fractionated many ceric earths; in the 50th fraction obtained. they recorded some X-ray spectra without, however, finding the presence of element 61. In 1925, Gerald J. Druce and Frederick H. Loring looked for it in preparations of manganese but without success.[235] Although the X-ray spectroscopy of the 1920s was more reliable than the technique of 10 years earlier, and the law of Moseley could be used to distinguish one element from another, the pathway that chemists set out on in search of element 61 was tortuous indeed.

IV.9.1. FLORENTIUM, THE METAL OF THE FLORENTINES

Element 61 was also the subject of research in Italy. In November 1919, the first anniversary of the end of World War I and after four interminable years of struggle and an epidemic of influenza (known as the "Spanish flu") that left more dead than those who fell on the battlefields, Europe sought to rise again. This was the year of the peace conference at Versailles, of Prohibition in the United States, and of the repudiation by that nation of the creation wished for by its own President Woodrow Wilson (1856–1924): the League of Nations. Italy in particular was wounded by deep social tensions. Those were the years in which research suffered from the loss of many scientists, either called to the front or engaged in the war effort. Slowly, the universities of the Kingdom of Italy resumed their research activities.

In that same year, shortly before the transformation of the Institute for Practical Higher Studies and Specialization into the University of Florence, the superintendent, the Marquis Filippo Torrigiani (1851–1924), called 37-year-old Luigi Rolla, already famous as an inorganic chemist, to assume the chair of general chemistry.

Luigi Rolla, born in Genoa, on May 21, 1882, had been a student of chemists Jacobus Henricus Van't Hoff (1852–1911) and Walther Nernst (1864–1941) at the Prussian Academy of Sciences in Berlin during the 2 years preceding World War I. He was one of the first chemists to master physics, and he was famous for telling his students that "I had the honor of hearing from the very mouth of my teacher of the discovery of the third law of thermodynamics,[236] and I was the only one of his students who understood it!"

Prematurely deaf and bald, with stern features held rigid in a perennial frown, Rolla was an autocrat who stood out in the scientific community both on account of his great height and for his indisputable scientific stature. He was one of the scientists who had dominated Italian chemistry for the entire period between the two World Wars. In 1921, the young Giorgio Piccardi returned from the front, brilliantly finished his studies in chemistry, and remained at the university as a volunteer assistant to Rolla. Figures IV.06 and IV.07 are formal portraits of Rolla and a departmental colleague, Angelo Angeli.

After the war, Luigi Rolla reestablished contacts with his German colleagues and followed with great attention the development of atomic physics; he was, in fact, among the first—and the first in Italy—to conceive of a link between ionization energy and the various atomic species belonging to the same group. With Piccardi's assistance, he performed

FIGURE IV.06. Luigi Rolla (1882–1960), Professor of Chemistry, University of Florence. With Lorenzo Fernandes, he carried on a massive but fruitless search for element number 61, calling it *florentium*. Courtesy, Galileo Galilei Museum, Florence, Italy.

FIGURE IV.07. Angelo Angeli (1864–1931), Professor of Organic Chemistry at the Newly Formed University of Florence. In the late 1920s, he played a role in the unfortunate epilogue to the discovery of *florentium*, aligning himself against his colleague, Luigi Rolla. Courtesy, Galileo Galilei Museum, Florence, Italy.

experiments able to measure the first ionization potentials of the various elements. Eighty-six substances were known in 1919. Six boxes remained vacant in the periodic table: atomic numbers 43, 61, 72, 75, 85, and 87.

The work of separation and chemical purification of the elements and the X-ray check for the purity of the rare earths required a lot of time and needed a good deal of manpower. Professor Rolla had at his disposition the entire first floor of the Institute of Chemistry at the university and two new graduates: Giovanni Canneri (1897–1964) and the very young Lorenzo Fernandes, who was extremely interested in the fractionation of the rare earths. Fernandes was born in Florence, on May 24, 1902, and received his degree in chemistry with high marks and great praise on July 11, 1924. He was immediately Luigi Rolla's favorite student, and, from that day until 1930, he held the post as assistant on the permanent staff of the "maestro."

When the work of purification was at a good point with respect to the X-ray spectra of samarium and neodymium, Fernandes perceived some unknown lines in the K series. Rolla was at first skeptical about assigning these lines to a new element. He well knew the law of Moseley formulated in 1913, and he knew that an element not yet discovered had to fall between Nd (60) and Sm (62). For the moment, Rolla thought of ending his studies on the ionization potential of the elements, but the idea of the possible discovery of element 61 insinuated itself in his mind.

When the aforementioned work was finished, 400 g of ceric earths remained, containing Gd, Ce, Nd, Sm, and Pr that had been acquired from the de Haen company.[237] The work of chemical purification and spectroscopic purity checking of the samples began again in search of element 61. From the beginning, the researchers supposed that the element would be contained in Brazilian monazite sands[238] in amounts so small that extraction would be impossible. For two years, Rolla worked on these sands. Finally, in the spring of 1924, Rolla and Fernandes announced that they had photographed the "fingerprints" (for those trained in the work "the characteristic X-ray spectrum") of element 61. The hunt was over, yet, instead of rejoicing, Rolla was assailed with doubt, conscious of the fact that many scientists had fallen into the fatal error of announcing a discovery shown later to be false. What to do? Wait for more confirmation? To temporize in science often meant running the risk of blowing a discovery. Rolla had to resolve the dilemma: either make a premature announcement or postpone the discovery. Figure IV.08 is a view of Rolla's laboratory, financed with public and private funding, for the separation of element number 61 from the monazite sands brought from Brazil.

The weeks passed and Rolla hesitated: he didn't want to act imprudently, but he knew time was of the essence. If he announced the isolation of the new element, he would be the first and only Italian to make such a discovery. In the end, the prospect of success and the prestige he would derive from it drove him to put aside his last reservations.

Rolla was by nature cautious and so, in announcing his work to the scientific community, he opted perhaps for the less compromising course. In June 1924, he sent a sealed packet to the *Accademia dei Lincei* containing a sample of the presumed element and the results of his analyses.[239] The packet would remain secret until he or other chemists had succeeded in proving the existence of element 61. In this way, he could defend the priority of his discovery without exposing himself unduly. It was a solution of compromise that satisfied no one. Rolla and Fernandes returned to their studies and, with the help of three other young chemists—Giorgio Piccardi, Giovanni Canneri, and Luigi Mazza (1898–1978)—sought to extract the elusive 61st element.

FIGURE IV.08. Luigi Rolla's Laboratory, University of Florence. Courtesy, Chemical Heritage of the Department of Chemistry of the University of Florence, Italy.

In those years, it was thought that the problem of isolating element 61 consisted only of finding a large enough quantity of raw material and in conducting a large enough number of fractional crystallizations. As to the matter of the element Rolla was seeking, a Genoese industrialist, Senator Felice Bensa (1878–1963), became so impassioned that he gave a million lire to the University of Florence to buy the necessary instruments and a sufficiently large amount of monazite sand to do the work. A large amount of this mineral was sent to the university and from it was extracted a ton of impure *didymium*.[240] On this material, in late autumn of 1925, they began the work of isolating the missing element. The first floor of the laboratory of chemistry at the university began to look like an industrial laboratory, such was the amount of material that came under treatment there.[241]

Unfortunately, the work that should have concentrated the mysterious element was not without its difficulties: some of the laboratory workers were overcome by a strange malaise, and they also complained about the death of one of their co-workers. Later, it was discovered that the cause of these illnesses was due to an abundant use of bromates[242] in the fractional crystallizations that, upon heating, released elemental bromine. Through successive fractional crystallizations of the ceric earths, many rare earth elements were obtained in purities never before achieved.

When the chemist Georges Urbain told the Academy of Sciences of Paris that he had completed about 15,000 fractional crystallizations to isolate element 71 (lutetium), the assembly was impressed. We know that at Florence, between 1925 and 1942, first Fernandes and then Piccardi accomplished a total of 56,142 fractional crystallizations on 1,200 kg of oxalates[243] obtained by treating the original monazite. The porcelain evaporating dishes were made especially for the purpose, and the largest of these had a diameter of a meter; the filters had a capacity of 5 L. During the work of isolating this most elusive metal, Piccardi and his co-workers obtained remarkable quantities of spectroscopically pure samarium, cerium, gadolinium, neodymium, and praseodymium. Later, a certain quantity of gadolinium was given by Rolla to Enrico Fermi for his studies on neutron-induced radioactivity.[244] In addition, samples of Nd, Pr, and Sm were given to Emilio Segrè who tried to obtain *florentium* by bombardment with α particles.[245]

Because Professor Rolla and his assistants did not succeed in isolating element 61 even after so many crystallizations from enriched samples, Rolla decided to send the material to Professor Rita Brunetti (1890–1942) at Arcetri, the location of the University of Florence's observatory and department of physics, where Brunetti was chair.

Rita Brunetti, born at Ferrara, on June 23, 1890, had come to Florence before World War I as an aide to the physicist Antonio Garbasso (1871–1933). When the latter was called to the front, Brunetti assumed not only his teaching duties but also the work of advancing his scientific research. On June 1, 1927, Rita Brunetti became the first woman in Italy to occupy a chair of physics.

Rolla hoped that Brunetti, being a spectroscopist, would be able to resolve the dilemma of the existence of the presumed unknown element in his preparations.[246] The intensity of some spectral lines would be taken as proof of the existence of the new element; the intensity of the lines obtained by Brunetti were not as weak as those recorded 2 years earlier, and as more convincing proof, they became even more intense in the later fractions of Sm, those designated by the numbers 2677 and 2682, that ought to have been enriched in element 61.[247] Rita Brunetti did not limit herself to a study of the roentgenographic emissions, but also carried out a study on the discontinuities of X-ray absorption.[248] It was the first time that this method was utilized in the search for a missing element in the periodic table. The results in this case were viewed as positive.

IV.9.2. THE AMERICANS DISCOVER ILLINIUM

The years passed in this long and drawn out manipulation of the Brazilian earths when, in 1926, like a lightning bolt out of a blue sky, a group of American chemists announced the discovery of element 61. Shortly before Rolla would have given notice of the partial results of the separation and concentration of the new metal, pointing out his new method of fractional crystallization based on the double nitrates of thallium, U.S. chemists B Smith Hopkins,[249] J. Allen Harris (1901–72), and Leonard Yntema (1892–1976) announced their discovery of element 61[250] (Figure IV.09).

The team from the University of Illinois had worked on the same material as Rolla and had arrived at the same results. B Smith Hopkins, the principal investigator in the discovery, decided to call the new element *illinium* in honor of the state and university of the discovery. Simultaneously, he proposed the symbol Il for it.

B Smith Hopkins was born at Owosso, Michigan, on September 1, 1873. He took his doctorate in chemistry at The Johns Hopkins University in 1906. He occupied various academic positions before coming to the University of Illinois in 1912, where he started a long series of researches on beryllium, yttrium, tantalum, and, finally, on the rare earths. From 1923 to 1941, he was professor of inorganic chemistry there.

While the scientific world was congratulating the American scientists for their discovery, the existence of the presumed *illinium* was confirmed by groups of Anglo-Saxon and German researchers.[251] The dismay in Florence was very great. After an initial period of bewilderment, Rolla hastened to Rome and asked that the Accademia dei Lincei break the seals on the packet that he had deposited there 2 years earlier.

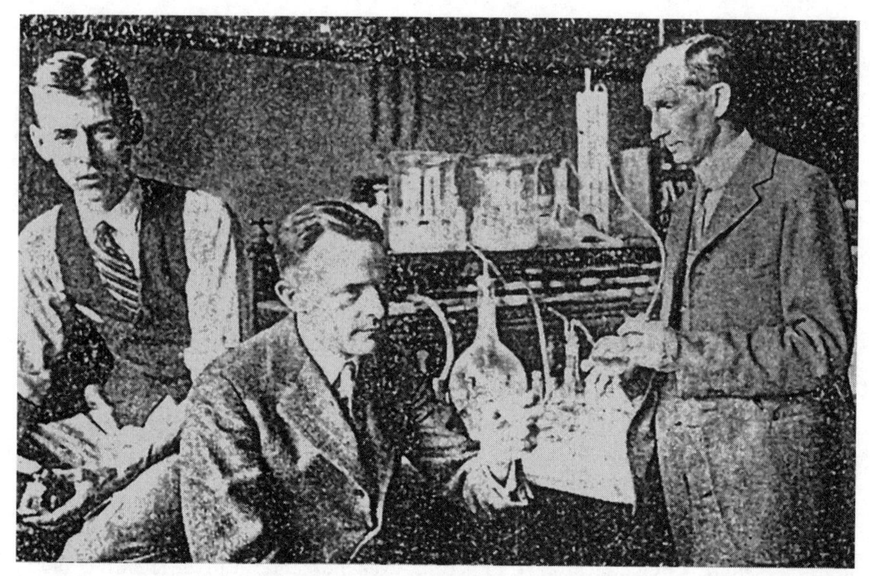

FIGURE IV.09. From left to right, Allen Harris (1901–72), predoctoral student; Leonard Yntema (1892–1976), spectroscopist; and Professor B Smith Hopkins (1873–1952). The U.S. research group that, simultaneously with Luigi Rolla and Lorenzo Fernandes (1902–77), worked on tracking down element number 61, which they called *illinium* in honor of the state of Illinois. Gift of B Smith Hopkins, Jr.

The Accademia, founded in 1603, brought together the most famous scientists and men of culture in Italy, and, from the platform of these meetings, discoveries were announced, disagreements were raised, and prophecies were launched. Rolla, during his October 1926 address to Accademia members placed a simple but dramatic problem before the group. He reviewed the great scientific events of the past 2 years, including his own work, then culminated with the sensational announcement of his discovery of a new element: *florentium*. The message of this discovery was launched out into the world in a fever of showmanship: "the element searched for in vain for such a long time, the rarest of the rare earth elements, ought to take its name from that of the most Italian of all Italian cities where with Dante the spirit of our noble lineage is expressed. For this element, with atomic number 61, we propose the name of florentium and the symbol Fr."

In Italy, the notice was wrapped in a climax of nationalism, and laurels for the two discoverers were not slow in arriving: on July 14, 1926, Rolla was elected to the Accademia dei Lincei, the greatest scientific authority in the country. In the same year, at only 24 years of age, Lorenzo Fernandes was named professor at the Royal University of Florence. Between the two shores of the Atlantic Ocean, however, a bitter polemic broke out to establish who had effectively discovered the 61st element. What alarmed the presumed discoverers most was not the content of the research, but the name proposed for the new metal: would it be called *florentium* or *illinium*?

Rolla did not lose heart. He had every intention of claiming what, according to him, he was entitled to and sent a letter to the journal *Nature*[252] in which he claimed priority and made note that the name *florentium* had been proposed a good 18 months before *illinium* saw the light. What followed was a long academic diatribe aimed at establishing who should receive the palm of victory. To assign recognition to one or the other of the research groups was not easy, considering that not only was the prestige of individual scientists and their respective universities at stake, but also that no one in either country, Italy or the United States, had until now discovered an element. This was a year in which relationships between Italy and the United States were particularly tense against a backdrop of renewed nationalistic posturing: in this xenophobic climate, two Italian anarchists, Nicola Sacco (1891–1927) and Bartolomeo Vanzetti (1888–1927), were waiting to be executed; and on May 29, 1926, the U.S. polar explorer Richard E. Byrd (1888–1957) was the first to arrive at the North Pole, while the Italian expedition of Umberto Nobile (1885–1978) ended tragically. Byrd clinched the acclaimed record by way of a Broadway ticker-tape parade before being received by President Calvin Coolidge (1872–1933).

As if things weren't complicated enough, Brauner offered his opinion on the sensitive issue of who should get credit for the discovery of element 61. He congratulated his American colleagues but dismissed the discovery as simply a matter of technique. He believed that the discriminating factor was who had predicted the existence of this new element before the law of Moseley, and that factor was himself. In fact, in 1902, Brauner had published a periodic table of the elements on which he reported "the existence of the missing element [element number 61] was predicted by me in 1902."

For a long time, Luigi Rolla maintained a written correspondence with his colleague and rival across the ocean. The relationship between the two, apparently cordial, concealed a mutual lack of faith even though reciprocal help and cooperation were promised. But Rolla went much further. Alarmed by the astonishing notices from his colleague asserting that he had isolated traces of *illinium*, when his own research seemed unshakable, in 1927, Rolla set sail for New York to see with his own eyes what progress was being made in the isolation of element 61. In these circumstances, Rolla met William Albert Noyes (1857–1941), head of the Department of Chemistry at the University of Illinois where Hopkins worked.

Fortunately, the progress claimed by Hopkins and his collaborator Harris was not enough to alarm Rolla. On his return from America, always carrying with him his samples of the presumed *florentium*, he stopped off at the Institute of Physics directed by Niels Bohr at Copenhagen. There, he subjected one of his enriched samples to a scrupulous spectroscopic examination, much more accurate than the one done a few months earlier by Rita Brunetti. If Brunetti's response had been largely positive, that of Bohr left no room for doubt.

In a fiery letter addressed to Rita Brunetti, Rolla wrote and underlined twice: "Dear Professor Brunetti,. . . in the samples analyzed by you, where you assured me that there was element 61, there is nothing."[253] Rita Brunetti is shown in Figure IV.10 with some illustrious colleagues. In keeping with his character, Rolla, unable to openly accuse his colleague at Arcetri, instead vented his anger against his transatlantic rivals and his subordinates in Florence.[254]

In 1922, the American Leonard Yntema investigated the X-ray emission spectra of samples of monazite, excluding from them the presence of a new element. Four years later, in collaboration with Hopkins and Harris, he retracted the conclusions of his article and claimed priority for the discovery of *illinium*. Rolla emphasized this aspect of the American research, but in his heart he knew that none of this was enough to salvage his discovery, but only to gain some time. If he wanted to win, he would have to isolate the new element first.

The years passed without anyone being able to extract the element in macroscopic quantities. It occurred to the Florentine chemists that their discovery could be false, but the applause of the Lincei academics that had filled the great hall of the Corsini palace at

FIGURE IV.10. Arcetri, Institute of Physics, 1925. Enrico Fermi (1901–54), left, certainly the greatest Italian physicist of the 20th century, announced the false discovery of the first transuranium elements to which the names *ausonium* and *hesperium* were suggested by his colleague, Franco Rasetti (1901–2001), far right. In the center is Nello Carrara (1900–93), physicist, and fellow student of Fermi and Rasetti; he is best-known for having coined the word "microwave." In the second row, Rita Brunetti (1890–1942), the first Italian woman to achieve a top-level academic position. With Luigi Rolla and Lorenzo Fernandes, she was involved in the "element 61 affair." Her experimental measurements supplied the proof of the existence of *florentium* (later shown to be erroneous). Courtesy, Chemical Heritage of the Department of Chemistry of the University of Florence, Italy.

the moment of Rolla's sensational announcement obscured their doubts. With the passing of the years, the relationship of esteem and collaboration between the two discoverers deteriorated and, as we shall see later, Lorenzo Fernandes left the scene at the beginning of the 1930s.

In 1926, Walter Noddack and his wife Ida Tacke Noddack, who had announced in 1925 the discovery of *masurium* (Z = 43) and rhenium (Z = 75), suggested that *illinium* might be related to samarium the way that radium was related to radon; that is, it could be a gaseous emanation produced by a type of radioactive decay as yet unknown.[255] This statement was almost immediately accepted as fanciful speculation later shown to be groundless, but, by sheer prescient intuition Ida Tacke Noddack was looking in the right direction: element 61 might be radioactive!

Noddack and Tacke Noddack tested minerals that could contain *illinium*. After working on 100 kg of rare earths but finding no trace of element 61, they announced that if the American chemists actually had identified *illinium*, they would have been able to isolate it even if it were 10 million times rarer than Sm and Nd. The hypotheses formulated to explain the failed attempt at identification could be explained admitting that (1) element

61 was so rare that the investigative techniques used were not able to identify it; and (2) the minerals analyzed were the wrong ones.

The geochemists opposed the second hypothesis because the abundance of all the known rare earths was more or less the same. There was no reason why *illinium* would be an exception. They suggested that *illinium* be looked for in minerals of Ca and Sr because all the rare earths were trivalent, but some of them showed a valence of 2 or 4. Perhaps *illinium* could be found in some natural minerals of strontium. The Noddacks processed many minerals rich in the alkaline earth elements, but the search for *illinium* seemed to have arrived at a dead end. At this point, the story of element 61 is bound indissolubly to that of element 43, technetium. In accordance with the laws of Josef Mattauch (1895–1976), technetium could not exist because it did not have any stable isotopes. The same law prohibits the existence of stable isotopes of element 61. *Illinium* and *florentium* were dead even before they were born, but element 61 survived.

Ida Tacke Noddack suggested to her husband that element 61 could have once existed on the earth but was highly radioactive and with a very short half-life, so that it could have decayed away long ago.

Later, S. Tackvorian, working with Maurice Curie (1855–1941), the brother-in-law of Marie Curie, established, after a long and meticulous work of crystallization, that some fractions of radioactive ceric earths between Nd and Sm behaved anomalously.[256] Later, J. K. Marsh[257] suggested that, under certain conditions, actinium and bismuth could crystallize into fractions in which they held that element 61 could be found.

The physicists first had the idea that, in order to obtain *illinium*, one would need to synthesize it artificially using nuclear methods, which is more or less what happened in 1937 with technetium.

IV.9.3. INTEGRITY COMES WITH A PRICE TAG

If any one story of the discovery of an element can encapsulate within itself the entire history of chemistry, the search for element 61 would take the prize. A highly complex discovery, a twisted pathway of numerous claimants[258] and almost as many names, one that unfolded against the backdrop of a growing understanding of the theoretical power of the periodic table and the problem of accommodating the rare-earths within it, a phenomenological discovery (X-rays by W. Roentgen) and the use to which it could be put (atomic numbers by Moseley), and the development of new analytical methods (ion exchange chromatography)—all of these ingredients and more are part of the story of this problematic element.

But one more ingredient, rarely referred to, is moral integrity. If Charles James (1880–1928), who stands at the center of this story, had known Oprah Winfrey, he could have either taken a cue from her or handed her this line: "Real integrity is doing the right thing, knowing that nobody's going to know whether you did it or not." James did the right thing. His reward is that the scientific world at large barely knew about it or, what's worse, hardly cared. Although recent literature[259] has drawn attention to his story and the American Chemical Society has designated a National Historic Chemical Landmark[260] at the University of New Hampshire to commemorate his work, Charles James's contributions have been largely ignored.

James was born on April 27, 1880, at Earls Barton, Northamptonshire, England. He studied under William Ramsay at University College, London, prior to emigrating to the

United States, where he became a professor in the chemistry department of what is now the University of New Hampshire at Durham. Over the years, he became the world's recognized expert on the separation and purification of the rare earths, amassing a collection of more than 200 rare earth specimens and publishing the results of his research on their compounds and atomic weights in more than 60 papers. He supplied Moseley with the sample of terbium that was used to determine its atomic number. Although at the time that Georges Urbain announced the discovery of lutetium James had accumulated a large amount of lutetium oxide, he withdrew his paper and made no further claim to its discovery. Today, however, he is often recognized as a co-discoverer.

Here is the story of how James "did the right thing." According to R. F. Gould, writing in *Chemical and Engineering News*,[261]

> when [B Smith] Hopkins made his announcements in March 1926, James and [Heman C.] Fogg of the University of New Hampshire had just completed their fractionation of ytterspar and had sent the 61-rich concentrate to [James M.] Cork at the University of Michigan for X-ray analysis. The results were reported in December, but by this time the controversies over the other three claims were in full swing, and the fourth entry went almost unnoticed in spite of the fact that the evidence was perhaps better than that of any other claimant. Probably contributing to this neglect was the fact that the announcement was published in a relatively obscure journal [*Proceedings of the National Academy of Sciences, PNAS*[262]]. . . to date, no other X-ray spectrogram of element 61 has been published, and while James' work has never been successfully repeated, neither has it been denied or repudiated.

According to Murphy,[263] the situation was much more complicated and revolves around the fact that James published in the *PNAS* rather than in the *Journal of the American Chemical Society (JACS)*, a much more widely read journal, where almost all of his other many publications resided. Why? James had been carrying on extensive research on element 61 for a number of years. Examination of scrapbooks and letters in the University of New Hampshire archives show that James was about to submit his paper to *JACS* when he received a request from the editor of that journal to referee two papers from Smith Hopkins claiming discovery of *illinium*. He quickly withdrew his own paper, gave positive reviews to the two Illinois papers, then submitted his own paper to the PNAS to avoid any conflict of interest—and thus, in the opinion of some, consigned his own work to oblivion.[264] Later evaluation of this work indicates that the six spectral lines reported by James and his co-workers came uncannily close to those reportedly taken on an actual sample of element 61 at Oak Ridge National Laboratories in 1949.[265]

In 1926, James began a study of uranium, refining it by his own methodology. In 1927, the college (by then a university, as of 1923), awarded him an honorary doctorate in science. This was remarkable because the university did not give such degrees to active faculty members, and he was very proud of this honor. This was also the year in which James persuaded the university to build a new chemistry building, which he helped design.

In 1928, he obtained a discarded greenhouse from the university and attached it to a new garage (although he had never owned a car). The greenhouse was destined to house his plant-growing ambitions: he and his wife were avid gardeners. The construction of the important new chemistry building had begun the preceding autumn. A glowing future seemed to lie ahead, but James began to suffer from increasing stomach pain. In early

December, he entered Boston Deaconness Hospital where he underwent cancer surgery. The surgeons realized that nothing could be done, and James died on December 10, 1928, at the age of 48.[266]

Although his legacy as a luminary in rare earth research has been neglected or forgotten, he is remembered by his department as a dedicated and meticulous chemist who, at the same time, was a sympathetic mentor to the many students whom he developed into accomplished research professionals.

IV.9.4. FLORENTIUM ENDS UP IN COURT

After smoldering for a long time, a running dispute marked by a vicious diatribe between Lorenzo Fernandes and Luigi Rolla broke out into the open at the beginning of 1930; it was the second act in a long-running drama that tormented the consciences and pitted the colleagues against one another. Called in to arbitrate between the opposing parties were the Dean of Pharmacy Guido Pellizzari (1858–1938), Giusto Coronedi (1863–1941), and the mild and impartial professor Angelo Angeli (1864–1931). However, despite the strength Angeli might have drawn from his Alpine ancestors, he could not hold up and on May 31, 1931, a fatal heart attack removed him from this unfortunate episode.

After the various events relative to the identification of *florentium* already narrated, in April 1928, Rolla decided to send Fernandes to Fribourg to gain practical knowledge of the more recent advances in X-ray spectroscopy. On his return, the young man prepared the apparatus and, after about a month, the first frames were recorded. They were all sharp and rich in spectral lines, but none of them was identifiable with element 61. After so much work and sacrifice, discouragement overtook Lorenzo.[267]

During the summer holiday months of 1928, at first verbally and later in writing, he tried to convince his teacher to publish a retraction of the discovery of *florentium*. In reply, Rolla forbade the young man to speak about the negative results to anyone. The first disagreements between disciple and teacher were noticed in the Institute, and they grew with time to culminate in Fernandes's dismissal on March 5, 1930. Rolla accused his ex-student of negligence, working on behalf of third parties while exploiting the goods and services of the university, damage to the X-ray equipment through his evident lack of skill, obstructionism and sabotage, and scientific dishonesty. Rolla essentially accused Fernandes of faking experimental data.[268]

Fernandes wasted no time in going directly to the university rector, Enrico Burci (1862–1933), a far from prudent choice. Burci was an iron-fisted fascist who subscribed wholeheartedly to the Mussolini regime: Fernandes was sacked on the spot despite having on his side influential persons like Pellizzari, Angeli, Coronedi, Senator Salvatore Gatti (1879–1951), Canneri, and many others.

Research on *florentium* was then entrusted to Giorgio Piccardi and Leo Cavallaro, who appeared to be in accord with the director. They assured Burci that, with Fernandes gone, they would succeed in isolating *florentium* in a very short time.

Fernandes looked for other work in chemistry but was blackballed by Rolla everywhere he went.[269] He then decided to bring suit against his professor, and *florentium* became an object—not of the chemistry bench where it was never extracted, but of the courtroom. Rolla's defense briefs were handled by the lawyer Piero Calamandrei (1889–1956), the future rector of the University of Florence. He accused Fernandes of overarching ambition, negligence, and utter disrespect of his mentor.[270]

Fernandes's response was to paint Rolla as a charlatan who sought to keep the failure of his alleged discovery hidden from the scientific world. Rivers of ink flowed; a lot of sweat was poured out and later many tears, but in the end the verdict was in favor of the old chairman, who was well integrated into the academic ambit. It appears curious that the national scientific community, still in the dark a year later about the sad event at Florence, voted to confer on Rolla the Cannizzaro prize for his research on element 61.

IV.9.5. CYCLONIUM

During the last years of his tenure at Florence, Rolla witnessed enormous changes in the methods of chemistry and physics. The experiments were complicated, the first accelerators entered the scene, and the epoch in which the presumed discovery of element 61 took place within the walls of the ex-stables of the Tuscan Grand Duchy (site of the Institute of Chemistry) was drawing to a close and the focus of scientific research was inexorably moving elsewhere.

In 1938, two physicists from Ohio State University conducted the first experiments on the synthesis of element 61.[271] A target of Nd was bombarded with beams of fast deuterons, D+.[272] Their hope was to obtain an isotope of *illinium*:

$$Nd + D^{+} \longrightarrow Il + n \qquad \text{(Eq. IV.13)}$$

Their results were inconclusive, and yet they obtained an isotope with a mass of 144 of a new element with a half-life of 12.5 hours.

The neodymium oxide, Nd_2O_3, that they used showed, by arc spectrum analysis, no other rare earths present, but to eliminate the effects of other contaminants easily activated by deuterons, they chemically separated the neodymium oxide by dissolving it and precipitating it with oxalic acid. The spectrum that they obtained from the bombarded sample showed the same lines observed by Hopkins, Yntema, and Harris in 1926. No chemistry was done in this work, and the nature of this mysterious radioactive element was never clarified. From 1938 on, many types of particles were used as projectiles, and many rare earth elements as targets; at the same time, the techniques of radiometric measurement to determine activity were greatly improved.

Reports on isotopes of *illinium* began to appear in the scientific journals. Element 61 became a reality, even though created artificially. The credit for the revival of this search went to Lawrence Larkin Quill (1901–89). He was born on February 24, 1901. He had studied chemistry at the University of Illinois, and in 1928, he received his research doctorate under the mentorship of B Smith Hopkins while working on concentrating *illinium*. He had not been on the team that made the presumed discovery of illinium, although a similar "honor" had touched his contemporary, Harris; this fact had left a bitter taste in the mouth of the young Quill. This exclusion tormented him for years. Quill in the meantime got married, had two daughters, and worked in a physics laboratory where it was possible to synthesize element 61.

In 1941, M. L. Pool (1900–82), H. B. Law, J. B. Kurbatov, and L. L. Quill[273] bombarded a target of Sm with 5 MeV protons and 10 MeV deuterons and discovered two isotopes of element 61. The team guided by Quill renamed the element *cyclonium* (symbol, Cy) because it was synthesized with a cyclotron. However, the symbol Cy did not stay long in box 61 in the periodic table. The researchers had measured the radioactive signal of Cy, but no one succeeded in extracting even a milligram of the new element, and, what is more, its spectrum had not been recorded. Unfortunately, they had obtained only indirect evidence of the existence of *cyclonium*. After the war, Quill shifted his research interests to other fields; he died on February 13, 1989, at the ripe old age of 88.

Samples of Pr, Nd, and Sm were donated by Rolla to Segrè that he might accomplish the same experiments. Also in this case the results were not clear, and the discoverer of Tc very prudently limited himself to reporting the half-life of about 100 days for an isotope of element 61—without mentioning the name of *florentium*.

IV.9.6. THE RETRACTION OF THE DISCOVERY OF FLORENTIUM

Finally, Rolla and Piccardi excluded the presence of element 61 from their neodymium- and samarium-bearing preparations, in agreement with the predictions of isotopic statistics (Mattauch's Law).[274] If the moment of triumph associated with the disclosure of the discovery of *florentium* had appeared in the more widely-read journals of Germany, England, and France,[275] the note of retraction appeared only in a minor journal of the Vatican State and was, for the most part, written in Latin. In September 1941, Rolla and Piccardi presented to the *Pontificia Academia Scientiarum* a long document divided into numerous parts regarding the identifications of the rare earth elements and in particular of the element that occupied box 61. The history of the material "neodymium," which was subjected to a very long process of fractional crystallization according to the "bromate" method, was briefly summarized. The authors described the spectroanalytical research on the resulting residue after 56,000 crystallizations. This research was conducted with the help of a "galvanic arc" for the visible and ultraviolet field. For the comparison made between the various fractions both of the substance itself and of very pure samples of the rare earths known to exist, in the fractions between neodymium and samarium, the presence of gadolinium became evident, and only gadolinium. No trace of any other element was found, neither known nor new, and above all, there was not even the slightest trace of element 61.

Luigi Rolla definitively abandoned every desire of priority for the discovery of element 61 by speaking of this metal no longer as *florentium*, but as *illinium*, as if he were ashamed of his work and wished to attribute to B Smith Hopkins all the "credit of failure."

Professor Rita Brunetti was unaware of these maneuvers because, in 1929, she had been transferred from the Royal University of Florence to the peripheral seat at Cagliari. Only in 1936 did she succeed in obtaining the assignment of a chair of physics at Pavia, not far from her native Milan, but she arrived there in too poor a state of health to be able to undertake new wide-ranging scientific research. Attacked by a malignant tumor, on June 28, 1942, five days after her 52nd birthday, Rita Brunetti closed her eyes forever. As a symbol of her adherence to fascism, she wanted to be laid out wearing the fascist uniform.[276]

As for the identification and isolation of the mysterious element 61, Rolla, like Moses, arrived within sight of the promised land without being able to enter. Rolla was around long enough to know about the synthesis by fission of the element that would take the name promethium, but he would never know of the existence of natural promethium in pitchblende,[277] a discovery that took place in 1968 through the work of the chemist Paul K. Kuroda (1917–2001).[278]

Luigi Rolla died on November 8, 1960, in Genoa, the city of his birth, where he had returned, embittered by the missed discovery of *florentium* and of the polemics that followed it. Hopkins's results, like those of Rolla, were also soon called into question and later refuted.[279] A communication in the *Quarterly Review* of the Chemical Society of London[280] asserted that the spectra observed by Hopkins could be reproduced by adding traces of neodymium to a gadolinium salt. This made the erroneous discovery of *illinium* highly probable due to the fact that neodymium was always found together with gadolinium in the fractional crystallizations done by the bromate method, which was the one employed by Hopkins.

IV.9.7. CONCLUSION

One of the greatest discoveries of the 20th century was the neutron fission of uranium in 1938 through the work of Lise Meitner and Otto Hahn. Thirty isotopes of elements ranging from zinc to gadolinium were produced by the fission of uranium-235. About 3% of the products consisted of a mixture of isotopes of element 61. The work of quantitatively extracting these was impossible using the techniques of the 1930s. A group of American chemists—Jacob Akiba Marinsky (1918–2005), Lawrence Elgin Glendenin (1918–2008), and headed up by Charles DuBois Coryell (1912–71)—developed new ion-exchange chromatographic techniques that they used to separate the fragments of uranium fission. At the bottom of this "sieve" were found two true treasures: the isotopes of mass 147 and 149 of the coveted element 61. In the end, element 61, after having changed its name from *illinium* to *florentium* and then to *cyclonium*, would receive a permanent name.[281] During a working supper, Mrs. Coryell proposed to the three researchers the name *prometheum* for this element. The three co-workers are shown in Figure IV.11 and one of them, J. A. Marinsky is shown in Figure IV.12 with one of the authors of this volume.

In ancient Greek mythology, Prometheus stole fire from heaven and gave it to mankind, and for this he was tortured by Zeus. The name *prometheum*, said its discoverers, was not only a symbol of the difficult and dramatic road taken to obtain this element in appreciable quantities through the difficulty of controlling the nuclear fission, but also served as a warning regarding the danger of nuclear war, represented by the eagle of Zeus. The name was accepted by the international commission, which modified only its spelling, transforming *prometheum* into promethium, but leaving unchanged the symbol, Pm, proposed by the discoverers. Pm was obtained in 1945, but only in 1947 did the first publication concerning it appear. In June 1948, participants at the Syracuse meeting of the American Chemical Society were among the first to see samples of promethium: three milligrams each of yellow $PmCl_3$ and pink $Pm(NO_3)_3$.

As far as anyone knew, in 1968, the entire world held no more than about 10 g of Pm. The law of Mattauch forbade the existence of natural promethium, but this law was not absolute and above all could not foresee that this element could be produced in nature by the spontaneous fission of uranium. From a recent estimate, it is believed that the total amount of promethium in the earth's crust amounted to about 780 mg; that is, practically nothing. A tremendous undertaking was tried to find naturally occurring Pm. Beginning in 1956, a group of American scientists headed up by Paul K. Kuroda, organized a gigantic

FIGURE IV.11. Charles DuBois Coryell (1912–71), Lawrence Elgin Glendenin (1918–2008), and Jacob Akiba Marinsky (1918–2005). The team that, during World War II, isolated element number 61 and, in 1947, proposed the name *prometheum* (later changed to promethium). Gift of Jacob Akiba Marinsky.

FIGURE IV.12. Jacob Akiba Marinsky. Marinsky (left) was primarily responsible for the discovery of promethium. 50 years after its isolation, he is seen here at the 1998 American Chemical Society meeting in Boston with Marco Fontani. Photograph by Marco Fontani.

task force to extract the natural promethium present in uranium-bearing deposits of pitchblende at Oklo in Gabon.[282] The mass of the natural isotope is 147.

As in the case of Tc, Pm also has two discovery dates; the first, 1945, is the date of its synthesis; the second, in 1968, was its identification in nature. This discovery is linked to the new capacities of chemical physics and to new methods of analysis, but its accomplishment remained purely theoretical because no one, to date, could even think of extracting natural promethium. The synthesis of Pm was not a true and proper synthesis like that of Tc because it was obtained by fission of uranium. This makes promethium a unique case among all the other synthetic elements.

IV.9.8. EPILOGUE

Unlike Rolla, B Smith Hopkins remained faithful to his discovery to his dying day. Made a widower in 1938 by his first wife Maude Sarah Child (1874–1938), in 1942, he married an ex-student, Dr. May Lee Whitsitt, a chemist at Southern Methodist University in Dallas, Texas. During a period of emergency following the Japanese attack on Pearl Harbor, Hopkins was recalled into service at the University of Illinois, where he remained until 1946 as a lecturer.[283]

Together with his second wife, he tramped the length and breadth of the United States and spent a considerable fortune in the vain attempt to salvage his *illinium* from oblivion. In 1948, he went to the American Chemical Society meeting in Syracuse, New York, and saw the first samples of promethium. Jacob A. Marinsky, to whom goes the major part of the credit for the discovery[284] of element 61, told of a very old professor, irate because he did not want to admit that he saw before his very eyes the element that he had looked for in vain for more than 20 years. After this sad interval, Hopkins returned to Urbana-Champaign, Illinois, and there died on August 27, 1952, not long before his 79th birthday.[285] Doctor Whitsitt continued to tenaciously defend *illinium* even after the death of her husband, and in a certain way "took up the battle. . . hoping the discovery of her husband would be vindicated. She had many of Hopkins's samples and she wanted to know if more modern techniques would help clear the matter up."[286] In 1970, May Whitsitt moved to Detroit, Michigan, to be near a niece. There, 5 years later, she died at approximately 84 years of age. Allen Harris,[287] the favorite student of B Smith Hopkins, died on February 6, 1972, at the age of 71. Finally, in 1976, Leonard Yntema passed from this life at the age of 84.

The last survivor of the discoveries of 1926 was Lorenzo Fernandes. He was forced to emigrate to France following the promulgation of the Italian racial laws of 1938. After the liberation of Florence in 1944, he returned to his home town and was one of the founders of the first Italian company to build radar units. This activity gave the unfortunate chemist a very good living. A shy and reserved person, he showed no interest in chemistry for the rest of his life, nor did he wish to recall those sad days of *florentium*. On Saturday June 25, 1977, toward noon, while having a friendly conversation with guests in the living room of his villa in the hills of Bellosguardo, he was felled by a fatal heart attack. He had just turned 75.

Rolla's disciple and successor at Florence, Giorgio Piccardi, spent many years in researching the rare earths and on the fractional crystallizations in search of the non-existent *florentium*; he was a man of exceptional intellectual honesty, and when his students asked him what he thought of all the work done searching for *florentium*, he replied courteously: "My dear boys, the great Poincaré defined science as the cemetery of hypotheses; if in it our own is also buried, I will be honored." Then, with a courtliness lacking any form of affectation, he took up his explanation again at the point where he had been interrupted.

Notes

221. Crookes, W. *Proc. Roy. Soc.* **1886**, *40*, 502.
222. Demarçay, E. A. *Compt. Rend. Chim.* **1886**, *102*, 1551; Demarçay, E. A., *Compt. Rend.*, 1887, *105*, 276.
223. Becquerel, H. *Compt. Rend. Chim.* **1887**, *104*, 777; Becquerel, H., *Compt. Rend. Chim.* **1887**, *104*, 1691.
224. Thompson, C. M. *Chem. News* **1887**, *55*, 227.
225. Kiesewetter, P.; Krüss, G. *Ber. Dtsch. Chem. Ges.* **1888**, *21*, 2310.
226. Krüss, G.; Nilson, L. F. *Ber. Dtsch. Chem. Ges.* **1887**, *20*, 2134.
227. Brauner, B. *Nature* **1926**, *118*, 84.
228. A recent book discusses six of Brauner's seven elements, substituting element 91 for 89 on Brauner's list. Please see Scerri, E. *A Tale of Seven Elements*; Oxford University Press: New York, 2013, pp. 176–179. The book paints a fascinating picture of chemical research—the wrong turns, missed opportunities, bitterly disputed claims, serendipitous findings, accusations of dishonesty—all leading finally to the thrill of discovery, a happy ending that has eluded most of the element seekers described in this volume.
229. Brauner, B. *Z. Anorg. Chem.* **1902**, *32*, 1–30.
230. Eder, J. *Sitzber. Akad. Wiss. Wien, Math. Natur.* **1917**, *Klasse IIa*, 125.
231. Brinton, P. H. M.; James, C. *J. Am. Chem. Soc.* **1921**, *43*, 1446.
232. Mineral having the chemical formula $(Ce,La)F_3$.
233. Hadding, A. *Z. Anorg. Allgem. Chem.* **1922**, *122*, 195.
234. Prandtl, W.; Grimm, A. *Z. Anorg. Allgem. Chem.*, **1924**, *136*, 283.
235. Druce, G.; Loring, F. *Chem. News* **1925**, *131*, 273.
236. This discovery was enough to win Nernst the Nobel Prize in Chemistry for 1920.
237. Rolla, L. *Atti Soc. It. Progresso Sci.* **1926**, *15*, 58.
238. Rolla, L.; Fernandes, L. *Le Terre Rare*; Zanichelli: Bologna, 1929.
239. Rolla, L.; Fernandes, L. *Gazz. Chim. It.* **1926**, *56*(7), 535.
240. This was the commercial name for a very crude mixture of Nd, Pr, and Sm.
241. Rolla, L.; Fernandes, L. *Gazz. Chim. It.* **1927**, *57*(9), 704.
242. Piccardi, G. *Atti Soc. It. Progresso Sci.* **1932**, 21 riunione, vol. II, 3.
243. Rolla, L. *La Ricerca scientifica* **1933**, anno IV, vol. II, N. 7–8, 3.
244. Amaldi, E.; D'Agostino, O.; Fermi, E.; Pontecorvo, B.; Rasetti, F.; Segrè, E. *Proc. R. Soc. London* **1935**, *149*, 522.
245. Segrè, E.; Wu, C. S. *Phys. Rev.* **1942**, *61*, 203.
246. Brunetti, R. *Gazz. Chim. It.* **1927**, *57* (5), 335.
247. Rolla, L.; Fernandes, L. *Gazz. Chim. It.* **1927**, *56*(7), 688.
248. Rolla, L.; Fernandes, L. *Atti Soc. It. Progresso. Sci.* **1926**, *15*, 60.
249. Although the "B" in Smith Hopkins's name is followed by a period in the literature, personal communications from the chemistry department at the University of Illinois assert that "B" is actually a full name. We have adhered to this practice.
250. Hopkins, B S.; Harris, J. A.; Yntema, L. *Nature* **1926**, *117*, 792; Hopkins, B S.; Harris, J. A.; Yntema, L. *Science* **1926**, *63*, 575; Hopkins, B S.; Harris, J. A.; Yntema, L. *J. Am. Chem. Soc.* **1926**, *48*, 1585; Hopkins, B S.; Harris, J. A.; Yntema, L. *J. Am. Chem. Soc.* **1926**, *48*, 1594.
251. James, C.; Cork, J. M.; Fogg, H. C. *Proc. Nat. Acad. Sci.* **1926**, *12*, 696; Meyer, R. J.; Schumacher, G.; Kotowski, A. *Naturwissenschaften* **1926**, *14*, 771; Dehlinger, U.; Glocker, R.; Kaupp, E. *Naturwissenschaften* **1926**, *14*, 772.

252. Rolla, L. *Nature* **1927**, *119*, 637.
253. Letter of Luigi Rolla to Rita Brunetti (1927); by the kind permission of Professor Michele Della Corte.
254. Rolla, L.; Fernandes, L. *Gazz. Chim. It.* **1927**, *57*(4), 290.
255. Noddack, W.; Tacke, I. *Metallborse* **1920**, *16*, 985.
256. Curie, M.; Takvorian, S. *Compt. Rend. Chim.* **1933**, *196*, 923; Tackvorian, S. *Compt. Rend. Chim.* **1931**, *192*, 1220.
257. Marsh, J. K. *Nature* **1946**, *158*, 134.
258. Here is the line-up of element-61 hunters; there may have been more: B Smith Hopkins (1873–1952), Allen Harris (1901–72) and Leonard Yntema (1892–1976); Luigi Rolla (1882–1960) and Lorenzo Fernandes (1902–77); Carl C. Kiess (1887–1967); James M. Cork (1894–1957), Heman C. Fogg (1895–1952), and Charles James; Kenneth Joseph Marsh (fl. 1921–52), Paul H. M. -P. Brinton (fl. 1911–43), Assar Robert Hadding (1886–1962), Emilio G. Segrè (1905–89), M. L. Pool (1900–82), H. B. Law, J. B. Kurbatov, and Lawrence Larkin Quill (1901–89), Lars Fredrik Nilson (1840–99), Bohuslav Brauner (1855–1935), C. M. Thompson, P. Kiesewetter, and G. Krüss. This is not to mention the real discoverers Jacob Akiba Marinsky (1918–2005), Lawrence Elgin Glendenin (1918–2008), and Charles DuBois Coryell (1912–71).
259. Murphy, C. J. Charles James, B Smith Hopkins, and the tangled web of element 61. *Bull. Hist. Chem.* **2006**, *31*(1), 9–18.
260. http://portal.acs.org/portal/PublicWebSite/education/whatischemistry/landmarks/earthelements/index.htm (accessed April 13, 2014). American Chemical Society National Historic Chemical Landmarks. Separation of Rare Earth Elements.
261. Gould, R. F. *Chem. Eng. News* **1949**, *25*, 2555.
262. Cork, J. M.; James, C.; Fogg, H. C. *Proc. Natl. Acad. Sci.* **1926**, *12*, 698.
263. Murphy, C. J., *art. cit.*, 10.
264. Although the PNAS is a very prestigious journal, its circulation did not match that of JACS in terms of dissemination of James' work.
265. Peed, W. F.; Sptizer, K. J.; Burkhart, L. E. The l spectrum of element 61. *Phys. Rev.* **1949**, *76*, 143–144.
266. http://unhmagazine.unh.edu/f10/charles_james.html (accessed April 13, 2014).
267. *Archivio Storico dell'Università di Firenze anno* **1930**, 10d./2908.
268. *Archivio Storico dell'Università di Firenze anno* **1930**, 10d./1930.
269. *Archivio Storico dell'Università di Firenze anno* **1930**, 10d./3389.
270. *Archivio Storico dell'Università di Firenze anno* **1930**, 10d./443.
271. Pool, M. L.; Quill, L. L. *Phys. Rev.* **1938**, *53*, 437.
272. Deuterons are deuterium nuclei, ^{2}H (heavy hydrogen).
273. Law, H. B.; Pool, M. L.; Kurbatov, J. D.; Quill, L. L. *Phys. Rev.* **1941**, *59*, 936; Kurbatov, J. D.; Pool, M. L. *Phys. Rev.* **1943**, *63*, 463; Pool, M. L.; Kurbatov, J. D.; Quill, L. L.; MacDonald, D. C. *Phys. Rev.* **1942**, *61*, 106.
274. Mattauch, J. *Naturwissenschaften* **1940**, *19*, 206.
275. Rolla, L.; Fernandes, L. *Chimie et Industrie* **1927**, 394.
276. Ollano, Z. *Nuovo cimento* **1942**, *19*(8), 225.
277. The colloidal varieties of the radioactive mineral uranite go under the name of pitchblende; it is one of the main natural sources of uranium The chemical formula is UO_2.
278. Attrep, M. Jr.; Kuroda, P. K. *J. Inorg. Nucl. Chem.* **1968**, *30* (3), 699.
279. Prandtl, W. *Angew. Chem.* **1926**, *39*, 897; Prandtl, W. *Angew. Chem.* **1926**, *39*, 1333; Aston, F. W. *Mass Spectra and Isotopes*; Longmans Green and Co.: New York, 1933; Mattauch, J. *Z. Physik* **1934**, *91*, 361; Noddack, W.; Tacke, I. *Angew. Chem.* **1934**, *47*, 301; Segrè, E. *Scientific Monthly* **1943**, *57*, 12.
280. Marsh, J. K. *Quart. Rev. Chem. Soc. London* **1947**, no. 1–2.

281. Marinsky, J. A.; Glendenin, L. E. *Chem. Eng. News* **1948**, *26*, 2346; Marinsky, J. A.; Glendenin, L. E. Coryell, C. D. *J. Am. Chem. Soc.* **1947**, *69*, 2781.
282. Marinsky, J. A.; Glendenin, L. E. *Chem. Eng. News* **1948**, *26*, 2346; Marinsky, J. A.; Glendenin, L. E.; Coryell, C. D. *J. Am. Chem. Soc.* **1947**, *69*, 2781.
283. Private communication of B Smith Hopkins Jr. to Marco Fontani, January 21, 1998.
284. Marinsky, J. A. The search for element 61. *Episodes from the History of the Rare Earth Elements*, Evans C. H., Ed.; Springer: Heidelberg, Germany, 1996, pp. 91–107.
285. An interesting conversation between two chemists, Gregory Girolami and Richard Perry, at the University of Illinois on the presumed discovery of *illinium* by a member of their department, B Smith Hopkins, can be viewed at http://www.youtube.com/watch?v=EyYk_mHXS-c (accessed April 13, 2014).
286. Wood, P. *The Sun,* March 16, 1986, C-1.
287. Harris took his degree in chemistry at the University of British Columbia; he was elected a Liberal member for South Okanagan, a position occupied formerly by J. W. Jones, minister of finance of the previous Conservative government. Shortly thereafter, he won a grant to study for his doctorate at the University of Illinois, publishing his dissertation in 1925. He worked until the early 1930s with Professor B Smith Hopkins in the attempt to isolate element 61 by chemical means. Later, his mentor sent him to George Urbain, in Paris, to become intimately familiar with the extraction and purification of the rare earth metals. When he returned to the United States in 1932, he was hired as an associate professor by the University of British Columbia. He remained in this position until he retired in 1966.

IV.10

MASURIUM: AN X-RAY MYSTERY

IV.10.1. THE DISCOVERY OF RHENIUM AND MASURIUM

There were many attempts made to isolate the element with atomic number 43; the most famous was that carried out in 1925 by the husband-wife team of Walter Noddack and Ida Tacke Noddack.

Walter Noddack, descendant of an ancient family originally from East Prussia, was born on August 17, 1893, at Berlin. Ida Eva Tacke was born February 26, 1896, in the small town of Wesel, on the lower courses of the Rhine. After having earned the degree of Doctor of Engineering in 1921, she completed her chemical education by work on the anhydrides of high-molecular-weight fatty acids, a research subject that she abandoned after having completed her research doctorate. Later, she obtained a position in the thriving German chemical industry at Berlin: first at Allgemeine Elektrizität Gesellschaft and later at Siemens-Halske. In 1925, she left this employment and joined the Physikalische Technische Reichsanstalt (Imperial Research Laboratory for Technical Physics) at Berlin. In this government organization, a chemistry laboratory was directed by the young Walter Noddack, her future husband.

The two chemists expressed interest in the missing elements in the periodic table. Up until the end of the 19th century, the elements had been discovered almost accidentally. In 1896, Dmitri Mendeleev, on the basis of his idea of periodicity, proposed to name the still missing elements from the seventh group *eka-manganese* and *dvi-manganese*, to which he gave the symbols Em and Dm. In 1913, the physicist Henry Gwyn J. Moseley, in formulating his law, confirmed Mendeleev's predictions: the two homologues of manganese, elements 43 and 61, were missing from the roll call. No known element with atomic weight higher or lower than theirs was radioactive or unstable, and for this reason it was believed that the two elements could exist in nature. Ida Tacke Noddack and Walter Noddack focused their attention on the mysterious missing elements, carrying out tedious systematic examinations of the chemical properties of the elements adjacent to the two they were seeking. They noticed a gradual change in the chemical properties of the transition metals belonging to a same group, such that the first and last elements did not resemble each other as much as Mendeleev had predicted.

This was one of the reasons why they succeeded in isolating the element with atomic number 75. The chemists who came before them had investigated minerals of manganese, convinced that they would find the elusive metal whose properties had been confirmed as being identical to the element with atomic number 25. All attempts resulted in failure.

A nearly forgotten French chemist, Gerber, hypothesized that elements 43 and 75 would have properties closer to those of molybdenum and tungsten than of manganese. During World War I, he carried out investigations along these lines and made note of the discovery of these two metals, which he called *neo-molybdenum* and *neo-tungsten*.[288]

Tacke Noddack and Noddack must have read Gerber's work because, soon after the end of the war, they affirmed the same ideas: that is, that the properties of element 43 and 75 would be more similar to those of the adjacent metals in the periodic table rather than to those of manganese. Consequently, they concentrated their efforts on the examination of deposits of metals like molybdenum, tungsten, ruthenium, and osmium. Treating minerals coming from different areas, they prepared more than 400 enriched samples that, in June 1925, they sent to Otto Berg, the spectroscopic specialist at Siemens-Halske and an ex-colleague of Tacke Noddack. Among the many trials accomplished, Berg found in Norwegian columbite two new elements.[289] The level of element 75 amounted to 5% whereas that of element 43 was 0.5%.

Later, traces of element 43 were found in gadolinite, fergusonite, and zircon, while the higher homologue was present in tantalite and tungstite. Berg was able to record three characteristic lines for element 43:

$K\alpha_1 = 0.672$ Å, $K\alpha_2 = 0.675$ Å, and $K\beta_1 = 0.601$ Å.

The names attributed to the two metals were *masurium* (symbol, Ma) and rhenium[290] (symbol, Re) to commemorate the birthplaces of the two discoverers, the Rhineland and the region of the Masuri lakes. World War I was hardly over and the scientific world was uneasy about the name *masurium*. British chemist John Newton Friend noted that the name rhenium was appropriate, whereas *masurium* was not a very friendly choice, representing blatant propaganda that mirrored the discontent, chauvinism, and revanchism (a political view that looked to regain losses due to war as a duty and a right to foment another war) inherent in the German nation. From September 6 to September 15, 1914, in the region of the Masurian lakes, the Kaiser's troops had inflicted a terrible defeat on the Russians, who left more than 125,000 men dead on the battlefield. Newton Friend condemned the choice of Noddack and Tacke Noddack (Berg had not taken part in the selection of the name) as "a stupid psychological blunder which no civilized scientist would make."[291] Numerous papers on rhenium and *masurium* appeared over a period of many years. While all three co-workers, Walter Noddack, Ida Tacke Noddack, and Otto Berg shared co-authorship, it was Ida Tacke Noddack's name alone that appeared on all of them, indicating the predominant role that the young woman had in the work of isolating the two metals.

In 1926, Ida Tacke married her boss, Walter Noddack, thus becoming a chemikerin (German for "woman chemist"). She held a subordinate position to her husband as an unpaid collaborator for a long time; she had no laboratory, no instruments, and no funds of her own for research. That same year, the quantity of rhenium isolated in the pure state was hardly 2 mg, but over the following 12 months, they extracted another 120 mg of the precious metal from molybdenite. Finally, in 1928, 660 kg of molybdenite were treated and Walter Noddack was proud to announce to the world at large "Die Herstellung von einem Gram Rhenium"[292]—the team had extracted the first gram of rhenium. The exact quantity was 1.04 g, a reserve that in 1929 increased to 3 g, thus enabling the Noddacks to establish the properties of the element and to study its compounds. The expense of extraction came to more than $180,000 in today's currency.

Between May 4 and 9, 1931, the 13th Chemical Industries Exposition was held in New York City. Its success surpassed all expectation: 360 exhibitors and more than 103,000 visitors. One of the objects that monopolized the interest of industrialists was one of the first models of a pH-meter and a jar about 10 cm high and 2.5 cm in diameter containing a sample of rhenium valued at around $3,000.

IV.10.2. NO MORE MENTION OF MASURIUM

As discussed, the Noddacks were chemists highly respected for the discovery of rhenium. In the same work, they had also announced the discovery of element 43, which they called *masurium*. The discovery of rhenium was quickly confirmed, and the element was soon prepared in macroscopic quantities, although there was no more mention made of *masurium*. In 1934, when Enrico Fermi charged Emilio Segrè with obtaining all the chemical elements so that he could irradiate them with neutrons, Segrè brought him a sample of rhenium, but not a grain of *masurium*.

The Noddacks, however, continued to claim that they had also discovered the element with atomic number 43. Starting out with a sample of Norwegian columbite dissolved in mineral acid, they concentrated small quantities of the sulfides (ReS and MaS). The X-radiographs finally confirmed the presence of about 0.001 g of rhenium and 0.2 mg of *masurium*. Furthermore, the authors were able to measure with great accuracy the $K\alpha_1$, $K\alpha_2$, and $K\beta_1$ of element 43 and the $K\alpha_1$, $K\alpha_2$, $K\beta_1$, $K\beta_2$, and $K\beta_3$ of element 75. They concluded their article[293] with an argumentative note directed at their colleagues Gerald J. F. Druce, Jaroslav Heyrovský, Vaclav Dolešek, and Wilhelm Prandtl who had all repeated the Noddacks' experiments in vain and had expressed doubts about the existence of *masurium*.

After this research, the Noddacks were held in very high regard in Germany and their opinions were highly valued. When they reported in Germany on the discovery of *illinium* on the part of U.S. chemist B Smith Hopkins;[294] their endorsement of the discovery brought about its acceptance by the entire German Chemical Society.

In 1935, Walter Noddack became professor of chemistry at the University of Freiburg, where he remained for 6 years. After the Nazi invasion of France and the consequent annexation of Alsace to the Third Reich, Noddack occupied the chair of chemistry at the newly established Reichsuniversität Straβburg.

The influential analytical chemist Wilhelm Prandtl rejected the discovery of *masurium*, advancing the hypothesis that both the Noddacks and their spectroscopist Otto Berg had been deceived by the presence of trace amounts of impurities of zinc and tungsten in their samples, whose spectral lines would have led to an erroneous conclusion.[295]

IV.10.3. PANORMIUM AND TRINACRIUM

While the young scholarship recipient Emilio Segrè spent a certain period of specialization under the guidance of the renowned Otto Stern (1888–1969) at Hamburg, his mentor never stopped singing the praises of Ernest Orlando Lawrence's cyclotron, foretelling a great future for this instrument. Segrè neither doubted nor forgot the words of his mentor. When he went to the United States in 1935, he made contact with Lawrence, and, once back in Rome, he spoke seriously with Fermi about the possibility of constructing a cyclotron of their own.[296] A few years later, he went to see personally the cyclotron at Berkeley; while visiting Lawrence's laboratory, he noticed some pieces of highly radioactive metals heaped up helter-skelter, and no one knew what they might contain. Segrè asked for some of these samples to bring back to his laboratory at Palermo, and Lawrence was extremely kind in giving him all the material, happy that they could be used to help a researcher at a poor university like Palermo.[297] After a tour of the United States, including a visit to Death Valley, Segrè returned to Palermo, where he began to study the radioactive products. They were found to contain phosphorus, silver, zinc, and cobalt.

In February 1937, Segrè received a letter from Lawrence that contained a small plate of molybdenum that had been part of the cyclotron's deflector. Lawrence was an engineer and did not know much about chemistry (or at least was not interested in it), but Segrè knew just enough to realize the magnitude of the gift given him by his American colleague. He suspected that the plate could contain isotopes of element 43.[298] Upon examining the plate, he found that the face that had been exposed to the beams was much more radioactive than the nonexposed face, which indicated to him that the nuclear reaction was due to a charged particle and not neutron activation. In dissolving the metal with acid, he had preferentially attacked the active surface, thus concentrating the products of the reactions (d,n) and (d,p)[299] and of their decays. Among them, Segrè believed, ought to be isotopes of element 43.

Segrè was sharp enough to know that the so-called *masurium*, announced in 1925 by the Noddacks, was nothing more than the result of an experimental error. Among other things, arguments from nuclear systematics[300] would make it highly unlikely that the element was present in nature. What remained now was to demonstrate that he had effectively observed a new element, created artificially and lacking any stable isotopes. For this work, Segrè collaborated with Carlo Perrier (1886–1948) and the radiochemist Nestore Bernardo Cacciapuoti (1913–79). The team found two isotopes of element 43: $^{95}43$ and $^{97}43$, both excited isomeric states produced by the bombardment of stable isotopes of molybdenum. Through their work, they discovered the first synthetically created element.[301]

Perrier and Segrè decided not to name this element. There was no lack of suggestions for names that celebrated fascism or Sicily, like *trinacrium*, or the university, *panormium*, but the discoverers did not like any of these. As well, to avoid polemics,[302] it was necessary to refute the Noddacks' discovery or to allow it to die on its own—which is what happened.[303] Segrè knew very well that the number of elements named exceeded the number of elements that truly existed or had been discovered.[304] It seemed to him that it would be smarter to show that he and Perrier were not in a hurry. Segrè wrote to George de Hevesy, who knew first-hand the work of the Noddacks on element 43. In a letter to Segrè he confirmed its groundlessness.

In 1937, after having read de Hevesy's letter carefully, Segrè decided to see the Noddacks' results with his own eyes. In September 1937, returning from a conference in Copenhagen, Segrè stopped at Freiburg, where the Noddacks had their laboratory. Professor Noddack kept Segrè waiting a long time before he received him. His wife, Ida Tacke Noddack, was not present. Segrè showed him a draft of his work on element 43, presented to the Lincei, and asked him if their work agreed with his. "*Yes*," he replied. Then Segrè asked him how much *masurium* was available, and he replied "about a milligram," which seemed highly unlikely to Segrè. Noddack shunted off other questions by adding that he could not show him the sample because he had sent it to F. W. Aston (1877–1945) for isotopic analysis. Segrè pressed Noddack, asking him if he had the X-ray photographic plates showing the characteristic spectrum of element 43, given that this had been their method of discovery. The response was also negative; the plates could not be shown because they had been broken some time earlier. When Segrè asked him "Why

FIGURE IV.13. Palermo, 1937. Carlo Perrier (1886–1948), left, and Emilio Segrè (1905–89), right, receive a visit from Walter Noddack (1893–1960) and his wife, Ida Tacke Noddack (1896–1978), center. In 1925, the Noddacks claimed to have discovered and isolated elements 75 and 43, to which they gave the names rhenium and *masurium*, respectively. *Masurium* was later shown to be an erroneous discovery. Not being present in the earth's crust, element 43 was synthesized artificially and then identified by Perrier and Segrè. After sorting through a number of possible names, including *panormium* and *trinacrium*, they finally settled on technetium.

haven't you made others?"[305] a long, embarrassed silence followed. Segrè believed that the Noddacks did not have in their hands the slightest trace of the much-proclaimed *masurium*, and furthermore were looking to gain time. He took his leave and returned to Palermo.

He was fairly surprised when, a few weeks later, Walter Noddack and a retinue of assistants presented themselves at the Institute of Physics at Palermo, where Segrè was happy to show all that he had of the elusive element number 43. Figure IV.13 commemorates the Noddacks' visit to Palermo.

After the war, in 1947, when nuclear reactors could supply macroscopic quantities of element 43, Segrè had the pleasure of determining that there had been no errors in his work. Segrè's pleasure would grow with the years and, in 1959, he would receive the Nobel Prize; Carlo Perrier, on the contrary, would die following a brief illness in 1948, at almost 62 years of age.

But perhaps the greatest pleasure that Segrè enjoyed was the surprise visit he received from his teacher, Enrico Fermi. He came into the Institute without announcing himself and, as soon as he saw Emilio Segrè, he greeted him with the words: "Your research on element 43 is the best that was done over the course of the past year!" Fermi did not make assertions like that lightly, and Segrè was very pleased with the compliment and gratified by the visit of his friend.

A review of the work,[306] colored by a bit of controversy,[307] on the detection limits possible to the Noddacks in their presumed discovery of technetium (*masurium*) as a fission product in nature was published in the *Journal of Chemical Education* in 2005. We know from Ida Tacke Noddack's own recollections[308] that neither she nor her husband used radioactivity measurements to detect its presence.

IV.10.4. THE IGNORED AND UNDERRATED "CHEMIKERIN" AND HER FISSION HYPOTHESIS

The confirmation of *masurium*'s existence in the Noddacks' samples had been repeatedly confirmed by means of X-ray investigations, beginning in 1925 by the spectroscopist Otto Berg.[309] As an X-ray specialist, Berg very much supported the work of Ida Tacke Noddack and Walter Noddack, and their names all appeared together on the announcement of the discovery of the two metals (*masurium* and rhenium). Furthermore, Berg's studies were the foundation on which the confirmation of the existence of *masurium* was placed. He seemed to have found the presence of this metal in at least 28 photographic plates on around 1,000 spectra, while another 70 cases remained uncertain.[310]

However, with the passing of the years, he distanced himself from this line of research, and the Noddacks preferred to consult other specialists, even though their work on *masurium* was also becoming increasingly reduced. Berg was born on November 23, 1873, in Berlin, where he studied inorganic chemistry; later, he specialized at Heidelberg and Freiburg.

Between 1902 and 1911, he had held the post of lecturer at Greifswald and, in 1911, he was transferred to Charlottenburg, in Berlin, to be hired by Siemens-Halske as a specialist in X-ray analysis. While at Siemens-Halske, he verified the discovery of rhenium and *masurium*. However, his contributions rapidly dried up and, after a few years, his name disappeared from the scientific literature. Otto Carl Berg died in 1939, at the age of 66.

Beginning in 1933, Ida Tacke Noddack developed her own line of research quite apart from that of her husband: intensive study of the periodic table. In her articles, she discussed the possibility of the discovery of the transuranium elements.[311] Ida Tacke Noddack's interests in this subject were motivated by the work that Enrico Fermi was carrying out at that time in Rome. As is well known, Fermi produced synthetic radioelements by neutron bombardment. When he came to irradiating the last known naturally occurring element, uranium, he believed that, following neutron capture, he could synthesize the first two transuranium elements.[312] He placed the first of the two substances under rhenium in the periodic table; this fact did not pass unobserved by the discoverer of rhenium. Tacke Noddack had studied at length the properties of the elements under manganese and, after reading Fermi's work, she wrote him a note in which she asserted that his experiments were too incomplete to arrive at the conclusions that he advanced. Fermi ought to have examined all of the elements in the periodic table before excluding their presence from among the neutron-irradiated products and claiming to have synthesized a new element. Tacke Noddack went even further:[313] "When heavy nuclei are bombarded with neutrons presumably it comes about that they break up into large fragments that are isotopes of known elements and nowhere near [in atomic weight] the bombarded targets." Thus, with these words, in advance of anyone else, Ida Tacke Noddack prophesied nuclear fission; however, she was not believed. On the contrary, in Rome, Fermi and his collaborators sneered at her and her work because it was effectively lacking in even minimal theoretical underpinnings. They alleged that Tacke Noddack's hypothesis was comparable to that of "shooting a rifle at an armored tank and watching the vehicle fall to pieces."[314]

Fermi seldom left anything to presumption, so, before he criticized her work, he sat down and did the calculations that led him to conclude that the probability of a scission of the uranium nucleus was extremely low. Thus, he repudiated Tacke Noddack's claim secure in the knowledge that his theory was right—but it was based on the wrong experimental information,[315] and this is what led to the downfall of *ausonium* and *hesperium*.

Fermi's experiments were repeated by Otto Hahn and his co-workers in Berlin. They confirmed the results obtained in Rome, and they published an extensive work about the properties and radiochemical separation of the presumed transuranium elements.[316] In 5 years of intensive research and after numerous publications, the results became so contradictory, however, that the concept of transuranium elements had to be abandoned. On January 6, 1939, Hahn and Fritz Strassmann (1902–80) wrote the famous sentence that is now accredited as the discovery of uranium fission: "In light of the facts, as chemists we ought to say that the new particles do not behave like radium, but in fact are reminiscent of barium; as nuclear physicists we cannot help but arrive at the conclusion that this is in conflict with all of our experience in nuclear physics."[317] At that time, Hahn was 60 years old and director of the prestigious Kaiser Wilhelm Institute of Chemistry. He was a scientist who had "arrived": his fame was assured in 1918, when he discovered *protoactinium* with Lise Meitner. But, by 1939, Hahn was mentally "ancient." He rejected the revolutionary idea that an atom of uranium could split into two large fragments. The mysterious motives that drove Otto Hahn to publish his results and snatch the discovery from Ida Tacke Noddack are, even today, being examined by historians of science.

When, between 1935 and 1936, Walter Noddack repeatedly suggested to Hahn that he could have referred in his many conferences and publications to Tacke Noddack's work and her criticisms of Fermi's work, Hahn replied that he did not wish to ridicule Noddack's wife about her absurd ideas about the fission of the uranium nucleus in front of the scientific community.[318]

Ida Tacke Noddack wrote a brief article in the same journal in which Hahn and Strassmann published. In it, she clinched the fact that, 5 years earlier, she had repudiated the hypothesis of the transuranium elements and predicted uranium fission. She concluded her remarks by regretting that Hahn had not acknowledged her with even a "thank you" nor with a simple citation of her work, even though in the preceding years there had been conversations on this subject between the two of them. The editor of the journal asked Hahn to comment on Tacke Noddack's remarks so that he could place them in the same issue, but Hahn indignantly refused. The editor was constrained to publish the following note: "Otto Hahn and Fritz Strassmann have informed us that they have neither the time nor the interest in responding to the criticisms leveled at them in the preceding note."[319]

Ida Tacke Noddack lacked the international and institutional support necessary for her work to be taken seriously. She published her works only in German journals because she did not know any other foreign language; in addition, the prevailing theory of matter seemed to contradict her. In the end, her status as a mere research associate in chemistry made her seem to professional physicists like a simple amateur. The controversial discovery of *masurium*, never confirmed, threw a shadow over her reputation as an inorganic chemist, so much so that it eclipsed the discovery of rhenium. In addition, the Noddacks, because of their limited knowledge of foreign languages, never went abroad to publish their successes. Nevertheless, Ida Tacke Noddack received three nominations for the Nobel Prize in Chemistry[320] in 1933, 1935, and 1937.

IV.10.5. DECLINING YEARS: SYMPATHY FOR NAZISM

In the meantime, their adherence to nazism produced its first fruits: the promotion of her husband to university professor at Freiburg in 1935 and an enormous influx of public money for research. Nevertheless, Ida remained in her husband's shadow as a research associate. The two chemists stayed at Freiburg for 6 years; then, with the annexation of Alsace to the Reich, Noddack, thanks to his espousal of National Socialism, was named professor of physical chemistry in the occupied city of Strasbourg. The scientific output of the Noddacks at Strasbourg was meager. Between 1940 and 1951, they published only one article: the obituary of the inorganic chemist Wilhelm Jander (1898–1942).[321] In 1944, Strasbourg was returned to France, and they had to make a rather hasty exit.

The French chemist Professor Jean-Pierre Adloff, one of the last students of Marguerite Perey at the University of Strasbourg, when questioned about what happened in those days, reported that nothing is known about the scientific work of the Noddacks during the period 1940–51.

When, after liberation the French chemists returned to the institute of chemistry in the Alsatian capital, the only trace of the Noddacks' work on *masurium* that could be found was the symbol "Ma" painted on the large periodic table in the main lecture hall.

After the war, Walter Noddack was brought to trial at the Denazification Court that, in the end and not without some strong objections, absolved him of wrongdoing. Ida was not prosecuted simply because she did not hold a high enough academic position. The trial's outcome was that the two chemists lost their jobs and moved to Turkey, where they lived for 12 twelve years. Nothing is known about the period of time that they spent abroad.

Meanwhile, Emilio Segrè and Carlo Perrier, at the suggestion[322] of Friedrich Adolf Paneth, were named the true discoverers of element 43 and thus were invited to propose a name for it: they prudently called it technetium (from the Greek, meaning "artificial").[323]

In 1956, the Noddacks returned to the Federal Republic of Germany and took jobs at the newly established Staatliche Forschungs Institut für Geochemie. Ida Tacke Noddack was interested in problems related to the rare earth elements. Walter Noddack's new employment was too brief to produce any significant scientific contributions; 4 years later, on December 7, 1960, he died in Bamberg, at 67 years of age. Ida continued her work in the same institute up until her retirement 6 years later, at the age of 70. In 1969, for her great contributions to inorganic chemistry with the discovery of rhenium, she was invited by the Soviet Academy of Sciences to participate in the centenary of the birth of Mendeleev. At that time, she was the last person still living to have discovered an element existing in nature. However, political reasons and poor health prevented her from being present at the ceremony. She sent a typed manuscript on the events surrounding the discoveries of rhenium and *masurium*. The manuscript was translated into Russian and read to the conference assembly.

Not having children or close relatives, she spent the last years of her life at Wohnstift Augustinum, a home at Bad Neuenahr in the neighborhood of Bonn. There, on September 24, 1978, Ida Eva Tacke Noddack closed her eyes forever at the age of 82.

The varied history of the Noddacks has fascinated many chemists. Some of them have sought to reevaluate their work related to *masurium*, but despite these attempts,[324] one can assert without a shadow of a doubt that, in the 1920s and with the means at their disposal, it was impossible to detect the infinitely tiny quantities of technetium[325] present in nature as a by-product of the spontaneous fission of uranium.

Notes

288. Gerber, M. *Mon. Sci.* **1917**, *5*, 219.
289. Noddack, W.; Tacke, I.; Berg, O. *Naturwissenschaften* **1925**, *13*, 567; Noddack, W.; Tacke, I.; Berg, O. *Z. f. techn. Phys.*, **1925**, *6*, 599.
290. Noddack, W.; Tacke, I.; Berg, O. *Naturwissenschaften,Sitzungsber. Preuss., Akad. Wiss. Berlin* **1925**, 400.
291. Newton Friend, J. *Men and Chemical Elements*, 2nd Ed.; Charles Griffin: London, 1961, p. 251.
292. Noddack, W. The production of a gram of rhenium. *Z. Anorg. Allg. Chem.* **1929**, *183* (1), 353.
293. Noddack, W.; Tacke, I. *Continental Met. & Chem. Eng.* **1926**, *1*, 109.
294. Noddack, W.; Tacke, I. *Chem. -Metall. Z.* **1926**, *16*, 985.
295. Kenna, B. T. *J. Chem. Educ.* **1962**, *39*, 436.
296. Fermi had asked for a 37-inch magnet like the one used in the cyclotron at Berkeley that was then in Coltano, near Pisa, a left over (leftover?) from the radio station of Guglielmo Marconi (1874–1937).
297. Segrè, E. *The Mind Always in Motion: The Autobiography of Emilio Segrè*; The University of California Press: Berkeley, 1993.
298. Artom, C.; Sarzana, G.; Terrier, C.; Santangelo, M.; Segrè, E. *Nature* **1937**, *139*, 836.
299. In the abbreviated, or Bethe, notation given here, the first letter within the parentheses indicates the incident particle, d, a deuterium nucleus, and the letter at the right indicates the particle expelled, either n, a neutron, or p, a proton. The atom subjected to bombardment is Mo (Z = 42) and the product is Tc (Z = 43); neither is indicated in the abbreviated notation.
300. Mattauch, J. *Z. Physik*, **1934**, *91*, 361.
301. Perrier, C.; Segrè, E. *Rendiconti Lincei* **1937**, *25*(6), 723; Perrier, C.; Segrè, E. *Rendiconti Lincei* **1937**, *27*(6), 579; Perrier, C.; Segrè, E. *J. Chem. Phys.* **1937**, *5*, 712; Perrier, C.; Segrè, E. *J. Chem. Phys.* **1939**, *7*, 155.
302. The recollection of the regrettable diatribe between the Americans and Italians over the attribution of the name for element 61—*illinium* vs. *florentium*—was still fresh in the minds of chemists and physicists in Italy.
303. For their entire lives, even when confronted with irrefutable spectroscopic evidence, neither Walter Noddack, nor his wife Ida Tacke Noddack, admitted their failure to isolate and identify *masurium* by spectroscopic means.
304. As a member of the Fermi group, he was also involved in the discovery of those elements that were thought to be the first transuranium elements: *ausonium* and *hesperium*.
305. Segrè, E. *The Mind Always in Motion: The Autobiography of Emilio Segrè*; The University of California Press: Berkeley, 1993, p. 209.
306. Zingales, R. From masurium to trinacrium: The troubled story of element 43. *J. Chem. Educ.* **2005**, *82*, 221–227.
307. Habashi, F. *J. Chem. Educ.* **2006**, *83*, 213.
308. Habashi, F. *Ida Noddack (1896–1978). Personal Recollections on the Occasion of the 80th Anniversary of the Discovery of Rhenium*; Laval University: Québec City, Canada, 2005, p. 59.
309. Berg, O. *Z. Techn. Physik* **1925**, *6*(11), 599.
310. Berg, O. *Z. Angew. Chem.* **1927**, *40*, 254.
311. Noddack, I. *Z. Angew. Chem.* **1934**, *47*, 301.
312. Fermi, E. *Nature* **1934**, *133*, 898.
313. Noddack, I. *Z. Angew. Chem.* **1934**, *47*, 653. See also Habashi, F. Ida Noddack: proposer of nuclear fission. In *A Devotion to Their Science: Pioneer Women of Radioactivity*; Rayner-Canham, M. F.; Rayner-Canham, G. W., Eds. Chemical Heritage Foundation: Philadelphia, PA, 1997, pp. 217–25.
314. D'Agostino, O. *Il chimico dei fantasmi*; Mephite: Atripalda (AV), 2002.

315. Rhodes, R. *The Making of the Atomic Bomb.* Simon and Schuster: New York, 1995, p. 231, reporting on the verbal reminiscences of Edward Teller (1908–2003). Rhodes offers four possible reasons for Fermi's intransigence: (1) Tacke Noddack's idea was ahead of her time, and there was no theoretical basis of support; (2) Bohr had not yet formulated his liquid drop model of the atomic nucleus, which would have furnished the needed support; (3) F. W. Aston's mis-measurement of the mass of helium introduced a systematic error into calculating the mass and energy of nuclei; (4) Fermi had relied too much on his own self-proclaimed "formidable intuition." Whatever the reasons, Ida Tacke Noddack's intervention was, seemingly misogynistically, shelved.
316. Shea, W. R. *Otto Hahn and the Rise of Nuclear Physics*; D. Reidel: Dordrecht, The Netherlands, 1983.
317. Hahn, O.; Strassmann, F. *Naturwissenschaften* **1939**, *27*, 11.
318. Jungk, R. *Brighter than a Thousand Suns*; Harcourt Brace, Jovanovich: New York, 1958, p. 62.
319. Noddack, I. *Naturwissenschaften* **1939**, *27*, 212.
320. Crawford, E.; Heilbron, J. L.; Ullrich, R. *The Nobel Population, 1901–1937*; Office for History of Science and Technology: UCB, Berkeley, CA, 1987.
321. Noddack, W. *Die Chemie* **1943**, *56*, 53.
322. Paneth, F. A. *Nature* **1947**, *159*, 8.
323. Perrier, C.; Segrè, E. *Nature* **1947**, *159*, 24.
324. Van Assche, P. H. M. *Nucl. Phys.* **1988**, *A480*, 205.
325. Herrmann, G. *Nucl. Phys.* **1989**, *A505*, 352.

IV.11

THE TWILIGHT OF THE NATURALLY OCCURRING ELEMENTS: MOLDAVIUM, SEQUANIUM, AND DOR

The more the vacant boxes in the periodic table diminished, the more scientists increased their efforts to identify the elements still missing. Although the techniques they used were increasingly sophisticated, the elements seemed more elusive and difficult to find. Despite the risk of reporting false discoveries, the number of announcements increased, and scientific journals received many papers [326] proposing fanciful names for elements 85, 87, and 93.

In the years in which physicists were successfully reassessing the great number of new discoveries that would lead to the synthesis of artificial elements, in Paris, two spectroscopists were looking for the presence in nature of precisely these missing elements. Yvette Cauchois (1908–99) was a famous woman of science who profoundly influenced the development of optics and X-ray spectroscopy. Born in Paris on December 19, 1908, she attended the Faculty of Physics at the Sorbonne, where, in 1933, she concluded her PhD study under the supervision of Jean Perrin. In the same year, she was appointed assistant researcher at the Centre National de la Recherche Scientifique (CNRS). Cauchois became associate researcher in 1937.

In 1934, Cauchois suggested the use of a curved crystal for the transmission of X-rays and for their more highly sensitive, higher resolution analysis.[327] Later, with the same equipment, she also focused weak-penetrating X-rays. With the aid of her curved crystal apparatus, she developed the diffraction imaging technique. In the 1930s and 1940s, she determined the inner transition energy levels of mono- and multiple-ionized atoms. Cauchois was also involved in searching for rare, naturally occurring elements such as radon and polonium. Along with her colleague, physicist Horia Hulubei, she developed a new survey method to deal with the study of the actinide elements.

Horia Hulubei was born in the northeastern Romanian town of Iaşi on November 15, 1896. He entered the university there in 1915, but within a year, he joined the Romanian army and took part in World War I as a volunteer. When Romania was invaded by the united Austro-German forces, he escaped to France, where he was enrolled as an aviator-fighter. At the end of the war, Hulubei was decorated with the Legion d'Honneur and, a few months later, he returned to Romania, where he served as a civilian pilot. He continued his academic studies and, in 1926, received his university degree magna cum laude. In 1927, Hulubei returned to France and took up research activity in Jean Perrin's laboratory of physical chemistry at the Sorbonne. There, he became acquainted with Yvette Cauchois. In 1933, he took his doctorate: he predicted and, later discovered, the multiple Compton effect. At the end of the 1930s, he collaborated with Cauchois, aiming their research at identifying the missing elements with atomic numbers 85, 87, and 93

FIGURE IV.14. Tomb of Yvette Cauchois (1908–99). Photograph by Marco Fontani.

and supposed by many to be present in nature in extremely minute trace amounts. Figure IV.14 is an image of the tomb of Yvette Cauchois in Romania and Figure IV.15 is a picture of Horia Hulubei, her collaborator.

IV.11.1. EKA-CÆSIUM: FROM RUSSIA TO MOLDAVIA, THROUGH VIRGINIA

An accurate and critical study confirms that, from the beginning of the 20th century until the early 1930s, many scientists searched for element 87 in nature, including three Nobel laureates. First, in 1903, T. W. Richards[328] employed the very suitable and classical method of determining the atomic weight of its neighboring elements. Using radiochemical analysis, Otto Hahn[329] and George de Hevesy[330] also looked for element 87. The hunt for this missing element utilized not only arc and flame spectra,[331,332] but also studies on physiological effects induced by alkali metals on frog heart.[333] The Dutch physician Hendrik Zwaardemaker (1857–1930) thought that he saw naturally occurring radioactive *eka-cæsium* in his samples.[334] Research on the missing elements, however, had to wait on one of the most effective techniques ever-developed—X-ray analysis—to receive fresh impetus.[335]

Gerald J. F. Druce and Frederick H. Loring[336] were among the first to utilize this investigative technique in search of element 87. Their work dragged on for many years, and

FIGURE IV.15. Horia Hulubei (1896–1972). Yvette Cauchois, French physicist, and Horia Hulubei, Romanian physicist, were both students of Jean Perrin. By means of spectroscopy, they sighted the elements with atomic numbers 85, 87, and 93, which they proposed be called *dor* ("burning desire" in Romanian), *moldavium* (after Moldavia, a region in Romania), and *sequanium* (after a people who occupied what is today modern France in the Roman era), respectively. All of these discoveries were shown to be baseless, but the careers of the two scientists did not seem to suffer: Cauchois occupied the chemical physics chair that had been Perrin's, and Hulubei became rector of the University of Bucharest.

they finally proposed (with great caution) the name of *alkalinium* for this element. In the years spanning the gap between the two world wars, some scientists who actively sought traces of element 87, provisionally called element 87 *eka-cæsium,* with the symbol Eka-Cs.

At the beginning of 1923, the future Nobel laureate in physics, Alfred Kastler (1902–84), as yet a young student at the Ecole Normale Superieure, was attending Professor Georges Urbain's lectures. Urbain offered him such a fascinating glimpse into inorganic chemistry that Kastler asked to be allowed to work in Urbain's laboratory. Urbain accepted, immediately giving Kastler a sample of pollucite (a mineral whose content was extremely rich in cesium) and suggesting that he start measuring the natural radioactivity of the rock. Many years later, on the occasion of the celebration of the 100th anniversary of Urbain's birth, Kastler wrote:[337] "Urbain had an unwavering idea: in the column of the periodic table containing the alkali metals, the box following cesium—that of element 87—remained maddeningly empty; this missing 'wedge' was close to that of radium, element 88. It could be supposed that element 87, still undiscovered, ought to be strongly radioactive."

Urbain's intuition was right on the mark: element 87 would be shown to be radioactive. The weak radioactivity coming from pollucite could not, however, be attributed to the presence of the missing element. Urbain, being a renowned expert in the chemistry of the elements, was certain that *eka-cesium* would not be present in his samples. A comparative examination of potassium salts and pollucite samples showed Urbain and Kastler that the radioactivity was due to the isotope ^{40}K.

In 1929, Kenneth Tompkins Bainbridge (1904–96) published an original paper on this subject.[338] He looked for element 87 with the aid of a high-sensitivity, high-resolution mass spectroscope developed by Arthur Jeffrey Dempster (1886–1950). Substantially, this amounted to chemically concentrating ores rich in the alkali metals, such as lepidolite and pollucite. Then, with the aid of a Dempster mass spectrometer, by the thermoionic effect, a current of positive ions of the elements K, Rb, and Cs, and of the presumed *eka-Cs* would be produced that previously had been deposited on a tungsten wire. By varying the ionization potential, a weak current would be produced, characteristic of each element. The idea, in and of itself, was original: it was possible to analyze a mixture of all the alkali elements without having to separate them. This method was indeed sensitive and would have bypassed the fact that element 87 might be radioactive. Unfortunately, the major criticism of Bainbridge's method lay precisely in the method: under the conditions fixed by the experiment, they would have also observed multiple ionizations of the alkali metals present.

A few years earlier, in 1925, the Russian chemist Dmitri Konstantinovich Dobroserdov (1876–1936) observed weak radioactivity in his potassium samples. Contrary to Urbain's idea, he thought that this observation could signal the presence of *eka-cæsium* in his sample.[339] He immediately named this presumptive element *russium*, after his fatherland. Before the outbreak of World War I, Dobroserdov was appointed professor of chemistry to the imperial University of Kazan. In the mid-1920s, when the civil war ended, he moved to Odessa Polytechnic. During the last years of his life, he devoted much time in chemical education and lost interest in the newly discovered element he had named *russium*.

In the 1930s, an American physicist, Fred Allison, and his assistant E. J. Murphy, claimed that they had found element 87 in lepidolite, a lithium ore, and pollucite, a mineral containing cesium. He named the presumed element *virginium*,[340] after his native state of Virginia. Soon afterward, Allison's magneto-optic effect turned out to be a non-discovery, and Nobel laureate Irving Langmuir (1881–1957) referred to it as an example of "pathological science."[341] (A more extensive discussion of Allison's work appears later in Part IV.)

While the magneto-optic effect was still a subject of criticism, chemists J. Papish and E. Wainer[342] hastened to make sure it was remembered that they had prepared the samples examined by Allison and Murphy. Furthermore, they emphasized that they had also been able to observe, using the more traditional technique of X-ray emission, the L lines characteristic of *eka-Cs* in other ores. They had worked on a sample of 10 kg of samarskite and arrived at the same results. Although they disagreed with Allison both with respect to the name given to *eka-Cs* and the priority of discovery, Papish and Wainer decided to wait for additional experimental confirmation before proposing their name for element 87—a confirmation that never came.

Five more years elapsed before talk of *eka-Cs* surfaced again. As many chemists had before him, Horia Hulubei, studying pollucite, an ore rich in cesium, believed that he found its higher homolog. He and his colleague, Yvette Cauchois, using their highly sensitive, high-resolution curved crystal X-ray apparatus, reported weak lines that they assumed were a doublet of element 87. Hulubei analyzed pollucite as Urbain had almost 10 years earlier. He found many characteristic L emission lines for *eka-cæsium*: $L\alpha 1 = 1032$ Xu, $L\alpha 2 = 1043$ Xu, including the secondary lines $L\beta$ and $L\gamma$, although they were not very intense. (Please see the next brief section for an explanation of Xu). He claimed that Cauchois's spectrometer had a sensitivity down to less than 1 part per 10 million level.[343] At the end of their work, they announced the discovery of *eka-cæsium*, and they suggested the name *moldavium* for this element.

The following year, in November 1937, in an article that contained an extensive bibliography and a critical comparison of the scientific work that preceded it, Hulubei announced that he had observed the lines Lβ1 = 838 Xu, Lβ2 = 856 Xu, and Lγ = 715 Xu[344] of element 87. Absolutely convinced that his samples contained element 87, he regretted that he was unable to determine with absolute precision the Lα1 and Lα2 lines on account of the extremely weak signals, but the fact that he had obtained the same lines with two different spectrographs, using different crystals (quartz and mica) with different reticular planes, made Horia Hulubei even more certain of the existence of moldavium. Such certainty induced him to publish a detailed work in which he enlarged on the reason for the proposed name: "For this element I propose the name of moldavium (Ml) in honor of Moldavia, a Romanian province, on the eastern borderland of the former Roman Empire." Among his unwritten reasons was certainly a love for his native land.

In 1939, Hulubei published his last paper on *moldavium*. He had fractionated more samples of pollucite; he digested the rock with hydrochloric acid and ethanol and extracted a tiny amount of *moldavium* chloride (MlCl). He also processed new minerals suspected to contain *moldavium*: lepidolite, beryl, and radioactive autunite-columbite from Bavaria.

In 1937, an American, F. R. Hirsh Jr.,[345] bitterly criticized Hulubei's methodological approach in searching for *moldavium*. He believed Hulubei suffered from a case of self-deception. Hirsh examined all the attempts his colleagues had made to find *eka-cæsium*, from the beginning up to Hulubei's most recent work, and he was deeply convinced that element 87 would not be found in nature. He suggested that Hulubei mistook mercury or bismuth X-ray lines for *moldavium* lines.

Just before the outbreak of World War II, Hulubei was appointed full professor at Iaşi University. During the war, he published little, just a half dozen papers, including the obituary of his former teacher, Jean Perrin. Few articles on the missing elements, such as those that refer to the discovery of *dor*, appeared during the waning days of the war[346,347] and soon after its conclusion.[348] War interrupted scientific communication between Romania and the United States, and Hulubei[349] learned of Hirsh's criticism[350] only in 1943. In March 1947, although element 87 had finally been discovered, he launched a scathing reply to Hirsh, asserting that his X-ray apparatus was so sensitive and that he had handled the samples so accurately that he excluded the presence of mercury or bismuth, even in traces, among his samples. He pointed out that he had predicted a stable isotope of element 87 as early as the middle of 1936, and he found Marguerite Perey's discovery of the radioactive isotope of *eka-cæsium* troubling. In fact, in January 1939, the young chemist, Marguerite Perey, announced the discovery of element 87, and she gave it the provisional name of actinium-K (Ac-K). In 1929, Perey had entered the Institut du Radium under the direct supervision of Marie Curie. When Marie Curie died, some problems arose regarding the leadership of the laboratory. Perey joined Curie's daughter, Irène Joliot-Curie, in her laboratory, but she was formally under the supervision of André Debierne, the new director. Perey observed the α decay of ^{227}Ac, which gave rise to the only naturally occurring isotope of element 87[351] as a decay product, and she soon informed Irène Joliot-Curie. This discovery was also an additional reason for anger and resentment between Debierne and Joliot-Curie. Debierne was regarded as Perey's supervisor, and he felt angry that he was ignored. In addition, he disliked being forced to accept Perey's annoying proposal for the name of element 87. Debierne refused to accept Joliot-Curie as a co-discoverer with Perey if he could not be considered co-discoverer as well. At the end

FIGURE IV.16. André Louis Debierne (1874–1949). Student and later co-worker with Pierre and Marie Curie, in 1899, he discovered actinium. After the death of Marie Curie (1934), he succeeded to the directorship of the Institut du Radium. His later scientific interests were in some respects odd and unconventional. He claimed to have discovered new physical phenomena at temperatures approaching 0 kelvin, as well as *néo-radium* and *néo-actinium*.

of this painful discussion, the two chemists recognized Perey as the only discoverer of element 87.[352] However, a new dilemma was appearing on the horizon: the Nobel laureate Jean Perrin, Hulubei's teacher and mentor, was asked to communicate Perey's discovery to the Académie des Sciences de Paris. Perrin was doubtful about her work; he chose to believe in his student's discovery of *moldavium*, and, in Perrin's mind the two discoveries could not be compatible. The more Perrin downplayed Perey's discovery, the more Perey was prudent in her criticism of Hulubei's work. Finally, under Debierne's influential recommendation, Marguerite Perey proposed the name *catium*[353] for *eka-cæsium*, but this proposal soon conflicted with those of Irène Joliot-Curie and her husband Frédéric Joliot, who sarcastically declared that the sound of this word would remind English-speaking chemists of the word "cat" instead of the wished-for name "cation." Finally, in 1946, Perey suggested the name francium and the symbol Fa[354] for this element. In 1951, the scientific community bestowed this name on element 87, but changed the symbol to the present one, Fr.[355] Figure IV.16 is an image of André Debierne.

IV.11.2. A DIGRESSION ON X-RAY WAVELENGTH: PRECISION, UNITS, AND CONVERSION FACTORS

Knowledge of the absolute values of the wavelengths of X-rays was a very confusing subject for a long time. This was due, in part, to the existence of three units of measure that were commonly used to designate the wavelengths of X-ray emission lines and the parameters of the standard crystal lattices on which they depended. Only one of these, the Ångstrom (Å) was an absolute unit (10^{-10} m). In 1919, Manne Siegbahn introduced the unit X (designated by Xu), and in 1959, Charles Thomson Rees Wilson (1869–1959) introduced the kilo Xu (kX). In the decades of the 1920s–1940s, these latter two units were widely used. One Xu was commonly taken to be equal to 10^{-3} Å, but it was implicitly defined from the value of 3029.04 Xu of the crystal lattice of calcite at 291 K.

With the passage of time, this definition was found to be unsatisfactory because relative measurements of X-ray wavelengths could not be made with absolute precision and because the parameters of the lattice varied from one crystal of calcite to the next. Therefore, today, the Ångstrom remains the only unit of measure for X-rays.

IV.11.3. EKA-RHENIUM: CUM CAESAR IN GALLIAM VENIT, ALTERIUS FACTIONIS PRINCIPES ERANT HAEDUI ALTERIUS SEQUANI. . .[356]

In 1934, it occurred to Enrico Fermi to use neutrons to produce radioactivity instead of α particles, which are repelled by the positive charge of the target nuclei. When Fermi's group reached the heaviest known element, uranium, they expected that neutron bombardment would produce some new elements heavier than uranium, with properties similar to rhenium and osmium (i.e., eka-rhenium and eka-osmium).[357] To the contrary, Hulubei and Cauchois hypothesized that element 93 would be present among the uranium ores. However, the two physicists were not the first to believe in this hypothesis. Soon after World War I, Richard Swinne (1885–1939) (Figure IV.17) empirically predicted some chemical and physical properties of the not yet discovered transuranium elements[358] on the basis of Bohr's theory. A few years later, in 1931, he believed that transuranium elements[359] could be present in the cosmic dust embedded in some Greenland glaciers and, with X-ray techniques, he identified the characteristic lines of element 108.

In 1934, the engineer Odolen Koblic (1897–1959), after processing pitchblende from Jàchymov, in Czechoslovakia, concluded that element 93 was present in it. He also predicted that the element would have an atomic weight of 240. Koblic, like Hulubei, went to

FIGURE IV.17. Richard Swinne (1885–1939). In the period between the two world wars, Swinne claimed to have observed the characteristic X-ray lines of the superheavy element with atomic number 108, which came from a mysterious interstellar dust entrapped in Arctic ice. Gift of Dr. Edgar Swinne, his son.

Paris, to the Sorbonne, where polonium, radium, and actinium were discovered, to finish his postdoctoral studies. Ironically, that laboratory influenced both men so much that they overenthusiastically and prematurely announced the discoveries of new elements. In the summer of 1934, Koblic published a brief communication with a very forthright title: *Bohemium.*[360] In it, Koblic concluded: "All the research that has been conducted bears witness to my success in isolating the presumed element number 93 to which I bestow the name of bohemium (Bo) in honor of my native land."

In the same year, another article about *bohemium* appeared by the same author. At the urging of Ida Tacke Noddack,[361] Koblic retracted his first report on *bohemium*, and admitted that his mistake was due to an unclear analytical error.[362,363] Meanwhile, in 1938, as Hulubei and Cauchois were examining and concentrating some minerals from Madagascar (tantalite, monazite, and betafite) suspected of containing element 93, they observed enhanced lines (L series) of element 93.[364] The accuracy of this measurement was even greater than the *moldavium* data. This quantity of data allowed them to hypothesize on the presence of element 93 in the ores under examination. Hulubei needed richer samples and therefore chemically processed much more raw material. A second article appeared a year later.[365] In it, studies on other minerals were also reported: tantalite, monazite, and betafite gave positive responses; whereas in molybdenite, gadolinite, and fergusonite not a trace of element 93 was found. Using Cauchois's spectrograph, which had proven extremely high sensitivity, Hulubei also looked for element 43 (at the time called *masurium*) that was supposed to be present in the materials examined. His and Cauchois's negative results further contributed to the deterioration of the scientific reputation of the presumed discoverers, the couple Walter Noddack and Ida Tacke Noddack.[366] In those days, Hulubei enjoyed the glory of being one of the most famous spectroscopists on the European continent. The Noddacks' reaction was not long in coming:[367] they were very skeptical about the validity of Hulubei's work.

After complete digestion of the ore with mineral acids, Cauchois and Hulubei removed element 93 using PtS as carrier. They observed a weak radioactivity in the sample, but they attributed it to traces of uranium. It is strange that neither Cauchois nor Hulubei were at all sure that element 93 was radioactive, despite the law of Mattauch,[368] and they eventually announced the discovery of this element: "If the existence of element 93 should be confirmed, we would like it to have the name of sequanium (Sq) in honor of the rich and talented civilization that flourished along the banks of the Seine."

If *moldavium* would have gratified Hulubei's fatherland, the name *sequanium* paid homage to Cauchois's native country. Sequani tribesmen, who settled along the River Siene, were first mentioned by Julius Caesar in his Gallic War memoirs. The outbreak of World War II forced the two scientists to interrupt their work. In 1940, at a time when the possibilities of finding new elements appeared to be exhausted, Edwin Mattison McMillan and Philip Abelson produced the first transuranium element[369] and thus extended the periodic system beyond the limits which, one might say, Nature seemed to have established.

IV.11.4. ALABAMINE AND VIRGINIUM

The short-lived case associated with the announcement of the discovery of *alabamine* and *virginium* turns out to be very difficult to interpret. The odd behavior of Professor Fred Allison, associated with the discovery of these two elements, is unjustifiable both on the human level and on the scientific because, as was written later, he came close to appearing intellectually dishonest.

In early 1930, Fred Allison, professor of physics at the Polytechnic of Alabama, published some scientific works on the search for elements 85 and 87. Allison was born in Virginia, on July 4, 1882. After having finished his studies in optics, he moved to Alabama, which at the time was one of the least advanced states in the Union.

In the first 40 years of the 20th century, scientists were anxiously searching for elements 85 and 87, whose existence had been predicted as far back as the second half of the 19th century. In those years, Fred Allison, together with his assistant, Edgar J. Murphy, developed an analytical method called the "magneto-optic method of chemical analysis," with which they were able to observe the presence of elements, dissolved in solution, in very small trace amounts.

The instrument that Allison and his colleague constructed made use of the physical effect first noted by Michael Faraday in 1845 and from him it took its name: if a beam of polarized light is made to pass through a liquid immersed in a magnetic field, one observes rotation of the plane of the polarized light. This effect is easily visible to the naked eye because the beam of light will appear more or less bright.

Allison's apparatus had two glass tubes placed in series and filled with the solutions under examination. The two cells were wrapped by spirals of copper wire, one in one direction and the other in the opposite, in order to guarantee magnetic fields with inverse directions. The light source was produced by striking an electric spark.

At the same instant that the electric current flowing in the copper wire created a magnetic field around the solution, Allison could observe the amount of rotation of the light simply by turning the second cell to compensate for the effect in the first cell. Later, he found that the amount of rotation depended on a second factor: the chemical composition of the substance dissolved in the tubes. Using a water-filled first cell as a blank, he read the values for a large number of substances dissolved (at different concentrations) in the second tube. What surprises us today is that Allison claimed that his apparatus could have a sensitivity of 1 part per 100 billion.

Allison had long since developed a scale for chlorides, nitrates, sulfates, and hydroxides. In addition, his method also allowed him, according to his claims, to identify a compound in the presence of a limited number of other substances. Thus, in the autumn of 1929, Allison used his magneto-optic technique to look for the possible presence of *eka-cesium* in nature. The laboratories of the General Electric Company furnished him with samples of pollucite and lepidolite, minerals rich in cesium, the lower homolog of element 87. Allison and colleagues found six minima in each of the compounds that they examined (chlorides, nitrates, sulfates, and hydroxides). After having meticulously repeated the measurements for 2 months and now sure of having eliminated whatever other element might be present except *eka-cesium*, they announced the discovery of element 87.[370]

Allison wanted to call it *virginium* (symbol, Va) in honor of the state of his birth.[371] Later, when the discovery of the last alkali metal and the magneto-optic technique were both shown to be nothing more than an enormous soap bubble, this symbol continued to appear in some periodic tables of the elements, although modified to Vi.

Because he had found six minima for every salt of *virginium*—VaCl, $VaNO_3$, Va_2SO_4, and VaOH—he asserted that *virginium* consisted of a mixture of six stable isotopes. Allison also found *virginium* present in other minerals, such as monazite, whether from

Brazil or from North Carolina; in pitchblende; in samarskite; and in the brackish waters of Lake Searles in California, as well as in ordinary seawater.

In 1932, the discovery of element 85, which he called *alabamine*,[372] was officially announced; it was named in honor of the state of Alabama, where the Polytechnic school in which he taught was located.[373] The search for *eka-iodine*, the last of the halogens, was initiated in the summer of 1930.[374] Allison used his instrument as he had for *virginium*; he determined a scale relative to the halogen salts already known: fluorides, chlorides, bromides, and iodides. In this way, he could extrapolate to the region where the minimum corresponding to the signal for element 85 would fall. Although the concentration of the new halogen was 1 part in 10^{11}; that is, at the limit of the sensitivity of his apparatus, Allison saw without a shadow of a doubt the presence of *alabamine* in his samples. He indicated the element with the letters Am, but later changed the symbol to Ab. The first samples of *alabamine* were extracted from 100 lbs (45 kg) of Brazilian monazite, but later Allison found it in trace amounts, never more than 1 part in 10^8, in many other minerals and in brackish water.

By means of the instrumentation at his disposal, he declared that he had observed in solution the whole series of the oxyacids of this halogen: HAmO, $HAmO_2$, $HAmO_3$, and $HAmO_4$. His studies quickly raised doubts in the academic world. Not much time passed before scientists realized that the magneto-optic effect was entirely nonexistent, as was the case with *virginium* and *alabamine* that Allison claimed to have isolated.

Allison's assertion that his apparatus, a relatively simple instrument, was able to distinguish between the different isotopes of the same element puzzled his scientific colleagues. This announcement shook many from their certainties because the chemical and physical means known at the time were very complex, required a great deal of time, and were very accurate. It was this last of Allison's claims that motivated the chemist Irving Langmuir, Nobel Laureate in Chemistry in 1932, to assert that the work of the Alabama physicist was a clear example of bad science.[375]

Allison became the target of a large part of the American scientific community. However, in the middle of the tempest raised by his discoveries, he published a work in which he asserted that he had discovered, aided by his magneto-optic technique, 16 isotopes of lead.

Irving Langmuir was at that time working at the University of California, hosted by the renowned Gilbert N. Lewis (1875–1946). Wendell M. Latimer (1893–1955), head of the local Department of Chemistry, was also on the scene. Talking among themselves, Latimer expressed himself in favor of these discoveries whereas his two colleagues remained strongly critical, so much so that Lewis bet $10 that the magneto-optic apparatus was nothing more than a simple hoax. Latimer was fascinated by the idea of discovering the isotope of hydrogen with a mass of 3, of whose existence physicists had been hypothesizing for a long time. It was with this idea in mind that he visited Allison in Alabama. He remained there 3 weeks and learned the techniques necessary. On his return, he constructed a model analogous to Allison's apparatus. He collected the necessary data and published a work in which he announced the discovery of tritium. Lewis paid off the $10 bet. A year later, Ernest Rutherford discovered tritium in his turn, using a completely different method. A curious fact is that the international scientific community recognized only Rutherford's discovery. Wendell Latimer was suspicious, but Langmuir was not surprised. He told his colleague that the methodology developed by Allison was able to deceive many scientists, all in good faith.

Allison was persuaded that he had found the two elements, but the international scientific community did not seem to notice. His eccentric work was tolerated for a few

FIGURE IV.18. Fred Allison (1882–1974). Allison announced the discovery of two elements to which he gave the names of *alabamine* and *virginium*, in honor of the states of Alabama and Virginia, making use of a method of his own invention, the magneto-optic technique. Both discoveries were shown to be in error. From the mid-1930s on, the American Chemical Society refused to publish any article referring to the magneto-optic technique, maintaining that it was fraudulent. Nevertheless, Professor Allison became chair of the Department of Physics at Auburn University in Alabama, and later, of the Department of Physics and Mathematics at the University of Texas. As professor emeritus, he taught until the age of 87, never backing off from his original position as discoverer. Courtesy, Auburn University Physics Department.

years by other scientists until, in 1934, the American Chemical Society (ACS)—substantially at Langmuir's insistence—forbade Allison to publish any articles relative to the magneto-optic effect in the journals published by the Society. Two years later, the American Physical Society (APS) banned the same types of works by Allison from its own publications. Langmuir thus became Allison's most strenuous adversary. On December 18, 1953, Langmuir held a conference with the title "Pathological Science" that had among its subjects Allison and his discoveries, true or presumed.[376]

It is a curious fact that from 1927 to 1935, a good 1,698 publications appeared on this subject in American scientific journals as testimony to the fact that Allison truly had a large group of supporters.

Allison's elements were quickly removed from the periodic table, although the symbols Va and Ab continued to appear in some American chemistry manuals and textbooks for the entire duration of World War II. For Allison, fate was even more favorable. He was made chair of the Department of Physics at Auburn University in Alabama from 1922 until 1953 (the physics building at Auburn is named in his honor); from 1953 to 1955, he was head of the Department of Physics and Mathematics at the University of Texas; and from 1956 to

1961, he chaired the Department of Physics and Mathematics at Huntington College. In 1961, he was named professor emeritus, but he continued to teach until 1969. At the time of his retirement from teaching, the laboratories of physics at the above-mentioned universities were named in his honor. Despite the bans by the ACS and the APS, Allison continued to publish various articles on the magneto-optic effect. The last publication that had this phenomenon as its subject appeared in 1966. Figure IV.18 is a portrait of Fred Allison that hangs in the physics hall named in his honor at Auburn University.

Allison was a tireless worker, spending almost every waking hour in his laboratory or at the university. He was gregarious, loved his students, and always welcomed them into his laboratory and office. He somehow found time to be an amateur beekeeper and loved to give friends and colleagues gifts of honey. He was esteemed and admired by both his students and faculty colleagues. Professor Allison survived all of his discoveries: he died of leukemia on August 2, 1974, at the age of 92.[377]

Spectroscopists today tend to be more indulgent than Langmuir was in judging Allison's work. They all agree that the magneto-optic apparatus could not have functioned as it was built by Allison. Theoretically, it would have been possible to obtain these measurements only under two conditions: that the spark would have a sufficiently short lifetime (of the order of nanoseconds) and that it would produce a light source very stable and coherent (laser light). Unfortunately the lifetime of the spark in Allison's apparatus was of the order of microseconds, but this was not known at the time, and, furthermore, the light from the spark could not in the least approach the characteristics of laser light. Finally, elements 85 and 87 are radioactive and not present in nature, a fact unknown to Allison and his contemporaries. Nevertheless, Allison never mentioned the possibility that both eka-iodine (*alabamine*) and eka-cesium (*virginium*) could be radioactive.

What put Allison into a bad light was not so much the double announcement of *alabamine* and *virginium* but his behavior: he began to work on this effect with Jessy W. Beams (1898–1977) when he was still at the University of Virginia. Beams began to slowly change his research interests until he abandoned completely the magneto-optic effect even before this line of research acquired the "odor of heresy." From a study of the chronological events, a strange coincidence emerges: Beams abandoned these investigations as Allison began to announce his claims ever more loudly. It is not clear if they both discovered their mutual error and acted in different ways. The fact remains that Beams made himself scarce and never openly accused his colleague of wrongdoing. Other students of this affair hypothesized that Beams was aware of the wrong hypothesis based on the theory of the magneto-optic effect, but said nothing in order not to damage the reputation of his ex-colleague who, in the meantime, was exposed as having exaggerated.

However, under the name *alabamine* (or *alabamium* [Am]), element 85 figured in textbooks and reference works until 1947. And surprisingly, Allison is still listed as the discoverer of astatine (i.e., *alabamine*) in the 1991 *Concise Columbia Encyclopaedia*.

IV.11.5. EKA-IODINE ASSUMES THE FANCIFUL NAME OF DOR

Finally, in 1939, Horia Hulubei and Yvette Cauchois observed unknown lines in the emission spectrum of radon, some of which could indicate the presence of *eka-iodine* among the disintegration products of this noble gas. They observed only the Kα1 line at 151.1 Xu, and they attributed it to element 85. Hulubei soon announced the discovery but waited

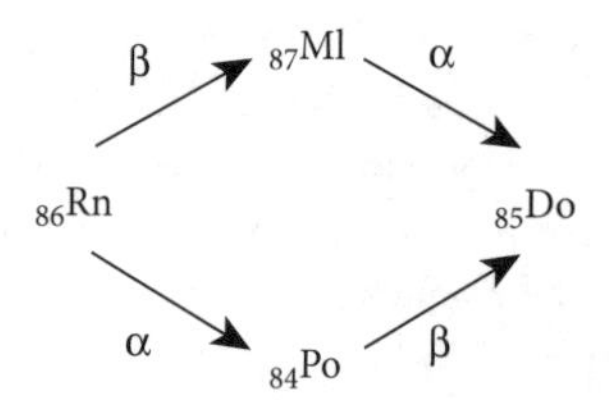

FIGURE IV.19. Decay Scheme of Radon Producing Dor (Do) by Two Different Pathways as Proposed by H. Hulubei.

for 5 years before he named this element *dor* (symbol = Do), meaning longing (for world peace). His exhaustive study of radon spectra allowed him to detect traces of element 85 as a product of the α-decay of *moldavium* (symbol = Ml), following the emission of one electron (β) from Rn. Figure IV.19 is Hulubei's proposed decay scheme of radon, producing both *moldavium* and *dor.*

Just a year later, a new claimant for *eka-iodine*, the Swiss physicist Walter Minder (1905–92), came into the limelight with an article that appeared in *Helvetica Chimica Acta.*[378] He was born in Scheuren, Switzerland, on August 6, 1905. In 1931, he was appointed professor of radiology at the Institut du Radium at the University of Bern, a position he held until his retirement in 1964. Minder observed an extremely weak β decay of RaA.[379] For this purpose, he connected two ionization chambers with an electrometer. The first chamber was placed in series with a second connected by a window. Because he was able to verify the simultaneous passage of current in both electrometers, he guessed that RaA followed a pathway of ß-decay. In fact, using other substances that were pure α-emitters,[380] no signal was observed in the second ionization chamber. As chemical proof of the existence of element 85, he used the fact that eka-iodine, formed by the decay, caused the solution in which the parent element (radon) was bubbling, to gel. The same behavior was observed for the preceding halogen, iodine. Chemical tests confirmed the analogy of this element with iodine. Minder named it *helvetium*, with the symbol Hv, after the Latin name for Switzerland. A question of priority rose between him and Hulubei:[381] "The chemical reactions attempted by Minder to support this interpretation cannot be and are not, even according to him, conclusive."

Their arguments soon became trifling and the proposed symbols became illegal squatters in the periodic table. Minder went on with his research and, 2 years later, with his colleague Alice Leigh-Smith, surprisingly repeated the announcement of the discovery of *eka-iodine.*[382] Minder and Leigh-Smith were influenced to repeat their measurements by Perey's work[383] on element 87 and by Louis Turner's speculations on naturally occurring isotopes and their distribution.[384] Minder and Leigh-Smith had, in fact, extracted a sample of 40 mg of ThA (radiothorium)[385] and characterized it by exploiting a characteristic of the halogen group: they sublimed the radioactive element on a conducting wire. This time, they accorded it the name *anglo-helvetium* with the symbol Ah. (Please see the next section for more details on this episode in chemical history.)

In the middle of World War II, a young physicist, Manuel Valadares (1904–82), was carrying on his research at the Istituto Superiore di Sanità in Rome. He repeated Hulubei's experiments with a large sample of radon (of the order of 600 millicuries) and observed new characteristic lines of element 85.[386,387,388] Except for this study, Hulubei's discoveries did not receive experimental confirmation outside of France. Then, in 1940, Dale R. Corson (1914–2012), Kenneth R. MacKenzie (1912–2002), and Emilio Segrè (1905–89), using the Berkeley 60-inch cyclotron, bombarded bismuth with helium ions[389] to discover element 85, which was later named astatine.[390,391] Today, it is known that some isotopes of astatine are present in uranium and thorium ores. The first experimental evidence of their presence was demonstrated by the Austrian radiochemists Berta Karlik (1904–90) and Traude Bernert (1915–98).[392] They were able to identify the isotopes $^{215}85$, $^{216}85$, and $^{218}85$, tentatively naming their discovery *viennium*, after the city and University of Vienna, where they worked. Karlik had a very successful career, becoming the first female member of the Austrian Academy of Sciences in 1973.

In October 1944, Hulubei, a professor of physics and (from 1941) rector of Bucharest University, reported to the Română Academiei Regale de Ştiinţe (Romanian Royal Academy of Science) his complete spectroscopic identification of element 85. It required many years of hard work: first he escaped from the Nazi conquest of Paris, and then he lost part of his equipment in the fire following the American bombardment of Bucharest on April 15, 1944:

> Now that we are virtually certain that our research and statements of 1939, on a natural element with atomic number 85, are correct, we would like to propose a name for this box in the periodic system, in case the confirmation of these experiments is finalized and the priority of our work recognized officially. We would like to call this element DOR (Do). It was identified during a period of terrible suffering for humanity. The name would, by its meaning in Romanian, recall a longing for the time when peace will bring an end to the most hateful war history has ever known.

In 1946, Hulubei spoke of element 85 for the last time. Horia Hulubei criticized the radiochemist F. A. Paneth, who wrote about the discovery of the missing chemical elements[393] without mentioning his work. The foundation of the Institute of Atomic Physics (IAP) in 1949 was the accomplishment of Hulubei's dream to build a modern, Western-type institution in his own country. He was removed from his directorship of IAP in 1968, and 4 years later, on November 22, 1972, he died at the age of 76.

Yvette Cauchois became an associate professor at the Sorbonne in 1945 and a full professor in 1951. She was the second woman, after Marie Curie, to be president of the French Society of Physical Chemistry. At the age of 90, Cauchois met a Romanian priest and embraced the Orthodox faith. According to her last will and testament, she wished to be buried in the monastery of Bârsana, Romania. She died at age 91, on November 19, 1999, following a bout with bronchitis acquired during a visit to northern Romania.[394,395]

IV.11.6. CONCLUSION

The work of Hulubei and Cauchois in the field of spectroscopy remains fundamental and innovative, and their attempts to identify very rare elements, some not present in nature, does not lessen their value. Examining the data in Table IV.3, we can see that, effectively, these two physicists may have been able to observe elements 85, 87, and 93.

By the end of the 1940s, solid confirmation of the existence of these elements by other workers bestowed on them their final names: astatine, francium, and neptunium, respectively. It is possible that minute amounts of element 87 exist in nature, but definitely not in the mineral samples analyzed by Cauchois and Hulubei. Naturally occurring traces of element 93 do not exist at all. And it might be hypothesized that the discovery of *moldavium*,

Table IV.3 Comparison of the Wavelengths of the X-ray Emission Lines of Elements 85, 87, and 93 Observed by Hulubei and Cauchois, with Current Values

Element	*Lα1 λ in Xu*	*Lα2 λ in Xu*	*Lβ1 λ in Xu*	*Lβ2 λ in Xu*	*Lγ λ in Xu*
Astatine[a]	1085.0	1096.6	893.6	904.3	872.0
Dor	1082.6	–	892.0	–	–
Francium[a]	1030.0	1042.1	840.0	858.0	824.8
Moldavium	1028.0	1043.0	838.0	856.0	715.0
Neptunium[a]	889.3	901.0	698.4	736.2	704.2
Sequanium	886.9	–	696.5	734.2	596.0

[a]Data taken from "International Tables for X-ray Crystallography," vol. IV. Kynoch Press, Birmingham, UK, 1974.

like the presumptive discovery of the first "transuranium element" harmoniously named *sequanium*, was the consequence of incorrect interpretation of experimental data. A different conclusion is possible for *dor*. Since it is now known that an isotope of element 85 is found as an occasional branch product among the decay products of radon, it is quite possible that some lines of its X-ray emission spectrum may be found in the radiation from radon sources. Nevertheless, it is very doubtful if such weak radiation could be detected by Hulubei and Cauchois, even with the focusing spectrograph they used.[396]

Notes

326. Costa, M.; Fontani, M.; Manzelli, P.; Papini, P. *Rendiconti dell'Accademia delle Scienze detta dei* XL **1997**, *115*, 21, 431; Fontani, M.; Costa, M. *Il Chimico Italiano* **1999**, *3*, 9, 26.
327. Hulubei, H.; Cauchois, Y. *Compt. Rend. Chim.* **1934**, *199*, 857.
328. Richards, T. W.; Archibald, E. H. *Proc. Am. Acad.* **1903**, *38*, 443.
329. Hahn, O. *Naturwissenschaften* **1926**, *14*, 158; Hahn, O.; Erbacher, O. *Phys. Z.* **1926**, *27*, 531.
330. de Hevesy, G. *Chem. Rev.* **1927**, *3*, 321.
331. Dennis, L. M.; Wyckoff, R. W. G. *J. Am. Chem. Soc.* **1920**, *42*, 985.
332. Baxter, G. P. *J. Am. Chem. Soc.* **1915**, *37*, 286.
333. Libbrech, H. *Nederlands Tijdschrift voor Geneeskunde* **1921**, *65*(II), 796.
334. Zwaardemaker, H.; Ringer, W. E.; Smith, E. *K. Akad. Amsterdam Proc.* **1923**, *26*, 575.
335. Hertzfinkiel, M. *Compt. Rend. Chim.* **1926**, *184*, 968.
336. Druce, J. G. F.; Loring, F. H. *Chem. News* **1925**, *131*, 289, 305, 337; Druce, J. G. F. *Chem. News*, **1925**, *131*, 273; Loring, F. H. *Chem. News* **1925**, *131*, 338; Loring, F. H. *Chem. News* **1926**, *132*, 101.
337. Kastler, A. *Rev. Chim. Miner.* **1973**, *10*, 1.
338. Bainbridge, K. T. *Phys. Rev.* **1929**, *34*, 752.
339. Dobroserov, D. *Ukrainskii Khem. Zhurnal* **1925**, *1*, 491; Dobroserov, D. *Chem. Zentr.* **1926**, *2*, 162.
340. Allison, F.; Murphy, E. J. *Phys. Rev.* **1930**, *35*, 285.
341. Langmuir, I. *Physics Today* **1989**, *42*(10), 36–50.
342. Papish, J.; Wainer, E. *J. Am. Chem. Soc.* **1931**, *53*, 3818.
343. The Cauchois spectrometer, still used today for studying the γ and high-energy X-ray regions, had a curved-crystal arrangement capable of greatly improving the resolution of heavy element spectra.
344. An old unit of wavelength, named for the Swedish physicist Manne Karl Siegbahn, corresponding to $1.002\,02 \times 10^{-13}$ m. Please see IV.11.2 for a discussion of units.

345. Hirsh, F. R., Jr. *Phys. Rev.* **1937**, *51*, 584–586.
346. Hulubei, H. *Bull. Soc. Roum. Phys.* **1944**, *45*, no. 82, 3.
347. Hulubei, H. *Bull. Acad. Roum.* **1945**, *27*, no. 3, 124.
348. Hulubei, H. *Journal de Chimie-Physique* **1947**, *44*, 225.
349. Hulubei, H. *Phys. Rev.* **1947**, *71*, 740.
350. Hirsh, F. R., Jr. *Phys. Rev.* **1943**, *63*, 93.
351. Perey, M. *Compt. Rend. Chim.* **1939**, *209*, 97.
352. Kauffman, G. B.; Adloff, J. -P. *Education in Chemistry* **1989**, *26*, 135.
353. Grinstein, L. S.; Rose, R. K.; Rafailovich, M. H., Eds. *Women in Chemistry and Physics*; Greenwood Press: Westport, CT, 1993, p. 470.
354. Perey, M. *J. Chim. Phys. Phys. -Chim. Biol.* **1946**, *43*, 155.
355. Crane, E. J. *Chem. Eng. News* **1949**, *51*, 3779; Perey, M. *Bull. Soc. Chim.* **1951**, *18*, 779.
356. "When Caesar arrived in Gaul, the leaders of one group were the Aedui; the Sequani were the leaders of the other," Julius Caesar, *de Bello Gallico*, book VI, chapter 12.
357. Fontani, M.; Costa, M. *Rendiconti dell'Accademia delle Scienze detta dei* XL **2001**, *117*, XXV, 433.
358. Swinne, R. *Z. Tech. Phys.* **1926**, *7*, 166.
359. Swinne, R. *Wiss. Veroffentlich. Siemens-Konzern* **1931**, *10*(4), 137.
360. Koblic, O. *Chem. Obzor* **1934**, *9*, 129; Koblic, O. *Chem. Ztg.* **1934**, *58*, 581; Koblic, O. *Osterr. Chem. Ztg.* **1934**, *37*, 140.
361. Ida Tacke Noddack had received some samples of material from Koblic with the request to search for traces of element 93. Both the chemical and spectroscopic analyses yielded negative results. To Koblic's great dismay, the samples analyzed by Tacke Noddack contained a mixture of silver, thallium vanadate, and tungsten salts. Following a personal communication from Tacke Noddack to this effect, Koblic published his retraction of *bohemium*, but in doing so he did not mention the presence of vanadium or silver in his samples. Although Tacke Noddack had communicated the results of her investigations to Odolen Koblic by letter, he preferred to report that he had fallen into an unspecified analytical error.
362. Koblic, O. *Chem. Obzor.* **1934**, *9*, 146.
363. Noddack, I. *Angew. Chem.* **1934**, *47*, 653.
364. Hulubei, H.; Cauchois, Y. *Compt. Rend. Chim.* **1938**, *207*, 333.
365. Hulubei, H.; Cauchois, Y. *Compt. Rend. Chim.* **1939**, *209*, 476.
366. A third person is numbered among the discoverers of rhenium and *masurium*: Otto Berg, who was the X-ray specialist. He belonged to the world of industry, working at the Siemens-Halske company in Berlin. Because of this fact and also his advanced age, he did not take part in the ongoing controversy following the isolation of the hypothetical milligram of masurium.
367. Noddack, I. *Trav. congr. jubilaire Mendeleev* **1937**, *2*, 371.
368. Josef Mattauch formulated the following empirical law: "With the exception of only the rarest of cases, two isobaric nuclei that differ in their electronic charge by one unit, cannot both be stable at the same time. One of the two will be stable, and the other will be unstable, that is, radioactive."
369. McMillan, E.; Abelson, P. H. *Phys. Rev.* **1940**, *57*, 1185.
370. Allison, F.; Murphy, E. J. *Phys. Review* **1930**, *35*, 285.
371. Allison, F.; Murphy, E. J.; Bishop, E. R.; Sommer, A. L. *Phys. Rev.*, **1931**, *37*, 1178.
372. At times, this element was called *alabamium* by other authors.
373. Allison, F.; Bishop, E. R.; Sommer, A. L.; Christiansen, J. H. *J. Am. Chem. Soc.*, **1932**, *54*, 613; Allison, F.; Bishop, E. R.; Sommer, A. L. *J. Am. Chem. Soc.*, **1932**, *54*, 616.
374. Allison, F.; Murphy, E. J. *J. Am. Chem. Soc.*, **1930**, *52*, 3796.
375. Langmuir, I. *Physics Today* **1989**, *42*(10), 36–50.
376. The full text of Langmuir's speech can be found on Princeton University faculty member Ken Steiglitz's website, http://www.cs.princeton.edu/~ken/ (accessed April 13, 2014). During the question-and-answer period, someone mentions that, with respect to Allison's magneto-optic

effect, Berkeley physicist Raymond T. Birge (who did not believe it) made the cleverest remark: he reportedly called the effect "Allison Wonderland."

377. Kauffman, G. B.; Adloff, J. -P. *The Chemical Educator* **2008**, *13*, 358–64.
378. Minder, W. *Helv. Phys. Acta* **1940**, *13*, 144.
379. In the years preceding World War II, ^{218}polonium was called Radium A or RaA.
380. The mean free path of an α particle in air is much less than that of an electron.
381. Hulubei, H.; Cauchois, Y. *Compt. Rend. Chim.* **1940**, *210*, 696.
382. Minder, W.; Leigh-Smith, A. *Nature* **1942**, *150*, 767.
383. Perey, M. *J. Phys. Rad.* **1939**, *10*, 435.
384. Turner, L. A. *Phys. Rev.* **1940**, *57*, 950.
385. ^{216}Polonium came to be called *radiothorium*.
386. Valadares, M. *Rendiconti dell'Istituto di Sanità Pubblica* **1940**, *3*, 953; Valadares, M. *Rendiconti dell'Istituto di Sanità Pubblica* **1941**, *4*, 713.
387. Valadares, M. *Rendiconti della Reale Accademia d'Italia* **1940**, *2*, 351.
388. Valadares, M. *Rendiconti della Reale Accademia d'Italia* **1941**, *2*, 1049; Valadares, M. *Rendiconti dell'Istituto di Sanita Pubblica* **1940**, *3*, 953.
389. Corson, D. R.; Mackenzie, K. R.; Segrè, E. G. *Phys. Rev.* **1940**, *57*, 459; Corson, D. R.; Mackenzie, K. R.; Segrè, E. G. *Phys. Rev.* **1940**, *57*, 1087, Corson, D. R.; Mackenzie, K. R.; Segrè, E. G. *Phys. Rev.* **1940**, *58*, 672.
390. Corson, D. R.; Mackenzie, K. R.; Segrè, E. G. *Nature* **1947**, *159*, 24.
391. *Chem. Eng. News* **1949**, *51*, 2996.
392. Karlik, B.; Bernert, T. *Naturwissenschaften* **1943**, *31*, 289; Karlik, B.; Bernert, T. *Sitzber. Akad. Wien, Math. -Naturw. Klasse* **1943**, *152*, 103.
393. Paneth, F. A. *Nature* **1947**, *159*, 8.
394. Dana Aldea Archive-1999; *Dana Aldea* is a francophone newspaper available in Romania.
395. Bonnelle, C. Yvette Cauchois. *Physics Today* **2001**, *54*(4), 88.
396. *Supplement to Mellor's, A Comprehensive Treatise on Inorganic and Theoretical Chemistry*; Longmans Green: London, 1956, p. 1067.

IV.12

A COCKTAIL OF CHEMISTRY AND ESPIONAGE: HELVETIUM, ANGLO-HELVETIUM, AND A PAIR OF INDIAN ELEMENTS

In 1937, at an Indian university, an unknown radiochemist published the discovery of a pair of elements found in the mineral monazite, one of which was presumably eka-iodine, long sought by chemists all over the world. This discovery, which could have signaled the presence of really fine chemists in India, was published in an obscure journal of the University of Dacca and thus passed into the chemical literature unnoticed by the larger international chemical societies. But, in 1956, almost 20 years later, this discovery made news and was reported in the supplement to Mellor's *A Comprehensive Treatise on Inorganic and Theoretical Chemistry*[397] dedicated to astatine. The work of this unknown scientist, Rajendralal De, had fallen into oblivion, since the discovery noted by Mellor had previously gone unnoticed and therefore had never been challenged. Three years passed, when a Swiss physicist, at the end of his ingenious radiochemical preparations, confirmed his isolation of an isotope of element 85. Finally, in 1942, this young man, Walter Minder, by name, burst onto the restless and somewhat elitist international scientific scene with a second note on the 85th element. For this second publication, he depended on the collaboration of a somewhat improbable British spy disguised as a lovely female research scientist, Alice Leigh-Smith. Neither publication passed unnoticed. Apart from what had happened because of De's announcement, they were the subject of acrimonious criticisms and passionate hostility on the part of many European scientists.

IV.12.1. RAJENDRALAL DE AND HIS TWIN ELEMENTS: GOURIUM AND DAKIN

The hunt for element 85, eka-iodine, involved many scientists over the course of many years. As we have already seen, the British chemist John Albert Newton Friend [398] betook himself to the Holy Land not as a pilgrim but to find dissolved salts of eka-iodine and eka-cesium in the waters of the Dead Sea, to no avail.[399] Subsequently, in 1928, the American chemist Samuel Coleville Lind (1879–1965) had maintained that elements 85 and 87 would be radioactive and best searched for using radiochemical methods.[400]

In those years, on the shores of the Gulf of Bengal, a young and unknown chemist was beginning his scientific career: information on the life of Rajendralal De seems to have vanished with his person, swallowed up in the histories of millions of inhabitants of the Indian subcontinent. De was a radiochemist who began publishing his work in 1916; he ended his very long scientific career 60 years later. His last labor, accomplished when he was already a very old man, was a 1976 article on the organometallic complexes of the lanthanides.[401]

De's first work as a young chemist was co-authored by the celebrated chemist, educator, and visionary Indian nationalist Prafulla Ray (1861–1944)[402] who played a great role in the education of many generations of young men in India. On the other hand, articles written in De's mature years are more associated with the physician, chemist, and revolutionary Ashtoush Das (1888–1941). De and Das worked together on the elements of the uranium family.[403] Lacking precise dates, we estimate that De was born in the last or next-to-last decade of the 19th century and that he died after 1976.

Following his first work in inorganic chemistry (valence and ionization potentials of the elements), De turned his interests to radiochemistry and mineral chemistry. He sought ways of concentrating uranium-X and measuring its percentage in minerals. At the University of Dacca, now in Bangladesh, De collaborated with the celebrated physicist Satyendranath Bose (1894–1974)[404] who constructed for his colleague various instruments that De needed for his radiochemical experiments. He carried on lively correspondence with Otto Hahn at the Kaiser Wilhelm Institute in Dalhem, near Berlin. In 1936, he perfected the method of preparing a neutral solution of ferric hydroxide in the presence of rare earth ions for the purpose of finding a selective technique for the precipitation of various metallic cations.[405] In the following year, while investigating products of the thorium decay series, he made a double announcement:[406] the discovery of a new element, eka-iodine, and Th-F (in other words, an isotope not yet known, ^{208}Po).

What is amazing about De's work is the quantity of material that he used, which had to be quite large. Th-F hydride (i.e., polonium hydride) turned out to be volatile. The chloride of this same element, obtained by placing the hydride in HCl, precipitated out as green crystals. With HBr, he obtained a pink compound; in the presence of KOH, a green precipitate that became pink with oxidation of the metal. Metallic Th-F electrochemically deposited on an aluminum wire had a grayish color and was radioactive, with a half-life of about 1,000 years. De proposed the name *gourium* for this isotope.

During this same work of separation of the elements present in his samples of monazite, he isolated and characterized the 85th element and reported the following properties: its compounds with oxygen, bromine, and iodine were volatile; its halides thermally decomposed, leaving a black deposit (with the exception of the chloride). The most curious property, which he associated with the halogens, seemed to be the capacity of the oxide to react with aluminum in an alkaline medium. De bestowed the name *dakin* on this element.

Two years later, he announced that he had found in his monazite samples a radioactive element that was a weak α emitter.[407] Electroscopic and photographic evidence led to the identification of the same *gourium* that he had announced in 1937. In this publication, he determined the half-life of the isotope with greater precision: 1,020 years. In January 1947, as India neared independence from Great Britain, De again took up his first work on eka-iodine and the fractionation of monazite sands,[408] now more than 10 years old.

From his initial treatment of monazite sands with concentrated sulfuric acid, De obtained a residue. He concentrated his efforts on characterizing the new elements. The insoluble fraction was placed in an electrolytic cell composed of two flasks containing a mixture of sulfuric and nitric acids: the first concentrated and the second dilute. The two solutions were kept separate by a porcelain membrane. The residue containing the sought-after elements was placed in the flask with the dilute solution and connected to the anode. Initially, white fumes developed near the cathode, attributed by the author to *gourium* and *dakin* (^{208}Po and ^{211}At), that were then collected in a solution of glacial acetic acid. The volatile compounds were then analyzed and, to his great amazement, De discovered that they contained, in addition to elements 84 and 85, also sulfur, chlorine, and oxygen.

After two or three days of electrolysis, dense brown fumes, attributed to the presence of only halogenous material, were collected in a trap with glacial acetic acid. Subsequently, De abandoned this extraction procedure for *dakin* from monazite in favor of another, based on chemical fractionation. The sandy mineral was melted, taken up again with water, and the solution was treated with sulfuric acid. The compound that precipitated out was oxidized with nitric acid, and the solution was made alkaline with ammonia until a crystalline precipitate was obtained that, in its turn, was converted to a sodium salt with a solution of NaOH. The hypothetical sodium *dakinide* was heated in the presence of HBr and HI, liberating eka-iodine in the elemental state. A closely analogous procedure was used for the separation of Th-F, or *gourium*. De calculated that Th-F was present in monazite in the amount of 4.07×10^{-4} g per gram of monazite sand.

At the conclusion of his work, De changed his mind about the absorption spectrum of the rare earth fraction and even about the presence of element 61. Moreover, he experienced in himself some evidence of the toxicity of the compounds of *dakin* and *gourium*. In 1962, many years after the artificial synthesis of astatine (eka-iodine) by Emilio Segrè and his colleagues Dale R. Corson and Kenneth R. MacKenzie,[409] De felt the need to confirm his discoveries for a third time. In an atmosphere of complete indifference on the part of the scientific community, a brief four-page report on all of his preceding work appeared, including a name change to *dekhine*.[410]

In the second half of the 1930s, other names attributed to this elusive element appeared.[411] A very strange case that deserves mention follows. The difficulty of tracking down the original articles must have added to the spread of the error about the "discovery."[412] Such was the case of the hypothetical *dacinium*, changed from *dakin*, and cited erroneously by N. A. Figurovskii in his book,[413] *Discovery of the Elements and the Origin of Their Names*.[414] Not a single article about this element exists, nor has any scientist ever claimed its discovery. Figurovskii introduced this element in his book out of the blue, and the error has been perpetuated by others.[415] Figurovskii probably believed that *dakin* was derived from the region of Dakov in Romania and, for some reason, he arbitrarily changed the name to *dacinium*. In fact, the Roman province of Dacia corresponds roughly to the territory of present-day Romania. Another hypothesis could be tied to the fact that Figurovskii knew about the presumed discoveries of elements 87, 93, and 85 by Horia Hulubei. The results of the discoveries of the first two elements (*moldavium* and *sequanium*) were published in *Comptes Rendus*,[416] while the announcement of the discovery and the naming of element 85 appeared in two very obscure Romanian journals.[417] It may be that Figurovskii knew of them but did not know the name that Hulubei had given to eka-iodine and, appropriately enough, he may have imagined it could be called *dacinium*, seeing that Hulubei was himself Romanian.

IV.12.2. WALTER MINDER AND HELVETIUM

Walter Minder was born on August 6, 1905, and received his degree in chemistry in Bern, in 1930, with a dissertation on mineralogy. He quickly became interested in understanding the radioactive decay series of the thorium and uranium families. During 1936, he traveled to Berlin, where he came into contact with most of the German atomic physicists of the time: Hans Bethe, Siegfried Flügge (1912–97), and Carl Freidrich von Weizsäcker (1912–2007). In 1938, he was named assistant to professor Adolf Liechti (1898–1946) at the local Radium Institute of the hospital in Bern. That year, he published his first article, in which he hypothesized on the existence in nature of eka-iodine and eka-cesium.

Minder devised a graph in which he plotted the neutron-to-proton ratio as a function of atomic number, Z, for the elements between lead and thorium. A discontinuity in the three known pathways of fragmentation impelled him to hypothesize a new pathway that might lead to the formation of elements 87 and 85.

A detailed report on the decay of Ra-A (^{218}Po) appeared on March 13, 1940. In addition to the already known α decay of this isotope, he believed that he also observed β decay. For this purpose, he passed some radon into an ionization chamber connected to an electrometer. This chamber had a double window beyond which, 5 mm away, a second ionization chamber was placed, and it too had a window that was permeable to the passage of charged particles. The charged particle current was measured, but in the second chamber (placed in series with the first) only the β radiation arrived because of the stopping effect on the α particles by the windows placed in both chambers. Strangely enough, the radiation in the second chamber was more intense than that measured in the first. Minder explained these results by assuming that Ra-A was decaying into element 85, and that, in its turn, it was emitting electrons. He also conducted other tests to characterize halogens, after which, convinced that he had discovered eka-iodine,[418] he wrote: "The beta-decay of Ra-A leads us certainly to hypothesize the formation of element 85. For this reason we suggest the name *helvetium*. Chemical tests to verify the nature and existence of this element continue."

It quickly became clear that Minder (Figure IV.20), not yet 35 years old, was involved in a game much larger than he could imagine. From the pages of *Comptes Rendus*, the Romanian physicist Horia Hulubei criticized the way in which Minder had announced his discovery. Hulubei said that, 2 years earlier, he had observed some lines in the emission spectrum of X in a concentrated preparation of radon.[419] His judgment of the young radiochemist was harsh, but he did not discuss the discovery. In the following autumn, Emilio Segrè and his colleagues at Berkeley synthesized the first isotope of this element, throwing Minder into a panic.

Meanwhile, in Switzerland, the local press had reported the discovery of *helvetium*, and this had given rise to a certain amount of national pride. The clamor created embarrassment for Minder's superiors, who did not seem to share the young researcher's enthusiasm. Minder found himself thrust into the limelight; his colleagues hastened to publish some articles in which they weakly welcomed the news of the discovery as if they were not completely convinced of the quality of Minder's work,[420] giving major credit instead to the atomic synthesis of Segrè: "The discovery of isotope 218 of element 85 still remains an open question, and it is possible to cite with certainly only the synthesis of isotope 211 of element 85, created artificially from bismuth, which is well-known."

IV.12.3. ALICE LEIGH-SMITH AND ANGLO-HELVETIUM

In 1942, Minder became acquainted with a young and beautiful English physicist, Alice Leigh-Smith, née Prebil. She had begun her career at the end of the 1920s under

FIGURE IV.20. Walter Minder (1905–92), Swiss Physicist and Radiologist. In 1940, Minder provided experimental evidence for the element with atomic number 85, which he called *helvetium*. Two years later, in collaboration with Alice Leigh-Smith, he revisited the same discovery, proposing a new name for the element: *anglo-helvetium*. This decision was the cause of a great deal of skepticism on the part of many of his European colleagues. Courtesy, Jubliläum 25 Jahre Schweizerische Gesellschaft für Strahlenbiologie und Medizinische Physik (Zürich 1989).

the mentorship of Nobel Laureate Owen William Richardson (1879–1959). In Paris, in 1933, she married Philipp Leigh-Smith (1892–1967), an employee at the British Embassy in Bern. The young physicist followed her husband all over Europe, to France, Greece, Switzerland, and, after the war, to Italy. One can easily follow her "grand tour" around Europe by reading the series of journals in which she published.

Alice Leigh-Smith approached Walter Minder and sought to convince him to relocate so that he could work with her German friends and colleagues in Berlin, with the purpose of picking up important information on the status of atomic research in Germany.[421] An analogous plan for the construction of an atomic bomb was advancing in great secrecy in the United States. The British, not having the same means of allying themselves with Americans, preferred to send, with the promise of enormous remuneration, a spy to Berlin to discover the enemy's plans. Minder was not exactly enchanted with this proposition and refused to accompany Leigh-Smith. However, a strong bond grew between them nevertheless, culminating in their joint publication on their work in radiochemistry on December 26, 1942.[422] Figure IV.21 is a rare photograph of Alice Leigh-Smith in a meeting with Irène Joliot-Curie.

Leigh-Smith and Minder had at their disposal about 40 mg of Ra-Th, and from this material they sought to extract element 85 that was formed through radioactive decay. To do this, they tried to pass "emanation of thorium" (i.e., ^{224}Ra) through two copper foils, one charged positively, the other negatively. The gas, with a median half-life of only 54 seconds, forced them to do multiple passes for 20 minutes in order to enrich the deposit of element 85 at the copper foil connected to the negative pole. The hypothesis—or rather

FIGURE IV.21. Irène Joliot–Curie (1897-1956) and Alice Leigh-Smith (1907–87). Leigh-Smith (right) is known for having perceived glimpses of element number 85 in nature. She collaborated with Irène Curie (left) in Paris. She worked at the Radium Institute, Berne, with Walter Minder, and at the Istituto Superiore di Sanità in Rome. Gift of Christopher Leigh-Smith.

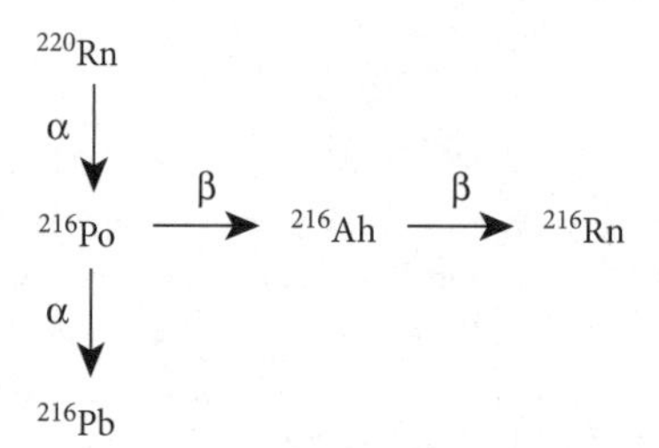

FIGURE IV.22. Branching Decay of Po-216 to Produce Anglo-Helvetium (Eka-Iodine; Z = 85) Proposed by W. Minder and A. Leigh-Smith.

their hopes—on which they based their entire experiment was that element 85 had a half-life greater than that of the "emanation of thorium" and that the nuclear reaction would proceed according to the sequence shown in Figure IV.22.

Eka-iodine could then be sublimed from the copper foil and heated to 180 °C on a silver wire kept at a lower temperature. After 10 minutes, the silver wire was placed in a Wilson cloud chamber.[423] The wire was kept at a certain distance in such a way that α particles could not reach it (and thus generate artificial radioactivity in the silver), nor could it be contaminated by the initial Ra-Th, nor by that produced by the other disintegration pathway (namely, α) from ^{216}Po.

Following this procedure, Minder and Leigh-Smith were convinced that the silver wire would only be reached by the sublimed halogen atoms (eka-iodine). They photographed the tracks of the α and β particles in the cloud chamber, attributed only to the decay products of eka-iodine. At the conclusion of their work, they both expressed a desire to name the 85th element, "as a tribute to the scientific work of our two countries," *anglo-helvetium*.[424]

Why Minder wanted to repeat his experiment from 2 years earlier is not clear. The fact that, in his work with Alice Leigh-Smith, he makes no mention of his previous publication on *helvetium* gives credence to the hypothesis that his former work could be erroneous and incomplete, as attested to by the skepticism of his superiors and his willingness to change the name to *anglo-helvetium*.

A few years later, Minder's scientific focus changed. Under the guidance of Adolf Liechti, director of the Röntgen Institute in Bern, Minder began research on the therapeutic uses of radium and other radioisotopes, especially for the treatment of neoplasms. During World War II, he began a series of publications on dosimetry and on the effects of ionizing radiation. Minder is remembered above all for this work.[425]

IV.12.4. C. W. MARTIN AND THE "ELUSIVE" PARENTHESES OF LEPTINE

A letter to the editor appeared in the pages of *Nature* on March 13, 1943. Because it was a somewhat bizarre and exaggerated fantasy, it is worth lingering here a moment to describe this curious episode. C. W. Martin, the letter's author, was an instructor at the King Edward's Grammar School in Birmingham. He vehemently flung himself against his two adversaries, Minder and Leigh-Smith, claiming that the name *anglo-helvetium* was simply ridiculous. Martin acknowledged that the periodic table contained other elements with compound names such as neodymium, praseodymium, and dysprosium, but that these were nothing in comparison to the unfortunate choice of *anglo-helvetium*. He continued sarcastically: "Assuming its existence to be confirmed and the chemistry of this element to be worked out, are we to talk of hydroanglo-helvetic acid [formula HAh perhaps] and the peranglo-helvetates? By comparison with the possibilities which might be made of anglo-helvetium, we may come to regard dysprosium and praseodymium as old friends."[426] His criticism, certain aspects of which were amusing, against these "modern" scientists ended with the following sentence: "The more science has been divorced from the humanities the more has mankind been afflicted by unpleasing words."

In the end, Martin proposed the name *leptine*[427] for eka-iodine because its ending was analogous to the names of all the other halogens. Its root, derived from the Greek *leptos*, which means "subtle, elusive," was acceptable to Emilio Segrè[428] and his colleagues at the time they were about to name element 85.

The war ended and the world forgot C. W. Martin's humorous taunt, swallowed up as it was by the enormous problems of reconstruction following almost 6 years of death and devastation. *Leptine* was never used as a name for a chemical element, but some years later its Greek root was used by physicists to name an entire family of subatomic particles: the leptons.[429]

IV.12.5. ACADEMIC CONFLICTS WITH HULUBEI, PANETH, AND KARLIK

In the winter of 1943, two Bern newspapers, *Der Bund* and *Neue Zücher Zeitung*, published two articles on element 85 written by Professor A. Liechti,[430] Minder's superior, and by Paul Scherrer (1890–1969).[431] The very next day the *Berner Tagblatt* carried a long article by Alice Leigh-Smith.[432] These reports, appearing as they did in the local press, were the fruit of an agreed-upon counteroffensive between the director of the Radium Institute, Walter Minder, and Alice Leigh-Smith in order to confront the criticisms that had pelted their work. As a matter of fact, first from France, and then from Romania, Horia Hulubei thundered against the Swiss physicists, accusing them of having ignored his work on element 85 going back to 1938. Adolf Liechti's intervention was made to salvage Minder's work: his work on element 85 needed a bit of rectifying, but the priority of Minder's discovery ought not be placed in doubt. Liechti explained Minder's work in detail but skipped over the reports of other scientists: he limited himself to referring to a few spectroscopic lines of element 85 that had been observed by "some Romanian physicists."[433] Minder, in the meantime (1940), had been awarded the Jubiläumspreis der schweizerischen Roentgengesellschaft (the Swiss Roentgen Society Jubilee Prize) for his discovery of *helvetium*. Another criticism arrived from the theoretical physicist Louis A. Turner[434] who strongly doubted the validity of the discovery since the observed radiation pathway would be incompatible with the energy associated with the β decay of Ra-A.

Finally, another criticism arrived from England, this one in the form of a letter from the radiochemist F. A. Paneth. On May 23, 1942, he wrote:[435] "There is so far no trustworthy indication of a branching of any of the main radioactive series leading to an element 85. Nor has a stable form of this element been found."

Paneth's second criticism relative to the existence of stable isotopes of element 85 was addressed to Hulubei's spectroscopic work, and this led to the only correct interpretation. Shortly after Paneth's intervention, the Viennese radiochemist Berta Karlik succeeded in discovering the only natural isotope of element 85, but it was not the same one "identified" by Minder.

The experiments of Minder and Leigh-Smith were repeated by Karlik and her colleague Traude Bernert,[436] but they did not observe the weak β radiation that Minder and Leigh-Smith claimed was characteristic of *anglo-helvetium*. Karlik considered the work of the two Bern physicists as the height of error. Using a methodology totally different from that used to "discover" *anglo-helvetium*, in 1943, Berta Karlik discovered the short-lived natural isotope of eka-iodine (^{218}At), with a half-life of about 2 minutes.

Minder never replied to Berta Karlik's criticisms, as he did to those of Emilio Segrè and Horia Hulubei. He continued his own radiological research until retirement in 1964. A confirmed pacifist, he regretted for the rest of his life that the atom bomb was dropped on Hiroshima on his 40th birthday. In 1960, when the Swiss Lower Chamber discussed the possibility of purchasing nuclear weapons and of deploying them to the Swiss Army, he was so opposed that he participated in two pacifist demonstrations. He died on April 1, 1992, at almost 87 years of age.

After the war, Leigh-Smith went on to Italy and settled in Rome. For some curious circumstance, she found herself working in the physics laboratory of the Institute for Advanced Medicine as the colleague of Neapolitan chemist Oscar D'Agostino.[437] He, after a brief stay in Marie Curie's laboratory, had joined Enrico Fermi's research group as the only chemist. This was the glorious period when Fermi and his team were irradiating elements with slow neutrons. He remained involved in this research on the transuranium elements—presumably obtaining some samples of uranium by nuclear bombardment[438] that would have a certain negative effect on his future career.[439] After the erroneous announcement of the discovery of the first two elements beyond uranium, elements 93 and 94, named by the Fermi group *ausonium* and *hesperium*, respectively, D'Agostino moved on to the radiochemistry laboratory at the Institute for Advanced Medicine. Here, he met Alice Leigh-Smith who encouraged him to repeat the experiments that she had carried out in Bern 5 years earlier. And so the saga of *helvetium* and *anglo-helvetium* finally concluded in Rome in 1947.

D'Agostino and Leigh-Smith published a report of their work and once again confirmed their discovery of element 85.[440] This time, they did not propose a third name for element 85; this did not help to save *anglo-helvetium* from ending up on the list of erroneous discoveries. After this publication, the last of 12 that took up a fair amount of time, we lose track of the beautiful British spy. We do not know if she continued to look for more missing elements or became involved in synthesizing transuranium elements. What we do know is that she did not publish scientific papers ever again. Her name, and information on her less honorable profession as a spy, came up many years later when, in 1981, Minder published a book on the history of radioactivity full of anecdotes and personal reminiscences.

IV.12.6. CONCLUSION

The groundlessness of the existence of *helvetium* and of *anglo-helvetium* was quickly proven. Berta Karlik's scrupulous work demolished Minder's. Although we can continue to speak enthusiastically about Karlik's work, of Walter Minder, Alice Leigh-Smith, and their fantasy elements there is hardly a trace.

The announcement of the separation of a macroscopic amount of element 85 from the monazite sands of Travancore, as claimed by Rajendralal De, could not be believed for the simple reason that the estimated total amount of this element in the Earth's crust would not exceed 30 g. It therefore seems impossible that he could have collected a macroscopic quantity of astatine, amounting to a few kilograms, from monazite. Furthermore, De's published data are so disconnected that it is impossible to ascertain the nature of the substance that he obtained and attempted to characterize.

In their excellent summary[441] of the work on the discovery of element 85, Thornton and Burdette discuss the ambiguity of discovery using this element as a sort of case study. They set out the present-day criteria for claiming credit for a discovery, among which are timing, instrumental verification, reproducibility, chemical verification when possible, and the ability to convince one's scientific peers of the experiment's success. They argue that there are three defendable "discoveries" of element 85 based on these criteria: the X-ray emission lines for $^{218}85$ reported by Cauchois and Hulubei (1934–39); the cyclotron production of $^{211}85$ at Berkeley, followed by chemical characterization (1940); and the detection of naturally occurring $^{218}85$ by Karlik and Bernert (1942). Any one of these groups, at various times, might have been deemed the true discoverers of eka-iodine.

There are obviously differences of opinion about priority at the frontiers of science. Two excellent reviews discuss this issue from the sociological[442] and scientific[443] points of view and cite numerous examples in which, due to what one author calls a "pathogenic" culture, scientists were led to deviant behavior and relativization of values.

Eka-cesium was the first of the two elements, 85 and 87, to be discovered: in 1939, Marguerite Perey, one of Marie Curie's last students at the *Institut du Radium* at Paris found this element as one of the decay products of actinium[444] (at first she called this element Ac-K, actinium-K). After the end of World War II, she proposed the name francium in honor of her native land.[445] The IUPAC accepted her proposal but not its symbol, Fa. The present symbol of francium is Fr.

The year following the announcement of the discovery of francium, Emilio Gino Segrè, Kenneth R. MacKenzie (1912–2002), and Dale R. Corson (1914–2012) synthesized the isotope $^{211}85$ by bombarding bismuth with α particles. These three scientists were so prudent that, only in 1947, pressured by the famous radiochemist Friedrich Adolf Paneth, did they advance the proposal of naming this new element astatine[446] from the Greek αστατοζ (astatos), that is, "unstable."

Notes

397. *Supplement to Mellor's A Comprehensive Treatise on Inorganic and Theoretical Chemistry*; Longmans Green: London, 1956.
398. John Albert Newton Friend was head of the department of chemistry at the Technical College of Birmingham until his retirement in 1946. He was the author of many books on chemistry and on the history of the elements. The reprinting of his book, *Man and the Chemical Elements*, in 1961, was his last publication.
399. Newton Friend, J. *Nature* **1926**, *117*, 789.
400. Lind, S. C. *Chem. Bull.* **1928**, *15*, 319.
401. Dutt, N. K.; De, R. *Indian J. Chem., Sect. A: Inorg., Phys., Theor. Anal.* **1976**, *14A*(7), 498.
402. Ray, P. C. ; De, R. *J. Chem. Soc., Abstracts* **1916**, *109*, 122.
403. De, R. ; Das, A. *J. Chim. Phys. Phys.-Chim. Biol.* **1937**, *34*, 386.
404. Indian physicist; his most famous works were in the area of statistical mechanics. Albert Einstein's extension of his work led to the formulation of the so-called Bose-Einstein statistics for subatomic particles with integral spin, now called "bosons" in his honor.
405. Das, A.; De, R. *J. Indian Chem. Soc.* **1936**, *13*, 197.
406. De, R. *Separate* (Bani Press, Dacca) **1937**, 18.
407. De, R. *Indian J. Phys.* **1939**, *13*, 407.
408. De, R. *Separate* (Univ. Dacca) **1947**, 21.
409. Corson, D. R.; Mackenzie, K. R.; Segrè, E. G. *Phys. Rev.* **1940**, *57*, 459; Corson, D. R.; Mackenzie, K. R.; Segrè, E. G. *Phys. Rev.* **1940**, *57*, 1087; Corson, D. R.; Mackenzie, K. R.; Segrè, E. G. *Phys. Rev.* **1940**, *58*, 672.
410. De, R. *Separate* (Bani Press, Dacca) **1962**, 4.
411. Pérez-Bustamante, J. A. *Fresenius' Journal of Analytical Chemistry* **1997**, *357*(2), 162–72.
412. Karpenko, V. *Ambix* **1980**, *27*, 77.
413. Figurovskii, N. A. *Discovery of the Elements and the Origin of Their Names*; Nauk: Moscow, 1970.
414. *Dakin*, transliterated into *dacinum*, is derived from the ancient region in Central Europe that was occupied by the Dacians, a group who formed part of the Gothic peoples.
415. Trifonov, D. N.; Trifonov, V. D. *Chemical Elements and How They Were Discovered*; MIR Publisher: Moscow, 1982.
416. Hulubei, H. *Compt. Rend. Chim.* **1936,** *202*, 1927; Hulubei, H. *Compt. Rend. Chim.* **1937**, *205*, 854; Hulubei, H.; Cauchois, Y. *Compt. Rend. Chim.* **1939**, *209*, 476.
417. Hulubei, H. *Bull. Soc. Roum. Phys.* **1944**, *45*, no. 82, 3; Hulubei, H. *Bull. Acad. Roum.* **1945**, *27*, no. 3, 124.
418. Minder, W., *Helvetica Physica Acta*, **1940**, *13*, 144.
419. Hulubei, H.; Cauchois, Y. *Compt. Rend. Chim.* **1940**, *210*, 696.
420. von Labhart, H.; Medicus, H. *Tatung der Schweizerischen Physikalischen Gesellschaft*, **1943**, 225.
421. Minder, W. *Geschichte der Radioaktivität*; Springer: Heidelberg, 1981.
422. Leigh-Smith, A.; Minder, W. *Nature* **1942**, *150*, 767.

423. Charles Thomson Rees Wilson, Scottish physicist, who devoted much of his energy to the study of fog and clouds. He tried to reproduce them artificially in the laboratory. He determined that the presence of dust or electrically charged particles was favorable to the formation of tiny droplets of water and hence of clouds or fog. Wilson prepared air free of dust, so that when it was humidified, the water droplets would have been prevented from condensing due to the absence of dust that could serve as condensation nuclei. When a charged particle was made to cross this humidified chamber, while at the same time the chamber was expanded adiabatically, little drops of water were formed around the ions produced by the passage of the particle so that one could observe not only the charged particle's passage, but also the route that it took through the chamber. If the cloud chamber were placed in a magnetic field, the curvature of the trajectory of the particle indicated the nature of its electrical charge, furnishing information about its mass. Wilson perfected his cloud chamber in 1911, and it soon became an important instrument in nuclear research. For this work, Wilson received the Nobel Prize for physics in 1927.
424. Leigh-Smith, A.; Minder, W. *Nature* **1942**, *150*, 767.
425. See, for example, Minder, W. *Manual on Radiation Protection in Hospitals and General Practice*, vol. 5. World Health Organization: Geneva, Switzerland, 1980.
426. Martin, C. W. *Nature* **1943** *151*, 309.
427. Leptine was suggested by analogy with the other halogens: chlorine, bromine, and iodine.
428. Corson, D. R.; MacKenzie, K. R.; Segrè, E. *Nature* **1947**, *159*, 24.
429. Leptons, those particles that feel the effects only of gravitational, electromagnetic, and weak interactions, are the electron, particle μ, particle τ, neutrinos, and their corresponding antiparticles.
430. Liechti, A.; Wegelin, C. *Der Bund* **1943**, January 13.
431. Scherrer, P. *Neue Zucher Zeitung* **1943**, January 13.
432. Leigh-Smith, A. *Berner Tagblatt* **1943**, January 14.
433. In reality, the work to which Liechti referred was signed by a Romanian, H. Hulubei, and by a Frenchwoman, Y. Cauchois.
434. Turner, L. A. *Phys. Rev.* **1940**, *57*, 950.
435. Paneth, F. A. *Nature* **1942**, *149*, 565.
436. Karlik, B.; Bernert, T. *Naturwissenschaften* **1942**, *30*, 685.
437. D'Agostino, O. *Il chimico dei fantasmi*;Mephite: Atripalda (AV), 2002; reprint.
438. D'Agostino, O. *Gazzetta Chimica Italiana* **1934**, *64*, fasc. XI; D'Agostino, O. *Gazzetta Chimica Italiana* **1935**, *65*, fasc. X.
439. D'Agostino was the only one of the "boys of Via Panisperna" to not become a university professor.
440. Leigh-Smith, A.; D'Agostino, O. *Rendiconti Istituto Superiore di Sanita* **1947**, *10*, 523.
441. Thornton, B. F.; Burdette, S. C. Finding eka-iodine: Discovery priority in modern times. *Bull. Hist. Chem.* **2010,** *35*(2), 86–96.
442. Merton, R. K. Priorities in scientific discovery: A chapter in the sociology of science. *American Sociological Review* **1957**, *22*(6), 635–59.
443. Gross, A. G. Do disputes over priority tell us anything about science? *Science in Context* **1998**, *11*, 161–79.
444. Perey, M. *Compt. Rend. Chim.* **1939**, *97*, 209; Perey, M. *J. Phys. Rad.* **1939**, *10*, 435.
445. Perey, M. *Bull. Soc. Chim. Fr.* **1951**, *18*, 779.
446. Segrè, E. G.; Corson, D. R.; Mackenzie, K. R. *Nature*, **1947**, *159*, 24.

IV.13

IS FAILURE A SEVERE MASTER?

Research on elements 85 and 87 did not seem to halt. Dozens of scientists passed their most productive and creative years and invested enormous financial resources in investigating these two elements, among the most elusive in nature. In the course of a few decades of research, it is possible to document a long list of personal and group failures, often with ruinous consequences to otherwise up-and-coming careers.

The case of the following two erroneous discoveries is rather remarkable: in fact, it is not absolutely certain who proposed their names, nor is the original work known. At any rate, these substances succeeded in making their way into the literature and, furthermore, a large number of compounds of these elements are mentioned in detail, even though the elements themselves do not exist in nature!

IV.13.1. ELINE

The element that goes by the name *eline* corresponds to eka-iodine, element number 85, today known as astatine. The name *eline*[447] appears—without any other documentation or etymology—in *Hackh's Chemical Dictionary* of 1946 and in successive editions up to 1969. The proposed symbol was El. However, neither the name of the discoverer nor the date of the supposed discovery are given.

Eline would be a halogen with an uncertain atomic weight, but most probably around 218. The element would be abundant in nature (the exact opposite of what we now know to be true), especially in the deserts of the American Southwest, and would be found in elemental form as a solid metal or in elongated brownish crystals. The mystery associated with this element becomes even greater when, in the chemical dictionary cited, a list of some of the properties of its compounds is given. *Eline* chloride would have an indefinite stoichiometry; a white, waxy appearance; and be soluble in water, alcohol, and ether. The nitrate, with the formula $ElNO_3$, would be a yellowish solid soluble in water or in CCl_4. Eline's only other compound mentioned was the sulfate, $El(SO_4)_2$, and this heavy halogen would have a valence of 4. It would be a white, hygroscopic solid soluble in the same solvents as the chloride.

IV.13.2. VERIUM

Hackh's Chemical Dictionary seems to have even less information on this element, with atomic number 87. The name, fairly unusual, was given as *verium*, and its atomic mass was estimated to be around 224. This alkali metal would appear in the elemental state as a liquid similar to mercury. It would also be expected to be the most electropositive element in the periodic table. What is remarkable is the absolute certainty that the author has with respect to experimental data. *Verium* would not be radioactive, and neither would it have a visible spectrum. It would be precipitated upon the addition of tungsto-silicic acid. The abundance of this element could not be compared to that of *eline*, but it would be expected to be found in macroscopic quantities. Deposits would be found in lithic clay sediments and alkaline deposits in the southwestern United States. It would also be possible to find *verium* as a monovalent cation in various minerals and in the oceans. The dictionary reports only one compound, the silico-tungstate, the properties of which are stated as "white rhomboidal prisms, insoluble in water." Furthermore, the unknown author reported *verium*'s symbol as Ve.

One supposes that Julius Grant (1901–91),[448] the author of *Hackh's Chemical Dictionary*, was familiar with Fred Allison's work on the presumed identification and concentration of elements 85 and 87, and of his naming them, respectively, *alabamine*[449] and *virginium*.[450] Grant[451] was born in London, on October 19, 1901. Educated first at Strand School and then at King's College, he completed his studies at the University of London, with a doctorate in chemistry in 1931. At first, he was assistant to the organic chemist, Alfred Chaston Chapman (1869–1932);[452] then he was successively a consultant (until 1950) on behalf of some paper mills, as well as an expert witness in questions of medical law. Along the way in his long career, he also worked for a period of time as a forensic chemist. He died in London, on July 5, 1991, almost 90 years of age.

Getting back to the two elements: one might say that, over time, the name *alabamine* morphed into *eline*, just as *virginium* could have been corrupted into *verium*. Grant, in his dictionary, describes all four elements, but his entries for *alabamine* and *virginium* contain much more detail. Perhaps we find ourselves confronted with a somewhat eccentric attribution due to nonexistent discoveries. . . or to a little playful fiction?

Notes

447. *Hackh's Chemical Dictionary*; J & A Churchill Ltd.: London, 1946, pp. 84, 303.
448. Julius Grant was a prolific author. Numerous books and articles on chemistry appeared under his name, from the field of textiles to that of paper, from analytical chemistry to spectroscopy. More generally, he published various textbooks and laboratory manuals, all during almost a half-century, from 1924 to 1969.
449. Allison, F.; Bishop, E. R.; Sommer, A. L. *J. Am. Chem. Soc.* **1932**, *54*, 616.
450. Allison, F.; Bishop, E. R.; Sommer, A. L. *J. Am. Chem. Soc.* **1932**, *54*, 613; Papish, J.; Wainer, E. *J. Am. Chem. Soc.* **1931**, *53*, 3818.
451. Anon. *Science & Justice* **1991**, *31*(3), 397.
452. B. D. *Biochem. J.* **1932**, *26*(6), 1715.

PART V

1939–Present: Beyond Uranium, to the Stars

Fascination with the Unknown surpasses everything.
—Homer

PROLOGUE TO PART V

In Part V, we have collected discoveries done almost exclusively by physicists, with the notable exception of Glenn T. Seaborg and his team. The time span discussed begins with the invention of the first particle accelerators and ends in the present. The sway that chemists held over the discovery of the elements ended with the discovery of the last naturally occurring element, francium, in 1939. From then on, it was a case of creating new elements, feats largely inaccessible to chemists due to either lack of instrumentation or competence or both. We also report here discoveries that are not true and proper ones, but rather predictions based on atomic theory and not yet experimentally verifiable.

But, in almost every chapter, we find situations that epitomize some of the most distressing in the annals of modern science. Starting with the United States' domination of the field of nuclear chemistry by unilaterally imposing the names of many of the transuranium elements in the midst of the Cold War, through dramatic turns of events worthy of Ian Fleming (1908–64), to sensational about-faces, this part concludes with the admonitory dismissal of Victor Ninov, an American, and coincides with the rise of two new and expert teams on the superheavy element scene, one German and the other Russian.

V.1

THE OBSESSION OF PHYSICISTS WITH THE FRONTIER: THE CASE OF AUSONIUM AND HESPERIUM, LITTORIUM AND MUSSOLINIUM

The attempt to find the first synthetic transuranium elements[1] occurred via investigations completely different from anything that one could imagine. They were conducted in Rome by the renowned team of "the boys of Via Panisperna,"[2] led by the young Enrico Fermi, affectionately called "the Pope" by his colleagues because, like the Supreme Pontiff, he was considered infallible. Nevertheless, this presumed infallibility in every area of the experimental sciences ought not stray into radiochemistry. Such hubris led to a spot on an otherwise splendid record: a clumsy interpretation of data that led to the doubtful attribution of the discovery of two transuranium elements. The hasty attempt to first name, and then retract, the two radioelements, would tarnish the prestigious and somewhat controversial figure of Enrico Fermi. On the other hand, this nonexistent discovery also sped the Roman professor to Stockholm, to receive the 1938 Nobel prize in physics.

On March 25, 1934, Enrico Fermi announced the observation of neutron-induced radiation in samples of aluminum and fluorine. This brilliant experiment was the culmination of preceding discoveries: that of the neutron and that of artificial radioactivity (produced by means of α particles, deuterons, and protons). The following October, a second and crucial discovery was announced: the braking effect of hydrogenous substances on the radioactivity induced by neutrons, the first step toward the utilization of nuclear energy. The year 1934, thanks to Fermi's research, was one of great expectations for the rebirth of Italian physics, an area that for centuries had remained in the backwater compared to the United States and the great countries of Europe. At the beginning of the 1930s, the members of Fermi's team had explained the theory of ß decay and, after 1934, with their induced radioactivity experiments, had also laid down the guidelines for research on the physics of neutrons. Rome became a reference point for nuclear research on the international level. The project of the director of the Rome Physics Institute, Senator Orso Mario Corbino (1876–1937), was nearly accomplished, a project that, from the end of the 1920s, Corbino had believed in and had not spared any expense to realize, investing all of his resources in the youthful Fermi, who was called to occupy the first chair in theoretical physics in Italy, created especially for him, when he was only 25 years of age.

Enrico Fermi (Figure V.01) was born in Rome, on September 29, 1901, and from his earliest youth he was distinguished for his extraordinary talent in matters scientific. Soon after becoming a tenured professor, on March 18, 1929, Fermi was in the first group to be enrolled in the Royal Academy of Italy and, some weeks later, he joined the National Fascist Party during the period in which Benito Mussolini (1883–1945) was celebrating the triumph of his concordance with the Holy See.

FIGURE V.01. Enrico Fermi (1901–54). In the 1930s, Fermi and his team explained the theory of β decay and laid down the guidelines for research on the physics of neutrons.

In 1933, the Fermi team was not numerous, but counted on the fact of being homogeneous in both age and talent. In addition to Fermi, the group consisted of Franco Rasetti (1901–2001), nicknamed the "Cardinal Vicar" because he was the chief's spokesperson; Emilio Segrè, alias "the Basilisk";[3] Edoardo Amaldi (1908–89); and the young Bruno Pontecorvo (1913–93), alias "the greenhorn." On the recommendation of Giulio Cesare Trabacchi (1884–1959), alias "Divine Providence," the chemist Oscar D'Agostino was added to Fermi's group.

Fermi was convinced that physics had come of age: knowledge of the atom was in large part complete, but what remained ripe for investigation was the components of the nucleus. In the year in which Irène and Frédéric Joliot-Curie announced the discovery of artificial radioactivity, he decided to radically change his area of research from theoretical to experimental physics. This was no small decision—he succeeded again and again in making his research world-class in scope and depth.

The Joliot-Curie discovery of artificial radioactivity had a great impact on scientific research worldwide, and Fermi was among the first to understand its enormous importance. He decided to attack the atom with neutrons instead of with α particles, but he did not have sufficient irradiated material like the Joliot-Curies at Paris. (Figure V.02 is a photograph of Frédéric Joliot-Curie's cyclotron, one of the first in Europe.) In January 1934, D'Agostino was sent to Marie Curie, by now mortally ill, to learn all the radiochemical techniques necessary for Fermi to conduct his research. D'Agostino received a respectful welcome and set himself to studying the methods for purifying polonium with the Ukrainian radiochemist Moïse N. Haïssinsky (1898–1976).

FIGURE V.02. The Cyclotron of the Collège de France. Frédéric Joliot-Curie had this instrument built in 1937. With it, he and Irène Joliot-Curie worked on nuclear transmutation. It was capable of a deuteron beam output of 7 MeV, which was remarkable for its time. Musée des Arts et Métiers, Paris. Photograph by Mary Virginia Orna.

Meanwhile, in Rome, Fermi procured from Trabacchi a very precious treasure, 1.6 g of radium chloride from which he could extract *emanation* (or radon) that would be necessary for the production of neutrons. Fermi's clever idea was to use these bodies, lacking any electric charge, so that they would not be repulsed by the charge on the nucleus. These projectiles, unlike the α particles (helium nuclei) used by the Joliot-Curie group, were not spontaneously emitted by radioactive materials. To obtain them, it was necessary to resort to bombarding lighter elements (like beryllium) with α particles emitted from natural substances. In this way, Fermi obtained one neutron per every 100,000 α particles emitted. The very low yield of neutrons made this method of production doubtful, but Fermi decided to try it. Having obtained a source of neutrons from "Divine Providence," he personally constructed, with Amaldi's help, the detectors for counting atomic disintegrations. A short time later, they began the bombardment experiments: first hydrogen, then lithium, then boron, carbon, nitrogen, and oxygen. These targets did not exhibit any induced radioactivity.

Success came later, when they began to irradiate the next element in turn, fluorine. As soon as the target was placed near a Geiger-Mueller counter, the physicists listened with absolute astonishment to the crackling that indicated that fluorine had become radioactive. After that, the number of atomic nuclei that became radioactive by neutron bombardment slowly grew as atoms of higher atomic number were irradiated. Fermi required the help of a chemist to characterize the new radioactive elements.

In March 1934, Marie Curie closed the Institut du Radium for the Easter holidays, and D'Agostino returned to Rome. On the day after Easter, he visited the Institute of Physics to greet his colleagues and found an astonishing scene: almost every one of the physicists was working feverishly. D'Agostino was immediately co-opted and never returned to Paris. In April of that same year, the first work on induced radioactivity by neutrons on fluorine and aluminum was published,[4] and many other elements were quickly added. The following June 3, during a solemn session of the newly formed Accademia d'Italia, Corbino, as director of the Institute of Physics and in the presence of King Victor Emmanuel III (1869–1947), gave an address that stirred up a true and proper hornets' nest. Corbino, in a highly polished discourse, spoke of his "boys" with unusual warmth. The public did not grasp the strictly scientific part of the subject; what struck them was that Fermi's group had succeeded in discovering a new element, the first transuranium element, one with atomic number 93.

The national press spoke immediately of a "fascist victory,"[5] but beyond the borders of Italy many scientists expressed grave doubts. Fermi, in his interviews, spoke of "prudence" and of "new and delicate tests." The controversy was a drawn-out affair. Further work by Fermi and his collaborators seemed to actually point to the discovery of two new elements,[6] with atomic numbers 93 and 94. In 1934, Fermi and his team discovered a new property of uranium-238 when irradiated with neutrons: it absorbed the neutrons and was changed into an isotope, uranium-239. Because the latter had an excess of neutrons, it should have shown the tendency to emit β-particles. The reaction should have been:

$$^{239}U \rightarrow {}^{239}93 + \beta \qquad \text{(Eq. V.1)}$$

The verification of the new transuranium elements was done by means of radiochemical techniques. It was demonstrated that the activity induced by the neutrons in uranium apparently did not belong to any of the elements that came before it in the periodic system. Element 93 seemed to have the properties of manganese.

Fermi extracted two β-active substances from the uranium target irradiated with neutrons. Element 93 became transformed into the successive element, number 94. Initially, Fermi and the chemists Otto Hahn and Fritz Strassmann believed that the transuranium elements would be the higher homologs of rhenium and iridium and therefore ought to be placed in the seventh period of the periodic table.

The arrogance demonstrated by Corbino in his speech turned out to be harmful to the group. Not long afterward, some journalists claimed that Fermi had been enamored of the idea of naming element 93 *mussolinium*, even though this idea never crossed the minds of anyone in the group.[7] Benito Mussolini (1883–1945) kept an eye on the work of the young physicist, above all for the prestige that he brought to Italy. The dictator, who held Fermi in such high esteem that he named him a member of the Academy of Italy, hoped that the new element could be named *littorium*. But physicist and political animal Orso Mario Corbino, demonstrating a marked sense of humor, pointed out that the half-life of the new radioactive element was very short and it would augur ill for any regime to be associated with it! In reality, the results of the team's experiments were not at all clear and were also very badly interpreted.

The relationship between Fermi and Mussolini was and remained cordial up until the promulgation of the racial laws[8] in 1938, the year that coincided with the discovery of nuclear fission (by Otto Hahn and Lise Meitner) and the conferral of the Nobel Prize in Physics (on Fermi).

The story of Fermi and his group's error is also somewhat the story of the discovery of uranium fission. The research conducted for almost 4 years (from 1934 to 1938) led the "boys of Via Panisperna" to hypothesize that the neutron bombardment of uranium brought about the formation of one or more new elements that did not exist in nature, elements with atomic numbers greater than 92.

The question raised a sharp controversy in scientific circles. Two chemists at the University of Fribourg, Ida Tacke Noddack and her husband Walter Noddack, in a highly controversial article,[9] placed in doubt the transuranic nature of the synthetic elements that Fermi had obtained. Their opinion, which the community of physicists branded with scornful and ill-concealed superiority as ridiculous, was a description of nuclear fission.

Rasetti, as soon as he read the article, burst his sides laughing; even Fermi shook his head. Fermi's faith was firmly placed in the incomplete nuclear theories of the time; he contended that the nucleus was like an "armored tank" and that a slow neutron was on the level of a small-caliber bullet. So, even though Segrè was irritated, Fermi was worried about the Noddacks' criticisms. If they were correct, their hypothesis would tarnish his reputation. At that moment, he received some highly critical comments from Aristid V. Von Grosse (1905–85), an American chemist, originally German, and a specialist in the chemistry of protactinium. Fermi decided to seek the opinion of the Nobel laureate Niels Bohr. The response that arrived in Rome from Copenhagen was a masterpiece of diplomacy. It basically said that it was impossible to hypothesize about inexact experimental data: that perhaps everything was possible, as perhaps everything was impossible.

The issue remained unresolved. In the following year, Otto Hahn and Lise Meitner, working in Berlin, repeated Fermi's experiments with facilities much better than those available in Rome and confirmed in great detail the data found by Fermi. Furthermore, according to them, they were also able to observe traces of elements 95, 96, and 97, which they provisionally called *eka-iridium, eka-platinum*, and *eka-aurum,* respectively.[10]

These were the confirmations that Fermi was waiting for. When Rasetti, the "Cardinal Vicar," arriving at the Institute to solemnly declare to Fermi that element 93 ought to be called *ausonium* and element 94 *hesperium*, two ancient names for Italy,[11] Fermi accepted this unusual proposition. The news was then sent to Corbino[12] in a communication obscured by nationalism and rhetoric. To fully understand Fermi's actions, we have to look back to the year 1935, when Irène Joliot-Curie, Hans von Halban, and Pierre Preiswerk (1907–72) published some conclusive notes on the artificial radioactivity of thorium.[13] Their conclusions did not agree with the possibility of a nuclear reaction that would lead to elements with atomic numbers greater than thorium, but, at the same time, they suggested the idea of the possibility of splitting the thorium nucleus. They arrived at an analogous conclusion for uranium. Otto Hahn and Lise Meitner, unlike Fermi, wanted to evaluate these assertions and redid their experiments on uranium, effectively finding that, in uranium, under the experimental conditions of the time, they did indeed split the nucleus instead of causing the nuclear reaction described by Fermi.[14] Fermi's shortsightedness and Rasetti's arrogance had made them lose the opportunity to discover atomic fission. A totally new scenario had opened up for humankind. Fermi had involuntarily lit the fuse and made true the fearful suspicion expressed by German chemist Walther Nernst 20 years earlier: "We live on an island of guncotton, but thanks be to God that we have not yet found the fuse."[15]

By 1938, the racial laws promulgated in Italy, mimicking its German ally, brought about a massive brain drain through flight. The research atmosphere was not what it was in 1933. Within a short period of time, the University of Göttingen, where Fermi had also studied, lost almost its entire teaching faculty; no one was left except the renowned mathematician David Hilbert (1862–1943), by now very old. The answer he gave to the minister of culture, Bernhard Rust (1883–1945), when he visited the university makes clear the situation of the time. "Is it true, Professor, that your institution is suffering very much from the expulsion of the subversive and the Jewish members of the faculty?"

"It is not suffering at all," replied Hilbert calmly. "My institution no longer exists."[16]

Fermi, as indicated earlier, had married a Jewish woman and watched the future with a certain amount of fear even though, on the threshold of World War II, he had reached an enviable social position. And so it was that Fermi, in September 1938, sent job applications to four American universities: all responded, and Fermi accepted the offer of a tenured position at Columbia University in New York City. He prepared to depart, telling the fascist authorities that he would stay in the United States for only 6 months. To ease his transition, the awarding of the Nobel Prize "providentially" came to him along with a hefty amount of ready cash. With his wife and two children, he left Italy on December 4 for Stockholm, and from there for New York. He only ever spoke of *ausonium* and *hesperium* once, during the address he gave following the receipt of the Nobel Prize. During the ceremony, Professor Pleijel, president of the Nobel Committee for Physics of the Royal Academy of Swedish Sciences, explained to the king of Sweden and to those present the scientific merits for which Fermi deserved the prize. He used these words: "Fermi's researches on Uranium made it most probable that a series of new elements could be found, which exist beyond the element up to now held to be the heaviest, namely Uranium with rank number 92. Fermi even succeeded in producing two new elements, 93 and 94 in rank number. These new elements he called Ausonium and Hesperium."

On that occasion, and for the first time, *ausonium* and *hesperium* were officially named: Fermi described the series of nuclear reactions thus:

$$^{238}U + n \rightarrow {}^{239}U \rightarrow {}^{239}Ao + \beta \rightarrow {}^{239}Hs + \beta \qquad \text{(Eq. V.2)}$$

He could not have chosen a worse moment to make his announcement. A few days later, Otto Hahn and Fritz Strassmann discovered uranium fission.[17] They thus realized that the products that Fermi obtained by the bombardment of uranium were not the elements Ao and Hs, but fragments of uranium nuclei. *Ausonium* and *hesperium* lasted only the space of a morning.

Elements 93 and 94 were verifiably prepared by nuclear reactions in 1940. The first was synthesized by E. M. McMillan and P. H. Abelson.[18] They called element 93 neptunium, after the planet Neptune. The second, plutonium, was discovered as the ^{238}Pu isotope by Glenn T. Seaborg, Arthur G. Wahl, and Joseph W. Kennedy. They named it after the planet Pluto, following the tradition used to name uranium and neptunium.[19]

With Fermi's departure, the Roman group fragmented like a uranium nucleus under neutron bombardment. Franco Rasetti and Emilio Segrè (the latter was Jewish and had married a German Jewish woman) emigrated to Canada and the United States, respectively. Bruno Pontecorvo went to France, where he became involved with the Communist Party; after the war he, made a sensational escape to Russia. Figure V.03 is a picture of the "Boys of Panisperna," Fermi's research group, before this difficult but inevitable breakup.

In the years to come, Fermi became involved with the construction of the first atomic bomb, thus alienating himself from his close friend Franco Rasetti. He returned to Italy a few times between 1949 and 1954 to hold seminars or conferences. During his last visit, in the summer of 1954, he was diagnosed with an advanced-stage malignant stomach tumor. He died in Chicago on November 29 of that same year, 2 months after his 53rd birthday.

FIGURE V.03. The Boys of Panisperna. From left: Oscar D'Agostino (1901–75), Emilio Segrè, Edoardo Amaldi, Franco Rasetti, Enrico Fermi. D'Agostino was a Neapolitan chemist who, at a very young age, joined "The Boys of Panisperna," the research group run by Enrico Fermi. He put forth experimental evidence for the existence of *ausonium* and *hesperium*. During World War II, he transferred to the Istituto Superiore di Sanità, where he had Alice Leigh-Smith as one of his colleagues.

Notes

1. Unlike Fermi, Richard Swinne in Germany and Odolen Koblic searched for element 93 in nature in vain.
2. Via Panisperna, located in the Monti area of Rome, is one of the most ancient streets, and its very name signifies this because it means "bread and ham," two ancient major food staples. It was here that the Institute of Physics of the University of Rome was found at the time of Enrico Fermi.
3. It was known throughout the university that Fermi and his colleagues used to call each other by ecclesiastical titles to make fun of the hierarchical pompousness of the nearby Holy See. Segrè, number three in the group, was the only one who earned his strange name on account of his character: a mere trifle would make him throw a childish tantrum.
4. Fermi, E.; Amaldi, E.; Segrè, E.; D'Agostino, O. *Ricerca Scientifica* **1934**, 5(1), 330; Amaldi, E.; D'Agostino, O.; Fermi, E.; Rasetti, F.; Segrè, E. *Ricerca Scientifica* **1934**, 5(1), 452; Amaldi, E.;

D'Agostino, O.; Fermi, E., Rasetti, F.; Segrè, E. *Ricerca Scientifica* **1934**, *5*(1), 467; Amaldi E.; D'Agostino O.; Fermi E.; Rasetti F.; Segrè, E. *Ricerca Scientifica* **1934**, *5*(2), 21.
5. *Il Giornale d'Italia,* 5 giugno 1934, anno XII.
6. Fermi, E.; Rasetti, F.; D'Agostino, O. *Ricerca Scientifica* **1934**, *6*(1), 9; Amaldi, E., D'Agostino, O.; Fermi, E.; Pontecorvo, B.; Rasetti, F. *Ricerca Scientifica* **1934**, *6*(1), 435.
7. D'Agostino, O. *Il chimico dei fantasmi*, Acocella, G., Ed.; Mephite: Atripalda (AV), Italy, 2002.
8. In 1928, Fermi married a Jewish woman, Laura Capon (1907–77).
9. Noddack, W.; Tacke Noddack, I. *Angew. Chem.* **1934**, *47*, 653.
10. Hahn, O.; Meitner, L.; Strassmann, F. *Nature* **1938**, *23*, 37.
11. *Ausonium* is taken from Ausonia, an ancient poetic name for Italy; *Hesperium* is taken from Hesperia, a Greek name for Italy meaning "land of the West."
12. Corbino, O. M. *La Nuova Antologia* **1935**, 16 December.
13. Joliot-Curie, I.; von Halban, H.; Preiswerk, P. *Compt. Rend. Acad. Sci.* **1935**, *200*, 2079.
14. Hahn, O.; Strassmann, F. *Naturwissenschaften* **1939**, *27*(1), 11; Meitner, L.; Frisch, O. R. *Nature*, **1939**, *143*, 239; Frisch, O. R. *Nature* **1939**, *143*, 276.
15. Engelhardt, H. T., Jr. *Scientific Controversies, Case Studies in the Resolution and Closure of Disputes in Science and Technology*; Cambridge University Press: Cambridge, UK, 1987, p. 532.
16. Reid, C. *Hilbert*; Springer: New York, 1996, p. 205.
17. Powers, T. *Heisenberg's War: The Secret History of the German Bomb*; Jonathan Cape: New York, 1993.
18. McMillan, E.; Abelson, P. H. *Phys. Rev.* **1940**, *57*, 1185.
19. Seaborg, G. T.; Wahl, A. C.; Kennedy, J. W. *Phys. Rev.* **1946**, *69*, 367; Seaborg, G. T.; Wahl, A. C. *J. Am. Chem. Soc.* **1948**, *70*, 1128.

V.2

FINIS MATERIAE

It is actually due to the physicists that the number of elements grew beyond the 92 naturally occurring ones hunted down by chemists in the previous 200 years. The physicists, one might say, united the elements discovered by the chemists to obtain new ones.[20]

Why are these super-heavy elements sought with such eagerness at Lawrence Berkeley National Laboratory (LBNL); at the Joint Institute for Nuclear Research in Dubna, Russia; and at the Gesellschaft für Schwerionenforschung (GSI) in Darmstadt, Germany, with each nation investing millions of dollars in research that, within the span of our lifetimes, will certainly not have useful technological fallout? The answer lies in the motive that drives the scientist to do research: a journey toward the unknown that lasts all of one's life, curiosity, economic interests, and—why not?—parochialism and national pride.

The declared goal of the physicists is to synthesize new superheavy elements in the attempt to arrive at that element with a nucleus composed of a magic number of nucleons that would give it a half-life of years and not a few fractions of a second. Also, according to theory, those elements with a number of nucleons close to a magic number would be more stable; so, for this reason as well, the search for superheavy elements is moving forward.[21]

Someone has said that every society celebrates itself with the construction of works of art or works of other kinds; in this case also, the names of the latest elements discovered—meitnerium, rutherfordium, dubnium, seaborgium, flerovium, and bohrium—are examples of auto-celebration within the physical sciences.[22]

At the beginning of the 19th century, and also in the preceding century, the discovery of new elements made chemists pose this question: is there a finite or infinite number of elements in the earth's crust? Following the publication of the periodic table of the elements in 1869, it was widely accepted that this question seemed to have a simple answer: there was a finite number of chemical elements. The elements not yet discovered were called "missing elements," and each of them had an empty box in the periodic system. Uranium, discovered in 1789, was shown to have the highest atomic number ($Z = 92$) until 1940. Therefore, looking for a missing element was a little like fishing in a barrel whose limits were represented by uranium.[23]

Already in the decade following the formulation of the concept of the periodic table, the discovery of many rare earth elements created more than a few problems in positioning them correctly within the periodic system. As an example, no chemist or physicist at the beginning of the 20th century could say how many rare earth elements there were. Also, since these elements could not be placed in the periodic system, was there an infinite number of them?

Radioactivity and radioactive isotopes (each one at first treated like a distinct element) further complicated things. The 1913 Law of Moseley limited the number of elements and smoothed the road toward the concept of isotopes. Finally, the quantum studies of Niels Bohr in 1923 laid to rest a long controversy that placed him at odds with the last great classical chemist, Georges Urbain: the number of the rare earth elements was not 15, as Urbain had supposed, but 14, as clarified by Bohr.

Physicists, with Emilio Segrè, from 1937 on (the date of the discovery of technetium), began to replace chemists in the discovery of new elements.[24] Already in 1925, the German physicist Richard Swinne, in the light of knowledge of radioactivity, hypothesized that there could be traces of transuranium elements in the stellar dust trapped in the ice mountains of Greenland. He said that he would be able to identify element 108 by its X-ray diffraction spectrum on a sample of this Arctic dust. At any rate, the certainty that the element with the highest atomic number was uranium, the certainty of which some chemists counted as a reason for pride, was crumbling with the rise of atomic physics.

The discovery of the proton and the hypothesis of the neutron through the work of the great experimental physicist, Ernest Rutherford; the discovery of artificial radioactivity on the part of the husband-wife team of Frédéric and Irène Joliot-Curie; and of slow neutrons thanks to Enrico Fermi and Ernest O. Lawrence's invention of the cyclotron were the milestones that led to this rapid evolution of scientific thought.

Shortly after the discovery of technetium, during World War II, American physicists discovered astatine, neptunium, plutonium, americium, and curium.[25] Meanwhile, chemists had discovered francium (1939) and promethium (1945), but it was in vain that chemists sought to set themselves up against the overwhelming invasion of the physical scientists into research on the elements.[26]

After World War II, the synthesis of new elements was due to the cyclotron, with increasingly larger machines better able to accelerate particles and nuclei with greater masses. The invention of the cyclotron can easily be compared to the invention of the voltaic pile by Alessandro Volta, an invention that led to the discovery of many alkali and alkaline-earth elements, or to spectroscopy, that led to the discovery of cesium, rubidium, gallium, indium, thallium, and the identification of some rare earth elements, not to mention helium and other noble gases. The analogies do not end here. In fact, neither Volta nor Lawrence personally utilized their discoveries for practical reasons, but left these kinds of investigations respectively to Humphry Davy and Glenn Seaborg.

1949 marked the birth of berkelium; in 1950, californium was born; in 1952, einsteinium was discovered; and a year later, it was fermium's turn. In 1955 and 1958, mendelevium and nobelium, respectively, were synthesized by nuclear reactions. All of these elements were discovered by Seaborg, Albert Ghiorso (1915–2010), and their co-workers at LBNL. In that same year, 1958, LBNL mourned the loss of their director and founder, Ernest Orlando Lawrence, inventor of the cyclotron that had led to the discovery of so many new elements. In 1961, on the occasion of the discovery of element 103, Seaborg and Ghiorso proposed that it be named lawrencium (Lw, later changed to Lr in 1963 by the International Union of Pure and Applied Chemistry (IUPAC)) in honor of their mentor.[27]

In 1957 research scientists in the laboratory at Dubna announced the discovery of element 104. At first, they believed that they had obtained the element by the following reaction:

$$^{242}\mathrm{Pu}\left(^{22}\mathrm{Ne},4\mathrm{n}\right)^{260}104$$

(Eq. V.3)[28]

but their error soon became apparent. The product they had obtained and that had a half-life of 14 milliseconds was ^{242}Am. News of the synthesis of this element returned to the limelight in 1964, when the Russians announced that they had synthesized various isotopes of element 104, including $^{261}104$, which was quite stable with a half-life of 1 minute. Some chemical tests were conducted on this sample to ascertain its nature. The name proposed for it was *kurchatovium* with the symbol Ku, but in 1969, Ghiorso laid claim to this discovery and wanted to call the element rutherfordium, the name presently accepted for this element, and to give it the symbol Rf.[29] In February 1970, the Russians made another attempt: starting with an americium target, they synthesized element 105 and quickly called it *nielsbohrium* (Ns). The reaction, according to Hans Bethe's shorthand notation was:

$$^{243}\mathrm{Am}\left(^{22}\mathrm{Ne},5\mathrm{n}\right)^{260}\mathrm{Ns} \qquad \text{(Eq. V.4)}$$

Two months later, the American group at LBNL bombarded a californium target with a beam of nitrogen ions to obtain:

$$^{249}\mathrm{Cf}\left(^{15}\mathrm{N},4\mathrm{n}\right)^{260}\mathrm{Ha} \qquad \text{(Eq. V.5)}$$

the element with atomic number 105 that they called *hahnium*. This element is presently called dubnium and has the symbol Db.

Up until 1974, the most effective method for producing the transuranium elements consisted in irradiating targets of heavy elements with beams of neutrons or lighter elements (up to $Z = 8$). In this way, the Seaborg-Ghiorso team at LBNL continued in progression to obtain element number 106. Not having transuranium isotopes available in sufficient quantity to use them as targets, the Russian Yuri Ts. Oganessian (b. 1933) and his co-worker A. Demin exploited a different mechanism, one based on the method of *fusion-evaporation* at low excitation energy. This method was a matter of employing less heavy, but at the same time more stable targets, such as lead ($Z = 82$) or bismuth ($Z = 83$). The compound nucleus resulting from the fusion of the target and projectile was produced with an excitation energy as weak as possible or rather with a correspondingly lower temperature. The fusion of the accelerated ions with the target was followed by expulsion of neutrons from the composite nuclei. The nuclear reactions in question were carried out I Dubna, the first in 1974 and the second in 1976, with chromium ions inside an accelerator:

$$^{207}\mathrm{Pb} + {}^{54}\mathrm{Cr} \rightarrow {}^{259}\mathrm{Sg} + 2\mathrm{n} \qquad \text{(Eq. V.6)}$$

$$^{209}\mathrm{Bi} + {}^{54}\mathrm{Cr} \rightarrow {}^{261}\mathrm{Bh} + 2\mathrm{n} \qquad \text{(Eq. V.7)}$$

A short time prior to the Russian syntheses, Ghiorso and his co-workers succeeded in synthesizing element 106, which they wanted to call seaborgium with the symbol Sg after their leader Glenn Seaborg. The synthesis of seaborgium is the last example of a synthesis of a transuranium element based on bombardment with light ions on a target having an atomic number slightly less than that sought in the reaction.

Working with the low excitation energy fusion-evaporation method, the German scientist Peter Armbruster (b. 1931), at GSI in Darmstadt, identified element 107 (bohrium) in 1981 and in 1982 discovered element 109, which he called meitnerium after the Austrian physicist Lise Meitner. Two years later, Armbruster synthesized element 108, calling it hassium.[30]

In May 1994, Ghiorso recounted that, in 1991 at LBNL, he glimpsed the existence of element 110 with a mass of 267 ($^{267}110$), but the official discovery of this element was attributed to Armbruster and Sigurd Hofmann, who, on November 9, 1994, synthesized it unequivocally, bombarding a target of ^{208}Pb with ^{62}Ni and obtaining the $^{269}110$ isotope. On November 23 of the same year, they obtained the $^{271}110$ isotope by the following reaction:

$$^{208}\mathrm{Pb}\left(^{64}\mathrm{Ni,n}\right)^{271}110 \quad \text{(Eq. V.8)}$$

A month later, on December 8, they also synthesized element $^{271}111$:

$$^{209}\mathrm{Bi}\left(^{64}\mathrm{Ni,n}\right)^{272}111 \quad \text{(Eq. V.9)}$$

Finally, in the first weeks of 1996, the GSI team at Darmstadt created element number 112 with a mass of 265, refusing to propose a name until the IUPAC definitively approved the name of hassium,[31] from the ancient Latin name of the region where the GSI is situated, Hesse.

The creation of three new elements, 114 at Dubna and 116 and 118 at LBNL, is certainly the scientific event of recent years, at least in the field of the superheavy elements and atomic physics.

The element with atomic number 114 was synthesized by means of a nuclear fusion reaction by Yuri Ts. Oganessian at Dubna, bombarding a target of ^{242}Pu and ^{244}Pu with the ion ^{48}Ca:

$$^{242}\mathrm{Pu}\left(^{48}\mathrm{Ca,3n}\right)^{287}114 \quad \text{(Eq. V.10)}$$

and

$$^{244}\mathrm{Pu}\left(^{48}\mathrm{Ca,3n}\right)^{289}114 \quad \text{(Eq. V.11)}$$

The advantage of this reaction lies in the fact that the projectile, ^{48}Ca, is an isotope rich in neutrons (N = 28). The products of this reaction are the element $^{A}Z = {}^{289}114$ and $^{A}Z = {}^{287}114$ (that is, with a Z that is a magic number). The magic number of neutrons N = A-Z from theory turns out to be 184. In the two isotopes produced at Dubna, N is 175 ($^{289}114$) and 173 ($^{287}114$); very far from the center of the nuclear stability island $^{A}Z = {}^{298}114$, although these nuclides are stable enough to have a half-life of almost a minute.

In the last experiment done at LBNL, a target of ^{208}Pb was used: this isotope has a closed proton (Z = 82) and neutron (N = 126) shell (the filling of the nucleons happens in the nucleus analogously to the filling of electronic orbitals) that confers extraordinary stability on it. The excess binding energy for this system of doubly closed shells leads to a *colder* nuclear system that necessitates the evaporation of only one neutron to avoid spontaneous fission (SF). Because the projectile that was accelerated, ^{86}Kr, was incapable of obtaining Z = 114 directly, the fusion reaction had as its product element 118:

$$^{208}\mathrm{Pb}\left(^{86}\mathrm{Kr,n}\right)^{293}118 \qquad \text{(Eq. V.12)}$$

that, by successive α decay generates first element 116:

$$^{293}118 \rightarrow {}^{289}116 + \alpha + \gamma \qquad \text{(Eq. V.13)}$$

and then 114:

$$^{289}116 \rightarrow {}^{285}114 + \alpha + \gamma \qquad \text{(Eq. V.14)}$$

The stability of the isotope with mass 285 of element 114 discovered at LBNL has a half-life four orders of magnitude less than that of $^{289}114$ discovered at Dubna. Thus, the researchers at Dubna, although they had not discovered any elements with higher atomic numbers, made a discovery much more important: they were the first to set their feet on the shores of the island of nuclear stability.

The synthesis of the transuranium elements reached its apotheosis in the 1990s. A rapid calculation tells us that in the 1940s there were seven elements discovered (five transuranium among them); five elements in the 1950s (all of them transuranium); and in the 1960s, two; in the 1970s, two; in the 1980s, three; and, finally, in the 1990s, four.

None of these artificial elements has stable nuclei because they emit α or β particles or undergo SF. The half-lives of these nuclides are tied to a relationship of inverse proportionality to their atomic number. The higher the Z, the lower the half-life of the element. The superheavy elements have extremely short half-lives (e.g., $^{269}110$ has a half-life of 170 ms); moreover, the events (collisions) used that lead to the fusion of the accelerated nuclei are very few.

V.2.1. THE ISLAND OF NUCLEAR STABILITY

The stability of a nucleus correlates directly to its binding energy, that is, to the difference between its mass and that of its components.[32] At first, the nucleus was treated like a statistical grouping of neutrons and protons. This model, in which the nucleus was considered analogous to a charged liquid drop, was very good at explaining many nuclear properties well. However, there was also strong experimental evidence of a shell structure analogous to the electronic shell structure in the extranuclear part of the atom, although not so prominent. The fact that a certain number of neutrons and protons led to the formation of particularly stable configurations was observed by Walter M. Elsasser (1904–91) in 1934. However, with the exception of the attempt to explain the well-known special stability of the light elements with values of Z and N equal to 2, 8, and 20 (respectively: ^{4}He, ^{16}O, and ^{40}Ca), the subject of nuclear shell structure was not taken seriously until 1948, when Maria Goeppert Mayer (1906–72) demonstrated the existence of additional magic numbers, 50 and 82 for protons, and 50, 82, and 126 for neutrons. Since then, 28 has been recognized as a magic number for both protons and neutrons.

Finally, quantum mechanics has given much greater information than hitherto obtained on the energies of the orbitals that nucleons occupy in the nucleus. A magic number of nucleons is discerned when they complete the levels just below the discontinuity in the shell. The compression of the energies of the levels that each nucleon occupies can lead to a significant increase in the total binding energy and, consequently, in stability. This shell effect persists beyond a certain radius of the magic number of nucleons. This stability is highest if either the protons or the neutrons are magic numbers, and it decreases if we distance ourselves from either one or both. The results of theoretical calculations indicate that the next magic numbers ought to be Z = 114 and N = 184, leading to the predicted island of nuclear stability for nuclei with values of Z between 104 and 124.[33] In the middle of the 1960s, Adam Sobiczewski (b. 1931) reported two new magic numbers, the results of his theoretical calculations: Z = 114 and N = 284.

V.2.2. UNFORTUNATE EPISODES IN THE ATTRIBUTION OF THE NAMES OF THE ELEMENTS BETWEEN 101 AND 109

The attribution of names for the elements that come between 101 and 109 constitutes an example of how scientists, able to overcome the obstacles that Nature has placed before them, are utterly incapable of overcoming national and academic pride. Intrigues and plots took place from the end of the 1950s, when mendelevium and nobelium were discovered, until the IUPAC conference at Guilford in 1995. During this time, the various research groups exchanged harsh letters via pages in the journal *Nature*. Why did the IUPAC Commission wait so long to assign names to elements that were discovered 40 years earlier? According to some, the Commission met only once there was a sufficient number of new elements waiting for names. According to other sources, however, the Commission waited so as not to run into the error of legitimizing false discoveries. But the two hypotheses do not seem very plausible. In fact, during the years of the Cold War, the IUPAC Commission took no initiative and only after the dissolution of the Soviet Union, on December 31, 1991, did it show any desire to settle the controversy surrounding the names of the elements between atomic numbers 101 and 109. Unfortunately, the meeting that led to the assignment of names to these superheavy elements assumed the likeness of the 1878 Berlin Congress, in which the African continent was divided up among the major colonial powers of the era. The teams that were competing inch by inch for the names of the elements were the Americans of the LBNL, the Russians of the institute at Dubna, the Germans of the GSI, and the Swedes of the Academy of Physics of the Nobel Institute.

The error discovered regarding nobelium on the part of the Swedish Academy was resolved in this way: the name nobelium was kept and ratified even though the Swedes and later the Americans, who accepted the proposed name, were in error, and that only the Russians at Dubna had correctly identified and characterized the element.

More unfortunate was the episode regarding assignment of names to elements 103, 104, 105, and 106. These were claimed by the Americans (with the names lawrencium, rutherfordium, *hahnium*, and seaborgium) and by the Russians (with the names *kurchatovium, nielsbohrium*, dubnium, and *joliotium*).

The discovery of element 107 was claimed by the American, Russian, and German groups, whereas elements 108 and 109 were claimed only by the Germans.

To emphasize the dubious motivation that led to the assignment of the names of these elements at Guilford in the summer of 1995, it is sufficient to look at the composition of the examining commission: one-third of the group's members were American, as was the group's president; another one-third of the group's members were Germans. The remaining third was composed of members from other countries, but there was no national representative for the Russians. It is not difficult to extrapolate the consequences of this situation: out of 10 elements, six took the names proposed by the Americans. Nobelium was also placed among these elements, and the Americans who were later recognized as the discoverers of this element wanted to keep the name proposed by the Swedish physicists. Three were named according to the German group's proposal: bohrium, hassium, and meitnerium. Finally, element 105 took the name of dubnium after the place of the Russian group's laboratory, Dubna.

Finally, it should be observed that LBNL had willfully ignored the prohibition of naming element 106 seaborgium. In fact, a rule in the statutes of the IUPAC forbids naming an element after a living person and, in 1994, Glenn Seaborg was very much alive. In addition, the same body made note that element 106 should be called rutherfordium, and they recommended this in order that elements 104 and 105 might take the names of *joliotium* and dubnium.

Through the pages of the journal *Nature*, the powerful combined lobby of the LBNL and the American Chemical Society (ACS) publicly intimated to the IUPAC Commission that they should modify the rule that prevented naming element 106 seaborgium by threatening not to recognize any decision that the Commission might make that differed from their own. In 1995, with Seaborg still living, element 106 was dubbed seaborgium, with the symbol Sg.

So that these deplorable incidents not be repeated, it would be fairer if, in the future, that the IUPAC Commission assign the names to every new element rapidly and recognize the priority of the discovering group by means of unambiguous official documentation. In a world where scientific news and information is transmitted in real time, it is unthinkable to have to wait decades before a discovery is recognized.[34]

V.2.3. FROM ATOMS TO THE STARS

To return to the question that chemists of Berzelius's era posed—that is, how many elements are there?—is there a limit? We can try to answer this question in light of present knowledge. For certain, we can say that 117 elements were known as of October 16, 2006. The element with the highest atomic number is 118; the element that was last ratified with a name and symbol was element 116, livermorium.

How far will we be able to go in the synthesis of the superheavy elements? Some day, it may be possible to accelerate a beam of uranium ions and bombard a target of the same element with it. Or it may be possible to cause three beams of uranium ions to collide to obtain a super atom made up of the sum of the three masses of uranium nuclei. All these ideas are fascinating but, for the moment, fantasy.

In fact, the problem essentially comes down to two concepts: the first is the half-life of the synthetic elements; some isotopes of elements 110, 111, and 112 have half-lives of less than 10^{-6} s, and even shorter half-lives are predicted for elements with higher atomic numbers, with the exception of those with a magic number of nucleons.

The second concept lies in the definition of the fusion of two nuclei. If we consider the two elements A and B that, on collision, produce element C, we have before us two cases: the first case is that C, obviously unstable, would decay according to the classical mechanisms of radioactive decay. The second case that could happen is that C would reverse direction and revert again to A and B (the phenomenon called *scattering*).

From the laws of atomic physics, we know that the minimum time to verify an interaction between two nucleons is around 10^{-11} s. Consequently, one can suppose that an element with a half-life of less than 10^{-11} s cannot exist. If we observe therefore a scattering phenomenon with times of the order of less than 10^{-11} s, we will not be able to say if it has led to the formation of a new element.

Ultimately, we have proven that the number of elements that can be synthesized is finite. In this case, the limit that we can give to matter is not expressed directly in Z but is imposed by the time of interaction of particles that obey the laws of strong nuclear force and, consequently, to the half-life of the element synthesized.

Atoms consisting of up to about 240 nucleons obey the laws of strong nuclear force (i.e., the force that holds the nuclear components together), but atomic structures containing 10^{57} nucleons would also feel the gravitational force that is notably weaker than the strong nuclear force but acts over a greater distance.[35] We come then to ask ourselves what kind of mass could an object have that is composed of 10^{57} nucleons, and above all, if it could possibly exist. The first answer tells us that it would have a mass of around 10^{33} g, roughly the mass of the sun; the second answer is yes, on condition that all, or almost all, of the nucleons would be neutrons. Objects with these characteristics are the so-called neutron stars. The term "neutron star" was coined in 1934, by Walter Baade (1893–1960) and Fritz Zwicky (1898–1974) who suggested that the energy dissipated by supernovae comes from the condensation of a star with solar characteristics in a neutron star.[36]

The neutron stars consist of about 99.0–99.9% neutrons, with the remaining particles being protons (and electrons). They have a radius of 3–10 km. Neutron stars derive, according to a model of stellar evolution, from stars with masses between $0.4 \leq M_S \leq 4$, where M_S is the solar mass. The density (ρ) of a neutron star is easily obtained from the following relationship:

$$\rho = 10^{33} \frac{g}{\left[\frac{4}{3}\pi r_S{}^3\right]} \qquad \text{(Eq. V.15)}$$

where r_S (the radius of a neutron star with a mass equal to the sun) = 4.4×10^5 cm, from which we calculate that $\rho \sim 10^{16}$ g/cm^3.

Densities of this magnitude were totally unexplainable at the beginning of the 20th century; solids and liquids were thought to be substantially incompressible because it was held that the atoms would touch one another (the maximum density is achieved by osmium, which is 22.5 g/cm^3). But in 1911, Rutherford's experiments on the scattering of α particles on a sheet of gold demonstrated that atoms are essentially empty space with a very tiny solid kernel, the nucleus, and atomic nuclei have densities precisely of the order of 10^{15} g/cm^3.

Can we hypothesize that a neutron star might be a superheavy transuranium element? According to this conclusion, one could say that in two distinct domains stable "elements" do exist. These regions, expressed as mass numbers, that for a qualitative calculation one assumes are equal to the number of nucleons, are included between $10^0 \leq A \leq 2.38 \times 10^2$ and $10^{56} \leq A \leq 10^{58}$.

Notes

20. Rowley, N. *Nature* **1999**, *400*, 209.
21. Friedlander, G., Kennedy, J. W., Macias, E. S., Miller, J. M. *Nuclear and Radiochemistry*, 3rd Ed.; John Wiley: New York, 1981.
22. Verrall, M. *Nature* **1994**, *372*, 306; Lehrman, S. *Nature* **1994**, *371*, 639; News in Brief, *Nature* **1996**, *379*, 762; News in Brief, *Nature* **1997**, *389*, 10.
23. Trifonov, D. N., Trifonov, V. D. *Chemical Elements, How They Were Discovered*; MIR Publishers: Moscow, Russia, 1982.
24. Perrier, C.; Segrè, E. G. *J. Chem. Phys.* **1937**, *5*, 712.
25. Seaborg, G. T.; Segrè, E. G. *Nature* **1947**, *159*, 863.
26. Marinsky, J. A.; Glendenin, L. E.; Coryell, C. D. *J. Am. Chem. Soc.* **1947**, *69*, 2781.
27. James, L. K., Ed.; *Nobel Laureates in Chemistry 1901–1992*, American Chemical Society: Washington, DC and Chemical Heritage Foundation: Philadelphia, PA, 1993.
28. For a certain period of time, this element was called *joliotium* (Jl).
29. Trubert, D. et al. *Compt. Rend.* **1998**, *1*(10), Serie II, 642.
30. Armbruster, P., et al. *Ann. Rev. Nucl. Part. Sci.* **1985**, *35*, 135.
31. Hofmann, S.; Ninov, V.; Hessberger, F. P.; Armbruster, P.; Folge, H.; Munzenberg, G.; Schott, H. J.; Popeko, A. G.; Yeremin, A. V.; Andreyev, A. N.; Saro, S.; Janik, R.; Leino, M. *Z. Physik* **1995**, *A 350-4*, 281; Hofmann, S.; Ninov, V.; Hessberger, F. P.; Armbruster, P.; Folger, H.; Munzenberg, G.; Schott, H. J.; Popeko, A. G.; Yeremin, A. V.; Saro, S.; Janik, R.; Leino, M. *Z. Physik* **1996**, *A 354*, 229.
32. The mass of every nucleus is always less than the sum of the individual nucleons it contains (i.e., the protons and neutrons). The mass loss that accompanies the formation of nuclei is called the *mass defect*, and it correlates with the nuclear binding energy via Einstein's equation, $E = mc^2$.
33. Sobiczewski, A. *Phys. Particles Nuclei* **1993**, *25*, 259.
34. Inorganic Chemistry Division, Commission on Nomenclature of Inorganic Chemistry, "Names and Symbols of Transfermium Elements (IUPAC recommendations 1997)," *Pure and Appl. Chem.* **1997**, *69*, no. 12, 2471.
35. Fontani, M.; Costa, M. *Atti del IX Convegno Nazionale di Storia e Fondamenti della Chimica, Modena 25–27 Ottobre 2001*, p. 444; vol. 119, Memorie di Scienze Fisiche e Naturali, Rendiconti dell'Accademia Nazionale delle Scienze detta dei XL, serie V, vol. XXV, parte II, tomo II 2001.
36. Braccesi, A. *Esplorando l'Universo*; Le Ellissi-Zanichelli: Bologna, 1988.

V.3

THE SEARCH FOR PRIMORDIAL SUPERHEAVY ELEMENTS: BETWEEN SCIENTIFIC RIGOR AND ATOMIC FANTASY

In the 1970s, inorganic chemistry had largely become reconciled with its original mother, physical chemistry, and indeed it was sometimes confused with physics. The search for new transuranium elements was the business almost exclusively of physicists (with some outstanding exceptions) and particularly of the centers at Dubna, Russia, where Georgy Nikolayevich Flerov (1913–90) was director, and at Berkeley, California, where the research group headed by Glenn T. Seaborg and Albert Ghiorso was working.

Element 105 was discovered in 1970, 106 in 1974, and 3 years later the Soviets maintained that they had synthesized 107. Announcements and retractions piled up, and members of both countries' scientific communities began quarreling over the names to assign these new elements. The last free spaces in the periodic table had been filled up in the 1940s, but some researchers still believed in the possibility of finding new superheavy elements in nature.

In the intervening years between the two world wars, there was a handful of chemists searching for elements 93 and 94. However, even 40 years later, research was continuing, stubbornly, on the identification of naturally occurring superheavy elements—but this time they turned their attention to an exogenous, or better, extraterrestrial source: meteorites.

On the night of February 8, 1969, at 1:05 AM, near the city of Pueblito de Allende, in the state of Chihuahua, Mexico, a very large meteorite crashed into the earth. Two tons of fragments were extracted from the crater; they seemed to have had their origin on the red planet, Mars.[37] The Mexican authorities sent samples to the United States, where they were examined by an expert in the field, Edward Anders.

Anders was born of a Jewish family in Liepaja, Latvia, on June 21, 1926. Originally, the family name was Alperovitch. During World War II, he lost his father and 23 other members of his family, killed by the Nazis, but fortunately, he and his mother survived by hiding in Germany until, in 1949, he emigrated to the United States. He changed his last name and in 1954 received his doctorate from Columbia University. He began research on meteorites, specifically their dating and chemical composition, becoming very soon an international authority in this area.

In the Enrico Fermi Laboratory in the Department of Chemistry at the University of Chicago, the Allende meteorite slowly began to reveal its secrets. It was composed of carbonaceous chondrites with inclusions, *chondrules*—one of the most primitive forms of material aggregation known. Anders hypothesized that these "cosmic rocks deprived of geological evolution" were generated by the explosion of a supernova about 4.6 billion years ago.

As in other meteorites as well as in the Allende fragment, an abnormally large amount of two isotopes of xenon, ^{131}Xe and ^{136}Xe, were found. Their origin was a mystery. For Anders, the only explanation for this experimental evidence was the spontaneous fission of an as yet unknown transuranium element. In fact, none of the known transuranium elements at that time gave xenon nuclei as fission products. Anders and his colleague Dieter Heymann (b. 1927) further hypothesized that the ancestor of xenon should be an element between atomic numbers 112 and 119. Three years later, in 1972, he reviewed his hypothesis[38] and set the range of the "superheavies" between 111 and 116; finally, in December 1975, he reduced the number of elements to three:[39] 115, 114, and 113. Anders's next step was to study the distribution of xenon in some other meteorites and compare their relative and absolute abundances with other elements present in trace amounts. The work presented some original features, whereas other logical routes were purely speculative: the distribution of the elements present in the meteorite led them to exclude certain transuranium elements. An example is element 119: because its lower homologs, the alkali metals, are quite rare in meteorites, Anders supposed that the parent element could not be an alkali metal. Using similar reasoning, he also excluded elements 118, 117, 112, and 111. He studied in detail six samples of meteorites looking for traces of 26 elements present in nature. The superheavy element would seem to lie in minerals like the chromites present in as much as 0.04% of the entire mass. Traces of the sulfates of some toxic metals like Pb, Tl, and Bi, plus bromine and some noble gases like Ar, Kr, and Xe accompany those already listed. Evidence of the temperatures at which these elements condense, about 400–500 K, like those present in nebulas, led Anders to maintain that the only superheavy elements that could possibly be present at the moment of meteorite formation were elements 113, 114, and 115. The estimate that he gave of their half-lives was quite high, about 10^8 years; this would not be enough for them to last down to our own times but, in his opinion, enough to leave measurable traces of their existence. Even though Anders set the half-life of an element like 114 arbitrarily, there is no doubt that his work rests on a solid theoretical basis and on rigorous scientific investigation.

Interest in research on the transuranium elements quickly infected other investigators scattered around the world. On July 5, 1976, Robert V. Gentry (b. 1933), a chemist working in the United States, announced that he had found evidence of primordial superheavy elements. Gentry's work followed shortly after that of Anders, but it did not have the same rigor or originality.[40] Having come into possession of a biotite from Madagascar, Gentry believed that he had identified elements 126, 116, 124, and 127 (listed in order of decreasing abundance). He examined some pleochroic haloes in biotitic mica, present as inclusions in microcrystalline Malagasy monazite. It was noted that, over time, the structural damage induced by α radiation produced by radioactive decay could generate spherical pleochroic haloes when the radioactive elements were contained as inclusions in transparent material like mica. He used the technique of characteristic X-ray emission induced by the bombardment of the haloes with a proton beam of appropriate energy. The technique, although more developed, went back to the classical spectroscopy of Moseley. One comparative study of the characteristic wavelengths, $L\alpha$ and $L\beta$, convinced Gentry of the presence of the superheavies in his samples, and indeed, at the conclusion of his work, he even ventured to assert that some of them could have concentrations of as much as 10^{-10} g per gram of monazite.

The theoretical physicist Cheuk-Yin Wong (b. 1941) was also enthusiastic on the subject. Two months after Gentry's publication, on September 13, 1976, he published the results of his own research on the half-life of a hypothetical primordial element[41] with atomic number 126. Wong was born in Guangdong in 1941 and grew up in Hong Kong. In 1966, he received his doctorate at Princeton and, in the same year, began work at Oak Ridge National Laboratory in Tennessee, where he became acquainted with Gentry. Wong's specialty was theoretical nuclear physics and, starting from Gentry's assumptions, he published a paper on the element with atomic number 126 hypothesized by Gentry to be in biotite. The model of the nucleus of this superheavy would not be spherical, and, consequently, the unstable configuration would not allow the nucleus to have a very long half-life. For this reason, Gentry's data had to be interpreted in another way. Far from taking into consideration the possibility that his colleague might have committed an error, he hypothesized that the nucleus of element 126 would have a toroidal form, but he asked to see further study and confirmation of such a hypothesis, inviting Gentry to determine the mass number of the hypothetical element—a confirmation that never happened.

Almost contemporaneously, on October 18, Robert Wolke came to the attention of the scientific community by proposing, in the pages of *Physical Review Letters*, an experiment able to corroborate the existence of element 116 in nature.[42] His bizarre idea was that elements like 116 and 118 can enter into and become part of the metabolism of some invertebrates and fish. He used as his starting point the following hypothesis: elements 116 and 118 are the higher homologs of polonium and radon (*eka-polonium* and *eka-radon* in Mendeleev's terminology) and therefore could have similar chemical characteristics. The isotopes ^{210}Po and ^{222}Rn, even though they have very short half-lives (138.4 and 3.82 days, respectively), are continually generated by the decay of uranium, and their level in surface marine waters is 10^{-2} pCi/L. Browsing through some specialized journals, Wolke found that in the hepatopancreas (a kind of liver) of some marine invertebrates, and in a type of pelagic fish, the content of Po and Rn is 10^{6} times greater than in nature. Could elements 116 and 118 also be present at similarly higher levels in similar organisms? In his opinion, it would be easier to find traces of elements 116 and 118 in biological tissues than in biotites. If one were to dry up the organs containing Po and Rn and then treat them with advanced radiochemical techniques, it might be possible to concentrate elements 116 and 118 by more than a factor of 10^{14}. Wolke wrote in the prestigious pages of *Physical Review Letters*, hoping that someone would pick up on his idea and put it into practice, worrying only about the fact that elements 116 and 118 might have properties similar to Po and Rn and not about their probable nonexistence. His urgings were not acted on, and the proposed experiment remains a dead letter.

On December 6, 1976, the French physicist Claude Stéphan of the Institut de Physique Nucléaire in Orsay, near Paris, published an exhaustive study highly critical of the content of the work done by Gentry.[43] Stéphan examined some samples of monazite taken from the same geological formation studied by Gentry. He used an apparatus with which, by neutron bombardment, he could study all the nuclear fragments in the region between 294 and 361 mass numbers. His instrument was a hundred times more sensitive than Gentry's (10^{-12} g/g), but nevertheless he found no evidence of superheavy elements.

On December 27 of the same year, Bruce Hubert Ketelle (1914–2003), a colleague of Gentry's at Oak Ridge National Laboratory, published an article mildly critical of the studies on the biotites and superheavy elements.[44] This article had as its goal more or less to reappraise Gentry's work. Gentry's scientific orthodoxy rapidly declined and in 1982, he was obliged to leave the Oak Ridge laboratory after 13 years of employment.

From then on, he was occupied with spreading his creationistic ideas by participating in conferences, writing articles, publishing books, and founding a society which, in certain respects, looked like a religious sect.

Then, in late 1983, another article on the superheavy elements appeared, authored by Anders. It was a retraction[45] of his hypothesis on the existence of primordial superheavy elements entrapped in the Allende meteorite, ancestors of the anomalous isotopes ^{131}Xe and ^{136}Xe. He had found an alternative to the unexplainable abundance of these two nuclei through a new, quite different, hypothesis: the two heavy isotopes of xenon with masses of 131 and 136 would have been formed through nucleosynthesis as a result of neutron capture and would have remained entrapped in the meteorite from time immemorial. The solution to the mysterious abundance of xenon was possible by positing synthesis starting from lighter elements through a thermonuclear process present in supernovae. Anders, with a providential change of direction, managed to salvage his academic authority, something that neither Gentry nor Wolke had been able to do.

In 2003, a large group of Russian scientists, headed by Yuri Oganessian, conducting experiments on synthetic nuclei, had been thus able to send a second and much truer message regarding the island of nuclear stability. Already in 2001, after having succeeded in synthesizing element 114 and thanks to a sensational scientific fraud in the ranks of the U.S. scientists, they remained the only competitors in the field of synthesizing superheavy elements.

The young Victor Ninov (b. 1959), a naturalized American citizen originally from Bulgaria, was blamed for having created his results on the synthesis of element 118 out of thin air. He was first suspected of this (November 2001), and then fired (May 2002), effectively killing the superheavy element program at Berkeley.[46] Ninov had actually suggested that the supposed element 118 be named *ghiorsium* in honor of the elderly nuclear chemist. For now, the Russians remain virtually the absolute masters of this research field. Between 2001 and 2003, they announced the discovery of another three superheavies: 116, 115, and 113. This synthetic method is relatively simple on paper: fusion of two nuclei by impact of a light nucleus with a heavy nucleus being used as the target.[47] The reactions are the following:

$$^{248}Cm + ^{48}Ca \rightarrow ^{296}116 \qquad \text{(Eq. V.16)}$$

and

$$^{243}Am + ^{48}Ca \rightarrow ^{291-x}115 + xn \qquad \text{(Eq. V.17)}$$

$$^{291-x}115 \rightarrow ^{289-x}113 + \alpha \qquad \text{(Eq. V.18)}$$

The synthesis of nuclei with odd atomic number will afford us new knowledge in the field of nuclear stability even if is too early to say very much about what direction it will take.[48] The world of the transuranium elements is in continuous evolution,[49] and there is no doubt that academic controversies will arise to establish who was the first to synthesize this or that element.[50] Nevertheless, the exploration of a world so distant from our own previous experience cannot help but continue to fascinate us.

Notes

37. Anders, E.; Heymann, D. *Science* **1969**, *164*, 821.
38. Anders, E.; Latimer, J. W. *Science* **1972**, *175*, 981.
39. Anders, E.; Higuchi, H.; Gros, J.; Takahashi, H.; Morgan, J. W. *Science* **1975**, *190*, 1262.
40. Gentry, R. V.; Cahill, T. A.; Flocchini, R. G. *Phys. Rev. Lett.* **1975**, *37*, 11.
41. Wong, C. Y. *Phys. Rev. Lett.* **1976**, *37*, 664.
42. Wolke, R. *Phys. Rev. Lett.* **1976**, *37*, 1098.
43. Stéphan, C.; Epherre, M.; Cieślak, E.; Sowiński, M.; Tys, J. *Phys. Rev. Lett.* **1976**, *37*, 1534.
44. Ketelle, B. H.; O'Kelley, G. D.; Stoughton, R. W.; Halperin, J. *Phys. Rev. Lett.* **1976**, *37*, 1734.
45. Lewis, R. S.; Anders, E.; Shimamura, T.; Lugmair, G. W. *Science* **1983**, *222*, 1013.
46. *The Chronicle of Higher Education; Section Research & Publishing*, 2002, August 16.
47. Fontani, M.; Costa, M. *Il Chimico Italiano* **2003**, *3/4*, 39.
48. Fontani, M.; Costa, M.; Cinquantini, A. *La Chimica e l'Industria* **2003**, *85*, 65.
49. Fontani, M.; Costa, M.; Manzelli, P. *Il Chimico Italiano* **2002**, *4*, Luglio/Ottobre, 34.
50. Karol, P. J.; Nakahara, H.; Petley, B. W.; Vogt, E. *Pure and Appl. Chem.* **2003**, *75*, 1601; Fontani, M.; Costa, M.; Manzelli, P. *Memorie di Scienze Fisiche e Naturali. Accademia delle Scienze detta dei XL*, 433, vol. XXV, serie V, parte II, tomo II, 2001.

V.4

NAMES, NAMES, AND NAMES AGAIN: FROM A TO ZUNZENIUM

Because many decades have passed since the discovery of the first transuranium elements, the chapter relative to the synthesis of the superheavy elements cannot be said to be closed. It is likewise true that all of the elements discovered—or rather, synthesized—up until now do not have a definitive name, and those that have received names have a fascinating history behind them, one made up of contrived compromises, of names accepted with good will or imposed by force, of complex academic controversies, and of questionable scientific rivalries. But what strikes the imagination of the reader and astounds the scholar is the fact that some elements not yet discovered—and we do not know if they ever will be created artificially—have received names. This is the case of the elements with atomic numbers of 145 and 243, called *hawkingium* and *zunzenium*.

V.4.1. THE ELEMENTS FROM NEPTUNIUM TO MENDELEVIUM SEEN FROM BOTH SIDES OF THE IRON CURTAIN

The synthesis of the first transuranium elements witnessed the virtually absolute domination of American scientists, even though the beginning of this fascinating chapter in science was born in Italy. After Enrico Fermi and his co-workers announced the discovery of the first two transuranium elements by neutron irradiation in 1934, it seemed a natural consequence that this discovery would be credited to Italian science.[51] Otto Hahn and his young assistants were in a position to reproduce the same data and, curiously enough, were also able to identify elements 95 and 96. During a press conference, Orso Mario Corbino, formally the chief of nuclear physics research in Rome, let fall the names of the two elements just discovered, *ausonium* and *hesperium*, giving rise to great disappointment on the part of Fermi. The years passed and in 1938, Enrico Fermi received the Nobel Prize in physics. Even though the reason for the coveted award made reference to the synthesis of the first transuranium elements, with the discovery of uranium fission, Fermi's research was first disputed and then refuted. Even stranger was how news of the prize leaked out: it was communicated to Fermi by Niels Bohr, well in advance of the communication from the Swedish Academy. In fact, Fermi had found among his neutron radiation products traces of both transuranium elements and lighter atoms. Wanting to determine at all costs whether he had actually discovered the first transuranium elements might have ended by backfiring against his career. Fermi never rectified his position because no one explicitly asked him to do so. He had barely received the Nobel Prize when he emigrated to the United States; he set foot in his native country again only years later. Even in the following years, no amendment came from his colleagues who had carried out the irradiation of uranium with neutrons.[52] World War II was at the gates, and Fermi was certainly preoccupied with a good many other problems: waiting to change his citizenship and wanting to downplay his past with the fascist regime that, in 1941, had declared war against the United States. He threw himself totally into the American war effort and never laid claim—not for himself and not for his colleagues—to the discovery of elements 93 and 94. There might have been an occasion to claim this discovery if international conditions had been different, but in those terrible years, many scientists were overwhelmed by events beyond their control. Witness the bitter fate reserved for Lise Meitner and her lack of recognition for the discovery of nuclear fission. In the synthesis of new elements, the baton passed to young American chemists and physicists headed, on the one hand, by Edwin M. McMillan and Philip H. Abelson and, on the other, by Arthur C. Wahl and Glenn T. Seaborg. The only Italian physicist from Fermi's old group who continued this type of research was Emilio Segrè who synthesized astatine.[53] In the spring of 1940, McMillan and Abelson[54] caught a glimpse of element 93 and realized that its properties were not entirely attributable to rhenium, as had been previously thought, but to its near neighbor, uranium. In the following autumn, McMillan spoke of his experiments to Seaborg, convincing him to collaborate in the attempt to separate the product of β decay of element 93. For a variety of reasons probably related to the approach of the war, McMillan soon left the group and toned down his interests in research. However, during the following January, McMillan, Seaborg, Kennedy, and Wahl succeeded in sending to Washington their first report that contained the results of their research on element 94, but they had to wait until 1946 to publish them.[55] For national security reasons, during World War II, elements 93 and 94 were called by their code names: silver and copper. When the time came to give them a definitive name, the U.S. nuclear chemists proposed *extremium* and *ultimum*, thinking that they had arrived at the extreme limit of the periodic table. McMillan arrogated to himself the decision to name element 93, and he called it neptunium[56] after the planet Neptune. The other chemists found themselves facing a fait accompli and inevitably followed their chief's example. The name chosen for the 94th element was plutonium, after the planet Pluto; at first, they thought of the name *plutium*, but this decision was soon abandoned because it was not euphonious. Then followed a long debate on the symbol, whether it should be Pl or Pu (in this case, it was decided that "p" and "u" had the better sound). In later years, Seaborg told how he had loved the idea of calling element 94 *chronium* (for the ancient Greek god called Saturn by the Romans) or *minervium* (after the goddess derived by the combination of two divinities, the ancient Roman Minerva, patron of the arts, and the Minerva of the Etruscans, goddess of war). Hand in hand with the work done by Seaborg and his colleagues in synthesizing new elements, they got very good at proposing new names, a fact in itself unique in the history of the discovery of the elements. As a result, we witnessed a rapid increase in bizarre proposals and outlandish suggestions. On November 11, 1945, during a radio program called *Adventure in Science*, Seaborg made a brief appearance and let slip the news of the discovery of elements 95 and 96. Seaborg told how one of his laboratory co-workers, Tom Morgan, referred to this pair of elements by the names *pandemonium* and *delirium*, even to the point of seriously considering proposing these names to the IUPAC Commission. On December 15, Seaborg again appeared on the same radio program and read a long, long list of names that friends, acquaintances, scholars, researchers, or simply the curious had suggested to him. J. D. Boon of the Department of Physics at Southern Methodist University proposed a nomenclature system that would have covered the names of all the

Table V.1 List of element names suggested to G. Seaborg, December 15, 1945

Element 95	*Etymology*	*Element 96*	*Etymology*
Proximogravum	Proximus gravissimus (L)	*Gravum*	Gravissimus (L)
Alium	Another (L)	*Novium*	New (L)
Quintium	Fifth (L)	*Sextium*	Sixth (L)
Solium or Solonium	Sun (L)	*Lunium*	Moon (L)
Sunonium	Sun (E)	*Moononium*	Moon (E)
Solium—Sunian	Sun (L)	*Nebulium*	Cloudy (L)
Big Dipperian	–	*Big Bearianen*	–
Dipperium	–	*Cometium*	Comet (L)
Stellanium	Star (L)	*Astronium*	Star (L)
Bolidium	Heavenly body (L)	*Asteroidium*	Asteroid (L)
Transneptunium	Beyond neptunium (L)	*Universum*	Universe (L)
Siderium	Sidereal (L)	*Stellium*	Star (L)
Astralium	Star (L)	*Cosmium*	Cosmos (G)
Draconium	Constellation Drago	*Leonite*	Constellation Leo
Sirium	Sirius (very bright star)	*Canopium*	Very bright star
Deimos	Moon of Mars	*Phobos*	Moon of Mars
Virgonium	Virgo (Zodiac)	*Ariesium*	Aries (Zodiac)
Terrium	Earth (L)	*Finium or Ultimum*	Last (L)
Amerium	America	*Artificium or Artifician*	Artificial (L)
Cyclo-Europium	Cyclotron	*Cyclo-Gadolinium*	Cyclotron
Mechanicum	Artificial (G)	*Scientium*	Science
Alphonium	α Particle	*Cosmonium*	Cosmic rays (G)
Neutronium	Neutron	*Alphanium*	Alpha particle
Splittium	Split (E)	*Fissium*	Fission (L)
Fermium	Enrico Fermi	*Bohrium*	Niels Bohr
Becquerelium	Henri Becquerel	*Rutherfordium*	Ernest Rutherford
Curium	Pierre and Marie Curie	*Einstenium*	Albert Einstein
Einsteinium	Albert Einstein	*Rooseveltium (FDR)*	Franklin D. Roosevelt
Washingtonium	George W. Washington	*Roosium*	Franklin D. Roosevelt
Vulcanium	Vulcan	*Herculium*	Hercules (L)
Zeusium	Zeus	*Venusium*	Venus (L)
Apollium	Apollo	*Martium*	Mars (L)
Unonium	United Nations Org.	*Paximum*	Peace (L)
Mondium	World (L)	*Worldliness*	Transience (E)

(Continued)

Table V.1 Continued

Element 95	*Etymology*	*Element 96*	*Etymology*
Eternium	Eternity	*Futurium*	Future (E, L)
Seaburnium	–	*Nutronium*	–
Nonagintium	–	*Ytunium*	–
Xtinium	–	*Curium*	Pierre and Marie Curie
Unicalium	University of California	*Bordium*	Glenn Seaborg
Seadium	Glenn Seaborg	*Bastardium*	Pluto raped Persephone
Persephonium	Persephone (goddess)		

L = Latin; E = English; G = Greek.

elements up to 100: *pentonium* (95), *sextonium* (96), *septonium* (97), *octonium* (98), *novanium* (99), and *centurium* (100). The complete list of the other names is given in Table V.1.

As is well-known, on March 5, 1946, Seaborg abandoned his reserve and, at the meeting of the Heavy Isotopes Group at the Metallurgical Laboratory, proposed the name americium (from America) and curium (for Pierre and Marie Curie) for the elements with atomic numbers 95 and 96. In this way, he linked himself to the homologous names of the rare earths: one name that honored a continent (europium) and the other a chemist (Gadolin). Seaborg had judged rightly. In fact, the actinide family was very probably similar to the lanthanides. Ironically, we saw how many years and how many workers it took to isolate the rare earth elements and how few were necessary to complete the transuranium element group. The next step was the discovery of elements 97 and 98, which today carry the names of berkelium and californium. They were discovered very close in time to one another, one at the end of 1949 and the other at the beginning of 1950, so that their discoveries were reported simultaneously.[57] As for the preceding elements, the choice of names was rather difficult. The names proposed are given in Table V.2.

Other proposals were also advanced to name the elements with atomic numbers 97, 98, 99, and 100 *universitum, offium, californium,* and *berkelium*, respectively, but Seaborg and Ghiorso preferred not to go beyond the elements discovered. Consequently, with respect to the selection of the last two names, they declined to specify the spelling of element 98. Would the name berkelium or *berklium* be more euphonious? And which symbol should be selected, Bk or Bm? In the midst of the uproar generated by the continuous discoveries of the Seaborg team and during the many imaginative decisions on names to attribute to new arrivals in the periodic table, two scientists in the Soviet Union, A. P. Znoyko and V. I. Semishin, claimed the discovery of element 97 and proposed to call it mendelevium[58] with the symbol Md, in recognition of the great Russian chemist who was father of the periodic table. Their claim did not have a solid basis and was quickly rejected, but it signalled that Americans were not alone in this field of research. Theirs would no longer be an uncontested domination, even though for many years they remained tops in this field of research. The problem was initially ignored until, 5 years later, the Berkeley scientists revisited the reasons advanced by the Russians and called the 100th element mendelevium.

Table V.2 Names proposed for element 98

Name of Element 98	*Etymology*
Lewisium	G. N. Lewis
Cyclotronium	Cyclotron
Cyclonium	Cyclotron
Euprosium	Greek: *eu*, good and *prósopon* (person)
Nonactinium	Ninety-eighth (Latin)
Ennactinium or *Enactinium*	Ninety-eighth (Greek)
Lawrencium	E. O. Lawrence
Radlabium	Radiation Laboratory
Praedicium	Foretold
Accretium	Increased
Colonium	After the city of Cologne (Colonia Agrippa)

That the subject of the discovery of new elements thrilled the American press is witnessed to by the fact that many newspapers and magazines reported the news at every possible opportunity. For example, the *New Yorker*, convinced that the game would be played strictly by Americans, ventured out on a limb by entertaining the idea of giving names to the not-yet-discovered elements 99 and 100: "we are already at work in our office laboratories on '*newium*' and '*yorkium*.' So far we just have the names."

Elements 99 and 100 were actually synthesized in a much more dramatic way than their predecessors: they were found among the products of detonation from the first thermonuclear weapon in history. Samples from the ground contaminated by the event were sent to both Berkeley and to the Argonne National Laboratory near Chicago. One month after the explosion, at the end of 1952, the California team headed by Seaborg was ready to announce the first results. A few days before Christmas, Seaborg gathered his colleagues and drew up a memo in which he reported his version of the facts surrounding the discovery of elements 99 and 100. This created friction between him and his colleagues at Argonne, who had laid claim to the discovery. Later, it became evident that the credit for the discovery would not go only to Seaborg and his co-workers, but also to colleagues at laboratories in competition with them. At first glance, the positions of both teams seemed irreconcilable. In 1955, Seaborg asked Ghiorso to mediate the difficulties between the groups. On that occasion, the team at Los Alamos withdrew their proposal for the name *losalium* (after Los Alamos) for element 99, hitherto strenuously advocated. In fact, while publications relative to the purification and characterization of the two elements proceeded at a sustained pace both at Berkeley and Argonne, at the beginning of 1954, Nobel laureate Manne Siegbahn, president of the Nobel Foundation, sent a letter to Seaborg in which he reminded him—newly a Nobel laureate himself[59]—that element 100 had been first synthesized in Sweden. Siegbahn had a rather aggressive temperament, coupled with the fact that he was not very prudent. In his younger days, he had heaped abuse on the spectroscopic work of Alexandre Dauvillier and later on the Austrian exile Lise Meitner,[60] and in his later years he engaged in a lively controversy with Seaborg.[61] He emphasized how Hugo Atterling and co-workers[62] of the Nobel Institute of Physics in Stockholm had discovered the first isotope of element 100 by bombarding, in their cyclotron, a sample of ^{238}U with projectiles of ^{16}O. Siegbahn exerted strong pressure on the International Commission to accept the name nobelium for element 100, but Seaborg didn't seem to care. In August 1955, on the occasion of the first Atoms for Peace Congress held in Geneva, Seaborg announced that he and his team wanted to call these elements by the names of einsteinium and fermium, respectively. Those present cheered on hearing this impartial decision. News of the wish to honor the memories of Enrico Fermi and Albert Einstein sent a very strong message. They were indeed two American citizens, but they were also born in two countries that, a dozen years earlier, had declared war on the United States. Seaborg and his colleagues could have selected any number of native-born American scientists, but they did not do so, perhaps because of the unassailable prestige of Fermi and Einstein.

In the autumn of 1954, Ghiorso made the decision to honor Enrico Fermi, now mortally ill with stomach cancer, by dedicating the name of the 100th element to him.[63] Before the official approval of einsteinium and fermium, these two elements had many alternative names sustained by many picturesque proposals that came out of the woodwork. The obvious name, *centurium*, for element 100 was given serious consideration. The scientists at Los Alamos seriously considered names that referred to their laboratory: *losalium, losalamium, losalamosium, alamosium, laslium,* or *laslucium*. The scientists at the University of California who had played a decisive role in analyzing the radioactive material following the first thermonuclear explosion proposed the name *uclasium* (acronym of the University of California at Los Angeles, UCLA). The scientists at Argonne claimed the right to give a name to at least one of the new elements and proposed *phoenicium* (perhaps from the Latin *phoeno*, "light"). Scientists at the Materials Test Reactor (MTR) in Idaho also threw themselves into the contest and, believing that they had synthesized element 100 through neutron bombardment before any of their colleagues, proposed the name *arconium*, after the city of Arco, Idaho, where their laboratory is located. They said that to respect the lanthanides'[64] homologous tradition, the 100th element ought to take the name of a city. Following this proposal, many others made their appearance: *ucalium*, by the researchers at the University of California, again *losalium* from Los Alamos, and *anlium* from Argonne National Laboratory (with the acronym ANL from which one can derive the proposed name of *anl*-ium).

The story of the names *athenium* (Z = 99) and *centurium* (Z = 100) is very unusual. They mysteriously appeared in the literature of the 1950s[65] as the result of a sensational misinterpretation of what Luis Alvarez (1911–88) had reported at a conference held at Oxford in 1950. In reality, Alvarez simply limited himself to announcing the possibility of synthesizing elements 99 and 100 by way of certain nuclear reactions (the discoveries came later), but somehow the news got out in a remarkably distorted form to the press. The newspapers reported that he had actually discovered elements 99 and 100, and because they remained nameless, someone coined the names *athenium* and *centurium*, names that were taken up by the Spanish, French, and Russian press. In the same year, in a letter addressed to the editor of *Physical Review*, a peculiar correspondent wrote: "I stated very plainly. . . a new atomic theory which named element 99 ninetynineum, symbol Nn, and element 100 centurium, symbol Ct."[66]

In 1955, Albert Ghiorso, after having announced at the Geneva Conference that he wanted to name these two new elements einsteinium and fermium, published at the same time a very short article on the subject that he then sent to his colleagues at Berkeley, Argonne, and Los Alamos.[67] This act put the official seal on the discovery of elements 99 and 100. The next element was discovered in 1955, by Ghiorso, Bernard Harvey, Gregory

Choppin (b. 1927), Stanley Thompson (1912–76), and Glenn T. Seaborg, who produced only 17 atoms of this element.[68] Shortly after the official announcement of the discovery of element 101, the magazine *Daily Cal* reported a story woven of pure fantasy in which a young man, barely 15 years old, by the name of Leonardo da Vinci, had discovered elements 100 and 101 at the Nuclear Metaphysical Laboratories of the University of California and that he had named them *centium* and *percentium*. This insignificant story, from the scientific point of view, struck the imagination of scientists so much so that Ghiorso made mention of it in his memoirs relative to the discovery of mendelevium.[69] The name mendelevium was pondered for at least a year before its discovery, and it was subsequently conferred on the new element. However, these were the years of the Cold War, and to go fishing for a name for the new element among the Russians seemed, at the very least, to be out of place. Seaborg spoke to Ernest O. Lawrence, a first-rate experimental physicist and Nobel laureate in physics but also a fearful reactionary who ran his research center like a despot. Contrary to every expectation, Lawrence voiced no objection to the idea, and so Seaborg proposed the name mendelevium for element 101. Some time later, at the Atoms for Peace Conference in Geneva, the French chemist Moïse Haïssinsky approached Seaborg and, with a certain show of affection, confided that the choice he had made (that is, to honor a Russian scientist) had done more for international relations than everything that the U.S. secretary of state had managed to do in his entire career!

V.4.2. THE STEP LONGER THAN ITS LEG: NOBELIUM

The synthesis of element 101 brought to light the necessity of finding projectiles heavier than helium if scientists wished to pursue the synthesis of the superheavy elements. In fact, it was taken as a certainty that it was impossible to find a nuclear target with an atomic number greater than 99, so it was necessary to increase the mass of the bombarding nucleus to achieve the desired effect. In 1957, there were three particle accelerators in the whole world capable of accelerating heavy ions. At Berkeley, where U.S. scientists were developing a new instrument; at Moscow, at the Kurchatov Institute; and at Stockholm, at the Nobel Institute for Physics. All three cyclotrons were at work trying to overcome this barrier. The Nobel Institute had constructed a really good accelerator. Its president, Manne Siegbahn, had traveled to the United States to learn as much as he could in the field of particle physics. At great expense, the Nobel Institute was founded during and after World War II. However, its scientific results, despite the great outpouring of funds, were late in being realized; Siegbahn urged his co-workers to "accelerate" time: his team was the first to engage in the difficult synthesis of element 102, and, in 1957, B. Aström[70] and his colleagues announced its discovery. In making the announcement, Aström, with great enthusiasm, allowed "nobelium" to escape his lips as the name of the new element. The name would have been very gratifying to the great Swedish philanthropist and chemist, Alfred Nobel, who was also the founder and benefactor of the annual award that bears his name.

Unfortunately, some years later, the Berkeley scientists showed that the data and related chemical analyses that the Swedish scientists published on the presumed new element did not match their experimental observations.[71] Siegbahn's obstinate determination in holding that the future of physics research lay in the synthesis of the transuranium elements had convinced him to throw himself headlong into a discipline new to him. Although he was starting out in a disadvantageous position, he spurred his colleagues into a race in which his American and Russian competitors were much more expert. Failure was almost inevitable, and the Americans and Russians found themselves in a position of claiming the discovery of the new transuranium element. The instruments at the disposal of physicists and chemists in 1957 were not able to synthesize or analyze an element with an atomic number greater than 101.

It was necessary, both in Russia at Dubna and in the United States at Berkeley and Argonne, to develop a new technology: on the one hand, increasingly powerful accelerators, and on the other, physicochemical instruments capable of analyzing for smaller and smaller amounts. When in 1959 Ghiorso and his colleagues unequivocally clarified the properties of element 102, the scientific community waited for the Swedish research group to retract the discovery of the presumed element and assumed that the name nobelium would disappear from the list of elements in the periodic table. However, the Nobel Institute group, led by Aström and with the tacit approval of Siegbahn, refused to recognize its error. The pressure that Seaborg experienced from Stockholm was very insistent, making note of the fact that he had recently received the Nobel Prize in chemistry and that he had distant Swedish ancestry. If the Swedes were not able to salvage the attribution of the discovery, they at least tried to salvage the name nobelium. A decade passed, and new tests seemed to partially confirm Aström's work of 1957. In 1967, the Americans, with Ghiorso, and the Swedes, with Torbjørn Sikkeland, reached an agreement; they published an article in which they both confirmed that they had no wish to change the name nobelium. Their reasons were twofold: (1) the name was already very much used in the literature, and it would be counterproductive to change it because there were already dozens of articles using the name; and (2) the name was well-recognized and also pleasantly euphonious.

Meanwhile, back in the Soviet Union, after their first erroneous attempts[72] in 1957, the scientists who worked at Dubna under the leadership of Georgy Nikolaevich Flerov from 1963 to 1966, had discovered many new nuclides of element 102 with atomic masses between 251 and 259. For this reason, they arrogated to themselves the discovery of the element. The interval between their first work (1957) and the year that they resumed research (1963) was due to the transfer of the nuclear laboratories (the Kurchatov Institute at Moscow) to Dubna, a village on the banks of the Volga about 100 km from Moscow. The new research center (the United Institutes for Nuclear Research), surrounded by a beautiful birch forest, was the Soviet answer to CERN at Geneva.

The important dates regarding the discovery of nobelium are as follows:

- 1957, September: The Nobel Institute at Stockholm announces the discovery of nobelium;[73]
- 1957, December: A few months after the announcement of the discovery, the group at Berkeley shows that the results arrived at by their European colleagues are in error;[74]
- 1957, December: At Moscow, at the Kurchatov Institute, the first attempt to produce isotopes of element 102 are undertaken, but the data are conflicting;[75]
- 1958: The discovery of the first isotope of 102 is made at Berkeley: $^{254}102$;
- 1959: U.S. scientists synthesize the second isotope, $^{252}102$, and in 1961, the third, $^{257}102$;
- 1964: The Soviet group synthesizes isotopes $^{256}102$, $^{255}102$, and $^{253}102$; they show that the first isotope discovered by the U.S. scientists ($^{254}102$) is unreliable;

- 1967: The Americans and Swedes come to an agreement to keep the name nobelium for 102;
- 1997: The IUPAC Commission confirms the name and symbol (No) for nobelium.

The events summarized above are narrated in great detail in a 1992 review article by G. N. Flerov et al. in *Radiochimica Acta* entitled "A History and Analysis of the Discovery of Element 102." This note also contains a response from the Berkeley group.[76]

V.4.3. CHAOS SURROUNDS LAWRENCIUM, RUTHERFORDIUM, DUBNIUM, AND SEABORGIUM

The history of lawrencium lacks the regrettable episodes we have just witnessed, nor is it surrounded by the controversies that would plague the discoveries of rutherfordium, dubnium, and seaborgium. The fact that lawrencium has only changed its symbol once over the course of the years while those near it in the periodic table have changed names and discoverers, certainly must make one smile at its rather uneventful journey. Prepared for the first time in 1961 by Albert Ghiorso, Torbjørn Sikkeland, Almon E. Larch, and R. M. Latimer at the Berkeley Laboratory of the University of California, the isotopes of lawrencium were created by bombarding a californium target with boron ions.[77] For element 103, Ghiorso and his colleagues suggested the name lawrencium with the chemical symbol Lw, which subsequently was changed to Lr, in honor of E. O. Lawrence.[78] The name lawrencium and its symbol (Lr) were ratified[79] by the IUPAC Commission during its meeting at Geneva in August 1997.

The story of the discovery of elements 104, 105, and 106, that later took the names of rutherfordium (Rf), dubnium (Db), and seaborgium (Sg), respectively, is rightfully recalled as encompassing the greatest controversy over elemental discoveries ever recorded, one that makes the controversy over *celtium* and hafnium pale by comparison. The extremely long life of the controversy (1960–97) was due to the principal personage of the Russian faction, Georgy N. Flerov, who died before the IUPAC Commission officially made a decision on the names of these elements.

According to American sources, element 104 was discovered by a group headed by Ghiorso[80] during experiments made in 1969 and 1970. On that occasion, the isotopes with masses of 257, 259, and 261 were synthesized. In 1974, an American ad hoc committee[81] rejected the discovery of ^{260}104 by Yuri Oganessian and G. N. Flerov dating back to 1964, claiming that the Russians had erroneously interpreted their experimental data. In this way, the Americans opened the way for the recognition of the discovery of 104 by the California team. The Russians did not willingly accept the American pronouncement, repeating on various occasions that the American commission was too partisan. The Dubna group proposed the name *kurchatovium*[82] for the new element, with its associated symbol of Ku, in honor of Igor Vasilyevich Kurchatov (1903–60). Even to the present, they tend to maintain that rutherfordium (named in honor of Ernest Rutherford) was synthesized for the first time in 1964, at Dubna.[83] The research team bombarded plutonium with neon ions accelerated to 133–115 MeV and maintained that they had found traces of nuclear fission on a special type of glass using a modified microscope. In 1969, the Berkeley group synthesized the element by subjecting californium-249 and carbon-12 to high-energy collisions. The group also reported that they had not been able to reproduce the method used by the Soviets. This fact led to a controversy regarding the element's name. Because the Soviets asserted that they had synthesized the element at Dubna, they proposed the name *kurchatovium* (Ku); on the other hand, the U.S. scientists proposed the name rutherfordium (Rf) in honor of the famous New Zealand physicist. Both IUPAC and the International Union of Pure and Applied Physics (IUPAP) temporarily adopted the name unnilquadium (Unq) until, in 1997, the dispute was resolved with the adoption of the name rutherfordium.

Dubnium, according to reports in the literature, was synthesized in 1967 at the United Institute for Nuclear Research in Dubna, Russia. It was produced in the form of two isotopes, $^{260}105$ and $^{261}105$, starting with the bombardment of ^{243}Am with ^{22}Ne.[84] Toward the end of April 1970, U.S. scientists at the University of California under the leadership of Ghiorso also identified element 105. The American team accomplished this by bombarding a target of ^{249}Cf with a beam of nitrogen nuclei. They used a linear accelerator that allowed them to produce $^{260}105$ with a half-life of 1.6 seconds.[85] Atoms of element 105 were conclusively identified on March 5, 1970, although some experimental evidence suggested that they actually had produced the element 1 year earlier, during instrument testing at Berkeley. The Berkeley scientists subsequently failed to confirm the Soviet results using the latter's methods. Consequently, Ghiorso and his co-workers proposed that the element be named *hahnium* (symbol Ha) in honor of the German chemist Otto Hahn. Subsequently, this name became so widespread among American and European scientists that American scientific journals seemed to ignore the IUPAC decision to call element 105 dubnium.

Element 106 was synthesized by Ghiorso's group in 1974,[86] but it had to wait another 20 years before it was officially called seaborgium.[87] It was only a few days later that a similar paper by Yuri Oganessian appeared, claiming the same discovery.[88] Both teams claimed the same discovery, but they also indicated that they had no wish to raise a controversy similar to the one that tainted the discovery of the preceding two elements. Because their methods of production of element 106 were substantially different, they decided that if both groups were correct, they would decide jointly on a name for the element. Ghiorso, at this point, dragged out a story which, if it had been accepted, would have put his Russian competitors in great difficulty. In 1971, a good 3 years before the discovery of the disputed element 106, Ghiorso's team, in its attempt to synthesize element 105, ran unexpectedly into element 106. They did not understand the discovery at the time, but a review of the experimental plans spoke very well in their favor. As Ghiorso expressed it: "Wow! Do you mean that we found element 106 on January 24, 1971 and didn't report it?"[89]

Understandably, the Russians were not pleased. They claimed that it was a plot to discredit them, and they refused to accept Ghiorso's explanation. They noted that even if the experiment was correct and element 106 had actually been discovered, the fact remained that the California team did not understand that they had made the discovery. So, they were back to square one in their relationship, and the stress levels began to mount dangerously in the two groups. The tension due to the Cold War was getting more acute, and the rivalry between the two research groups made them feel it even more. In June 1974, Flerov and some of his co-workers visited the United States to meet with their California colleagues. They saw the instrumentation for the synthesis of new superheavy elements and, according to some present, Flerov was very impressed by the many advanced techniques available to the Americans. However, neither group had the others' apparatus and therefore no one was able to repeat the others' tests to determine if they were correct. Thus, the two teams were held to be the legitimate discoverers of element 106 until, in 1984, the Russian physicist A. G. Demin published a note in which, very timidly, he criticized the 1974 experiment conducted by Flerov and his co-workers. In short, Demin took issue with the form of Flerov's article, not with its substance. Flerov had written that he had observed the spontaneous fission of at least two nuclei of element 106; in realty, Demin asserted that the Dubna group had observed the fission of element 104 after radioactive decay. In substance, the Russians had synthesized element 106, but they had not observed it. What they did observe was an α particle and element 104, which very shortly thereafter underwent nuclear fission. The fragments of element 106 had been observed (the α particle and element 104), but the Russians preferred to pass over these facts, leaving the Americans, who were apprised of the news, to wage a smear campaign against them. A few years later, the Germans also entered the lists and rechecked both American and Russian results, finding even more "wormholes" in the work of the latter.

At this point, the Americans were the only champions in the field, and Ghiorso began to seriously consider naming element 106. According to his judgment, the most appropriate name was *alvarezium,* in honor of the great physicist Luis Walter Alvarez who had strongly advocated for the development and understanding of many nuclear phenomena.[90] However, many of Ghiorso's co-workers did not agree with this proposal and suggested *joliotium* after Frédéric Joliot, the son-in-law of Marie Curie. *Joliotium* would have been a compromise name. The Soviets had proposed it for element 102, at the time of the regrettable incident over the false discovery of nobelium by the Swedes, and this would have paid them back for the frustration they experienced by having the privilege of naming it snatched away from them. There were many other names proposed: *newtonium* (after Sir Isaac Newton), *edisonium* (after Thomas Edison), *davincium* or *vincium* (after Leonardo da Vinci), *columbium* (after Christopher Columbus), *magellanium* (after Ferdinand Magellan), *ulyssium* (after Ulysses), *washingtonium* (after George Washington), *kapitzium* (after Peter Kapitza), *sacharovium* (after Andrei Sacharov), and *finlandium* (after Finland). The scientists did not succeed in resolving the dilemma but only in localizing it: from a conflict between the U.S. and Russia, it was reduced to an internal controversy. Table V.3 summarizes the names finally recommended for elements 101–109.

When in 1994 the IUPAC Commission confirmed and attributed the discovery to Ghiorso's group, he received a telephone call from a reporter on the staff of the *New York Times* who, after congratulating him, completely surprised him by asking: "What are you going to name element 106—ghiorsium?" This was in no way a new idea. Ghiorso had participated in the discovery of many elements and, in 1957, Glenn T. Seaborg, on the occasion of the ACS Division of Nuclear Chemistry pre-Christmas dinner, had given him a bottle of wine with a label that read: "A weightless sample of ^{299}Gr, Ghiorsium." Ghiorso laughed both at the remembrance of Seaborg's joke and at the reporter's remark, but it made him think: why not call the element *seaborgium*? After all, the names of elements 99 and 100, einsteinium and fermium, were proposed while Einstein and Fermi were still living. Ghiorso felt that seaborgium would be the most appropriate name, and he proposed just that to Seaborg himself. At first, Seaborg seemed somewhat undecided, but in the end, he acceded to the idea.

Meanwhile, the ad hoc committee created in 1974 to resolve the controversies arising over the naming of elements 104 and 105 never met again, and its highly criticized

Table V.3 The recommended names for elements 101–109 reached by the Joint Commission of the International Union of Pure and Applied Chemistry (IUPAC) and the International Union of Pure and Applied Physics (IUPAP) in 1994

Element	*Name*	*Symbol*
101	Mendelevium	Md
102	Nobelium	No
103	Lawrencium	Lr
104	Dubnium	Db
105	*Joliotium*	Jl
106	Rutherfordium	Rf
107	Bohrium	Bh
108	*Hahnium*	Ha
109	Meitnerium	Mt

decisions were never acted upon. It was clearly better to postpone any action until a more opportune time. This American "creature" survived until 1984, when it became clear that its functioning could no longer benefit the situation: in the succeeding years, the discoveries of elements 106 through 109 suffered from the same problems of attribution as had the discoveries of elements 104 and 105.

In the 1970s, 1980s, and 1990s, each research group continued to use its own adopted elemental names: rutherfordium (Rf) and *hahnium* (Ha) at Berkeley, and *kurchatovium* (Ku) and *nielsbohrium* (Nb) at Dubna. After the discovery of elements 107, 108, and 109 at the GSI in Hamburg, Germany, Peter Armbruster suggested that all the discoveries be reconsidered on a more solid basis and by means of their isotopic identification.[91] The response to Armbruster's suggestion was the creation of the Transfermium Working Group (TWG), but the controversies relative to the naming of the elements, far from dying down, only grew more contentious. Geoffrey Wilkinson (1921–96), Nobel Laureate in Chemistry in 1973, placed in the unhappy position of international arbitration supervisor, proposed to the American and German scientists that if the symbol of Kt (as a compromise on the dispute over kurchatovium) would be satisfactory to them, it would be understood that the attribution of the discovery could go to the Russians. The response of both groups was decidedly negative.

On October 24, 1990, Oganessian visited Seaborg and Ghiorso at Berkeley in an attempt to reach an accord on the names of elements 102–106. Oganessian prudently recognized the priority of the Americans in the discovery of elements 102 and 103, but he firmly defended the name *kurchatovium* for element 104. Ghiorso was open to this possibility, but Seaborg was adamantly opposed, asserting that the Russians had not in fact discovered element 104 and therefore should not have the right to name it. Another meeting took place the following afternoon, attended also by Darleane C. Hoffman (b. 1926), and an agreement seemed to have been reached. They also discussed the naming of element 106, and Oganessian suggested that it might be called *flerovium* in honor of Georgy Nikolaevich Flerov,[92] the Russian physicist who had long been head of the Russian laboratory for the synthesis of the transuranium elements. Yet, once again, this difficult agreement vanished. At the New York meeting of the ACS in August 1991, he incorrectly interpreted the agreement that had been reached with Oganessian the year before and said that *kurchatovium* would be taken into consideration as the name of element 106 and that all the other elements would be named following the U.S. suggestions. Oganessian reacted very badly to what seemed to him an about-face on Seaborg's part and, on August 10, 1992, in revenge for Seaborg's insult, he met with Peter Armbruster to agree on what names to give the other elements from 102 to 109.[93] They decided on the following list:

Z = 102, *joliotium* (Jt)
Z = 103, lawrencium (Lr)
Z = 104, meitnerium (Mt)
Z = 105, *kurchatovium* (Ku)
Z = 106, rutherfordium (Rf)
Z = 107, *nielsbohrium* (Ns)
Z = 108, hassium (Hs)
Z = 109, *hahnium* (Ha)

Oganessian asserted that because his group had not received any credit for the discovery of element 107, Armbruster ought to have given him the privilege of bestowing a name of his liking to honor the discovery of the technique called "cold fusion" that had permitted the discovery of elements with Zs of greater than 106. As a compromise, he withdrew the name of *flerovium* as a candidate.

The day after this announcement, Ghiorso and Seaborg wrote to Armbruster deploring his decision to draw up a list of names with the Russians: they did not want any element named after the father of the Soviet atomic bomb (i.e., Kurchatov). They recognized that the Russians were the only ones with the right to name element 105, and they pushed the Soviets to name it *gamowium*, *goldanskium*, or *landauvium* for the names of three great Russian physicists, respectively: George Gamow (1904–68), Vitalii Iosifovich Goldanskii (1923–2001), and Lev Davidovich Landau (1908–68).

In August 1994, the IUPAC Commission meeting at Antwerp essentially matched the Russo-German idea for the names of elements 101–109, although they inverted some of them and also changed the spelling of the symbols for lawrencium (Lr reratified in 1994) and *joliotium*. The Commission also asserted that the choice of names would be definitive and that there would be no appeal.[94]

In November of the same year, the Committee on Chemical Nomenclature, expressly created by the ACS), rejected the international decision and, on the strength of the fact that almost half of the publications in the field were the exclusive property of the ACS, proposed alternative names. Table V.4 summarizes the recommendations of the Antwerp meeting.

The Russians, via the then-president of IUPAC, Karol I. Zamarev, protested vehemently and requested a new and urgent international meeting, which was held in Guilford, England, in August 1995. On this occasion, the participants looked for a new compromise that was reached by sacrificing the name nobelium, which everybody thought was an erroneous Swedish discovery and that the Americans did not seem to want to risk very much to save. In addition, the commission removed rutherfordium from the list and replaced it with seaborgium. An outcry arose among English journalists who criticized the arrogance of the ACS, which had imposed by force the name seaborgium in defiance of the ban on endowing an element with the name of a living person. Table V.5 lists the 1995 Guilford scheme.

Table V.4 The International Union of Pure and Applied Chemistry (IUPAC) recommendations given at the Antwerp meeting, August 10–11, 1994.

Element	*Name*	*Symbol*
101	Mendelevium	Md
102	Flerovium	Fl
103	Lawrencium	Lr
104	Dubnium	Db
105	*Joliotium*	Jl
106	Seaborgium	Sg
107	*Nielsbohrium*	Ns
108	*Hahnium*	Ha
109	Meitnerium	Mt

The committee made some changes to the Russo-German choice of names.

Table V.5 The 1995 Guilford (UK) scheme

Element	*Name*	*Symbol*
104	Rutherfordium	Rf
105	*Hahnium*	Ha
106	Seaborgium	Sg
107	*Nielsbohrium*	Ns
108	Hassium	Hs
109	Meitnerium	Mt

Nevertheless, the Americans were less than satisfied and, with the not inconsequential support of the Chinese and Japanese chemical and nuclear societies, requested that the controversy be reopened. A new meeting of the IUPAC Commission for the naming of the elements was held in Geneva in August 1997. The Commission published a new table of names for elements 101–109. The major consequence was a notable reduction in the Russian proposals, with the cancellation of the names *flerovium* and *joliotium*, and with the substitution in their place of rutherfordium and nobelium.[95]

Many other superheavy elements were discovered in the meantime, but for none of these did any faction feel they had to resort to subterfuge or gross international blackmail to impose a preferred name.

Notes

51. Procopio, M. *Chimica nell'Industria* **1939**, *15*, 803.
52. Guerra, F.; Robotti, N. *Atti dell' XI Convegno Nazionale di Storia e Fondamenti della Chimica, Accademia Nazionale delle Scienze detta dei XL*, 2005, 295.
53. Segrè, E. G.; Corson, D. R.; Mackenzie, K. R. *Nature* **1947**, *159*, 24.
54. McMillan, E. M.; Abelson, P. H. *Phys. Rev.* **1940**, *57*, 1185.
55. Seaborg, G. T.; McMillan, E. M.; Kennedy, J. W.; Wahl, A. C. *Phys. Rev.* **1946**, *69*, 366.

56. The first element following uranium was named after the first planet following Uranus, Neptune. The name of this planet was derived in its turn from the Roman god of the sea, Neptune.
57. Thompson, S. G.; Ghiorso, A.; Seaborg, G. T. *Phys. Rev.* **1950**, *77*, 838; Thompson, S. G., Street K. A. Jr.; Ghiorso, A.; Seaborg, G. T. *Phys. Rev.* **1950**, *78*, 298.
58. Znoyko, A. P.; Semishin, V. I. *Novaya Seriya* **1950**, *5*, 917.
59. Glenn T. Seaborg was of Swedish descent and only 3 years earlier had received the Nobel Prize.
60. Friedman, R. M. "Remembering Miss Meitner," in Shepherd-Barr, Kirsten (ed.), *Science on Stage: From "Dr. Faustus" to "Copenhagen";* Princeton University Press: Princeton, 2006.
61. Siegbahn, a master at designing experiments, had achieved a dominant position in the field of physics in Sweden. He had contributed to the perfection of X-ray spectroscopy, providing increasingly exact measurements on the behavior of electrons. These measurements were found to be essential in the development of the new quantum physics. The 1924 Nobel Prize in Physics went to him largely due to the pressure exerted by those colleagues of his from the University of Uppsala who were members of the narrowly constituted Nobel Committee. None, or nearly none of his international physics colleagues regarded him to be at the level of the prize, and furthermore, some members of the Nobel Committee had observed that Siegbahn's candidacy was in conflict with the rules of the Prize. He had not made any significant discovery, and he had not invented or designed a new instrument. The Rockefeller Foundation, to which he had applied for a significant amount of funding during World War II, observed that Siegbahn's work was excellent with respect to exactitude and precision, but it was very limited with respect to new ideas. Soon after the war, however, the Swedish government financed the setting up of the Nobel Institute for Experimental Physics, and Manne Siegbahn was named director of the new institute.
62. Atterling, H.; Forsling, W.; Holm, L. W.; Melander, L.; Åström B. *Phys. Rev.* **1954**, *95*, 585.
63. Segré, E. *Enrico Fermi, Fisico. Una biografia scientifica*, 2nd Ed.; Zanichelli: Bologna, 1987.
64. The homologue of element 100 is erbium, whose name is derived from the Swedish town of Ytterby.
65. Acera, L. H. *Met. y elec.* (Spain) **1951**, *15* (no. 164), 36.
66. Thompson, S. G. et al. *Phys. Rev.* **1950**, *77*, 838–39.
67. Ghiorso, A.; Thompson, S. G.; Higgins, G. H.; Seaborg, G. T.; Studier, M. H.; Fields, P. R.; Fried, S. M.; Diamond, H.; Mech, J. F.; Pyle, G. L.; Huizenga, J. R.; Hirsch, A.; Manning, W. M.; Browne, C. I.; Smith, H. L.; Spence, R. W. *Phys. Rev.* **1955**, *99*, 1048.
68. Ghiorso, A.; Harvey, B.; Choppin, G.; Thompson, S.; Seaborg, G. T. *Phys. Rev.* **1955**, *98*, 1518.
69. Ghiorso, A. In *Adventures in Experimental Physics*, Maglich, B., Ed.; World Science Communications: Princeton, NJ, 1972, *2*, p. 245.
70. Fields, P. R.; Friedman, A. M.; Milsted, J.; Atterling, H.; Forsling, W. Holm, L. W.; Åström, B. *Phys. Rev.* **1957**, *107*, 1460.
71. Maly, J., et al. *Science* **1968**, *160*, 1114.
72. Flerov, G. N. et al. English translation in *Sov. Phys. Dokl.* **1958**, *3*, 546.
73. Fields, P. R. et al. *Phys. Rev.* **1957**, *107*, 1460.
74. Ghiorso, A. et al. *Phys. Rev. Lett.* **1958**, *1*, 18. This work was performed in late 1957, but the publication was not received nor issued until mid-1958.
75. Flerov, G. N. et al. *Dokl. Akad. Nauk SSSR* **1958**, *120*. 73; Flerov, G. A. et al. *JETP* **1960**, *38*, 82.
76. Flerov, G. N. et al. *Radiochimica Acta* **1992**, *56*, 111-124; Ghiorso, A.; Seaborg, G. T. *Radiochimica Acta* **1992**, *56*, 125.
77. Ghiorso, A. et al. *Phys. Rev. Lett.* **1967**, *18*, 401.
78. Ernest Orlando Lawrence, American physicist. His invention of the cyclotron opened the way to the production of artificial radioisotopes. He was professor of physics at the University of California at Berkeley and, from 1930 to 1936, he was the director of the radiation laboratory that developed into a great nuclear physics research center. He received the Nobel Prize in Physics in 1939.
79. Lw was changed to Lr in 1963. The symbol was ratified by the IUPAC in 1997.

80. Ghiorso, A. et al. *Phys. Rev. Lett.* **1969**, *22*, 1317; Ghiorso, A. et al. *Phys. Lett.* **1970**, *32B*, 95.
81. Hyde, E. K.; Hoffman, D. C.; Keller, Jr. O. L. *Radiochimica Acta* **1987**, *42*, 57.
82. Igor Vasilevich Kurchatov was head of Soviet nuclear research and father of the first Russian thermonuclear bomb.
83. Flerov, G. N. et al. *At. Energ.* **1964**, *17*, 310; Flerov, G. N. et al. *Phys. Lett.* **1964**, *13*, 73.
84. Flerov, G. N. et al. *Joint Institute of Nuclear Research* 1968, Preprint P7-3808, Dubna.
85. Ghiorso, A. et al. *Phys. Rev. Lett.* **1970**, *24*, 1498.
86. Ghiorso, A. et al. *Phys. Rev. Lett.* **1974**, *33*, 1490.
87. Three years after the assignment of this name, the IUPAC and IUPAP Commissions decided to change it and call it seaborgium.
88. Oganessian, Yu. T. et al. *JEPT Lett.* **1974**, *20*, 265.
89. Hoffman, D.; Ghiorso, A.; Seaborg, G. T. *Transuranium People: The Inside Story;* Imperial College Press: London, 2000, p. 309.
90. The American physicist Luis Alvarez was a pioneer in the construction of bubble chambers. He directed the construction of the first modern proton accelerator. He won the Nobel Prize in 1968.
91. Armbruster, P. *Ann. Rev. Nucl. Part. Sci.* **1985**, *35*, 235.
92. Georgy Nikolaevich Flerov, terminally ill, passed away after a short illness on November 20, 1990, at the age of 77.
93. Seaborg, G. T. *Personal Journal*, August 10, 1992, as cited in Hoffman, D. C., Ghiorso, A., Seaborg, G. T. *The Transuranium People: The Inside Story*, Imperial College Press: London, 2000, p. 384.
94. Sargeson, A. M. et al. *Pure & Applied Chem.* **1994**, *66*, 2419.
95. Inorganic Chemistry Division Commission on Nomenclature of Inorganic Chemistry, *Pure & Applied Chem.* **1997**, *69*, 2471.

V.5

DO WE HAVE TO LIVE WITH FANTASY? HAWKINGIUM AND ZUNZENIUM

The idea of placing an "upper limit on the atomic number" was documented for the first time in 1936, in the work of French physicist Georges Fournier (1881–1954), who was very active between the two world wars. After having begun his career in 1923, working beside Irène Joliot-Curie studying the γ-ray emission of radium-D and radium-E,[96] in the 1930s, Fournier began to work with great interest on the classification of atomic nuclei. He proposed a theory about their origin related to their radioactive disintegration. After the discovery of the neutron[97,98] by James Chadwick (1891–1974), Fournier developed a personal theory based on the geometry of α particles, protons, and neutrons.[99] He suggested that both neutrons and protons should be considered like tetrahedra and α particles like octahedra. Any atomic nucleus could be made up from an assemblage of these particles, and Fournier carried his theory to the extreme. Using it,[100] he showed that the highest achievable atomic number was $Z = 137$, and for atomic weight, the result he arrived at was $A = 360$. He said that he arrived at the same results when starting from the relativistic approach proposed by Niels Bohr, following the theory of Paul A. M. Dirac (1902–84). Unfortunately, among the many things that disappeared in the wake of World War II was Fournier's curious theory about the extreme limit of atomic dimensions.

In 1972, the dilemma about where to place the last box in the periodic table was taken up again by Professor Tang Wah Kow of the New Method College in Hong Kong.[101] He proposed a very bizarre form for the periodic table, one simultaneously octagonal and prismatic. Up to this point, there would not have been anything particularly confusing or innovative about his ideas: many other chemists and physicists before him had been involved in "acrobatic speculations" about the form of the periodic table or the positioning of the elements inside it. However, Kow went further by introducing a complex network of definite laws: the rule of series, the rule of triads, and the rule of octaves, whose names must have elicited a profound déjà-vu feeling among his readers. At the conclusion of his article, Kow listed the three brief consequences that would follow from acceptance of his system. First, he emphasized that the law of triads would be best illustrated by basing it on the nucleonic configuration of the elements and not only on their electronic configuration. As a corollary to this law, Kow proposed that all atomic nuclei be classified into eight nucleonic typologies. Finally, for his third point, Kow ventured that because the elements with $Z = 244$ and $Z = 245$ had no place in his periodic table, and seeing that they would have run into the limits imposed by the law of triads, no element could have an atomic number greater than 243. Kow did not bother to explain how his law questioned the validity of the quantum theory, but, on the contrary, he dedicated the last paragraph of his article to justifying the name and symbol of the 243rd element in this way: "there is a suggestion offered to the supposable future founder of element 243 that it may be called zunzenium (symbol Zz). It is deduced from Chinese idiom. The name stands behind Zun Zen, who (Zun Zen) comes last on the list of successful candidates in a royal examination."

Obviously, after this inconsequential publication, nothing more was heard of the fanciful *zunzenium*. Perhaps we should wait for the synthesis of element 243 to see if the candidacy of this improbable element is accepted, but, fortunately, this is not likely to happen in the foreseeable future.

The last scientist (at the time of writing) to develop a theory capable of predicting some of the properties of the as-yet-unknown transuranium elements is an all-but-unknown Macedonian physicist. In February 2004, Petar K. Anastasovski of the Faculty of Physics at the University of Saints Cyril and Methodius in Skopje attended a convention of the American Institute of Physics (AIP) in Albuquerque, New Mexico. The subjects discussed at this conference were rather special: thermophysics and microgravity, space travel for civilian and commercial purposes, propulsion in space using nuclear fuel, space exploration, and human colonization of other heavenly bodies. Anastasovski participated in the thermophysics and microgravity session as a theoretical physicist. A passionate student of the most recent, but least orthodox, theories in physics, he carried with him some rather eclectic cultural baggage. He began his career in the 1970s with the practical problem of eliminating interference signals in the helium-neon laser,[102] but, with the passing of the years, he turned his scientific curiosity to other areas of physics. In 2001, he was very interested in superluminal (speed faster than light) theory and its effects.[103] A year earlier, he published a paper at the Swedish Royal Academy of Sciences on how to extract energy from a vacuum.[104]

But it was without a doubt his last work, published in the *Acts* of the conference, that could be marked as the most controversial of all of his scientific ideas. In the article put together from Anastasovski's oral communication delivered on February 11, 2004, the author conjoined two apparently disparate areas of physics: antigravity and the superheavy elements.[105] In conformity with the spirit of the conference, Anastasovski opened his discourse by referring to the concept of propulsion. In a few sentences, he changed direction and began to speak of gravity and antigravity, maintaining that the essence of any concept of propulsion lay in overcoming gravity. Antigravity, he said, would be the most natural means of accomplishing this goal. Therefore, the technology that exploited antigravity through the use of the superheavy elements would be the first to supply the world with a new method of propulsion. According to him, the theory of superluminal relativity furnished a hypothesis on the existence of elements with atomic numbers up to $Z = 145$, and this indicated that some of these atomic nuclei could have antigravitational properties.

Anastasovski reaffirmed the existence of the space–time curve; he showed that gravitational and antigravitational properties acted not only around the nuclei, but also inside them. He extracted from the theory the idea that two groups of elements (the first with $Z < 64$ and the second with $63 < Z < 145$) seemed to have these properties. The nuclei belonging to the first group of elements had masses that allowed for only gravitational properties and therefore would in no way be useful for his purposes. On the other hand, the nuclei of the elements in the second group seemed to have masses suitable for both gravitational and antigravitational properties.

Drawing always on his antigravity theory, Anastasovski ascertained the properties of the heaviest element belonging to the second group ($Z = 145$). This hypothetical element would be the only one, of all the elements taken into consideration, with antigravitational properties. At the end of this paper, Anastasovski suggested that this element be called *hawkingium* in honor of the renowned English physicist and cosmologist, Stephen W. Hawking.[106]

In conclusion, it would be useful to ponder the following question: if one day some scientists actually succeeded in synthesizing an element with the atomic number 137, or 145, or even 243, would they willingly give up the right to propose a name that they liked?

Notes

96. Curie, I.; Fournier, G. *Compt. Rend.* **1923**, *176*, 1301.
97. Fournier was a very talented physicist whose name, unjustly, has been forgotten. He had intuited the existence of the neutron 2 years before Chadwick and published an article to that effect: see *Journal de Physique*, **1930**, *1*, 194
98. Chadwick, J. *Proc. Roy. Soc.* A **1932**, *136*, 692.
99. Fournier, G. *Compt. Rend.* **1936**, *203*, 1138.
100. Fournier, G. *Compt. Rend.* **1936**, *203*, 1495.
101. Kow, T. W. *J. Chem. Educ.* **1972**, *49*, 59.
102. Anastasovski, P.; Pop-Janev, S. *Optical and Quantum Electronics* **1975**, *7*(4), 331.
103. Anastasovski, P. K.; Hamilton, D. B. *Advances in Chemical Physics* **2001**, *119* (Pt. 3, Modern Nonlinear Optics, 2nd Edition), 655.
104. Anastasovski, P. K.; Bearden, T. E.; Ciubotariu, C.; Coffey, W. T.; Crowell, L. B.; Evans, G. J.; Evans, M. W.; Flower, R.; Jeffers, S.; Labounsky, A.; Lehnert, B.; Meszaros, M.; Molnar, P. R.; Vigier, J. P.; Roy, S. *Physica Scripta* **2000**, *61*(5), 513.
105. Anastasovski, P. K. *AIP Conference Proceedings* **2004**, *699* (Space Technology and Applications International Forum—STAIF 2004), 1230.
106. Stephen Hawking is considered one of the greatest theoretical physicists now living. He is the Lucasian professor of mathematics at Cambridge.

V.6

NAMING THE LAST FIVE ARRIVALS IN THE GREAT "FAMILY OF THE TRANSURANIUM ELEMENTS"

In recent years, the IUPAC Commission officially assigned names to elements 110, 111, and 112: respectively, darmstadtium (Ds), roentgenium (Rg), and copernicium (Cn). And on March 30, 2012, it officially approved the name flerovium (Fl) for element 114 and livermorium (Lv) for element 116. The pathway leading to these two approvals is dealt with later in this chapter.

Darmstadtium, synthesized for the first time in 1994 by Sigurd Hofmann, Viktor Ninov, Fritz Peter Heßberger, Peter Armbruster, H. Folger, Gottfried Münzenberg (b. 1940), H. J. Schött (Gesellschaft für Schwerionenforschung, Darmstadt, Germany), A. G. Popeko, A. Vladimirovich Yeremin, A. N. Andreyev (Flerov Laboratory for Nuclear Reactions, Dubna, Russia), S. Saro, R. Janik (Univerzita Komenského, Bratislava, Slovakia), and M. Leino (b. 1949) (University of Jyväskylän, Finland), first had the systematic IUPAC name of ununnilium (symbol Unn).

In discovering darmstadtium, the team headed by the Germans Hofmann and Armbruster had observed a cascade of nuclear reactions arising from the fragmentation of isotope $^{269}110$. In their reactor they had created the new superactinide according to the reaction,

$$^{62}\mathrm{Ni} + {}^{208}\mathrm{Pb} \rightarrow {}^{269}110 + \mathrm{n} \quad \text{(Eq. V.19)}$$

during which nickel ions, suitably accelerated, were directed at a lead target. The Darmstadt physicists observed a chain of four α decays. An accurate study of the daughter elements, from $Z = 108$ to $Z = 102$, allowed them to assign the mass to the element 110 thus created.[107] Six years later, the IUPAC–IUPAP Joint Working Party (JWP) confirmed the discovery and recognized the priority of the German-Russian-Slovak-Finnish team.[108] In January 2003, the JWP released a communication highly recommending that the scientific community adopt the name darmstadtium with the symbol Ds. The commission used the utmost caution in making this recommendation because its assignment of names to elements 103–109 in 1997 raised a veritable wasps' nest of protests. Among the reasons the commission gave for this recommendation was the fact that there was already a solid tradition for deriving an element name from the city in which it had been synthesized or discovered. Some examples are elements 67 (holmium for Stockholm) and 71 (lutetium for Paris).

The team headed by Hofmann had discovered elements 108, 109, 110, 111, and 112. Number 108, hassium, was named after Hesse, the German region where Darmstadt is located. Number 109 was named after Lise Meitner, the great Austrian physicist who, unfortunately, was ignored on account of racial persecution. (At the end of 1939, correctly interpreting the experimental results of her German colleague, Otto Hahn, she discovered uranium nuclear fission.) The stories of the discovery of the other two "Darmstadt" elements follow.

Element 111, roentgenium, was also discovered by the Hofmann team on December 8, 1994. Produced by the technique of cold fusion (nuclear fusion at low energy) between nickel ions and a bismuth target in a linear accelerator, only three atoms of $^{272}111$ were observed. The reaction was as follows:

$$^{209}\mathrm{Bi} + {}^{64}\mathrm{Ni} \rightarrow {}^{272}111 + \mathrm{n} \qquad \text{(Eq. V.20)}$$

In 2001, the JWP felt there was insufficient evidence to confirm the discovery, but 2 years later, after the Darmstadt (GSI) team had repeated the experiment and collected a few more atoms, the commission awarded them the discovery. The GSI group proposed the name roentgenium (symbol Rg) in honor of the German physicist who had discovered X-rays, and this name was accepted as permanent on November 1, 2004.[109]

The GSI team first created copernicium on February 9, 1996, by firing accelerated zinc-70 nuclei at a target of lead-208. A single atom of element 112 was produced by the reaction:

$$^{70}\mathrm{Zn} + {}^{208}\mathrm{Pb} \rightarrow {}^{277}112 + \mathrm{n} \qquad \text{(Eq. V.21)}$$

In May 2000, the GSI successfully repeated the experiment to synthesize a further atom of copernicium-277. This reaction was repeated at RIKEN in 2004 to synthesize two further atoms and confirm the decay data reported by the GSI team. However, the JWP found still insufficient evidence to support the claim, relating mainly to contradictory decay data for two isotopes of rutherfordium, which was subsequently cleared up so that, in May 2009, the GSI team was officially recognized as its discoverers. The IUPAC then asked the discovery team to propose a permanent name for element 112, heretofore referred to as *ununbium*; on July 14, 2009, they proposed copernicium with the element symbol Cp in honor of Nicolaus Copernicus, the great Polish scientist who literally turned our worldview inside-out. On February 10, 2010, on Copernicus's 573rd birthday, the name was officially recognized, with the change of symbol from Cp to Cn because of previous use of Cp for *cassiopeium*, now known as lutetium, as well as its use to abbreviate the cyclopentadienyl ligand.[110]

A summary of the proposals and outcomes for all the elements from 103 to 112 is given in Table V.6.

In the following years, the discoveries of numerous other superheavy elements have been reported with atomic numbers 113, 114, 115, 116, 118, and even 122. The naming of these elements may become even more complicated than those already dealt with earlier in this part of the book. In fact, the names proposed by the discoverers of element 113 had been given by S. N. Dmitriev of the laboratory at Dubna and by Kenji Morita from the group working at RIKEN in Japan. The JWP has not yet made a decision regarding 113, although two names already exist: *japonium* (symbol Jp), after the country of discovery, and *rikenium* (symbol Rk), after the RIKEN Institute in Japan.[111] As of 2011, the IUPAC

Table V.6 Summary of various proposals for elements 103–112 and the final International Union of Pure and Applied Chemistry (IUPAC) decisions

Z	*Systematic IUPAC designation*	*American proposals*	*Russian proposals*	*German proposals*	*IUPAC recommendations 1994*	*IUPAC 1997–2010 and present names*
103	Unt unniltrium	Lw lawrencium	–	–	Lr lawrencium	Lr lawrencium
104	Unq unnilquadium	Rf rutherfordium	Ku kurchatovium	–	Db dubnium	Rf rutherfordium
105	Unp unnilpentium	Ha hahnium	Ns nielsbohrium	–	Jl joliotium	Db dubnium
106	Unh unnilhexium	Sg seaborgium	–	–	Rf rutherfordium	Sg seaborgium
107	Uns unnilseptium	Ns nielsbohrium	–	Ns nielsbohrium	Bh bohrium	Bh bohrium
108	Uno unniloctium	Hs hassium	–	Hs hassium	Ha hahnium	Hs hassium
109	Une unnilennium	Mt meitnerium	–	Mt meitnerium	Mt meitnerium	Mt meitnerium
110	Uun ununnilium	–	–	Da Darmstadtium	–	Ds darmstadtium
111	Uuu unununium	–	–	Rg roentgenium	–	Rg roentgenium
112	Uub ununbium	–	–	Cp copernicium	–	Cn copernicium

Table V.7 Proposals of names and symbols rejected by the International Union of Pure and Applied Chemistry (IUPAC)-International Union of Pure and Applied Physics (IUPAC) for elements 114, 115, and 116

Z	*Name*	*Symbol*	*Reason*
114	*Russium*	Rs	Name already used for an unconfirmed discovery; furthermore, element 44, ruthenium, has a similar name
114	*Kurchatovium*	Ku	Name already used for the unconfirmed discovery of element 104
115	*Russium*	Rs	Name already used for an unconfirmed discovery; furthermore, element 44, ruthenium, has a similar name
115	*Kurchatovium*	Ku	Name already used for the unconfirmed discovery of element 104
116	*Leosium*	Ls	Name already used for the unconfirmed discovery of element 43
116	*Kurchatovium*	Ku	Name already used for the unconfirmed discovery of element 104
116	*Flerovium*	Fl	Name already used for the unconfirmed discovery of element 102

conclusion is that the RIKEN experiments did not meet their criteria for discovery, but the RIKEN team has put forward claims to the discovery of ununtrium in any case.

There are other cases in which the IUPAC Commission has rejected names proposed by discoverers on the basis of questionable assertions or of norms already in force, for example, the norm that forbids reusing names already proposed in the past by other persons or groups. In proposing the name *nipponium* (symbol Np), RIKEN experienced double jeopardy: *nipponium* had already been proposed for the discovery of element 72, and Np was already the symbol for neptunium. Also, the group at Dubna experienced a similar rejection: their proposal of the name *russium* (symbol Rs) had already been used in a false discovery of element 43.

Other plausible symbols and names were bandied about in the scientific community, but without acceptance. Among these were the RIKEN proposal of *nihonium* (symbol, Nh), after a Japanese name for that country, and another was proposed by Dubna, *becquerelium* (symbol, Bq) after Henri Becquerel, the discoverer of radioactivity. Table V. 7 shows the names and symbols rejected by the IUPAC-IUPAP joint commission for the elements with atomic numbers 114, 115, and 116.

In Table V.8 are reported names that were previously contenders for elements with atomic numbers 114, 116, and 118. These names were proposed more or less by their presumed discoverers.

Following the recommendations of a JWP of experts drawn from IUPAC and IUPAP, the IUPAC has officially approved the name flerovium, with symbol Fl, for the element of atomic number 114 and the name livermorium, with symbol Lv, for the element of atomic number 116.

Table V.8 Names and symbols proposed for elements 114, 116, and 118

Element	*Proposed Name*	*Symbol*	*Derivation*
114	*Atlantisium*	An	Atlantis; reference to the island of nuclear stability
114	*Lazarevium*	Lz	Yuri Lazarev (1946–96), leader of the Russian research group
114	*Oganessium*	Og	Yuri Oganessian, leader of the Russian research group
116	Flerovium	Fl, Fv	Georgy Flerov, head of the Russian research group
116	*Butlerovium*	Bu, Bv	Aleksandr Butlerov (1828–86), Russian chemist, but with the "defect" of having been an organic chemist
116	*Rossijium*	Ro, Rs	Rossija; transliteration of the word "Russia" from Russian
116	*Taldomskium*	–	Taldomsky; Russian district where the Dubna research center is
118	Flerovium	Fl, Fv	Georgy Flerov, head of the Russian research group
118	*Dubnabium*	Dn	Very similar to the name of element 105, dubnium
118	*Moscovium*	–	After Moscow. Variation: *moscowium*

In accordance with agreed-upon criteria, the Commission assigned priority for these discoveries to the collaboration between the Joint Institute for Nuclear Research (Dubna, Russia) and the Lawrence Livermore National Laboratory (Livermore, California). The collaborating teams proposed the names flerovium and livermorium, which have now been accepted and formally approved by IUPAC, thus contravening the IUPAC's own rules for not reusing the names and symbols of unsuccessful past candidates (see Table V.8). The choice of flerovium for element 114 is curious because once a name has been proposed for an element, the name gets only one shot at appearing in the periodic table. If the evidence for the element falls apart, or if the international governing body of chemistry (IUPAC) rules against an element's name, it is blacklisted. This might feel satisfying in the case of Otto Hahn, but it also means that no one can ever name an element "joliotium" after Irène or Frédéric Joliot-Curie, since "joliotium" was once an official candidate name for element 105. It is unclear why flerovium got another shot at the periodic table.

Flerovium honors Georgy N. Flerov, an appropriate choice because the element was synthesized in 1991 in the laboratory that bears his name. Livermorium honors the heavy element research group at Lawrence Livermore National Laboratory which, over the years, has made important contributions to nuclear science.[112,113,114] A new JWP has taken up the task of assigning priority for the discoveries of elements 113, 115, 117, and 118 and any heavier elements for which claims may be submitted.

Finally, a few words on the superheavy element 122. In this case, we are not dealing with a discovery, but with a rediscovery in nature. On April 24, 2008, a research group at the Racah Institute of Hebrew University in Jerusalem, led by Professor Emeritus Amnon Marinov, asserted that it had found "single atoms" of unbibium in samples of thorium-232 in concentrations between 10^{-11} and 10^{-12} g. The Israeli researchers placed in evidence a superheavy nucleus marked by a mass number of A = 292 and having a Z = 122.[115] The discovery of Marinov and his colleagues was immediately criticized by the scientific community. Copies of the manuscripts sent simultaneously to the journals *Nature* and *Nature Physics* were returned without having been taken into serious consideration by the editors. Seeing name of Robert Gentry listed among the authors of this paper could not but have evoked a certain skepticism with respect to this communication. In fact, Gentry had already, during the 1970s, publicly claimed to have discovered primordial superheavy elements, discoveries that we know today were not true.

Notes

107. Hofmann, S. et al. *Z. Phys.* **1995**, *A350*, 277.
108. Karol, P. J. et al. *Pure Appl. Chem.* **2001**, *73*, 959.
109. Corish, J.; Rosenblatt, G. M. et al. *Pure Appl. Chem.* **2004**, *76*(12), 2101–03.
110. Tatsumi, K.; Corish, J. *Pure Appl. Chem.* **2010**, *82*(3), 753–55.
111. RIKEN NEWS, November 2004. The proposal was withdrawn on February 9, 2008.
112. Group Report, *Pure and Appl. Chem.* **2011**, *83*(7) 1485–98.
113. Recommendations, Pure and Appl. Chem. **2012**, *84*(7), 1669–72.
114. http://www.iupac.org/news/news-detail/article/element-114-is-named-flerovium-and-element-116-is-named-livermorium.html (accessed April 13, 2014).
115. Marinov, A.; Rodushkin, I.; Kolb, D.; Pape, A.; Kashiv, Y.; Brandt, R.; Gentry R. V.; Miller H. W. e-Print Archive, *Nuclear Experiment* **2008**, 1–14.

PART VI

No Place for Them in the Periodic Table: Bizarre Elements

THOSE ARE MY PRINCIPLES.
IF YOU DON'T LIKE THEM I HAVE OTHERS.
GROUCHO MARX (1890–1977)

PROLOGUE TO PART VI

The sixth part of this volume is, from a certain point of view, the most bizarre as well as the most diverse. As the subtitle, "without a place in the periodic table" indicates, these discoveries have only one thing in common: their arrangement in chronological order. It begins in the first year of the 19th century and draws this period of numerous failures to a close around the middle of the following century. We range from fantastic theories propounded by renowned university professors, such as the aged Mendeleev and the stubborn Harkins, to a self-proclaimed bishop of the Reformed Catholic Church and an occultist by hobby who casually skipped from one eccentric interest to another. The lively and spirited practitioners of these "periodic arts" come alive in these pages.

VI.1

INORGANIC EVOLUTION: FROM PROTO-ELEMENTS TO EXTINCT ELEMENTS

Pyotr Nikolaevich Chirvinsky (1880–1955), the eminent Russian geologist, is best known as the founder of the science of meteorology. In the 1920s, Chirvinsky became the director of the Donskoi Polytechnic at Novochercassk. He spent a great deal of time as a consultant for the mines scattered throughout the Russian empire: along the Donets Basin,[1] on the Kola[2] and Crimean[3] peninsulas, on the northeastern slopes of the Caucasus,[4] and in the enormously rich mineral deposits of the Urals.[5] His major objective in this work was to establish connections between the chemical composition of terrestrial minerals and meteorites by studying the quantity of a mineral present in a given sample of rock and the physicochemical conditions leading to its formation. He insisted that meteorites be considered legitimate objects of study in petrology, and because they had been formed in heavenly bodies and not on earth, they might provide clues regarding the formation of elements from primal material. Chirvinsky had predecessors in this way of thinking, as we shall see.

VI.1.1. A STEP BACKWARD: PRIME MATTER, *ANDRONIA*, AND *THELYKE*

The concept of *prime matter* is very old, coming before the definition of a chemical element, but connected to the idea of the elements. Raymond Lull (ca. 1235–1315), in his book, *De Materia*, defined the concept of prime matter as an element *in potentia* in all possible substances. The idea was very acceptable to many alchemists up until the end of the 19th century.

In 1800, Jakob Joseph Winterl[6] (1732?–1809) was a famous physician and professor at the University of Nagyszombat, in present-day Hungary. He developed a vitalistic and dualistic concept that was, from a certain point of view, anti-Enlightenment, according to which all of the chemical elements would have originated from two immaterial principles:[7] one male, *andronia*, and the other female, *thelyke*.[8] Although Winterl's speculations may have been based on doubtful or misinterpreted experimental evidence, many German chemists accepted his theory. The physicist Heinrich Pfaff (1773–1852) embraced Winterl's theory with enthusiasm, as did the pharmacist Johann Friedrich Westrumb (1751–1819) who propagated the concepts of *thelyke* and *andronia*.

The first problems occurred when Winterl was unsuccessful in experimentally proving his theory. To complicate this already difficult scenario, he found himself in open hostility to Antoine Laurent Lavoisier, refusing to believe in the theory of oxidation and going to great lengths to support the idea that no acid contained oxygen. He sought to explain all chemical phenomena by way of the dualistic principle of *andronia* and *thelyke*. The weakest point in his theoretical edifice lay in his claim to having discovered substances even simpler than the elements themselves: *andronia* and *thelyke* identified with the acid principle and the basic principle, respectively. Winterl said that he had isolated the acid principle, *andronia*, as a white substance with peculiar properties: with water and with oxygen it gave rise to nitric oxide or nitric acid (depending on the proportions of the reactants); with hydrogen, milk or egg white were formed; combining it with gypsum would yield marble.

The preparation of the two principles, *andronia* and *thelyke*, looks ridiculous, if not downright absurd, to chemists of the 21st century. However, Winterl described in detail the process for extracting these two "principles." To get *andronia*, it was necessary to mix one part charcoal with four parts saltpeter, after which the "principle" formed from these could be removed by cooling everything down with three parts snow and one part salt or, lacking this, with carbon dioxide. *Thelyke* was obtained by dissolving some marble or material from a stalactite in hydrochloric acid. Adding ammonia to the resulting solution yielded a precipitate; the filtrate was washed and redissolved, and reprecipitating with potassium carbonate would yield pure *thelyke*.

The chemists Adolf Ferdinand Gehlen and Wilhelm August Lampadius tried unsuccessfully to obtain the same results as Winterl, and for this reason they were disparaged by another chemist, Karl W. Gottlieb Kastner (1783–1857), who asserted that he had succeeded in his attempt. In 1804–05, at the University of Jena, Kastner proudly held a series of lectures on the two *principles*.

Winterl himself tried to extract more *andronia* because his colleagues were requesting more and more samples of it. The first scientist to openly line up in opposition to the two principles was the renowned French chemist Louis Bernard Guyton de Morveau.[9] In 1807, worried about the growing skepticism surrounding his elements, Winterl asked Gehlen, who was a famous experimentalist, to verify their existence. Gehlen passed the request on to the German pharmacist Christian Friedrich Buchholz (1770–1818), asking him to repeat Winterl's experiments. A short time later, Buchholz published a work[10] with a hopelessly damning conclusion: "no trace of the problematic *andronia*." Winterl rejected the findings. Collecting his last amount of *andronia*, he sealed it in a bottle and sent it to the highest authority on matter, the Academy of Sciences at Paris, with the request that the chemists there analyze it and give their opinion. The eminent chemists Claude Louis Berthollet, Antoine François de Fourcroy (1755–1809), Louis Nicholas Vauquelin, and Guyton de Morveau found that the bottle contained nothing more than clay, plaster, potash, and a trace of iron.[11] Their conclusion was indeed a gloomy verdict for Winterl, whose scientific reputation was irreparably demolished:

> We will therefore conclude this report by saying that the alleged andronia does not exist at all. . .; that the theory that he has propounded on andronia is a hypothesis devoid of any type of fundamental principle and that his way of thinking is likely to make science go backward instead of advancing it.

Winterl died on November 23, 1809, in the city of Buda, in the same year that the crushing refutation of his work was published, but no one knows if he was ever aware of it.

A few years later, in 1817, Johannes L. G. Meinecke (1781–1823), a professor at the University of Halle, Germany, revived the concept of prime matter[12] using the word *urstoff* (i.e., element).

VI.1.2. PANTOGEN

Gustavus Detlef Hinrichs (1836–1924) was born at Lunden in Holstein, then a region of Denmark, but presently in Germany. He was a very successful student at the University of Copenhagen. From his earliest youth, he was a prolific writer, and, by the time he reached the age of 20, he already had numerous articles and one book to his name. During the years 1855–57, Hinrichs developed a bizarre theory on the unity of matter[13] based on a single universal element that he called *pantogen*.[14] His thought spanned the ideas of the ancient Greeks, of Raymond Lull and his prime matter, and of Johannes L. G. Meinecke and others. He took his degree in 1860, and in 1861, he emigrated to the United States right at the beginning of the American Civil War, settling at the University of Iowa. His interests, in addition to his teaching, could not have been more numerous and varied: influenced by the scientific eclecticism of Michael Faraday, whom he admired greatly, these ranged from dielectrics, to magnetism, geomagnetism, astronomy, physics, chemistry, and meteorites. A prolific writer even into late old age, he published more than 300 articles. He later moved to a chemistry position at the College of Pharmacy in Saint Louis. He retired at the age of 71, but continued to publish prolifically. On December 2, 1923, apparently in good health, he celebrated his 87th birthday, but died suddenly the following February.

Among his many interests, one must certainly number his cosmological dead-end. On the model of Hesiod's "Theogony," a large-scale synthesis of a vast amount of Greek traditions and ideas, Hinrichs identified four stages in the creation of the universe:[15] in the first stage, prime matter, or *urstoff*, gave rise to the chemical elements; the second stage was the development of the heavenly bodies; the third "era" resulted in the subsequent "cooling off" and formation of geological structures; and the fourth and last stage corresponded to the present era.

In the field of chemistry, Hinrichs was greatly influenced by the work of classification of the elements advanced in France first by Jean-Baptiste Dumas[16] and later by Alexandre Émile Beguyer de Chancourtois, inspector general of the French mines. Chancourtois, in 1862, before Newlands announced his Law of Octaves and Mendeleev had described his Periodic System, presented a paper to the French Academy of Sciences in which he described a spiral-shaped periodic table on which the elements were arranged around a central "parent" element, *pantogen*. If we exclude the concept of a pan-element or *pantogen*, his attempt at classifying the elements according to their chemical properties (groups) cannot be rejected out of hand.

Hinrichs, in a brief article that appeared in *Chemical and Metallurgical Engineering*, postulated that weighable matter was not chemically active except for a very small special part of it that would take up, at most, a hundredth part of it. It is curious to note that his analogy between electrons, atomic size, and chemical reactivity was purely coincidental. As a corollary, he asserted that matter was one, and therefore that the chemical elements could not be simple substances but a complex combination of a single substance, *pantogen*. He promised his readers that he would furnish proof of his assertions in the next issue, but this never happened.[17] Hinrichs concluded another series of articles by maintaining that the hypothetical element *pantogen*, from which all the other elements were formed, would have an atomic weight 1/128 that of hydrogen.[18] He arrived at the conclusion that if, from a liter of *pantogen* weighing 0.697 mg, one were to subtract the observed experimental weights of 1 liter of O, H, N, and C (gases), the new atomic weights of these elements would turn out to be the whole numbers of 16, 1, 14, and 12, respectively. The atomic weight of hydrogen, estimated at the time to be 1.007813, would be shown to be equal to (128+1)/128, where 1/128 would be nothing other than the atomic weight of *pantogen*. By means of this unnecessarily complex system, Hinrichs maintained that, for example, oxygen was composed of 16 × 128 atoms of *pantogen*.[19] Hinrichs was the author and practically the only user of this very artificial construction. Because very few read him, no one contradicted him.

As an aside: in 1894, when he was almost 60 years of age, Hinrichs proposed an international subscription with the purpose of collecting funds to erect a statue in honor of Lavoisier.[20] His initiative was greeted with enthusiasm, but in certain respects it summed up the fate of all of Hinrichs's ideas. The irony was that the funds were found, the bronze statue was forged by E. Barris of the French Institute of Fine Arts, and the statue was erected in Paris at the Place de la Madeleine with great pomp and circumstance on July 27, 1900—only to be removed and melted down by Nazi troops during the Occupation of 1942.

Meanwhile, in Manchuria, on the other side of the world, the Russian physicist Nikolai Morozov (1854–1946) was following developments in the maturing discipline of atomic physics from a very unusual point of view. He taught at the Russo-Chinese Polytechnic Institute of the Far East in Harbin, Manchuria. The city and its surroundings had undergone many different changes in government. This university, founded in 1899, was the easternmost one in imperial Russia.

Calling on the most recent discoveries of cathode and anode rays, Morozov developed a personal theory of *pseudoelements* based on three simple substances that he called *anodium, cathodium*, and *archonium*.[21] However, due to his extremely remote location, far from the more advanced research centers where his ideas could be received and propagated, but perhaps also because of the scanty basis for his hypothesis, Morozov (Figure VI.01) and his three elements were soon forgotten.

VI.1.3. PROTYLE

Almost 30 years after Hinrichs put forth his hypothesis, in 1886, Sir William Crookes raised the concept of *protyle*,[22] that is, matter *in potentia*, not organized. According to its discoverer, it would be intangible, unable to be perceived by humans, and "probably" not subject to the law of gravity: "Protyle is a word analogous to protoplasm, to express the idea of the original primal matter before the evolution of the chemical elements. The word I have ventured to use for this purpose is compounded of a Greek word 'earlier than,' and 'the stuff of which things are made."'[23]

Crookes's idea, unlike the scientific positivism of the French, was pervaded with a "pagan neo-mysticism." As a spiritualist and a believer in almost every aspect of the occult world, he was convinced that all forms of observable matter represented different stages of growth in complexity of one unique element or form of matter, a hypothesis suggested by William Prout in 1816, who claimed that hydrogen could be the fundamental unit from which all matter was made. Because experimental atomic weights did not conform to this view, Crookes suggested that his *protyle* was not hydrogen, but perhaps a half or a fourth part of hydrogen or other particle of low atomic weight—the discovery of isotopes 30 years later soon did away with the necessity for this hypothesis. If the admirable conceptual effort ended up absorbing Crookes in a morass of sterile and inconclusive investigations, his obstinate attempts to have it accepted made him quite unpopular with his European colleagues. Nevertheless *protyle* did not fall into oblivion immediately. In fact,

FIGURE VI.01. Nikolai Aleksandrovich Morozov (1854–1946). A Soviet revolutionary, scientist, and scholar *sui generis*, he proposed that matter was composed of positive and negative atoms of electricity that he called *anodium* and *cathodium*.

on the occasion of the meeting of the British Association for the Advancement of Science held at Portsmouth in 1911, the youthful John William Nicholson (1881–1955) used the work as a basis for a discourse on the theory of the structure of the chemical elements.[24]

Nicholson was a mathematical physicist who had the good fortune of being at Cambridge in the years that Ernest Rutherford was the leading physicist there. In a series of articles between 1911 and 1912, Nicholson interpreted certain lines observed in stellar spectra as evidence of "transverse oscillations in the rotational orbits of the electrons around the nucleus." Nicholson's calculations, in addition to leading him to hypothesize somewhat heterodox ideas on the presence of new elements, also led him to discover the quantization of angular momentum of electrons that performed transverse oscillations, which Bohr himself later recognized. Despite the fact that Nicholson was associated with a teacher like Rutherford, his own personal structural theory was a synthesis of old and new theories with very little basis in fact.

From observations of spectra from the Orion nebula made by Henry Bourget (1864–1921), Charles Fabry (1867–1945), and Henri Buisson[25] (1873–1944), Nicholson built up a theory based on the existence of elements lighter than helium. He also modified their system for the periodic table.[26] The lines of hydrogen, helium, and of other presumed elements present in nebulas were explained with the aid of dynamic vibrations of simple atoms or *primary substances*. These atoms represented "nodules" of positive electricity surrounded by one or more rings of electrons. Nicholson differentiated the "nodules" using the following terminology: +e, +2e, +3e, +4e, +5e, +6e, and so forth. He resurrected the term *protyle*, subdividing this principal generator of all matter initially into four *primary substances* and later into many more. The following are some primary substances having a name and some physical characteristics determined by Nicholson:

- *Coronium*, the simplest *primary substance*, consists of two electrons rotating around a central positive nucleus.
- *Hydrogen*, the second *primary substance*, is composed of an electrically neutral atom around whose central nucleus three electrons rotated.
- The third *primary substance* was identified as *nebulium*, which was impalpable, and had four electrons rotating around a positive central nucleus; its atomic weight was equal to 1.31.
- The fourth *primary substance* was recognized as *proto-fluorine*, whose name, according to the author, was provisional while waiting for a better one. This element had five electrons and, contrary to what one might expect from its name, it bore no resemblance to fluorine but, like *coronium*, was discovered in the spectrum of the solar corona.
- The last named *primary substance* was *archonium*, composed of a ring of six electrons rotating around a positive "nodule." Nicholson calculated that *archonium* had an atomic weight of 2.945.

Nicholson's theoretical studies were based on Rutherford's primitive atomic model and consequently were incapable of giving the hoped-for interpretation of the origin of the spectral lines of the hydrogen atom. In the following year, Niels Bohr, having a correct model of the atom at his disposition, succeeded in formulating what Nicholson had failed to do. Furthermore, Nicholson had unwisely dragged in nonexistent *primary substances* on which he based part of his theoretical work. Consequently, his scientific credibility was greatly diminished and his academic career, to use an astronomical term, was as brief as a meteor. After a series of articles about the many different aspects of quantum mechanics,[27] about which no one could say he was incompetent, Nicholson exited definitively from the scene.

John Nicholson was a mathematics teacher at Balliol College, Oxford. After the events just outlined, in the 1920s, he began to nurse a growing resentment toward the founding fathers of quantum mechanics. He maintained that he had been the victim of a conspiracy for not having received adequate recognition for his contributions to physics, and he pointed the finger at the most influential people in the discipline. He found consolation at the university tavern so much so that, by 1930, he was no longer able to accomplish his academic duties, and he lost his job. Nicholson fell into a profound state of depression and alcoholism; he passed the last 25 years of his life, practically forgotten, in the hospital at Warneford.

VI.1.4. OTHER THEORIES OF CHEMICAL EVOLUTION

Toward the beginning of the 19th century, Jöns Jacob Berzelius attempted to explain the relative positions of the elements in an electrochemical series by the assumption that each atom carries charges of positive and negative electricity, the preponderance of one or the other serving to determine the chemical character of the substance.

Fifty years later, in 1885, Thomas Carnelley (1854–90), in spite of his short life, did fundamental research on the relationship between the physical properties of the elements and their compounds and their position in the periodic table (Carnelley's rules).[28] He put forward the idea that these substances are not strictly simple or elemental but are compound radicals made up of at least two simple elements *A* and *B*. Element *A* was supposed to be identical with carbon, whereas *B* was associated with a negative weight of −2, and it was suggested that it might be the elusive ether of space. The concept of a negative weight has never been acceptable, and the hypothesis dropped out of sight.

According to another suggestion, by Charles Skeele Palmer (1858–1939), one can assume the existence of two subelements, which he named *kalidium* and *oxidium*.[29] Palmer studied under Ira Remsen (1846–1927) at Johns Hopkins, graduating in 1886. Shortly afterward, he did his postdoctoral work at the University of Leipzig under Wilhelm Friedrich Ostwald (1853–1932). In 1894, he was elected president of the Colorado Scientific Society, and, in 1900, he became head of the Chemistry Department of Colorado Academy of Science. His work includes many articles on chemistry, mineralogy, and meteorites. However, his passion concerned theoretical chemistry and the composition of matter. His articles appeared in obscure journals and consequently went unnoticed[30,31].

Palmer soon discarded the hypothesis that hydrogen is the proximate ingredient of the elements because the atomic weights were not found to be exact multiples of unity and because hydrogen is inherently basic; although it might be looked upon as the prototype of base-forming elements, it could not be the origin of the acid-forming elements.

Palmer developed his own hypothesis and suggested that hydrogen could possibly be a member of a completely independent series of elements as yet unknown. He thought he would call the last element of this series *prefluorine*. From a certain point of view, Palmer seemed to have anticipated some of Mendeleev's ideas on elements lighter than hydrogen. As for *kalidium* and *oxidium*, the two hypothetical components of all the elements, Palmer did not regard them as being forms of matter that could be isolated but merely as representing antithetic qualities that are jointly responsible for the properties of the elements as we know them.

The investigations on the discharge of electricity through gases, carried on especially by J. J. Thomson (1856–1940) and his school, and the consequent incomplete development of a corpuscular theory of matter, seemed not completely in disagreement with Palmer's hypothesis regarding the constitution of matter, the elements, and their periodic relation to atomic weight. However, when the nature of subatomic particles was clarified, his theory was no longer tenable.

In 1896, a cautious report from the Harvard College Observatory described six lines in a peculiar star spectrum that formed a rhythmical series similar to hydrogen and were interpreted as "apparently. . . due to some element not yet found in other stars or on the Earth."[32] In a paper appearing the following year, the author retracted the idea of an unknown element and hypothesized that the peculiar spectrum was more than likely due to the presence of hydrogen under not yet achieved conditions of temperature and pressure.[33]

Then, in 1900, Norman Lockyer observed that *Fraunhofer lines* (dark lines in the visible spectrum) have greater intensity in the spark than in the arc spectrum. He called these "enhanced lines," and, to the vapors producing them, he affixed the prefix proto-, giving rise to a series of *proto-metals* such as *proto-iron* and *proto-magnesium*, suggesting that a finer form of the element developed from them.[34]

Finally, in 1910, Morozov hypothesized on the inorganic evolution of the chemical elements in the stars and other heavenly bodies, giving the name *protohydrogenium* to the primary substance with an atomic weight of 0.0818.[35]

Surely, there was no lack of imagination during this century of theorizing on the nature of the elements, with ideas drawn from earthly and heavenly measurements. These ideas were soon channeled into the great edifice of quantum mechanics early in the following century.

VI.1.5. THE ASTEROID ELEMENTS

The second year of civil war was going on in Russia when Pyotr Nikolaevich Chirvinsky developed a hypothesis of asteroid elements.[36] As we have briefly seen, other ideas of a similar nature had been put forth, and these had been woven together using circumstantial evidence and at times even supported by gross manipulation of raw data. One so-called "metaphysical" scientist and a font of bizarre ideas, was the eclectic Sir William Crookes whose reasoning seems to have been greatly influenced by the ideas of Charles Darwin. Crookes maintained that the elements would have originated through the condensation of a primitive form of matter and that the different elements, at the moment of their appearance, would have evolved by way of a rigorous "selection" process. According to him, the elements would have behaved like living organisms, undergoing a true and proper struggle for their survival, and those that were not in "harmony with their own development" would have disappeared. For these, Crookes coined the phrase *extinct elements.*[37] In Crookes's complex chemical world, other categories also existed, such as *common elements* and *scarce elements*, with the growth and diffusion of the latter limited by adverse evolutionary conditions. Within the category of *extinct elements*, Crookes discerned the presence of a subcategory: *asteroid elements*. These were created in remote past time on a par with all the others, but on Earth they did not succeed in their competition with the other elements and therefore became extinct. In synthesis, this was the fantastic theory of a Victorian chemist. And if this were not enough, in his hypothesis, also lay a hidden possibility that outside of our planet some "extinct elements" may have "survived," although in "evolutionary niches" and on a very limited scale. Crookes thought that these "evolutionary niches" would be found in meteorites. At the end of the 1880s, as Crookes was expounding his idea of "inorganic evolution," the scientific community had not yet taken a position against it. In fact, a 1907 paper[38] seems to embrace it wholeheartedly, postulating four "protons," the earliest forms of matter existing in nebulae, two of which are already known (viz., hydrogen and helium) and two that the authors put forth to explain their observations (viz. *proto-beryllium [proto-glucinium]* and *proto-boron*). The authors of this paper traced how the process of direct (same valency) evolution and indirect (different valency) evolution of these four "protons" gave rise ultimately to all of the elements in the periodic table and also solved the problem of atomic weight pair-inversion (of atomic weights such as tellurium-iodine and potassium-argon) that so troubled Mendeleev. The authors, in testing their hypothesis, found that the elements would exhibit higher atomic weights than those normally found—giving rise to a further hypothesis of a disturbing influence, what they termed "devolution," to lower atomic weights, for example, through radioactive decay.

Meanwhile, Bohuslav Brauner, in a 1902 paper[39] spoke of *asteroid elements* in the context of the condensation of a primordial *Ursubstanz* during the formation of the rare earth elements. One might suppose that he had taken his original idea from Crookes because they were frequent correspondents since their days together in Manchester. However, Brauner seems to have been totally unaware of Crookes's lectures on this topic. Although Brauner is credited with the *asteroid* hypothesis with respect to a methodology of accommodating the puzzling rare earth group in the periodic table (pigeonholing all of them in the same box!), the prior idea seems to have originated in a paper by Jan Willem Retgers,[40] in which he mentions a group of planetoids between Mars and Jupiter that occupied the orbit of one planet. So, not surprisingly, he put La, Ce, Di, Sm, Er, and Yb all in the same place in the periodic table, and this became the forerunner of a whole series of papers appearing into the early 20th century, all of them adhering to the same *asteroid* hypothesis and claiming this as the solution to the troubling rare earth problem.

Thirty years after Crookes's trilogy of papers on the process of the genesis of the chemical elements, Chirvinsky appeared on the scientific scene. The ideas of the scientific community had changed and, although he worked in a country on the periphery of the great scientific centers, he should have used more caution in reviving concepts even vaguely referable to Sir William Crookes.

Chirvinsky observed that the chemical elements, whether coming from rocks in the Earth's crust or from meteorites, had some analogous relationships to the classification of organic compounds. As a further step, he advanced the hypothesis that rock formation would have taken place in different zones of a hypothetical "primordial gaseous sphere." The heavier components would have formed in the lower zones where the pressure was greater; the isomorphous elements and isotopes, on the other hand, would have formed in a similar zone of the "gaseous sphere." The mean chemical composition of eruptive rocks from the Earth's crust showed the elemental percentages of Si, 20.5%; metals, 19.5%; and O, 60.5% that would give rise to the empirical formula of a metasilicate, $MSiO_3$ (with M standing for a generic metal or pseudo-element that he called *crustaterrium*). According to Chirvinsky, *crustaterrium* would have an atomic weight of 20.56. In the same way, he determined the following progressive series of pseudo-elements: *terrium*, or *primordial matter*, was an element that represented the median weight of the entire earth; its atomic weight was estimated at 39.98. Then came *chondrium*,[41] which corresponded to the formula $M_2SiO_4 + MSiO_3$ and with the generic M, to an atomic weight of 24.36; *pallasium*,[42] $M_2SiO_4 + 3M$, with an atomic weight of M = 30.90; and finally, *siderium* that had a weight of 55.72.

For some years, Chirvinsky sought a law that would explain the formation of the chemical elements in the universe.[43] In 1924, he finished collecting the analytical data on meteorites that had fallen to Earth between 1492 and his current work. A statistical study of the material led him to formulate the following hypothesis: all the meteorites had the median composition of $M_2SiO_4 + MSiO_3 + M$, where with the letter M indicated the "collective metal" having an atomic weight of 40.05. This would be composed of 50% iron and other similar metals and the remaining 50% of Mg, Ca, and other alkaline earth elements. He called the pseudo-element, M, *cosmium*.[44]

Over the following 2 years, Chirvinsky analyzed the composition of the lithic meteorites. He discovered that the median weight of the atomic weights of the elements present in this type of meteorite corresponded to 24.36, equivalent to the imaginary "collective element" that he called *chondrium*.[45] He determined the chemical composition of the meteorites using *chondrium* as a basis, but this did not add anything substantially new to his work of 1919.

VI.1.6. THE PAINFUL FINALE

Pyotr Nikolaevich Chirvinsky (or Chirvinskii) was a very famous expert in mineralogy, petrology, and meteorites when, in 1931, the Soviet Secret Police[46] arrested him on the accusation of having sabotaged Soviet science. After a mock trial with a rapidly foreseen verdict of condemnation, he was exiled to a remote region of the Kola Peninsula. Although his discoveries were not completely credible and his hypotheses off track, the punishment inflicted on him was certainly unjust and far greater than he deserved for his eccentric thinking. Chirvinsky remained imprisoned until 1938, but he didn't completely abandon his research in mineralogy. In 1938, at Yukspor on the tundra of Khibina, where he was confined, he discovered a new sorosilicate. After having determined its formula, $Ca_3Si_2O_7H_2O$, he proposed that it be called foshallasite. After his release, he had to wait for the end of World War II before he could be officially "rehabilitated"; in that same year, he moved his activities to the University of Perm.

In 1953, by now very old and ill, he published his last article. It was a long (70 pages) and detailed study of the similarity of the chemical composition of meteorites then known and other heavenly bodies.[47] His data were derived from knowledge of the average composition of meteorites, from the crust and the center of the Earth, from the sun and from the stars. The resemblance of the composition of all of the meteorites was described in terms of "pseudo-elements"—*cosmium, siderium, pallasium, chondrium, terrium,* and *crustaterrium*—a concept he introduced in 1919. The significance of the atomic weight, density, heat capacity, metallic and metalloid content, and other properties of these substances were reported in detail. At the conclusion of his analysis, Chirvinsky dealt with the problem of the formation of three of the pseudo elements, namely, *cosmium, chondrium* and *terrium*. Although this was not new work—he had already dealt with this problem in the 1920s—according to him it was very important: *cosmium, chondrium,* and *terrium* would have been generated in the cosmos following the cooling and crystallization of an "improbable" *gaseous magma*.

Chirvinsky died 2 years later, in 1955, at the age of 75. With his exit from the scientific scene, one could definitively place the final word in the chapter of the *asteroid elements*—and by this late date, the place of the rare earths in the periodic table had also been resolved.

Notes

1. Chirvinsky, P. N. *Bull. North Caucasian Meliorations, Novocherkassk* **1919** no. *12*, 193; Chirvinsky, P. N. *Compt. rend. Acad. sci. U. R. S. S.* **1928**, 367.
2. Chirvinsky, P. N. *Mem. Soc. Russe Mineral.* **1935**, *64*, 315; Chirvinsky, P. N. *Mem. Soc. Russe Mineral.* **1936**, *65*, 163.
3. Chirvinsky, P. N. *Bull. Inst. Polytechn. Don* **1916**, *5*(Sec. 2), 88.
4. Chirvinsky, P. N. *Neues Jahrb. Mineral. Geol., Beil. Bd.* **1931**, *A64*, 649; Chirvinsky, P. N. *Centr. Mineral. Geol.* **1929**, *A*, 366; Chirvinsky, P. N. *Centr. Mineral. Geol.* **1926**, *A*, 344.
5. Chirvinsky, P. N. *Bull. Soc. Naturalistes Moscou, Sect. geol.* **1938**, *16* (no. 1), 3.
6. Jakob Joseph Winterl was both a chemist and a botanist. The date of his birth is uncertain, and various sources give three dates: 1731, 1732, or 1739. In addition to his interests cited in the text, he also was involved with the areas of hydrology, water analysis, and pharmacy.
7. Winterl, J. J. *Prolusiones ad chemiam saeculii decimi noni*; Typis ac sumptibus Typographiae Regiae Universitatis Pestinensis: Buda, Hungary, 1800.

8. Winterl, J. J. *Gehlen Allgem. Journ der Chemie* **1805**, *2(III)*, 336; Winterl J. J. *Gehlen's J. für Chem. u. Phys.* **1807**, *3*, 135; Winterl, J. J. *Sein system d. dualist Chemie*; Jena, 1804; Snelders, H. A. M. "The Influence of the Dualistic System of Jakob Joseph Winterl (1732–1809) on the German Romantic Era."*Isis* **1970**, *61*, 231–40.
9. de Morveau, G. *Ann. Chim.* **1802**, *47*, 312; de Morveau, G. *Ann. Phys.* **1803**, *15*, 496.
10. Buchholz, C. F. *J. Chem. Phys.* **1807**, *3*, 342.
11. de Serres, M. *Ann. Chim.* **1809**, *71*, 225.
12. Meinecke, J. L. G. *Schweigger's Journ.* **1817**, *22*, 138.
13. Kauffman, G. B. *Voprosy Istorii Estestvoznaniya i Tekhniki* **1969**, no. *4*, 36; Nielsen, H. T. *Dansk Kemi* **1982**, *63*(5), 144.
14. Hinrichs, G. D. *Programm der Atom-Mechanik, oder die Chemie eine Mechanik der Pan-Atome*; Augustus Hageboek: Iowa City, 1867; Hinrichs, G. D. *Am. Journ. Science* **1861**, *32*, 350; Hinrichs, G. D. *Amer. Assoc. Rep.* **1869**, *18*, 112; Ewing, R. *The Lattice* **2002**, *18*, 24.
15. Hinrichs, G. D. *Am. J. Sci.*, 1866**, 42**, 350.
16. Dumas, J. *Liebig's Annalen* **1858**, *195*, 74.
17. Hinrichs, G. D. *Chemical and Metallurgical Engineering* **1910**, *8*, 138.
18. Hinrichs, G. D. *Rev. Gen. Chim.* **1911**, *13*, 351; Hinrichs, G. D. *Rev. Gen. Chim.* **1911**, *13*, 377; Hinrichs, G. D. *St. Louis Met. Chem. Eng.* **1910**, *8*, 200; Hanssen, C. J. T. *Chem. News.* **1909**, *99*, 229.
19. Hinrichs, G. D. *Compt. Rend.* **1909**, *147*, 797.
20. Williams, W. D. *Bull. Hist. Chem.* **1999**, *23*, 47.
21. Perez-Bustamante, J. A. *Fresenius' Journal of Analytical Chemistry* **1997**, *357*(2), 162.
22. Crookes, W. *Chem. News* **1885**, *54*, 117.
23. Crookes, W. *Chem. News* **1886**, *55*, 95.
24. Nicholson, J. W. *Phil. Mag.* **1912**,S. 6, *22*, 864.
25. Bourget, H.; Fabry, C.; Buisson, H. *Compt. Rend.* **1914**, *158*, 1017.
26. Nicholson, J. W. *Compt. Rend.* **1914**, *158*, 1322.
27. Nicholson, J. W. *Proc. Phys. Soc.* **1918**, *30*, 65; Nicholson, J. W. *J Chem. Soc. Trans.* **1919**, *115*, 855; Nicholson, J. W. *Phil. Mag.* **1922**, *44*, 193; Nicholson, J. W. *Nature* **1922**, *110*, 37; Nicholson, J. W. *Phil. Mag.* **1923**, *25*, 801.
28. Wisniak, J. *Educ. quím.*, **2012**,*23*(4), 465–73.
29. Tilden, W. A. *The Elements: Speculations as to Their Nature and Origin*. Harper Brothers, London, 1910, pp. 93–94.
30. Marple, H. A. *Industrial and Engineering Chemistry, News Edition*. **1934**, *12*, 305.
31. Palmer, C. S. The Nature of the Chemical Elements (fourth paper). *Proceedings of the Colorado Scientific Society* **1893**, *4*, 56–74.
32. Pickering, E. C. *The Astrophysical Journal* **1896**, *4*, 369–70.
33. Pickering, E. C. *The Astrophysical Journal* **1897**, *5*, 92.
34. Lockyer, N. *Inorganic Evolution as Studied by Spectrum Analysis*; Macmillan: London, 1900.
35. Morozov (or Morosoff), N. A. *Die Evolution der Materie auf den Himmelskörpern*; Theodor Steinkopff: Dresden, Germany, 1910.
36. Chirvinsky, P. N. *Bull. Inst. Polytechn. Don* **1919**, *7*(Sect. 2), 94.
37. Mellor, J. W. *A Comprehensive Treatise on Inorganic and Theoretical Chemistry*; Longmans-Green: London and New York, 1931; vol. V, p. 3.
38. Jessup, A. C.; Jessup, A. E. The Evolution and Devolution of the Elements. *Phil. Mag.* **1907**, *15*(VI), 21–55.
39. Brauner, B. *Z. anorg. Chem.* **1902**, *32*, 1.
40. Retgers, J. W. Über einige Änderungen im periodischen System der Elemente. *Z. phys. Chem.* **1895**, *16*, 644.
41. This name derives from "chondrite." The chondrites were a type of rocky meteorite principally composed of iron and magnesium silicate. They have remained virtually unchanged

(geologically speaking) from the dawn of the solar system 4.5 billion years ago. About 86% of all the meteorites that fall to Earth are chondrites, rendering them the most common type of all meteorites. They arise from asteroids that have never been melted or undergone differentiation except for the loss of elements like hydrogen and helium. It is therefore thought that they have the same elemental composition as the primitive nebulas that gave rise to the solar system.

42. The term "pallasite" designates one of the two classes into which the ferrolithic meteorites are subdivided (the other class is called "mesosiderite"). The term is derived from the last name of Peter Simon Pallas (1741–1811), the German naturalist who discovered them in Siberia in 1772. Chemical analysis of the meteorite, done several decades later, showed the presence of a great deal of iron. Not knowing how to distinguish this iron from iron of terrestrial origin, and erroneously thinking that they were different, chemists of the era called it "iron of Pallas."
43. Chirvinsky, P. N. *Bull. Inst. Sci. P. F. Lesgaft.* **1920,** *1*, 46; Chirvinsky, P. N. *Izv. Don. Polytech. Inst.* **1923**, *8*, 68.
44. Chirvinsky, P. N. *Astronomische Nachrichten* **1924**, *222*(5311), 103.
45. Chirvinsky, P. N. *Centr. Mineral. Geol.* **1926**, *A*, 246.
46. During the 1930s, the secret police, also known by the initials CEKA, instead of carrying out intelligence services under Joseph Stalin (1879–1953), were responsible for terrorizing the entire country.
47. Chirvinsky, P. N. *Publ. Kiev. Astron. Observatorii* **1953**, no. *5*, 105.

VI.2

DAZZLING TRACES OF FALSE SUNS

VI.2.1. THE MIRAGE OF THE SOURCE OF STELLAR ENERGY

The astronomer Tycho Brahe (1546–1601) was the first Western scientist to observe a "nova" (in reality, a supernova, today indicated by the alphanumeric designation of SN 1572) in the constellation Cassiopeia. He described it in his book *De Stella Nova*, thereby originating the term *nova* or new star. The nature of novas remained a mystery for physicists for many centuries, but the history of these heavenly bodies took root in times even more remote, when Europe was immersed in the pitch black intellectual darkness of the Middle Ages.

Whatever remained of science and culture after the disastrous fall of the Roman Empire and the upheavals that followed it lay buried in the oblivion of monastic libraries. The blind, undisputed acceptance of the word of the ancient teachers snuffed out every wish to do research, and knowledge was reduced to merely perpetuating tradition. Thus it was that students of astronomy knew the *Almagestum* of Claudius Ptolemaeus (ca. 100–175) perfectly, but rarely lifted their eyes to look at the sky, fearing change more than error. Perhaps it was for this reason that no European document makes mention of an exceptional event that the Chinese astronomers described in detail in their chronicles: in July 1054, in the constellation Taurus, there emerged, apparently from nothing, a star so bright that it was visible in broad daylight. After some weeks of increasing brightness, it began to decline such that, toward the middle of April 1056, the naked eye could no longer perceive it. But the traces remained, even if hidden from the unaided eye, such that, having found it again, today we know that what remains is still more marvelous than the extraordinary appearance described by the ancient Eastern astronomers. The history of the observation of this fascinating object, begun in the East, was taken up again in Europe about 700 years later. In 1731, when the telescope was already widespread among students and enthusiasts of astronomy, a Scottish amateur discovered a curious cloudiness in the constellation Taurus. A few decades later, Charles Messier (1730–1817) started his famous star catalogue precisely with this object that we now call the "Crab Nebula." It was a nebula with a diffuse background and a set of widely branched filaments radiating out from a central area. In 1921, Knut Lundmark (1889–1958) observed that this object occupied precisely the same region in space in which, according to Eastern documents, the extraordinary star of 1054 appeared. Others discovered, almost simultaneously, that the nebula was expanding (documented by small changes shown in photographs taken years apart), and they calculated that to arrive at its present-day dimensions, it had to have been expanding for 900 years. It was enough, at this point, to consider these two pieces of data to understand the relationship between *novas* and nebulas.

$2(1^2) = 2$	atomic number of helium, He
$2(1^2 + 2^2) = 10$	atomic number of neon, Ne
$2(1^2 + 2^2 + 2^2) = 18$	atomic number of argon, Ar
$2(1^2 + 2^2 + 2^2 + 3^2) = 36$	atomic number of krypton, Kr
$2(1^2 + 2^2 + 2^2 + 3^2 + 3^2) = 54$	atomic number of xenon, Xe
$2(1^2 + 2^2 + 2^2 + 3^2 + 3^2 + 4^2) = 86$	atomic number of radon, Rn
$2(1^2 + 2^2 + 2^2 + 3^2 + 3^2 + 4^2 + 4^2) = 118$	atomic number of element 118

In 1928, in a totally different context, W. S. Andrews of General Electric published an essay in which he proposed the existence of a new element accountable for being the source of stellar energy, the nature of which was completely unknown.[48] At that time, physicists knew that the energy radiating from stars was in the form of heat and light, but for novas, things were not that clear. Some people thought that novas were the result of huge conflagrations of heavenly bodies, whereas others thought that they were stars exploding under very special circumstances. Andrews tried to explain which element would have been responsible for the energy emitted by novas, but his theory had many holes in it. To do this, he described a very odd numerical series (displayed above) that resulted in the atomic numbers of all the noble gases known up until 1928.

So far, we cannot but admire his original work, in which he discovered the numeric series that regulates the atomic number of all the noble gases and predicted that the one with Z = 118 would belong to that family. The oddness of his reasoning was based on the fact that he believed that the 118th element was responsible for the energy emitted from stellar novas. Why precisely the 118th and not the 93rd, 94th, 95th, etc.? A response to this question does not exist in Andrews's work.

W. S. Andrews began a long, speculative, and fantastic rationalization. Starting from his knowledge of radioactivity and radioactive families, he hypothesized that the 118th element could be the first in a family of transuranium elements present under certain circumstances in the sun and other heavenly bodies. He even immortalized this element: "Let us name this element 'hypon' for future reference, and let us give it, if you please, 'a place in the sun.'"

Hypon would have to be radioactive and disintegrate into lighter fragments, thus liberating energy, but the frightfully large gravitational pressure of the stars would have prevented this phenomenon. It was precisely this pressure that would have caused the "decomposition" of *hypon* to be slowed down so that it would "persist" for millions or even billions of years; that is, for the lifetime of a star. After having lost energy (and therefore, mass) for a sufficiently long period of time, the star would have reached a minimum of gravitational pressure, beyond which all the *hypon* present in the star would decay[49] in a very short period of time, giving rise to a very violent explosion of cosmic proportions. Nothing of the star would remain but dust, incandescent gas, heat, and electromagnetic radiation, hurled out in every direction into space. After many light years had passed, the light would reach our eyes as a witness to what Andrews called a "celestial cataclysm." "So it was that we watched the gradual development to its maximum brightness and beauty and then the slow decline of this wonderful and mysterious apparition in the sky which was called a nova, and which, if our hypothesis be the truth, was but the manifestation of atomic number 118, the seventh of our number series, hypon!"

Regarding this alleged discovery, so far-reaching, dramatic, and at times delirious, no scientist or amateur posed the question: did Andrews really believe this, or was he just pulling our leg? We will never be able to refute or deny his assertions: no trace remains of Andrews's life and work except those already cited.

VI.2.2. THE CURIOUS APPEARANCE OF KOSMIUM AND NEOKOSMIUM

In the 1880s, a German chemical journal published an article that contained a curious announcement. The author reported that he had succeeded in determining the existence of two new elements to which he gave the somewhat strange names of *kosmium* and *neokosmium*.[50]

At that time, the discovery of new elements had turned into a "mass phenomenon" that created serious problems for the chemical sciences. The announcement of new elements alternated like a seesaw between true and false discoveries. The names of these new simple substances were also getting increasingly more bizarre, narrow in viewpoint, provincial, and at times in bad taste. Some scientists, such as E. Demarçay, P. E. Lecoq de Boisbaudran, C. Auer von Welsbach, and others did not even take the trouble to propose a name for their "newborn" elements, but called them simply by Greek letters or Roman numerals.

The announcement made by H. Kosmann was immediately suspect. It occurred to some in the scientific community that the discoverer of *kosmium* and *neokosmium* was making fun of his colleagues and of the climate that permeated the academic community in those years, one that fostered an "epidemic of discoveries." But it is possible that the article had another, more serious purpose as well. The author was Bernhard Hans Kosmann (1840–1921). He was born on February 4, 1840, in the German city of Lobsens (Łobżenica, in present-day Poland). After having obtained his doctorate in engineering in 1870, he became inspector of weights and measures, and later *Bergmeister* (superintendent) of the mines in Koenigsulte, Charlottenburg, Berlin, and Joachimstadt. On June 9, 1896, Kosmann claimed that he had isolated the oxide of a new metal that he called *kosmiumoxyd*. A few weeks later, on July 25, he announced that he had discovered another element, which he called *neokosmium*. Furthermore, on November 16, 1896, Kosmann made a patent application for the preparation of *kosmium* and *neokosmium*.[51] Because the fixed price (of the patent process) was very high, one could not immediately dismiss Kosmann's claims as an ironic joke, one that made a mockery of the element-naming process by using his own name as the etymological source. Perhaps a clue to his motivation is the title of the patent he filed in England (number 18915), "Separation of certain rare earths, and the manufacture therefrom of fabrics for use in incandescent gas-lighting."[52] So it is quite possible that Kosmann went to all this trouble and expense to circumvent or counteract von Welsbach's patents on the use of rare earths in incandescent lamps by setting himself up as a possible competitor.

Hans Bernhard Kosmann's work was incorrect, and the names he chose for these so-called elements were flippant, but they were not entirely a joke. He survived World War I long enough to suffer the defeat of his homeland, passing away in 1921 at the age of 81.

Although the precise number of false names given to the elements will never be known, a good approximation is something over 200. The names *kosmium* and *neokosmium* appear in this great cemetery of elements like a touch of merriment in the middle of a sea of failed and delusional discoveries.

Notes

48. Andrews, W. S. *The Scientific Monthly* **1928**, *27*(6), 535.
49. The term used by Andrews was "dematerialized," referring to Albert Einstein's well-known equation, $E = mc^2$.
50. Kosmann, H. B. *Z. Elektrochem.* **1896**, *3*, 279; Kosmann, H. B. *Berg. u. H.* **1896**, *50*, 225.
51. Anon. *La Revue Scientifique* **1897**, *35*, 259.
52. See *Patents for Inventions*; Great Britain Patent Office: London, 1899, p. 149; there is an extensive summary of this English patent in German: Herzfeld, J; Korn, O. *Chemie der seltenen Erden*; Springer: Heidelberg, Germany, 1901, p. 72.

VI.3

FROM THE NONEXISTENT ELEMENTS OF MENDELEEV TO THE PUZZLE OF THE EXISTENCE OF THE ETHER

It probably is not well-known that Mendeleev had predicted the existence of more than 10 elements. Their discoveries were sometimes the result of lucky guesses (like the famous cases of gallium, germanium, and scandium), and at other times, they were erroneous. Historiography has kindly passed over the latter, forgetting about the long line of imaginary elements that Mendeleev had proposed, among which were two with atomic weights lower than that of hydrogen, *newtonium* (A = 0.17) and *coronium* (A = 0.4). He also proposed the existence of six new elements between hydrogen and lithium whose existences were false.

To his credit, Mendeleev discovered the periodic law of the elements while he was still young, and he also publicized his periodic table better than his colleagues. During the years in which the first noble gases were discovered, he was already in his 60, and he was older still when radioactivity was discovered. All of these discoveries greatly upset his "cast-iron" certainties. At first, he was very much opposed to the discovery of the noble gases, saying that argon was nothing more than N_3. In a famous telegram sent to William Ramsay, the discoverer of argon, Mendeleev sarcastically wrote: "[I'm] delighted about the discovery of argon. [I] think the molecules contain three nitrogens bonded by heat."[53]

Among other things strongly supported by Dmitri Mendeleev was the concept of the *ether*, which physicists had created in order to explain the propagation of electromagnetic waves. Mendeleev recognized in the *ether* the ability to penetrate all substances, as postulated by physicists, but he went further, asserting that this form of matter would be characterized by its inability to form any stable chemical compound with ordinary atoms and therefore would be unable to chemically bond. The *ether* could be likened in this case to helium or argon.

In 1904, Mendeleev published a fascinating small work[54] in which he expressed his concept of the *ether*:

> The ether may be said to be a gas, like helium or argon, incapable of chemical combination. . . This point lies at the basis of our investigation into the chemical nature of ether, and includes the following two fundamental propositions: (1) that the ether is the lightest (in this respect ultimate) gas, and is endowed with a high penetrating power, which signifies that its particles have, relative to other gases, small weight and extremely high velocity, and (2) that ether is a simple body (element) incapable of entering into combination or reaction with other elements or compounds, although capable of penetrating their substance, just as helium, argon, and their analogues are soluble in water and other liquids.

In 1869, in the act of bringing the periodic table to birth—continued Mendeleev—an element like the *ether* was not even remotely conceivable, but in the fact that his predictions were shown to be reliable, they corroborated the entire periodic system of the elements, so much so that he claimed, without much modesty, that he had discovered an "absolute" law comparable to Newton's. Following this assertion, Mendeleev ventured to make some additional remarks about the *ether*, the element lighter than hydrogen. He treated the "ether gas" as an interstellar atmosphere composed of at least two lighter-than-hydrogen elements. He stated that these gases originated due to violent bombardments internal to stars, with the sun being the most prolific source of such gases. According to Mendeleev's booklet, the interstellar atmosphere was probably composed of several additional elemental species.

There were two considerations that determined Mendeleev's position with respect to the *ether*. In the first place, he did not think he had long to live. Mendeleev was actually obsessed with death. In 1905, he sent a sealed packet to the prime minister, Sergei Witte (1849–1915) with instructions to follow in case of his death and how to provide for the support of his wife and his young children who were still living at home. In the second place, in his later years, Mendeleev had heard a lot of discussion about the subdivision of atoms into electrons. These were imaginative speculations that rested on the existence of the first real subatomic particle, the electron. At first, when Mendeleev feared that the electron could compromise the survival of the periodic system, these ideas, from his point of view, were all smoke and mirrors, but later he welcomed them, although with the hidden intention of adapting and inserting them into his periodic edifice. Hence his desire to define any notion about the *ether* on the basis of its physical properties and intermolecular forces to replace the vague ideas that were being bandied about on its chemical nature.[55]

To Mendeleev, it seemed that the right moment had arrived to speak about the chemical nature of the *ether* because no one had done so before. Since the *ether* was not "weighable," one could extrapolate its atomic weight from the periodic table. The periodic law supplied the upper limit of an element (indicated by the letter x) belonging to the group 0 and the period 0 ($x \leq 0.17$, taking H = 1). Mendeleev assured himself of the lower limit of the atomic weight of the *ether* by resorting to the kinetic theory of gases, from which he calculated the mass of a particle light enough to escape from the atmosphere of the heaviest known star at that time. From these considerations, he showed that the atomic weight of the *ether* would lie between $5.3 \times 10^{-11} \leq x \leq 0.17$, taking the atomic weight of hydrogen as 1.

After having expounded on how to extrapolate the atomic weight of the new element, Mendeleev concluded his speculations on the *ether* with these words: "I consider the majority of phenomena are sufficiently explained by the fact that the particles and atoms of the lightest element x capable of moving freely everywhere throughout the universe have an atomic weight nearly one millionth that of hydrogen, and travel with a velocity of about 2,250 kilometers per second."

Mendeleev was overtly hostile toward all those phenomena that did not fit in with his atomic edifice and might for this reason be able to damage or destroy it. He acquired a growing conservative attitude with the passing of the years, and not even the phenomenon of radioactivity escaped his merciless criticism. When, in 1902, he went to Paris to visit the Curies, he was shocked to hear of an element that transmuted itself into another by emitting helium. In his eyes, this presented the risk that his periodic table would lose its centrality and importance. When he was shown the newly discovered radioactive elements, eka-barium and eka-tellurium, instead of being happy about it, he remarked to a friend:[56] "Tell me! How many grams of radium salts are there on the Earth? A couple of grams? And with this ridiculous amount they want to destroy my idea of the nature of matter?"

In his last years, Mendeleev referred more and more often to Sir Isaac Newton (1642–1727) as one of his precursors. It might seem strange that Mendeleev considered himself the rightful follower of Newton, a physicist, and not of Lavoisier, a chemist, but on closer examination, he seemed to be on target. Lavoisier's fame grew exponentially only after about a century following his death, whereas Newton became instantly famous after his formulation of the law of gravity, much like Mendeleev who received fame and honor at only 35 years of age after his publication of the periodic law. Furthermore, Newton's laws predicted discoveries (like Halley's Comet and the planets Uranus and Neptune), analogously to the periodic law that had predicted the discovery of the eka-elements. Lavoisier had not done anything like this. Mendeleev considered that the discoveries of an entire group of noble gas elements (from 1894), of radioactivity (1896), and of the electron (1897) were violent attacks on his periodic law. He became very discouraged and nurtured the idea that great changes in the periodic table were imminent. As a first step in avoiding what he considered a "death knell" for his system, he decided to incorporate the entire family of the noble gases into his periodic system. But he still maintained the existence of the two gases,[57] the *ether* and *coronium* (with a mass of 0.4, taking hydrogen as a base).

Mendeleev took to heart his project of expounding the chemistry of *ether* because of his own longing to be recognized as the rightful successor of Newton. In his article on the *ether* as a chemical, he added a brief footnote: "I would like preliminarily to call it newtonium—in honor of the immortal Newton." In the preceding draft of the article, Mendeleev scribbled his second consideration regarding the ether, but it was almost illegible and never published: "[The ether is] the lightest elementary gas which penetrates everything [Row 0, Group 0] which I would like to preliminarily call newtonium, since the thoughts of Newton penetrate all parts of mechanics, physics, and chemistry."

On February 2 (January 30 according to the old Julian Calendar), 1907, Dmitri Ivanovich Mendeleev, often hailed as the most renowned of all Russian chemists, died in Saint Petersburg at the age of 73, still tormented by the obsession that his periodic law could be overshadowed by recent discoveries and full of regret that he had, in his old age, espoused false assumptions and postulated the existence of some elements that did not exist. The man died; the myth was born.

VI.3.1. CORONIUM AND ITS AFTERMATH

In 1897, Stanislao Cannizzaro proposed that the Reale Accademia dei Lincei (the national academy of Italy, embracing both literature and science among its concerns) finance research into the nature of the gaseous emanations to be found in geologically active areas around Italy. This large-scale study had as its purpose the discovery of pockets of noble gases, such as argon and helium, as well as the presence of possible new elements. The work was assigned to three men: Raffaello Nasini (1854–1931), a Sicilian chemist and ex-student of Cannizzaro's; Francesco Anderlini (1844–1933), a chemist and Italian patriot who led an adventurous life; and Roberto Salvadori (1873–1940), a young student of Nasini's. The places selected for analysis were the active sulfur beds of Pozzuoli, the boric acid fumaroles of Tuscany, and some other fumaroles in the Apennines of Tuscany and Emilia Romagna.

In the spectra of the collected gases, the three chemists discerned a rather bright line whose wavelength was identical to the *coronium* proposed by Mendeleev, an element lighter than hydrogen and whose existence had apparently been observed in the sun's corona, but never on Earth. In addition to this "element," Nasini and his co-workers also found three very bright lines that did not belong to either helium or argon. Strangely, they seemed to come from iron, potassium, or titanium, elements whose presence in these gases was highly improbable. This caused the three chemists to propound the idea that they had in their hands one or more new gaseous elements which, for the sake of prudence, they did not dare assign even provisional names.[58]

The Reale Accademia dei Lincei did not refinance the project the following year, and so the research on these mysterious gases was terminated abruptly. It is quite possible that Nasini, who was a very fine chemist, was aware of the erroneous interpretation of the observed spectra or that he had discovered a source of contamination in the collected gases. In any event, he seems to have voluntarily abandoned his research into these hypothetical gaseous elements; mention of them never appeared in the literature again.

VI.3.2. THE GEOCORONIUM HYPOTHESIS

Molecular nitrogen was long believed to be the principal element whose lines composed the spectrum of the aurora borealis. However, a completely unknown green line appeared in the spectrum, one whose origin was hotly debated even as late as 1920. In 1911, Alfred Wegener attributed this line to a hypothetical gas[59] called *geocoronium*, with a mass of 0.4, whose existence had been advanced years before by Mendeleev. In his contribution, coming at the time when modern meteorology was developing, Wegener arrived at the conclusion that the atmosphere was composed of four strata:

- The troposphere, the innermost region, where the clouds are located and storms develop; its height is about 11 km depending on the latitude; its temperature is inversely proportional to altitude;
- The stratosphere, which extends to a height of about 70 km above Earth and has a roughly constant temperature of –55 °C;
- The hydrogenous sphere, which extends up to about 220 km and is characterized by extremely low pressure (ca. 0.01 mmHg); and
- The *geocoronium* sphere, which has an approximate extension to about 500 km above Earth.

The stratum formed by *geocoronium*, which has an atomic weight of 0.4, would not be neatly separated from the hydrogenous sphere as the other spheres are from one another.

In the following year, 1912, Wegener confirmed the existence of this last stratum in the Earth's atmosphere[60] by identifying the presence of *geocoronium* in the aurora borealis. If we attribute to Mendeleev the doubtful credit of having hypothesized the existence of this gas, Wegener was the first who showed us how to find it in nature. Furthermore, he said that *geocoronium* was responsible for the luminescence of the aurora.

Alfred Lothar Wegener (1880–1930) was born in Berlin on November 1, 1880. He was an interdisciplinary scientist who was interested in the many-faceted aspects of the physical sciences, particularly meteorology, but his name will always be associated with his theory of continental drift (*Kontinentalverschiebung*). In 1904, Wegener took a doctorate in astronomy at the University of Berlin, but he immediately became interested in meteorology, making many balloon ascents with the purpose of tracing the routes of air currents.

Wegener also took part in a number of expeditions to Greenland to study the circulation of polar air. On his last study voyage, he and his assistant Rasmus Villumsen (1909–30) disappeared among the ice fields of Northern Greenland. His body was found 6 months later reverently buried and in perfect condition; it is speculated that he died of heart failure from overexertion in a hostile climate.

Always opposed to the existence of *geocoronium*, the Norwegian scientist Lars Vegard (1880–1963) tried on many occasions to find the origin of the unknown line in the spectrum of aurora borealis. Vegard experimentally reproduced the conditions of the atmosphere at high altitude in the laboratory: he bombarded a crystal of solid nitrogen[61] with cosmic rays and obtained a spectrum virtually identical to the aurora borealis without having recourse to the hypothesis of *geocoronium*. However, Vegard's work was disputed by the Canadian team of John McLennan (1821–93) and Gordon Shrum who showed that the mysterious green line did not arise from solid nitrogen but from a "forbidden" transition of atomic oxygen.[62] Their explanation stands to this day, as documented in H. Kragh's comprehensive review article on the subject.[63]

A sidebar to the search for the origin of the green line led a philosopher who had worked as an assistant at the Harvard College Observatory, Charles Sanders Peirce (1839–1914), to hypothesize the existence of a hitherto unknown element in the auroral gas that he called, appropriately enough, *aurorium*. This hypothetical element appeared for the second time in the chemical literature in 1923, when it was mentioned by B Smith Hopkins in his book on the rarer elements.[64]

Aurorium's first appearance came in 1867 when Jonas Anders Ångström (1814–74) observed a particular line in the spectrum of the aurora borealis that led him to hypothesize on the existence of a new element. Ångström was born in Lögdö, Sweden, on August 13, 1814. He became a very well-known physicist and is rightly considered one of the founding fathers of the science of spectroscopy. In 1843, he was appointed director of the Astronomical Observatory of the University of Uppsala, and 15 years later, he became professor of physics there. Combining the spectroscope with photography for the study of the solar system enabled him to prove that the sun contains hydrogen. At the age of 44, he published his research on the solar spectrum in a volume[65] that includes detailed measurements of more than 1,000 spectral lines. He was the first to examine the spectrum of the aurora borealis, and he identified and measured the characteristic bright line in its yellow-green region. These studies led him to mistakenly assume the presence of a new element that he named *aurorium*.[66] Not yet 70, Ångström passed away at Uppsala on June 21, 1874.

As late as 1918, the existence of *aurorium* was considered possible. During solar eclipses, lines attributed to either *coronium* or *aurorium* were observed. Astronomers asked themselves if the presence of the *coronium* line in the sun's corona could be regarded as an electromagnetic phenomenon similar to what they observed for *aurorium* in the aurora borealis.

VI.3.3. *ETHERIUM*: ELEMENTARY GAS OR SUBATOMIC PARTICLE?

The *ether* of Mendeleev could have been the discovery of Charles Brush who, in August 1898, in a paper read before the American Association for the Advancement of Science

(AAAS), claimed the discovery of *etherion* or *etherium*, a new elementary gas in the atmosphere.[67]

Brush, at the time about 50 years old, was an American physicist who enjoyed a solid reputation in his home country. On August 24, 1898, at the end of the AAAS conference, readers of the *Cleveland Plain Dealer* and the *Boston Evening Transcript* would have learned of Brush's presumed discovery. The existence of *etherium* was published on the following day in the *Cleveland Leader*,[68] and the news continued to appear in the columns of daily local papers in large city newspapers nationwide.[69] Finally, after the discovery was reported by the *New York Commercial*[70] on September 2 and by the *New York Sun* [71] on September 6, and publicized in specialized journals as well,[72] the news hopped across the ocean and appeared in the pages of many European scientific journals[73] up until the end of 1898.

Before the end of the year, an exhaustive work on *etherium* appeared in the *Journal of the American Chemical Society*.[74] In this publication, Brush discussed the discovery of the new gas, provisionally called *etherion*, with a clear religious reference to "high in the Heavens." According to the author, this gas was a constituent of the atmosphere and would be contained in many substances. At the time of publication, Brush had only determined what he thought was the principal property of this gas, that is, its high conductivity of heat at very low pressures. At the conclusion of his article, he published a detailed comparison of the thermal conductivity of *etherium* with other known gases.

Brush thought that the new gas was lighter than air and even of hydrogen and could be separated from the atmosphere by successive diffusion techniques. He devoted himself to this end for many months by carrying out complex experiments. He obtained his best results by filtering the air with porous porcelain. By bringing a porcelain tube to a pressure of 1.3 mmHg, Brush was able to collect by diffusion about 19 cubic centimeters per hour of the new gas. A strange result of his work was his observation of the chemical properties of *etherium*. He observed that phosphorus pentoxide and lime were able to absorb the new gas, two clues that would have made a chemist suspicious. He, on the other hand, felt that *etherium* was so light that it could easily penetrate ordinary matter. The answer to this puzzle was even simpler in that he did not need to postulate the existence of a subatomic gas, and it was supplied, not even 2 years later, by one of his Polish colleagues. Brush was a competent physicist with the interests of a polymath, and he started looking for a practical application for his discovery. Certainly, he sought to produce new ways of verifying his hypothesis on the existence of *etherium*, but never succeeded.

In 1900, a young Polish physicist named Marian Ritter von Smolan Smoluchowski harshly refuted Brush's discovery.[75] This young and little-known Pole attacked the renowned American physicist on a matter totally distinct from *etherium*. He reminded his colleague that, years ago, James Clerk Maxwell had proven that the thermal conductivity of a gas was independent of the pressure except at much reduced pressures or in the vicinity of a solid–gas interface. The supposed existence of the new gas was indeed not justified, and Brush's presumed new element was probably nothing more than water vapor.

Smolan Smoluchowski was a pioneer in statistical physics. A fervently nationalistic Pole, he studied in Vienna, then the capital of the Habsburgs. In 1913, he obtained the chair of experimental physics at Cracow, at that time still part of the Austro-Hungarian Empire. When he was only 45 years old, he died of dysentery in 1917, during World War I.

The events that characterized the life of Charles Brush were certainly a bit happier than those of his Polish colleague. After the unfortunate *etherium* fiasco, Brush wrote many articles between 1910 and 1929 in which he made known his personal views on the "kinetic theory of gravitation" based on a type of new electromagnetic wave. Despite the fact that in his late old age he had become a scientist who, in certain respects, was swimming against the tide and harbored bizarre ideas, he continued to be recognized by everyone as a skillful inventor and an accomplished physicist, not to mention a generous philanthropist. He died on June 15, 1929, at 80 years of age.

Notes

53. Gordin, M. D. *A Well-Ordered Thing: Dmitrii Ivanovich Mendeleev and the Shadow of the Periodic Table*; Basic Books: New York, 2004, p. 210.
54. Mendeleev, D. *An Attempt towards a Chemical Conception of the Ether*; Longmans Green & Co.: London, 1904. This work is now available in a modern rendition in Jensen. W. B., Ed. *Mendeleev on the Periodic Law: Selected Writings, 1869–1905;* Dover Publications: Mineola, New York, 2005.
55. Bensaude-Vincent, B. L'éther, élément chimique: un essai malheureux de Mendéleev en 1904. *Brit. J. Hist. Sci.* ***1982***, *15,* 183–188.
56. Morozov, N. *D. I. Mendeleev i znachenie ego periodicheskoi sistemy dlia khimii budushchago*; I. D. Sytin: Mosca, Russia, 1908.
57. Mendeleev, D. *Vesnik i Biblioteca Samoobrazovanii* **1903**, *1–4*, 25; 83; 113; 161.
58. Nasini, R.; Anderlini, F.; Salvadori, R. *Rendiconti della Regia Accademia dei Lincei* **1898**-II (5) 7, 73.
59. Wegener, A. *Physik. Z.* **1911**, *12,* 170; Wegener, A. *Physik. Z.* **1911**, *12*, 214.
60. Wegener, A. *Z. anorg. Chem. B.* **1912**, *75*, 107.
61. Vegard, L. *Nature* **1924**, *113*, 716.
62. McLennan, J.C.; Shrum, G. *Proc. Roy. Soc. London* **1925**, *108*, 501.
63. Kragh, H. *Bull. Hist. Chem.* **2010**, *35 (2)*, 97.
64. Hopkins, B S. *Chemistry of the Rarer Elements*; D.C. Heath & Co.: Boston, 1923.
65. Ångström, J. A. *Recherches sur le Spectre solaire*; W. Schultz: Uppsala, Sweden, 1868.
66. Ångström, J. A. *Nova Acta Uppsala Sci.* **1867**, 9(3), 29.
67. Brush C. F. "New Gas in the Atmosphere," presented at the AAAS Meeting of 1898.
68. "Etherion Is the Name of the New Gas," *Cleveland World*, Aug. 24, 1898; "The New Gas," *Cleveland Leader*, Aug. 25, 1898.
69. "A New Gas," *New York Evening Post*, Aug. 24, 1898; "A New Gas," *Cleveland Plain Dealer*, Aug. 24, 1898; "New Gas in Atmosphere," *Boston Evening Transcript*, Aug. 24, 1898; "The New Element," *Cleveland Plain Dealer*, Oct. 4, 1898.
70. "The Composition of Air," *New York Commercial*, Sept. 12, 1898.
71. *New York Sun*, Nov. 6, 1898; "A New Gas," *New York Sun*, Nov. 8, 1898.
72. *Electrical Engineer*, Sept. 16, 1898.
73. Brush, C. *Chem. News* (October 21, 1898), 197; Anon. *Compt. Rend.* **1898**, 767; Anon., *Bulletin de la Sovele Des Sciences Naturelles*, Nov. 11, 1898.
74. Brush, C. F. *J. Am. Chem. Soc.* **1898**, *20*(12), 899.
75. von Smolan Smoluchowski, M. R. *Oesterr. Chem. Zeit.* **1900**, *2*, 385.

VI.4

ANODIUM AND CATHODIUM

A curious episode, almost totally forgotten, is the one associated with the nebula hypothesis of two subelements, and it deserves to be treated, even if marginally, in this connection. In 1907, while Mendeleev was dying in Saint Petersburg, in Moscow an obscure Russian physicist was printing a monograph in which he expounded his ambitious theory on how chemical bonds are formed between elements.[76] The author, Nikolai Aleksandrovich Morozov (1854–1946), had been a friend of Mendeleev, but he was also a prominent anti-Tsarist revolutionary. He had written this monumental work while imprisoned in Schlisselburg Fortress and in other prisons of Tsarist Russia.

Morozov was a scientist and scholar entirely original in his thinking, given the fact that, considering the 20 years of his virtual isolation in prison, he was unable to keep up to date with developments taking place in chemistry and physics. In addition to his opus magnum, he also authored many essays in the fields of the social sciences, chemistry, physics, biology, astronomy, cosmology, botany, geophysics, meteorology, and aeronautics; he was also an esteemed poet and philosopher. After pursuing his initial interests in electrotechnology, he was imprisoned in 1874 for his anti-Tsarist activities and was freed only after the revolution of 1905. From then on, he devoted many years to the publication of his works, many of which were inevitably out-of-date.

In his book, Morozov expressed his very personal point of view with respect to atomic structure and the nature of the chemical bond. According to him, matter was composed of atoms of negative and positive electricity that he called *anodium* and *cathodium*. His theory required that all the chemical bonds in inorganic compounds, as well as in organic compounds in some cases, have an electrolytic character. They were formed by means of the "*cathodium* valence" of one atom and the "*anodium* valence" of the other. However, things did not seem to work out correctly using these terms, and the first person to recognize this was Morozov himself, who was unable to explain the nature of the carbon–carbon bond. For this reason, Morozov hypothesized that the C–C bond had a different nature from all the other chemical bonds formed by the other atoms. In short, he did not think his theory was erroneous, but rather that nature had made an exception with the C–C bond. In this case, each atom of carbon would use its own "*cathodium* valence," thus doubling the total positive charge. To explain why two positive charges did not repel one another, Morozov postulated that charges of the same sign, placed in very close contact with one another, would cease to repel each other and would form a "common field."

Although Morozov continued for many years to propound his *anodium-cathodium* theory, his contemporaries associated his name with the unfortunate hypothesis of like charges forming a common field, and thus he became scientifically marginalized. However, a year after his book appeared, Johannes Stark (1874–1957) also came up with an original, limited, but more orthodox theory on the nature of the chemical bond in organic compounds.[77]

A curious fact emerges from all this. Morozov could be held up as an example of "controlled convergent evolution" because his original ideas and his discoveries, which lay outside the context of the scientific community, developed in a completely independent manner, but, the historical and scientific context being sufficiently mature, other colleagues scattered throughout the world came up with the same ideas.

Years later, when the communists had consolidated their power following the October Revolution of 1917, they elected Morozov an honorary member of the Academy of Sciences of the Soviet Union. In his old age, Morozov became director of the P. S. Lesgaft Institute of Natural Sciences. The Soviet government decorated him with two medals of the Order of (Vladimir) Lenin (1870–1924) and one of the Order of the Red Flag of Labor. But the sign of honor that must have impressed him most as patriot and revolutionary was what the government awarded him as a type of "ransom" for his long stay in prison. An industrial complex in the area of Leningrad, not far from Schlisselburg Fortress, was named after him.

In the 1920s, also on Lenin's initiative, Morozov was given the extensive property of Borok, the village where he was born, and there he died on July 30, 1946. On his tomb, the sculptor Georgiy Ivanovich Motovilov (1884–1963) erected a statue of Morozov seated with a book in his hands, an open allusion to the long years of prison and of forced idleness spent in study and thinking.

Notes

76. Morozov, N. A. *The periodic System of the Structure of Substances. Theory of Formation of Chemical Elements*; Sytin Publishers: Moscow, Russia, 1907.
77. Stark, J. *Phys. Z.* **1908**, 9, 85; Stark, J. *Jahrb. d. Radiakt. und Elektr.* **1908**, 5, 124.

VI.5

THE EXOTIC DAMARIUM

The last and probably the most fantastic and least known of all the ethereal elements is certainly *damarium*. What befell *damarium* was bizarre, an event marginally associated with chemistry, but having more the character of an adventure novel. These are the words of Fried (or Fraenkel) Much[78] as he defined it: "The world of chemical processes is similar to the scene of a drama on which takes place the wonderful story of the discovery of damarium."

At the height of the European Colonial era, mining engineer Karl Lauer, a subject of the Kaiser, was in Namibia (then under the imperial German crown) for the purpose of finding new mineral deposits. His work in Africa began in 1888, but it was only on April 2, 1890 that he and his friend, a chemist named Paul Antsch, sent their vivid account to the weekly *Chemiker Zeitung*.[79]

Lauer's area of investigation was Damaraland, an arid and mountainous region of Namibia, squeezed between the Namib and Kalahari deserts. An expert geologist, he noted that the area was mostly red sandstone whose formation could be dated to more than 150 million years in the past. What really impressed him was finding, at the top of a plateau about 2 km square (about 0.8 square miles), 17 rocky depressions in the form of funnels with perfectly circular circumferences. Their diameters were between 0.2 and 0.8 m (about 8 inches to 2.5 feet) and in none of the craters examined was the depth greater than 2 m (about 6.5 feet). Lauer noticed that the edges of the craters were composed of a material different from that of the surrounding terrain. A fact yet stranger was that inside these small craters were the bodies of many dead animals (mammals, birds, reptiles, and insects). He also saw that the bodies of other dead animals, perfectly preserved, lay round about them, which caused him to suppose that a prevailing wind would have carried the poisonous vapors spewing from the craters far enough to kill these animals as well.

During the 20 minutes that he and his bearers had inspected the small craters, they noticed with astonishment that their blue uniforms, their beards, and their hair had become white. Other sensory impressions that warned the explorers to leave the area were nausea, malaise, and a rotting odor that one associates with sulfur compounds. Overcome with the heat from a tropical sun, they removed their sun helmets and mosquito netting in order to take a drop of cognac. Having removed their headgear, they could hear the sound emitted by these natural funnels, somewhat like a boiling teakettle.

Lauer had another moment of astonishment when he found that the taste of the cognac was changed, and he put that down to the gaseous fumes as well. Heedless of the danger, after preparing the necessary equipment, he lowered himself into one of the fumaroles carrying three bottles with him to collect the gas. When he came out, he felt very ill, but, indefatigable, he told his co-worker to seal the bottles with wax. When his colleague lit a flame in order to heat the wax, one of the small fumaroles nearby immediately caught fire. Lauer immediately extinguished it with his sun helmet.

FIGURE VI.02. Volcanic Region in Damaraland (Namibia). In these craters, at the end of the 19th century, the German colonial official Karl Lauer and the chemist Paul Antsch collected a gas that seemed to have completely unknown properties; they called it *damarium*. Later, the gas was shown to be a mixture of hydrogen and hydrogen sulfide. Photograph by Mariagrazia Costa.

When they returned to camp, upset by what had happened and still feeling ill from the effects of the gas, Lauer wrote a long report and sent the three sealed bottles containing the mysterious inflammable gas to his friend Paul Antsch, a chemist in Capstadt, Germany, so that he could analyze them. His response came at the beginning of 1889.

Antsch pointed out immediately the reducing character of the gas and reported his bewilderment over its inconsistent properties. After having bubbled 10 cm^3 of the gas into a buret filled with water, he began to study its reactivity. With oxygen it formed an explosive mixture that yielded an oxide. Three cm^3 of this oxide were electrolyzed: at the positive electrode, Antsch collected oxygen, whereas at the negative electrode a mysterious elemental gas was emitted. From the ratio of the two volumes collected, Antsch calculated that the atomic weight of the gas was surprisingly small: 0.5, that is, only half the weight of hydrogen. Not at all upset by this bizarre result, he named the new element: "With 'D' ['Damarium'] I name this strange new gaseous element, which is present in the elemental state in the gaseous emissions [in the region] of Damara." Figure VI.02 is an image of the fumaroles that emitted the mysterious elemental gas.

The formula that he proposed for the oxide was D_4O. *Damarium* would be the element with the smallest atomic weight. According to Antsch, the formula for the oxide of *damarium* would make it necessary to review the valence of oxygen and the other elements. His reasoning was based on the theory of the tetrahedral atom of Jacobus Henricus van't Hoff and Joseph Achille Le Bel (1847–1930).

In addition, according to Antsch, *damarium* would also be the element with the highest reduction potential, able to reduce to the elemental state salts of platinum, gold, silver, copper, and even lead. It was also able to reduce sulfur dioxide to elemental sulfur. Indigo solutions were decolorized by the gas. And *damarium* reacted with the oxides of phosphorus, forming the following acids:

$$P_2O_5 + D_4O \rightarrow 2PD_2O_3 \qquad \text{(Eq. VI.1)}$$

$$P_2O_3 + 3D_4O \rightarrow 2PD_6O_3 \qquad \text{(Eq. VI.2)}$$

Additional work confirmed that *damarium* was also able to reduce carbon dioxide to carbon monoxide. Antsch did not try to react the gas with either chlorine or hydrochloric acid for fear of an explosion.

Antsch went on to explain to Lauer why he had not detected the classic odor of cognac when he was trying to taste it in the vicinity of the fumarole. His explanation was based on the fact that gaseous *damarium* would have immediately reacted with the ethyl alcohol to produce *damarium* oxide with its characteristic rotten odor. In trying to resolve the puzzle of the cognac's odor and taste, Antsch proposed unwittingly the empirical formula for gaseous *damarium*, D_4, but he didn't seem to care much about the consequences to which theory this might lead.

News of the discovery of the new element was taken up by some specialized journals,[80] but either their lack of widespread distribution or the impossibility of any chemist going to fetch another sample of the gas led to the fact that no one seemed to be interested in the mysterious element, nor could they refute the bizarre properties that were described. In all probability, *damarium* was nothing more than a mixture of hydrogen and hydrogen sulfide: the first gas would have been the cause of its remarkable reducing properties and the second of its unpleasant odor.

Notes

78. Much, F. *La Revue Scientifique—Science*, 1925, 259.
79. Antsch, P.; Lauer, K. *Chemiker Ztg.* **1890**, *14*(27), 435.
80. *Proc. Am. Pharm. Assn.* **1890**, 571; *Chem. News* **1890**, 217; *Pharmazeutische Zentralhalle fur Deutschland* **1890**, 231 & 252; *The Pharmaceutical Journal and Transactions of Pharmaceutical Society of Great Britain*, **1890**, 893.

VI.6

SUBTLE IS THE AIR: THE CASE OF *ASTERIUM*

During the solar eclipse of August 18, 1868, Joseph Norman Lockyer (1836–1920) observed a brilliant orange-yellow line in the spectrum of the solar photosphere.[81] At first, the yellow line was confused with the sodium-D line. Later, Pierre Jules César Janssen (1824–1907) established the separate identity of this line,[82] which formed a trio together with the sodium-D doublet. Because of this chance proximity to the other two lines, it was designated as D_3.

At first, Lockyer felt that the substance responsible for the emission of line D_3 could be an allotropic form of hydrogen or a product of "dissociation."[83] Lockyer's assumption was dismissed almost immediately: Father Angelo Secchi (1818–78) of the Vatican Observatory realized that this line could not possibly be attributed to hydrogen,[84] so the hypothesis of a new element began to make its way into the minds of the three spectroscopists.

Although in the past Lockyer had put forth hypotheses that were somewhat bizarre, he was the first who recognized the new element and to call it helium. The next step was made by another Italian. In 1881, Luigi Palmieri (1807–96) rediscovered the mysterious D_3 line in the spectrum of gases being emitted from Vesuvius.[85] This observation was useful for establishing the presence of the new element not only in the sun, but also on Earth, but it did not after all throw new light on the chemical nature of this elusive element.

After the discovery of argon, Sir William Ramsay, with his assistants John Norman Collie and Morris William Travers, decided to repeat some experiments of William Francis Hillebrand (1853–1925). Hillebrand had observed that minerals like uranite or clevite gave off nitrogen when they were treated with sulfuric acid, or if they were melted with a mixture of alkaline carbonates.[86] In 1895, Ramsay extracted some of this supposed nitrogen from the minerals in question, and Sir William Crookes observed that the spectrum showed the unmistakable presence of line D_3: what had rashly been assumed to be nitrogen was none other than helium.[87]

With the help of a more sophisticated instrument, Aristarkh Bélopolsky (1853–1934) showed that the line in question was in reality a doublet.[88] A short time later, other spectroscopists, among them Louis Karl Heinrich Friedrich Paschen[89] (1865–1947), Sir William Huggins[90] (1824–1910), George Ellery Hale[91] (1868–1938), and Henri Alexandre Deslandres[92] (1853–1948) supplied sufficient data to establish that terrestrial helium and the element present in the solar corona were identical.

In that same period, physicist Carl David Tolmé Runge (1856–1927) together with his colleague Heinrich Gustav Johannes Kayser (1853–1940) were intent on examining spectroscopically a large range of chemical elements[93] and came across Paschen's work. Because Paschen and Runge worked at the same university and their research interests matched, they began a very fruitful collaborative effort. They combined their theoretical and experimental expertise to map out a complete spectrum of helium. The recorded spectrum ranged from the ultraviolet to the infrared regions and showed three repetitive arrangements of doublets, as well as many other single lines. They decided to publish their results in the English journal *Philosophical Magazine* because they thought the propagation of their work would be easier, but also to somewhat challenge the discoverer of helium, J. Norman Lockyer,[94] who often announced his more sensational discoveries in that journal.

The "true" helium was supposed to be the one that gave rise to the first series of doublets; the second group of lines was attributed to a foreign element that they provisionally called *co-helium*.The authors, with much care and attention, mapped out and recorded all the lines found in the spectrum of the gas extracted from cleveite. With the help of a bolometer, used to record infrared rays, they scanned the infrared region, which had remained at that time totally unexplored, and recorded six new series of lines that they considered belonged to two different atomic systems. Furthermore, compared with the spectra already known, they arrived at establishing the presence of a constituent of higher atomic weight, a result confirmed via a simple practical observation: the gas was allowed to enter through an asbestos window into a Geissler tube kept under vacuum. The lines attributed to "true" helium did not reach their maximum intensity rapidly. Based on this experimental observation, the co-workers assigned atomic weights of 5 and 3 to the two probable constituents.

With the passing of time, Runge and Paschen felt ever more secure about their observations and asserted that helium was composed of a mixture of two gases: *orthohelium* (or the helium already known) and *parahelium*, the first characterized by a marked yellow spectral line, the other by a green line.

But it fell to the authoritative Lockyer to have the last word; he wanted to call the two elements by different names. Since he was still enamored of a name that he had proposed many years before, he decided to call the first element helium again, whereas for the second gas the choice fell on *asterium*, from the Greek *astros*, meaning star. Lockyer put forth this name not only to satisfy his own passion for astronomy, but also because he discovered the same lines, not only in the gas produced by cleveite and in the solar corona, but also in the hottest stars.[95]

The truth was far from being reached because there was no unanimous opinion on these experimental observations. As an example, in 1895, the American E. A. Hill interpreted the spectrum of helium published a short time before by Crookes in a completely different manner. In his opinion, it was quite clear that the spectrum contained evidence of the presence of 15 new elements, as many as the new lines observed![96]

William Ramsay and John Norman Collie asserted that some clarification was needed if science were to make any progress in this particular area. They therefore decided the riddle of the spectrum for helium and *asterium* once and for all. Through the process of separating a gaseous mixture obtained by diffusion through a porous membrane (atmolysis), they claimed that they had succeeded in separating helium into two components, one portion with a lower density and one with a greater density. If the separation operation seemed crowned with great success, spectroscopic observation left the physicists greatly perplexed: both fractions had an identical spectrum. The meteorologist William Jackson Humphreys[97](1862–1949) and the physicist Joseph Sweetman Ames[98] (1864–1943) arrived independently at the same conclusion.

No scientist was able to unravel the mystery, and so a hypothesis was put forward that satisfied very few: Ramsay had separated a mixture of helium into two identical constituents but with different weights. If he were correct, the path he was following would have led, about 20 years ahead of its time, to the concept of the isotope, but the assumptions were wrong and the times were not ready. Working on the process of atmolysis, Ramsay was unwittingly moving farther away from a solution to the entire problem.

The riddle of *asterium* was in fact solved by Runge and Paschen and only later by Ramsay's assistant, Travers. The two German physicists showed that the complexity of a spectrum is not in and of itself proof for establishing if an element is pure, in a mixture, or chemically combined with other elements. Ramsay and Travers proved that the color of helium's spectral line depended on the gas pressure. *Asterium* did not exist, simply because the mysterious lines that appeared in the spectrum were those of helium recorded, time after time, under different experimental conditions.

It was thus that this new gaseous element, after a fleeting appearance, disappeared from the list of simple substances, but no one seemed to feel its passing. Within a few years, Ramsay and his colleagues enchanted the world with their discoveries of argon, neon, krypton, and xenon. And a few years later, between 1899 and 1900, an international team of scientists[99] discovered the last and heaviest of the noble gases, radon, or *niton*.[100]

We turn back now to the two physicists who were the unwitting discoverers of *asterium*. Lockyer was born at Rugby, in England, on May 17, 1836. After what has already been related here, he partially changed his scientific interests: he was the first to be convinced that Stonehenge and other similar stone circles scattered in the southern parts of Great Britain were built for astronomic purposes. His ideas brought about the first crude archaeological research efforts in these areas. For this reason, he has been called the Father of Archaeoastronomy. Before retiring to a private life in 1911, Lockyer established an observatory near his house at Salcombe Regis, in Devon, where, on August 16, 1920, he died at the age of 84.

Runge, born on August 30, 1856, was much younger than Lockyer. After having left school at the age of 19, he spent 6 months traveling with his mother around the cultural centers of Italy. When he got back to Germany, he enrolled at the University of Munich to study literature, but abandoned this idea after he heard a lecture by the renowned mathematician Carl Weiersstrass (1815–97). He wrote to his mother that at that moment he changed his course of study and intended to take his degree in mathematics.

After having succeeded brilliantly in all his examinations and beginning study for his doctorate, he devised a procedure for the numerical solution of algebraic equations, invented a method for approximating the solution to differential equations, and became interested in spectroscopy. Runge was moved to seek a mathematical relationship among the wavelengths and spectral lines of all the elements. J. J. Balmer (1825–98) had found a similar empirical law, but it was only good for the spectral lines of the simplest element, hydrogen.

Thanks to his great scientific ability, Runge was appointed to a professorship at the University of Hannover and there he remained for 18 years. In 1904, he was persuaded to accept a professorship in mathematics at the University of Göttingen, and this he held until the year of his retirement in 1925. Runge was an imposing and energetic man. He remained active and gifted with extraordinary vitality until the end of his life. It is said that on the occasion of his 70th birthday, he entertained his young grandchildren by walking on his hands upside-down. But a few months later he suffered a fatal heart attack and died on January 3, 1927.

Notes

81. Lockyer, J. N. *Phil. Mag.* **1869**, *37*(4), 143.
82. Janssen, J. P. C. *Compt. Rend.* **1869**, *68*, 93; Janssen, J. P. C. *Compt. Rend.* **1869**, *68*, 181; Janssen, J. P. C. *Compt. Rend.* **1869**, *68*, 367.
83. Today we would say ionization.
84. Secchi, A. *Compt. Rend.* **1869**, *68*, 237.
85. Palmieri, L. *Rendiconti dell'Accademia di Napoli* **1882**, *20*, 233.
86. Hillebrand, W. F. *Bull. U. S. Geol. Sur.* **1891**, 78.
87. Ramsay, W.; Collie, J. N.; Travers, M. W. *Ann. Chim. Phys.* **1895**, *58*, 81; Ramsay, W.; Collie, J. N.; Travers, M. W. *J. Chem. Soc.* **1895**, *67*, 684; Ramsay, W.; Collie, J. N.; Travers, M. W. *Nature* **1895**, *52*, 7; Ramsay, W.; Collie, J. N.; Travers, M. W. *Nature* **1895**, *52*, 306; Ramsay, W.; Collie, J. N.; Travers, M. W. *Nature* **1895**, *52*, 331; Crookes, W. *Chem. News* **1898**, *78*, 221.
88. Bélopolsky, A. *Mem. Soc. Spettr. Ital.* **1895**, *23*, 89.
89. Born on January 22, 1865, he was known for his work on electrical discharge in gases and for the series of lines, named after him, in the infrared region of the hydrogen spectrum, which he observed for the first time in 1908. He became professor at the Polytechnic University of Hamburg in 1893, and 8 years later moved to take up the chair of physics at the University of Tübingen. He was also the president of the Physikalisch-Technischen Reichsanstalt in the 10-year interval of 1924–33. Made honorary professor at the University of Berlin in 1925, he continued to teach until his death, which took place at Potsdam on February 25, 1947.
90. Huggins, W. *Chem. News* **1895**, *71*, 283
91. Hale, G. H. *Astrophys. Journ.* **1895**, *2*, 165.
92. Deslandres, H. *Compt. Rend.* **1895**, *120,* 1112; *Deslandres, H. Compt. Rend.* **1895**, *120*, 1331.
93. Kayser was a German physicist born in Bingen am Rhein on March 16, 1853. He carried out his first scientific investigations on sound waves. He discovered the presence of helium in Earth's atmosphere. In 1905, he expounded his personal theory about electrons. The unit of measure in the CGS system called the *kayser*, related to wavelength, was named in his honor. He died in Bonn, at the age of 87, on October 14, 1940.
94. Runge, C.; Paschen, F. *Phil. Mag.* **1895**, *42*(2), 297.
95. Lockyer, J. N. *Proc. Roy. Soc. London* **1897**, *62*, 52.
96. Runge, C.; Paschen, F. *Nature* **1896**, *53*, 245.
97. Humphreys, W. H.; Ames, J. S. *Astrophys. Journ.* **1897**, *5*, 97.
98. He was known, among other things, for having been one of the founders of National Advisory Committee for Aeronautics (NACA), the government agency responsible for programs in air and space research that preceded the creation of NASA.
99. Among these were the renowned names of Pierre Curie, Marie Skłodowska-Curie, Ernest Rutherford, Robert Bowie Owens (1870–1940), Friedrich Ernst Dorn (1848–1916), and André-Louis Debierne.
100. This element was discovery independently by Ramsay and by Robert Whytlaw-Gray, who called it *niton* (from the Latin *nitens*, shining) to emphasize its characteristic of inducing phosphorescence in certain materials.

VI.7

CLAIRVOYANCE AS A MEANS OF INVESTIGATING SOME "OCCULT ELEMENTS"

The person behind research into the strange group of chemical elements carrying the fanciful name of "occult" was the Englishman Charles Webster Leadbeater (1847–1934), who also coined the term. When he was a little boy, he met the famous exoticist Edward Bulwer-Lytton (1803–73) who called his attention to the literature of the occult. When he was 13, Leadbeater emigrated with his family to Brazil, where at Bahia, during a rebellion, the rebels tried to make him, his father, and his younger brother Gerald trample on a cross. Refusing, Gerald was killed and Charles and his father were tortured. Returning to England, he studied at Oxford and, in 1878, he was ordained a priest of the Church of England.

His faith helped him very much in overcoming his childhood trauma, but his psyche was troubled all of his life. Leadbeater was known for his restlessness; his soul seemed to reach out in a continuous search for contact with the world of the dead. In 1883, after having met Alfred Percy Sinnett (1840–1921) and Helena Petrovna Blavatsky (1831–91), he joined the London Theosophical Society.[101] He quickly became known as a clairvoyant and wrote many books based on his extrasensory experiences. After leaving Europe, he went to Colombo, on the island of Ceylon (present-day Sri Lanka), where he was converted to Buddhism under the guidance of High Priest Waligama (or Weligama) Sri Sumangala. In 1884, he settled in Adyar, India, where he succeeded Damodar K. Mavalankar (1857–after 1930) as secretary of the Theosophical Society.

In 1889, he returned to England with the 14-year-old Curuppumullage Jinarajadasa (1875–1953) who would become the fourth president of the Theosophical Society and would play a large part in the study of the chemistry of the "occult elements." Thanks to Jinarajadasa's supernatural gifts, Leadbeater advanced in the development of his own clairvoyance. In 1890, he met Annie Besant (1847–1933), whom he taught how to develop clairvoyance, and he worked with her for more than 40 years. In 1893, they held their first public presentation on clairvoyant phenomena in a period characterized by considerable materialistic skepticism.

VI.7.1. A CLAIRVOYANT INVESTIGATES THE STRUCTURE OF NEW AND OLD ATOMS AND THEIR POSITION IN THE PERIODIC TABLE

Occult Chemistry: Investigations by Clairvoyant Magnification into the Structure of the Atoms of the Periodic Table and Some Compounds is a book written by Annie Besant, Charles Webster Leadbeater, and Curuppumullage Jinarajadasa, members of the Theosophical Society, with their center in Adyar, India. The first edition of the book[102] came out in 1908, but some "scientific" observations could place it at the end of the preceding century. The book was reprinted in 1919[103] and a third time in 1951,[104] when only one of the three authors was still living. Right from its first edition, the book aroused great curiosity and stirred up heated discussions that persist up to the present. The book proposed a strange view of the microscopic world, asserting that the structure of the chemical elements could be deduced by clairvoyant observations. Their esoteric investigations of matter were carried out over a long period of time: from August 1895 and at more or less regular intervals until October 1933. The book was composed of descriptions both of the presumed ethereal counterparts of ordinary atoms and of chemical elements unknown at the time of its publication, as well as speculations in the fanciful field of "occult physics." The academic world at the time was violently critical of Besant and Leadbeater's ideas, and such harsh criticism is still to be found today.[105]

The book's principal researchers, Besant and Leadbeater, were both very famous clairvoyants. Their research method was unique and difficult for a 21st-century person to comprehend. Because the meaning of the word *clairvoyance* ("to see clearly") implies knowledge of sounds and sights not perceived by ordinary persons, the way in which they used this cognitive capacity in the service of pseudoscientific investigations is completely indescribable.

The two researchers affirmed that the entire atomic universe, composed of atoms and molecules, was laid out before their very eyes. Everything that they observed and wrote down was not subjective—that is, the results of their imagination—but as it actually appeared in the submicroscopic world.

The object examined, whether it be an atom or a molecule, was observed exactly as it was, that is, not subjected to any perturbing forces like an electric or magnetic field or by the action of heat. Because all the objects in this submicroscopic world are in rapid motion, the only force applied to them was a special "force of will" that made the atomic movements slow enough to allow the clairvoyants to observe all the particulars in detail.

Their initial investigations were conducted in England in 1895, and the first atoms observed were four gases present in the air: hydrogen, oxygen, nitrogen, and a fourth gas (with an atomic weight of 3) not yet discovered by chemists. The problem for the clairvoyants was the identification of the elements themselves since, as Leadbeater and his colleagues said, "the atoms did not have labels on them." The most reactive of the four gases was the one that they decided was oxygen. The gas that seemed to them to be more "lethargic" they thought might be nitrogen. The smallest atoms of the four gases they identified as hydrogen.

Only after a more complete examination of the constituents of the gases was conducted successfully were Besant and Leadbeater able to assert that the atoms were not the ultimate building blocks of matter but themselves were composed of even smaller particles: they said that hydrogen was composed of 18 subparticles, nitrogen of 261, and oxygen of 290. Finally, they found that the fourth gas was composed of 54 of these units. The fourth gas, with an atomic weight of 3, was at first thought to be helium, about which much was said in 1894—both in the scientific journals and in the daily papers—following its discovery by Sir William Ramsay. When it was clear that the atomic weight of helium was 4, the gas observed by the clairvoyants, with its atomic weight of 3, was thought to be a new constituent in the atmosphere. Therefore Besant and Leadbeater (Figure VI.03 and Figure VI.04) coined for it the name of *occultum*.

More detailed descriptions of the internal structure of the atoms of hydrogen, oxygen, and nitrogen, and of the ultimate constituents of these atoms, which the two clairvoyants

FIGURE VI.03. Annie Besant (1847–1933), Chemist and British Occultist. Image Courtesy of the Theosophical Society in America Archives.

called *anu*, were published at London[106] in 1895. Their work was suspended for about 12 years and was only taken up again in 1907, the year in which they observed a good 59 elements by means of clairvoyance. When the elements could be obtained in sufficient purity, as for example, sulfur, iron, and mercury, the two investigators had no trouble in identifying them and discerning their relative structures. The difficulties arose in the case of lithium and some other elements. It was then that they requested very pure samples of these elements from Sir William Crookes, a friend of both as well as a member, for a time, of the Theosophical Society.

As their investigations progressed, many other atoms were examined, but the work became ever more exacting and stressful. So the two researchers decided to take a break, spending their summer vacation at Weisser-Hirsch, near Dresden, Germany. However, their principal occupation remained the cataloguing of the elements and their organization according to complex diagrams.

In the city of Dresden, Besant and Leadbeater found a museum with an exceptionally well-endowed section dedicated to minerals. After having made a list of the elements they were looking for, they went to the museum and took the corresponding minerals from their glass cases. Leadbeater examined them rapidly and obtained a rather complex image of the composition of the minerals.

When they returned to Weisser-Hirsch, Leadbeater, with the help of clairvoyance, calmly conjured up the images he had seen at Dresden. Then, exerting his power on a

FIGURE VI.04. Charles Webster Leadbeater (1847–1934), a Bishop of the Catholic Liberal Church. Together with Besant, he became interested in "occult" chemistry, carrying out bizarre investigations into atomic structure and the elements in the periodic table. The two scholars, using the help of a clairvoyant, claimed to have discovered many elements. Those that stand out on account of their originality are *kalon, adyarium,* and *occultum*. They were certainly the first to "observe with their minds" and to later name a nonexistent subatomic particle: the *anu*. Image Courtesy of the Theosophical Society in America Archives.

"molecule of mineral," he succeeded in uncovering its complex structure. He observed every atom in all its parts and determined that all the atoms were formed from additional units. While each atom was being examined, his collaborator Annie Besant sketched an approximate model; she counted the number of units and divided them by 18, which was the number of units (*anu*) present in an atom of hydrogen.

Fifty-nine elements from the investigations at Weisser-Hirsch were drawn by Annie Besant and later by Curuppumullage Jinarajadasa. Their results were printed month after month in the journal *Theosophist*,[107] edited at Adyar, a suburb of Madras, beginning in January 1908.

In 1907, they recorded the presence of three elements that had never before been observed by chemists, to which they gave the provisional names of *occultum, kalon*, and *platinum-B*.

The descriptions for all of these elements were drawn by their assistant, Curuppumullage Jinarajadasa, and appeared in the first edition of the book *Occult Chemistry*,[108] published in 1909, as well an article on the *ether* in space.[109] In that same year, Leadbeater again took up his work at the center of the Theosophical Society in Adyar, where he and his co-workers made a detailed study of an additional 20 elements.

In 1919, Leadbeater was in Sydney to investigate clairvoyantly the composition of dissolved salts in water. A second edition of *Occult Chemistry* was published in that same year, but without any additional unedited or supplementary material.[110] The study of water was taken up again in 1922, and the results were published between March 1924 and August 1926. Also in 1919, he took up the study of diamond, whereas hafnium[111] was described in 1928 and rhenium 3 years later.

After Charles Webster Leadbeater moved to Adyar in 1930, the last elements in the periodic table that had not been studied previously were an object of examination. Between 1932 and 1933, new material was published in the journal *Theosophist*, including a description of elements 85, 87, and 91 and an updated list of atomic weights. In 1932, they announced the discovery of a new element with an atomic weight of 2 and called it *adyarium*[112] because it had been discovered in the city of Adyar.

The third edition of *Occult Chemistry* saw a complete revision of the text, with the results of more recent research incorporated. From this material, the following conclusions emerged:

- From 1895 on, Leadbeater and Besant replaced the atom as the ultimate unit of matter with the subatomic particle termed *anu*.[113]
- According to the researchers, there was no way to determine the dimensions of *anu*. They could only specify that *anu* existed in two forms: one positive and one negative. The negative particles united to form a helix that was the mirror image of the one made up of positive particles. No further investigation in this respect was carried out.
- The research of 1907 allowed them to identify a new "neutral gas" called *kalon*, heavier than xenon but lighter than radon.
- Two other elements, *adyarium* and *occultum*, took their place in the periodic table between hydrogen and helium. (The description of *occultum* was drawn up in 1896 and published in 1909.)

Years later, Leadbeater discovered that, among the rare earths, a group of three elements existed that would form a new interperiodic grouping. The first element was found in 1909 in the pitchblende sample that Curuppumullage Jinarajadasa had sent to Leadbeater from the United States. Earlier, Leadbeater had discovered the nonexistent fourth member of the platinum group that he called *platinum-B*.

According to the three clairvoyants, the isotopes of many elements would have been seen and described by 1907, well before Frederick Soddy realized their existence and coined the name in 1913. In fact, in 1907, during the clairvoyant investigations at Weisser-Hirsch, some isotopes were found. The researchers used the term "meta" to designate the "second variety of element." And it was thus that these substances made their appearance: *meta-neon* and *meta-kalon*. According to Leadbeater, every type of *meta*-element or isotope had 42 *anu* more than the element that gave it its name. The first refutation of this empirical law was argon. It had an isotope that was lighter and so, for this element, Leadbeater proposed the name *proto-argon*.

The power of clairvoyance allowed the three researchers to observe the forms of the elements, which had definite shapes. With few exceptions, all the elements fell into seven groupings according to shape: point, linear, tetrahedral, cubic, octahedral, X-shaped, and star-shaped.

The valence of every element was also revised based on their observations of *anu*. They could be subdivided; that is, an atom with a valence of 1 could be divided into two halves, each with an effective valence of one-half. The valence of hydrogen could be composed of two or six parts, each one of which could have 1/2 or 1/6 of the valence. Similarly, other elements with valences of 2, 3, or 4 had the ability to subdivide these valences. According to Leadbeater, valence was closely connected to the form of an atom. He gave as an example that the divalent elements would be, for the most part, tetrahedral.

Leadbeater and Besant also had a way of revising the periodic law. Whether out of patriotism or based on the spiritism associated with the Theosophical Society, they declared the atomic arrangement proposed by Sir William Crookes "better than any other periodic table and corresponding most with the truth." This construction, today totally forgotten, described the atoms as the oscillation of a pendulum. The results were read during a conference at the Royal Institution, in London, on February 18, 1887, and later published.

As previously mentioned, many years after Leadbeater's death, Jinarajadasa edited the last edition of *Occult Chemistry*. In this version, he spoke about the action of the Demiurge and Supreme Surveyor. In Jinarajadasa's universal plan, one can discern the influence of Crookes and his thesis of "Genesis of the Elements" by the work of a "divine mind."

In his theoretical processing of their findings, Jinarajadasa found new elements with atomic weights of 185, 187, 189, 191, and 193. As a result of 55 years of reflection and study in the field of occult chemistry, Jinarajadasa made his own the words of the English physicist Sir James Jeans (1877–1946) who affirmed in his book, *The Mysterious Universe*, that "from the intrinsic evidence of his creation, the Great Architect of the Universe now begins to appear as a pure mathematician."[114] Even after more than 50 years have elapsed since this prophecy, we cannot say if the Demiurge imagined by Jinarajadasa has been descried by scientists, but what is certain is that mathematics, physics, and chemistry have begun to thrust themselves ever more profoundly into the "kingdom" of the life sciences.

VI.7.2. THE LAST YEARS OF THE THREE CLAIRVOYANTS

Leadbeater's publications transported his readers immediately into his world, but his ideas on chemistry and physics, borrowed from some of the most influential persons of his epoch, were always generic and evasive. Those who have a vague notion of the English mindset might take this as a sort of very British reticence to assert things categorically, but not in this case: Leadbeater was obliged express himself in that manner. In the crowded world of "occult chemistry," there never was, nor could there ever be, space for scientific clarity.

In 1908, he moved to Adyar, and he finally retired to The Manor in Sydney, Australia. In his travels around the world, he invariably sought more modest lodgings and, if they existed, the possibility of arrangements more suited to his age. While he was in England, he hobnobbed with some very influential personages and was often a guest in English country manor houses, places rarely accessible to most people. It is difficult to say how or why he could pass in one night from a castle to a hovel, or eat one day off a silver plate and another from a pauper's bowl.

In 1916, he was consecrated bishop in the Liberal Catholic Church and, in 1930, he returned to Adyar to be at the bedside of Annie Besant[115] who was suffering from a serious illness. She passed away in September 1933. Old Bishop Leadbeater wished at all costs to return to Sydney for a visit. While he was aboard ship, he fell gravely ill and disembarked at Perth, in Western Australia, in search of medical treatment. He was in the hospital for 16 days, alternating between moments of recovery and of worsening health. On February 26, 1934, a recurrence of his illness made him realize that there was no hope of a cure. He died a few days later, on March 1.

Meanwhile, Jinarajadasa became president of the Theosophical Society in Adyar. From 1945 until he passed away on June 18, 1953, he was also president of the International Theosophical Society.

Notes

101. The theosophical term derived from the Greek θεός, "God," and σοφία, "knowledge or wisdom." It specifies various mystical and philosophical doctrines that have occurred historically and that refer to each other. In neo-Platonic philosophy, *theosophy* means "divine wisdom," which man can approach only by way of a supposed mystical experience. Theosophy was taken up again as a philosophical doctrine in the 17th century and maintains that all religions have a single origin. Later, in the 19th century, this esoteric religious movement developed into a syncretic system with Christian, Eastern, and philosophical elements; it admits the possibility of direct contact with the divine and preaches reincarnation (literally passing from one body to another). To sum up, the Theosophical Society founded in 1875 in New York by Helena Petrovna Blavatsky advocated disclosing the theosophical doctrine: "all religions derive from one unique divine Truth. This Truth has been passed down over the course of history by a very small circle of initiates who have revealed only those aspects of it conforming to the historic period in which they found themselves."
102. Besant, A.; Leadbeater, C. W. *Occult Chemistry. A Series of Clairvoyant Observations on the Chemical Elements,* 1st ed.; Theosophical Publishing Society: London and Benares City, 1909.
103. Besant, A.; Leadbeater, C. W. *Occult Chemistry. Clairvoyant Observations on the Chemical Elements,* 2nd ed., A. P. Sinnett, Ed.; Theosophical Publishing House: London, 1919.
104. Besant, A.; Leadbeater, W. C.; Jinarajadasa, C. *Occult Chemistry. Investigations by Clairvoyant Magnification into the Structure of the Atoms of the Periodic Table and of Some Compounds,* 3rd ed., C. Jinarajadasa, Ed.; Theosophical Publishing House: Adyar, Madras, India, 1951.
105. Morrisson, M. *Modern Alchemy: Occultism and the Emergence of Atomic Theory*; Oxford University Press: New York, 2007.
106. Besant, A. *Lucifer* **1895**, November, 211.
107. Besant, A.; Leadbeater, C. *Theosophist* **1908**, *I*, 347; Besant, A.; Leadbeater, C. *Theosophist* **1908**, *II*, 437; Besant, A.; Leadbeater, C. *Theosophist* **1908**, *III*, 531; Besant, A.; Leadbeater, C. *Theosophist* **1908**, *IV*, 625; Besant, A.; Leadbeater, C. *Theosophist* **1908**, *V*, 729; Besant, A.; Leadbeater, C. *Theosophist* **1908**, *VI*, 841; Besant, A.; Leadbeater, C. *Theosophist* **1908**, *VII*, 929; Besant, A.; Leadbeater, C. *Theosophist* **1908**, *VIII*, 1019; Besant, A.; Leadbeater, C. *Theosophist* **1908**, *IX*, 1111; Besant, A.; Leadbeater, C. *Theosophist* **1908**, *X*, 43; Besant, A.; Leadbeater, C. *Theosophist* **1908**, *XI*, 166; Besant, A.; Leadbeater, C. *Theosophist* **1908**, *XII*, 253; Besant, A.; Leadbeater, C. *Theosophist* **1908**, *XII*, 258; Besant, A.; Leadbeater, C. *Theosophist* **1909**, *I*, 386; Besant, A.; Leadbeater, C. *Theosophist* **1909**, *VI*, 355; Besant, A.; Leadbeater, C. *Theosophist* **1909**, *VII*, 455; Besant, A.; Leadbeater, C. *Theosophist* **1909**, *VII*, 458; Besant, A.; Leadbeater, C. *Theosophist* **1909**, *VII*, 468; Besant, A.; Leadbeater, C. *Theosophist* **1909**, *IX*, 721.

108. Besant, A.; Leadbeater, C. W. *Occult Chemistry. A Series of Clairvoyant Observations on the Chemical Elements,* 1st ed.; Theosophical Publishing Society: London and Benares City, 1909.
109. Besant, A.; Leadbeater, C. W. *Theosophist* **1908**, *VI*, 823.
110. Besant, A.; Leadbeater, C. W. *Occult Chemistry. Clairvoyant Observations on the Chemical Elements,* 2nd ed., A.P. Sinnett, Ed.; Theosophical Publishing House: London, 1919.
111. Jinarajadasa, C.; Leadbeater, C. W. *Theosophist* **1928**, *V*, 193.
112. Jinarajadasa, C.; Leadbeater, C. W. *Theosophist* **1932**, *XII*, 361.
113. The term was borrowed from the Sanskrit; *anu* means "atom, molecule, or fundamental particle."
114. Jeans, J. *The Mysterious Universe*; Cambridge University Press: Cambridge, 1930, p. 122.
115. Annie Besant was born in London of an Irish family. She married the Reverend Besant, with whom she had two children. Her political interests and her feminism drove her to abandon her religion and become an atheist. In 1872, she left her husband and took up a friendship with Charles Bradlaugh (1833–91), a Mason and antireligious propagandist. She became an organizer in the feminist movement and the struggle for social justice and participated in framing the constitution of the English Labour Party. In 1889, she met Helena Blavatsky and was converted to the cause of theosophy, on which she left a quasi-mystical and quasi-Christian stamp that would engender much discussion and lead to a schism in the ranks. Having become president of the Theosophical Society, she lived for a long while in India, where she participated in the struggle for liberation from English domination.

VI.8

WILLIAM HARKINS'S ELEMENT ZERO: NEUTRONIUM

Having completed the identification of the naturally occurring chemical elements, scientists seemed to find themselves confronting the only alternative to go further: to synthesize new elements with ever-increasing atomic masses.

William Draper Harkins (1873–1951) was a scientist who swam against the tide, and, as such, he had the idea of moving in the opposite direction from that of his colleagues: he looked for a chemical element lighter than hydrogen, the *neutronium*. Based on the knowledge of the time, it would be composed of an aggregate of neutrons and therefore lack a nuclear charge, thus going against the definition of "element," at least in the narrower sense of the term. It might seem strange, but Harkins's absurd idea, although it has never had any reliable experimental confirmation, has not fallen into oblivion and even today has some supporters.

At the beginning of the 20th century, scientists finally began to travel the pathway that would lead to the solution of the problem relative to the source of solar and stellar energy. The English physicist Arthur Stanley Eddington (1882–1944) pointed out that the probable source lay in the transmutation of radioactive elements. At that time, the only source of nuclear energy known was that produced by uranium and thorium decay. Could the stars draw energy from these substances? The reply, as we know today, was negative. The explanation was supplied by spectroscopy that at that time was capable of accurately showing the composition of the sun and the other stars. By examining Fraunhofer's lines, one could determine not only the types of chemical elements present in the sun, but one could also rule out the idea that our star was just an enormous ball of uranium and thorium.

In 1915, Harkins presented his own theory, according to which many types of nuclear transformations could produce energy.[116] He showed how an extremely large amount of energy could be produced when four hydrogen nuclei came together to form a nucleus of helium; he put forth the hypothesis that this reaction was the source of solar energy.

Harkins, born in Titusville, Pennsylvania, began his teaching career at the University of Montana in 1900, but only in 1908 did he receive his doctorate from Stanford. He remained in Montana for another 4 years, and, after a brief period of study with Fritz Haber, he moved to the University of Chicago, where he remained for the next 39 years. He was a physical chemist with many and varied interests, and one in particular attracted his attention: nuclear structure. His vision of science was ahead of its time (e.g., his hypotheses on the existence of the neutron and stellar nucleosynthesis). However, his speculative talents also led him to hypothesize other properties of the neutron later shown to be inexact. His idea of searching for the lightest element present in nature, the hypothetical free neutron, was not in fact new.

VI.8.1. A PLACE IN THE PERIODIC TABLE FOR THE ELEMENT WITHOUT A NUCLEAR CHARGE

On the other side of the Atlantic, between 1925 and 1926, Professor Andreas von Antropoff published a strange periodic table[117] and suggested the existence of a new form of matter composed entirely of neutrons.[118] This hypothetical element, the *neutronium*, with atomic number 0 (formally, one would not treat it as an element since it lacked a proton) was placed at the head of a new version of a table of the elements. Later, it was inserted into the group of the noble gases and also appeared in some spiral-form classifications of the periodic system.[119] At the same time, Prince Louis de Broglie (1892–1987) tried to justify the insertion of another subatomic particle, the neutrino,[120] into the periodic table together with the *neutronium*, or *neuton*: the *neutronium* (symbol = Nn) was placed in box 0, whereas for the *neutrino*, a box was created that preceded it, indicated by 00.

Andreas von Antropoff developed in an original way an idea that was not his own. This idea had been put forth by the eminent physical chemist Walther Nernst who, in 1903, had postulated that the ether, the hypothetical rarefied gas present around Earth, might be composed of "neutrons," weightless particles derived by the mutual annihilation of negative and positive electrons. According to this theory, the "neutrons" would be able to react chemically with ordinary elements.[121] The idea and the name "neutron," understood as paired positive and negative electrons, had been coined a short time earlier[122] by William Sutherland (1859–1911). However, the neutral particle hypothesized by Nernst and conceptualized by Antropoff was, of its very nature, completely different from the neutron.

Von Antropoff was born in Reval,[123] on August 16, 1878, in the era of Imperial Russia. He soon moved to Germany, where he pursued his scientific and university careers. While in his 50s, he enrolled in the Nazi Party and was one of its most active members at the University of Bonn. Immediately after Adolf Hitler (1889–1945) rose to power in January of 1933, he distinguished himself by the singular gesture of raising the Nazi swastika to the highest point of the university building. At the end of World War II, Antropoff was tried for his past collaboration with the Nazi Party. In a 1945 report, the committee charged with examining his behavior under the Nazi regime was opposed to his readmission to an academic position at the university. He appealed the decision, but the university was unyielding in denying him permission to return to his former position. He passed away in Bonn on June 2, 1956.

VI.8.2. FROM THE NUCLEAR "ALPHABET" TO THE HYPOTHESIS OF NEUTRONIUM

In the 1920s, physicists had made great advances in the field of atomic physics. The time was ripe for the discovery of the neutron (1932) and the subsequent reinterpretation of the structure of the atomic nucleus. However, during the brief interval between these two events, Harkins and von Antropoff succeeded in introducing and developing their own ideas, formulated by erroneously interpreting the data obtained by experimental physicists of the time.

In 1919, at the end of World War I, Harkins published an article on atomic structure and a new periodic system that he had developed.[124] According to Harkins, the periodic system was not related to the distribution of extranuclear electrons but was connected to the actual structure of the atomic nuclei. The system was composed of only two periods: the first contained the elements with even atomic numbers, and the second those with odd atomic numbers.

The first category (called the "helium system") contained those nuclei that Harkins imagined were composed of multiple α particles; the second category (called the "hydrogen system") consisted of those nuclei made up of α particles plus one or three nuclei of hydrogen. The principal difference between the two classifications lay in the greater stability of atomic nuclei in the first category with respect to the second. Harkins belonged to the "old school" and did not seem to be aware of the fundamental discoveries taking place at that time, especially in Europe, that would change the face of physics. Thus, in the following year, he published an extensive monograph on his reinterpretation of periodic classification starting from atomic nuclei.[125] Using this theory, Harkins was able to explain a great many atomic phenomena, such as radioactivity. It was further possible to establish which elements would have isotopes and which would not, and which radioactive elements would undergo isomeric transitions and which would not. It was, in short, an all-encompassing atomic theory that would have been undoubtedly useful had it been correct.

The data on which his whole theory of nuclear structure rested were mass numbers and atomic numbers—that Harkins assumed were equal to the nuclear charges—and atomic stability: in the case of the radioactive elements, the half-lives had to be considered. From these data, Harkins arrived at his first assumption that all nuclei would be composed of positive electrons (nuclei of hydrogen and indicated as "n+" or alternatively as particles η) and by negative electrons (β−). Almost all nuclei could be seen as a set of α particles or helium nuclei α++ = n4 + β2−; other nuclei could be grouped together as particles of μ (μ = η2+ β2−). Both particles α and μ carry a net charge equal to 0; a third class of particles indicated by the Greek letter ν had an excess positive charge, ν = (η3+ β2−)+. This class would include the major part of the atoms with uneven atomic numbers. Furthermore, according to Harkins, pairs of electrons would have been present in all the heavy nuclei and would serve to "cement" the additional α particles. According to his reasoning, only the "cementing" electrons would be emitted during the disintegration of the radioactive nuclei. From a further deepening of his theory, Harkins was able to assert that chlorine, silicon, magnesium, neon, nickel, and all the elements following up to atomic number 80 (Hg) would be composed of isotopic mixtures.

Among the many academic duties that crowded Harkins's day, one might suppose that this line of research would quickly be abandoned, but such was not the case.[126] In the years following these first publications, he devoted much of his energy to the study of nuclear phenomena, without making significant progress or receiving much appreciation for his efforts from his scientific colleagues.

In a year devoid of any great events on the political, cultural, or artistic scene, two fundamental discoveries took place in the scientific area: 1932 was the *annus mirabilis* for physics. James Chadwick made concrete the intuitional idea of his mentor Ernest Rutherford by discovering the long-awaited neutron.[127] And in September of that same year, young Carl D. Anderson (1905–91) introduced antimatter onto the stage of science by discovering an electron that carried a positive charge, the positron,[128] thus opening up issues that are still present in the conceptual framework of the microcosm.

Harkins wasted no time in making contributions to this changed world of nuclear structure. After Chadwick's discovery, Harkins dusted off his old (1920) hypothesis that a neutral particle could exist in isolation.[129] Rutherford came up with this idea at the same time: he, like Harkins, was a mine of new ideas, but the difference between them was that Rutherford always tried to confirm his ideas through experiment.

At the beginning of 1933, Harkins was a very disappointed scientist for not having discovered anything significant in almost 20 years of studying atomic physics. Now approaching 60, the discovery of *neutronium* seemed to offer him one last chance. He therefore threw out the very improbable hypothesis that neutrons could form bonds among themselves to form more massive particles without electric charge. His article began with a rather controversial introduction: he arrogated to himself the idea of the existence of the neutron by bringing up his 1920 hypothesis. At the time, he had said that a neutral particle inside the nucleus could better explain the emission of α particles from radioactive nuclei: according to the laws of classical physics, a positively charged particle would escape with more difficulty from a nucleus where there was an equal number of positive and negative charges. Unfortunately, things were not as Harkins had presented them. For the most part, the laws he used to support his hypothesis could not be applied to the microscopic world with the casualness he used to explain his reasoning.

At the conclusion of his presentation, Harkins took into consideration the existence of an aggregate of two neutrons. These aggregates could be present in deep space and perhaps be concentrated through gravitational effects around a planet or even better, a star. He could not have found a better solution to salvage his theory: by basing the fundamentals of his theory on something that could not be determined by experiment, no one would be able to refute them. Harkins never reported any experimental data in support of his theories except the confirmation of the mass of the neutron, as already measured experimentally by Chadwick. Harkins remarked that the chemical elements could be classified on the basis of their atomic number (numbers of nuclear charges). The neutron, he noted, was a particle devoid of charge; it could therefore be designated as the element with atomic number 0. He thus proposed a name for this particle, although the real discoverer, Chadwick, had already done so. In trying to snatch part of the discovery for himself, Harkins clumsily proposed as many as four alternative names for the neutron: *neutronium, neutronon, neuteron,* or *neuton*."[130]

VI.8.3. WILLIAM DRAPER HARKINS: A VERSATILE AND OBSTINATE CHEMIST

Around 1937, William Draper Harkins changed his research interests. He began to study the absorbance of gases on dusts. He retired in 1939, but when the United States entered World War II, he accepted a position as a member of the Defense Research Committee. To his credit, he was the first to show, through his work on the structure of the atomic nucleus, that in the then-hypothetical nuclear fusion of hydrogen to produce helium there was an enormous mass loss, theorizing that this was the source of stellar energy. He also was able to show that atomic nuclei with even-numbered masses were more stable than nuclei with odd-numbered masses.[131] Harkins was a popular and prolific writer. In all, he published 271 scientific articles on a variety of subjects. His scintillating prose continued to appear uninterruptedly in specialty journals right up until the time of his death.[132] Harkins died suddenly on March 7, 1951, following a coronary thrombosis.

Harkins's greatest deficiency was certainly that his simultaneous interests in so many aspects of chemistry and physics dispersed his efforts. He never succeeded in creating a reference center for the simple reason that his interests changed too rapidly.

He was educated as a chemist with a solid cultural background. However, his postdoctoral experience with Haber in Germany did not contribute to developing his ideas of modern physical theory. On his return to the United States, Harkins was probably lacking in the mathematical expertise indispensable for solidifying his understanding of atomic theory.

Some of Harkins's ideas have survived in altered form. The free neutron was discovered, but it was different from what Harkins had theorized in that it is not stable, having a half-life of about 15 minutes. It decays by emitting a β particle and is thus transformed into a proton or hydrogen nucleus.

Physicists have succeeded in verifying the existence of dineutron (^{2}n) as well as tetraneutron (^{4}n), although numerous experiments were required to do so.[133] However, it was not possible arrive at an unambiguous conclusion for the simple fact that not all physicists recognize the validity of such experiments. Theoretical calculations have yielded totally inconsistent results: some data report the absolute impossibility of forming a neutron–neutron bond;[134] in other cases, the transient stability of an n–n bond could possibly be permitted only in a cluster of three neutrons,[135] that is, ^{3}n; in other calculations, the bond would be formed only in complexes of very high mass,[136,137] possibly as high as 100 or even 1,000! Because, in 1977, experimental physicist Claude Détraz (b. 1938) asserted that he has detected neutron aggregates[138] with masses between ^{2}n and ^{10}n, one cannot say that this discussion is by any means concluded.

Notes

116. Harkins, W. D.; Wilson, E. D. *Proceedings of the National Academy of Sciences of the United States of America* **1915**, *1*(5), 276.
117. von Antropoff, A. *Z. angew. Chem.* **1926**, *39*, 722.
118. von Antropoff, A. *Z. angew. Chem.* **1925**, *38*, 971.
119. Ternstrom, T. *J. Chem. Educ.* **1964**, *41*, 190.
120. de Broglie, L. *Théorie générale des particules à spin*, 2nd ed.; Gauthier-Villars: Paris, 1954, p. 94.
121. Nernst, W. *Theoretische Chemie*, 4th ed.; F. Enke: Stuttgart, Germany, 1903, p. 139.
122. Sutherland, W. *Phil. Mag.* **1899**, *47*, 269.
123. Before 1918, the city carried the German name of Reval, but in the year in which Estonia became independent, the name of Tallinna was adopted; in the early 1920s, it became the present-day Tallinn.
124. Harkins, W. D. *Science* **1919**, *50*, 577.
125. Harkins, W. D. *Phys. Rev.* **1920**, *15*, 73.
126. Harkins, W. D. *J. Am. Chem. Soc.* **1920**, *42*, 1964; Harkins, W. D. *J. Am. Chem. Soc.* **1920**, *42*, 1996; Harkins, W. D. *Nature* **1921**, *107*, 202; Harkins, W. D. *Phil. Mag.* **1921**, *42*, 305; Harkins, W. D. *Journal of the Franklin Institute* **1923**, *195*, 553; Harkins, W. D. *Phys. Rev.* **1931**, *38*, 1270.
127. Chadwick, J. *Nature* **1932**, *129*, 312; Chadwick, J. *Proc. Roy. Soc.* **1932**, Series *A 136*, 692.
128. Anderson, C. *Science* **1932**, 328; Anderson, C. *Phys. Rev.* **1932**, *42*, 145.
129. Harkins, W. D. *Physical Letters* **1920**, *15*, 73.
130. Harkins, W. D. *Nature* **1933**, *131*, 23.
131. Coffey, P. *Cathedrals of Science: The Personalities and Rivalries That Make Modern Science.* Oxford University Press: New York, 2008, p. 131. This same reference documents Harkins's proclivity for making dubious priority claims, so much so that he earned the nickname "Priority" Harkins. See also Kauffman, G. B. *J. Chem. Educ.* **1985**, *62*, 758–61.
132. Mulliken, R. S. *Nat. Acad. Sci. Biogr. Mem.* **1975**, *47*, 49.

133. Fiarman, S.; Hanna, S. S. *Nuclear Physics* **1975**, *A251*, 1; Fiarman, S.; Meyerhof, W. E. *Nuclear Physics* **1973**, *A206*, 1; Cerny, J. et al. *Physical Letters* **1974**, *53B*, 247.
134. Goldanskii, V. I. *Journal of Experimental and Theoretical Physics Letters* **1973**, *17*, 41.
135. Mitra, A. N.; Bhasin, V. S. *Phys. Rev. Lett.* **1966**, *16*, 523.
136. Baz, A. I.; Bragin, V. N. *Physical Letters* **1972**, *39B*, 599.
137. Antonchenko, V. Ya.; Bragin, V. N.; Simenog, I. V. *Journal of Experimental and Theoretical Physics Letters* **1974**, *19*, 314.
138. Détraz, C. *Physical Letters* **1977**, *66B*, 333.

PART VII

Modern Alchemy: The Dream to Transmute the Elements Has Always Been with Us

Some Experiments in Physics Work Best with the Instrument turned off
—G. N. Flerov (1913–90), Russian Nuclear Chemist

PROLOGUE TO PART VII: ALCHEMY THEN AND NOW

Alchemy, more than a pseudoscience, is one of the most fascinating areas in the history of literature. It has coursed through the dreams and emotions of ancient and modern people like an underground stream, running deeply and quietly at times, and at other times bursting noisily forth. This "stream of consciousness" is made of secrets, both esoteric and exoteric, and also of poetry, science, technology, and above all, symbols. It has concentrated within itself both the spiritual and material. It has nourished for centuries three specific dreams: perfect health, eternal life, and the transmutation of matter. It has been exploited by the powerful figures of the past[1] and has known crises[2] and persecutions.[3] Alchemy's main thrust has always been to get to the root of things by reflection on the infinity, variety, and transformations that take place in matter as an alternative to the conventional wisdom posed by philosophical and scientific thought. Above all, alchemy had as its goal the reproduction of Creation in the laboratory, putting all matter under the control and at the service of humankind. In other words, alchemy is the promethean dream of *homo faber.*

The most ancient alchemical prescriptions date back to the age of Constantine (280–337 AD). They are practices that aim at the creation of imitation gold and silver, precious stones, and crimson dye. In later times, the first alchemical symbols were developed, as well as the first rudimentary apparatus for distillation, as described by Maria the Jewess's treatise on furnaces, now lost. But it was in the Middle Ages that alchemy blossomed into a kind of "mystical chemistry." A chief proponent was Duns Scotus (ca. 1180–1236)[4] who was among the first to recognize the contributions of the Arab civilizations; among their writings, mention of the transmutation of lead into gold appeared for the first time.[5]

With the rise and diffusion of sciences such as chemistry and physics, the influence of alchemists declined rapidly, hastened also by the fact that its practice was banned from academies and universities. Nevertheless, alchemy, although sidelined, has not been extirpated completely from the human consciousness. And while the evolution from alchemy to science took place, the idea of the transmutation of metals remained latent in the dreams and minds even of illustrious scientists.[6] One of the more famous of these modern "alchemists" was Ernest Rutherford who, working at Cambridge, was a pioneer in transmutation by bombardment of atoms with subatomic particles. He was the first to discover that, by bombarding nitrogen with α particles, it was possible to transform it into oxygen and hydrogen.[7] With Rutherford's amazing discovery, the ever-smoldering dream of the alchemists returned to the limelight. The new physics would supply the impetus to achieve a complete understanding of the process and theory of transmutation of the elements.

Some of these attempted transmutations actually gave rise to particles that were sometimes mistaken for new elements that were given spurious elemental names. And even before the periodic table was "completed" with the last element in Period 7, number 118, some scientists and pseudoscientists began to try their hand at transmuting known elements into others using a variety of means, some scientifically sound, others the product of fantasy. Hence, this last part of our search for the "lost elements," a section on modern alchemy, deserves a place in this book.

VII.1

A PIECE OF RESEARCH GONE UP IN SMOKE: DECOMPOSITION OF TUNGSTEN INTO HELIUM

This rather improbable research project had its start long ago when, in 1912, J. J. Thomson, the famous Cambridge physicist, caught a glimpse of a particle with an atomic weight of 3 in his discharge tubes. He labeled it with the symbol X_3, but decided to finish other higher priority studies before analyzing this phenomenon in detail. In the meantime, the American physicist William Duane[8] (1872–1935) and the young Gerald Louis Wendt became interested in this remarkable discovery and decided to repeat Thomson's experiments. They began by investigating the properties of hydrogen after exposure to α particles. Their intense activity continued throughout World War I and, eventually, in September 1919, Wendt published a paper dealing with the synthesis of ozone by hydrogen excited with α particles.[9] Most probably, trace amounts of oxygen present in the discharge tube were converted into ozone as a result of collisions with the charged particles. At that time, it was not very clear what happened in the subatomic world, and 3 years passed in unproductive attempts to fuse hydrogen nuclei to form ozone.

In April 1922, Wendt and Clarence E. Irion of the University of Chicago reported their "Experimental Attempts to Decompose Tungsten at High Temperatures" to a meeting of the American Chemical Society. They claimed to have completely disintegrated tungsten wire into helium by means of a high-voltage discharge in glass bulbs. Their work was viewed with suspicion at the time and, today, cognizant physicists have commented that their experimental design was faulty. Unfortunately, the Associated Press widely published an exaggerated account of the "transmutation" experiment, based on their oral presentation. In a footnote to their published article, the scientists emphasized that "this report is preliminary, and that nothing is proven beyond the importance of the problem and the promise of this method."[10]

Wendt and Irion had planned a complete analysis of the gas they collected, but the sample was lost in an accident. Two years later, S. K. Allison and William Harkins reported inconclusive results from their version of the experiment.[11] The harsh criticism of Harkins and Allison was a hard blow for Wendt, one that eventually interrupted his research activities. Wendt's passion for the astonishing developments of nuclear chemistry was never extinguished, however, and, in the years following World War II, he became deeply involved in promoting the public understanding of science. Gerald Louis Wendt passed away in 1973, at the age of 82.

Notes

8. William Duane was born in Philadelphia and graduated from Harvard in 1893, later obtaining a PhD at the University of Berlin. He taught at the University of Colorado from 1898 to 1907. He spent 5 years at the Laboratoire du Radium in Paris, and in 1913, he returned to the United States and held the first chair of biophysics at Harvard (1917–34). He is mostly known for his research in the field of radioactivity, X-rays, and their application to radiotherapy.

9. Wendt, G. L. *Proc. Acad. Science* **1919**, *5*, 518.
10. Wendt, G. L.; Irion C. E. *J. Am. Chem. Soc.* **1922**, *44* (9), 1887; Wendt G. L. *Science* **1922**, *55*, 567.
11. Allison, S. K.; Harkins, W. D. *J. Am. Chem. Soc.* **1924**, *46*, 814.

VII.2

TRANSMUTATIONS OF MERCURY INTO GOLD

In March 1924, Professor Hantaro Nagaoka[12] of the Tokyo Imperial University described his group's studies "on the isotopes of mercury and bismuth revealed in the satellites of their spectral lines" and of gold in particular. In May 1925, they reported subjecting paraffin oil to high voltage and detecting Au in the viscous residue.[13,14] Nagaoka stated that when a discharge was passed through drops of Hg falling between iron electrodes, the formation of silver and other elements was observed. Considerations of the satellites of the spectral lines of Hg led Nagaoka to the conclusion that a proton is "slightly detached" from the nucleus of Hg and that it can be removed: If his assumption was valid, he could perhaps realize the dream of alchemists by striking out a proton from the nucleus by α particles or some other powerful means of disruption to produce Au from Hg.[15]

At about the same time, Professor Adolf Miethe of the Photochemical Department at the Berlin Technical High School found that a sooty deposit often formed in mercury vapor lamps contained gold. Subsequently, he and his assistant, Hans Stammreich,[16] were issued German Patent Specification No. 233,715 (May 8, 1924) for "Improvements in or Relating to the Extraction of Precious Metals." This news was widely discussed in scientific circles.[17]

Adolf Miethe was born in Potsdam on April 25, 1862. Even as a child, he was interested in the new science of photography and in optical instruments. He eventually became a professor in Charlottenburg as successor to Hermann Wilhelm Vogel (1834–98), the discoverer of the sensitizing action of dyes on photographic emulsions. Miethe was responsible for teaching scientific and practical photography in all its branches: photomechanical methods, spectral analysis, optics, and astronomy. He was also well versed in botany, mineralogy, and other subjects.

In July 1924, Miethe and Stammreich announced that they had changed mercury into gold in a high-tension mercury vapor lamp, producing the equivalent of 1 euro of gold at a cost of 60,000 euros of electricity.[18] Otto Honigschmid (1878–1945) and Eduard Zintl (1898–1941) determined the atomic weight of Miethe's mercuric Au via potentiometric titration of the auric salt with $TiCl_2$. It was found to be 197.26, which is heavier than ordinary Au (197.20). For a mass spectrographic analysis,[19] they sent samples to Frederick Soddy, who suggested that such a change might occur by the collapse of an electron into the mercury nucleus,[20] but F. W. Aston argued strongly against this possibility.[21]

In December 1924, *Scientific American* announced that it would arrange for a comprehensive and exact test of the Miethe experiment at New York University by Professor H. H. Sheldon and Roger Estey. The negative results of their three experiments established a strong probability that the transmutation announced by Professor Miethe could not be confirmed.[22,23]

Scientific American published another report of "More Mercuric Gold from Germany" in April 1926, announcing that a 10,000-fold increase in yield had been obtained in the production of the mercuric-gold process.[24,25] Other researchers were not so optimistic. In 1925, two Italian scientists, Arnaldo Piutti (1857–1928) and Enrico Boggio-Lera (1862–1956) vainly tried to confirm the experiments of Nagaoka and Miethe.[26] Erich Tiede and his colleagues reported that transmutation of Hg into Au was theoretically but not practically possible.[27] Milan Garrett at Oxford published completely negative results of his repeated attempts to reproduce the Hg–Au transmutation experiment by several methods.[28] Even Fritz Haber made careful attempts to repeat the work of Nagaoka and Miethe. Mercury, in which no Au could be detected, was subjected to six different treatments, but no Au was formed.[29]

Miethe and Stammreich defended themselves against accusations of amateurism, asserting that the formation of gold from mercury depended on the application of intermittent electric discharges. Alois Gaschler attempted to reverse the Miethe-Nagaoka experiment by treating gold with high-speed hydrogen nuclei. He assumed that one of them might penetrate deeply into the electron shells of Au, be held by the innermost shells as a "paranucleus," and form a "*Tiefenverbindung*." After 30 hours of bombardment, the spectrum of the tube began to show Hg lines that steadily increased in intensity, causing Gaschler to postulate that Hg is a gold hydrogen "compound."[30] The scientific community gave a fair and thorough review of the claims of Miethe, Stammreich, and Nagaoka (who also skillfully managed the criticism).

The "conventional" transmutation of mercury to gold was achieved only in 1941. Using fast neutrons and a mercury target, with the aid of an atomic particle accelerator, Rubby Sherr (1913–2013), Kenneth Tompkins Bainbridge, and H. H. Anderson of Harvard synthesized three radioactive isotopes of gold, all of them with a short half-life.[31]

The history of the German alchemist Franz Tausend—who performed transmutation of mercury into gold in the same period—is much more intriguing and, although it is not really a topic in chemistry, deserves to be told. Franz Tausend (1884–1942) began to produce gold from mercury in the 1920s, eventually working in association with General Erich Friedrich Wilhelm Ludendorff (1865–1937) in 1925 to produce artificial gold, based on "Kabbalistic principles," for the Nazis. General Ludendorff must have been both an ambitious and a not very intelligent man to be misled by such an absurd and grotesque charlatan. He was, moreover, a fervent nationalist, blinded by his desire for Germany to once again become a great nation. The war reparations that Germany was obliged to pay in gold marks gave rise to tremendous inflation, and the economy was on the edge of chaos. Thus, the ingenuous General entrusted himself to this dishonest adventurer.

Tausend's process captured the interest of the Nazi Party, and a group of particularly influential Nazis even allowed him to meet Adolf Hitler (1889–1945) who gave him plenty of money and equipment, but all in vain. In the meantime, Tausend had embezzled the equivalent of approximately 10 million euros earmarked to set up five laboratories and a research institute. Instead, he hid this remarkable fortune and swindled his many investors.

While the investors lost their money, Tausend became extremely rich, but German civil authorities discovered the deceit and he just managed to escape across the Italian border. He was eventually arrested and extradited to Germany but, even when confronted with the evidence of his scientific and financial fraud, he continued to claim that his formula was correct. Tausend never changed his story. At the end of his trial, he was sentenced to many years in prison for embezzlement.[32] After 4 years of detainment, he was released, but, in 1937, he was arrested again for fraud. The Nazi prison regime was much harsher than he had experienced during his years of detention under the Weimar Republic. It is not completely clear when he died, although it seems likely that it was while he was still in prison, in 1942.

Finally, in chronological order, one should mention Roger Caro (1911–92), alias Kamala-Jnana, of the French alchemical school of the Temple of Ajunta, and Roberto Monti, for years involved in the search for nuclear reactions at extremely low energy (the so-called cold fusion process) and in the redefinition of theories concerning atomic structure. Roberto Monti and his wife, the epistemologist Gerardina Cesarano Monti (b. 1963), pursued this work for many years.

During the 1960s, Roger Caro published some booklets concerning alchemy. The methodologies that he described, based on classical alchemical language, were nevertheless clearly described. Monti was convinced that the 21st century would herald the rebirth of alchemy.[33] We do not believe in such forecasts and we therefore limit ourselves to the hope that, if cold fusion is a practical road, then the rebirth of alchemy may remain only a bad dream.

Notes

12. Hantaro Nagaoka was born in Omura on August 15, 1865. He became the foremost Japanese professor of modern physics during the early 20th century, and, between 1893 and 1896, he moved to the Cavendish Laboratory in Cambridge and to the University of Berlin, Vienna, and Munich. Inspired by the work of James Maxwell on the "rings of Saturn," he postulated his Saturnian model of the atom in 1903. His quasi-planetary model for the atom was superior to Jean Perrin's semiqualitative model, but, in 1911, it was replaced by that of Sir Ernest Rutherford. In his later years, his former pupil Hideki Yukawa won the Nobel prize. Nagaoka died on December 11, 1950.
13. Nagaoka, H. *Nature* **1924**, *114*, 197; Nagaoka, H. *Nature* **1926**, *117*, 604; Nagaoka, H. *Nature* **1926**, *117*, 758.
14. Nagaoka, H. *Die Naturwissenschaften* **1925**, *13*(29): 635–37; Nagaoka, H. *Die Naturwissenschaften* **1925**, *13*(31), 682–84; Nagaoka, H. *Die Naturwissenschaften* **1926**, *14*, 85; Nagaoka, H. *Chem. Abstracts* **1925**, *19*, 3209; Nagaoka, H. *Nature* **1925**, *116*(2907), 95.
15. Nagaoka, H. *Journal de Physique et la Radium* **1925**, *6*, 209.
16. Hans Stammreich was born in Remschield, Germany, on July 6, 1902. In 1933, he fled Nazi Germany for Paris. When France was conquered by the Germans in 1940, he fled to the University of Rio de Janeiro to seek protection from the Nazis. He died in São Paulo, Brazil, on March 6, 1969.
17. *Literary Digest* (March 14, 1925); "Attempts at Artificial Au"; *Literary Digest* (December 12, 1925); "Negative Evidence in the Hg-Au Case." *Literary Digest* (February 6, 1926).
18. Miethe, A. *Nature* **1924**, *114*, 197; Miethe, A. *Nature* **1926**, *117*, 758–60.
19. Honigschmid, O.; Zintl, E. *Die Naturwissenschaften* **1925**, *13*(29), 644.
20. Soddy, F. "The Reported Transmutation of Hg into Au." *Nature* **1924**, *114*, 244.
21. Aston, F. W. *Nature* **1924**, *114*, 197; Aston, F. W. *Nature* **1926**, *117*, 758–60.
22. Anon. *Sci. Amer.* **1924**, *131*, 389; Anon. *Sci. Amer.* **1925**, *132*, 157; Anon. *Sci. Amer.* **1925**, *133*, 296.
23. Sheldon, H.; Estey, R. S. *Phys. Rev.iew;* **1926**, *27*(2), 515.
24. Anon. *Sci. Am.* **1926**, *135*, 90; Anon. Sci. Am. **1928**, *138*, 208.
25. Siemens & Halske Akt.-Ges., German Patent Spec. No. 243,670 [Cl. 39(i) & 82 (i)]; "Treating Hg."
26. Piutti, A.; Boggio-Lera, E. *Giornale di Chimica Applicata* **1925**, *8*, 59.
27. Tiede, E. et al. "The Formation of Au from Hg." *Ber. Dtsch. Chem. Ges.* **1926**, *59*, 1629–41.

28. Garrett, M. W. **1926**, *Nature, 118*, 2959.
29. Haber, F. et al. *Z. Anorg Allg. Chem.* **1926**, *153*, 153–83.
30. Gaschler, A. *Zeit. Elektrochem.* **1926**, *32*, 186–87; Manley, J. J. *Nature* **1925**, *114*, 861; Manley, J. J. *Nature* **1925**, *115*, 337.
31. Sherr, R.; Bainbridge, K. T.; Anderson, H. H. *Physical Review* **1941**, *60*, 473.
32. Schleff, H. *Der Goldmacher Franz Tausend der größte Abenteurer der Gegenwart*; Jos. C. Hubers Verlag: München, Deutschland, 1929.
33. Certosi, P. *Alla ricerca della pietra filosofale*; Newton & Compton: Rome, Italy, 2002, p. 274.

VII.3

TRANSMUTATIONS OF SILVER INTO GOLD

The manufacture of gold from other elements can be achieved by several methods. For alchemists, the *Ars Magna* was not the art of imitating "nature" but improving upon it. With this understanding, transmutation should not be an unattainable "chimaera." Martinus Rulandus the Elder (1531–1602), a follower of Paracelsus (1493–1541), compiled an alchemical dictionary that was published posthumously. His son, Martinus Rulandus the Younger (1569–ca. 1611), professor at Regensburg and later at Prague under the patronage of the Habsburg Emperor, Rudolph II (1552–1612), eventually became a member of the imperial court. Rulandus convinced the ingenuous emperor that it was possible to transmute silver into gold in his laboratory.

During the first half of the 19th century, new charlatans and unscrupulous people returned to cast the shadows of mysticism and alchemy on the Western scientific panorama. Most of these experimenters used a variety of "wet" techniques with nitric acid or "dry" transmutations with catalytic alloys (especially arsenic) in a furnace.

In the years between 1854 and 1855, Cyprien-Théodore Tiffereau (1819–after 1898) submitted six memoirs to the French Académie des Sciences concerning transmutation of silver to gold, eventually published under the title, *Les Metaux sont des Corps Composés*[34] in 1855. He later joined the French Expedition Corps that Emperor Napoleon III (1807–70) sent to Mexico.

Tiffereau conducted his experiments at considerable expense while supporting himself financially by making daguerreotypes (the first photographic process) in Mexico. He claimed that Mexican silver possesses peculiar qualities that favor its transmutation into gold, attributing the production of gold in the earth to the action of the "microbe of gold." This was confirmed in the 1980s by the discovery that placer gold nuggets form around a nucleus of *Bacillus cereus*. Tiffereau, considered by many to be one of the fathers of modern alchemy, died, presumably in his 80s, on the eve of the 20th century.

On the other side of the Atlantic, in 1908, Sir Henry Baskerville mentioned a contemporary claim for the production of artificial gold by a Mr. R. M. Hunter of Philadelphia who claimed he had perfected a process to produce it. He forwarded samples of silver in which the gold was "growing" and some was already "grown-up," said to have been produced by his secret process. Sir Henry did not analyze the samples.[35]

A few years later, the well-known occultist Arthur E. Waite[36] (1857–1943) wrote "A Collection of Alchemical Processes," which includes a part entitled "Silver Transmuted into Gold by the Action of Light."[37] His processes were limited to simple manipulations of silver nitrate with hydrochloric acid and the consequent formation of silver chloride (or other silver halides).

In this brief panorama of neo-alchemy, the name of Fulcanelli should not be allowed to disappear. Behind this name, which one presumes is the pseudonym of an author of alchemical books of the 20th century, may be hidden the French alchemist Jean Julien

Champagne (1877–1932) or actually René Adolphe Schwaller de Lubicz (1887–1961). Whoever Fulcanelli[38] happened to be, he published an extensive description of the transmutation of silver in *Les Demeures Philosophales*.[39] Other popular candidates for the title of Fulcanelli are Pierre Dujols (1862–1926) and Eugène Canseliet (1899–1982).

François Jollivet-Castelot (1874–1937) was the secretary general and later president of the Alchemical Society of France (founded in 1896). He also edited the Society's journal *L'Hyperchemie*. He authored several books and articles on alchemy and "hyperchemistry," a system of nonoccult chemical methods of transmutation.[40] In 1920, he published *La Fabrication Chimique de L'Or* to report his successes using both "wet" and "dry" methods of transmutation of metal.

The case of Stephen Henry Emmens is most curious: did he actually find the key to the dreams of the medieval alchemists or was he a clever impostor? The question remains unanswered. But there is no doubt that he did produce gold from some source, which he sold to the U.S. Mint.

Stephen Emmens was a scientist with an international reputation, an author of numerous books, and a member of some prestigious professional societies. Although he attended regular chemistry courses at King's College London and signed his name "Dr. Emmens," it is not known if he ever obtained that academic title. In the 1880s, he emigrated to the United States and founded the Emmens Chemicals and Explosives Co., the Emmens Metal Co., and the Argentaurum Laboratory. He was an authority in the field of nickel and zinc metallurgy,[41] but he tried to inflate his reputation by citing false reviews of scientists who had never met him or read his papers.[42]

The most ambitious enterprise of Emmens's career was *The Argentaurum Papers No. 1*, a book that, despite the title, did not merely deal with silver, gold, or transmutation: It was also a crackpot attempt to demolish accepted scientific theories.[43]

In 1895, Emmens, while conducting geological studies, noticed that gold is found in greenstone that has made its way from the interior of the earth under conditions that permit very slow cooling. He also observed that gold is not found in ordinary lava flows where the heat has been quickly dissipated. Because lava and greenstone are composed of similar elements, he decided that subjecting a nonauriferous limestone to the same treatment as an auriferous greenstone could produce gold by transmutation. Likewise, he suggested that, in the course of natural chemical evolution, silver becomes transmuted into gold, or gold into silver, "or that a third substance exists which changes partly into gold and partly into silver." This third intermediate substance he called *argentaurum*.

He started experiments in his New York laboratory, and, several years later, he claimed that he had produced *argentaurum* by a secret method, although he never revealed his process. He used as his starting material Mexican silver dollars, certified as containing less than 1/10,000 part of gold. Using an apparatus he called a "force-engine," he announced the discovery of a new element to fill the "vacant space existing in the sub-group of Group I,"[44] and which he thought to be the intermediate matter from which silver and gold are formed.[45] In 1897, Emmens produced more than 660 ounces (almost 20 kg) of gold from silver and sold it to the U.S. Assay Office. He revealed a few historical and technical details of his transmutation process[46] in his book, *Argentaurum Papers No. 1: Some Remarks Concerning Gravitation*.

Rumors of Emmens's alchemy circulated throughout the scientific world before it reached the public. In May 1897, Sir William Crookes wrote to Emmens inquiring about his experiments, and their correspondence continued for about a year. Almost from the beginning, however, the personalities of the two men came into conflict, and their relationship ended in bitterness and controversy. Sir William questioned the theory of *argentaurum* as an intermediate substance between silver and gold. In reply, Dr. Emmens outlined his method only in general terms, never revealing all the details of his process. Crookes tried to duplicate Emmens's method twice, but with mixed results, finding in a sample sent by Emmens only common well-known elements.[47] A year later, Emmens published a book entitled *Argentaurana, or Some Contributions to the History of Science.* It contained a general outline of his methods, together with his correspondence on the subject with Sir William Crookes. Shortly after, he exhibited his process at the Greater Britain Exhibition.

Did Emmens actually create artificial gold? In one assay report of "*argentaurum gold*" made by the government, it was stated that the ingots contained impurities of a kind "constantly present in old jewelry." In referring to this report some years ago, the British writer Rupert T. Gould (1890–1948) stated that this "was as neat a way of calling Emmens a 'fence' as could be imagined." However, the same impurities—traces of copper, platinum, lead, zinc, and iron—can also be found in coined Mexican dollars.[48] Emmens died shortly after the turn of the 20th century, and his secret died with him. No evidence of fraud has ever been found to discredit him. And his mysterious *argentaurum* gold, in coins and in bars, remains buried below Fort Knox.

The last member of this group of chemists and alchemists involved in transmuting silver into other elements is Matthew Carey Lea. In 1889, while studying the reduction of silver nitrate, he discovered the preparation of allotropic silver,[49] his best-known discovery.[50] Matthew Carey Lea was born in Philadelphia on August 18, 1823; he belonged to a well-to-do Irish Catholic family with deep interests in matters scientific. His father, Isaac Lea (1792–1886), was a renowned naturalist; his brother, Henry Charles Lea (1825–1909), was a publisher of medical texts. Due to poor health, Matthew Carey Lea hardly ever left his home, but he had a home laboratory where he carried on his own experiments.[51] Due to his pioneering experiments of 1866, he came to be known as the father of chemical photography and of mechanico-chemistry (i.e., the chemical effects of mechanical action).[52] Unfortunately, his notebooks were destroyed according to his will,[53] limiting the information about his work to the contents of his published papers. Some allotropic silver samples prepared by Lea are preserved in the library of the Franklin Institute, in Philadelphia.[54] What Lea considered solutions of allotropic silver are in fact colloids, but that became clear only many years later. In fact, his results were cited in the Nobel Award address of Richard Zsigmondy (1865–1929) in 1925 "for his demonstration of the heterogeneous nature of colloids," among them alloptropic silver, by the use of the ultramicroscope.[55] Matthew Carey Lea died on March 15, 1897, aged 75.

Notes

34. Tiffereau, T. *Les Metaux Sont Des Corps Composes*; Vaugirard: Paris, 1855; Tiffereau, T. *L'Or et le Transmutation des Metaux*; Chacornac: Paris, 1889; Tiffereau, T. *Compt. Rend. Acad. Sci. Paris* **1854**, *38*, 383; Tiffereau, T. *Compt. Rend. Acad. Sci. Paris* **1854**, *38*, 792: Tiffereau, T. *Compt. Rend. Acad. Sci. Paris* **1854**, *38*, 942; Tiffereau, T. *Compt. Rend. Acad. Sci. Paris* **1854**, *39*, 374; Tiffereau, T. *Compt. Rend. Acad. Sci. Paris* **1854**, *39*, 642; Tiffereau, T. *Compt. Rend. Acad. Sci. Paris* **1854**, *39*, 743; Tiffereau, T. *Compt. Rend. Acad. Sci. Paris* **1854**, *39*, 1205; Tiffereau, T. *Compt. Rend. Acad. Sci. Paris* **1855**, *40*, 1317; Tiffereau, T. *Compt. Rend. Acad. Sci. Paris* **1855**, *41*, 647; Tiffereau, T. *Compt. Rend. Acad. Sci. Paris* **1896**, *123*, 1097.

35. Baskerville, C. *Popular Science Monthly* **1908**, *72*, 46–51.
36. Arthur E. Waite was a scholarly mystic who wrote extensively on occult and esoteric matters, and he was the co-creator of the Rider-Waite Tarot deck. As his biographer, R. A. Gilbert described him, "Waite's name has survived because he was the first to attempt a systematic study of the history of western occultism—viewed as a spiritual tradition rather than as aspects of proto-science or as the pathology of religion."
37. Waite, A. E. *A Collection of Alchymical Processes*; S. Weiser: New York, 1987.
38. According to tradition, he was born in 1839. He died (or better he "disappeared") either in 1923 or 1924. It is possible that he died in the vicinity of Paris. It has also been reported that he may not have really existed.
39. Fulcanelli. *Les Demeures Philosophales, Vol. 1*; J. Pauvert: Paris, 1964, 184–200.
40. Jollivet-Castelot, F. *Chimie et Alchimie*; E. Noury: Paris, 1928.
41. Trimble, R. F. *The Hexagon* **1981**, *71*, 41.
42. Gould, R. T. *Enigmas*; Philip Allan & Co.: London, 1929.
43. Emmens, S. H. *Argentaurum Papers #1: Some Remarks Concerning Gravitation*; Plain Citizen Publishing Co.: New York, 1896.
44. *New York Press*, 8 August, **1896**; *Evening Sun*, 10 August, **1896**; *New York Journal*, 16 August, **1896**.
45. Emmens, S. H. *Science* **1897**, *5*(112), 314; Emmens, S. H. *Science* **1897**, *5*(113), 343; Emmens, S. H. "The Argentaurum Papers No. 1, Some Remarks Concerning Gravitation," *Argentaurana*, Du Boistel, G., Ed., Bristol, 1899; Emmens, S. H. *Science*, **1898**, *9*, 386; Emmens, S. H. *Arcanae Naturae*, Pamphlet, Paris, **1897**.
46. Emmens, S. H. *Chem. News* **1897**, *76*, 117; Emmens, S. H. *The Engineering & Mining Journal* **1896**, *62*(10), 221; Emmens, S. H. *The Engineering & Mining Journal* **1896**, *62*(11), 243; Emmens, S. H. *The Engineering & Mining Journal* **1896**, *62*(14), 315
47. Between May 8, 1897 and May 12, 1898, Emmens wrote 34 letters to Crookes. The latter replied only 14 times.
48. Woodward, R. S. *Science*, **1897**, *5*(112), 343.
49. Lea, M. C. *Amer. J. Sci.* **1889**, *37*(series 3), 476; Lea, M. C. *Amer. J. Sci.* **1889**, *38*, 47; Lea, M. C. *Amer. J. Sci.* **1889**, *38*, 129; Lea, M. C. *Amer. J. Sci.* **1889**, *38*, 237; Lea, M. C. *Amer. J. Sci.* **1891**, ***41***, 179; Lea, M. C. *Amer. J. Sci.* **1891**, *42*, 312; Lea, M. C. *Amer. J. Sci.* **1894**, *48*, 343; Lea, M. C. *Amer. J. Sci.* **1896**, *51*, 259; Lea, M. C. *Amer. J. Sci.* 1896, *51*, 282.
50. Lea, M. C. *Zeit. Anorg. Allgem. Chem.* **1894**, *7*, 340; Lea, M. C. *Am. J. Sci.* **1889**, *38*(series 3), 237.
51. Lea was educated at home by a tutor, Eugenius Nulty, a native of Ireland who also taught his older brother. He devoted himself chiefly to the chemistry of photography, to which he made a number of important contributions. His publications include numerous papers on the chemical action of light and an excellent *Manual of Photography*.
52. Takacs, L. *Bull. Hist. Chem.* **2003**, *28(1)*, 26; Takacs, L. *J. Material Science* **2004**, *39*, 4987.
53. Smith, E. F. *M. Carey Lea, Chemist*; University of Pennsylvania Press: Philadelphia, PA, 1923.
54. Lea, M. C. *Photogenic Collection on the Properties of Allotropic Silver*, 1891, Library of the Franklin Institute, Philadelphia.
55. Soederbaum, H. G. *Nobel Lectures, Chemistry, 1922–1941*; Elsevier: Amsterdam, 41, 1966.

VII.4

TRANSMUTATION OF ORES

The synthesis of elements by high-energy bombardment of other elements is common knowledge and practice among nuclear physicists. In their way, modern physicists have accomplished one of the goals of alchemy: the production of artificial gold. However, the yields are low, and the product is unstable and very expensive. Such nuclides find only limited use in medicine and chemistry. We review some examples here.

In 1972, Soviet physicists at a nuclear research facility near Lake Baikal in Siberia accidentally discovered a fusion reaction for turning lead into gold when they found the lead shielding of one of their experimental reactors had changed to gold.[56]

In 1980, a group of researchers at Lawrence Berkeley National Laboratory, headed by Nobel laureate Glenn Seaborg, reported the production of a few billion atoms of gold as the "side product" of an experiment with a Bevalac accelerator. A bismuth target was bombarded with a "relativistic projectile" that chipped some protons from bismuth nuclei, thus forming gold. The experiment produced less than one-billionth of a cent worth of gold.[57]

However, some researchers in the 20th century have reported their methods of producing profitable amounts of noble metals from base metals and low-grade ores without the use of nuclear reactors. Some of the methods were reported to be genuine low-energy alchemical transmutations.

In early 1931, newspapers in Europe and the United States frequently reported stories about Zbigniew Dunikowski, a Polish engineer who claimed to have a secret formula enabling him to produce artificial gold from ordinary sand and rocks by the action of mysterious rays he called "Z-rays." Although he was very soon nicknamed the "Polish alchemist," his vain promises attracted the attention of financiers and even of some European political leaders. After a few years of futile experiments, he was sued by his impatient financial backers, arrested, and imprisoned. It remains unclear whether Dunikowski was truly convinced that his formula for making gold could work or if he was a simple swindler. He claimed he was accused of fraud by bankers who feared that his method would undermine the status quo of world's economy.

Zbigniew Dunikowski later founded *Metallex, Société Anonyme* with Belgian stockholders and established a factory on Lake Neuchâtel. Soon after the beginning of World War II, he reported that Franco-British authorities had asked him to transfer his work to southern England and continue his experiments on the transmutation of silica into gold, to support the Allies in the war. Nothing more is known about this ridiculous affair because all subsequent proceedings were kept secret.[58]

There is some doubt about his date of birth, but he was probably born in 1889; in the early 1950s, he ended up in the United States as a political refugee, where he died on March 15, 1964.

In this same period, an analogous case that involved an Austrian chemist named Adalbert Klobasa came to light. In 1937, Klobasa claimed that he had produced gold using an electromagnet and an induction coil with which he treated a mixture of base metal salts, claiming a 1% yield of gold. He published a booklet[59] in which he cryptically reported the advancement of his research. The chaos caused by World War II canceled both the work and any trace of this mysterious man.

In 1950, Thomas H. Moray investigated the possibility of improving the extraction of uranium ores. The Moray Research Institute (MRI) proceeded by bombarding the ore in an "environment" with X-rays as high as 24 MeV before attempting to extract any metals. The average ore contained 0.23% uranium oxide. After irradiation, the ore yielded from 7.0–7.5% uranium oxide!

Thomas Henry Moray was an American inventor born in Salt Lake City, Utah, on August 28, 1892. Moray graduated from the Latter Day Saints Business College, and he studied electrical engineering through an international correspondence course. He later received a PhD in electrical engineering from the University of Uppsala, Sweden. Moray developed what he termed the "Moray Valve"—a device for extracting "radiant energy" from the "energy waves of the universe," which he thought to be an inexhaustible energy source freely available in the environment.

In 1953, the MRI proposed that the Atomic Energy Commission (AEC) investigate a project for the "aging" of atomic ores by a "breeding type reaction with high-energy particles or X-rays in the presence of a proper environment." The AEC declined to grant a contract.

The X-ray tube used for bombardment in the Moray process was developed by Moray himself and became the fulcrum of the equipment that he called "Electrotype-Therapeutic" (U.S. patent number 2,460,707, Cl. 128-421; February 1, 1949). Moray claimed yields of 50–100 ounces (1.4–2.8 kg) of gold per ton (909 kg) of ore[60] via his "therapeutic" bombardment process.

In 1963, some scientists decided to investigate Moray's process and found that the presence of gold in the solutions treated by Moray's apparatus had a scientific explanation quite different from the transmutation of heavier metals. They proposed that colloidal gold dispersed in Moray's mysterious "environmental solution" was concentrated by the action of the X-rays.

Later in life, Moray reported that, during the 1930s, he and his family had received death threats on several occasions and that his laboratory had been ransacked by "mysterious" government agents who had sabotaged his prototype instruments in order to stop his research. Embittered by the hostility of the establishment, Moray withdrew to private life; he died at age 82 in 1974.

Another "modern alchemist," Arnold Conrad, claimed that he knew of a simple method of transmutation which, as he said, "ripens green ores" (volcanic sulfides, pyrites, or tellurides). He learned the process from a German scientist whose name he refused to reveal. The technique balances the ore's electropositive charge with 10–150 volts DC. The precious metals produced are removed by electroplate refining.[61]

In the 1980s, David Hudson discovered the existence of Orbitally Rearranged Monoatomic Elements (ORMEs), which are virtually undetectable by conventional means (except for a distinct infrared doublet located between 1,400 and 1,600 cm^{-1}) because they lack a *d*-electron. Hudson and associates developed a method to recover ORMEs and convert them into their metallic forms. Although not a transmutation of one element into another (but rather the conversion of an allotrope into the common visible form of the element), the extraction and conversion of ORMEs to metal may explain the claims of some other experimenters.

In the 1990s, Joe Champion announced a variety of methods for the transmutation of black sands by thermal burns, melts, and kinetic methods. He was convicted of fraud in Arizona after being accused by an irate investor who failed to achieve satisfactory results. Other researchers validated his processes, however, so the question remains open. The process was developed from a method of "growing gold" in an electrolytic cell that was originally developed by Walter Lussage, a Czech geologist who revealed his process to a Jack Keller, who taught it to Joe Champion in 1989. Champion subsequently developed the method further. The necessary "parental" isotopes needed for the "transmutation" process were cobalt, iron, manganese, nickel, and calcium. Other methods "championed" by Champion included using the kinetic energy of a ball mill with 40 kg carbon steel balls, microwave digestion, and the use of "dimensional" chemical reactions.[62]

After such affirmations, the ephemeral contact that Champion may have had with scientific truth vanished. Very often, modern authors of alchemic texts report innovative ideas that they simply copy from the latest discoveries in the field of physics, genuine discoveries that are less well known to the general public. These authors then adapt the real discoveries for their own personal gain, masking in incomprehensible jargon a pseudoscientific "get rich quick" scheme capable of deceiving gullible clients.

Notes

56. Melchanov, A. *Chicago Elite*, January, **1980**.
57. *American Business* **1980**, April 16; Garretson, F. *Oakland Tribune* **1980**, Sat. March 22, A-7; *Star*, **1980**, February 12. However, the first person to accomplish this synthesis was Geoffrey Wilkinson in 1949 (Wilkinson, G. *Phys. Rev.* **1949**, *75*, 1019).
58. Doberer, K. K. *The Goldmakers*, Nicholson & Watson: London, 1948; Nelson, R. A. *Adept Alchemy*; Rex Research: Jean, Nevada, 1998.
59. Klobasa, A. *Künstliches Gold. Versuch und Erfolg in der Goldsynthese.* Hartleben Publishers: Vienna, Austria, 1937, p. 47.
60. Moray, T. H. *Recovery of Minerals from Low-Grade Ore by High Energy Bombardment*; 68th National Western Mining Conf., Denver, CO, February 4, 1965; Hooper, W. J., *Startling Possibilities in Artificial Transmutation*, 68th National Western Mining Conf., Denver, CO, February 4, 1965.
61. Conrad, A. *California Mining Journal* **1973**, February, 13.
62. Champion, J. *Producing Precious Metals at Home*; Discover Publishing, Westboro WI, 1994; see also: Bockris, J. *Fusion Technology* **1994**, *26*, 261.

VII.5

OTHER TRANSMUTATIONS

The history of nonconventional science is marked by much credible documentation that provides evidence for many types of transmutations accomplished without particle accelerators. Long before the discovery of "cold fusion" by Stanley Pons (b. 1943) and Martin Fleischmann (1927–2012), other scientists found evidence of nonradioactive, low-energy transmutation of light elements in plants, animals, and minerals. These reactions have come to be known as "biological transmutations" or "nuclido-biological reactions." Many scientists and nonscientists alike believe that this class of transmutations is of great importance to the progress of human knowledge in the fields of physics, cosmology, biology, geology, ecology, medicine, nutrition, and agriculture. The exact mechanisms of biological transmutations remain unknown, although a few theories have been proposed to explain them. Many think that biological transmutations cannot be denied and that they are essential for living organisms, which could not function without them. This considerable literature consists mostly of fanciful theories based on absurd hypotheses—mostly, but, surprisingly, not all.

VII.6

BIOLOGICAL TRANSMUTATION

The study of biological transmutation began in the 17th century with the famous experiment by Johann Baptista von Helmont (1579–1644), who grew a willow tree in a clay vase with 200 pounds (91 kg) of soil.[63] After 5 years, he dried the soil and found that its weight had decreased by only 2 ounces (0.06 kg): "Water alone had, therefore, been sufficient to produce 160 (73 kg) pounds of wood, bark and roots" (plus fallen leaves which he did not weigh). Presumably, he claimed, there were minerals in the water he fed to the tree. Today, we know that plants form carbohydrates from atmospheric carbon dioxide, but their mineral content is derived from soil, not air. Lacking controls, it is difficult to presume the origin of the necessary mineral content for plant growth.

However, numerous reports have been made over the past few centuries regarding possible nuclear transmutations taking place in plants, animals, and microorganisms, of both the fusion and fission varieties. Here are some examples:

In 1799, Louis-Nicholas Vauquelin, famous for having discovered chromium (1797) and beryllium (1798), found that hens excreted five times more Ca than they ingested. Although he was a follower of Lavoisier's thesis, he was forced to conclude that lime had been created, but he could not understand how it happened.

J. J. Berzelius reported on several experiments involving evidence that plants contained minerals not previously found in their seeds, although care was taken to eliminate the possibility of their admission to the system.[64]

Albrecht von Herzeele (b. 1821) noted the weight variation of magnesium in plants, and Pierre Baranger, professor of organic chemistry at the Ecole Polytechnique in Paris, noted the same thing for calcium and phosphorus. He concluded that plants can perform low-energy transmutations that we cannot do without resorting to high-energy physics.[65]

In 1946, Henri Spindler, director of the Laboratoire Maritime de Dinard, investigated the origin of iodine in seaweed and found that the algae *Laminaria* manufactured iodine out of water that did not contain the element.[66]

Among the many examples of biological transmutations that can be cited, the work most well-developed and well-known is that of Louis C. Kervran (1901–83), a French scientist who was also a member of the New York Academy of Sciences. Kervran presented the idea that sodium, potassium, and dozens of other elements change into each other under certain natural conditions in the mineral, vegetable, and animal kingdoms.[67] Biological transmutations have been demonstrated, crucial experiments replicated, and the theoretical principles verified by many scientists who are finding new industrial, medical, and agricultural applications for these discoveries.

Inspired by Kervran's pioneering work, George (Ryogji) Ohsawa sought to transmute sodium into potassium in vitro. Inspired by a symbolic dream, Ohsawa and his co-worker, Michio Kushi, constructed an experimental electric discharge tube with copper (Yin) and iron (Yang) electrodes and a valve through which to create a vacuum or admit oxygen. The first transmutation with this equipment was achieved on June 21, 1964. After applying 60 watts of electricity for 30 minutes, the heat produced transformed the sodium into a plasma state. A molar equivalent of oxygen was then introduced. Viewed with a spectroscope, the orange band of sodium gave way to the blue band of potassium, which was formed according to the reaction:

$$^{23}Na + ^{16}O = ^{39}K \qquad \text{(Eq. VII.1)}$$

Analysis of the reaction product confirmed the result and revealed an unexpected bonus: a trace of gold was produced by the combination of Na, O, and K with the Cu and Fe electrodes. Several different metals were tested as electrode materials. Neon and argon atmospheres were found to enhance the yield of potassium and other elements. External heating of the reaction tube also served to ionize the sodium.

Since the initial experiments conducted by George Ohsawa[68] and Michio Kushi[69] (b. 1926) in the 1970s, several other researchers have reported similar results but in more detail, thanks to modern analytical equipment, computers, and communication. Notable among them is the work of Solomon Goldfein of the U.S. Army Mobility Equipment Research and Development Command (MERADCOM) at Fort Belvoir, Virginia. He proposed, granted that such transmutations existed, a mechanism in which magnesium adenosine triphosphate (Mg-ATP), in its part-by-part disintegration, played the role of a molecular cyclotron.[70]

The world of low-energy transmutations seems to have become much more accessible through such work. New discoveries are being reported at an increasing rate in the scientific literature, particularly cold fusion and biological transmutations. Perhaps within a few decades we shall see the mass production of elements on demand.[71]

Although the idea of biological transmutations has its adherents, it lies outside the realm of conventional physics and chemistry for the following reasons:

- There has been little or no consideration of the magnitude of energy changes that must take place in a nuclear transmutation such that, if they take place in vivo, the living subject could be incinerated by the amount of energy released.
- There has been little or no consideration of the many other pathways via "normal" or conventional chemical reactions that could be taken to reach the same result.
- The thesis flies in the face of Lavoisierian chemistry, which has as its foundation that elements do not change—they retain their identities (with the Curie corollary: unless they are unstable).

Nevertheless, there is a considerable body of literature on the subject, and it is not without its many adherents and admirers who feel that this new direction could lead to knowledge that may not only explain many anomalous biological observations, but also solve the energy problems of modern society.[72]

Notes

63. Biberian, J.-P. "Biological Transmutations: Historical Perspectives." *J. Condensed Matter Nucl. Sci.* **2012**, *7*, 11–25. This review article shows how, over the past two centuries, a number of experimentalists have questioned the Law of Conservation of Mass.

64. Berzelius, J. J. *Treatise on Mineral, Plant & Animal Chemistry*; Chez Didot Frères, Libraires: Paris, France, 1849.
65. Report of January 27, 1959, to the Institut Génévois, as cited in Kervran, L. C. *Une Histoire des Transmutations Biologiques*. http://fr.wikibooks.org/wiki/Une_histoire_des_transmutations_biologiques/Louis_Corentin_Kervran_et_son_%C3%A9poque (accessed April 12, 2014).
66. Spindler, H. *Bull. Lab. Maritime Dinard* **1946**, December; Spindler, H. *Bull. Lab. Maritime Dinard* **1948**, June 15.
67. Kervran, L. C. *Biological Transmutations*; Happiness Press: Asheville, NC, 1989.
68. George Ohsawa, born Nyoichi Sakurazawa into a poor Samurai family on October 18, 1893, died on April 23, 1966. He was the founder of the journal *Macrobiotic Diet and Philosophy*. While living in France, he also used the French name Georges, and his name is sometimes given this spelling.
69. In 1987, Michio Kushi, his wife, and a team of macrobiotic medical associates were invited by the government of the Republic of the Congo to visit West Africa and give a symposium on the macrobiotic approach to AIDS in Brazzaville.
70. Goldfein, S. "Energy Development from Elemental Transmutations in Biological Systems." U.S. Army MERADCOM Report 2247, May 1978 http://research.whnlive.com/BiologicalFusion/USArmyGoldfeinReport.htm (accessed April 13, 2014).
71. Anonymous *The Order of the Universe* **1975**, 3(10), 12–17; Gardiner, B. *East-West Journal* **1975**, February, 15; Grotz, T. *Fulcrum* **1996**, *4*(3), 6–10; Harris, P.M. Unpublished lab notes (March **1965**); Jovivitsch, M. Z. *Monatschrift f. Chemie* **1908**, *29*, 1–14; Kervran, L. *Transmutations A Faible Energie*; Librerie Maloine: Paris, 1968; Kervran, L. *Preuves Relatives A l'Existence de Transmutations Biologiques*, Librairie Maloine: Paris, France, 1968; Kervran, L. *Transmutations Biologiques en Agronomie*, Librairie Maloine: Paris, France, 1970; Kervran, L. *Biological Transmutations*, Swan House: New York, 1972; Kushi, M. *East-West Journal* **1975**, February, 22–26; Mallove, E. *Infinite Energy* **1996**, March–April, no. 7; McKibben, J. L. *Infinite Energy* **1996**, November–December, no. *11*, 37; Ohmori, T.; Enyo, M. *J. New Energy*, **1996**, *1*(1), 15; Singh, M., et al. *Fusion Technology* **1994**, *26*, 266; Sundaresan, R.; Bockris, J. *Fusion Technology* **1994**, *26*, 261.
72. Komaki, H. "Observations on the Biological Cold Fusion or the Biological Transmutation of Elements," *Frontiers of Cold Fusion* (Proceedings of the Third International Conference on Cold Fusion, Nagoya, October, **1992**); H. Ikegami, ed., Universal Academy Press: Tokyo, 1993, 555.

VII.7

THE TRANSMUTATION OF HYDROGEN INTO HELIUM AND NEON

Dozens of scientific papers were published between 1905 and 1927 concerning the mysterious appearance of hydrogen, helium, and neon in vacuum tubes. Eventually, when the phenomena could not be reliably reproduced, most scientists concluded that the results were due to contamination. The first report, written by Clarence Skinner, was published in *The Physical Review* in July 1905: While making an experimental study of the cathode potential of various metals in helium, it was observed that no matter how carefully the gas was purified, hydrogen radiation, observed spectroscopically, persistently appeared in the cathode glow. Skinner eventually located its source in the cathode.[73].

In 1912, Sir William Ramsay published a paper entitled "The Presence of Helium in the Gas from the Interior of an X-Ray Tube," and the following year, J. J. Thomson published an article "On the Appearance of Helium and Neon in Vacuum Tubes." Thomson was investigating a new gas called *X3* that he determined to have an atomic weight of 3. This heavy isotope of hydrogen is now called tritium, but at that time he believed it was a polymerized form of hydrogen. He did not fully comprehend the discovery that could have led him to the concept of isotopes before Frederick Soddy.[74] Despite every precaution, he and John N. Collie and Hubert S. Patterson, working independently, repeatedly obtained traces of helium and neon, despite the fact that they performed numerous blank experiments to exclude the possibility of contamination from various sources. It appeared that neon was formed by a union of helium and oxygen.[75].

John William Strutt, 3rd Baron Rayleigh, and other workers found no helium in their experiments.[76]

In 1914, Collie published an article entitled "Attempts to Produce the Rare Gases by Electric Discharge." In it, he discounted the presence of the He and Ne due to air leaks because other elements, such as nitrogen and argon, would also have been evident but were not.[77]. Collie used every precaution but could not explain his results.[78]

The issue lay dormant for several years, but research was resumed after World War I. In 1926, Fritz Paneth and K. Peters determined that palladium had caused the transmutation of hydrogen to helium in their experiments. Excluding every possible source of error, Paneth and Peters absorbed H in colloidal Pd and subsequently detected the main spectral lines of He. No He production was observed with Pd preparations that had not absorbed hydrogen. The experiment was repeated three times with the same results.[79] However, no trace was detected of any energy liberated during the transformation, either as heat or radiation.

Many of their American colleagues didn't believe their results.[80] No one had considered the energetic balance of the transmutation reaction, which would have to appear as radiant heat. The well-known American physicist Richard C. Tolman (1881–1948) had advanced a similar hypothesis[81] in 1922.

In Italy, soon after the end of World War I and at the periphery of the international scientific stage, the Italian chemist Arnaldo Piutti was disappointed by not being able to reproduce the transmutation experiments carried out by his long-standing friend, William Ramsay. He felt that he was not as good a chemist as Ramsay, but in reality, his experiments confirmed that transmutation was not possible and that Ramsay was wrong. In fact, Piutti published his negative results but blamed himself, not Ramsay, for his lack of success. Inadvertently, Piutti had laid the basis for the confutation of all these experiments.[82] Some years later, an important figure in German chemistry, Paul Walden (1863–1957), in referring to these transmutation experiments, sarcastically asked if modern chemists had something still to learn from the ancient alchemists.[83]

Notes

73. Skinner, C. A. *Phys. Rev.* **1905**, *21*(1), 1–15.
74. Ramsay, W. *Nature* **1912**, *89*(2229), 502; Ramsay, W. *Proc. Chem. Soc.* **1913**, *29*(410), 21; Thomson, J. J. *Nature* **1913**, *90*(2259), 645–47; Reprinted in *Science*, **1913**, *37*(949), 360–64; reprinted in *Scientific American Supplement* **1913**, *75*(1940), 150; Thomson, J. J. *Nature*, **1913**, *91*(2774), 333–37; Thomson, J. J. *Proc. Royal Soc. London* **1922**, *101-A*(711), 290–99.
75. Collie, J. N.; Ramsay, W. *Proc. Royal Soc. London* **1896**, *59*, 257–70; 356; Collie, J. N.; Patterson, H. *Proc. Chem. Soc.* **1913**, *29*, 271; Patterson, H. S. *Proc. Chem. Soc.* **1913**, 233
76. Egerton, A. C. G. *Proc. Royal Soc. London* **1915**, *91-A*(627), 180–89; Merton, T. R. *Proc. Royal Soc. London* **1914**, *90-A*(621), 549–53; Strutt, R. J. *Proc. Royal Soc. London* **1914**, *89-A*(613), 499–506.
77. Collie, J. N. *et al. Proc. Royal Soc. London* **1914**, *91-A*(623), 30–45.
78. Masson, I. *Proc. Chem. Soc.* **1913**, *29*(417), 233.
79. Paneth, F. *Science* **1926**, *64*(1661), 409–417; Paneth, F. *Nature* **1927**, *119*(3002), 706; Paneth, F.; Peters, K. *Ber. Deutschen Chem. Ges.* **1927**, *59*, 2039; Paneth, F. *Ber. Deutschen Chem. Ges.* **1927**, *60*, 808.
80. Allison, S. K.; Harkins, W. D. *J. Am. Chem. Soc.* **1924**, *464*, 814–824; Harkins, W. D.; Wilson, E. F. *The London, Edinburgh & Dublin Philosophical Magazine & Journal of Science*, **1915**, *30*(179), 723–34.
81. Tolman, R. C. *J. Am. Chem. Soc.* **1922**, *44*(9), 1902–08.
82. Piutti A. *Gazzetta Chimica Italiana* **1920**, *50*, 5.
83. Walden, P. *Science*, **1927**, *66*(1714), 407.

VII.8

RADIOCHEMISTRY: A CHILD OF BOTH PHYSICS AND CHEMISTRY

Soon after the death of Pierre Curie, his widow, Marie Skłodowska Curie, began to assume a prominent role in their laboratory, a role that she presumably could not have aspired if her husband had survived. Within 3 years of her husband's fatal accident (1906), the number of researchers in the small laboratory in Paris' Rue Cuvier grew from seven to 24. Marie Curie had a managerial approach to running the laboratory and soon increased her international prestige, gaining supremacy in the field of radioactivity. [84] Her scientific authority was evident while Pierre was still living, but it came into prominence internationally when she criticized two claims—one of which was substantially erroneous—of a German and of an English colleague.

However, before we can pronounce on the disagreements that arose, we must look at the state of confusion that came in the wake of the discovery of radioactivity, slowly recognized by many scientists to actually be the alchemists' transmutational dream—with a hook. Once scientists realized that one of their major articles of faith—the immutability of the atom—was demolished by the radioactive decay phenomenon, they found themselves sailing on an uncharted sea. Physicists could deal with the study of rays, the measurement of energy, and the eventual necessity of half-life measurement. But they also found that the decays gave rise to new products whose separation from each other required the expertise of chemists. Some physicists, like Marie Curie, became expert at these separations; others came to rely on chemists to untangle the many decay sequences that would eventually lead to the key principles necessary for understanding the phenomenon.[85]

For some decades, the tried-and-true way of identifying and characterizing a new simple substance depended on the determination of its atomic weight. These new substances were so fleeting and present in such infinitesimally small quantities that new methods had to be invented to measure them, such as electrochemical and conservation of momentum techniques. One clever method that relied on chemical similarity was the use of a so-called carrier. The radioelement was placed in solution along with a salt of a known nonradioactive element, and a reagent was added to precipitate the carrier. If the active species could be found in the precipitate, then it had reacted like the known element, to which it must be chemically similar. Then, by trial and error and repetition of the process, a separation could hopefully be effected. But chemists soon encountered some stubborn mixtures that defied separation, and when these cases continued to multiply, so did the chaos and confusion accompanying them.

The first well-documented problem was the case of radiolead. In 1900, Karl Andreas Hofmann (1870–1940)[86] and his co-worker Eduard Strauss reported on the isolation of radioactive lead sulfate from a variety of uranium minerals. In the following year, they extracted from those same minerals a type of radioactive lead that behaved exactly like ordinary lead in every respect, including its spectroscopic characteristics and its approximate atomic weight, but they were convinced that it was a different element.[87,88,89]

This discovery of *radiolead* would draw many others into a series of fruitless attempts to separate it from its nonradioactive counterpart. Furthermore, many similar radioactive species made their appearance on the radiochemical stage, so much so that scientists worried about the lack of space for them in the periodic table. Within the decade, Bruno Keetman (1883–1918) hazarded a suggestion that perhaps more than one species could occupy the same space in the table.[90] Things moved quickly after that: Willy Marckwald concluded in 1910 that, having unsuccessfully tried to separate radium from mesothorium, the two were chemically completely similar—which amounts to the same thing as saying they were the same thing![91] When Frederick Soddy dared to step over the line to proclaim the chemical identity of these inseparable pairs, he was on the road to a Nobel Prize.[92]

With his 1911 hypothesis, Frederick Soddy crossed the line, admitting that the doctrine of atomic weight was no longer the identifying criterion for an element. But as we shall see, radiochemistry had a long way to go to reach this point.

VII.8.1. WILLY MARCKWALD MAKES HIS MARK: THE POLONIUM CONTROVERSY

Now we flash back to 1902, to take up the first of two controversies that embroiled Marie Curie. In that year, Wilhelm (Willy) Marckwald (1864–1942), professor of chemistry at the University of Berlin, reported that he had extracted a radioactive substance along with bismuth from pitchblende residues, although it could not possibly have been bismuth because metallic bismuth displaced it from solution. This was interesting because, in their original 1898 note on the discovery of polonium from pitchblende, and for nearly the next 10 years, the Curies maintained steadfastly (although, in retrospect, incorrectly) that they were unable to separate polonium from bismuth. Marckwald also found that this substance's radioactivity did not diminish with time (unlike Marie Curie's polonium, which decayed markedly over a period of about a year), that it apparently emitted α rays, and that chemically it resembled tellurium. Hence, he provisionally named his new find *radiotellurium*.[93] Marie Curie read Marckwald's article and was convinced that the "new element" was really polonium, which she had already described, despite the apparent chemical differences. She published a retort in a German journal, the better to reach Marckwald's accustomed audience. She emphasized that she was responding to Marckwald's communication to reaffirm that she had no doubts about the existence of polonium,[94] even though she had not been able to isolate it yet.[95] Moreover, "the new polonium of Marckwald"—Marie concluded—"was identical to hers."

However, Marckwald continued to proclaim the validity his work: in his opinion *radiotellurium* was different from polonium.[96] Thus began the great polonium controversy, fueled most likely by confusion due to the fact that nobody yet understood that the radioactive materials with which they were working changed with time and behaved differently after each transmutation (radioactive decay).

Eventually, Frederick Soddy and Ernest Rutherford concluded that Marie Curie was right:[97] the substance that Marckwald had called *radiotellurium* was indeed polonium. The controversy subsided when Marckwald reported in January 1905 that *radiotellurium* decayed with a half-life of 139.8 days; 1 year later, Marie Curie announced with great satisfaction that she had determined a more accurate value of 140 days for the half-life of polonium, and she claimed that Marckwald's substance was her polonium.[98]

Marckwald, lacking any substantial argument in reply, had to admit his error.[99] First he cited a few lines of *Romeo and Juliet*: "'What's in a name? That which we call a rose, by any other name would smell as sweet'. . . I propose in the future to replace the name of 'radiotellurium' by 'polonium.'"[100]

In hindsight, we might say that Marie Curie and Willy Marckwald were both right and both wrong. Without a clear vision of what they were dealing with, they could not see all the ramifications. Marckwald claimed that his *radiotellurium*—"polonium"—did not lose activity over time; it actually did, and he soon found out that he had to wait about 5 months to observe it. However, he also recognized its great chemical similarity to tellurium, thus pointing to its rightful place in the periodic table. On the other hand, Marie Curie recognized its physical properties (half-life,[101] α emission) but was stymied regarding its chemical properties, early on maintaining incorrectly that it was chemically related to bismuth—and possibly not a new element at all.

Marckwald was actually an organic chemist and after this adventure in the land of radiochemistry, he rapidly returned to his original interests. He concluded his scientific activity and academic career on a very sad note. In 1932, he fled from Nazi Germany and ended up dying in Rolândia, Brazil, in 1942.

VII.8.2. WILLIAM RAMSAY "OUT OF HIS ELEMENT"

Marie Curie's second challenge took place in 1907 with the publication of an article entitled "The Chemical Action of Radium Emanation. Part II. On Solutions Containing Copper and Lead,"[102] which claimed that contact with radium emanation could induce the radioactive disintegration of copper.

The cast of this drama included, as always, Marie Curie. In this case, she was pitted against the most respected chemist of the time, Sir William Ramsay. In addition to his discovery of the noble gases, in 1903, he also established that helium was continuously produced by the natural decay of radioactive substances, an important confirmation of the theory of nuclear disintegration proposed by Ernest Rutherford.

Later, Ramsay made the unfortunate decision to continue research on radioactivity on his own. Although he was considered a skilled experimenter, he was a neophyte in the field of radioactivity, and many experienced scientists incredulously observed Ramsay's bizarre experimental results. He asserted that radium, upon radioactive decay, not only produced helium, but also neon and argon. He and his co-worker Alexander Thomas Cameron (1882–1947) even asserted that copper and neon could be transformed into lithium!

In 1907, a young Norwegian, Ellen Gleditsch (1879–1968), arrived in Paris to follow courses in chemistry, mineralogy, and radioactivity at the Sorbonne. By 1912, while working as Marie Curie's personal assistant, she completed her *licence ès sciences* (equivalent to the bachelor's degree). She was proud of her work[103] on lithium in radioactive minerals,[104] spurred by Rutherford and Soddy's radioactive transformation theory. Soddy joined Ramsay's laboratory in 1903 at University College London. Soon after his arrival, they reported that they had detected helium in condensed gaseous radium emanation (radon). Ramsay took this to be evidence for the transformation of one element into another, in contrast to Rutherford's view that "the production of helium was due to accumulated α particles expelled from the radioactive matter."

In those years, Gleditsch also participated in the controversy regarding the refutation by Curie and herself of Ramsay and Cameron's claim that copper could be transformed into lithium. From their experiments, they hypothesized that, in the presence of radium emanation, copper could be transformed into lithium, assuming that there was a genetic relationship between copper and lithium, with lithium being the lowest element in the series or group.

Ramsay received the Nobel Prize for Chemistry in 1904 for his work on the noble gases, but now his work was met with general disbelief.

The news from Ramsay's laboratory traveled around the world. Bertram Borden Boltwood (1870–1927), chair of chemistry at Yale, wrote to his friend Rutherford in England: "I imagine then the excitement and surprise when Ramsay announced in *Nature* [July 18, 1907][105] that if some emanation of radium was mixed with water there appeared neon, with only a trace of helium; but that if the emanation was mixed with a solution of copper sulphate there appeared no helium at all but only argon, while the copper gave rise to lithium." Here at last was the transmutation of elements with a vengeance! These statements, backed by Ramsay's deserved prestige and supported by his known skill in handling small quantities of gas and in using the spectroscope, produced a mixture of admiration, astonishment, and bewilderment.

Marie Curie had a sufficient amount of radium to test Ramsay's results. She and Gleditsch repeated Ramsay and Cameron's experiments and concluded that the traces of lithium they had detected probably came from the glass beakers they had used. Moreover, they suggested that the copper salts used by Ramsay and Cameron might have contained tiny amounts of lithium. Ramsay's biographer, Morris William Travers, has suggested a more prosaic explanation of Ramsay's observation of lithium: that Ramsay, a chain-smoker, had contaminated his experiments with tobacco ash, known to be rich in lithium.[106]

In October, Boltwood cackled in a letter to Rutherford that Ramsay had "entered the field exhibiting false credentials."[107] Boltwood, who was a chauvinist and anti-Semite,[108] normally refused to give credit to Marie Curie, but on this occasion, blinded by his resentment of Ramsay, he recognized that she was right. His greatest scientific discovery was the identification of a radioelement (an isotope of thorium), which he erroneously thought to be a new element,[109] *ionium*. When Marie Curie pointed out his blatant error, perhaps not too politely, he felt so injured that he developed a fierce hostility toward her and, on many occasions, said so to Rutherford. [110] Boltwood was affected by mental disorders. His difficult personality was clouded increasingly by periods of depression, culminating, at the age of 57, in his suicide in the summer of 1927.[111]

Conversely, the life of Sir William Ramsay was filled with joy and happiness until the end. On July 23, 1916 at the age of 64, he died of cancer. He had achieved the peak of his scientific reputation a dozen years earlier with the discovery of the noble gases, and the unpleasant incident of lithium transmutation did not blemish it in the least.[112]

VII.8.3. TELLURIUM X

When the science of radioactivity was still in embryonic form, in March 1902, Giovanni Pellini (1874–1926) entered the field hoping to be among the first to break new ground.

His research up until that time had concentrated on the dilemma regarding the inversion of the atomic weights of three pairs of elements: nickel-cobalt, potassium-argon, and tellurium-iodine. For tellurium, he maintained that, according to an unconfirmed hypothesis of Mendeleev, it might contain within itself a hitherto unknown element with similar chemical properties. Another idea, more pragmatic but less revolutionary, was being bruited about at that time by the Florentine chemist, Augusto Piccini. He felt that human error in the determination of tellurium's atomic weight might be the problem. No one could imagine at that time the concept of the isotope, which would open the way toward a resolution of the problem.

Pellini did not accept either of these ideas. After having read the pioneering work of Becquerel and the Curies, as well as Willy Marckwald's work on radiotellurium, he became convinced that this mineral could contain a radioactive substance with a higher atomic weight. So he began to analyze tellurium samples, both in the elemental form and as TeO_2. After fractionating a quantity of the raw material, Pellini felt that this substance, which he called *tellurium X*,[113] was also present in extremely minute amounts and that it was the higher homologue of tellurium, namely, polonium. He found that its atomic weight was 212, fairly close to today's accepted value of 209 for polonium.

Although Pellini's idea was confirmed by Marckwald's work, his natural reserve kept him from contesting the latter. He simply continued to work for a long time on tellurium without mentioning or claiming the discovery of element 84. Some years later he determined with great precision that the atomic weight of tellurium was 127.6; today's accepted value is 127.60. He also investigated tellurium's isomorphism with selenium, serving to confirm its position (despite its anomalous atomic weight) in the periodic table. He also wrote authoritative articles about tellurium for several encyclopedias.

Giovanni Pellini was born in Meina on Lake Maggiore on August 14, 1874. He studied at Padua and eventually took a permanent position at the University of Palermo. His interests were very broad, especially on the practical side of chemistry: fragrances, vegetable extracts, explosives, and chemical warfare. With respect to the latter, he worked for the Italian Ministry of War on the synthesis and properties of new war gases and even tried them out on himself! This may very well be the reason for his sudden and premature death on January 26, 1926.

Notes

84. Please see an excellent review, Adloff, J. -P.; MacCordick, H. J. "The Dawn of Radiochemistry," *Radiochimica Acta* **1995**, *70/71*, 13–22.
85. Malley, M. C. *Radioactivity: A History of a Mysterious Science*; Oxford University Press: New York, 2011.
86. Hofmann, K. A.; Strauss, E. *Ber. Dtsch. Chem. Ges.* **1900**, *33*, 3126–31.
87. Hofmann, K. A.; Strauss, E. *Ber. Dtsch. Chem. Ges.* **1901**, *34*, 8–11.
88. Hofmann, K. A.; Strauss, E. *Ber. Dtsch. Chem. Ges.* **1901**, *34*, 907–13.
89. Hofmann, K. A.; Strauss, E. *Ber. Dtsch. Chem. Ges.* **1901**, *34*, 3033–39.
90. Keetman, B. *Über die Auffindung des Joniums, einer neuen radioaktiven Erde in Uranerzen*; Schade: Berlin, Germany, 1909; Keetman, B. "Über Ionium," *Jahrb. Radioaktiv.* **1909**, *6*: 265–74.
91. Marckwald, W. *Ber. Deutsch. Chem. Ges.* **1910**, *43*, 3420.
92. Soddy, F. "The Origins of the Conceptions of Isotopes." Nobel Lecture, December 12, 1922.
93. Marckwald, W. *Ber. Deutsch. Chem. Ges.* **1903**, *36*, 2662.

94. The Curies had certainly done so with their declaration that polonium "is a kind of active bismuth; it has not been proven that it contains a new element, it does not emit radiations deviated by a magnetic field and does not produce induced radioactivity." Curie, P.; Curie, M. *Compt. Rend. Acad. Sc. Paris* **1902**, *134*, 85.
95. Curie, Frau. *Phys. Z.* **1902–03**, *4*, 234–235.
96. Marckwald, W. *Ber. Deutsch. Chem. Ges.* **1905**, *38*, 591.
97. Soddy, F. *Nature* **1904**, *69*, 347.
98. Curie, M. *Compt. Rend. Acad. Sc. Paris* **1906**, *142*, 273.
99. Marckwald, W. *Physikal. Zeit.* **1906**, *7*, 369.
100. Translation from Romer, A. *Classics of Science: Radiochemistry and the Discovery of Isotopes*; Dover Publications: New York, 1977, p. 105.
101. In 1900, Ernest Rutherford introduced the concept of the half-life when examining an "emanation" emitted from thorium compounds; his mathematical analysis of the rate of decay of the radioactive species became a universal criterion for its characterization. It was quickly adopted by Marie Curie, Marckwald, and others. Rutherford, E. *Phil. Mag.* **1900**, *49*, 1–14.
102. Ramsay, W. *J. Chem. Soc.* **1907**, *91*, 1593.
103. Lykknes, A.; Kragh H.; Kvittingen, L. *Phys. Perspect.* **2004**, *6*, 126.
104. Curie, M.; Gleditsch, E. *Compt. Rend.* **1908**, *147*, 345.
105. Ramsay, W. *Nature* **1907**, *76*, 269.
106. Travers W. M. *The Life of Sir William Ramsay*; Arnold: London, 1956.
107. Boltwood to Rutherford, Badash, L. *Dictionary of Scientific Biography*, **1969**, 190.
108. Boltwood was very pleased to learn that Yale did not confer an honorary doctorate on Einstein because he was Jewish; Boltwood to Rutherford, July 14, 1921, Badash, L. **1969**, 346.
109. Boltwood B. B. *Amer. J. Sci.* **1908**, *25*, 365.
110. Boltwood to Rutherford, December 5, 1911; Badash, L. *Dictionary of Scientific Biography*, **1969**, 260.
111. Rutherford E. *Nature* **1928**, *121*(3049), 535.
112. *Chem. News* **1916**, August 4, 60.
113. Pellini, G. *Gazz. Chim. It.* **1903**, *33* (II), 35.

VII.9

TRANSMUTATION OF LEAD INTO MERCURY

In 1924, Arthur Smits and his assistant A. Karssen, at the University of Amsterdam, published astonishing reports of their alleged transmutation of lead into mercury and thallium.[114] Their work was inspired by that of Adolf Miethe, who claimed to have transformed mercury into gold in a modified Jaenicke mercury ultraviolet lamp.[115]

The experiment, carried out in a quartz-lead lamp, was monitored with a quartz spectroscope. After a current of 30–35 A/8 V had been passed through the system for 6 hours, a few mercury lines began to appear in the spectrum. After 10 hours, the entire series of mercury lines, plus those of thallium, were apparent in the visible and ultraviolet spectrum.

In 1926, Smits and Karssen reported further developments of their experimental protocol. The lamp was redesigned, and all the equipment was examined spectroscopically to make certain it was free from mercury and thallium. The researchers also conducted experiments in a nitrogen atmosphere at various pressures and a liquid dielectric (carbon disulfide) with 100 kV/2 mA for 12 hours. The mercury was chemically detected as the iodide. Similar results were obtained with 160 kV/10–20 mA. In six such experiments, 0.1–0.2 mg of mercury was recovered. The researchers suspected that the CS_2 contained a trace of some organic mercury compound. Positive results were still obtained, however, even after it had been thoroughly purified. Smits offered this explanation for the transmutation:

$$_{82}\mathrm{Pb} \rightarrow {}_{80}\mathrm{Hg} + {}_{2}\mathrm{He} \qquad \text{(Eq. VII.2)}$$

In the case of the transmutation of lead into thallium, he assumed a cumbersome and unrealistic process:

$$_{82}\mathrm{Pb} + {}_{1}\mathrm{p} \rightarrow {}_{81}\mathrm{Tl} + {}_{2}\mathrm{He} \qquad \text{(Eq. VII.3)}$$

However, soon after these experiments, new problems were reported that showed that the phenomena taking place in the quartz-lead lamp was at the least bizarre, depending on unknown factors, and that transmutation was not so easy to reproduce as he had expected.

In 1926, Frank Horton (1878–1957) and Ann Catherine Davies[116] reported that they had been unsuccessful in their attempts to replicate the Smits-Karssen experiments. More than half of Horton's 44 published papers concerned the characteristic X-ray emission from certain elements and were mainly in collaboration with students, in particular with Ann Catherine Davies, whom he later married.

Scientific papers concerning transmutation of lead into mercury had become the fashion[117] in the mid-1920s, but, for some unknown reason, these experiments were not continued, and the issue disappeared from the scientific literature after 1927.[118] This line of research remains open to exploration, since the questions it raised remain unanswered to this day. Most probably the Dutch researchers were deceived by tiny quantities of impurities of mercury and thallium present in their samples and instruments.

Notes

114. Smits, A.; Karssen, A. *Scientific American* **1925**, *133*(4), 230–31; Smits, A.; Karssen, A. *Scientific American* **1926**, *134*(2), 80–81; Smits, A. *Nature* **1924**, *114*(2869), 609–10; Smits, A. *Nature*, **1926**, *117*(1931), 13–15; Smits, A. *Nature*, **1926**, *117*(1948), 620; Smits, A. *Nature*, **1927**, *120*(3022), 475–76; Smits, A.; Karssen, A. *Die Naturwissenschaften* **1925**, *13*(32), 699.
115. Anon. "Attempts at Artificial Au," *Literary Digest* **1925**, March 14; Anon. "Negative Evidence in the Hg-Au Case," *Literary Digest*, **1926** February 6; Anon. *Nature* **1924**, *114*, 197; Anon. "Transmutation of Hg into Au," *Nature* **1926**, *117*(2952), 604; Anon. "The Present Position of the Transmutation Controversy" *Nature* **1926**, *117*(2952), 758–60.
116. Davies, A. C.; Horton, F. *Nature*, **1926**, *117*(2935), 152; Whiddington, R. *Biographical Memoirs of Fellows of the Royal Society* **1958**, *4*, 117–27.
117. Anon. *Science-Supplement* **1925**, *62*(1602), 14; Anon. *Science-Supplement* **1926**, *63*(1623), 10; Anon. *Nature* **1927**, *117*(2952), 758–60.
118. Thomassen, L. *Nature* **1927**, *119*(3005), 813.

VII.10

SOME LIKE IT "COLD"

Cold fusion is the concept of nuclear fusion in conditions close to room temperature, in contrast to the conditions for the well-understood fusion reactions, such as those inside stars and high-energy experiments.

The ability of palladium to absorb hydrogen was recognized as early as the 19th century by Thomas Graham (1805–69). In the late 1920s, two Austrian-born scientists, Friedrich Paneth and Kurt Karl Peters (1897–1978), reported the transformation of hydrogen into helium by spontaneous nuclear catalysis when hydrogen was absorbed on finely divided palladium at room temperature. However, the authors later retracted the report, acknowledging that the helium they measured was due to that present in the air.[119]

In 1927, Swedish scientist Johan G. Tandberg stated that he had fused hydrogen into helium in an electrolytic cell with palladium electrodes. On the basis of his work, he applied for a Swedish patent for "a method to produce helium and useful reaction energy." After Harold Urey (1893–1981) discovered deuterium in 1932, Tandberg continued his experiments with heavy water,[120] but his patent application was eventually denied.

Interest in the field increased dramatically after nuclear fusion was reported in a table-top experiment involving electrolysis of heavy water on a palladium (Pd) electrode by Martin Fleischmann,[121] an electrochemist, and the physicist Stanley Pons[122] in 1989. They reported anomalous heat production ("excess heat") of a magnitude they asserted would defy explanation except in terms of nuclear processes. They further reported measuring small amounts of nuclear reaction by-products, including neutrons and tritium. These reports raised hopes for a cheap and abundant source of energy.

Enthusiasm turned to skepticism as failure to replicate the results was explained by (1) several theoretical reasons why cold fusion is not likely to occur, (2) the discovery of possible sources of experimental error, and (3) the discovery that Fleischmann and Pons had not actually detected nuclear reaction by-products. By late 1989, most scientists considered cold fusion claims dead, and cold fusion subsequently gained a reputation as pathological science. However, cold fusion continued to be investigated, and some positive results have been reported at mainstream conferences and in peer-reviewed journals. Cold fusion research sometimes is referred to as *low-energy nuclear reaction (LENR) studies* in order to avoid the negative connotations associated with earlier projects.

Fleischmann and Pons moved their laboratory to France with a grant from the Toyota Motor Corporation. The laboratory, IMRA, was closed in 1998 after spending £12 million on cold fusion work. Between 1992 and 1997, Japan's Ministry of International Trade and Industry sponsored a "New Hydrogen Energy Program" of US$20 million to conduct research on cold fusion. Announcing the end of the program in 1997, the director and one-time proponent of cold fusion research, Hideo Ikegami,[123] acknowledged its failure.

Notes

119. Paneth, F.; Peters, K. *Naturwissenschaften* **1926**, *14*(43), 956.
120. Tandberg, J. *Teknisk Tidskrift* **1939**, *69*, 9.
121. A British chemist, born in 1927 in Karlsbad, Czechoslovakia, Fleischmann and his family fled to Britain in 1939. He was educated at Imperial College London, where he earned his PhD in 1951. He taught chemistry at the University of Durham and at Newcastle University. In 1967, he was appointed professor of electrochemistry at the University of Southampton.
122. Stanley Pons was born in 1943, in Valdese, North Carolina. He is an electrochemist known for his work with Martin Fleischmann on cold fusion in the 1980s and '90s. He met Fleischmann while he was a graduate student in Professor Alan Bewick's group at the University of Southampton, where he obtained his PhD degree in 1978.
123. Voss, D. *Science* **1999**, *284*, 1252.

VII.11

IS COLD FUSION HOT AGAIN?

Soon after the Fleischmann-Pons announcement, Francesco Scaramuzzi (b. 1929) created and headed the Cold Fusion Research Project at the ENEA Laboratories in Frascati, Italy, from 1989 until his retirement. He continues his work in Frascati to this day, where he is actively involved in this field.

In February 2002, U.S. Navy researchers at the Space and Naval Warfare Systems Center in San Diego, California, who had been studying cold fusion continuously since 1989, released a two-volume report entitled "Thermal and Nuclear Aspects of the Pd/D_2O System" with a plea for funding.

In 2005, Italy embarked again on a new wave of cold fusion research. The Italian Cold Fusion research program was supported by the Ministry of Education. The team was led by Vittorio Violante who had established a unique level of mastery of the metallurgy of palladium foils, essential for success in this field. He is also familiar with calorimetry, also important for the experiments, and has performed successful heat-producing "fusion" experiments. However, the project has been bitterly criticized.

A demonstration in Bangalore by Japanese researcher Yoshiaki Arata (b. 1924) in 2008 revived interest for cold fusion research in India. Projects have commenced at several centers, such as the Bhabha Atomic Research Centre, and the National Institute of Advanced Studies has also recommended that the Indian government revive this research.

EPILOGUE

Three of the great advances that revealed the true relationship of chemical elements to one another were Mendeleev's doctrine of the periodic table, Moseley's law that conferred a number and an identity on every element by virtue of its number of nuclear protons, and Soddy's discovery that more than one type of atom could occupy the same place in the periodic table as long as that all-essential proton number were the same.

Even after these three stepping stones were laid as markers on the trail, many, and perhaps too many, scientists continued to make conceptual, absurd, and even ridiculous errors. In these pages, we have seen that some of these wrong turns were the results of experimental errors of the grossest sort, such as sample contamination, simple carelessness, or misuse of a new scientific technique, whereas others arose from incompetence, scientific fraud, unorthodox beliefs, a misplaced nationalism, and just plain obstinacy.

Some highly respected scientists, including some Nobel laureates, fell into error when they moved out of their area of expertise into another where they were rank amateurs. Others, not so well-known and out to make a name for themselves or to ingratiate themselves with their superiors, found the path of no return: oblivion. Sadly, another conclusion we have come to is that, at times, renowned and highly prolific scientists have, late in their careers, deviated from orthodox science because they did not keep pace with its evolution.

It is evident from the collection of stories in this volume that the discoveries of the genuine elements are inextricably bound up with the "discoveries" of the false. And so, in some instances, it took many decades to distinguish between the false and true element, with the false sometimes even outliving its discoverer.

As we have seen, the discoverers of the false elements fall into three categories. The first group of scientists had the good fortune to outlive their false discoveries, and some even received special recognition for them. The second group includes those chemists or physicists who had the misfortune to see their discovery turn out to be a "nondiscovery," and, in some rare cases, they were marginalized by official science. The last and most controversial group includes a few scientists like Fred Allison, inventor of the "magneto-optic" technique—who were considered examples of practitioners of "pathological science." Despite this label, they enjoyed great prestige in the scientific community and were allowed to continue their research, publish their misleading ideas, and disseminate these ideas to their students. This situation is perhaps one of the most distressing in the history of science.

Some scientists chose to recognize and retract their false discoveries immediately and publish their errors in the scientific literature; for example, Odolen Koblic's admission of his error in the discovery of *bohemium*. Others chose to retract somewhat slyly by publishing in obscure journals and little-known and even dead languages, as in Luigi Rolla's 1942 retraction of *florentium*, published partially in Latin in the commentaries of the Pontifical Academy of Sciences. Still others, and by far the largest group, chose not to retract at all, but to stand their ground even in the face of obvious error; for example, Enrico Fermi's proven false "discovery" of the first transuranium elements, *ausonium* and *hesperium*.

But, whatever the error, in good faith or not, there are lessons to be learned and remembered. One such lesson is that the process of doing science, of testing and revising our picture of nature, is the only part that does not change. We sincerely hope that these tales and their documentation will contribute to more scholarship in this fascinating area.

Notes

1. Crisciani, C. *Il Papa e l'alchimia*; Viella: Rome, 2002.
2. Paravicini Bagliani, A. *Le crisi dell'alchimia*; Sismel Edizioni del Galluzzo: Florence, Italy, 1996.
3. Pope John XXII issued a decree, *Spondent pariter*, condemning a frequent fraudulent alchemical practice, that of claiming to transform base metals into precious ones. See http://www.alchemy-website.com/Papal_bull.html (accessed April 12, 2014).
4. He is known mostly for being cited in Dante's *Inferno*, Canto XX, in the Eighth Circle of Hell among the soothsayers for prophesying the location of the death of Frederick II.
5. Crisciani, C.; Paravicini Bagliani, A. *Alchimia e medicina nel medioevo*; Sismel Edizioni del Galluzzo: Florence, Italy, 2003.
6. Gamow, G. *Biography of Physics*; Dover Publications: New York, 1988.
7. Rutherford E. *Phil. Mag.*, 6th series **1919**, *37*, 571.

APPENDIX

Chronological Finder's Guide for the Lost Elements

Date	*Element*	*Discoverer*	*Reference*	*Page*
1777	Terra nobilis Siderum	T. O. Bergman	Klaproth, M. H. *Crell's Annalen*, **1784**, *1*, 390.	3
1777–78	Hydrosiderum (wassereisen)	J. K. F. Meyer	Meyer, J. K. F. *Schriften der Gesellsch. naturf. Freunde*, **1780**, *2*, 334.	3
1783	Metallum problematicum	F. Müller von Reichenstein	Müller von Reichenstein, F. *Physik. Arbeiten der einträchtigen Freunde in Wien* **1783**, *1*(1), 57.	12
1786	Saturnum Saturnit	A. G. Monnet	Monnet, A. G. *J. Physique* **1786**, *28*, 168.	43
1788	Terra adamantina Diamanthspatherde	M. H. Klaproth	Klaproth, M. *Beschaft. Ges. Nat. Fr.* Berlin, **1788**, *8*, 4.	10
1789	Caloric	A. L. Lavoisier	Lavoisier, A.-L. *Traité élémentaire de chimie*; Chez Cuchet: Paris, 1789.	64
1790	Apulium Borbonium Austrium Parthenium Bornium Hydrosideron	A. Ruprecht; M. Tondi	Ruprecht, A. Tondi, M. *Ann. Chim.* **1791**, *8*, 3. Klaproth, M. H. *Ann. Chim.* **1791**, *10*, 275.	27
1790	Sydneium Australium Austral sand Terra australis	J. Wedgwood	Wedgwood, J. *Phil. Trans.* **1790**, *80*, 306.	4
1799	Thermoxygen	L. Brugnatelli	Brugnatelli, L. V. *Ann. Chim. Phys.* **1799**, *17*, 29.	62

(Continued)

Date	*Element*	*Discoverer*	*Reference*	*Page*
1800	Agusterde	J. B. Trommsdorff	Trommsdorff, J. B. *Almanach der Fortschritte in Wissenschaften, Künsten, Manufakturen und Handwerken* **1800**, *5*, 65.	65
1800	Andronia Thelyke	J. J. Winterl	Winterl, J. J., *Prolusiones ad chemiam saeculii decimi noni*; Typis ac sumptibus Typographiae Regiae Universitatis Pestinensis: Buda, Hungary, 1800.	403
1801	Pneum-alkali	C. F. S. Hahnemann	Hahnemann, C. F. S. *Scherer's Journal of Chemistry* **1801**, *5*, 665.	6
1802	Silene Silenium	J. L. Proust	Proust, J. -L. *Journ. de Phys.* **1802**, *55*, 457.	43
1803	Gahnium Nitricium	J. Berzelius	Jorpes, J. E. *Jac. Berzelius. His Life and Work*; Almqvist Wiksell: Stockholm, Sweden, 1966, p. 29.	73 24
1804	Klaprothium	J. F. John	Mellor, J. W. *A Comprehensive Treatise on Inorganic and Theoretical Chemistry*; Longmans Green: London and New York, 1946, p. 404.	44
1808	Niccolanum	J. B. Richter	Richter, J. B. *Gehlen's Jour.*, **1808**, *4*, 392.	53
1810	Vestium (vestalium, vestaeium)	A. Śniadecki	Marshall, J. L.; Marshall, V. R. "The Curious Case of 'Vestium.'" *The Hexagon* **2011**, Summer, 20–24.	14
1817	Urstoff	J. L. G. Meinecke	Meinecke, J. L. G. *Schweigger's Journ.* **1817**, *22*, 138.	404
1818	Crodonium	J. B. Trommsdorff	Trommsdorff, J. B. *Ann. der Physik* **1820**, *36*, 208.	65
1818	Melinum (melinium)	K. J. B. Karsten	Karsten, K. J. B. *Ann. der Physik* **1818**, *29*, 104.	61
1820	Apyre (apyrit)	G. Brugnatelli	Brugnatelli, G. *Brugnatelli Giorn. Fis.* **1820**, *3*, 2.	64
1820	Aurum millium	"Mr." Mills	Silliman, B., Ed. *American Journal of Science and Arts*, Vol. II, S. Converse: New Haven, CT, 1820, p. 363.	67

Date	*Element*	*Discoverer*	*Reference*	*Page*
1820	Wodanium	F. Stromeyer	Stromeyer, F. *Taschenbuch für die gesammte Mineralogie*; J. C. Hermann: Frankfurt am Main, Germany, 1822, p. 225; Anon. *Journal de Pharmacie et des Sciences Accessoires* **1820**, *6*, 397.	66
1825	Ostranium	J. F. A. Breithaupt	Breithaupt, A. *Pogg. Ann.* **1825**, *5*, 377.	111
1828	Pluranium Polinium	G. Osann	Osann, G. *Pogg. Ann.* **1828**, *13*, 283.	73
1836	Donium	T. Richardson	Richardson, T. *Record of General Science* **1836**, *3*, 426.	77
1836	Treenium	S. H. Boase	Boase, S. H. *Record of General Science* **1836**, *4*, 20.	78
1842	Didymium	C. G. Mosander	Mosander, C. G. *Pogg. Ann.* **1842**, *56*, 503.	172
1844	Pelopium	H. Rose	Rose, H. *Compt. Rend. Chim.* **1844**, *19*, 1275.	46
1845	Norium	L. F. Svanberg	Svanberg, L. F. *Pogg. Ann.* **1845**, *65*, 317.	111
1850	Aridium	C. Ullgren	Ullgren, C. *Öfversigt af Kongl.vetenskaps- akademiens förhandlingar* **1850**, no. *3*, 55.	44
1851	Donarium	C. W. Bergemann	Bergemann, C. W. *Ann. Chim. Phys.* **1852**, 235.	70
1852	Thalium	D. D. Owen	Owen, D. D. *Silliman's Amer. Jour.* **1852**, *13*, 4.	82
1857	Sulphurium	J. Jones	Jones, J. *Mining J.* **1857**, *27, July 14.*	87
1858	Junonium Vestium Neptunium Astaeum Hebeium	J. F. W. Herschel	Herschel, J. F. W. *British Association for the Advancement of Science Reprints. Part 2*, **1858**, 41.	92
1860	Dianium	W. F. von Kobell	von Kobell, F. *Bull. d. K. Bayr. Ak. d. Wissen. München*, (II Classe), Sitzung, (1860); *Ann. Ch. Pharm.* **1860**, *114*, 837.	47
1862	Wasium	J. F. Bahr	Bahr, J. F. *Stockholm Ak. Handl.* **1862**, *19*, 8.	138
1867	Aurorium	J. A. Ångström	Ångström, J. A. *Nova Acta Uppsala Sci.* **1867**, *9*(3), 29.	423
1869	Jargonium	H. C. Sorby	Sorby, H. C. *Chem. News* **1869**, *17*, 511.	112

(*Continued*)

Date	*Element*	*Discoverer*	*Reference*	*Page*
1869	Nigrium	H. A. Church	Church, H. A. *Chem. News* **1869**, *19*, 121.	112
1869	Ouralium Uralium Udalium	A. Guyard	Guyard, A. *Monit. Scientif.* **1879**, *21*, 795;	84
1874	Ilmenium	R. Hermann	Hermann, R. *J. prakt. Chem.* **1846**, *38*, 91.	47
1877	Davyum (davyium, devium)	S. Kern	Kern, S. *Compt. Rend. Chim.* **1877**, *87*, 72.	129
1877	Lavœsium	J. -P. Prat	Prat, J. -P. *Le monde pharmaceutique* **1877**, *8*, 4.	128
1877	Mosandrum (mosandrium)	J. L. Smith	Smith, J. L. *Compt. Rend. Chim.* **1877**, *87*, 148.	121
1877	Neptunium	R. Hermann	Hermann, R. *J. prakt. Chem.* **1877**, *2*, 15; 105.	47, 93
1878	Decipium	M. Delafontaine	Delafontaine, M. *Compt. Rend. Chim.* **1878**, *87*, 632.	122
1879	Rogerium Columbium	J. L. Smith	Smith, J. L. *Am. Chem. Journ.* **1883**, *5*, 73.	124
1879	Norwegium	T. Dahll	Dahll, T. *Vid. Selsk. Forth.* **1879**, *21*, 4.	136
1880	Comesium	H. Kämmerer	Anon. *Chem. -Ztg.* **1880**, *17*, 273.	83
1881	Phipsonium	J. Cawley	Cawley, J. *Chem. News* **1881**, *44*, 167.	149
1884	Idunium	M. Websky	Websky, M. *Sitzungsberichte Berliner Akademie* **1884**, 331.	85
1885	A	T. Carnelley	Wisniak, J. *Educ. quím.*, **2012**, *23*(4), 465–73.	409
1885	B	T. Carnelley	Wisniak, J. *Educ. quím.*, **2012**, *23*(4), 465–73.	409
1885	Elements Zα, Zβ, Zγ, Zδ, Zε, Zζ	P. E. Lecoq de Boisbaudran	Lecoq de Boisbaudran, P. E. *Compt. Rend. Chim.* **1885**, 100, 1437, etc.	210
1886	Austrium	E. Linnemann	Linnemann, E. *Monatshefte für Chemie* **1886**, *7*(1), 121.	36
1886	Protyle	W. Crookes	Crookes, W. *Chem. News* **1885**, *54*, 117.	406
1886	Elements Gα, Gβ, Gδ, Gζ, Gη; meta-elements; extinct elements	W. Crookes	Crookes, W. *Proc. Roy. Soc.* **1886**, *40*, 502.	204–209
1886	Polymnestum Erebodium Gadenium Hesperisium	A. Pringle	Pringle A. *Chem. News* **1886**, *54*, 167.	156

Date	Element	Discoverer	Reference	Page
1887	Erbium α, β Xα, Xβ, Xγ, Xδ, Xε, Xζ, Xη; Tmα, Tmβ; Smα, Smβ Diα, Diβ, Diγ, Diδ, Diε, Diζ, Diη, Diθ, Diι, Di	Krüss, G.; Nilson, L.	Krüss, G.; Nilson, L. F. *Ber. Dtsch. Chem. Ges.* **1887**, *20*, 2134.	123
1887	Neo-erbium	G. Krüss	Krüss, G. *Ber. Dtsch. Chem. Ges.* **1887**, *30*, 2143.	215
1889	Austriacum (Austrium)	B. Brauner	Brauner, B. *Chem. News* **1889**, *29*, 295.	38
1889	Gnomium	G. Krüss; F. W. Schmidt	Schunck, E. *Memoirs of the Manchester Literary and Philosophical Society* **1890**, *4*(3), 170.	67
1890	Damarium	P. Antsch; K. Lauer	Antsch, P.; Lauer, K. *Chemiker Ztg.* **1890**, *14*(27), 435.	428
1892	Masrium	H. D. Richmond; H. Off	Richmond, H. D.; Off, H. *J. Chem. Soc. Trans.* **1892**, *61*, 491	158
1893	Kalidium Oxidium Prefluorine	C. S. Palmer	Palmer, C. S. *Proceedings of the Colorado Scientific Society* **1893**, *4*, 56–74.	409
1894	Bauxium	M. Bayer	Bayer, M. *Bull. Soc. Chim. Fr.* **1894**, *11* [3], 534.	201
1894	Demonium	H. A. Rowland	Rowland, H. A., *Chem. News* **1894**, *70*, 68.	163
1894	Constitutive substances of erbium, yttrium and cerium	H. A. Rowland	Rowland, H. A., *Chem. News* **1894**, *70*, 68.	163
1895	Co-helium Orthohelium Parahelium	C. Runge; F. Paschen	Runge, C.; Paschen, F. *Phil. Mag.* **1895**, *42*(2), 297.	432
1895	Metacerium	B. Brauner	Brauner, B. *Chem. News* **1895**, *71*, 283.	40
1895	Infra-elements	G. J. Stoney	Stoney, G. J. Argon: A Suggestion. *Chem. News.* **1895**, *71*, 67–68.	182
1895	Supra-elements	L. W. Andrews	Andrews, L. W. *Chem. News.* **1895**, *71*, 235.	182
1896	"Hydrogen"	E. C. Pickering	Pickering, E. C. *The Astrophysical Journal* **1897**, *5*, 92.	409

(*Continued*)

Date	Element	Discoverer	Reference	Page
1896	Actinium (Zn)	T. L. Phipson	Phipson, T. L. *Chem. News* **1896**, *74*, 260.	147
1896	Argentaurum	S. H. Emmens	Emmens, S. H., *Argentaurum Papers No. 1: Some Remarks Concerning Gravitation*; Plain Citizen Publishing Co.: New York, 1896.	458
1896	Kosmium Neo-kosmium	H. B. Kosmann	Kosmann, H. B. *Z. Elektrochem.* **1896**, *3*, 279; Kosmann, H. B. *Berg. u. H.* **1896**, *50*, 225.	417
1896	Lucium (metal A)	P. Barrière	Barrière, P. *Chem. News* **1896**, *74*, 213.	165
1897	Anglium Scotium Hibernium	W. Ramsay	Ramsay, W. *Nature* **1897**, *56*, 378.	179
1897	Bythium δ	T. Gross	Gross, T. *Elektrochem. Ztschr.* **1897**, *4*, 1–8.	95
1897	Glaucodymium Glaucodidymium Russium	K. D. von Chrustchoff	von Chrustchoff, K. D. *Journ. Russ. Phys. Chem. Soc.* **1897**, *29*, 206.	174
1898	Etherium Etherion	C. F. Brush	Brush C. F. "New Gas in the Atmosphere," presented at the AAAS Meeting of 1898.	423
1898	Monium	W. Crookes	Crookes, W. *Proc. R. Soc. London: Report of the Meeting of the British Association for the Advancement of Science*, **1898**, 3–38.	202
1898	Victorium	W. Crookes	Crookes, W. *Proc. R. Soc. London* **1900**, *65*, 237.	202
1898	Metargon (metaargon)	W. Ramsay; M. W. Travers	Ramsay, W.; Travers, M. W. *Compt. Rend. Chim.* **1898**, *126*, 1610.	180
1900	Proto-metals Proto-elements	N. Lockyer	Lockyer, N. *Inorganic Evolution as Stuided by Spectrum Analysis*; Macmillan: London, 1900.	409
1900	Elements Σ, Γ, Δ, Ω and Θ	E. A. Demarçay	Marshall, J. L.; Marshall, V. R. *The Hexagon* **2003**, Summer, 19; Demarçay, E. -A. *Compt. Rend. Chim.* **1900**, *130*, 1019.	210
1900	Krypton II	A. Ladenburg	Ladenburg, A.; Kruegel, C. *Sitzungsber. K. Preuss. Akad. Wiss.* **1900**, 212–17.	187
1900	Radiolead	K. A. Hofmann; E. Strauss	Hofmann, K. A.; Strauss, E. *Ber. Dtsch. Chem. Ges.* **1901**, *34*, 8–11.	470

Date	Element	Discoverer	Reference	Page
1900	Thorium-α Thorium-β	B. Brauner	Brauner, B. *Proc. Chem. Soc.* **1900**, *17*, 67.	41
1901	Berzelium Carolinium	C. Baskerville	Baskerville C. *J. Am. Chem. Soc.* **1904**, *26*, 922.	192
1901	Euxenium	K. A. Hofmann; W. Prandtl	Hofmann, K. A.; Prandtl, W. *Ber. Dtsch. Chem. Ges.* **1901**,*34*, 1064–69.	113
1901?	Amarillium	M. W. Courtis	Courtis, M. W. *Trans. Am. Inst. Min., Metall., Pet. Eng., Soc. Min. Eng. AIME* **1901**, *31*, 1080.	198
1902	Ursubstanz	B. Brauner	Brauner, B. *Z. anorg. Chem.* **1902**, *32*, 1.	410
1903	Brillium	Unknown	*Washington Post*, 18 November 1903.	97
1903	Newtonium Coronium	D. Mendeleev	Mendeleev, D. *Vesnik i Biblioteca Samoobrazovanii* **1903**, *1–4*, 25; 83; 113; 161.	419
1903	Radium foil	Baskerville, C.; Kunz, C.	Baskerville, C.; Kunz, G. F. *Am. J. Sci.* **1904**, *18*(4), 25–28.	104
1904	Ether	D. Mendeleev	Mendeleef, D. *An Attempt towards a Chemical Conception of the Ether*; Longmans Green & Co.: London, 1904.	419
1904	Radiomercurium	S. M. Losanitsch	Losanitsch, S. M. *Ber. Dtsch. Chem. Ges.* **1904**, *37*, 2904.	105
1906	Ionium Incognitum	W. Crookes	Crookes W., *Proc. Roy. Soc., London* **1886**, *40*, 7.	202
1907	Anodium Cathodium	N. A. Morozov	Morozov N. A. *The periodic System of the Structure of Substances. Theory of Formation of Chemical Elements*; Sytin Publishers: Moscow, Russia, 1907.	406
1907	Proto-glucinium Proto-boron	A. C. Jessup; A. E. Jessup	Jessup, A. C.; Jessup, A. E. The Evolution and Devolution of the Elements. *Phil. Mag.* **1907**, *15*(VI), 21–55.	410
1909	Occultum Kalon Platinum-B Anu Proto-argon	A. Besant; C. W. Leadbeater	Besant, A.; Leadbeater, C. W. *Occult Chemistry. A Series of Clairvoyant Observations on the Chemical Elements,* 1st ed.; Theosophical Publishing Society: London and Benares City, 1909.	436

(Continued)

Date	*Element*	*Discoverer*	*Reference*	*Page*
1909	Satellite Nitron "Helium"	F. H. Loring	Loring, F. H. *Chem. News* **1909**, *100*, 281.	248
1909–10	Primary substances Zoikon Sub-element X	J. Moir	Moir, J. *J. Chem. Soc. Trans.* **1909**, *95*, 1752; Moir, J. *Proc. R. Soc. London* **1910**, *25*, 213.	246
1910	Protohydrogenium Pseudoelements Archonium	N. Morozov	Morozov (or Morosoff), N. A. *Die Evolution der Materie auf den Himmelskörpern*; Theodor Steinkopff: Dresden, Germany, 1910.	406
1910	Element E or X_2	Exner, F.; Haschek, E.	Exner, F.; Haschek, E. *Sitzungsber. Akad. Wiss. Wien* **1910**, *119*, 771.	207
1911	Canadium	A. G. French	*Glasgow Herald*, 5 December 1911; Rayner-Canham, G. W. *Canadian Chem. Educ.* **1973**, *8*(3), 10–11.	224
1911	Coronium "Hydrogen" Nebulium Proto-fluorine Archonium	J. W. Nicholson	Nicholson, J. W. *Phil. Mag.* **1912**, S. 6, *22*, 864.	408
1911	Geocoronium	A. Wegener	Wegener, A. *Physik. Z.* **1911**, *12*, 170.	422
1911	Neo-holmium	J. M. Eder; E. Valenta	Eder, J. M.; Valenta, E. *Sitz. Akad. Wiss. Wien.* **1911**, *119*, 32.	215
1911	Pantogen	G. D. Hinrichs	Hinrichs, G. D. *Rev. Gen. Chim.* **1911**, *13*, 351.	405
1911	Thulium I, II, III	C. Auer von Welsbach	Auer von Welsbach, C. *Monatshefte für Chemie und Verwandte Teile Anderer Wissenschaften* **1911**, *32*, 373.	235
1912	Josephinium	T. A. Eastick	Eastick, T. A. *Chem. News* **1912**, *105*, 36.	198
1914	Asium	V. I. Vernadsky	Vernadsky, V. I. *Bull. Acad. Sci. Petrograd.* **1914**, 1353.	114
1916	Denebium Neo-thulium Dubhium	J. M. Eder	Eder, J. M. *Sitzungsber. K. K. Akad. Wiss. Vienna IIa* **1916**, *125*, 1467.	280
1917	Euro-samarium Welsium	J. M. Eder	Eder, J. M. *Sitzungsber. K. K. Akad. Wiss. Vienna IIa* **1917**, *126*, 473.	281
1917	Néo-molybdenum Néo-tungsten	M. Gerber	Gerber, M. *Le Moniteur Scientifique Quesneville* **1917**, *7*, 73; 121; 169; 219.	310

Date	*Element*	*Discoverer*	*Reference*	*Page*
1919	Asteroid elements Crustaterrium, Primordial matter Terrium Chondrium, Pallasium Siderium Cosmium	P. N. Chirvinsky	Chirvinskii, P. N. *Bull. Inst. Polytechn. Don* **1919**, *7*(Sect. 2), 94.	411
1919	"Helium system" "Hydrogen" system	W. D. Harkins	Harkins,W. D. *Science* **1919**, *50*, 577.	445
1921	Emilium	P. Loisel	Loisel, P. *Compt. Rend. Chim.* **1921**, *173*, 1098.	284
1922	Hibernium	J. Joly	Joly, J. *Proc. Roy Soc.* A **1922**, *102*, 682.	270
1923	Oceanium	A. Scott	Scott, A. *J. Chem. Soc.* **1923**, *38*, 311.	116
1925	Neutronium Neuton Neutronon	A. von Antropoff	von Antropoff, A., *Z. angew. Chem.***1925**, *38*, 971.	444
1933	Element Z = zero	W. D. Harkins	Harkins, W. D. *Nature* **1933**, *131*, 23.	445
1925	Masurium	W. Noddack; I. Tacke; O. Berg	Zingales, R. "From masurium to trinacrium: The troubled story of element 43," *J. Chem. Educ*. **2005**, *82*, 221–27.	310
1925	Pragium	G. Druce	Karpenko,V. *Ambix* **1980**, *27*, 77; Ref. 44a.	250
1925	Dvi-manganese	Dolejšek, J.; Heyrovský, J.	Dolejšek, J.; Heyrovský, J. *Nature* **1925**, *116*, 782.	250
1926	Illinium	B S. Hopkins, *et al.*	Hopkins, B S. *Nature* **1926**, *117*, 792	296
1927	Florentium	L. Rolla, *et al.*	Rolla, L. *Nature*, **1927**, *119*, 637	296
1928	Hypon	W. S. Andrews	Andrews, W. S. *The Scientific Monthly* **1928**, *27*(6), 535.	416
1930	Alkalinium	F. H. Loring	Loring F. H. *Chem. News J. Ind. Sci.* **1930**, *140*, 178.	253
1931	Virginium (verium)	F. Allison	Allison, F.; Murphy, E. J.; Bishop, E. R.; Sommer, A. L. *Phys. Rev.*, **1931**, *37*, 1178.	323
1931	Element 108	R. Swinne	Swinne, R. *Wiss. Veroffentlich. Siemens-Konzern* **1931**, *10*(No. 4), 137.	326
1932	Adyarium Meta-Elements	Jinarajadasa, C.; C. W. Leadbeater	Jinarajadasa, C.; Leadbeater, C. W. *Theosophist* **1932**, *XII*, 361.	439

(*Continued*)

Date	*Element*	*Discoverer*	*Reference*	*Page*
1932	Alabamine (alabamium, eline)	F. Allison	Allison, F. *et al. J. Am. Chem. Soc.* **1932**, *54*, 613.	328
1933	Néo-actinium Néo-radium Néo-elements	A. Debierne	Debierne, A. *Compt. Rend. Chim.* **1933**, *196*, 770.	151
1934	Ausonium Hesperium	E. Fermi and co-workers	Fermi, E.; Rasetti, F.; D'Agostino, O. *Ricerca Scientifica* **1934**, *6*(1), 9.	316
1934	Bohemium	O. Koblic	Koblic, O. *Chem. Obzor* **1934**, *9*, 129.	327
1937	Eka-iodine Th-F; Gourium Dakin (Dacinum), Dekhine	R. De	De, R. *Separate* (Bani Press, Dacca) **1937**, 18.	338
1937	Moldavium	H. Hulubei	Hulubei, H. *Compt. Rend. Chim.* **1937**, *205*, 854.	323
1938	Sequanium	H. Hulubei; Y. Cauchois	Hulubei, H.; Cauchois, Y. *Compt. Rend. Chim.* **1938**, *207*, 333.	320
1939	Dor	H. Hulubei; Y. Cauchois	Hulubei, H. *Bull. Soc. Roum. Phys.* **1944**, *45*, no. 82, 3; Hulubei, H. *Bull. Acad. Roum.* **1945**, *27*, no. 3, 124.	331
1940	Helvetium	W. Minder	Minder, W. *Helv. Phys. Acta* **1940**, *13*, 144.	340
1942	Anglo-helvetium	W. Minder, A. Leigh-Smith	Minder, W.; Leigh-Smith, A. *Nature* **1942**, *150*, 767.	342
1963	Sulfénium	M. Duchaine	Duchaine, M. P. J. *French Demande* (May 4, 1973) 4 pp., CODEN: FRXXBL FR 2149300.	88
1972	T. W. Kow	Zunzenium	Kow, T. W., *J. Chem. Educ.* **1972**, *49*, 59.	392
1997	Quebecium	P. Demers	Demers, P. *Le Nouveau Système des Elements: Le Système du Quebecium*; Presses universitaires: Montreal, Canada, 1997.	225
2004	Hawkingium	Anastasovski, P. K.	Anastasovski, P. K. *AIP Conference Proceedings* **2004**, *699* (Space Technology and Applications International Forum—STAIF 2004), 1230.	393

BIBLIOGRAPHY

Websites

A popular website devoted to the elements
http://elements.vanderkrogt.net/ (accessed April 12, 2014)
Interactive in Latin (and once in, you can choose many other languages)
http://www.ptable.com/?lang=la (accessed April 12, 2014)
A collection of 90 classic papers in chemistry and more
http://www.chemteam.info/Chem-History/Classic-Papers-Menu.html (accessed April 12, 2014)
Selected classic papers including primary sources on elements (nature, number and discovery), atoms, and the periodic table
http://web.lemoyne.edu/~giunta/papers.html (accessed April 12, 2014)
The internet database of periodic tables
http://www.meta-synthesis.com/webbook/35_pt/pt_database.php?Button=Reviews+and+Books. (accessed April 12, 2014)
Key information about the periodic table and the elements
www.webelements.com (accessed April 12, 2014)

Books

Adunka, R. *Carl Auer von Welsbach: Entdecker—Erfinder—Firmengründer*; Kärntner Landesarchivs: Klagenfurt, Austria, 2013.

Aldersey-Williams, H. *Periodic Tales. The Curious Lives of the Elements*; Penguin Books: London, 2011.

Ångström, J. A. *Recherches sur le Spectre solaire*; W. Schultz: Uppsala, Sweden, 1868.

Anon. *Annual Scientific Discovery or Year-Book Facts in Science and Art for 1863*; David A. Wells: Boston, MA, 1865.

Anon. *Handwörterbuch der reinen und angewandten Chemie*; F. Vieweg und Sohn: Braunschweig, Germany, 1859.

Anon. *Retrospect of Philosophical, Mechanical and Agricultural Discoveries*; J. Wyatt: London, 1806.

Aristotle. *Metaphysics*, Book I, Part 4. W. D. Ross, tr. ebooks@Adelaide, 2007.

Atkins, P. W. *The Periodic Kingdom: A Journey into the Land of the Chemical Elements*; Basic Books: New York, 1997.

Baird, D.; Scerri, E.; McIntyre, L. *Philosophy of Chemistry: Synthesis of a New Discipline*; Springer: Dordrecht, 2006.

Beebee, H.; Sabbarton-Leary, N. *The Semantics and Metaphysics of Natural Kinds*; Routledge: New York, 2010.

Bensaude-Vincent, B.; Simon, J. *Chemistry—The Impure Science*, 2nd ed.; Imperial College Press: London, 2012.

Berzelius, J. J. *The Use of the Blowpipe in Chemistry and Mineralogy*, 4th ed.; William D. Ticknor: Boston, MA, 1845.

Berzelius, J. J. *Treatise on Mineral, Plant & Animal Chemistry*; Chez Didot Frères, Libraires: Paris, 1849.

Besant, A.; Leadbeater, C. W. *Occult Chemistry. A Series of Clairvoyant Observations on the Chemical Elements*, 1st ed.; Theosophical Publishing Society: London and Benares City, 1909.

Besant, A.; Leadbeater, C. W. *Occult Chemistry. Clairvoyant Observations on the Chemical Elements*, 2nd ed.; A. P. Sinnett, Ed.; Theosophical Publishing House: London, 1919.

Besant, A.; Leadbeater, W. C.; Jinarajadasa, C. *Occult Chemistry. Investigations by Clairvoyant Magnification into the Structure of the Atoms of the Periodic Table and of Some Compounds*, 3rd ed.; C. Jinarajadasa, Ed.; Theosophical Publishing House: Adyar, Madras, India, 1951.

Beudant, F. S. *Traité élémentaire de Minéralogie*; Carilian: Paris, 1837.

Boyle, R. *The Sceptical Chymist*. Project Gutenberg ebook, 2007 (prepared from the 1661 first edition). http://www.gutenberg.org/files/22914/22914-h/22914-h.htm (accessed April 9, 2014)

Braccesi, A. *Esplorando l'Universo*; Le Ellissi-Zanichelli: Bologna, 1988.

Bradford, T. L. *The Life and Letters of Dr. Samuel Hahnemann*; Boericke & Tafel: Philadelphia, PA, 1895.

Brock, W. H. *The Norton History of Chemistry*; W. W. Norton & Co.: New York, 1993.

Campbell, J. *Scientist Supreme*; AAS Publications: Christchurch, New Zealand, 1999.

Carter, H., Ed. *The Tomb of Tut-Ankh-Amon*; George H. Doran Co.: New York, 1927.

Certosi, P. *Alla ricerca della pietra filosofale*; Newton & Compton: Rome, 2002.

Champion, J., *Producing Precious Metals at Home*; Discover Publishing: Westboro, WI, 1994.

Coffey, P. *Cathedrals of Science: The Personalities and Rivalries That Made Modern Chemistry*; Oxford University Press: New York, 2008

Cook, T. M. *Samuel Hahnemann: The Founder of Homeopathic Medicine*; Thorsons: Wellingborough, U.K., 1981.

Cotton, F. A.; Wilkinson, G.; Murillo, C.; Bochmann, M. *Advanced Inorganic Chemistry*; John Wiley & Sons: New York, 1999.

Crawford, E.; Heilbron, J. L.; Ullrich, R. *The Nobel Population, 1901–1937*; Office for History of Science and Technology: UCB Berkeley, CA, 1987.

Crisciani, C. *Il Papa e l'alchimia*; Viella: Rome, 2002.

Crisciani, C.; Paravicini Bagliani, A. *Alchimia e medicina nel medioevo*; Sismel Edizioni del Galluzzo: Florence, 2003.

Cronstedt, A. F. *An Essay Towards a System of Mineralogy*; E. and C. Dilly: London, 1788.

Crosland, M. P. *Historical Studies in the Language of Chemistry*; Harvard University Press: Cambridge, MA, 1962.

D'Agostino, O. *Il chimico dei fantasmi*; Mephite: Atripalda (AV), 2002; reprint.

de Broglie, L. *Théorie générale des particules à spin*, 2nd ed.; Gauthier-Villars: Paris, 1954.

de Chancourtois, A. E. B. *Vis Tellurique: Classement des Corps Simples ou Radicaux, Obtenu au Moyen d'un Système de Classification Hélicoïdal et Numérique*; Mallet-Bachelier: Paris, 1863.

De Ment, J.; Dake, H. C. *Uranium and Atomic Power*; Chemical Publishing Co.: Gloucester, MA, 1945.

Delamétherie, J.-C. *Théorie de la terre*; Chez Maradan Publisher: Paris, 1797; available from Kessinger Publishers LLC: Whitefish, MT, 2001.

Demers, P. *Le Nouveau Système des Elements: Le Système du Quebecium*; Presses universitaires: Montreal, 1997.

Doberer, K. K. *The Goldmakers*; Nicholson & Watson: London, 1948.

Donovan, A. *Antoine Lavoisier: Science, Administration, and Revolution*; The University Press: Cambridge, UK, 1993.

Duché, M. T. *Elements of Chemical Philosophy*; Corey & Fairbank: Cincinnati, OH, 1832.

Ede, A. *The Chemical Element: A Historical Perspective*; Greenwood Publishing Group: Westport, CT, 2006.

Emmens, S. H. *Argentaurana, or Some Contributions to the History of Science*; G. Du Boistel: Bristol, 1899.

Emmens, S. H. *Argentaurum Papers No. 1: Some Remarks Concerning Gravitation*; Plain Citizen Publishing Co.: New York, 1896.

Emsley, J. *Nature's Building Blocks, An A–Z Guide to the Elements, New Edition*; Oxford University Press: New York, 2011.

Engelhardt, H. T., Jr. *Scientific Controversies, Case Studies in The Resolution and Closure of Disputes in Science and Technology*; Cambridge University Press: Cambridge, UK, 1987.

Evans, C. H., Ed. *Episodes from the History of the Rare Earth Elements*; Springer: Heidelberg, 1996.

Exner, F.; Hascheck, E. *Die Spektren der Elemente bei normalem Druck*; F. Deuticke: Wien, 1911.

Figurovskii, N. A. *Discovery of the Elements and the Origin of their Names*; Science Ed.: Moscow, 1970.

Fownes, G. *Elementary Chemistry, Theoretical and Practical*; Blanchard and Loeb: Camp Hill, PA, 1855.

Friedlander, G., Kennedy, J. W., Macias, E. S., Miller, J. M. *Nuclear and Radiochemistry*, 3rd Ed.; John Wiley: New York, 1981.

Friend, J. N. *Man and the Chemical Elements*; Griffin: London, 1951.

Fulcanelli: *Les Demeures Philosophales*; J. Pauvert: Paris, 1964.

Gamow, G. *Biography of Physics*; Dover Publications: New York, 1988.

Goldschmidt, B. *Atomic Rivals*; Rutgers University Press: New Brunswick, NJ, 1990.

Goldschmidt, V. M. *Geochemistry*; Clarendon Press: Oxford, UK, 1954.

Gordin, M. D. *A Well-Ordered Thing: Dmitrii Ivanovich Mendeleev and the Shadow of the Periodic Table*; Basic Books: New York, 2004.

Graesse, J. G. Th. *Orbis Latinus: Lexikon lateinischer geographischer Namen des Mittelalters und der Neuzeit*; Richard Carl Schmidt & Co.: Berlin, 1909.

Gray, T. *The Elements: A Visual Exploration of Every Known Atom in the Universe*; Black Dog & Leventhal Publishers: New York, 2009.

Greenwood, N. N.; Earnshaw, A. *Chemistry of the Elements*; Pergamon Press: New York, 1984.

Griffin, J. J. *A Practical Treatise on the Use of the Blowpipe in Chemical and Mineral Analysis*; R. Griffin & Co.: Glasgow, 1827.

Grinstein, L. S.; Rose, R. K.; Rafailovich, M. H. *Women in Chemistry and Physics: A Biobibliographic Sourcebook*; Greenwood Press: London, 1993.

Habashi, F. *Ida Noddack (1896–1978). Personal Recollections on the Occasion of the* 80th *Anniversary of the Discovery of Rhenium*; Laval University: Québec City, 2005.

Hackh's Chemical Dictionary; J & A Churchill Ltd.: London, 1946.

Haehl, R. *Samuel Hahnemann: His Life and Work*; Jain B. Publisher: New Delhi, 1995.

Heilbron, J. L. *H. G. J. Moseley: The Life and Letters of an English Physicist*; University of California Press: Berkeley, 1974.

Hendrickson, W. B. *David Dale Owen: Pioneer Geologist of the Middle West*; Indiana Historical Bureau: Indianapolis, 1943.

Herzfeld, J; Korn, O. *Chemie der seltenen Erden*; Springer: Heidelberg, 1901.

Hibben, J. G. *Inductive Logic*; Read Books: Alcester, UK, 2007.

Hinrichs, G. D. *Programm der Atom-Mechanik, oder die Chemie eine Mechanik der Pan-Atome*; Augustus Hageboek: Iowa City, 1867.

Hoffman, D.; Ghiorso, A.; Seaborg, G. T. *Transuranium People, The Inside Story;* Imperial College Press: London, 2000.

Hopkins, B S. *Chemistry of the Rarer Elements*; D.C. Heath & Co.: Boston, 1923.

Ihde, A. J. *The Development of Modern Chemistry*; Dover Publications: New York, 1984.

Jacobi, J., Ed. *Paracelsus: Selected Writings*; Princeton University Press: Princeton, NJ, 1951.

Jaffe, B. *Crucibles: The Story of Chemistry*, 4th revised edition; Dover Publications: New York, 1976.

James, L. K., Ed. *Nobel Laureates in Chemistry 1901–1992*; American Chemical Society: Washington, DC and Chemical Heritage Foundation: Philadelphia, PA, 1993.

Jeans, J. *The Mysterious Universe*; Cambridge University Press: Cambridge, UK, 1930.

Jensen, W. B., Ed. *Mendeleev on the Periodic Law: Selected Writings, 1869–1905*; Dover Publications: Mineola, NY, 2005.

Jollivet-Castelot, F. *Chimie et Alchimie*; E. Noury: Paris, 1928.

Joly, J. *The Surface History of the Earth;* The Clarendon Press: Oxford, 1925.

Jones, F. A. *New Mexico Mines and Minerals*; The New Mexican Printing Company: Santa Fe, 1904.

Jorpes, J. E. *Jac. Berzelius. His Life and Work*; Almqvist Wiksell: Stockholm, 1966.

Jungk, R. *Brighter than a Thousand Suns*; Harcourt Brace, Jovanovich: New York, 1958.

Kant, I. *Sämmtliche Werke*; Rosenkranz, K.; Schubert, F. W., Eds.; Leopold Voss: Leipzig, 1840.

Kean, S. *The Disappearing Spoon: And Other True Tales of Madness, Love and the History of the World from the Periodic Table of the Elements*; Little, Brown & Company: New York, 2010.

Kervran, C. L. *Biological Transmutations*; Happiness Press: Asheville, NC, 1989.

Kervran, L. *Biological Transmutations*; Swan House: New York, 1972.

Kervran, L. *Preuves Relatives A l'Existence de Transmutations Biologiques*; Librairie Maloine: Paris, 1968.

Kervran, L. *Transmutations A Faible Energie*; Librerie Maloine: Paris, 1968.

Kervran, L. *Transmutations Biologiques en Agronomie*; Librairie Maloine: Paris, 1970.

Klobasa, A. *Künstliches Gold. Versuch und Erfolg in der Goldsynthese*; Hartleben Publishers: Vienna, 1937.

Lavoisier, A.-L. *Traité élémentaire de chimie*; Chez Cuchet: Paris, 1789.

Leicester, H. M. *The Historical Background of Chemistry*; Dover Publications: New York, 1956.

LeNormand, Payen, *et al. New Universal Dictionary of Technology or Arts and Trades*; Tipografia Giuseppe Antonelle: Venezia, Italy, 1834.

Levere, T. *Transforming Matter: A History of Chemistry from Alchemy to the Buckyball*; The Johns Hopkins University Press: Baltimore, 2001.

Livio, M. *Brilliant Blunders*; Simon and Schuster: New York, 2013.

Lockyer, N. *Inorganic Evolution as Studied by Spectrum Analysis*; Macmillan: London, 1900.

Macorini, E. *Enciclopedia della Scienza e della Tecnica*, 7th ed.; Mondadori-McGraw-Hill: Milan, 1980.

Macquer, P. J. *Chymisches Wörterbuch*, 3rd Ed.; Weidmann: Liepzig, 1809.

Maglich, B., Ed. *Adventures in Experimental Physics*; World Science Communications: Princeton, NJ, 1972.

Malley, M. C. *Radioactivity: A History of a Mysterious Science*; Oxford University Press: New York, 2011.

Manutchehr-Danai, M. *Dictionary of Gems and Gemology*: 2nd ed.; Springer: Heidelberg, 2005.

Mazurs, E. G. *Graphic Representations of the Periodic System during One Hundred Years*; University of Alabama Press: Tuscaloosa, 1974.

Mellor, J. W. *A Comprehensive Treatise on Inorganic and Theoretical Chemistry*; Longmans Green: London and New York, 1946.

Mendeleef, D. *An Attempt towards a Chemical Conception of the Ether*; Longmans Green & Co.: London, 1904. This work is now available in a modern rendition in Jensen. W. B., Ed. *Mendeleev on the Periodic Law: Selected Writings, 1869–1905;* Dover Publications: Mineola, New York, 2005, pp. 227–52.

Mendeleev, D. *The Principles of Chemistry* (translated from 6th Russian edition by George Kamensky); Longmans, Green: London, 1897.

Minder, W. *Geschichte der Radioaktivität*; Springer: Heidelberg, 1981.

Minder, W. *Manual on Radiation Protection in Hospitals and General Practice*, vol. 5; World Health Organization: Geneva, 1980.

Moeller, T. *Qualitative Analysis;* McGraw-Hill: New York, 1958.

Moissan, H. *Einteilung der Elemente*; M. Krayn: Berlin, 1901.

Moissan, H. *The Electric Furnace*; Edward Arnold: London, 1904.
Moore, F. J. *A History of Chemistry*; McGraw-Hill: New York, 1939.
Morozov (or Morosoff), N. A. *Die Evolution der Materie auf den Himmelskörpern*; Theodor Steinkopff: Dresden, 1910.
Morozov (or Morosoff) N. A. *The Periodic System of the Structure of Substances. Theory of Formation of Chemical Elements*; Sytin Publishers: Moscow, 1907.
Morozov, N. D. I. *Mendeleev i znachenie ego periodicheskoi sistemy dlia khimii budushchago*; I. D. Sytin: Moscow, 1908.
Morrisson, M. *Modern Alchemy: Occultism and the Emergence of Atomic Theory*; Oxford University Press: New York, 2007.
Nelson, R. A. *Adept Alchemy*; Rex Research: Jean, NV, 1998.
Newman, J. R. *The World of Mathematics*, vol. 2; Simon and Schuster: New York, 1956.
Newton Friend, J. *Man and the Chemical Elements*, 2nd ed.; Scribners: New York, 1961.
Ohly, J. *Analysis, Detection and Commercial Value of the Rare Metals*, 3rd ed.; Mining Science Publishing Co.: Denver, CO, 1907.
Orna, M. V. *The Chemical History of Color*; Springer: New York and Heidelberg, 2013.
Pacciani, S., Ed. *Zibaldone di Padre Matteo Pinelli (1577–1669) priore di Cerliano (Notebook of Father Matteo Pinelli, prior of Cerliano)*; Gianpiero Pagnini Editore: Bagno a Ripoli, 1997.
Paracelsus. *The Hermetic and Alchemical Writings of Aureolus Philippus Theophrastus Bombast, of Hohenheim, called Paracelsus the Great*; J. Elliott and Company: London, 1894.
Paravicini Bagliani, A. *Le crisi dell'alchimia*; Sismel Edizioni del Galluzzo: Florence, 1996.
Partington, J. R. *A History of Chemistry*; Macmillan: London, 1962.
Phipson, T. L. *Confessions of a Violinist: Realities and Romance*; Chatto & Windus: London, 1902.
Phipson, T. L. *Famous Violinists and Famous Violins*; Chatto & Windus: London, 1903.
Phipson, T.L. *Biographical Sketches and Anecdotes of Celebrated Violinists since Lulli*; R. Bentley & Son: London, 1877.
Phipson, T. L. *Voice and Violin*; Chatto & Windus: London, 1898.
Powers, T. *Heisenberg's War: The Secret History of the German Bomb*; Jonathan Cape: New York, 1993.
Ramsay, W. *Chimica e chimici; saggi storici e critici*; Remo Sandron: Firenze, Italy, 1913.
Rayner-Canham, M. F.; Rayner-Canham, G. W., Eds. *A Devotion to Their Science: Pioneer Women of Radioactivity*; Chemical Heritage Foundation: Philadelphia, PA, 1997.
Reeves, R. *A Force of Nature: The Frontier Genius of Ernest Rutherford*; W.W. Norton & Co.: New York, 2008
Reid, C. *Hilbert*; Springer: New York, 1996.
Rhodes, R. *The Making of the Atomic Bomb*; Simon and Schuster: New York, 1995.
Rolla, L.; Fernandes, L. *Le Terre Rare*; Zanichelli: Bologna, 1929.
Romer, A. *Classics of Science: Radiochemistry and the Discovery of Isotopes*; Dover Publications: New York, 1977.
Ronan, C. A. *The Shorter Science and Civilisation in China: An Abridgement of Joseph Needham's Original Text*, Vol. 1; Cambridge University Press: Cambridge, UK, 1978.
Sacks, O. *Uncle Tungsten*; Alfred A. Knopf: New York, 2001.
Scerri, E. *A Tale of 7 Elements*; Oxford University Press: New York, 2013.
Scerri, E. *The Periodic Table: Its Story and Its Significance*; Oxford University Press: New York, 2007
Seaborg, G. T. *Elements of the Universe*; E. P. Dutton: New York, 1958.
Seaborg, G.; Seaborg, E. *Adventures in the Atomic Age: From Watts to Washington*; Farrar, Straus and Giroux: New York, 2001.
Segrè, E. *Autobiografia di un Fisico*; Il Mulino: Bologna, 1995.
Segré, E. *Enrico Fermi, Fisico. Una biografia scientifica*, 2nd Ed.; Zanichelli: Bologna, 1987.
Segrè, E. G. *A Mind Always in Motion: The Autobiography of Emilio Segrè*; The University of California Press, Berkeley, 1993.
Seligardi, R. *Lavoisier in Italy*; Leo S. Olschki: Firenze, Italy, 2002.

Shea, W. R. *Otto Hahn and the Rise of Nuclear Physics*; D. Reidel: Dordrecht, 1983.
Shepherd-Barr, Kirsten, Ed. *Science on Stage: From "Dr. Faustus" to "Copenhagen;"* Princeton University Press: Princeton, NJ, 2006.
Siegfried, R. *From Elements to Atoms: A History of Chemical Composition*; American Philosophical Society: Philadelphia, 2002.
Silliman, B., Ed. *American Journal of Science and Arts*, vol. II; S. Converse: New Haven, CT, 1820.
Smith, E. F. *M. Carey Lea, Chemist*; University of Pennsylvania Press: Philadelphia, 1923.
Soederbaum, H. G. *Nobel Lectures, Chemistry, 1922–1941*; Elsevier: Amsterdam, 1966.
van Spronsen, J. W. *The Periodic System of the Elements: A History of the First Hundred Years*; Elsevier: London, 1969.
Stock, J. T.; Orna, M. V., Eds. *The History and Preservation of Chemical Instrumentation*; D. Reidel Publishing Company: Dordrecht, 1986.
Strathern, P. *Mendeleyev's Dream. The Quest for the Elements*; St. Martin's Press: New York, 2000.
Stromeyer, F. *Taschenbuch für die gesammte Mineralogie*; J. C. Hermann: Frankfurt am Main, 1822.
Stwertka, A. *A Guide to the Elements*, 2nd ed.; Oxford University Press: New York, 2002.
Svanberg, L. F. *Öfversigt af Kongl.vetenskaps- akademiens förhandlingar;* P. A. Norstedt & Söner: Stockholm, 1845.
Szabadváry, F. *History of Analytical Chemistry*; Gordon & Breach Science Publishers: London, 1992.
Thénard, L. J. *An Essay on Chemical Analysis*; W. Phillips: London, 1819.
Thomson, T. *A System of Chemistry of Inorganic Bodies*, 7th ed., vol. 1; Baldwin & Cradock: London, 1831.
Thyssen, P; Binnemans, K. Accommodation of the Rare Earths in the Periodic Table: A Historical Analysis. In *Handbook on the Physics and Chemistry of Rare Earths*, vol. 41; Gschneidner, K. A., Jr.; Bünzli, J.-C. G.; Pecharsky, V. K., Eds.; North Holland (Elsevier): Amsterdam, 2011; 1–94.
Tiffereau, T. *L'Or et la Transmutation des Métaux*; Chacornac: Paris, 1889.
Tiffereau, T. *Les Métaux Sont Des Corps Composés*; Vaugirard: Paris, 1855.
Tilden, W. A. *The Elements: Speculations as to their Nature and Origin*; Harper Brothers: London, 1910.
Travers, M. W. *A Life of Sir William Ramsay*; Edward Arnold: London, 1956.
Treadwell, F. P.; Hall, W. T. *Analytical Chemistry, Vol. 1: Qualitative Analysis,* 7th ed.; D. Van Nostrand Co.: New York, 1930.
Trifonov, D. N.; Trifonov, V. D. *Chemical Elements: How They Were Discovered*; MIR Publishers: Moscow, 1982.
Turner, E. *Elements of Chemistry: Including the Recent Discoveries and Doctrines*, 5th ed.; Desilver, Thomas and Co.: Philadelphia, 1828.
Urbain, G. *Les notions fondamentales d'éléments chimiques et d'atome*. Gauthier-Villars: Paris, 1925.
Voronkov, M. G.; Abzaeva, K. A. *The Chemistry of Organic Germanium, Tin and Lead Compounds*; Rappoport, Z., Ed.; John Wiley and Sons: London and New York, 2002, pp. 1–130.
Waite, A. E. *A Collection of Alchymical Processes*; S. Weiser: New York, 1987.
Weeks, M. E. *Discovery of the Elements*: Reprinted from a series of articles published in the *Journal of Chemical Education*; Kessinger Publishing: Whitefish, MT, 2003.
Winterl, J. J. *Prolusiones ad chemiam saeculii decimi noni*; Typis ac sumptibus Typographiae Regiae Universitatis Pestinensis: Buda, Hungary, 1800.
Yorifuji, B. *Wonderful Life with the Elements: The Periodic Table Personified*; No Starch Press: San Francisco, 2012.
Zoellner, T. *Uranium*; Viking Press: New York, 2009.

ABOUT THE AUTHORS

Marco Fontani received his undergraduate degree in materials chemistry from the University of Florence and subsequently his doctorate in the chemical sciences at the University of Perugia. In 2003, he joined the Department of Organic Chemistry at the University of Florence. He is the author of over 100 papers in materials chemistry, organometallic chemistry, and electrochemistry, as well as in the history of chemistry.

Mariagrazia Costa received her degree in physical chemistry at the University of Florence and served in its Department of Chemistry all of her professional life until 1986 when she moved to the Laboratory of Educational Research in Chemical Education and Integrated Science. She has more than 250 print and 10 multimedia publications to her name. Her research activities have been in the areas of spectrochemistry, electrochemistry, and chemical education, as well as in the history of chemistry.

Mary Virginia Orna received her doctorate in analytical chemistry from Fordham University. She has served as Professor of Chemistry at the College of New Rochelle, New York, and as Editor of *Chemical Heritage* magazine and Director of Educational Services at the Chemical Heritage Foundation. She has won numerous national awards for excellence in chemical education; she was also a Fulbright Fellow for Israel in 1994–95. She has authored or edited 12 books and numerous papers in the areas of chemical education, color chemistry, and the history of chemistry.

NAME INDEX

Aartovaara, Gustaf Alfred (1863–1940), 257
Abelson, Philip H. (1913–2004), 91, 133, 327, 358, 376
Adloff, Jean-Pierre (b. 1930), 317
Agassiz, Jean Louis (1807–73), 119
Agricola, Georgius (1494–1555), 56, 272, 276
Allison, Fred (1882–1974), 256, 323, 327–31, 349, 481, 491, 492
Allison, S. K., 451
Alvarez, Luis (1911–88), 380, 385, 390
Amaldi, Edoardo (1908–89), 354, 355, 359
Ames, Joseph Sweetman (1864–1943), 432
Ampère, André-Marie (1775–1836), 56, 57
Anastasovski, Petar K., 392, 393, 492
Anderlini, Francesco (1844–1933), 421
Anders, Edward (b. 1926), 370–73
Anderson, Carl D. (1905–91), 445
Anderson, H. H., 454
Andrews, Lancelot Winchester (b. 1856), 182, 416, 417, 487
Andrews, W. S., 491
Andreyev, A. N., 394
Angeli, Angelo (1864–1931), 291, 301
Ångström, Jonas Anders (1814–74), 423, 485
Antropoff, Andreas von (1878–1956), 444, 491
Antsch, Paul, 428–30, 487
Arata, Yoshiaki (b. 1924), 480
Aristotle (382–322 BCE), XXV, XXVI, XXVIII, 24
Armbruster, Peter (b. 1931), 363, 364, 386, 387, 394
Armstrong, Henry Edward (1848–1937), 106
Arnaiz y Freg, Arturo (1915–82), 16
Artmann, Paul, 106
Aston, F. W. (1877–1945), 189, 313, 319, 453
Aström, B., 381, 382
Atterling, Hugo, 379
Azyr, Felix Vicq-d' (1748–94), 3

Baade, Walter (1893–1960), 368
Bahr, Johann Friedrich (Jön Fridrik) (1815–75), 45, 138, 139, 485
Bainbridge, Kenneth Tompkins (1904–96), 322, 323, 454
Balmer, Johann Jacob (1825–98), 433
Baly, Edward Charles Cyril (1871–1948), 185
Banks, Joseph (1743–1820), 4, 5
Baranger, Pierre, 465
Barrière, Prosper, 165–69, 174, 488
Baskerville, Charles (1870–1922), XXX, 97, 104, 192–95, 489
Baskerville, Sir Henry, 457
Baxter, Gregory Paul (1876–1953), 189
Bayer, Karl Josef (Bayer, M.) (1847–1904), 201, 202, 487
Beams, Jessy W. (1898–1977), 331
Beaumont, Jean-Baptiste Elie de (1798–1874), 121
Becquerel, Edmond (1788–1878), 121
Becquerel, Henri (1852–1908), XXVII, 268, 269, 377, 397, 474
Beilstein, Friedrich Konrad (1838–1906), 103
Bélopolsky, Aristarkh (1853–1934), 431
Berg, Otto (1874–1939), 287, 311, 312, 315, 335, 491
Bergemann, Carl Wilhelm (1804–84), 70, 71, 485
Bergman, Torbern Olof (1735–84), 3, 4, 12, 23, 25, 27, 272, 276, 277, 483
Berlin, Nils Johannes (1812–91), 71, 98, 108

Bernert, Traude (1915–98), 333, 344, 345
Bernoulli, Daniel (1700–82), 56
Bernoulli, Friedrich-Adolph (1835–1915), 275, 277
Berthelot, Marcellin Pierre Eugène (1827–1907), xxvii, 180, 189, 284, 285
Berthelot, Paul Alfred Daniel (1865–1927), 284
Berthier, Pierre (1782–1861), 216
Berthollet, Claude (1748–1822), 26, 29, 45, 404
Berzelius, Jöns Jacob (1779–1848), 13–15, 22, 24, 38, 44, 46, 54, 70–75, 111, 147, 172, 193–95, 233, 273, 367, 408, 465, 484
Besant, Annie (1847–1933), 435–42, 489
Bethe, Hans (1906–2005), 318, 340, 363
Beudant, François Sulpice (1787–1850), 73
Bindheim, Johann Jacob (1743–1822), 79
Bischof, Karl Gustav (1792–1870), 70, 72
Bischoff (or Bischof), Carl (1812–84), 101, 102
Blackett, Patrick Maynard Stuart (1897–1974), 254, 287
Blomstrand, Christian Wilhelm (1826–97), 48
Blood-Ryan, H. C., 249
Blumenbach, Johann Friedrich (1752–1840), 5
Boase, Henry Samuel (1799–1883), 78, 79, 485
Boggio-Lera, Enrico (1862–1956), 454
Bohr, Niels (1885–1962), 117, 118, 192, 220, 228, 236–40, 244, 270, 297, 319, 326, 357, 361, 362, 375, 377, 391, 407, 408
Boisbaudran, Paul Emile (*dit* François) Lecoq de (1838–1912), 37, 40, 123, 165, 168, 169, 173, 175, 179, 201, 207, 210–14, 216, 277, 279, 289, 417, 486
Bolton, Henry Carrington (1843–1903), 80, 132, 134
Boltwood, Bertram Borden (1870–1927), 473, 475
Boltzmann, Ludwig (1844–1906), 191, 208
Bonnier, Gaston (1853–1922), 128–32
Bordet, Jules (1870–1961), 242
Born, Ignaz Elder von (1742–91), 27, 28
Bosanquet, Claude H. (b. 1896), 286, 287
Bose, Satyendranath (1894–1974), 338, 346
Boucher, Gethen George (b. 1869), 104, 106, 107
Boudouard, Octave (1872–1923), 40
Bourget, Henry (1864–1921), 407
Brackebusch, Ludwig (1849–1906), 85
Bragg, William H. (1862–1942), 190, 286
Brahe, Tycho (1546–1601), 415
Brauner, Bohuslav (1855–1935), 38–42, 171, 174–77, 182, 189, 238, 241, 279, 290, 297, 307, 308, 410, 411, 487, 489
Breithaupt, Johann Friedrich August (1791–1873), 111, 485
Broglie, Duke Maurice de (1875–1960), 238, 244
Broglie, Prince Louis de (1892–1987), 238, 444
Brugnatelli, Gaspare (1795–1852), 63, 64, 484
Brugnatelli, Luigi Valentino (1859–1928), 64
Brugnatelli, Luigi Valentino Gasparo (1761–1818), 62, 483
Brugnatelli, Tullio (1825–1906), 64
Brunetti, Rita (1890–1942), 295, 297, 298, 303
Brush, Charles (1849–1929), 423–25, 488
Brush, George Jarvis (1831–1912), 82
Buchholz, Christian Friedrich (1770–1818), 404
Buisson, Henri (1873–1944), 407
Bunsen, Robert (1811–99), 24, 34, 39, 84, 91, 97, 99, 119, 125, 166, 173–75, 188, 196, 201, 204, 233, 279
Burci, Enrico (1862–1933), 301
Bussy, Antoine Alexandre-Brutus (1794–1882), 31, 36, 80
Butlerov, Aleksandr (1828–86), 398

Cabrera, Blas (1878–1945), 238
Cacciapuoti, Nestore Bernardo (1913–79), 313
Cameron, Alexander Thomas (1882–1947), 472, 473
Campo y Cerdán, Angel del (1881–1944), 107

Canneri, Giovanni (1897–1964), xviii, 293
Cannizzaro, Stanislao (1826–1910), xxviii, xxxi, xxxv, 302, 421
Canseliet, Eugène (1899–1982), 458
Caritat, Antoine Nicolas de, Marquis de Condorcet (1743–94), 3
Carius, Georg Ludwig (1829–75), 84
Carnelley, Thomas (1854–90), 409, 486
Caro, Roger (1911–92), 455
Carobbi, Guido (1900–83), 145
Carrara, Nello (1900–93), 337
Carter, Howard (1874–1939), 116
Cauchois, Yvette (1908–99), 320–23, 326, 327, 331, 333, 334, 345, 347, 492
Cawley, J., 149, 153, 486
Chadwick, James (1891–1974), 391, 393, 445, 446
Champagne, Jean Julien (1877–1932), 458
Champetier, Georges (1905–80), 241
Champion, Joe, 463
Chancourtois, Alexandre Emile Beguyer de (1820–86), xxxi, 405
Chandler, Charles Frederick (1836–1925), 99–102
Chandler, William Henry (1841–1906), 100
Chapman, Alfred Chaston (1869–1932), 349
Charles, Jacques Alexandre César (1746–1823), 50
Chelpin, Hans Karl Euler (1873–1964), 242
Chevreul, Michel Eugène (1786–1889), 37, 121
Chirvinsky, Pyotr Nikolaevich (1880–1955), 403, 410–12, 491
Choppin, Gregory Robert (b. 1927), 381
Chrustchoff, Konstantin Dimitrievic von (1852–1912), 174, 488
Church, Arthur Herbert (1834–1915), 112, 113, 486
Clark, Frank Wigglesworth (1847–1931), 162, 164
Clarke, Edward Daniel (1769–1822), 32, 49, 50, 52, 147
Claude, Georges (1870–1960), 188
Cleve, Per Theodor (1840–1905), 123, 165, 167, 173, 175, 280
Collet-Descotils, Hippolyte-Victor (1773–1815), 15, 69
Collie, John Norman (1859–1942), 183, 184, 190, 431, 432, 468
Conrad, Arnold, 462
Corbino, Orso Mario (1876–1937), 353, 356, 357, 375
Coronedi, Giusto (1863–1941), 301
Corson, Dale R. (1914–2012), 332, 339, 346
Coryell, Charles DuBois (1912–71), 304, 305, 308
Cossa, Alfonso (1833–1902), 144, 145
Coster, Dirk (1889–1950), 116–18, 236–43, 255
Courtis, William M. (1842–1920s?), 198, 199, 489
Cranston, John Arnold (1891–1972), 263, 265
Crawford, Adair (1748–95), 27
Crookes, Sir William (1832–1919), 25, 82, 83, 88, 91, 129, 131, 163, 165–69, 174, 183, 190, 201, 203–10, 216, 217, 249, 252, 279, 281, 289, 290, 406, 410, 411, 431, 432, 437, 440, 458–60, 486, 488, 489
Cruikshank, William (~1745–~1810), 27
Cullen, William (1710–90), 7
Curie, Marie Skłodowska (1867–1934), xxviii, 62, 150–54, 159, 160, 183, 189, 193, 195, 210, 219, 262, 299, 324, 325, 333, 344, 345, 354, 355, 378, 385, 420, 434, 470–75
Curie, Maurice (1855–1941), 299
Curie, Pierre (1859–1906), xxviii, 150, 151, 160, 183, 189, 195, 210, 434, 470

D'Agostino Oscar (1901–1975), 382–3, 392–4, 492, 385, 397–8, 532
Dahll, Tellef (1825–93), 136–40, 486
Dalton, John (1766–1844), xxvii, 14
Damour, Augustin Alexis (1808–1902), 48, 70, 71, 177
Darwin, Charles Robert (1809–82), xxiii, 5, 410
Darwin, Erasmus (1731–1802), 5
Das, Ashtoush (1888–1941), 338
Dauvillier, Alexandre (1892–1979), 117, 236–40, 244, 255, 379
Davies, Ann Catherine, 476

Davy, Sir Humphry (1778–1829), 24, 30–32, 36, 49, 50, 52, 56, 57, 60, 79, 80, 95, 129, 147, 188, 196, 362
De, Rajendralal, 337–39
Debierne, André (1874–1949), 147, 150–53, 184, 185, 189, 195, 262, 324, 325, 492
Delafontaine, Marc Abraham (1837–1911), 119–25, 139, 171, 173, 281, 486
Delambre, Jean Baptiste Joseph (1749–1822), 57
Delamétherie, Jean Claude (or De la Métherie or de La Métherie) (1743–1817), 43–45
Demarçay, Eugène (1852–1903), 201, 210–12, 216, 281, 289, 417, 488
Demers, Pierre (b. 1914), 225–29, 492
Demin, A. G., 363, 385
Dempster, Arthur Jeffrey (1886–1950), 323
Deslandres, Henri Alexandre (1853–1948), 431
Détraz, Claude (b. 1938), 447
Deville, Henri Etienne Sainte-Claire (1818–81), 48, 49, 136, 141
Dewar, Sir James (1842–1923), 115, 179, 181
Dirac, Paul Adrien Maurice (1902–84), 192, 391
Divers, Edward (1837–1912), 219, 222
Dixon, Henry Horatio (1869–1953), 271
Dmitriev, I. S., XXXI
Dmitriev, S. N., 395
Döbereiner, Johann Wolfgang (1780–1849), XXXI
Dobroserdov, Dimitrii Konstantinovich (1876–1936), 323
Dolejšek, Vaclav (1895–1945), 250, 251, 255, 491
Dolezalak, Friedrich (1873–1920), 194, 197
Dolomieu, Deodat de (1750–1801), 33
Donnan, Frederick G. (1870–1956), 186
Dorn, Friedrich Ernst (1848–1916), 184, 189, 195
Drossbach, Paul Gerard (1866–1903), 40
Druce, John Gerald Frederick (1894–1950), 131, 132, 134, 249–57, 291, 312, 321, 491
Duane, William (1872–1935), 451, 452
Duchaine, P. J. Marcel, 88, 89, 492
Dufrénoy, Armand (1792–1857), 121
Dujols, Pierre (1862–1926), 458
Dumas, Jean-Baptiste (1800–84), XXXI, 121, 405
Dunikowski, Zbigniew (1889?–1964), 461
Duns Scotus, John (ca. 1180–1236), 450
Dupré, August (1835–1907), 98, 99
Dupré, Friedrich Wilhelm (d. 1908), 98, 99
Dupuy, Gaston, 152

Eastick, T. A., 198, 490
Eastman, George (1854–1932), 282
Eddington, Arthur Stanley (1882–1944), 443
Eder, Josef Maria (1855–1944), 201, 215, 216, 278–82, 490
Edison, Thomas Alva (1847–1931), 191, 385
Eglestone, Thomas (1832–1900), 100
Einstein, Albert (1879–1955), XXIII, 346, 369, 377, 380, 385, 418, 475
Ekeberg, Anders Gustaf (1767–1813), 46, 233
Elhuyar y de Zubice, Don Fausto d' (1755–1833), 15, 17, 27, 272, 273, 277
Elhuyar y de Zubice, Juan José d' (1754–96), 17, 71, 272, 277
Elsasser, Walter M. (1904–91), 365
Emmens, Stephen Henry (b. 1844 or 1845), 458, 459, 488
Estey, Roger, 453
Euler, Leonhard (1707–83), 56
Evans, Clare de Brereton (b. ca. 1865), 105, 106
Exner, Franz (1849–1926), 207, 208, 280, 281, 490

Fabry, Charles (1867–1945), 407
Fajans, Kasimir (1887–1975), 260–65
Faraday, Michael (1791–1867), XXXVI, 38, 60, 115, 328, 405
Fehling, Hermann von (1812–85), 103
Fermi, Enrico (1901–54), XVIII, 192, 264, 295, 298, 312–19, 326, 344, 353–59, 362, 370, 375–77, 380, 385, 481, 492
Fernandes, Lorenzo (1902–77), XV, XVIII, 173, 292–94, 296, 298, 301, 302, 306, 308
Fernandez, D. Dominique Garcia, 97

Fischer, Emil (1852–1919), 187
Fischer, Ernst Gottfried (1754–1831), 53
Fittig, Wilhelm Rudolph (1835–1910), 186
Flawitzky, F., 179
Fleischmann, Martin (1927–2012), 464, 478–80
Flerov, Georgii Nikolayevich (1913–90), 370, 382, 383–86, 398, 449
Flügge, Siegfried (1912–97), 340
Folger, H., 394
Fourcroy, Antoine François de (1755–1809), xxvi, 69, 70, 80, 404
Fournier, Georges (1881-1954), 391, 393
French, Andrew Gordon, 198, 224, 225
Fresenius, Carl Remigius (1818–97), 23, 165–68
Fresnel, Augustin Jean (1788–1827), 226
Freud, Sigmund (1856–1939), 203
Friedel, Charles (1832–99), 125, 150, 235, 242, 277
Friend, John Newton (1881–1966), 251, 252, 311, 337, 346
Fulcanelli, 457, 458
Gadolin, Johan (1760–1852), 171, 233, 378
Gahn, Johan Gottlieb (1745–1818), 22, 30, 73
Galissard de Marignac, Jean Charles (1817–94), 48, 119, 120, 123, 167, 171, 172, 235
Galle, Johann Gottfried (1812–1910), 93
Gamow, George (1904–68), 387
Garbasso, Antonio (1871–1933), 295
Garrett, Milan, 454
Gaschler, Alois, 454
Gay-Lussac, Louis-Joseph (1778–1850), 30, 31, 56
Gehlen, Adolph Ferdinand (1775–1815), 13, 404
Genth, Frederick Augustus (1820–93), 48, 82, 97, 98, 100
Gentry, Robert V. (b. 1933), 371–73, 399
Gerber (Carl Ludwig?; b. 1854?) M., 274–77, 310, 311, 491
Gerland, William Balthasar (b. 1831), 102, 103, 122
Ghiorso, Albert (1915-2010), 362–64, 378–87
Gibbs, J. Willard (1839–1903), 191
Gibbs, Oliver Wolcott (1822–1908), XXXI, 162
Giesel, Friedrich Oskar (1852–1927), 184, 262
Gilbert, Ludwig Wilhelm (1769–1824), 44, 59–61, 67
Gleditsch, Ellen (1879–1968), 472, 473
Glendenin, Lawrence Elgin (1918-2008), 304, 305, 308
Gmelin, Leopold (1788–1853), XXXI
Göhring, Ostwald Helmuth (1889–1915?), 261, 262
Goldanskii, Vitalii Iosifovich (1923-2001), 387
Goldschmidt, Bertrand (1912–2002), 151, 152
Goldschmidt, Victor Moritz (1888–1947), 222, 223
Gorceix, Claude-Henri (1842–1919), 174–76
Graham, Thomas (1805–69), 478
Gramont, Antoine Arnaud Alfred Xavier Louis de (1861–1923), 274, 275, 277
Gramont, Armand A. Agenor, Count of (1879–1962), 277
Grant, Julius (1901–91), 349
Gregor, William (1761–1817), 27, 33
Grignard, Victor (1871–1935), 242
Griner, Georges, 202
Gross, Theodor (1860–1924), 95, 96
Grosse, Aristid Victor (1905–85), 263, 264, 357
Guillaume, Charles-Edouard (1861–1938), 242
Guyard, Antony (d. 1884), 84, 486
Guyton de Morveau, Louis Bernard (1737–1816), XXVI, XXXV, 30, 404

Haber, Fritz (1868–1934), 262, 443, 447, 454
Hadding, Assar Robert (1886–1962), 291, 308
Hahn, Otto (1879–1968), 262–64, 266, 304, 316, 321, 338, 356–58, 375, 384, 395, 398
Hahnemann, Samuel Christian Friedrich (1755–1843), 6–11, 484
Haidinger, Karl (1756–97), 5

Haïssinsky, Moïse N. (1898–1976), 354, 381
Halban, Hans von (1908–64), 226, 357
Hale, George Ellery (1868–1938), 431
Hall, Charles Martin (1863–1914), 201
Hamer, Richard, 286, 287
Hampson, William (1859–1926), 179
Hans, Alexander, 95
Harkins, William Draper (1873–1951), 443–46, 451, 491
Harris, J. Allen (1901–72), 295–97, 302, 306, 309
Harvey, Bernard, 380
Haschek, Eduard (1875–1947), 207, 280, 281, 490
Hatchett, Charles (1765–1847), 5, 6, 46, 47, 51
Haüy, René-Just (1743–1822), 33, 79
Hawking, Stephen W. (b. 1942), 393
Hehner, Otto (1853–1924), 158
Heisenberg, Werner (1901–76), 192
Helmholtz, Hermann L. von (1821–94), 161
Helmont, Johann Baptista von (1579–1644), 465
Hénouville, Théodore Baron de (1715–68), 30
Hermann, Carl Samuel (1765–1846), 59–61
Hermann, Hans Rudolph (1805–79), 47–49
Hermbstaedt, Sigismund Friedrich (1760–1833), 7
Héroult, Paul Louis Toussaint (1863–1914), 201
Herschel, Caroline Lucretia (1750–1848), 92
Herschel, Sir John (1792–1871), 91–94, 485
Herschel, Sir Frederick William (1738–1822), 92
Herschel, William James (1833–1917), 94
Hervilly, Marie Melanie d' (1800–78), 9
Herzeele, Albrecht von (1821–?), 465
Hess, Viktor F. (1883–1964), 217
Heßberger, Fritz Peter, 394
Hevesy, George de (1885–1966), 116–18, 220–22, 236–43, 255, 271, 279, 313, 321
Heymann, D. (b. 1927), 371
Heyrovský, Jaroslav (1890–1967), 134, 242, 250, 251, 312, 491
Hilbert, David (1862–1943), 357, 358
Hill, E. A., 432
Hillebrand, William Francis (1853–1925), 431
Hinrichs, Gustavus Detlef (1836–1924), 405, 406, 490
Hisinger, Wilhelm (1766–1852), 13, 14
Hjortdhal, Torstein (1839–1925), 136
Hoff, Jacobus Henricus Van't (1852–1911), 291, 429
Hoffman, Darleane C. (b. 1926), XIX, 386
Hofmann, Augustus Wilhelm von (1818–92), 192, 204
Hofmann, Karl Andreas (1870–1940), 113, 215, 470, 488, 489
Hofmann, Sigurd (b. 1944), 364, 394
Hohenheim, Auroleus Phillipus Theophrastus Bombastus von, aka Paracelsus (1493–1541), XXV
Holtz, H. C., 107
Homberg, Wilhelm (1652–1715), 30
Home, Daniel Douglas (1833–86), 204
Honigschmid, Otto (1878–1945), 453
Hopkins, B Smith (1873–1952), XIX, 279, 295, 297, 300, 302–04, 306–09, 312, 423, 491
Horton, Frank (1878–1957), 476
Hudson, David, 462
Huggins, Sir William (1824–1910), 431
Hulubei, Horia (1896–1972), 256, 320–27, 331–34, 339, 340, 343–45, 492
Humboldt, Alexander, Baron von (1769–1859), 15
Humphreys, William Jackson (1862–1949), 432

Inoue, Toshi (1894–1967), 222
Irion, Clarence E., 451

Jacquin, Joseph Franz Freiherr von (1766–1839), 29
James, Charles King (1880–1928), 291, 299–301
James, William (1842–1910), 203
Jander, Wilhelm (1898–1942), 317
Janet, Pierre (1859–1947), 203
Janik, R., 394
Janssen, Pierre Jules César (1824–1907), 178, 431

Jasper, Charles (1864–1906), 268
Jeans, Sir James (1877–1946), 440
Jinarajadas, Curuppumullage (1875–1953), 435, 438–41, 491
Joliot, Frédéric (1900–58), xxx, 151, 192, 226, 242, 264, 325, 354, 355, 362, 385, 398
Joliot-Curie, Irène (1897–1956), xxx, 151, 152, 192, 242, 264, 324, 325, 341, 354, 355, 357, 362, 391, 398
Jollivet-Castelot, François (1874–1937), 458
Joly, John (1857–1933), 267–71, 491
Jørgensen, Christian K. (1931-2001), 209
Joule, James Prescott (1818–89), 124
Joy, Charles Arad (1823–91), 99

Kamerlingh-Onnes, Heike (1853–1926), 179
Kämmerer, Hermann (1840–98), 83, 84, 486
Kankrin, Frantsevich (1775–1845), 74, 75
Kant, Immanuel (1724–1804), 27
Karlik, Berta (1904–90), 333, 344, 345
Karssen, A., 476
Karsten, Dietrich Ludwig Gustav (1768–1810), 7
Karsten, Karl Johann Bernhard (1782–1853), 60, 61, 484
Kastler, Alfred (1902–84), 322
Kastner, Karl W. Gottlieb (1783–1857), 404
Kayser, Heinrich Gustav Johannes (1853–1940), 431
Keeley, T. C. (b. 1894), 286
Keetman, Bruno (1883–1918), 471
Keir, James (1735–80), 5
Kekulé von Stradonitz, Friedrich August (1829–96), 37, 188
Kennedy, Joseph W. (1916–57), 147, 358, 376
Kern, Sergius, 129–32, 486
Kervran, Louis C. (1901–83), 465
Ketelle, Bruce Hubert (1914–2003), 372
Kiesewetter, P., 289, 308
Kimura, Kenjiro (1896–1988), 222
Kirchhoff, Gustav (1824–87), 24, 91, 196, 204
Kirwan, Richard (1733–1812), 41, 45, 51
Kitaibel, Pál (1757–1817), 41
Klaproth, Martin Heinrich (1743–1817), 4, 5, 7, 10, 13, 16, 23, 25, 27, 29, 30, 32, 33, 41, 43, 44, 59, 79, 80, 111, 233, 273, 483
Klaus, Karl Ernst Karlovich (1796–1864), 68, 74–76
Klobasa, Adalbert, 462
Kobell, Wolfgang Xavier Franz von (1803–82), 47, 48, 485
Koblic, Odolen (1897–1959), 326, 327, 333, 359, 481, 492
Koenig, George Augustus (1844–1913), 137
Kosmann, Bernhard Hans (1840–1921), 417, 488
Kowarski, Lew (1907–79), 226
Kremers, Edward (1865–1941), 173
Kruegel, Curt, 187, 488
Krüss, Gerhard (1859–95), 67, 68, 123, 162, 163, 174, 215, 289, 290, 308, 487
Kunz, George Frederick (1856–1932), 104, 105, 489
Kurbatov, J. B., 302, 308
Kurchatov, Igor Vasilyevich (1903–60), 383, 387, 390
Kuroda, Paul K. (1917–2001), 304
Kushi, Michio (b. 1926), 465–67

Ladenburg, Albert (1842–1911), 187, 188, 488
Lampadius, Wilhelm August (1772–1842), 66, 67, 404
Lamy, Claude-Auguste (1820–78), 25, 88, 131
Landau, Lev Davidovich (1908–68), 387
Langmuir, Irving (1881–1957), 323, 329–31, 335
Larch, Almon E., 383
Latimer, R. M., 383
Latimer, Wendell M. (1893–1955), xxxiv, 329
Lauer, Karl, 428–30, 487
Laurent, Auguste (1807–53), 274, 276
Lavoisier, Antoine Laurent (1743–94), xxi, xxv–xxviii, xxxv, 1, 13, 15, 19, 21, 23, 24, 27–29, 33, 34, 45, 51, 56, 57, 62, 64, 87, 128, 132, 133, 179, 403, 406, 421, 465, 483
Law, H. B., 302, 308
Lawrence, Ernest Orlando (1901–58), 191, 312, 313, 362, 379–83, 389

Lazarev, Yuri (1946–96), 398
Le Bel, Joseph Achille (1847–1930), 429
Lea, Matthew Carey (1823–97), 459
Leadbeater, Charles Webster (1847–1934), 435–41, 489, 491
Lecoq de Boisbaudran, Paul Emile (*dit* François) (1838–1912), 37, 40, 123, 165, 168, 169, 173, 175, 179, 201, 207, 210–14, 216, 277, 279, 289, 417, 486
Leduc, A., 189
Lehmann, Johann Gottlieb (1719–67), 272, 276
Leigh-Smith, Alice (1907–87), xviii, 332, 337, 340–45, 359, 492
Leino, M. (b. 1949), 394
Lembert, Max Ernst (1891–1925), 261
Leonhardi, Johann Gottfried (1746–1823), 44
Lewis, Gilbert N. (1875–1946), 329, 379
Lieben, Adolf (1836–1913), 39
Liebig, Justus von (1803–73), 33, 38, 84, 121
Liechti, Adolf (1898–1946), 340, 343, 344
Lind, Samuel Coleville (1879–1965), 337
Lindemann, Frederick Alexander (1886–1957), 286
Linnemann, Eduard (1841–86), 36, 37, 40, 486
Lippi, Carminantonio (1760–1823), 28
Lippmann, Gabriel (1845–1921), 268
Lockyer, Joseph Norman (1836–1920), 188, 409, 431–33, 488
Loew, Carl Benedict Oscar (1844–1941), 102
Loisel, Pierre, 284, 285, 491
Loring, Frederick Henry, 134, 246–58, 291, 321, 490, 491
Lortie, Léon (1902–85), 225, 229
Lubicz, René Adolphe Schwaller de (1887–1961), 458
Lucas, Alfred (1867–1945), 116
Lull, Raymond (ca. 1235–1315), 403, 405
Lundmark, Knut (1889–1958), 415
Lussage, Walter (d. 1977), 463

Mach, Ernst (1838–1916), 40, 208
MacKenzie, Kenneth R. (1912–2002), 332, 339, 346
Maclure, William (1763–1840), 82
Majima, Toshiyuki (1874–1963), 220
Mallet, John William (1832–1912), 130, 141, 192
Marcet, Alexander (1770–1822), 64
Marckwald, Wilhelm (1864–1942), 471, 472, 474, 475
Maria the Jewess (or Maria Prophetissima), 450
Marignac, Jean Charles Galissard de (1817–94), 119, 120, 123, 167, 171, 172, 235
Marinov, Amnon (1930–2011), 399
Marinsky, Jacob Akiba (1918–2005), xix, 304–06, 308
Marsh, Joseph Kenneth (b. 1895), 299, 308
Mattauch, Josef (1895–1976), 299, 303, 304, 327, 335
Maxwell, James Clark (1831–79), 162, 268, 424, 455
Mayer, Maria Goeppert (1906–72), 365
Mazza, Luigi (1898–1978), 293
McLennan, John (1821–93), 423
McMillan, Edwin M. (1907–91), 91, 133, 147, 327, 358, 376
Meinecke, Johannes L. G. (1781–1823), 404, 405, 484
Meissner, Wilhelm (1792–1853), 60
Meitner, Lise (1878–1968), 239, 262–64, 304, 316, 356, 357, 364, 376, 379, 395
Mellor, J. W. (1873–1938), 32, 159, 174, 337
Mendeleev, Dmitri I. (1834–1907), XXI, XXIV, XXVII, XXXI, XXXV, 14, 19, 21, 22, 38–40, 68, 93, 102, 109, 114, 129, 130, 138, 160, 175, 182, 195, 228, 229, 231, 248, 253, 274–76, 310, 317, 372, 401, 405, 409, 410, 419–23, 426, 474, 480, 489
Messier, Charles (1730–1817), 415
Meyer, Johann Karl Friedrich (1733–1811), 4, 483
Meyer, Julius Lothar (1830–95), XXXI, XXXII, XXXVII, 93
Meyer, Stefan (1872–1949), 262, 263
Miethe, Adolf (1862–1927), 453, 454, 476
Minder, Walter (1905–92), 332, 337, 339–45, 492
Moir, James R. (1874–1929), 246, 247, 490

Moissan, Henri (1852–1907), 32, 56, 88, 234
Monnet, Antoine Grimoald (1734–1817), 45, 483
Monti, Gerardina Cesarano (b. 1963), 455
Monti, Roberto, 455
Moore, Richard B. (1871–1931), 185, 221
Moray, Thomas H. (1892–1974), 462
Morita, Kenji, 395
Morozov, Nikolai Aleksandrovich (1854–1946), 406, 407, 410, 426, 427, 489, 490
Mosander, Carl Gustav (1797–1858), 122, 171, 172, 215, 233, 235, 241, 485
Moseley, Henry G. J. (1887–1915), XXI, XXVII, XXXII, 14, 109, 171, 200, 201, 215, 231, 232, 235–37, 279, 287, 290, 291, 293, 297, 299, 300, 310, 361, 371, 480
Much, Fried (or Fraenkel), 428
Münzenberg, Gottfried (b. 1940), 394
Muthmann, Friedrich W. (1861–1913), 174

Nagaoka, Hantaro (1865–1950), 453–55
Nasini, Raffaello (1854–1931), 421, 422
Nenadkevich, Konstantin Avtonomovich (1880–1963), 114, 115, 117
Nernst, Walther (1864–1941), 291, 307, 357, 444
Neumann, Caspar (1648–1715), 39
Neumann, Karel Augustin (1771–1866), 39
Newlands, John Alexander Reina (1837–98), XXXI, XXXII, 39, 42, 160, 405
Newton, Sir Isaac (1642–1727), 385, 420, 421
Nicholson, John William (1881–1955), 407, 408, 490
Nicholson, William (1753–1815), 5, 23
Nicklès, Jérôme F. J. (1820–69), 139
Nilson, Lars Fredrik (1840–99), 123, 174, 290, 308, 487
Ninov, Victor (b. 1959), 226, 227, 351, 373, 394
Nobel, Alfred (1833–96), 381
Noddack, Ida Tacke (1896–1978), 104, 132, 221–23, 250, 251, 255, 279, 287, 298, 299, 310–19, 327, 335, 357, 491
Noddack, Walter (1893–1960), 104, 132, 221–23, 250, 251, 255, 279, 287, 298, 299, 310–18, 357, 491
Nordenskjöld, Ivar (1877–1947), 270
Noyes, William Albert (1857–1941), 297
Nylander, William (1822–99), 101

Odling, William (1829–1921), xxxi
Oersted, Hans Christian (1777–1851), 30, 31
Oganessian, Yuri Ts. (b. 1933), 363, 364, 398, 373, 383, 384, 386, 387, 398
Ogawa, Masataka (1865–1930), 105, 106, 219–23, 275
Ohly, Julius, 131
Ohsawa, George (Ryogji) (1893–1966), 465–67
Orfilia, Mathieu (Mateu) (1787–1853), 121
Osann, Gottfried Wilhelm (1796–1866), 74–76, 485
Ostwald, Wilhelm Friedrich (1853–1932), xxvii, 261, 409
Owen, C. T., 67, 68
Owen, David Dale (1807–60), 82, 485
Owens, Robert Bowie (1870–1940), 184, 434

Pallas, Peter Simon (1741–1811), 414
Palmer, Charles Skeele (1858–1939), 409, 487
Palmieri, Luigi (1807–96), 431
Paneth, Friedrich Adolf (1887–1958), xxxv, 16, 237, 287, 317, 333, 344, 346, 468, 478
Paracelsus (Auroleus Phillipus Theophrastus Bombastus von Hohenheim) (1493–1541), xxv
Pascal, Blaise (1622–1663), 43
Paschen, Louis Karl Heinrich Friedrich (1865–1947), 431–33, 487
Patterson, Hubert S., 468
Pauli, Wolfgang (1900–58), 192
Pauling, Linus (1901–1994), xxiii
Pebal, Leopold von (1826–87), 37
Peirce, Charles Sanders (1839–1914), 423
Peligot, Eugène Melchior (1811–90), 25, 44
Pellettier, Bertrand (1761–97), 29
Pellini, Giovanni (1874–1926), 473, 474
Pellizzari, Guido (1858–1938), 301
Pelouze, Théophile-Jules (1807–67), 121
Percy, John (1817–89), 101
Perey, Marguerite (1909–75), 151, 152, 317, 324, 325, 332, 345

Perkin Jr., William H. (1860–1929), 190
Perrier, Carlo (1886–1948), xviii, xxx, 317
Perrin, Jean (1870–1942), 184, 242, 320, 322, 324, 325, 455
Peters, Kurt Karl (1897–1978), 468, 478
Pfaff, Heinrich (1773–1852), 403
Phipson, Thomas Lambe (1833–1908), 147–50, 153, 488
Piccardi, Giorgio (1895–1972), 227, 278, 291, 293, 294, 301, 303, 306
Piccini, Augusto (1854–1905), 40, 474
Piutti, Arnaldo (1857–1928), 454, 469
Pizzighelli, G. (1849–1912), 278
Pons, Stanley (b. 1943), 478–80
Pontecorvo, Bruno (1913–93), 354, 358
Pool, M. L. (1900–82), 302, 308
Popeko, A. G., 394
Poulliet, Claude Servais (1791–1868), 121
Prandtl, Wilhelm (1878–1956), 113, 250, 251, 291, 312, 489
Prat, Jean-Pierre (b. 1834), 128–30, 133, 173, 486
Preiswerk, Pierre (1907–72), 357
Pribram, Richard (1847–1928), 37, 38
Priestley, Joseph (1733–1804), 5, 45, 51
Prigogine, Ilya (1917–2003), 227
Pringle, Alexander, 155–57, 486
Prochazka, George A. (1855–1933), 137–40
Proust, Louis Joseph (1754–1826), 43, 44, 50, 54, 484
Prout, William (1785–1850), xxxvii, 209, 406

Quill, Lawrence Larkin (1901–89), 302, 303, 308
Quin, Frederick Foster Hervey (1799–1878), 11

Ramsay, Sir William (1852–1916), xvi, xxvii, 105–07, 150, 178–90, 213, 219, 220, 222, 235, 246, 248, 250, 299, 419, 431–34, 436, 468, 469, 472, 473, 488
Rasetti, Franco (1901–2001), xix, 298, 357–59, 492
Ratan, Dahr Nil (1892–1987), 243
Ray, Prafulla (1861–1944), 338
Regnault, Henri Victor (1810–78), 25
Reich, Ferdinand (1799–1882), 22, 71
Reichenstein, Ferenc Müller von (1740–1825), 12, 13, 27, 41, 483
Remmler, Hugh, 68
Remsen, Ira (1846–1927), 409
Retgers, Jan Willem (1856–96), 411
Richards, Theodore William (1868–1928), 189, 321
Richardson, Owen William (1879–1959), 341
Richardson, Thomas, 77–80, 485
Richmond, Henry Droop (1867–1931), 98, 158–60, 487
Richter, Jeremias Benjamin (1762–1807), 53, 54, 484
Richter, Theodor Hieronymus (1824–98), 22, 71, 93
Rinman, Sven (1720–92), 4
Rio, Andrés Manuel del (1764–1849), 14–16, 71
Robinson, Robert H. (1886–1975), 287
Rogers, William Barton (1804–82), 124, 125
Rolla, Luigi (1882–1960), xviii, 173, 271, 279, 291–97, 301–04, 308, 481, 491
Roscoe, Henry Enfield (1833–1915), 16, 40, 174, 175
Rose, Gustav (1798–1873), 85
Rose, Heinrich (1795–1864), 23, 46, 71
Rosseland, Svein (1894–1985), 270, 271
Rowland III, Henry Augustus (1848–1901), 161–64, 279, 487
Rubies, Santiago Piña de, 107
Ruddock, F. G., 104–07
Rulandus, Martinus, the Elder (1531–1602), 457
Rulandus, Martinus, the Younger (1569–ca. 1611), 457
Runge, Carl David Tolmé (1856–1927), 431–33, 487
Ruprecht, Leopold Anton von (1748–1814), 12, 27–33, 36, 37, 483
Russell, Alexander S. (1888–1972), 260
Rutherford, Ernest (1871–1937), xxvii, 153, 183, 184, 189, 191, 192, 201, 219, 221, 231, 237, 238, 247, 254, 260, 268, 287, 329, 362, 368, 377, 383, 407, 408, 434, 445, 446, 450, 455, 471–73, 475
Ryan, H. C. Blood, 249

Sabatier, Paul (1854–1941), 242

Sacks, Oliver (b. 1933), 287, 288
Salvadori, Roberto (1873–1940), 421
Samarskij-Byhovec, Vasilij Evgrafovič (1803–70), 123, 226
Saro, S., 394
Savaresi, Andrea (1762–1810), 28–30, 32
Scacchi, Arcangelo (1810–93), 143–46
Scaramuzzi, Francesco (b. 1929), 480
Scheele, Carl Wilhelm (1742–86), 4, 27, 30, 56, 272, 273, 276, 277
Scherer, Alexander Nicolaus (1771–1824), 7
Scherrer, Paul (1890–1969), 343
Schiff, Hugo (1834–1915), 130, 277
Schmidt, Friedrich Wilhelm (1829–1903), 67, 68, 487
Schmidt, Gerhard (1865–1949), 183
Schmidt, Rudolf, 185, 186
Schönbein, Christian Friedrich (1799–1868), 24
Schött, H. J., 394
Schützenberger, Paul (1829–97), 40, 165, 167, 168, 170, 171
Schwarzenberg, Prince Karl Philipp zu (1771–1820), 8
Scott, Alexander (1853–1947), 115–17, 491
Seaborg, Glenn T. (1912–99), xxxiv, 147, 351, 358, 362, 363, 367, 370, 376, 377–86, 389, 461
Secchi, Angelo (1818–78), 431
Sefström, Nils Gabriel (1787–1845), 15, 16
Segrè, Emilio G. (1905–89), xxx, 162, 267, 295, 303, 308, 312, 314, 317, 332, 339, 340, 343, 344, 346, 354, 357–59, 362, 376
Shapleigh, Waldron, 167, 168
Sheldon, H. H., 453
Sherr, Rubby (1913–2013), 454
Shrum, Gordon, 423
Siegbahn, Karl Manne Georg (1886–1978), 222, 237, 256, 286, 325, 334, 379, 381, 382, 389
Sikkeland, Torbjørn (b. 1923), 382, 383
Silliman, Benjamin (1779–1864), 67, 484
Sjögren, Carl Anton Hjalmar (1822–93), 98
Sjögren, Sten Anders Hjalmar (1856–1922), 98
Skinner, Clarence, 468
Skrobal, A., 106
Smith, John Lawrence (1818–83), 82, 103, 119, 121–24, 133, 486
Smits, Arthur, 476
Smolan Smoluchowski, Marian Ritter von (1872–1917), 424
Śniadecki, Andreas (or Jedrzej) (1768–1838), 13, 14, 68, 69, 73, 484
Sobiczewski, Adam (b. 1931), 366
Soddy, Frederick (1877–1956), xxi, xxvii, 150, 183, 189, 209, 219, 231, 232, 260, 263, 265, 287, 288, 439, 453, 468, 471, 480
Sommerfeld, Arnold (1868–1951), 192
Sorby, Henry Clifton (1826–1908), 112, 113, 485
Soret, Jacques Louis (1827–90), 119–24, 208
Spindler, Henri, 465
Stammreich, Hans (1902–69), 453–55
Stark, Johannes (1874–1957), 426
Stéphan, Claude, 372
Stern, Otto (1888–1969), 239, 312
Stolba, Frantisek (1839–1910), 39
Stoney, George Johnstone (1826–1911), 182, 487
Stradonitz, Friedrich August Kekulé von (1829–96), 37, 188
Strassmann, Fritz (1902–80), 316, 356, 358
Stromeyer, Friedrich (1776–1835), 59–61, 66, 485
Strutt, John William, 3rd Baron Rayleigh (1842–1919), 178, 468
Sutherland, William (1859–1911), 444
Suzuki, Umetaro (1874–1943), 102
Svanberg, Lars Fredrik (1805–78), 111, 112, 485
Swinne, Richard (1885–1939), 326, 359, 362, 491

Tacke Noddack, Ida (1896–1978), 104, 132, 221–23, 250, 251, 255, 279, 287, 298, 299, 310–19, 327, 335, 357, 491
Tackvorian, S., 299
Tandberg, Johan G., 478
Tausend, Franz (1884–1942), 454
Teller, Edward (1908–2003), 319
Tennant, Smithson (1761–1815), 69, 70

Thalén, Robert Tobias (1827–1905), 125
Thénard, Louis-Jacques (1777–1857), 30, 31, 56
Thompson, Claude Metford (1855–1933), 174, 289, 308
Thompson, Stanley (1912–76), 381
Thomsen, Julius (1826–1909), 113, 178, 179, 181, 244
Thomson, Joseph John (1856–1940), xxvii, 190, 232, 409, 451, 468
Thomson, Thomas (1773–1852), 50, 57, 60, 61, 77
Thomson, William, Lord Kelvin (1824–1907), 257, 268
Tiede, Erich, 454
Tiffereau, Cyprien-Théodore (1819–after 1898), 457
Tolman, Richard C. (1881–1948), 469
Tondi, Matteo (1762–1835), 27–34, 36, 483
Trabacchi, Giulio Cesare (1884–1959), 354, 355
Travers, Morris William (1872–1961), 178–80, 183, 189, 431, 433, 473, 488
Trommsdorff, Christian Wilhelm Hermann (1811–84), 66
Trommsdorff, Ernst (1905–96), 66
Trommsdorff, Hieronymus Jacob (1708–68), 65
Trommsdorff, Johann Bartholomäeus (1770–1837), 7, 31, 65, 484
Trommsdorff, Wilhelm Bernhard (1738–82), 65
Troost, Louis J. (1825–1911), 49
Turner, Louis A., 332, 344

Ullgren, Clemens (1811–68), 44, 45, 485
Urbain, Georges (1872–1938), 40, 107, 117, 125, 126, 133, 171, 175, 185, 189, 200–11, 213–17, 225, 233–45, 249, 255, 279–82, 294, 300, 309, 322, 323, 362
Urbain, Pierre (1895–1968), 242
Urey, Harold (1893–1981), 478

Valadares, Manuel (1904–82), 332
Valenta, Hofrat Eduard (1857–1937), 201, 215, 279, 282, 490
Vallarta, Manuel Salvador (1899–1977), 16
Van't Hoff, Jacobus Henricus (1852–1911), 291, 429
Vauquelin, Louis Nicolas (1763–1829), 15, 65, 69, 70, 79–81, 233, 404, 465
Vegard, Lars (1880–1963), 423
Vernadsky, Vladimir Ivanovich (1863–1945), 114, 490
Verneuil, Auguste Victor Louis (1856–1913), 40
Vest, Lorenz Chrysanth von (1776–1840), 60, 61, 68
Vinci, Leonardo da (1452–1519), 226, 381
Violante, Vittorio, 480
Vogel, Hermann Wilhelm (1834–98), 453
Volta, Alessandro (1745–1827), 23, 62, 362

Wahl, Arthur C. (1917–2006), 147, 358, 376
Waite, Arthur E. (1857–1943), 457, 460
Walden, Paul (1863–1957), 469
Watt, James (1736–1819), 5
Watts, Oliver Patterson (1865–1953), 32
Websky, Friedrich Martin (1824–86), 85, 486
Wedgwood, Josiah (1730–95), 4–6, 10, 483
Weeks, Mary Elvira (1892–1975), xxiv, 32, 125
Wegener, Alfred Lothar (1880–1930), 422, 423, 490
Weiersstrass, Carl (1815–97), 433
Weizsäcker, Carl Friedrich von (1912–2007), 340
Welsbach, Carl Auer Freiherr von (1858–1929), 39, 40, 123, 165, 169, 170, 173–76, 214–16, 233–35, 240–43, 279–82, 289, 417, 490
Wendt, Gerald Louis (1891–1973), 451
Wenzel, Carl Friedrich (ca. 1740–93), 54
Werner, Alfred (1866–1919), xxxiv, 242
Westrumb, Johann Friedrich (1751–1819), 403
Whitsitt, May Lee (1891–1975), 306
Whytlaw-Gray, Robert (1877–1959), 184–86, 434
Wilkinson, Geoffrey (1921–96), 386, 463
Williamson, Alexander William (1824–1904), 222
Wilm, Theodor Eduard (1845–93), 103, 104

Wilson, Charles Thomson Rees (1869–1959), 325, 347
Winkler, Clemens A. (1838–1904), xxxii, 93
Winterl, Jakob Joseph (1732?–1809), 403, 404, 412, 484
Winthrop, John (1609–76), 46
Withering, William (1741–99), 5
Wöhler, Friedrich (1800–82), 15, 16, 30, 31, 75, 80, 172
Wolke, Robert (b. 1928), 372, 373
Wollaston, William Hyde (1766–1828), 14, 46, 50, 69
Wong, Cheuk-Yin (b. 1940), 371, 372
Woulfe, Peter (1727–ca. 1805), 272
Wurtz, Charles-Adolphe (1817–84), 188
Wynne, W. P. (1861–1950), 240
Wyrouboff, Gregoire (1843–1913), 40

Yeremin, A. Vladimirovich, 394
Yntema, Leonard (1892–1976), 295–97, 302, 306, 308

Zamarev, Karol I., 387
Zambonini, Ferrùccio (1880–1932), 145
Zeeman, Pieter (1865–1943), 162
Zeppelin, Ferdinand von (1838–1917), 95
Zintl, Eduard (1898–1941), 453
Zsigmondy, Richard (1865–1929), 459
Zwaardemaker, Hendrik (1857–1930), 321
Zwicky, Fritz (1898–1974), 368

LOST ELEMENT NAME INDEX

**—other names, presumed and otherwise, for true elements; §—proposed names for transuranium elements; #—names for radioactive isotopes of a known element

A, 409
accretium§, 379
actineon (Rn)**#, 184
actinium, 147, 488
actinon (Rn)**#, 184
adyarium, 438–9, 491
agusterde (Be)**, 65, 484
akton (Rn)**#, 184
alabamine (At)**, 191, 196, 327–31, 349, 492
alabamium (At)**, 331, 492
alamosium§, 380
aldebaranium (Yb)**, 207, 213–14, 233–34, 240, 279–82
aldebaranium-thulium I, II, III, 280
alium§, 377
alkalinium (Fr)**, 134, 252, 256–57, 322, 491
alphanium§, 377
alphonium§, 377
alvarezium§, 385
amarillium, 198–99, 489
amerium§, 377
andronia, 403–04, 484
anglium (Ar)**, 179, 182, 488
anglo-helvetium (At)**, XVIII, 332, 337, 340–45
anlium§, 380
anodium, 406–07, 426, 489
anu, 437–42, 489
apollium§, 377
apulium, 27–28, 31, 483
apyre (apyrit), 63–64, 484
archonium (He?)**, 406, 408, 490
argentaurum, 458–59, 488
aridium, 43–45, 485
ariesium§, 377
artificium or artifician§, 377
asium, 111, 114–15, 490
astaeum, 92, 485
asterium (He)**, 431–33
asteroidium§, 377
astralium§, 377
astronium§, 377
athenium§, 380
atlantisium§, 377
aurorium, 423, 485
aurum millium, 67, 484
ausonium, 226, 264, 298, 316, 318, 344, 353, 357–60, 375, 481, 492
austral sand, 5, 6
australium, 4, 5, 483
austriacum, 38–41, 487
austrium, 27, 28, 31, 32, 36–38, 41, 483, 486, 487

B, 409
barcenium, 135, 141
barote (Ba)**, 30
baryta (Ba)**, 29, 30, 49, 50
bastardium§, 377
bauxium, 201, 202, 217, 487
becquerelium§, 377
berklium (Bk)**, 378
berzelium, 192–94, 489
big bearianen§, 377
big dipperian§, 377
bohemium, 327, 335, 481, 492
bolidium§, 377
boracium (B)**, 31

borbonium (Ba)**, 27, 28, 31, 483
bordium§, 378
bornium, 27, 28, 31, 483
brevium (Pa)**#, 260–65
brillium, 97, 489
butlerovium§, 377
bythium, 95, 488
caloric, 25, 62, 64, 483
canadium, 198, 224, 225, 228, 490
canopium§, 377
carolinium, 191–95, 489
cassiopeium (Lu)**, 207, 213–14, 233–34, 240, 279–80, 395
cathodium, 406, 407, 426, 489
catium (Fr)**, XIX, 325
celtium (Hf)**, XVIII, 51, 111, 117, 118, 200, 203, 215, 225, 233, 235–41, 255–56, 264, 383
centium§, 381
centurium§, 378, 380
ceresium (Pd)**, 12–14, 17
chondrium, 411–12, 491
co-helium, 432, 487
colonium§, 379
columbium**, 46–48, 51, 119, 124, 135, 385, 486
comesium, 83, 84, 486
cometium§, 377
coronium, xix, 408, 419, 421–23, 489, 490
cosmium§, 377
cosmonium§, 377
crodonium, 65, 66, 484
crustaterrium, 411, 412, 491
cyclo-europium§, 377
cyclo-gadolinium§, 377
cyclonium§, 379
cyclotronium§, 379, 95, 488

dacinium, δ, 95, 488
dacinum, 339, 492
dakin, 337–39, 346, 492
damarium, 428–30, 487
danium (Hf)**, 117, 118, 238, 239
davincium§, 385
davyum (davyium), 128–35, 486
decipium (Sm)**, 119, 120, 122, 123, 135, 173, 249, 486
deimos§, 377
dekhine, 339, 492
delirium§, 376
demonium, 163, 487
denebium, 201, 280, 490
devium, 131, 486
diamanthspatherde, 10, 11, 483
dianium, 43, 47, 48, 485
didymium, 39, 40, 103, 122–25, 130, 138-39, 167, 171-76, 214, 234–35, 241, 289-90, 294
dipperium§, 377
donarium, 70, 71, 485
donium, 77–80, 485
dor, 320–24, 331–34, 492
draconium§, 377
dubhium, 201, 280, 490
dubnabium (or dubnadium)§, 398
dwi (or dvi)-manganese (Tc)**, 222, 250, 274–76, 310, 491
dwi (or dvi)-tellurium (Po)**, 38, 39

edisonium§, 385
eka-aurum§, 357
eka-barium (Ra)**, 160, 420
eka-cadmium (Ge)**, 93
eka-cesium (Fr)**, 252–56, 322, 328, 331, 337, 340, 345
eka-iodine (At)**, 253, 329–32, 337–45, 348, 492
eka-iridium§, 357
eka-manganese (Tc)**, 132, 222, 225, 250, 251, 258, 274, 275, 310
eka-osmium§, 326
eka-platinum§, 357
eka-polonium§, 372
eka-radon§, 372
eka-rhenium§, 326
eka-silicon (Ge)**, 93
eka-tantalum (Pa)**#, 264
eka-tellurium (Po)**, 420
element 108, 326, 491
element, atomic number zero, 443–44
element E, 207, 490
element of Chandler, 99
element of De Brereton Evans, 105
element of Dupré, 98
element of Fernandez, 97
element of Genth, 97

element of Holtz, 107
element of Loew, 102
element of Wilm, 103
element Sδ (Eu)**, 290
element X (Ho)**, 119, 123, 125, 135, 207, 208, 246, 280
element X_2, 207, 490
elements, asteroid, 411, 491
elements, extinct, 410, 486
elements Gα, Gβ, Gδ, Gζ, Gη, 204–09, 290
elements of Boucher and Ruddock, 104
elements of Gerland, 102
elements of Nylander and Bischoff, 101
elements, occult, 435–42, 489
elements Xα, Xβ, Xγ, Xδ, Xε, Xζ, Xη, 123, 201, 487
elements Zα, Zβ, Zγ, Zδ, Zε, Zζ, 201, 210, 486
elements Σ, Γ, Δ, Ω, Θ, 201, 210, 488
eline, 348, 492
emanation (Ac, Rn, or Th)**#, 184–86, 284, 285
emanium (Ac)**, 262
emilium, 284, 285, 491
enactinium (ennactinium)§, 379
eosium (Kr)**, 180, 181
Erα, 123
Erβ, 123
Erebodium, 156, 157, 486
erythronium (V)**, 14–16
eternium§, 378
ether, 419, 489
etherion (etherium), 423, 424, 488
euprosium§, 379
euro-samarium, 281
euxenium, 113, 117, 489
exactinium (Rn)**#, 184
extremium§, 376
exradium (Rn)**#, 184
exthorium (Rn)**#, 184

finium§, 377
finlandium§, 385
fissium§, 377
florentium (Pm)**, 226, 271, 289, 291, 292, 295–301, 303–06, 318, 481, 491
fluore, fluorure (F)**, 57
futurium§, 378

gadenium, 156, 157, 486
gahnium, 73, 484
gamowium§, 387
geocoronium, 422–23, 490
ghiorsium§, 373
glaucodymium, glaucodidymium, 173–74, 488
glucine, glucinum, glucium (Be)**, 79
glucinium (Be)**, 51, 79, 80, 160, 248
gnomium, 67, 68, 487
goldanskium§, 387
gourium, 337–39, 492
gravum§, 377

hahnium§, 239, 363, 366, 384, 386–88, 396
hawkingium, 375, 391, 492
hebeium, 92, 485
"helium", 248, 490
"helium system", 445, 491
helvetium, 332, 337, 339–45, 492
herculium§, 377
hesperisium, 156, 157, 486
hesperium, XIX, 226, 264, 298, 316, 318, 344, 353, 357, 375
hibernium (Ar)**, 179, 182, 270, 271, 488
"hydrogen system", 445, 491
hydrosideron, 29, 32, 483
hydrosiderum, 4, 483
hypon, 416, 491

idunium (or idumium), 85, 486
illinium (Pm)**, XIX, 175, 191, 196, 290, 295–309, 312, 318, 491
ilmenium, 47–49, 486
incognitum (or incognitium) (Gd)**, 83, 202, 205, 206, 489
infra-elements, 182, 487
ionium (Gd)**, 201–06, 473, 474, 489

japonium§, 395
jargonium, 111–13, 485
joliotium§, 366–69, 385–89, 398
josephinium, 198, 199, 490
junonium, 60, 61, 92, 485
kadmium (Cd)**, 59, 61

kalidium, 409, 487
kalon, 438, 439, 489
kapitzium§, 385
klaprothium, 13, 44, 59, 484
kosmium, 417, 488
krypton II, 187, 488
kurchatovium§, 363, 366, 383–87, 397

landauvium§, 387
lapis ponderosus (W)**, 272
laslium§, 380
lasluciium§, 380
lavoesium, 128–33, 135, 486
lazarevium§, 398
leonite§, 377
leosium§, 377
leptine (or leptin) (At)**, 343, 347
lewisium§, 379
lisonium, lisottonium (Pa)**, 239, 262
littorium§, 226, 353, 356
losalamium§, 380
losalamosium§, 380
losalium§, 379, 380
lucium (Y)**, 165–69, 174, 286, 488
lunium§, 377
lutecium (Lu)**, xviii, 114, 117, 126, 203, 233–35, 240, 241

magellanium§, 385
magnium (Mg)**, 31, 36
manganesium (Mg)**, 36
martium§, 377
masrium, 98, 158–60, 487
masurium (Tc)**, XVIII, 132, 221, 258, 286, 287, 298, 310–18, 327, 335, 491
mechanicum§, 377
melinum (or melinium) (Cd)**, 61, 484
menachite (or menachin) (Ti)**, 33
metacerium, 40, 41, 487
meta-elements, 204–09, 486
meta-kalon, 439
meta-neon, 439
metal A (Y)**, 166, 488
metallum problematicum, 12, 483
metargon, 180, 181, 488
minervium§, 376
moldavium (Fr)**, 256, 320, 322–25, 327, 332–34, 339, 492
mondium§, 377
monium, 83, 201–04, 488
moononium§, 377
mosandrium (or mosandrum), 103, 119, 121–24, 132–35, 486
moscovium, moscowium§, 398
moseleyum (Tc)**, 286
murium (Cl)**, 24
mussolinium§, 353, 356

nebulium§, xix, 377, 408, 490
néo-actinium or néoactinium, 151, 152, 325, 492
neo-celtium (Hf)**, 235–36
neo-didymium (Nd)**, 241, 289
neo-erbium, 215, 487
neo-holmium, 215, 490
neo-kosmium, 417, 488
néo-molibdène or néomolybdenum, 272, 275, 276, 310, 491
néo-radium (or néoradium), 151, 152, 325, 492
neo-thulium, 201, 490
neo-tungsten or néotungsten, 272, 276, 310, 491
neo-ytterbium (Yb)**, XVIII, 126, 200, 203, 206, 213, 215, 225, 233–35, 240, 280
neptunium, 43, 47–49, 91–93, 132–35, 147, 333, 334, 358, 362, 375, 376, 397, 485, 486
neuton, 491, 444
neutronium§, 377, 443, 444, 446, 491
newium§, 379
newtonium, XIX, 385, 419, 421, 489
niccolanum (Ni)**, 53, 54, 484
nielsbohrium§, 363, 366, 386, 387, 388
nigrium, 111–13, 486
nihonium§, 397
ninetynineum§, 380
ninovium§, 227
nipponium (Re)**, XVIII, 105–08, 220–22, 226, 249, 275, 286, 397
niton (Rn)#**, 183, 186, 189, 433, 434
nitricium, 24, 484
nitron, 248, 249, 490
nonactinium§, 379

nonagintium§, 378
norium, 111, 112, 485
norwegium, 136–40, 486
novanium§, 378
novum (Ne)**, 180
nutronium§, 378

occultum, 436, 438, 439, 489
oceanium, 111, 115, 116, 491
ochroite (Ce)**, 12, 13
octonium§, 378
offium§, 378
oganessium§, 398
orthohelium, 432, 487
ostranium, 111, 485
ouralium, 84, 135, 486
oxidium, 409, 487

pallasium, 411, 412, 491
panchromium (V)**, 12–15
pandemonium§, 376
panormium (Tc)**, 312–14
pantogen, 405, 406, 490
parahelium, 432, 487
parthenium, 27, 31, 483
paximum§, 377
pelopium, 46, 47, 485
pentonium§, 378
percentium§, 381
persephonium§, 378
philippium (Ho)**, 119–25, 135
phipsonium, 486
phobos§, 377
phoenicium§, 380
phtore (F)**, 56, 57
pilsum, 42
platinum-B, 438, 439, 489
pluranium (Ru)**, 73–76, 485
plutium§, 376
plutonium (Ba)**, 49, 50, 147
pneum-alkali, 6–9, 484
polinium (Ir)**, 73–76, 485
polymnestum, 155–57, 486
praedicium§, 379
pragium, 250, 251, 491
praseodidymium (Pr)**, 173, 214
prefluorine, 409, 487
primordial matter, XXXVII, 35, 411, 491
prometheum (Pm)**, 304–05
proto-argon, 439, 489
proto-beryllium, 410
proto-boron, 410, 489
proto-elements, 403–13, 488
proto-fluorine, 408, 490
proto-glucinium, 410, 489
protohydrogenium, 410, 490
proto-metals, 409, 488
protyle, 406, 408, 486
proximogravum§, 377
pseudoelements, 406, 490
ptene (or ptène) (Os)**, 69

quebecium, 225, 227–29, 492
quintium§, 377

radio-brevium (Pa)#, 264
radiolead (Pb)#, 470, 471, 488
radiomercurium, 105, 489
radion (Rn)#, 189
radioneon (Rn), 189
radiotellurium (Po)**, 471–74
radiothorium (Po)**#, 332, 336
radium emanation (Rn)**#, 184, 186, 472, 473
radium foil, 104, 105, 489
radlabium§, 379
rikenium§, 395
rogerium, 124, 135, 486
rooseveltium§, 377
roosium§, 377
rossijium§, 398
russium§, (Fr)**, 174, 323, 397, 488

sacharovium§, 385
Smα, Smβ, 124, 487
satellite, 248, 249, 490
saturnit, saturnite, 45, 483
saturnum, 45, 46, 51, 483
scheelium (W)**, 273
schwerspatherde (Ba)**, 30
scientium§, 377
scotium (Ar)**, 179, 182, 488
seaburnium§, 378
seadium§, 378

septonium§, 378
sequanium, 321–27, 334, 335, 339, 492
sextium§, 377
sextonium§, 378
siderium§, 377, 411, 412, 491
siderum, 3, 4, 483
silene, 43, 44, 484
silenium, 44, 484
sirium§, 60, 61, 377
solium§, 377
substances, constitutive, 162, 487
substances, primary, 407, 408, 490
splittium§, 377
stellanium§, 377
stellium§, 377
sub-element X, 247, 490
subelements, 246, 247, 409, 426
sulfénium, 88, 89, 492
sulphurium, 88, 89, 485
sunian§, 377
sunonium§, 377
supra-elements, 182, 487
sydneia, sydneium, 5, 6, 483
sylvanite, 41

talcium (Mg)**, 36
taldomskium§, 398
tellurium X (Po)**, 473, 474
terbium-I, -III (Gd)**, 213, 215
terbium-II (Dy)**, 213, 215
thermoxygen, 62, 63, 483
terra adamantina, 10, 483
terra australis, 5, 483
terra nobilis, 3, 483
terrium§, 411, 412, 491
thalium, 82, 83, 88, 485
thelyke, 403, 404, 484
thoreon (Rn)**#, 184
thorium-α, thorium-β, 41, 489
thorium emanation (Rn)**#, 184
Th-F, 338, 339, 492
thoron (Rn)**#, 184–86, 189, 197
thulium I, II, III, 235, 280, 490
Tmα, Tmβ, 124, 487
transneptunium§, 377
treenium, 78–80, 485
trinacrium (Tc)**, 312, 314, 318

uclasium§, 380
udalium, 84, 486
ultimum (or ultimium)§, 376
unicalium§, 378
universitum (or universitium)§, 378
universum§, 377
unonium§, 377
uralium, 84, 486
urstoff, ursubstanz, 404–05, 484
UX1 (Th)**#, 260, 261
UX2 (Pa)**#, 260, 261

venusium§, 377
verium, 348, 349, 491
vesbium (V)**, xviii, 135, 143–45
vestium, vestaeium, vestalium, 14, 60, 61, 484
victorium, 83, 201, 202, 204–06, 488
viennium (At)**, 333
vincium§, 385
virginium (Fr)**, 191, 196, 256, 323, 327–31, 349, 491
virgonium§, 377
volfram (W)**, 272
vulcanium§, 377

washingtonium§, 377
wasium, 45, 138, 139, 485
wassereisen, 4, 483
welsium, 201, 281, 282, 490
wodanium, 54, 66, 67, 485
worldliness§, 377
X (Xα, Xβ, Xγ, X, Xε, Xζ, Xη), 123, 487
Xtinium§, 378
Yorkium§, 379
Ytunium§, 378
Zeusium§, 377
Zoïkon, 246, 247, 490
Zunzenium, 392, 492

SUBJECT INDEX

Page numbers followed by n indicate notes. Page numbers followed by *f* or *t* indicate figures or tables, respectively.

Aberdonia, 77–81
Académie des Sciences, 274
Academy of Physics, 366
Academy of Saint Petersburg, 14
Academy of Sciences, Vienna (*Akademie der Wissenschaften*), 29, 234, 279
Accademia dei Lincei, 331, 333–35, 351
Accademia d'Italia, 394
Accademia Mineraria de Amadén, 52
ACS. *See* American Chemical Society
Adam Hilger Ltd. Company, 252–54, 258n93
Adventure in Science, 376
Alchemical Society of France, 458
alchemy, 449–50, 461, 481n3
Allison Wonderland effect, 336n376
allotropic silver, 459
aluminum, XXVI, XXVIII, 28–30
American Association for the Advancement of Science (AAAS), 423–24
American Chemical Society (ACS), 100, 304, 306, 330, 367, 386–87
American Chemist, 100
American Journal of Mining, 113
American Museum of Natural History (AMNH), 105
American Physical Society (APS), 330
americium, 362, 378
AMNH. *See* American Museum of Natural History
Analyst, 159
analytical methodology, 21–26. *See also specific methods*
ancienne chimie (traditional chemistry), 45–46
Ångstrom (Å), 325–26
ANL. *See* Argonne National Laboratory
Annalen der Physik, 44
Annalen (Liebig), XXXI
Annales de Chimie, 28–29
Anschluß, 282, 283n197
apophorometers, 271n142
APS. *See* American Physical Society
Ares (Αρης), 45
Argentaurana, or Some Contributions to the History of Science (Emmens), 459
argentaurum gold, 459
Argentaurum Laboratory, 458
The Argentaurum Papers No. 1 (Emmens), 458
argon, 178–79, 181–82, 188n250, 189n262
Argonne National Laboratory (ANL), 379–80
argyrodite, 93
Ars Magna, 457
arsenic, 59, 157t, 457
artificial elements, XXX, 365
Associated Press, 451
associations, 252
astatine, 42n76, 331, 333–34, 334*t*, 346, 362, 376
asteroid elements, 410–12, 491
atomic bomb, 358
Atomic Energy Commission (AEC), 462
atomic myth, 231–32
atomic theory, 246–47
Atoms for Peace, 380–81

Australia, 4–5
Avogadro's Law, 182
Avogadro's number, 151

Bacillus cereus, 457
Băile Herculane, Romania, 284
Bakerian Lecture, 30–31
barium, 28–31, 35n60, 49, 147
Bayer Process, 201–2
bementite, 137, 141n118
Berner Tagblatt, 343–44
beryllium, 51n107, 79–80, 160n183, 248, 465
betafite, 327
Bhabha Atomic Research Centre, 480
biological transmutation, 464, 465–67
biotite, 371
bismuth, xviii–xix, 38
bizarre elements, 401
blowpipe analysis, 21–22
blue diamonds, 88–89
bohrium, 363–64, 367
bombs, 358, 390n82
Bonzenfrei group (Das Bonzenfreie Kolloquium), 238, 239*f*
boron, 28–31
Boston Evening Transcript, 424
Boys of Panisperna, 358, 359*f,* 359n3
Der Brandner Kaspar und das ewig Leben ("Kaspar Brandner and eternal life") (Kobell), 48, 52n116
British Society for the Advancement of Science, 202–3
bubble chambers, 390n90
Der Bund, 343–44

cadmium, 59–61, 61*t*
Cairo Museum, 159
calcium, 28–31, 35n60
californium, 191, 362, 378
carbide diamonds, 88–89
Carnelley's rules, 409
carriers, 470
Cassiopeia, 415
cathodic phosphorescence, 205–6
Cauchois spectrometer, 334n343
Centre National de la Recherche Scientifique (CNRS), 320
Centre Quebécois de la Couleur, 226
Ceres, 13, 17n32, 32n55
cerium, 12–13, 16n25, 40, 60, 61*t,* 73, 162, 194
cesium, 24–25, 254
Chanson d'automne (Urbain), 242
Chemical and Metallurgical Engineering, 405
Chemical Educator, 169
chemical elements. *See* element(s); *specific elements*
chemical evolution
inorganic, 183, 410
theories of, 408–10
Chemical Heritage Foundation, 134n96
Chemical Industries Exposition, 311–12
Chemical News, 32, 38, 67, 103, 149, 157n180, 165–68, 169n213, 193–94, 204, 240, 249, 253–55
Chemical Society of London, 304
chemical splitting, 40
Chemiker Zeitung, 428
Chemistry and Industry, 237, 241
Chemistry International, 287
Chemists' Club, 100
chlorine, xxv–xxvi, 24
chondrules, 370
chromium, 15, 16n25, 23, 81n225, 465
"Chronic Illnesses" (Hahnemann), 8
clairvoyance, 435–42
Cleveland Plain Dealer, 424
code names, 376
cold fusion, 387, 395, 478–79, 480
Cold Fusion Research Project (ENEA Laboratories), 480
Cold War, 381, 384
collective elements, 411
Collège de France, 354, 355*f*
Colonial era, 428
coloring photographs, 278
Columbia School of Mines (later Columbia University), 100, 264, 358, 370
Communism, 358
A Comprehensive Treatise on Inorganic and Theoretical Chemistry (Mellor), 337
Comptes Rendus de l'Académie des Sciences de Paris, 180, 185, 205, 339

controlled convergent evolution, 427
copernicium, 132, 394–95
copper (code name), 376
copper, transformation of, 473–74
Crab Nebula, 415
Crodo, 65
Cromer Fund, 282
crystallization, fractional, 171
cupellation, 21
curium, 362, 378
cyclotrons, 354, 355*f*, 362, 381, 389n78
Cygnus, 280

Dacians, 346n414
dahllite, 142n132
Daily Cal, 380–81
Damaraland (Namibia), 428, 429*f*
darmstadtium, 132, 134n100, 394
dating of the earth, 267–70
davidsonite, 77–78
Dead Sea, 251–52
de Haen company, 293
dematerialization, 418n49
Deneb, 280
deviant science, 226
diamond(s), 88–89, 439
Discovery of the Elements and the Origin of Their Names (Figurovskii), 339
distillation, 21
Donar (Thor), 70
Dubna, Russia. *See* Joint Institute for Nuclear Research (Dubna, Russia)
dubnium, 282n192, 363, 366–67, 383–88
dynamization, 8
dysprosium, 210–11, 215

early elements, 1–2
early errors, 1–11, 19
earth(s), 1–2, 267–70
earth acids, 47
Easter Rebellion, 267
Eastman Kodak Research Laboratory, 282
einsteinium, 362, 380, 385
electrolysis, 21, 23–24
"Electrotype-Therapeutic" equipment, 462
element(s). *See also specific elements*
  alternative names for, 12–17
  artificial elements, XXX, 365
  asteroid elements, 410–12, 491
  bizarre elements, 401
  code names for, 376
  collective elements, 411
  concept of, XXIV–XXVIII, XXXVn23, 21
  discoveries of, XVIII–XXX
  early errors and elements, 1–11, 19
  extinct elements, 410, 486
  false discoveries, XXX, 1–11, 19, 91, 192–93, 415–18, 481, 483–92
  false names for, 417–18
  first errors, 19
  gaseous, 2, 412
  half-lives of, 367, 475n101
  harmonization of, 247–49
  identifying criteria for, 471
  inert elements, 247–49
  infra-elements, 182, 487
  of Kingdom of Naples, 27–35
  Lavoisier's notion of, XXVI
  lost elements, 483–92
  meta-elements, 183, 201, 205, 208–10, 216n321, 281, 486, 491
  missing, 361
  names of, 1–2, 375–90
  néo-elements, 151, 492
  not yet discovered, 375–90
  number of, 367–69
  occult elements, 435–42
  patented, 165–66
  periodic table of, XXX–XXXIV, XXXIII*f*, 130, 130*f*, 195, 228–29, 228*f*
  photochemical, 91–92
  presumed discovered in 1877, 132, 133*t*
  proto-elements, 403–14, 487
  pseudoelements, 406, 412, 490
  québecium system, 228–29, 228*f*
  radioelements, 151
  rare earths, 161–64, 192–93, 216n308, 233, 362
  solar, 415–17
  superheavy, 361, 367, 370–75
  supra-elements, 182, 487
  synthetic, 367, 375–76
  that breathe, 6–7

element(s) (*Cont.*)
trace, XIX
transuranium, 37, 333–34, 356, 364–65, 375–76, 394–99
unnamed, 97–108
element 43. *See* technetium
element 61, 289–309
element 75, 249–51
element 86, 183–84, 184*t*
element 87, 251–52, 256–57
element 93, 256, 356–57
element 110, 364
element 111, 364
element 112, 364
element 114, 364
element 116, 364
element 118, 364–65
elementality, 24
element hunters, 279
element zero, 443–48, 491
emerald, 79
emission spectroscopy, 24–25
Emmens Chemicals and Explosives Co., 458
Emmens Metal Co., 458
ENEA Laboratories, 480
energy, radiant, 462
Enrico Fermi Laboratory (University of Chicago), 370
erbium, 123–24, 162, 167, 233, 279–80, 389n64
ercinite, 126n39
Erebus, 156, 157n178
espionage, 340–43
ethical evolution, 208
europium, 210–11, 280–81, 378
evolution
controlled convergent, 427
ethical, 208
inorganic, 183, 410
theories of, 408–10
experimentalism, 464, 465–67, 466n63
extinct elements, 410, 486
extraction procedures, 29

La Fabrication Chimique de L'Or (Jollivet-Castelot), 458
false discoveries, XXX, 1–11, 19, 91, 192–93, 481, 483–92
false names, 417–18
false suns, 415–18
fascism, 356
Federal Advanced School of Graphic Arts (Höhere Graphische Bundes Lehr), 278–79, 282
fer cassant a froid, 3–4
fergusonite, 127n67
fermium, 362, 380, 385
ferricyanide, 278
flerovium, 132, 386–88, 394, 397–98
fluorine, 56, 58n131, 355
fluorium, 56
force-engine, 458
Fort Belvoir (Virginia), 466
fractional crystallization, 171
francium, 150–52, 325, 333–34, 334*t*, 345–46, 362
Frasch Process, 88
fraud, 480, 481n3
Fraunhofer lines, 409
frigadréaction, 152–53
fusion, 368, 455
fusion-evaporation, 363

gadolinite, 126n50, 233
gadolinium, 146n138, 200, 206, 215, 295
Gallia, 169
gallium, 146n138, 169, 212
gaseous elements, 2, 412
gasometric analysis, 23
Gazeta de México, 15
Gazzetta chimica italiana, 82n507, 438n385
General Electric Company, 328
germanium, XXXII, 71n169, 157t
Glasgow Herald, 224
Glimpse, 112
gods, pagan, 43
gold, XXVIII
argentaurum gold, 459
growing, 463
side products, 461
transmutation of mercury into, 453–56
transmutation of silver into, 457–60
GSI, 366, 395

Hackh's Chemical Dictionary, 348–49
hafnium, 51n107, 117, 118n30, 237–41, 256, 439
half-life, 367, 475n101
haloes, pleochroic, 269
Harvard University, 454
hassium, 364, 367, 394–95
heavy water, 478
Hebrew University, 398–99
helium, 178, 180, 188nn248–49, 246, 248
decomposition of tungsten into, 451–52
transmutation of hydrogen into, 468–69
helium system, 445, 491
heresy, 331
Hesse crucible, 34n39
Adam Hilger Ltd. Company, 252–54, 258n93
Höhere Graphische Bundes Lehr (Federal Advanced School of Graphic Arts), 278–79, 282
holmium, 119–21, 123, 126n60, 135, 394
homeopathy, 6–9
hydrated sodium carbonate, 10n17
hydrofluoric acid, 56–57
hydrogen: transmutation into helium and neon, 468–69
hydrogen system, 445, 491
hyoscine, 188
L'Hyperchemie, 458

IAP. *See* Institute of Atomic Physics
illumination, 165–66
IMRA, 478
indium, 22, 71n169, 226
Industrial School of Nuremberg, 83
inert elements, 247–49
influenza (Spanish flu), 291
infra-elements, 182, 487
inorganic evolution, 183, 410
inorganic respiration, 6–7
Institute of Atomic Physics (IAP), 333
Institute of Physics, 359n2
Institute of Theoretical Physics, 238
International Commission for Inorganic Chemical Nomenclature, 282n187, 282n192, 366–67, 387–88
International Commission on Atomic Weights, 117
International Commission on Weights and Measures, 241
International Committee on the Chemical Elements, 189n271
International Union of Pure and Applied Chemistry (IUPAC), 47, 80, 227, 234, 282n187, 282n192, 346, 362
1878 Berlin Congress, 366
1995 Guilford conference, 366–67
1995 Guilford scheme for elements 101–109, 387, 388*t*
IUPAC-IUPAP Joint Working Party (JWP), 394–95, 398
names and symbols rejected for elements 114, 115, and 116, 395, 397*t*
recommended names for elements 101–109, 387, 388*t*
Red Book, 276n166
International Union of Pure and Applied Physics (IUPAP), 384
IUPAC-IUPAP Joint Working Party (JWP), 394–95, 398
names and symbols rejected for elements 114, 115, and 116, 395, 397*t*
inverted microscopy, 126n40
ionizing radiation, 343
Ireland, Republic of, 267
iridium, 70
Irish Citizen Army, 267
Irish Republican Brotherhood, 267
iron, XXVIII
fer cassant a froid, 3–4
of Pallas, 414n42
iron phosphide, 34n40
isotopes, xxvii, 232, 260, 288n220
isotungstic acid, 276n177
IUPAC. *See* International Union of Pure and Applied Chemistry
IUPAC-IUPAP Joint Working Party (JWP), 394–95, 398
IUPAP. *See* International Union of Pure and Applied Physics

JACS. See Journal of the American Chemical Society
James Moir Medal (South African Chemical Society), 247
Japan, 219–23, 478
Japanese Chemical Society, 220
Joint Institute for Nuclear Research (Dubna, Russia), 364–67, 382–84, 397–98
Joly Process, 268
Journal de Physique, 43–44
Journal of Chemical Education, 314–15
Journal of Medicine, 149
Journal of the American Chemical Society (JACS), 100, 192, 194, 300, 308n264, 424
Journal of the Royal Society of Chemistry, 159

Kabbalism, 454
kenotime, 73
kilo Xu, 325
kobellite, 48
kobolds, 68
Kodak Company, 282
krypton, 178–79, 187–88
kunzite, 105
Kurchatov Institute, 382

Laminaria, 465
lanthanides, XXXII–XXXIV, 13
Law of Conservation of Mass, 27, 33n24, 466n63
Law of Mattauch, 303–4
Law of Moseley, XXVII, 236, 361, 480
Law of Soddy or of Chemical Displacement, 189n269
Law of "the optimum of cathodic phosphorescence in binary systems," 205–6
Lawrence Berkeley National Laboratory (LBNL), 225–26, 362–67, 370, 373, 379, 383, 389n78, 461
Lawrence Livermore National Laboratory (Livermore, California), 398
lawrencium, 362, 366, 383–88
LBNL. *See* Lawrence Berkeley National Laboratory
lead, XXVIII, 476–77
LENR studies. *See* low-energy nuclear reaction studies
livermorium, 132, 394, 397–98
lost elements, 483–92
low-energy nuclear reaction (LENR) studies, 478
luminescence
illumination, 165–66
induced, 149–50
Lunar Society of Birmingham, 4–5
lutetium, 117n19, 171, 200–201, 213–15, 234, 240, 279–80, 282n187, 394

Madagascar, 238, 371
Magagnose et Dyonisos (Urbain), 242
magnesium, xxvi, 31–32, 35n60
magneto-optic effects, 323, 328–29, 331, 481
manganese, 22, 31, 36
Manhattan Project, 264
masrite, 159
Materials Test Reactor (MTR), 380
Materia Medica Pura (Pure Medical Matter), 7–8
Mattauch's Law, 303–4
media relations, 379
The Medicine of Experience (Hahnemann), 8
meitnerium, 364, 367, 395
mendelevium, 362, 378, 381
mercury, XXVIII, 2
transmutation into gold, 453–56
transmutation of lead into, 476–77
mesosiderites, 414n42
meta-elements, 183, 201, 205, 208–10, 216n321, 281, 486, 491
Metallex, Société Anonyme, 461
meteorites, 370–71, 412
Méthode de Nomenclature chimique, XXVI
miasma, 8–9
microscopy, 112, 126n40
Minerva, 376
mining, 2, 40, 66, 125, 147n99, 211, 236
misogyny, 315–16, 319n315
missing elements, 361
modern alchemy, 449–50

James Moir Medal (South African Chemical Society), 247
molybdenum, 104, 157t, 275
monazite, 297, 327, 338–39
Moray Research Institute (MRI), 462
Moray Valve, 462
Moseley's Law, XXVII, 236, 361, 480
Mount Somma, 143
MRI. *See* Moray Research Institute
muriatic acid, 24
music, 242
The Mysterious Universe (Jeans), 440
mystical chemistry, 450
mysticism, 406–7, 460n36
mythology, 43

NACA. *See* National Advisory Committee for Aeronautics
names of elements, 1–2
- alternative names, 12–17
- code names, 376
- Committee on Chemical Nomenclature (ACS), 387
- criteria for, 398
- for element 98, 378, 379*t*
- for elements 95 and 96, 376–78, 377*t*–378*t*
- for elements 101–109, 366–67, 385, 386*t*, 387–88, 388*t*
- for elements 103–112, 395, 396*t*
- for elements 114, 115, and 116, 397, 397*t*
- for elements 114, 116, and 118, 397, 398*t*
- International Commission for Inorganic Chemical Nomenclature, 282n187, 282n192, 366–67, 387–88
- *Méthode de Nomenclature chimique*, XXVI
- nomenclature, XXVI, XXVn17
- for not yet discovered elements, 375–90
- pagan gods, 43
- three-letter designations, 37

Namibia, 428, 429*f*
Naples, 27–35, 34n34
National Advisory Committee for Aeronautics (NACA), 434n98
National Institute of Advanced Studies, 480
nationalism, 480
Nature, 16, 36–37, 180, 237–38, 241, 253, 265, 269, 287, 343, 366–67, 399, 473
Nature Physics, 399
Naturhistorische Gesellschaft (Natural History Society) of Nuremberg, 83
Nazism, 114, 242, 317, 444, 454–55
neodymium, 303
néo-elements, 151, 492
neo-mysticism, 406–7
neon, 178, 180, 188n254, 189n255, 468–69
Neues Allgemeines Journal der Chemie, 13
Neue Zücher Zeitung, 343–44
neutrino, 444
neutrons, slow, 362
neutron stars, 368–69
New Harmony, 82
New Hydrogen Energy Program (Japan), 478
New South Wales, Australia, 4–5
New York Commercial, 424
New Yorker, 379
New York Sun, 424
New York Times, 385
New York University, 453
nickel, 22, 61*t*
Niobe, 46, 51n105
niobium, 46–47, 157t
nitrogen, 247
Nobel Institute, 366, 379, 381, 389n61
nobelium, 362, 366–67, 379–83
Nobel Prize
- in Chemistry, 242–43, 243*t*, 260, 264, 314, 316, 473
- in Physics, 356, 358, 375, 389n61, 389n78

noble gases, XXXII, 178–80, 212
nomenclature, XXVI, XXXVn17
- Committee on Chemical Nomenclature (ACS), 387
- International Commission for Inorganic Chemical Nomenclature, 282n187, 282n192, 366–67, 387–88
- *Méthode de Nomenclature chimique*, xxvi

nondiscoveries, 481
Les notions fondamentales d'éléments chimiques et d'atome (Urbain), 242

novas, 415
nuclear fission
discovery of, 264, 315–16, 319n315, 356–58, 376, 395
low-energy nuclear reaction (LENR) studies, 478
nuclear formulas, 247
nuclear fusion, 226, 246n, 402, 433, 484, 516
Nuclear Metaphysical Laboratories (University of California), 381
nuclear stability, 365–66
nuclido-biological reactions, 464
Nuremberg Polytechnic Institute, 83

Oak Ridge National Laboratory, 300, 372
Occult Chemistry: Investigations by Clairvoyant Magnification into the Structure of the Atoms of the Periodic Table and Some Compounds (Besant, Leadbeater, and Jinarajadasa), 435–36, 438–40
occult elements, 435–42
Oceanus, 116
official science, 226–27
opera, 242
optimum of cathodic phosphorescence in binary systems, 205–6
Orbitally Rearranged Monoatomic Elements (ORMEs), 462–63
ores: transmutation of, 461–63
Organization for Vigilance and Repression of Anti-Fascism (OVRA), 285n201
The Organon of Rational Healing (Hahnemann), 8
The Organon of the Art of Healing (Hahnemann), 8
Orion, 407–8
osmium, 69–70, 264
OVRA. *See* Organization for Vigilance and Repression of Anti-Fascism
oxygen, 62–63

pacifism, 344
paganism, 43, 406–7
palladium, 12–14, 225
Pallas or Pallade, asteroid, 33n55, 52
paramelaconite, 137, 141n118
Paris, France, 27–28, 406
Paris Academy of Sciences, 14
patents, 165–66
pathological science, 323, 481
"Pathological Science," 329–30
Pelops, 46, 51n105
periodic table of elements, XXXIII*f*
development of, XXX–XXXIV, 195
of Hugo Schiff, 130, 130*f*
québecium system, 228–29, 228*f*
petalite, 50
petrographic microscopy, 112
Philosopher's Stone, XXIX
Philosophical Magazine, 432
Philosophical Transactions (Royal Society), 4–5
Phipson phenomenon, 149–50
phosphorescence, cathodic, 205–6
phosphorus, XVIII–XIX
photochemical elements, 91–92
photochemistry, 278–79
Photographic News, 91–92
photography
coloring photographs, 278
X-ray, 282
The Physical Review, 468
Physical Review Letters, 372, 380
pitchblende, 23, 154n167, 308n277
placebo effect, 8
platinum, 74–75, 84
pleochroic haloes, 269
PNAS. *See Proceedings of the National Academy of Sciences*
Poggendorff's Annalen, 75
polonium, xxix, 198, 336n379, 471–72, 475n95
Pontificia Academia Scientiarum, 303
porcelain, 10n10, 53
press relations, 379
primary substances, 407–8, 490
prime matter, 403–4
primordial matter, 411, 491
Proceedings of the National Academy of Sciences (PNAS), 300, 308n264
Prometheus, 304
protactinium, 260
proto-elements, 403–14, 487

proto-metals, 409, 488
protons, 410
provincialism, 195–96, 198–99
pseudoelements, 406, 412, 490
qualitative analysis, 21, 23
quantitative analysis, 21, 23
Quarterly Journal of Science, 204
Quarterly Review, 304
québecium system, 228–29, 228*f*
quinine, 7–8
Racah Institute, 398–99
racial persecution, 356–58, 395
radiant energy, 462
radiation, ionizing, 343
radioactivity, XXVI–XXIX, 183–86, 354–55
  dating of the earth by, 268–70
  discovery of, 362
radiochemistry, XXIX–XXX, 356, 470–75
radioelements, 151
radium, XXIX, 160n183, 198
radon, 178, 183–86, 189n271, 285n205, 331–32, 332*f*
rare earths, 161–64, 192–93, 216n308, 233, 362
Reale Accademia dei Lincei, 421–22
Red Book (IUPAC), 276n166
reductionism, 208
respiration, inorganic, 6–7
retractions, 481
rhenium, 73, 84, 221–22, 223n390, 250–51, 256, 264, 298, 310–12, 335n366, 439
rhodium, 14
RIKEN Institute, 395–97
Rockefeller Foundation, 389n61
roentgenium, 132, 287, 394–95
Rowland grating, 162
Royal Institution, 24, 50, 440
Royal Porcelain Works (Berlin), 53
Royal Society, 4–5, 30, 51*n101*, 79, 88, 112–113, 116, 164, 204, 223, 240, 286–87
rubidium, 24–25, 254
ruthenium, 72n183, 73–76, 76n210
rutherfordium, 366–67, 383–88
samarium, 123, 135, 146n138, 211, 226, 230n401, 271, 289
scandium, 123, 135, 146n138, 157t
scattering, 368
Schwanen Ring, 66
science
  definition of, 307
  deviant, 226
  official, 226–27
  pathological, 323, 329–30, 481
Science, 121
Scientific American, 453–54
scopolamine, 188
seaborgium, 363, 366–67, 383–88, 390n87
Seater, 65
sedative salt, 30–31
selenium, 73, 157t
silicon, XXVI, XXVIII
silver, XXVIII
  allotropic, 459
  transmutation into gold, 457–60
silver (code name), 376
simple bodies, XXV–XXVI
slow neutrons, 362
smelting, 21
Soddy's Law, 189n269
solar elements, 415–17
solids, 2
South African Chemical Society, 247
Space and Naval Warfare Systems Center (U.S. Navy), 480
Spanish flu, 291
spectral analysis, xix, 91–92, 207–8
spectroscopy, 21
Die Spektren der Elemente bei normalem Druck (Exner), 208
stars, neutron, 368–69
stauroscope, 48
Stockholm, 27, 44, 45, 138, 358
strontium, 16n25, 27
superheavy elements, 361, 367, 375
  element 98, 378, 379*t*
  element 122, 398–99
  elements 95 and 96, 376–78, 377*t*–378*t*
  elements 101–109, 385, 386*t*, 387, 388*t*
  elements 103–112, 395, 396*t*
  elements 113, 115, 117, and 118, 398
  elements 114, 115, and 116, 397, 397*t*
  elements 114, 116, and 118, 397, 398*t*
  search for, 370–74
supernovas, 415

supra-elements, 182, 487
"Sur la nature de quelques éléments et meta-éléments phosphorescent de Sir W. Crookes" (Urbain), 206
Sur l'herbe (Urbain), 242
Swedish Royal Academy, 3, 98, 366, 375, 392
Sydney Bay, Australia, 4–5
synthetic elements, 367, 375–76

tantalite, 327
tantalum, 46, 264
Tantalus, 46, 51n105
Taurus, 415
technetium, 221–22, 250–51, 258n96, 267, 314–15, 317
Technische Hochschule (Polytechnic Institute of Vienna), 279
tellurium, 12–13, 16n25, 27, 38–39, 41n73
Tenebrae, 156
terbium, 135, 205–6, 211, 233, 281
thallium, 24–25, 82, 88
Theosophical Society, 435, 441, 442n115
Theosophist, 438–39
Thor (Donar), 70
thorianite, 105–6, 222n381
thorium, 183–84, 196n291, 264
thulium, 123, 126n60, 135, 146n138, 279
titanium, 23, 27, 33n26
Toyota Motor Corporation, 478
traditional chemistry *(ancienne chimie)*, 45–46
Transfermium Working Group (TWG), 386
transmutation(s), 260–66, 449–50
  biological, 464, 465–67
  of hydrogen into helium and neon, 468–69
  of lead into mercury, 476–77
  of mercury into gold, 453–56
  of ores, 461–63
  of silver into gold, 457–60
  of tungsten into helium, 451
transuranium elements, 333–34, 356, 364–65
  last five arrivals, 394–99
  in the sun, 415–17
  synthesis of, 375–76
  three-letter designations, 37
tungsten, 17n37, 27, 71n169, 272, 275–76, 451–52

unbibium, 398–99
United Institutes for Nuclear Research, 382, 384
United States, 191–99
United States Army, 466
United States Mint, 458
United States Navy, 480
unit X, 325
University of California, 381
University of California at Los Angeles (UCLA), 380
University of Cambridge, 49, 264, 407, 450, 455*n*
University of Chicago, 370
University of Florence, 291, 293, 294*f*, 294–98, 303, 306
University of Göttingen, 357–58
University of Illinois, 295, 297, 302, 306, 307*n*249, 309*n*285, 287
University of Oxford, 183, 201, 235, 380, 408, 435, 454
University of Vienna, 207, 333, 455*n*
unnamed elements, 97–108
unnilquadium, 384
ununoctium, 227
uranium, 33n25, 260, 264, 304, 361
  discovery of, 16n25, 23, 25n12, 27, 44
  sources of, 308n277
uranium-239, 356
uranium fission, 316, 356–58

Vanadis, 15–16
vanadium, 12, 14–16, 71n169
Vatican Observatory, 431
A la veillée (Urbain), 242
Vesta, 14, 16n29
Vesuvius, 143, 164n201
Via Panisperna boys, 353, 356–58, 359*f*
vitalism, 9
"vitrified copper," 54

Wasa, 138–39
Washington Post, 97
wolfram, 17n37, 27, 264, 272–73, 276n166, 276nn155–56
women. *See also specific women by name in the Names Index*

ignored and underrated *chemikerin*, 315–16
World War II, 362, 376

xenon, 178, 180
xenotime, 73
X-ray analysis, xxxii, 169n209, 299, 321, 325–26
X-ray photography, 282
X-ray spectroscopy, 215–16
Xu, 325

Yale University, 191
ytterbium, 123, 135, 146n138, 225, 279, 282n187, 282n189
yttria, 233
yttric earths, 233
yttriotantalite, 48, 49*t*
yttrium, 162, 167, 171, 233

zinc oxide, 59
zirconium, 16n25, 23, 27, 111
Z-rays, 461

ENTE
CASSA DI RISPARMIO
DI FIRENZE